Metal Amide Chemistry

Metal Amide Chemistry

MICHAEL LAPPERT AND ANDREY PROTCHENKO

Department of Chemistry and Biochemistry, University of Sussex, UK

PHILIP POWER AND ALEXANDRA SEEBER

Department of Chemistry, University of California at Davis, USA

A John Wiley and Sons, Ltd, Publication

Registered office
John Wiley & Sons Ltd, The Atrium, Southern Gate, Chichester, West Sussex, PO19 8SQ, United Kingdom
For details of our global editorial offices, for customer services and for information about how to apply for permission to reuse the copyright material in this book please see our website at www.wiley.com.

Library of Congress Cataloging-in-Publication Data

Metal amide chemistry / Michael Lappert ... [et al.].
p. cm.
Includes bibliographical references and index.
ISBN 978-0-470-72184-1 (cloth : alk. paper) 1. Amides. 2. Alkaline earth metals. 3. Organometallic compounds. I. Lappert, M. F.
QD305.A7M48 2008
547'.042–dc22 2008044491

A catalogue record for this book is available from the British Library.

ISBN 9780470721841 (H/B)

Typeset in 10/12pt Times by Thomson Digital, Noida, India.

Printed and bound in Great Britain by CPI Antony Rowe, Chippenham, Wiltshire.

Contents

Biographies

Michael Lappert is Emeritus Research Professor of Chemistry at the University of Sussex. A graduate of Northern Polytechnic, his B.Sc. was followed by a Ph.D. (with W. Gerrard) to which in 1960 he added a D.Sc. (University of London). He has been at Sussex since 1964, having previously been at UMIST (1959–1964). He was the recipient of the first Chemical Society Award for Main Group Metal Chemistry (1970) and then of the Organometallic Award (1978). He won the ACS – F. S. Kipping Award for Organosilicon Chemistry (1976); with the RSC (he was its Dalton Division President, 1989–1991), he has been a Tilden (1972), Nyholm (1994), and Sir Edward Frankland (1998) Medallist and Lecturer. In 2008, he was awarded the GDCh Alfred-Stock Prize. He was elected FRS in 1979 and was awarded an honorary doctorate from Ludwig-Maximilians-Universität (München, 1989). With co-workers he has published 2 books, edited 4 others, and more than 780 papers and reviews and a few patents on various aspects of inorganic and organometallic chemistry.

Philip Power received a B. A. from Trinity College Dublin in 1974 and a D. Phil. under M. F. Lappert from the University of Sussex in 1977. After postdoctoral studies with R. H. Holm at Stanford University, he joined the faculty at the Department of Chemistry at the University of California at Davis in 1981, where he is a Distinguished Professor of Chemistry. His main interests lie in the exploratory synthesis of new main-group and transition-metal complexes. A major theme of his work has been the use of sterically crowded ligands to stabilize species with new types of bonding, low coordination numbers, and high reactivity. He has been the recipient of several awards including the A. P. Sloan Foundation Fellow (1985), Alexander von Humboldt Fellowship for Senior U. S. Scientists (1992), Ludwig Mond Medal of the Royal Society of Chemistry (2005), F.A. Cotton Award in Synthetic Inorganic Chemistry of the American Chemical Society (2005) and was elected Fellow of the Royal Society of London (FRS) (2005).

Andrey Protchenko was born in 1961 in Rostov-on-Don, Russia. A graduate of the Moscow State University (Russia) in 1983, he carried out his doctoral work at the Karpov Institute of Physical Chemistry (Moscow) and completed his Ph.D. in chemistry (1997) at the Razuvaev Institute of Organometallic Chemistry (Nizhny Novgorod) under the supervision of Prof. M. N. Bochkarev studying naphthalene complexes of the lanthanides. In 1998 he joined the group of Prof. M. F. Lappert at the University of Sussex (England) working on a series of projects (Royal Society, EPSRC and Leverhulme Trust grants) involving lanthanide redox chemistry based on cyclopentadienyl, N- and N,N′-centred ligands.

Alexandra Seeber, née Pickering, was born in 1979 in Lichfield, UK. After graduating in Chemistry at Keele University in 2001 and completing a Ph.D. in Inorganic Chemistry into the design and synthesis of ligands for the controlled self-assembly of transition metal complexes at the University of Glasgow (Scotland) in 2004, she carried out postdoctoral research into nanoparticle synthesis at the University of California, Davis with Prof. Philip Power under a Department of Energy grant for the design of high surface area materials as hydrogen storage vessels. Alexandra Seeber has worked since 2006 in Research and Development in the catalysis department at BASF in Ludwigshafen (Germany).

Preface

In 1980, Ellis Horwood Ltd (with John Wiley & Sons, Ltd) were the publishers of a book entitled *Metal and Metalloid Amides*, by M. F. Lappert, P. P. Power, A. R. Sanger and R. C. Srivastava. This comprehensive treatise, which gathered together for the first time information on this important subject, is still occasionally referenced: for example, there have been more than 10 citations per annum in the major American Chemical Society journals during 2003–2007. However, the intervening three decades (from 1980) have witnessed a massive interest in the area, as evident in the more than 3000 papers; and to bring the subject up to date was a substantial undertaking.

The present volume arose from discussions during the end of 2004 with the late Ellis Horwood, who urged us to take up this challenge. Two of us (MFL and PPP, with our respective research groups; we thank our able collaborators for their contributions) have continued to publish extensively in the area. MFL and PPP were fortunate to have dedicated colleagues AVP (at Sussex) and ALS (at Davis) to join in this enterprise and to have the encouragement of Paul Deards, Richard Davies and their colleagues especially Nicole Elliott of John Wiley & Sons, Ltd, Chichester.

The first drafts of Chapters 1, 4, 5, 9 and 10 were written by MFL and AVP, while Chapters 2, 3, 6, 7 and 8 owe their origin to PPP and ALS; however, there have been ongoing consultations between the Sussex and Davis groups.

M. F. Lappert
P. P. Power
A. V. Protchenko
A. L. Seeber

1

Introduction

1.1 Scope and Organisation of Subject Matter

Our principal focus is to describe the synthesis, structure, reactions and applications of metal amides and related compounds which are stable at ambient temperatures, and particularly those which are mono-, di- or oligonuclear. Thus, the main thrust is on compounds having one or more $\bar{N}(R)R'$ ligands attached to a metal M. Such species are now known for all the natural elements (and Np and Pu), except for the lighter rare gases. In the present treatise, M is restricted to an element of Groups 1, 2, 3 and a 4*f* metal, a lighter 5*f* metal, and a metal of Groups 4 to 11, 12, 13 (except B), 14 (excluding M^{IV} compounds as reagents), and 15 (excluding P); these are described in Chapters 2 to 10, respectively.

The amido ligand may be $\bar{N}H_2$, $\bar{N}(H)R$, and $\bar{N}(R)R'$, in which R and R′ are the same or different, and each is an alkyl, aryl, or silyl (particularly $SiMe_3$) group; selected NH_2-metal compounds feature mainly in Chapters 2, 3, 6, 7 and 8. The amido ligand may be bound to the metal in a terminal or bridging (double or single) fashion; examples of the latter are $[Sn(NMe_2)(\mu\text{-}NMe_2)]_2$ and $[Cu(\mu\text{-}NMe_2)]_4$. Within our scope for main group and *f*-elements are bi- and tri-dentate ligands, including not only *N,N′*- and *N,N′,N″*-centred species such as $1,2\text{-}C_6H_4(\bar{N}CH_2Bu^t)_2$ and $MeC\{Si(Me)_2\bar{N}Bu^t\}_3$ but also others with a single amido site such as **1**. Tetradentate encapsulating ligands are excluded; their more natural home is to be found for the most part in textbooks of organic and biological chemistry and, more recently, materials chemistry. Amide-free metal complexes containing some *N,N′*-bidentate ligands which are neutral (e.g. bipy) or monoanionic (e.g. amidinates, guanidinates, or β-diketiminates) are also outside our scope.

Imides are usually included for main group elements because bi- and oligonuclear metal imides are ubiquitous and were often developed in parallel with the amides. They generally feature trivalent nitrogen, as in $[As(Cl)(\mu\text{-}NBu^t)]_2$. Imido derivatives of the transition

Metal Amide Chemistry Michael Lappert, Andrey Protchenko, Philip Power and Alexandra Seeber

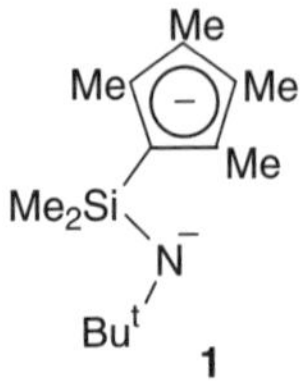

metals are not widely covered, as their chemistry is quite extensive and distinct from that of the amides. In general mononuclear metal imides are more frequent in *d*- and *f*-block chemistry; examples are $[Ti(=NC_6H_3Pr^i{}_2\text{-}2,6)(NMe_2)(NHMe_2)_3][B(C_6F_5)_4]$ or $[U\{(N(SiMe_2Bu^t)CH_2CH_2)_3N\}(=NSiMe_3)]$, but rare examples are also found in main group metal compounds having bulky ligands, as in $[In\{C_6H_3(C_6H_3Pr^i{}_2\text{-}2',6')_2\text{-}2,6\}\text{-}\{=NC_6H_3(C_6H_3Me_2\text{-}2',6')_2\text{-}2,6\}]$. The majority of metal amides and related compounds are neutral, but cationic and anionic complexes have featured; examples are $[Bi\{N(Bu^t)Si(Me)_2NBu^t][GaCl_4]$ and $[Li(OEt_2)]_2[Zr(NMe_2)_6]$. Compounds considered are mainly homometallic, but several are heterometallic such as the inverse crown complex $[Na_4Mn_2(tmp)_6(C_6H_4)]$ [tmp = $NC(Me)_2(CH_2)_3CMe_2$].

1.2 Developments and Perspectives

In 1980, metal and metalloid amides were already known for 57 elements, but then excluded were amides of the following elements: Sr, Ba, Tb, Dy, Er, Tm, Ru, Os, Rh, Ir and Pd (for several others, the first publications were post-1970: Sc, Y, La, Ce, Pr, Nd, Sm, Eu, Gd, Ho, Yb, Lu, Re, In and Tl). Since then, these lacunae have been filled. The variety of metal oxidation states for metal amides has also greatly expanded since 1980. The literature on metal amides has likewise burgeoned. For example, in the 1980 book there were 119 and 317 bibliographic citations for Group 1 and a conglomerate of Group 3 and 4*f*/5*f* and *d*-metal amides, respectively, whereas there are now 263 and 1110 for these two sections, respectively. Our aim has been to provide a comprehensive, but not exhaustive, coverage of the field of metal amides, with emphasis on the post-1980 developments through to the end of 2007. The total number of references for the succeeding Chapters 2 to 10 is in excess of 2500, while for these topics there were 923 citations in 1980.

Tables 1.1 and 1.2, laid out in the form of a Periodic Table, provide (i) a list of the elements, (ii) in brackets after each entry, the metal oxidation states for its amides, and (iii) the number of literature citations for each group of metals. Highlighted in bold are those metals for which amides were reported in the post-1979 period, while in italics are those metal oxidation states which likewise are of more recent date.

The variety of amido ligands now in use is very large. Many are bulky and often are free of β-hydrogen atoms. Among the newer monodentate amides are $\bar{N}Cy_2$ (Cy = cyclohexyl), $\bar{N}(SiHMe_2)_2$, $\bar{N}(SiMe_2Bu^t)_2$, $\bar{N}[(SiMe_3)C_6H_3Pr^i_2\text{-}2,6]$, $\bar{N}[(SiMe_3)\{C_6H_3(C_6H_2Me_3\text{-}2',4',6')_2\text{-}2,6\}]$, and $\bar{N}[Ad(C_6H_3Me_2\text{-}3,5)]$. Bi- and tridentate amides include not only those which are solely *N*-centred such as $\bar{N}(R)CH{=}CH\bar{N}R$ or $X\{CH_2CH_2\bar{N}Bu^t\}_3$ (X = $SiBu^t$ or N), but others in which the ligating atoms are not exclusively nitrogen: for illustrations of such ligands see **6–12** and **26–54** in Chapter 4.

Table 1.1 *List of metal amides of Group 1, 2, 12, 13 (not B), 14 (not M^{IV}) and 15 (not P) in various oxidation states (shown in brackets) and number of literature citations: new*[a] *metals and oxidation states are shown in bold and italics, respectively*

Group no.	1	2	12	13 (not B)	14 (subvalent)[b]	15 (not P)
Chapter no.	2	3	7	8	9	10
	Li (1)	Be (2)				
	Na (1)	Mg (2)		Al (*1*, 3)[c]	Si (*2*)[e]	
	K (1)	Ca (2)	Zn (2)	Ga (*1*, *2*, 3)[d]	Ge (2, 3)[f]	As (*1*, *2*, 3, 5)
	Rb (1)	**Sr** (2)	Cd (2)	In (*1*, *2*, 3)	Sn (2, 3)[f]	Sb (3, 5)
	Cs (1)	**Ba** (2)	Hg (2)	Tl (*1*, 3, *1/3*)	Pb (2)	Bi (3, *5*)
Total no. of references	**260**	**211**	**114**	**355**	**388**	**137**

[a]Since 1979
[b]M^{IV} compounds for Group 14 metal (M) amides are not discussed
[c]Also cluster compounds having metal oxidation states <1, *e.g.*, $[Al_{77}\{N(SiMe_3)_2\}_{20}]^{2-}$
[d]Also cluster compounds having metal oxidation states <1, *e.g.*, $[Ga_{84}\{N(SiMe_3)_2\}_{20}]^{4-}$
[e]Also oligomeric anions (*e.g.*, $[\{Si(NN)\}_4]^-$, $[\{Si(NN)\}_4]^{2-}$, $[\{Si(NN)\}_3]^{2-}$; $[Si(NN) \equiv Si\{(NR)_2C_6H_4\text{-}1,2\}$, R = Et, Bu^t])
[f]Also cluster compounds having metal oxidation states <1: $[Ge_8\{N(SiMe_3)_2\}_6]$, $[Sn_9\{Sn(N(C_6H_3iPr_2\text{-}2,6)SiMe_2R)\}_6]$ (R = Me, Ph)

Numerous amidometallates (several as cluster compounds) and, more rarely, cationic complexes, are well established. Representative examples of such salts are $[KMg(NPr^i_2)_3]_\infty$, $[Ti(NMe_2)_3(py)_2][BPh_4]$, $[Pt(\mu\text{-}Cl)\{SnCl(NR_2)_2\}(PEt_3)]_2$ (R = $SiMe_3$), $[Li(thf)_4][Ni(NPh_2)_3]$, $[U(NEt_2)_2(py)_2][BPh_4]_2$, $[As(C_5Me_5)(NMe_2)][AlCl_4]$ and $Li[Sb_3(\mu\text{-}NCy)_4(NMe_2)_2]$.

A feature of the post-1980 literature on mononuclear metal amides has been the discovery of numerous low-coordinate metal complexes. Significant examples among neutral compounds are the crystalline unicoordinate Group 13 metal and the dicoordinate Si^{II} and Group 15 metal amides $[Ga\{N(SiMe_3)C_6H_3Mes_2\text{-}2,6\}]$, $[\overline{Si\{N(Bu^t)CH{=}CHNBu^t}\}]$ and $[As\{N(H)Mes^*\}(=NMes^*)]$ (Mes = $C_6H_2Me_3$-2,4,6; Mes* = $C_6H_2Bu^t_3$-2,4,6). The highest metal coordination number recorded for a homo- or heteroleptic metal amide containing only monodentate ligands is six, in $[M(NMe_2)_6]$ (M = Mo or W), or seven, in $[U(NEt_2)_2(py)_5][BPh_4]_2$, respectively. The highest metal oxidation state of +6 is found not only in the Mo and W hexa(dimethylamides) but also for U, originally in *trans*-$[U(O)_2\{N(SiMe_3)_2\}_2(thf)_2]$; $U(NMe_2)_6$ was found to be thermally labile.

Bi-, tri- and oligonuclear compounds have ring, fused ring, or cluster core geometries, which feature $\bar{N}(R)R'$ as bridging ligands. For oligonuclear aggregates, the fused rings are often arranged as ladders or stacks, descriptions which were originally coined in the context of such alkali metal amides. The addition of a neutral ligand often fragments such an aggregate, a simple example being the conversion of $[Li\{\mu\text{-}N(SiMe_3)_2\}]_3$ into $3[Li\{N(SiMe_3)_2\}(tmeda)]$ (tmeda = $Me_2NCH_2CH_2NMe_2$).

A wider range of amido-containing complexes is of mixed metals, while others have imido bridges. Illustrations are provided by $[NaMg(NPr^i_2)_2(OBu^n)]_2$ (see Chapter 3, Figure 3.5) and $[Li_3(thf)_3Sb_6(NCy)_6(N)_3\cdot Li(N{=}NH)]$ (see **29** in Chapter 10). In the latter compound, each of the bridging cyclohexylimido or nitrido ligands has a coordination number ≥ 3 and hence may be regarded as a bis- or tris-(metallo)amide, respectively; the

Table 1.2 *List of Group 3–11 metal amides in various metal oxidation states (shown in brackets) and number of literature citations: new*[a] *metals and oxidation states are shown in bold and italics, respectively*

Sc (3)	Ti (2, 3, 4; *1/3*)	V (3, 4, 5; *3/5*)	Cr (2, 3, 4, *5*, *6*; *3/4*)	Mn (2, *3*)	Fe (2, 3; *2/3*)	Co (2, 3)	Ni (1, 2)	Cu (1)
Y (3)	Zr (4)	Nb (4, 5)	Mo (*1*, 2, 3, 4, 5, *6*; *4/5*)		**Ru** (*1*, 2)	**Rh** (1, *3*; *1/3*)	**Pd** (*2*)	Ag (1)
La (3)	Hf (4)	Ta (4, 5)	W (*2*, 3, 5, 6; *4/5*)	Re (*1*, *2*, *5*)	**Os** (*1*, *2*, *4*)	**Ir** (*2*, *3*)	Pt (2, 4)	Au (1, 3)
Ce (3, *4*, *3/4*)		Th (4)		**Total number of references:**				
Ln[b] (3)		U (*2*, *3*, 4, *5*, *6*; *3/4*)		1. Sc, Y, La and 4*f* elements (**Ch. 4**): **268**				
				2. Th, U (**Ch. 5**): **116**				
Ln[c] (*2*, 3)		**Np** (*3*)		3. Np, Pu (**Ch. 5**): **3**				
		Pu (*3*)		4. *d*-Block elements (**Ch. 6**): **692**				

[a]Since 1979
[b]Pr, Nd, Gd, **Tb**, **Dy**, Ho, **Er**, **Tm**, Lu
[c]Sm, Eu, Yb

diradicaloid compound $[SnCl(\mu\text{-}NSiMe_3)]_2$ is a simpler compound having bridging imido ligands in which the nitrogen atoms are three-coordinate.

Prior to 1980, 11 methods of synthesis of metal (M) amides were described and continue to be used. The salt elimination procedure $[L_nMX + M'\{N(R)R'\}]$ is the most general, usually with X = Hal and M′ = an alkali metal; whereas lithium is often the metal of choice, employing M′ = Na or K has the advantage of easier separation of the alkali metal halide from the target metal amide. More recently used leaving groups are tosylate, triflate, aryloxide, or a cyclopentadienyl (e.g. in Al^I or Pb^{II} chemistry). Newer sources of an amido ligand are $BrN(SiMe_3)_2$ and a *N,N′*-dihydrocarbyl-1,4-diazabuta-1,3-diene.

Molecular structure elucidation, principally by single crystal X-ray diffraction, has become almost routine and is now available for the majority of the metal amides presently discussed. In the 1980 book, however, such data were provided for just 112 compounds, 54 of which were for *d*- and *f*-block metals and 41 for the Group 13 metal amides. The contrast with the developing situation is illustrated by reference to Group 1 metal amides: from four X-ray data sets in 1980 there were more than 200 by the end of 2007.

Among other physicochemical techniques which now feature prominently are those based on vibrational, electronic, NMR (including ^2H, ^{13}C, ^6Li, ^7Li, ^9Be, ^{27}Al, ^{29}Si, ^{113}Cd, ^{119}Sn and ^{171}Yb) and EPR spectroscopy and magnetic measurements. Computational studies are increasingly significant, particularly for subvalent metal-nitrogen compounds. The nitrogen environment is almost invariably close to trigonal, or distorted trigonal, planar for homo- and heteroleptic metal amides having terminal $\bar{N}(R)R'$ ligands, respectively. This fact, as well as the generally relatively short M–N bond lengths (compared with the sum of the covalent radii of M and N) has often been interpreted in terms of the $\bar{N}(R)R'$ ligand being both a σ- and a π-donor. However, as has increasingly become evident, other factors should also be considered, as discussed, for example, in the context of transition metal amides and bis (amino)silylenes in Chapter 6 (Section 6.2.3.1) and Chapter 9 (Section 9.2.3), respectively.

Several metal amides are much employed as reagents, ligand transfer reagents or precursors for more complex molecules. Alkali metal, and mainly lithium, bulky amides, including $LiN(SiMe_3)_2$, $LiNPr^i_2$ and $LiNCy_2$, are important proton-abstractors for organic synthesis, as are recently the synergic couple $MN(R)R'/M'R_n$ (M′ = a Group 2, 12 or 13 metal). Magnesium, aluminium, tri(methyl)tin and, particularly, alkali metal amides are valuable ligand transfer reagents, being the most useful precursors to amides of other metals. The ready cleavage of the highly polar M–N(R)R′ bond, especially by protic reagents, make several metal amides important as (i) synthons for a wide range of that metal's compounds, and (ii) materials. Examples of (i) are $Sn(NMe_2)_2$ or $Sb(NMe_2)_3$ as the source of unusual Sn–N and Sn–P complexes such as the cubane $[\{Sn(\mu\text{-}NBu^t)\}_4]$ and $[Li_2Sn_2(PR)_4(tmeda)_2]$ (**53** in Chapter 9), or related Sb compounds, such as $[Li_4\{Sb_2(NCy)_4\}_2]$ (**30** in Chapter 10). Illustration of (ii) is provided by Tables 4.1 and 4.2 of Chapter 4, showing $[Ce\{N(SiMe_3)_2\}_3]$ as a starting material for a heteroleptic β-diketiminato compound and $[Y\{N(SiHMe_2)_2\}_3(thf)_2]$ as a substrate for the formation of yttrium-centred catalysts, usually immobilised, for various organic transformations. A valuable feature for (i) and (ii) is that the amine co-product, $NHMe_2$ or $NH(SiHMe_2)_2$, is volatile. Metal amide-catalysed hydroamination/cyclisation of *N*-unprotected alkenes to form heterocyclic nitrogen compounds is exemplified by the role of an optically active Sm^{III} amide as an enantioselective catalyst for such a process (e.g. see Equations (4.8) and (4.12) in Chapter 4). Various *d*- and 4*f*-block metal amides have been shown to be catalysts for some important processes, including olefin di-, oligo- or polymerisation, olefin hydroamination and hydrosilylation, copolymerisation of an oxirane and carbon dioxide, and ring-opening polymerisation of lactides or ε-caprolactam. Certain *d*- and 4*f*-block metal amides are unusual in their ability to 'activate' N_2 or N_2O, as in the conversion of [Mo{N(Ad)R}$_3$] into [Mo(X){N(Ad)R}$_3$], respectively (R = $C_6H_3Me_2$-3,5 and X = N or NO), or in the reactions of [Zr(**F**)Cl_2] with KC_8 and N_2 (**F** is a macrocyclic *P,P′,N,N′*-centred bis(amido) ligand) (see Schemes 6.3, 6.4 and 6.7 in Chapter 6).

2

Alkali Metal Amides

2.1 Introduction

Amido derivatives of the alkali metals enjoy the most widespread use of all metal amides. This is a result of their central importance as amido ligand transfer agents for the synthesis of other element amido derivatives throughout the Periodic Table. In addition, they are widely used in organic chemistry as powerful Bronsted bases and they are preferred bases for the formation of ketone enolates. Despite their importance in these fields, the primary focus of this chapter is on their synthesis and structures. In particular, the number of known structural types has undergone great expansion since 1980. It is instructive to compare knowledge of alkali metal diaorganoamides described in Chapter 2 of our 1980 book[1] with currently available data. At that time, a comprehensive listing of the known alkali metal amides, including the parent metal amides MNH_2 (M = Li – Cs), totalled less than 40 compounds with less than 100 citations from the primary literature. The detailed structures of just three compounds – $[\{LiN(SiMe_3)_2\}_3]$,[2,3] $[\{NaN(SiMe_3)_2\}_\infty]$[4] and $[\{(dioxane)_2KN(SiMe_3)_2\}_\infty]$[5] had been published. The lithium salt $[\{Li(tmp)\}_4]$ (tmp = 2,2,6,6,-tetramethylpiperidide), was also listed as structurally characterized but as unpublished work (see below).[1]

The intervening years have seen huge growth in the number of well-characterized compounds, the vast majority of which are lithium, sodium or potassium salts. Their structural chemistry has proven to be especially rich and the number of structures of alkali metal amides currently available exceeds 200. These involve a wide selection of structural motifs that were mostly unknown in 1980.

There have been several reviews on alkali metal amide structural chemistry.[6–12] Most of these deal with lithium amides but there is also significant coverage of the heavier elements[9–12] and other topics such as derivatives of primary amides and hydrazides.[12] Two comprehensive reviews[6,8] for lithium amides and related salts published in 1991 and 1995 include extensive tables of structural data. There have also been reviews of the solution behaviour[13–15] of lithium amides as well as of the extensive use of alkali metal

Metal Amide Chemistry Michael Lappert, Andrey Protchenko, Philip Power and Alexandra Seeber

amides in organic synthesis.[16–19] In addition, there have been a large number of related lithium complexes that incorporate chelating or multidentate amido ligands or ligands related to amides in which nitrogen is part of a delocalized array incorporating carbon, nitrogen, sulfur, or phosphorus atoms, exemplified by azallyl anions, guanidinates, benzamidinates, β-diketiminates, phospohoraneiminates, sulfaminides, and related ligands. However, these complexes lie outside our current scope and are thus not covered in this chapter. The present coverage focuses on the most important structural findings for alkali metal complexes of monodentate diorganoamido ligands with emphasis on the more recent developments.

2.2 Lithium Amides

2.2.1 Introduction

Lithium amides are the most important of the alkali metal amides. This is mainly due to the facility with which they can be prepared in solution by the simple reaction of the amine with commercially available $LiBu^n$. An analogous reaction with heavier metal alkyls is much more difficult due to the high reactivity of heavier alkali metal alkyls which attack many solvents. Another advantage of lithium amides is that they tend to be more soluble in hydrocarbons than their heavier element congeners. This is due to the small size (and hence greater polarizing power) of the lithium ion, which induces greater covalent character.

The parent lithium amide, $LiNH_2$, was discovered in 1894[20] but crystal data were not refined until 1972.[21] Unfortunately, the variation in NH_2 group orientation and the crystal packing of the ions in the unit cell made it impossible to fully determine the crystal structure. The crystal structure of $LiNH_2$ was reinvestigated more recently by powder neutron diffraction.[22] It was found to crystallize in the tetragonal space group I4. The distances between the nearest nitrogen and hydrogen atoms lie between 0.986 and 0.942 Å. The H–N–H bond angle was estimated to be 99.97°. These results differ significantly from those reported in the previous studies. The structure of the deuterated analogue $LiND_2$ also has the tetragonal space group *I*4 and a structure in which the lithium ions are coordinated by four amido (ND_2^-) ions in a distorted tetrahedral geometry (Li–N distances lie between 2.06 and 2.21 Å).[23] The LiN_4 tetrahedra are connected by edge-sharing interactions to form a three-dimensional network. The two N–D distances (av. 0.98 Å) are similar to those previously published. The crystal structure of a bromide ion-supported Li-NH_2 lattice in $Li_2Br(NH_2)$ has been determined.[24] It has an infinite array of $[Li_4(NH_2)_2]^{2+}$ rods held within a three-dimensional crystal lattice via coordinative interactions with the Br^- ions. The Li–N distances are between 2.01 and 2.17 Å, and the Li–Br distances lie between 2.50 and 2.63 Å.

In the vapor phase it is possible to obtain $LiNH_2$ as a monomer and its structure has been determined by millimeter/submillimeter – wave spectroscopy.[25] The r_0 structure of $LiNH_2$ has Li–N and NH distances of 1.736(3) Å and 1.022(3) Å as well as H–N–H angle of 106.9(1)°. The Li–N distance is the shortest measured for a lithium amide, and it is in good agreement with theoretical predictions.[26–29] There have also been theoretical investigations of the lower aggregates $(LiNH_2)_n$ (n = 2, 3, 4, etc.).[30–33] In recent work it has been shown that, with use of complexing agents at both lithium and boron, it is possible to stabilize monomeric $LiNH_2$ in the solid state.[34] The use of 12-crown-4 to complex lithium and of $B(C_6F_5)_3$ to coordinate nitrogen allowed the isolation of the unique [(12-crown-4)$LiNH_2$-{$B(C_6F_5)_3$}] which has an average Li–N bond length of 2.048(1) Å.

Parent lithium amide derivatives are also attracting attention due to their potential applications in hydrogen storage.[35–39] For example, the complex hydride system $LiNH_2$–LiH[40] can undergo reversible, thermally induced hydrogenation and dehydrogenation transformations and represents a material with a potential to provide over 10 wt.% hydrogen storage.[40–42] There is also interest in the reduction of the hydrogen desorption temperature by destabilization of complex hydrides via full or partial cationic substitutions with heterometallic atoms.[43–53] For example, a substitution of M = Li with M = $Li_{0.9}Mg_{0.1}$ results in a decrease of dehydrogenation temperature from 550 to 500 K.[54–56] Interest in the chemical hydrogen storage problem has also led to the isolation of several new complex hydrides. For example, $Li_2BH_4NH_2$ has a hexagonal array of distinct $(LiNH_2)_6$ clusters dispersed in a $LiBH_4$ matrix.[57] The lithium amide-borohydride system is dominated by a bcc compound of formula $Li_4BH_4(NH_2)_3$.[58]

2.2.2 Monomeric Lithium Amides

The simplest substituted lithium amide structures are mononuclear. Some of the earliest examples were isolated through the use of the multidentate ligand 12-crown-4 to complex the lithium and prevent oligomerization (Scheme 2.1). The reaction of lithium complexes $LiNR_2$ (where R = $SiMe_3$, Ph or $SiMePh_2$), obtained by treatment of the amine with $LiBu^n$, with 12-crown-4 forms the corresponding monomer [Li(12-crown-4)NR_2][59,60] with Li–N bond distances in the range 1.97 to 2.06 Å. Each monomer features a five-coordinate Li^+ ion in an approximately square pyramidal environment complexed by one nitrogen and four oxygen donor atoms. It is also possible to induce ion pair separation[61] by the addition of 12-crown-4 to lithium amide complexes $LiNR_2$ when R is very large, for example, [{Li(12-crown-4)$_2$}{N($SiPh_3$)$_2$}][62] and [{Li(12-crown-4)$_2$}{N{Si(Bu^t)$_2$F}$_2$}].[63] The crown

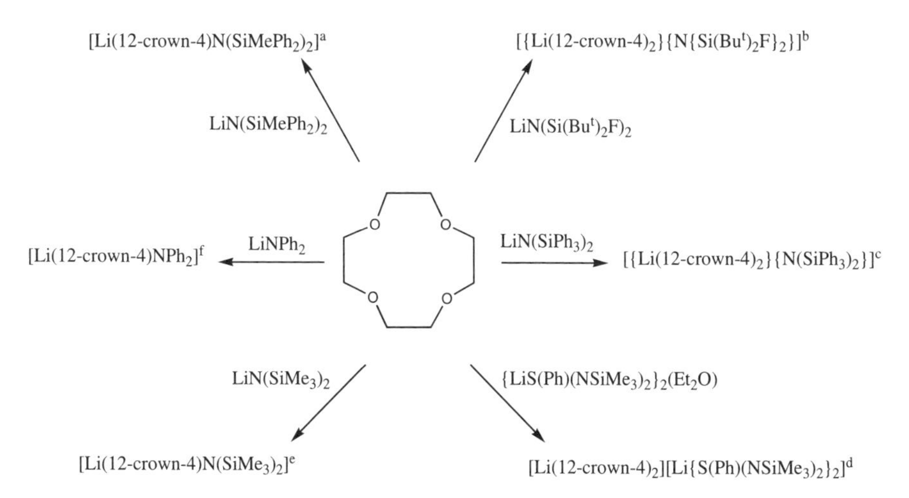

Scheme 2.1 *Lithium amido complexes formed upon addition of 12-crown-4: a,[60] b,[63] c,[62] d,[65] e,[59] f[60]*

ether acts as a stronger donor ligand to the lithium ion and ion pair separation allows minimization of the steric pressure within the bulky amido anion.[63] An interesting structure is observed upon addition of 12-crown-4 to the dimeric lithium amide [{LiS(Ph)(NSi-$Me_3)_2\}_2(OEt_2)]$,[64] which results in the ion pair separated compound [Li(12-crown-4)$_2$]-[Li{S(Ph)(NSiMe$_3$)$_2$}$_2$], where the anion and cation each contain a complexed lithium ion.[65] One of the lithium ions is sandwiched between two crown ether molecules to afford 8-coordinate Li^+ and the other is coordinated by the two chelating anions in a spirocyclic fashion. The Li–N bond lengths are relatively long (2.09 to 2.17 Å) due to the increased interelectronic repulsion within the anionic component.

Other monomeric silylamido complexes have also been obtained by the addition of tmeda (*N,N,N′N′*-tetramethylethylenediamine) or pmdeta (*N,N,N′,N″,N″* – pentamethyl-diethylenetriamine) to afford [(tmeda)LiN(SiMe$_3$)$_2$] which has three coordinate lithium Li–N(amide) = 1.893(3) Å and [(pmdeta)LiN(SiMe$_3$)$_2$]with 4-coordinate Li and Li–N (amide = 1.988(6) Å.[66] The use of very large substituents at the amido ligand can also afford monomeric structures. Examples include [Li(thf)$_3$N(SiMe$_2$Ph)$_2$],[60] [Li(thf)$_2$N(SiPh$_3$)$_2$],[60] [Li(thf)$_2$N(SiBu$^t_2$F)$_2$],[63] [Li(thf)$_2$N{(C$_6$F$_5$)(SiBu$^t_2$F)}][67] and [Li(thf)$_3${NMes*(SiPr$^i_2$X)}] (Mes* = C$_6$H$_2$Bu$^t_3$-2,4,6; X = F or Cl).[68] A more recent example is given by the monomeric [Li{N[Si(NCH$_2$Bu^t)$_2$C$_6$H$_4$-1,2-SiMe$_2$Ph][Si(NCH$_2$ = Bu^t)$_2$C$_6$H$_4$-1,2)SiMe$_3$](thf)].[69] The lithium ion is three-coordinate and is bound by two amido nitrogen atoms (average Li–N bond length is 1.96 Å) and one tetrahydrofuran donor molecule.

Monomeric structures are also observed in sterically crowded borylamides such as [(Et$_2$O)$_2$LiN(R)BMes$_2$] (R = Ph[70] or Mes[71]) or [(tmeda)LiN(Bu^t)BBu$^t_2$}[72] where B-N π bonding may lower the basicity of the nitrogen lone pair and disfavour aggregation.

Monomeric amido complexes were also obtained through the use of other large nitrogen substituents in combination with various donor ligands. For example, the mononuclear primary amide [Li(NHMes*)(tmeda)][73] formed by addition of tmeda to a toluene solution of the dimeric [{Li(NHMes*)(OEt$_2$)}$_2$].[74] The lithium ion in this structure has a very distorted trigonal planar coordination environment (Figure 2.1) because of an additional

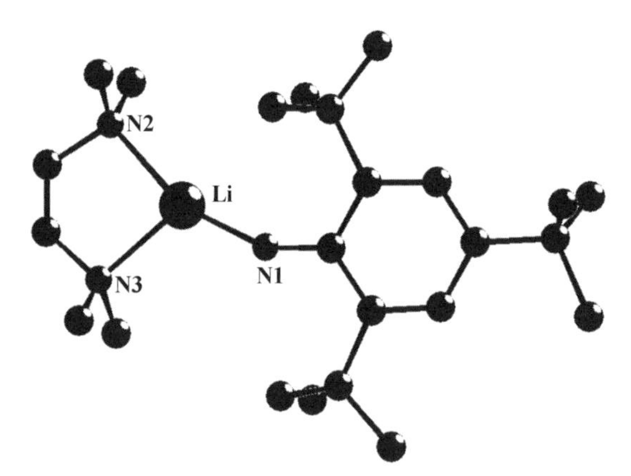

Figure 2.1 *Illustration of the monomeric primary amido complex [Li(NHMes*)(tmeda)]*[73] *showing the distorted trigonal planar geometry at the lithium ion. Selected bond lengths: Li-N1: 1.895(8) Å, Li-N2: 2.137(9) Å and Li-N3: 2.165(9) Å*

C-H· · ·Li interaction. The lithium ion has a much shorter bond to the amido nitrogen (Li–N = 1.897(5) Å) in comparison with the tmeda nitrogens (av. Li–N distance is 2.15 Å). This strategy has been further employed in the synthesis of the monomeric complexes [Li{NPh(naphthyl)}(tmeda)] and [Li{NPh(naphthyl)}(pmdeta)], which have a trigonal (Li–N bond lengths between 1.97 and 2.13 Å) and tetrahedral (Li–N distances between 2.00 and 2.22 Å) coordination of the lithium ion, respectively,[75] and [Li{N(CH(Me)Ph)-(CH_2Ph)}(pmdeta)],[76] which has tetrahedral lithium coordination with Li–N bond lengths between 1.96 and 2.31 Å. Each of the above complexes has shorter contacts to the nitrogen from the aryl group than to the amino groups of the multidentate donor ligands. Furthermore, reaction of NCBu^t with a pentane solution of [(tmeda)Li{$C_6H_7(SiMe_3)_2$}] at room temperature yielded the monomeric chelated lithium amido complex [(tmeda)LiN($SiMe_3$)-{C(Bu^t)$C_6H_7SiMe_3$}].[77] The presence of relatively bulky substituents at the amido ligand is an important factor in the preferential formation of monomers.[75–77] Smaller substituents on N facilitate higher aggregate species even in the presence of donor ligands, as exemplified by the formation of the dimeric species [{LiNMePh(tmeda)}$_2$][75] and (*R*)-[Li{N(CH_2Ph)-CH(Me)Ph}(pmdeta)].[76]

2.2.3 Dimeric Lithium Amides

Dimeric and higher aggregate lithium amides can generally be classified into the coordination motifs illustrated in Scheme 2.2. The four-membered $(LiN)_2$ ring is ubiquitous in lithium amide chemistry and is observed both in discrete dimeric structures in either planar (Scheme 2.2, A) or non-planar (Scheme 2.2, B) geometries as well as in oligomeric and polymeric (ladder) frameworks (Scheme 2.2, C). Trimeric six-membered $(LiN)_3$ ring

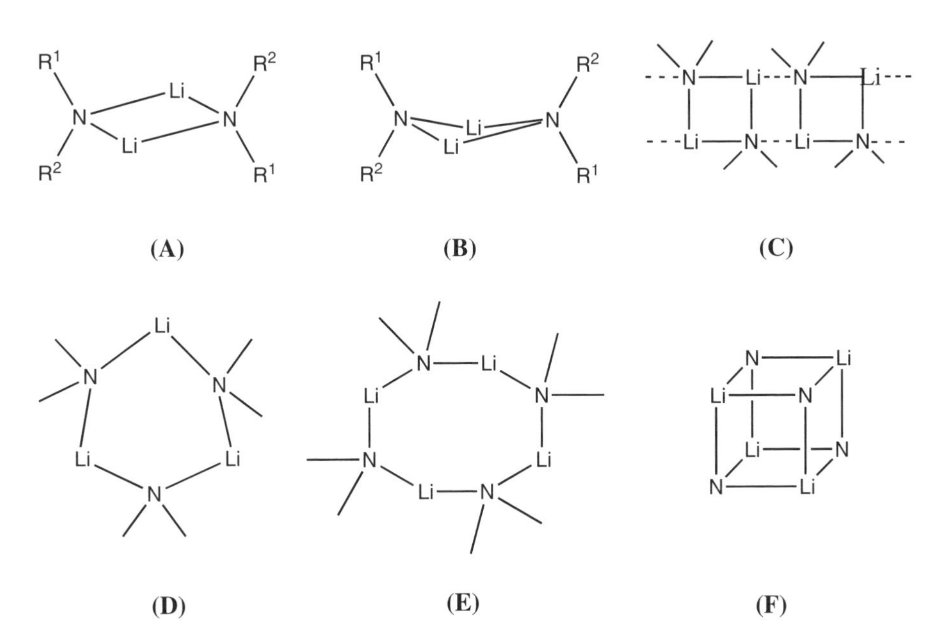

Scheme 2.2 *Common aggregation motifs of lithium amides*

(Scheme 2.2, D) and tetrameric eight-membered $(LiN)_4$ ring and cubane (Scheme 2.2, E and F) structures are also known. Comprehensive reviews of dimeric and other lithium amido structures were published in the 1990s. As a result only selected examples from the preceding literature are covered.[6,8]

The most common lithium amide structures are dimeric although in solution the dimers often exist in equilibrium with other structures.[13–15,78] The simplest are derived from monodentate, primary amido ligands, -NHR. For example, the solvated structures $[\{Li(OEt_2)NHR\}_2]$ (where R = Mes* or Dipp ($C_6H_3Pr^i{}_2$-2,6)) contain a planar Li_2N_2 framework (Structure A) with Li–N bonds of 1.987(5) and 2.041(6) Å for $[\{Li(OEt_2)NHMes^*\}_2]$,[74] and an average Li–N distance of 2.00 Å for $[\{Li(OEt_2)NHDipp\}_2]$. The lithiums are bridged by the two amido ligands, and the trigonal coordination geometry at the lithium atoms is completed by coordination of one (diethyl ether) molecule.

There are several simple dimeric structures of less crowded lithium derivatives of primary amides. Among these are $[\{Li(thf)_2NHPh\}_2]$,[79] and $[\{Li(thf)_2NHC_6F_5\}_2]$,[79] $[\{Li(py)_2NHPh\}_2]$[80] and $[\{Li(C_6H_4\text{-}NMe_2\text{-}4)_2NHPh\}_2]$,[80] the silylamides $[\{Li(tmeda)\text{-}NH(SiBu^t{}_2F)\}_2]$[81] and $[(LiNH(SiBu^t[OSi(F)\{N(SiMe_3)Pr^i\}_2)_2]$.[82] It has already been mentioned that $LiN(SiMe_2)_2$ is a trimer in the solid state.[2,3] However, in the vapor phase it has the dimeric formula $[\{LiN(SiMe_3)_2\}_2]$ (Structure A) with two coordinate Li and Li–N = 1.99(3) Å.[83] The more crowded $[\{LiN(SiMe_3)(SiMe_2Ph)\}_2]$[84] is a dimer in the solid state. The extensive use of the lithium amide $LiN(SiMe_3)_2$ has led to great interest in the structures of its complexes with various Lewis bases and the various factors that control its aggregation tendency. Several crystal structures of the general formula $[\{(L)LiN(SiMe_3)_2\}_2]$ (L = Et_2O,[85,86] thf,[87] NMe_2CH_2Ph,[66] and $MeOCH_2CH_2OMe$ (dme)[66] or 1,4,-dioxane[66]) are known. In addition, $LiN(SiMe_3)_2$ forms an unusual cationic species in the salt $[Li(\mu\text{-}NSiMe_3)(\mu\text{-tempo})(tempo)_2][Mg\{N(SiMe_3)_2\}_3]$ when it is treated with tempo (tetramethylpiperidine-N-oxide) and $Mg\{N(SiMe_3)_2\}_2$.[88] In addition, an unusual triple anion complex with three different types of anion shown in **1** was obtained from a solution containing a 1:1:1:2 ratio of LiBr, $LiN(SiMe_3)_2$, $LiOC(Bu^t)CH_2$ and tmeda.[89] The behaviour of lithium silylamides in solution in various solvents and in the presence of a range of donors has been studied by ^6Li and ^{15}N NMR spectroscopy.[13–15]

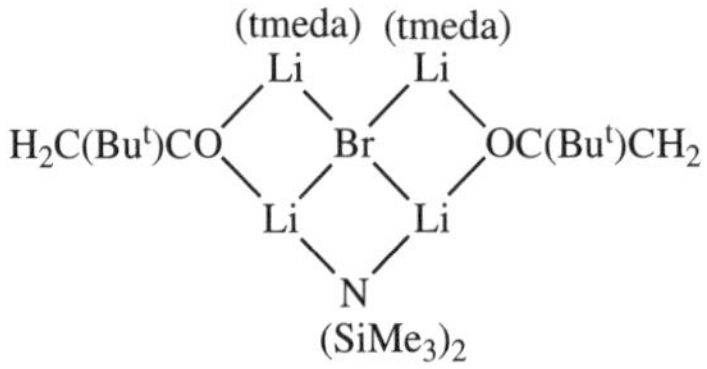

(tmeda) (tmeda)
Li Li
$H_2C(Bu^t)CO$ Br $OC(Bu^t)CH_2$
Li Li
N
$(SiMe_3)_2$

1

The borylamide derivative $[\{(Li(OEt_2)NHBMes_2)\}_2]$ also has a bridged dimeric structure despite the reduction in nitrogen lone pair character by B-N π-bonding.[90] A dimeric structure containing a crown ether moiety was formed upon monodeprotonation of 1,7-diaza-12-crown-4, $[\{1\text{-}Li\text{-}1,7\text{-}N\text{-}12\text{-}crown\text{-}4\}_2]$.[91] The lithium atom is coordinated to the macrocycle through four oxygen and one nitrogen donor (Li–N is 2.19 Å; Li–O = 2.23 and 2.26 Å). There is an additional, shorter bond to the deprotonated N atom (Li–N bond length, 2.00 Å). Two such units are dimerized through Li–N interactions to form a regular $(LiN)_2$ core.

Similar, isolated diamond-shaped $(LiN)_2$ frameworks are observed for many of the dimeric lithium amide complexes formed from monodentate,[75,76,92–105] bidentate,[102,106–113] or tridentate[114–118] amido ligands or ligands with multiple amido functions.[119–129] These ring systems are generally planar (Structure A), but puckering of the ring can occur (Structure B), as is seen in the compound $[(Li(OEt_2)$*c*-$\{NCH(n\text{-}Bu)CHCH\text{-}2,3\text{-}C_{10}H_6\})_2]$[130] formed by the reaction of 7,8-benzoquinoline with *n*-BuLi. The lithiums are three coordinate and bridge two benzoquinoline ligands and are coordinated to one diethyl ether ligand. The Li–N bond lengths are 2.038(3) and 2.104(2) Å. The lithium and nitrogen atoms in the $(LiN)_2$ ring in $[\{LiN(SiMe_3)(SiMe_2Ph)\}_2]$[84] alternate above and below the ring plane by approximately 0.033 Å, with Li–N bond lengths between 1.98 and 2.02 Å. Each lithium ion has an additional contact to the *ipso* carbon of the aryl group (2.45 Å). Likewise, the $(LiN)_2$ cores in the compounds $[Li\{1,2\text{-}(N(SiMe_3))_2C_{20}H_{12}\}(thf)_2]$ and $[Li\{1,2\text{-}(N(CH_2Bu^t))_2C_{20}H_{12}\}(thf)_2]$ are folded, and have Li–N distances of 1.85–2.25 and 1.92–2.24 Å, respectively.[131] Each lithium ion is four-coordinate with coordination by one tetrahydrofuran ligand (av. Li–O distance is 1.88 Å) and an additional short contact with an *ipso* carbon (Li–C contact, 2.24 Å).

An unusual variation of the $(LiN)_2$ core structure is observed in the lithium fluorosilylamide $[\{LiN(Bu^t)Si(Bu^t)_2F\}_2]$.[132] The fluorine atoms coordinate to the lithium atoms from the central $(LiN)_2$ ring in a boat-like conformation (Structure B, av. Li–N distance is 2.04 Å), which forms two four-membered Si–N–Li–F rings fused to the central $(LiN)_2$ core. Other examples of fluoride coordination to the $(LiN)_2$ core have been reported, for example in the complexes $[\{LiFSi(Bu^t)_2NSi(Pr^i)_2OSiMe_3\}_2]$[133] (**2**) and $[\{LiFSi(Pr^i)_2NSi(Bu^t)OSi(Me)\{N(SiMe_3)_2\}_2F\}_2]$.[134] In contrast, the fluorine atom in $[\{LiN(H)Si(Bu^t)_2F(tmeda)\}_2]$[81] does not display coordinative interactions, and a regular $(LiN)_2$ framework (Structure B) is observed with average Li–N distances of 2.06 Å for the fluorosilylamido nitrogens and significantly longer contacts with the bidentate tmeda ligands (average Li–N = 2.27 Å).

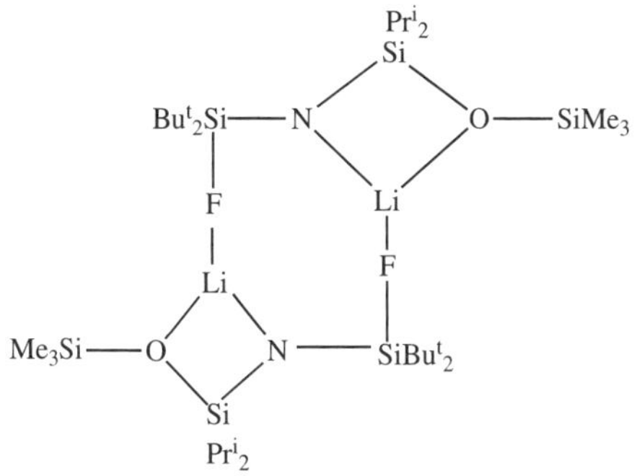

2

2.2.4 Trimeric Lithium Amides

The structure of $[\{LiN(SiMe_3)_2\}_3]$[2,3] (**1**), was originally reported in 1969;[2] it has a planar Li_3N_3 array and it represented the first substituted lithium amide to be structurally characterized. The Li–N bond lengths range from 1.98 to 2.02 Å (Structure D).[3] The germanium analogue [{LiN(GeMe$_3$)$_2$}$_3$][135] has an almost identical structure, with shorter Li–N contacts (average 1.96 Å), possibly as a result of lower steric crowding. The dibenzylamido derivative [{LiN(CH$_2$Ph)$_2$}$_3$](**2**)[90] also has a planar six-membered Li$_3$N$_3$ ring. The methylene linkers of the amido ligand effectively block delocalization, resulting in a low coordination number of two around the Li atoms (Li–N distances range between 1.91 and 2.04 Å). The trimer [{LiN(SiMe$_3$)(CH$_2$Ph)}$_3$] can be seen as a hybrid structure of compounds **1** and **2** as it contains a planar (LiN)$_3$ ring with two coordinate lithium atoms and short Li–N bond lengths (*ca.* 1.95 Å).[136] The internal ring angles at Li and N average 93.7 and 146.2° and the SiMe$_3$ and CH$_2$Ph substituents are disposed at opposite sides of the almost planar Li$_3$N$_3$ ring. Its treatment with OP(NMe$_2$)$_3$ yielded [{(Me$_3$N)$_3$POLiN(SiMe$_3$)-CH$_2$Ph}$_2$].[136] The trimetallic primary amido derivative [{(pmdeta)$_2$(LiNHPh)$_3$] features a central {Li(NHPh)$_3$}$^{2-}$ unit, one {Li(pmdeta}$^+$ cation coordinates to one of the NHPh nitrogens and the other bridges the two remaining nitrogens.[137] The two related complexes [{Bu^{t}Si(OSiMe$_2$N[Li]Bu^t)$_3$}$_3$] and [{Bu^{t}Si(OSiMe$_2$N[Li]SiMe$_3$)$_3$}$_3$] feature a trimetallic, six-membered ring in which the tripodal ligand coordinates to three lithium ions with average Li–N bond lengths of 1.99 and 1.98 Å, respectively.[138] Other trimeric lithium amide structures have also been reported, which feature chelating ligands.[139,140]

2.2.5 Tetrameric Lithium Amides

The most common structure observed for tetrameric lithium amides is the ladder structure C as shown in Scheme 2.2, in which edge bridged (Li$_2$N$_2$) units form a corrugated, generally linear, array. This is exemplified by the structures of [{LiNH{Si(NMe$_2$)$_3$}$_4$],[141,142] and [(LiNHPh)$_4$(py-Bu^t-4)$_6$],[80] chelated [LiNH(8-quinolyl)}$_4$(Et$_2$O)$_2$],[143] [{LiN(SiMe$_3$)(8-quinolyl)}$_4$(Et$_2$O)$_2$][143] and [{LiN(SiMe$_3$)(furfuryl)}$_4$].[144] A variation of the regular ladder structure is observed for the complex 2-Li(Me$_3$Si)NC$_6$H$_4$CH$_2$N(SiMe$_3$)Li and its thf solvate which displays an 'arched-ladder' motif probably as a result of the chelating effect to the amido ligand.[145] In contrast, the more sterically crowded [{LiNH(SiBu^{t}Me$_2$)}$_4$][146] has a Li$_4$N$_4$ cubane structure. Further increases in steric crowding can induce other structural changes. For example, the tetramer [{LiN(SiMe$_3$)Ph}$_4$][84] has the structure illustrated in **3** in which there is a dimeric

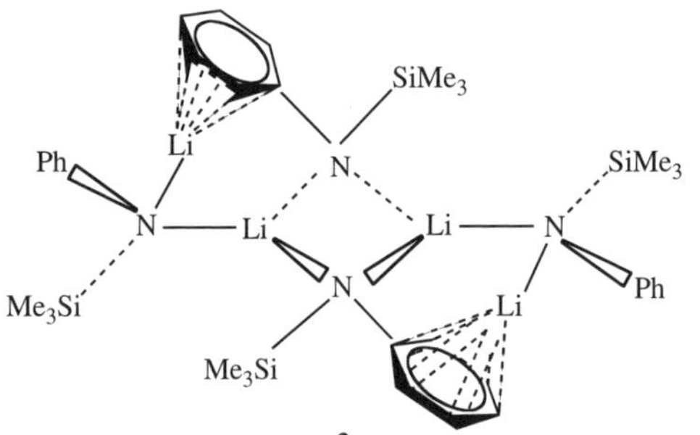

3

{LiN(SiMe$_3$)Ph}$_2$ core wherein each lithium is also bound to the nitrogen of another LiN-(SiMe$_3$)Ph unit whose lithium has η^6 by the phenyl contacts from the core {LiN(SiMe$_3$)-Ph}$_2$ dimer. The Li–N core distances are 2.057(9) and 2.183(9) Å and the 'terminal' Li–N distances are slightly shorter at 1.983(1) and 2.004(9) Å. There have also been several spectroscopic studies of lithium amide tetramers[13–15,146–151] in solution and an unusual tetrametallic amide/alkoxide adduct $[\{LiNPr^i_2\}_2\{LiOCH(CH_2NMe_2)\}_2]$ has been characterized.[149]

The pyrrolidide tetramer $[(tmeda)_2(LiN(CH_2)_3CH_2)_4]$ and hexamer $[(pmdeta)_2$-$(LiN(CH_2)_3CH_2)_6]$ have been investigated structurally, spectroscopically and by computation.[147] The tetramer has a Li_4N_4 ladder structure in which a lithium at each end of the ladder is attached to a tmeda donor. The hexamer also has an Li_4N_4 ladder structure but a lithium at each end of the ladder is bound to a further -$\bar{N}(CH_2)_3CH_2$. Calculations on model species showed that for $(LiNH_2)_4$ an 8-membered ring structure (see below) is preferred with ladder and cubane structures +7.4 and +11.4 kcal mol^{-1} less stable respectively. The ring/ladder order is reversed upon bis complexation of a base with the ladder structure $(LiNH_2)_4{\cdot}2H_2O$ being 7.2 kcal mol^{-1} more stable than the ring form. The larger sized piperidine ring also provides a tetrameric ladder structure in the form of the salt $[\{LiN(CH_2)_4CH_2{\cdot}HN(CH_2)_4CH_2\}_4]$.[152]

Uniquely, the tetrameric lithium amide $[\{Li(tmp)\}_4]$ (tmp = 2,2,6,6-tetramethyl piperidide)[153] has a planar, eight-membered $(LiN)_4$ ring with the coordinating piperidido ligands arranged around the ring alternating with the lithium atoms (Structure E). The Li–N distances within the ring are short (average 2.00 Å) with almost linear lithium geometry (average N–Li–N angle is 168.5°).

2.2.6 Higher Aggregate Lithium Amides

Treatment of 1,4–dimethylpiperazine (dmp) with lithium anilide LiNHPh gave $[(LiNHPh)_5(dmp)_2]$.[154] It has a distorted ladder structure with five Li–N rings. However, it is U-shaped rather than stepped as shown by **4** [N = NHPh; N(dmp) = 1,4-dimethylpiperidine] from which it can be seen that the bridgehead lithium is bound to

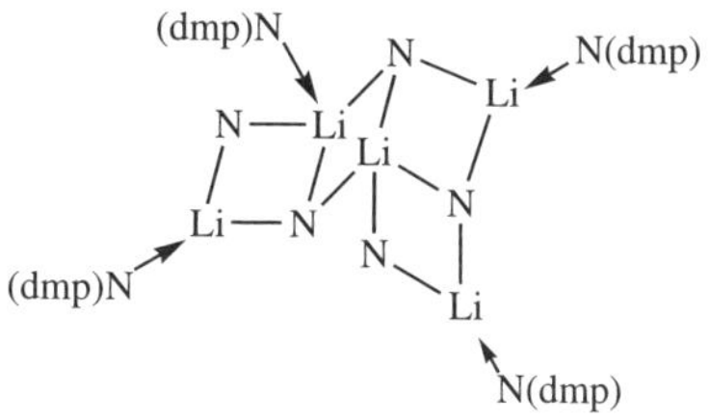

4

four anilido NHPh groups. Each pentamer is linked to four neighboring pentamers by four bridging dmp ligands to afford the 5:2 stoichiometry. Overall, the structure clearly demonstrates the importance of the neutral base in determining the structure. The stronger bases thf and py afford dimeric structures with LiNHPh.[79,80]

A number of hexameric lithium complexes are known.[103,155–158] For example, there is a lithium amide ladder motif in $[\{LiN(CH_2)_5CH_2\}_6]$[155] and the solvated $[(LiNHPh)_6$-$(thf)_8]$.[144] In contrast, the chelated $[(LiNHCH_2CH_2NMe_2)_6]$[158] has a hexagonal prismatic structure. The octameric lithium amide aggregate $[\{LiN(H)Bu^t\}_8]$[159] also exists as a ladder assembly (Li–N bond lengths between 2.03 and 2.08 Å) whereas the octameric lithium

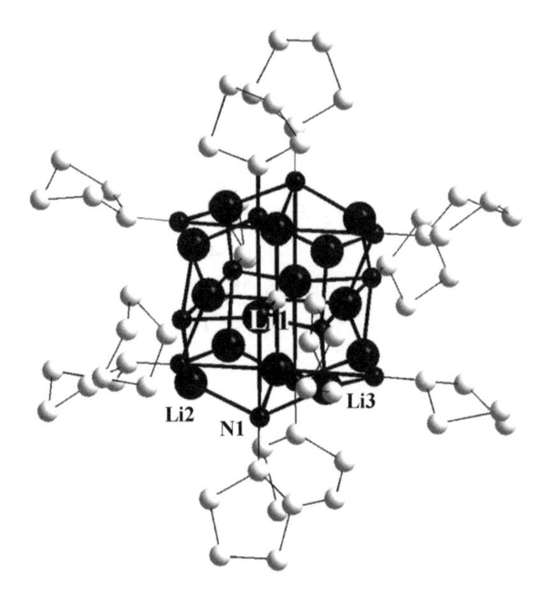

Figure 2.2 *Structure of crystalline* $[Li_{14}(O)(N(H)(C_5H_9))_{12}]$[161] *highlighting the central* $Li_{14}N_{12}O$ *metal-heteroatom unit in bold lines. Lithium atoms are shown as large black spheres, and nitrogen atoms are shown as small black spheres, the central oxygen atom and the carbon atoms are white spheres. Selected bond lengths: Li1-O1: 1.888(10) Å, Li1-N1: 2.080 (5) Å, Li2-N1: 2.062(8) Å, Li3-N1: 2.002(8) Å*

amide $[\{Li_2[N(SiMe_2(Bu^t))(2\text{-pyridylmethanidyl})]\}_4]$[160] has a large wheel-like skeleton with a diameter of approximately 50 Å and Li–N distances between 1.98 and 2.70 Å.

The largest known discrete lithium amide structure, the tetradecameric $[Li_{14}(O)(N(H)\text{-}(C_5H_9))_{12}]$ cage was synthesized in the presence of water; it has a central oxo-anion.[161] Fig. 2.2 highlights the $Li_{14}N_{12}O$ metal-heteroatom core.

Although lithium diisopropylamide (LDA), $LiNPr^i_2$, is one of the most widely used synthetic reagents because of its high deprotonating ability, its crystal structure was not solved until 1991[162] due to the difficulty in growing suitable single crystals. It represented the first polymeric lithium diorganoamide to be crystallographically characterized and comprises a helical arrangement of near-linear LiN_2 units, eight of which form a turn of a helix. It is the special steric requirements of the bulky diisopropylamido groups that prevent the formation of the common ladder-type (double stranded) framework. The ligand-ligand repulsions are minimized by the coiling of the single strands without disrupting the linearity of the $[LiN_2]$ units. LDA can be crystallized from tmeda/hexane as an infinite array of dimers linked by bridging tmeda ligands; ^6Li and ^{15}N NMR spectroscopy showed that in neat tmeda LDA is a dimer with an η^1 bound tmeda at each lithium.[163a] Another example of non-ladder polymeric chains of Structure A-type $(LiN)_2$ rings connected via bridging tmeda ligands has been observed in the complex $[Li_2(C_7H_{14}N)_2(tmeda)]_n$ (*inter*-cycle Li–N bond lengths are 2.038(3) and 2.050(3) Å, and the Li–N(tmeda) bond length is 2.161(2) Å).[163b] Similarly, the bridging mode of the bidentate tmeda ligand in the complexes $[\{Li\{N(Ph)\text{-}(SiMe_3)\}(tmeda)\}]_\infty$ and $[Li\{N(Ph)(CH_2Bu^t)\}(tmeda)]_\infty$ facilitates the formation of infinite chains of planar Structure A-type $(LiN)_2$ units.[164] The lithium ions are three-coordinate with nearly planar, distorted trigonal geometry. The *intra*-ring lithium-amido nitrogen distances (average 2.06 and 2.04 Å, respectively) are shorter than the Li–N bond lengths to the bridging tmeda groups (average 2.07 and 2.12 Å, respectively). A further example is provided by the structure of $[\{(1,4\text{-dioxane})LiN(SiMe_3)_2\}_\infty]$[66] in which 1,4-dioxane links the dimeric $\{LiN(SiMe_2)_2\}_2$ units into an infinite chain array (cf dimer structure for $[\{(1,4\text{-dioxane})_2LiN(SiMe_3)_2\}_2]$).[66]

2.2.7 Laddering

The dimeric $(LiN)_2$ structural motif (discussed in Section 2.2.3) is of central importance in a mode of lithium amide association known as 'laddering' (Structure C). The phenomenon was first described by Snaith and co-workers with the structure $[\{Li(\overline{N(CH_2)_3CH_2})\}_3\{pmdeta\}_2]$[165] (Figure 2.3). Such polymers form via lateral association of the self-assembled $(LiN)_2$ dimers (i.e. laddering), with the size and stereochemistry of the organo-substituents on the nitrogen as well as the solvating medium playing a key role in the extent of the association. It is generally accepted that a lower degree of solvation is crucial in the formation of polymeric $(LiN)_n$ ladder-type structures as opposed to isolated $(LiN)_2$ dimers.[7,10,166–170] A range of ladder structures is possible by varying the donor group, for example the corrugated assemblies $\{Li(\overline{N(CH_2)_3CH_2})\}_2(tmeda)]_2$ (Li–N bond lengths between 1.95 and 2.04 Å)[147] and $[\{Li(N(H)CH_2Ph)\}_2(thf)]_\infty$ (Li–N bond lengths between 2.05 and 2.10 Å).[166] Similar structures are observed in the polyamido compounds $[\{Li(C_6H_3Pr^i_2\text{-}2,6)NCH_2CH_2N(C_6H_3Pr^i_2\text{-}2,6)Li\}_2]$[124] and $[\{Me_2Si(LiNMes)_2\}_2]$,[124] which have average Li–N bond lengths of 2.04 and 2.01 Å, respectively. Polymeric lithium amide laddering is observed in the primary benzylamide complexes $[\{LiN(H)\text{-}CH_2Ph\}_2(H_2NCH_2Ph)\}_\infty]$[168] and $[\{LiN(H)CH_2Ph\}_2(thf)\}_\infty]$[166] which exhibit infinite,

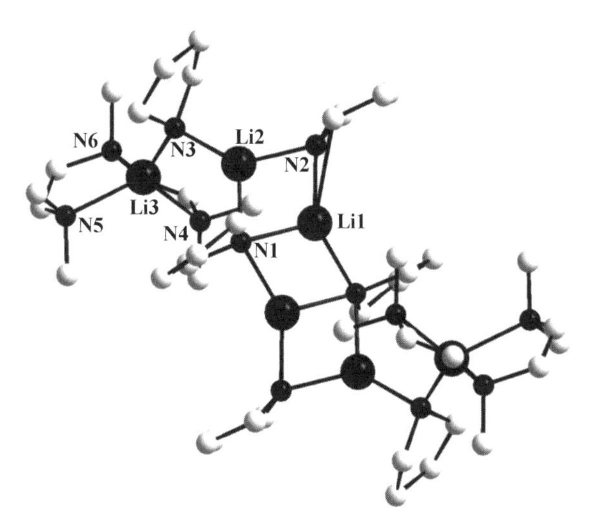

Figure 2.3 *Representation of the ladder structure of* $[\{Li(\overline{N(CH_2)_3CH_2})\}_3(pmdeta)_2]$.[165] *Selected bond lengths: Li1-N1: 1.93 Å, Li1-N2: 2.03 Å, Li2-N1: 2.12 Å, Li2-N2: 2.01 Å, Li2-N3: 2.01 Å, Li3-N3: 1.97 Å, Li3-N4: 2.19 Å, Li3-N5: 2.21 Å, Li3-N6: 2.17 Å*

twisted-ladder frameworks of fused $(LiN)_2$ rings. One edge of the ladder is solvated by donor molecules, while the other is donor-free. The chelating salt $[(LiNHCH_2CH_2NH_2)_\infty]$ is an example of a primary lithium amide with an infinite ladder structure.[170] Laddering of $LiNEt_2$ solvated by oxetane, thf or Et_2O was studied by ^6Li and ^{15}N NMR spectroscopy, identifying four-, five- and six-rung ladders.[169]

The ladder structures formed by lithium amides and their heavier group 15 analogues stand in contrast to those formed by the related lithium alkyls which generally prefer aggregates with three-dimensional or one-dimensional polymeric structures.

2.2.8 Heterometallic Derivatives

A very large number of heterometallic lithium amide complexes have been reported. Only a few examples are given here. Many of the heterometallic species will be discussed in subsequent chapters that deal with the hetero (derivatives of the other groups of the Periodic Table) metal. A listing of derivatives was published in a 1995 review.[8] The hetero species are derivatives of the parent amido ligand $-NH_2$. The earliest examples, $K_2Li(NH_2)_3$ and $RbLi(NH_2)_2$, were published in 1919 and resulted from the platinum catalysed reaction of the corresponding metals in liquid ammonia.[171] Three ternary amide systems of formulae $LiNa_2(NH_2)_3$, $Li_3Na(NH_2)_4$ and $Li_5Na(NH_2)_6$ were prepared from the reaction of metallic lithium and metallic sodium in supercritical ammonia, and some structural data have been reported.[172] Similar reactions with lithium and potassium with ammonia has led to the complexes $K_2Li(NH_2)_3$, $KLi(NH_2)_2$, $KLi_3(NH_2)_4$ and $KLi_7(NH_2)_8$,[173] the structures of which are dominated by layers of edge sharing anion-tetrahedra three-quarters occupied by lithium (Li$-$N distances are between 2.02 and 2.16 Å and Li$-$K distances are between 3.07 and 3.27 Å). Bimetallic compounds were also reported for caesium ions; the complexes $CsLi(NH_2)_2$ and $CsLi_2(NH_2)_3$ were shown to contain twisted tetrahedral chains of $[Li(NH_2)_2]^-$ ions held in hexagonal packing arrays by interactions with the caesium cations.[174] The cubic crystal structure of $Li_4BH_4(NH_2)_3$ was elucidated via a combination of high resolution synchrotron X-ray and neutron powder diffraction measurements.[175]

Interesting examples of simple substituted heterometallic species are $[(thf)M\{\mu\text{-}N(SiMe_3)_2\}_2M'(thf)_2]$ (M/M′ = Li/Na, Li/K and Na/K) which were obtained from solutions of 1:1 mixtures of the metal salts.[176] The unsolvated $[(\{LiN(SiMe_3)_3\}\{KN(SiMe_3)_2\})_\infty]$ has a polymeric, one-dimensional chain structure formed by alternating lithium and potassium centres bridged through the hexamethyldisilazides (Li$-$N = 1.935(1) Å and K$-$N = 2.86(1) Å) with linear coordination at K and almost linear coordination (N$-$Li = 176.4(3)°) at Li.[177]

An interesting crystal structure has been determined for the heterometallic complex $[\{[Bu^tN(H)]_4(Bu^tO)_4Li_4K_4{\cdot}(C_6H_6)_3\}(C_6H_6)]$.[178] It shows pairs of $(LiN)_2$ (Li$-$N between 2.00 and 2.04 Å) and $(KO)_2$ (K$-$O between 2.61 and 2.66 Å) rings fused in a 'carousel' arrangement via strong Li$-$O interactions (av. 1.87 Å, Figure 2.4).

Bimetallic complexes of lithium and group 2 metal amides (e.g. magnesium) can exhibit an 'inverse crown ether' structure when one type of amido ligand is present.[179–181] (See also Chapter 3). For example, the octagonal, mixed-metal $(LiMgN_2)_2$ ring system in $[(tmp)_4Li_2Mg_2(\mu_4\text{-}O)]$[179] contains three-coordinate lithium and magnesium atoms with bridging amido groups and at its core, square planar O^{2-} bound to mutually transoid $(Mg)_2$ and $(Li)_2$ ions. The ring is described as star-shaped, with the amido nitrogen atoms located at the apices. On the other hand, the mixed metal, mixed amido complex [{LiMg(tmp)(CH$_2$Si-

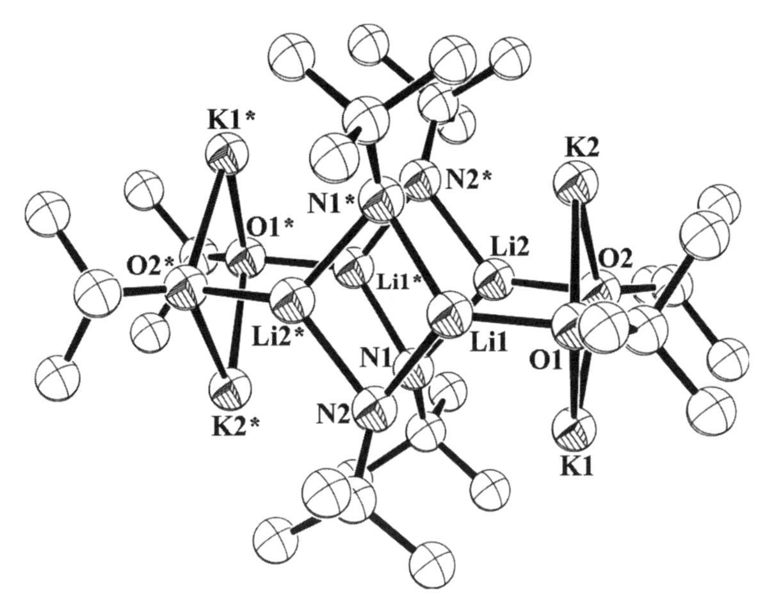

Figure 2.4 *Molecular structure of crystalline* $[\{[Bu^tN(H)]_4(Bu^tO)_4Li_4K_4\cdot(C_6H_6)_3\}(C_6H_6)]$[178]

$(Me_2)N(SiMe_3))\}_2]$[182] does not have an inverse crown ether structure, but rather a chain of five four-membered rings – a central $(MgC)_2$ core (Mg–C between 2.26 and 2.32 Å), two terminal planar (LiNMgN) rings (average Mg–N and Li–N bond lengths are 2.10 and 2.00 Å, respectively) and two bridging (MgNSiC) rings between them.

A variety of examples of discrete[183–187] and polymeric[188] lithium-group 13 metal amides have been reported (see Chapter 8). For example, addition of $LiBu^n$ to $[\{Me_2AlNH(Dipp)\}_2]$ yielded the heterometallic complex $[Me_2(Bu^t)Al(NHDipp)\{Li(thf)_3\}]$, (Li–N bond length is 2.171(21) Å).[189] A further lithium aluminium amide $[Me_2Al\{N(CH_2Ph)_2\}_2LiPy])$ was synthesized from $[(PhCH_2)_2NLi]$ and $[Me_2AlN(CH_2Ph)_2]$ in the presence of pyridine.[190] The structure is based on a central, non-planar $(LiAlN_2)$ framework. The endocyclic Li–N and Al–N bond lengths are 2.063(6) and 2.078(5) Å, and 1.904(2) and 1.934(2) Å respectively. The reaction of $2,6\text{-}Pr^i{}_2C_6H_3N(SiMe_3)Si(NH_2)_3$ with $MAlMe_4$ (M = Li, Na) in thf yielded the drum-shaped ternary metal amide-imide complexes **5** with an $M_2AlSi_2N_6$ core.[191]

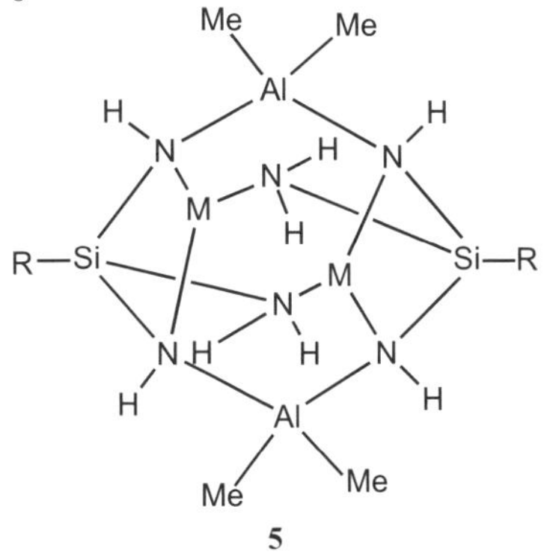

5

[M = Li, Na; R = $N(C_6H_3Pr^i{}_2\text{-}2,6)SiMe_3$]

Numerous transition metal lithium amides have also been reported, for example the reaction of $[\{(LiN(SiMe_3))_3SiMe\}_2]$ with $CpTiCl_3$ afforded $[Li\{Si(Ph)(NSiMe_3)_3Ti(Cl)\text{-}Cp\}]$,[192] which forms bridging coordinative interactions via the tridentate silicon-triamide and chloride ligands. The lithium atom is three coordinate being bound to two amido ligands (av. Li–N distance is 1.95 Å) and one (diethyl ether) solvent molecule. Unusually long Li–N interactions are found in $[Li(thf)\{N(SiMe_3)_2\}_2Mn\{N(SiMe_3)_2\}]$.[193] An interesting lithium coordination environment is shown in the trimetallic complex $[LiMn\text{-}\{MesNSiMe_2NMes\}_2MnN(SiMe_3)_2]$,[124] where the lithium ion forms a sandwich complex with the two η^6- aryl Mes groups, one of three-coordinate divalent manganese atoms is bonded to one $N(SiMe_3)_2$ ligand and two aryl $\{MesNSiMe_2NMes\}^{2-}$ ligands, and the other is bound to the nitrogen sites on these ligands. The triheterometallic lithium amide $[\{((Me_3Sn)(Me_3Ge)N)Li(OEt_2)\}_2]$ was obtained by reaction of the amine $(Me_3Sn)_2N\text{-}(GeMe_3)$ with methyllithium; it was used as a precursor for germanium(II), tin(II), and lead (II) imidocubanes possessing trimethylgermyl substituents.[194]

2.3 Sodium Amides

2.3.1 Introduction

The synthesis of sodium amide, $NaNH_2$ (or 'sodamide'), by passing ammonia over heated sodium metal, was first reported almost two centuries ago.[195] A number of studies have since been made of its properties,[196–198] but no crystal structure has been reported. Sodamide is used as a strong base in organic chemistry (often in liquid ammonia solution).[199] In contrast, sodium bis(trimethylsilyl)amide $NaN(SiMe_3)_2$ (or 'sodium hexamethyldisilazide', NaHMDS), whose crystal structure is discussed later, is widely used for deprotonation reactions or base catalysed reactions due to its solubility in a wide range of non-polar solvents. An overview of some of the types of chemical reactions in which NaHMDS is used is presented in Scheme 2.3.

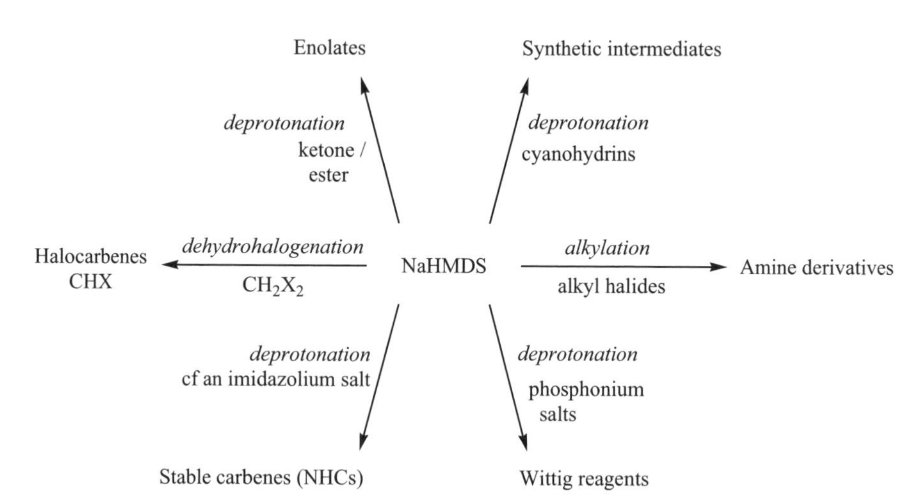

Scheme 2.3 *Illustration of some organic reactions using NaHMDS ($NaN(SiMe_3)_2$)*

A wide variety of crystal structures containing substituted sodium amides is now known. In addition, related, unusual sodium nitrogen derivatives have been reported such as $NaHCN_2$,[200] which has a three-dimensional array of hexagonally packed Na^+ and HCN_2^- layers (Na–N distances between 2.49 and 2.69 Å). The related disodium complex Na_2CN_2,[201] prepared from the reaction of $NaHCN_2$ with $NaNH_2$ at 200°C, contains linear CN_2^{2-} units coordinated to Na^+ ions in a square pyramidal geometry. The overall structure is a three-dimensional coordination network, with *inter*-sodium distances between 3.18 and 3.25 Å and Na–N bond lengths between 2.41 and 2.62 Å.

2.3.2 Monomeric and Dimeric Sodium Amides

Very little information is available for monomeric sodium amides. Addition of pmdeta to $NaN(CH_2Ph)(CHMePh)$ effects its transformation into the monomeric 1,3–diphenyl-2-azallyl salts $[(pmdeta)NaN(CHPh)_2]$ and [(pmdeta)NaN(CHPh)(CMePh)].[202] However, the sodium hydrazide salt $[(thf)_3NaN(Ph)N(SiMe_3)_2]$ crystallizes as monomers featuring four coordinate Na^+ ions.[203]

Two examples of dimeric primary amido derivatives are the $[\{(pmdeta)Na(NHPh)\}_2]$[137] and $[\{pmdeta)Na(NHC_6H_4\text{-}OPh\text{-}2)\}_2]$.[204] Structurally characterized secondary amido dimers are relatively common. In most of these two amido ligands bridge two sodiums which are usually coordinated to one or more Lewis bases. The most straightforward are derivatives of the ubiquitous $N(SiMe_3)_2$ ligand, as in $[\{(thf)NaN(SiMe_3)_2\}_2]$,[205] $[\{t\text{-}BuCNNaN(SiMe_3)_2\}_2]$,[206] and $[\{(tempo)NaN(SiMe_3)_2\}_2]$ $[\{RNCNaN(SiMe_3)_2\}_2]$ (R = 1-Ad or Bu^t)[104] or the related $[\{RNCNaN(SiMe_3)SiMe_2Ph\}_2]$.[104]

The formation of discrete frameworks is also possible via the chelation of multidentate multidentate ligands.[98,112] An example of this coordination mode is observed in the enamide $[Na\{N(H)(=CH_2)CPh\}(pmdeta)]_2$,[207] which has a planar (NaNNaN) core containing five-coordinate sodium ions with average distances of 2.41 and 2.55 Å for *endo*- and *exo*-cyclic Na–N bonds, respectively. The compounds $[Na(NPr^i{}_2)(tmeda)]_2$,[208] $[Na(NPr^iCy)\text{-}(tmeda)]_2$[208] and $[(Na\{N(SiMe_2Ph)(SiMe_3)\})_2(tmeda)]$[209] have similar dimeric rings and the chelation of tmeda or pmdeta ligands results in three or four-coordinate sodiums with average distances for *endo*-cyclic Na–N bonds of 2.44, 2.43 and 2.55 Å and 2.61, 2.60 and 2.63 Å, for *exo*-cyclic Na–N bonds. In contrast, $[\{Na\{N(Me)(PhCH_2)\}(tmeda)\}]_2$ has a buckled ring with short Na–N *endo*- (2.35–2.39 Å) and *exo*-cyclic (2.49–2.51 Å) contacts comparable to the planar structures described above.[210] The planar and folded $(NaN)_2$ frameworks of tmeda coordinated sodium amides are illustrated in Scheme 2.4.

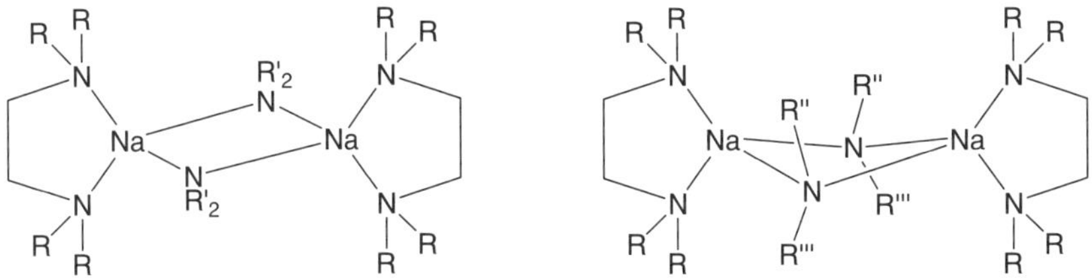

R and R''' = Me, R' = *e.g.* $\{NPr^i{}_2\}$, $\{NPr^iCy\}$ or $\{N(SiMe_2Ph)(SiMe_3)\}$, R'' = CH_2Ph

Scheme 2.4 *Examples of dimeric planar and folded $(NaN)_2$ frameworks with terminal tmeda ligands*[208,209]

Asymmetric coordination of the sodiums can also lead to planar, four-membered cyclic frameworks, such as the (NaNNaO) ring in the sodium ester enolate/sodium amide/tmeda mixed aggregate $[(C_8H_{15}O_2Na)\cdot(C_{16}H_{18}Si_2NNa)\cdot(C_6H_{16}N_2)_2]$ [obtained from $NaN(SiMe_3)_2 + Pr^iC(O)OBu^t + tmeda$],[211] wherein the planar NaONaO core has short Na–O (2.20–2.21 Å) and Na–N (2.44–2.46 Å) contacts. The sodium atoms are four-coordinate, with longer Na–N bond lengths between the cyclic sodium ions and the terminal ligands (2.50–2.58 Å). Addition of pmdeta or hmpa to phenyl (2-pyridyl) amidosodium yielded the corresponding dinuclear amido –(pmedeta) or (hmpa) adducts; in the former only 2 of the 3 nitrogen atoms of pmdeta are metal bound and in the latter the hmpa ligands bridges the two sodium ions.[212]

2.3.3 Higher Aggregate Sodium Amides

Larger ring systems have been observed, as illustrated by the trimeric complex $[\{Na(tmp)\}_3]$ (tmp = 2,2,6,6-tetramethylpiperidide),[213] the $(NaN)_3$ ring is planar with significantly different Na–N bond distances that range from 2.31 to 2.36 Å. Although the X-ray structure of $NaN(SiMe_3)_2$ was reported to be a one dimensional polymer when crystallized from mesitylene,[4] it can also be crystallized from hexane as the trimer $[\{NaN(SiMe_3)_2\}_3]$ with an almost planar six-membered $(NaN)_3$ ring (maximum deviation of 0.08 Å from the best plane).[203,214] In addition to coordination by the bridging $^-N(SiMe_3)_2$ amido ligands (Na–N bond lengths are between 2.36 and 2.40 Å), each sodium centre has close C–H contacts to two methyl groups from different neighbouring $^-N(SiMe_3)_2$ groups. An hexameric stacked structure has been observed for the hydrazide $[(NaN(H)NPh_2)_6]$.[203]

In contrast to metallation and subsequent fragmentation reactions of tmeda by organolithium compounds, which can proceed at room temperature and result in the metallation of terminal methyl groups, the remarkable high nuclearity sodium amido complexes $[Na_{10}(NMe_2)_{10}(tmeda)_4]$, $[Na_{12}(NMe_2)_{12}(tmeda)_4]$ and $[Na_{12}(NMe_2)_{10}(CH_2C_6H_4Me_4)_2(tmeda)_4]$ were formed in 30–50 % yields at room temperature by the treatment of tmeda with $NaBu^n$.[215] In this unusual synthesis, the fragmentation of the tmeda molecule (Scheme 2.5) resembles ether cleavage by organolithium reagents. These polynuclear complexes contain $(NaN)_2$ rings that are stacked into oligomeric coordination polymers.[215] Each sodium ion in each complex is four-coordinate, with Na–N bond lengths ranging from 2.35 to 2.60 Å.

The higher aggregation tendency of the sodium amides is illustrated by the fusion of the ubiquitous $(NaN)_2$ rings to form polymeric sodium amido complexes, as exemplified in the compound $[\{(Na\{N(SiMe_2Ph)(SiMe_3)\})_2\}_\infty]$.[209] The planar $(NaN)_2$ rings are connected via close Na–C contacts with phenyl and methyl substituent groups on adjacent $(NaN)_2$ rings. The structure of $[\{(1,4\text{-dioxane})_2NaN(SiMe_3)_2\}_\infty]$, like its potassium counterpart,[5] is composed of an infinite array of monomers linked by the dioxane ligands.[216] In contrast, the structure of $[\{NaN(SiHMe_2)_3\}_\infty]$[217] comprises an all-planar, infinite $(NaN)_2$ ladder

$$Me_2NCH_2\text{-}CH_2NMe_2 + NaBu^n \xrightarrow{-Bu^nH} Me_2NCH_2\text{-}CH(Na)NMe_2$$

$$\xrightarrow{\text{decomposition}} Me_2NCH_2{=}CH_2 + NaNMe_2$$

Scheme 2.5 *Decomposition of tmeda ligands to enable the formation of high nuclearity sodium amide complexes*[215]

arrangement. The Na–N bonds that form the rungs of the ladder are 2.456(2) Å in length, whereas the edges have alternating short and long Na–N bond lengths of 2.449(3) Å and 3.115(3) Å respectively. A variation on this ladder motif is observed in the infinite wave-like structure in $[\{[Na(N(H)Bu^t)]_3 \cdot H_2NBu^t\}_\infty]$ that has alternating long and short Na–N rung lengths (Na–N = 2.34 and 2.56 Å).[218]

The reaction of heavier alkali metal disilazides $MN(SiMe_3)_2$ (M = Na, K, Rb, Cs) with ferrocene yielded the unusual homologous series $[\{MN(SiMe_3)_2\}_2\{Fe(\eta^5\text{-}C_5H_5)_2\})_\infty]$. The $\{MN(SiMe_3)_2\}_2$ dimers aggregate through neutral ferrocene which interacts via η^5-cation-π interactions. The Rb and Cs compounds also display agostic M–C interactions and with toluene as the solvent, mixed toluene-ferrocene-$\{MN(SiMe_3)_2\}_2$ adducts were obtained.[219]

2.3.4 Heterometallic Sodium Amides

Several heterometallic complexes containing sodium amides have been reported. The crystal structure of $[Li_3Na(NH_2)_4]$[172] is related to the three-dimensional array seen in $LiNH_2$,[21] wherein the cations occupy half of the tetrahedral holes in an array of distorted cubic-packed amido ions. Li–N and Na–N distances are 2.066(3) Å and 2.486(4) Å respectively. The compound $[\{LiNa(N(CH_2Ph)_2)_2(OEt_2)\}_2]$[210] contains a chain of two terminal (LiNNaN) and one central $(LiN)_2$ ring, with Li–N bonds forming the internal rungs of the ladder and Na–N bonds the external rungs. The three-coordinate lithium atoms are bound to three nitrogen atoms of the ladder framework (bond lengths between 2.07 and 2.13 Å). The sodium atoms are also three-coordinate, interacting with two nitrogens within the ladder (Na–N between 2.39 and 2.54 Å) and an O atom in diethyl ether. A similar structure is observed in $[\{LiNa\{N(H)(Bu^t)\}_2(tmeda)\}_2]$,[219] with Li–N bond lengths ranging between 1.95 to 2.09 Å and Na–N distances between 2.31 and 2.57 Å. The tmeda molecules solvate the sodium atoms at the edges of the complex and prevent further polymerization of the ladder. The mixed Li/Na salt $[\{LiNa\{N(CH_2Ph)_2\}(OEt_2)\}_2]$[220] has a distorted corrugated ladder structure. There are two inner LiN and two outer NaN rungs and each sodium atom is also bound to an ether ligand.

The inclusion of a sodium hydroxide unit in the crystal structure of the tetralithium pentasodium alkoxide-amide complex $[Li_4Na_4(Bu^tO)_4\{PhN(H)\}_4(NaOH)NC_5H_4Me_4]$ (Figure 2.5),[221] forms a unique 17-vertex dome with an internally positioned OH^- ion. The dome is formed from a set of stacked ring structures, the largest of which is a basal $(NaN)_4$ macrocycle above which is a smaller $(LiO)_4$ ring and an apical sodium ion. The μ_5-oxygen atom from the hydroxide group is involved in the central $(LiO)_4$ ring and coordinates to all four lithium ions as well as the apical sodium ion.

The two puckered 12-membered $(N_6Na_4Mg_2)$ macrocycles in the compounds $[(tmp)_6Na_4Mg_2\{C_6H_3(CH_3)\}]$ (from toluene) and $[(tmp)_6Na_4Mg_2(C_6H_4)]$[222] (from benzene) host dideprotonated arene molecules, having interactions with two of the six carbon atoms in the aromatic rings via the six metals in the ring (av. Na–C distance is 2.69 Å and av. Mg–C distance is 2.19 Å). The mixed metal amides were synthesized from a 1:1:3 molar ratio of $NaBu^n$, $Mg\ Bu_2$, and the amine tmpH in the appropriate hydrocarbon solvent. Each metal ion is further coordinated by two macrocyclic nitrogen atoms in a distorted trigonal planar geometry with the metal ions being displaced toward the centre of the ring (Figure 2.6) (see also Chapter 3). Conversely, the sodium ions in the 8-membered macrocycle of $[Na_2Zn\{N(SiMe_3)_2\}_4(O)]$[223] (see also Chapter 4; obtained

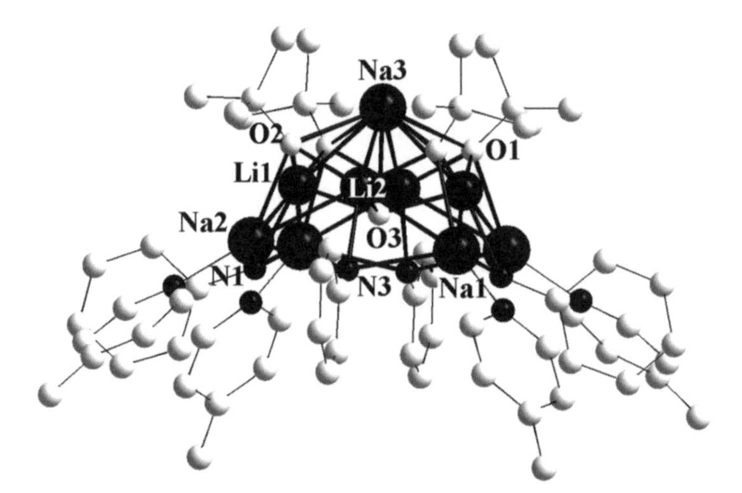

Figure 2.5 *Illustration of the dome-structure in $[Li_4Na_4(Bu^tO)_4\{PhN(H)\}_4(NaOH)NC_5H_4Me\text{-}4]$[221] highlighting the central coordination environment in bold lines. Sodium and lithium atoms are shown as large black spheres, nitrogen atoms as small black spheres, oxygen and carbon atoms as white spheres. Selected bond lengths: Na3–O1: 2.553(3) Å, Na3–O2: 2.560(3) Å, Na3-O3: 2.349(5) Å, Na2-O2: 2.408(3) Å, Na3-O2: 2.560(3) Å, Na3-O3: 2.396(4) Å, Li1-O3: 2.0187(7) Å, Li1-O1: 1.922(7) Å, Li1-O2: 1.922(8) Å, Li2-O1: 1.929(7) Å, Li2-O2 1.934(8) Å, Li2-O3: 2.011(6) Å, Na2-N1: 2.481(4) Å, Na1-N3: 2.507(4) Å, Na2-N1:2.518(4) Å, Na2-N3: 2.481(4) Å*

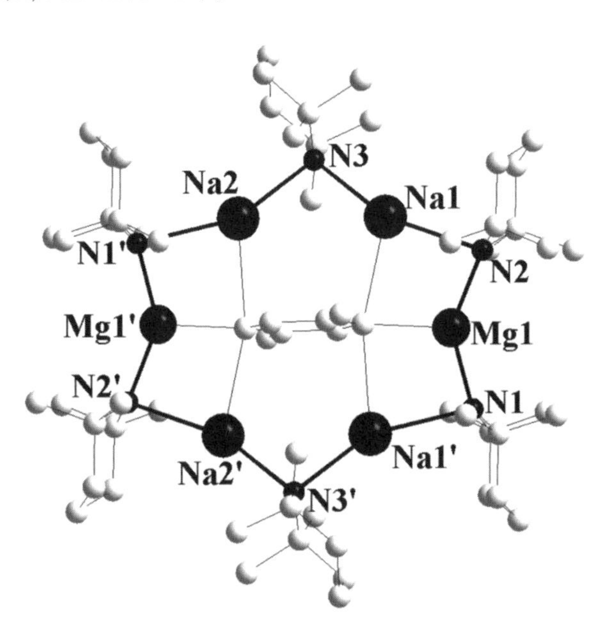

Figure 2.6 *Illustration of the macrocyclic ring structure in $[(tmp)_6Na_4Mg_2\{C_6H_3(CH_3)\}]$[222] showing guest dideprotonated toluene ligand. Sodium and magnesium atoms are shown as large black spheres, nitrogen atoms as small black spheres and carbon atoms as white spheres. Selected bond lengths: Na1-N2: 2.626(2) Å, Na1-N3: 2.393(2) Å, Na1-C28: 2.691(2) Å, Na2-N3: 2.350(2) Å, Na2-N1*: 2.596(2) Å, Na2-C28*: 2.682(2) Å, Mg1-N1: 2.048(2) Å, Mg1-N2: 2.051(2) Å, Mg1-C28: 2.200(2)*

from $NaN(SiMe_3)_2$, $ZnCl_2 + 2LiN(SiMe_3)$, are displaced away from the ring and the Zn ions are displaced towards the central oxide group. In this inverse crown ether complex has flanking organosilyl groups are attached to the corner nitrogen atoms, which coordinate to both the sodium (Na–N distances are 2.541(3) and 2.597(2) Å) and zinc (Zn–N distance is 1.984(2) Å) ions within the macrocycle.

2.4 Potassium Amides

2.4.1 Introduction

In the previous edition of this book[1] the structure of just one potassium amide, $[\{(1,4\text{-dioxane})_2KN(SiMe_3)_2\}_\infty]$, which has $KN(SiMe_3)_2$ units linked into an infinite array by 1,4-dioxane bridges to its neighbours, was known. Solvent free $KN(SiMe_3)_2$ crystallized from hexane has a centrosymmetric dimeric structure of formula $[\{KN(SiMe_3)_2\}]_2$with K–N distances of 2.770(3) and 2.803(3) Å with K_2N_2 core angles of 94.47(9)° (K) and 85.53(9)° (N).[224] When crystallized from the $C_6H_5Bu^t$ or mesitylene, the solvates $[\{(\eta^6\text{-ArH})KN(SiMe_3)_2\}_2]$ were obtained.[225]

The synthesis of the simplest potassium amide KNH_2 was first described in 1918,[198] and its crystal packing at low temperature was reported to show a deformed sodium chloride structure.[226] Recently, two studies have been conducted to investigate the hydrogen bonding in the compounds $KNH_2{\cdot}NH_3$[227] and $KNH_2{\cdot}2NH_3$.[228] The synthesis of several simple salts related to amides have been described, e.g. $KHCN_2$, obtained from the reaction of melamine $(C_3N_3(NH_2)_3)$ with KNH_2 in liquid ammonia,[229] and $K_5H(CN_2)_3$[230] and K_2CN_2,[231] synthesized by the reaction of $KHCN_2$ with metallic potassium or KNH_2 in liquid ammonia.[230,231] The K–N bond lengths in these complexes range between 2.81–3.49, 2.75–2.91 and 2.76–3.35 Å, respectively.

2.4.2 Potassium Parent Amides ($^-NH_2$ as Ligand)

A number of mixed metal systems containing non-substituted potassium amides are known. For example, the reaction of metallic potassium and lithium with ammonia yielded $K_2Li(NH_2)_3$,[232] $KLi_3(NH_2)_4$ and $KLi_7(NH_2)_8$,[174] which contain edge-sharing tetraamido-lithium tetrahedra. The K^+ ions in $K_2Li(NH_2)_3$ connect the tetrahedra into a three-dimensional network by coordination with nitrogen atoms (Li–N = 2.05–2.37 Å and K–N = 2.83–3.18 Å). A similar arrangement is seen in the compound $K_2Na(NH_2)_3$,[232] with almost identical K–N (2.85–3.11 Å) and shorter Na–N (2.33–2.53 Å) distances. The potassium-poor compounds $KLi_3(NH_2)_4$ and $KLi_7(NH_2)_8$ have square prismatic (K–N 3.07–4.03 Å, Li–N 2.02–4.18 Å) and eightfold (K–N 3.08–4.09 Å, Li–N 2.02–2.28 Å) coordination respectively.

The reaction of KNH_2 with metal oxides led to heterometallic potassium amides. For example, treatment of KNH_2 with GeO_2 gave $K_3GeO_3NH_2$ and $K_3GeO_3NH_2{\cdot}KNH_2$.[233] The germanium atoms have tetrahedral coordination and the tetrahedra are linked via hydrogen bonding into infinite, one-dimensional chains, which are further connected into a three-dimensional network by coordination of the potassium ions. The potassium ions have coordination numbers of either six or seven, with contacts to one (K–N = is 3.410 Å) and three (average K–N = 3.30 Å) nitrogen atoms, respectively. The function of potassium ions in connecting polyhedral units is also observed in $[KSn(NH_2)_3]$,[234] formed from KNH_2 with

$SnBr_2$ in liquid ammonia. The anionic trigonal pyramidal $[Sn(NH_2)_3]^-$ units are linked via interactions with the K^+ cations into a strongly distorted hexagonally close-packed array (K–N = 2.907(7) and 2.949(8) Å, and Sn–N = is 2.154(7) Å).

2.4.3 Potassium Primary and Secondary Amides

Potassium amides (NHR or NRR′) are generally produced via one of three synthetic routes: (a) transmetallation of lithium amides with $KOBu^t$, (b) reaction of $KN(SiMe_3)_2$ or (c) reaction of K, KH, KBu or KBz with an amine.

A large range of potassium amido complexes have been synthesized using KCH_2Ph (formed from the reaction of $KOBu^t$, $LiBu^n$ and toluene in hexane at −78°C). One example is the dimeric $[(K\{$*trans*$\text{-}N(SiMe_2Ph)(SiMe_3)\})_2]$, which has a planar $(KN)_2$ core and two-coordinate potassium ions (K–N bond lengths are 2.78 and 2.81 Å); additional interactions are between the potassium ions and the phenyl and methyl substituents (K–C distances are between 3.36 and 3.59 Å).[209] The tetrameric structure $[(K\{$*trans*$\text{-}N(Bu^t)(SiMe_3)\})_4(\eta\text{-}C_6H_6)_2]$[209] crystallizes as a ladder structure comprised of three rhomboidal planes; the potassium ions forming the innermost $(KN)_2$ ring are three coordinate (av. K–N bond length 2.73 Å) and the terminal potassium ions are coordinated by two amido ligands (av. K–N distances are 2.73 and 2.93 Å) and one η-phenyl group A. The compound $[(K\{$*trans*$\text{-}N(Ph)(2\text{-}C_5H_4N)\})_2(thf)_3]$ has a polymeric chain structure,[209] in which the potassium ions are connected via alternating bridging units of tetrahydrofuran solvent molecules (average K–O distance is 2.79 Å) and amido ligands (average K–N distance is 2.85 Å).[209] The three-dimensional polymeric complex $[(K\{N(2\text{-}C_5H_4N)_2\})_\infty]$ has a layered structure with six-coordinate potassium ions (K–N bond lengths are between 2.89 and 3.07 Å). Each $[N(2\text{-}C_5H_4N]^-$ ligand is hexadentate, with four nitrogen atoms from the μ^2-bridging pyridine atoms and two from the μ^2-amido groups.

The reaction of $KN(SiMe_3)_2$ with the silylene $Si[(NCH_2Bu^t)_2C_6H_6\text{-}1,2]$ in thf gave $[K\{N(SiMe_3)(Si[(NCH_2Bu)_2C_6H_4\text{-}1,2]SiMe_3)\}]$ has a single K–N bond (2.694(14) Å) and close K···C contacts with one methyl carbon (3.268(2) Å) and the $\eta^3\text{-}C_6H_4$ group (mean K–C distance is 3.17 Å).[235]

Metallation of a secondary amine with either potassium or potassium hydride has been used for the formation of $[\{K(\mu\text{-}$*trans*$\text{-}N(SiMe_2Ph)(SiMe_3))\}_2]_\infty$, $[(K\{\mu\text{-}$*trans*$\text{-}N(Bu^t)\text{-}(SiMe_3)\})_4(\eta\text{-}C_6H_6)_2]$,[209] $[(K\{\mu\text{-}$*trans*$\text{-}N(Ph)(C_5H_4N)\})_2(\mu\text{-}thf)_3\}$ and $[K\{\mu\text{-}N(C_5H_4N\text{-}2)\}]_\infty$.[209] The potassium atoms in the dimeric toluene solvate $[\{K(\mu\text{-}N(SiMe_3)_2{\cdot}PhMe\}_2]$, prepared by refluxing hexamethyldisilazane with potassium hydride in toluene followed by recrystallization from toluene, are two-coordinate K–N–K′ = 85.8°, N–K–N′ 94.2°, mean K–N bond length, 2.78 Å.[236] The structurally characterized complex $[(thf)\text{-}\{KN(SiMe_2H)(SiMe_3)\}_2]$ has one potassium bound to thf and the other interacts with the Si–H moiety.[237] The reaction of KH with carbazole HNC_8H_{10} afforded the carbazolate salt $[\{KNC_8H_{10}\}_\infty]$ which has an infinite helical structure.[238]

2.4.4 Heterometallic Potassium Amides

The triheterometallic compound $[\{(PhHN)_2(Bu^tO)LiNaK(tmeda)_2\}_2]$,[239] containing a 12-vertex $Li_2Na_2K_2N_4O_2$ cage structure (Figure 2.7), was synthesized in 78 % yield from lithium anilide, $NaOBu^t$, $KOBu^t$ and tmeda in hexane under ultrasonication.[235]

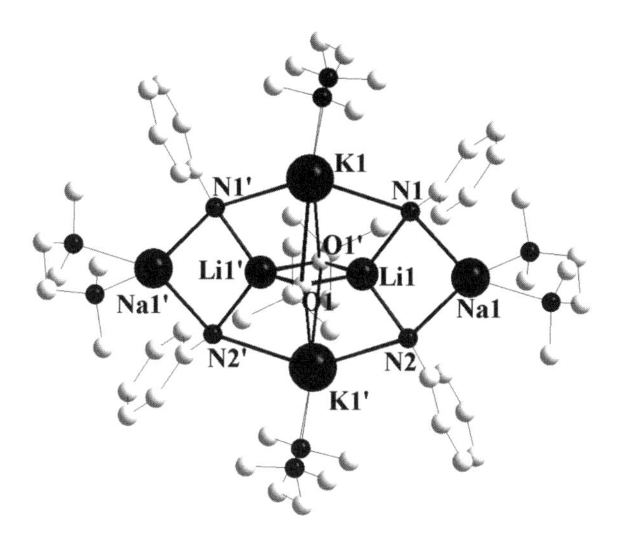

Figure 2.7 *Illustration of* $[\{(PhN(H))_2(Bu^tO)LiNaK(tmeda)_2\}_2]$[239] *highlighting the central* $Li_2Na_2K_2N_4O_2$ *cage. Potassium, sodium and lithium atoms are shown as large black spheres, nitrogen atoms as small black spheres, oxygen and carbon atoms are white spheres. Selected bond lengths: K1-O1: 2.925(2) Å, av. K-N: 2.93 Å, av. Na-N: 2.46 Å, Li1-O1: 1.949(5) Å, Li-N: 2.148(5) Å*

The inverse crown ether-type structure described previously for bis(alkali metal) compounds has also been observed for $[\{K_2Zn_2(N(SiMe_3)_2)_4(O_2)_x(O)_y\}_\infty]$,[223] formed from $KN(SiMe_3)_2$ with $ZnCl_2$ in the presence of $LiN(SiMe_3)_2$. The potassium and zinc ions form an octagonal framework with bridging amido ligands, and a third interaction with the oxide or peroxide guest molecule. The zinc ions are displaced towards the centre of the ring with and average Zn–N distance of 1.963(1) Å, whereas the potassium ions are displaced from the plane of the ring and form one short and one long interaction with the bridging amido nitrogens (2.797(1) and 2.838(1) Å). Each potassium atom forms a further interaction with a methylene group (K–C = 3.324(2) Å) to connect the rings into a linear polymer.

The complex $[\{((Me_3Si)_2N)_4K_2Mg_2(O_2)\}]$[240] obtained from KBu, Mg Bu_2 and $(SiMe_3)_2NH$ is similar to the related heterobimetallic compounds $[\{(Me_3Si)N\}_4Li_2Mg_2(O_2)_x(O)_y]$,[180] $[(tmp)_4Li_2Mg_2(O)]$[179] and $[\{(Me_3Si)_2N\}_4Na_2Mg_2(O_2)_x(O)_y]$[179] in that it is based upon an eight-membered $(MNMgN)_2$ ring system. But unlike the dilithium and disodium derivatives, the potassium complex polymerizes via $K–CH_3SiMe_2$ interactions to form chains of $(N_4K_2Mg_2)^{2+}$ octagonal rings, each surrounding a peroxo $(O_2)^{2-}$ core anion. The N–K–N and N–Mg–N bond angles are non-linear, with the potassium ions being displaced away, and the magnesium ions toward, the central peroxo group. The K–N bond lengths lie between 2.80 and 2.85 Å, and the average Mg–N distance is 2.04 Å.

Further examples of a cyclic heterometallic potassium amide, obtained from KBu, $MgBu_2$ and tmpH in C_6H_6 or PhMe, are the isostructural complexes $[(tmp)_{12}K_6Mg_6(C_6H_5)_6]$ (Figure 2.8) and $[(tmp)_{12}K_6Mg_6(C_6H_4CH_3)_6]$ comprised of 24-membered $N_{12}K_6Mg_6$ macrocycles that play host to six mono-deprotonated arene molecules.[241]

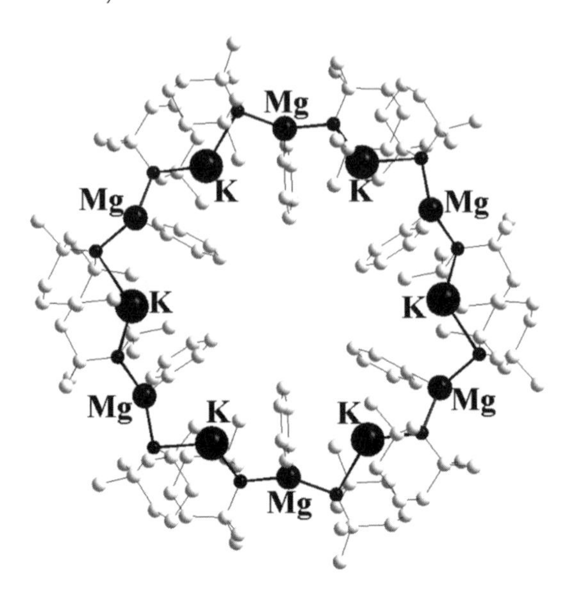

Figure 2.8 Molecular unit of $[(tmp)_{12}K_6Mg_6(C_6H_5)_6]$,[241] highlighting the $(KNMgN)_6$ macrocycle in bold lines. Potassium and magnesium atoms are shown as large black spheres, nitrogen atoms as small black spheres and carbon atoms as white spheres. Average Mg-C_{ipso} 2.19 Å, K···$C_6H_5(\eta^6)$ 3.35Å, Mg-N: 2.02 Å and K-N: 3.10 Å

The reaction of KCH_2Ph with $Me_3Si(Bu^t)NH$ in the presence of $[Mg\{N(SiMe_3)Bu^t\}_2]$ and $LiBu^n$ results in the trimetallic alkali metal amido complex $[Li_2K_2Mg_4\{N(Bu^t)SiMe_3\}_4\{N(Bu^t)Si(CH_2)Me_2\}_4]$,[242] which comprises a 16-membered $(KNMgNLiNMgN)_2$ macrocycle with four exocyclic, six-membered (MgCSiNLiN) rings. The shortest endocyclic K–K distance is 6.15 Å. The two-coordinate lithium (mean Li–N distance is 1.99 Å) and potassium (average K–N distance is 2.95 Å) ions are near-linear (average bond angles are 179.15° and 170.61° respectively), whereas the magnesium ions (mean Mg–N distance is 2.05 Å) have distorted trigonal planar geometries with wide N–Mg–N (mean 135.8°), intermediate C–Mg–N(K) (av. 115.58°) and narrow C–Mg–N(Li) (average 108.15°) bond angles. Each metal ion also forms two short contacts to aniomic $C(H)_2^-$ atoms (Li–C, K–C and Mg–C distances are 2.45–2.49 Å, 3.29–3.32 Å, and 2.20–2.21 Å respectively).

2.5 Rubidium Amides

Relatively few examples of rubidium amides have been reported. The simplest $RbNH_2$ ('rubidamide'), first characterized in 1965, was prepared from rubidium metal and gaseous ammonia in a nickel crucible.[226] The crystal structure showed it to have a deformed sodium chloride structure with the rubidium and nitrogen atoms in distorted octahedral environments lying along two mirror planes parallel to the plane of the crystallographic *a* and *c* axes (Rb–N = 3.14 and 3.36 Å). The deuterated analogue, $RbND_2$, has a similar structure with shorter Rb–N contacts (2.92–3.51 Å).[243]

Similarly, the reaction of metallic rubidium with another metal in ammonia led to the appropriate heterobimetallic amides, as in the reaction of Rb with Ca in supercritical ammonia at 573 K yielding $RbCa(NH_2)_3$,[244] which consists of one-dimensional infinite face-sharing anion-octahedra occupied by calcium and connected into a three-dimensional network by the rubidium ions (Rb–N contacts between 3.11 and 3.92 Å).

A more common synthesis of heterobimetallic rubidium amides involved the reaction of $RbNH_2$ with a metal or a metal complex. For example, $RbLi(NH_2)_2$, first reported in 1919[171] but not characterized until 2002,[245] was made from $RbNH_2$ and lithium metal in supercritical ammonia at 470 K. The structure comprises tetrahedra of amido ions occupied by half a lithium atom (Li–N contacts between 2.06 and 2.17 Å), resulting in zig-zag chains by sharing two vertices and two edges with four surrounding tetrahedra. The rubidium ions in the lattice are surrounded by eight nitrogen atoms (Rb–N distances between 2.95 and 3.45 Å), one lithium atom and ten hydrogen atoms, which form an unexpected hydrogen bridge bonding system between crystallographically independent amide ions.[245]

The reaction of $RbNH_2$ with $Cr(NH_2)_3$ in ammonia under a pressure of 3 kbar for 90 days gave $[Rb_4Cr_2(NH_2)_{10}]$,[246] which contains isolated, face-sharing N-octahedra around two Cr^{3+} ions. These $[Cr(NH_2)_3(NH_2)_{3/2}]_2{}^{3-}$ units form a hexagonal array, the connecting rubidium ions occupy 'tetrahedral' or 'octahedral' holes with contacts to ten amido groups ranging in length between 2.84 and 4.00 Å. Room temperature syntheses are also possible, as in the reaction between $RbNH_2$ and $Ni(SCN)_2{\cdot}4NH_3$ yielding $[Rb_6Ni_3(NH_2)_{12}{\cdot}(RbCN)_{1.0}{\cdot}(RbNH_2)_{0.3}]$.[247] The square planar $[Ni(NH_2)_4]^{2-}$ complex ions form positively charged cages that are partially occupied by CN^- ions. The rubidium cations cap the hexagonal prismatic proton cages with Rb–N distances between 2.97 and 3.55 Å.

The direct reaction for several days of rubidium metal with an excess of $HN(SiMe_3)_2$ and recrystallization from dioxane yielded $[\{RbN(SiMe_3)_2\}_2(1,4\text{-dioxane})_3]_\infty$.[216] The solvent-free $[\{Rb(N(SiMe_3)_2)\}_2]$ was obtained by recrystallization from toluene. It has Rb–N distances of 2.878(2) and 2.956(2) Å, with endocyclic angles of 92.0(1) (Rb) and 88.0(1)° (N).[248] The rubidium carbazolate $[(RbNC_{12}H_8)_\infty]$ has a polymeric structure very similar to that of its potassium analogue.[238] Treatment of $RbNC_{12}H_8$, prepared by metallation of $HNC_{12}H_8$ with $RbN(SiMe_3)_2$, with a crown ether yielded the dimer $[\{Rb(\mu\text{-}NC_{12}H_8)(15\text{-crown-}5)\}_2]$.[249]

Substituted rubidium amides can also be formed by the reaction of the $RbCH_2Ph$ with an amine, as for the dimeric complex [*trans*-$(Rb\{\mu\text{-}N(SiMe_2Ph)(SiMe_3)\})_2]_\infty$[209] which comprises a planar $(RbN)_2$ framework with Rb–N bond lengths ranging between 2.89 and 3.06 Å; the rubidium ions are also involved in an η^6-bonding mode with the phenyl substituent of the adjacent $(RbN)_2$ moiety (Rb–C contacts are between 3.34 and 3.72 Å). The rubidium ions in the compound $[Rb_8\{P_4N_6(NH)_4]\}(NH_2)_2]$ are integral to the crystal lattice as they connect the adamantane-like $[P_4N_6(NH)_4]^{6-}$ anions into a distorted hexagonal array.[250] The octameric rubidium complex is formed from the reaction of $6RbNH_2$ with P_3N_5 at 400°C, and contains six different Rb environments: three distorted octahedral, two distorted square pyramidal and one distorted trigonal prismatic geometries, with Rb–N distances between 2.79 and 3.44 Å.

Like its sodium and potassium analogues, $RbN(SiMe_3)_2$ when reacted with ferrocene forms an infinite array of formula $[(\{RbN(SiMe_3)_2\}_2\{Fe(\eta^5\text{-}C_5H_5)_2\})_\infty]$. Its structure resembles those of its sodium and potassium counterparts in that the $\{RbN(SiMe_3)_2\}_2$

dimers interact in the same way with ferrocene, although both the rubidium and caesium complexes also possess inter-chain M–C agostic interactions.[219]

2.6 Caesium Amides

$CsNH_2$ has been known since 1905,[251] its ammonia complex $CsNH_2{\cdot}NH_3$[227] contains a distorted hexagonal close packed arrangement of caesium ions with ammonia molecules on octahedral sites and amido anions on trigonal bipyramidal interstitial positions. The caesium ions are surrounded by ten crystallographically independent nitrogen atoms (Cs–N distances between 3.09 and 4.39 Å). The NH_3 groups form a hydrogen-bonded network via the three hydrogen atoms bridging two amido ions (the $(NH_3){\cdots}N(NH_2)$ distance is 3.19 Å) and a further $(N(NH_3){\cdots}N(NH_3)$ distance of 3.56 Å). Thermodynamic investigations have also been carried out on this compound.[252] Azido derivatives of caesium amide are known, the complexes $Cs_2(NH_2)N_3$[253] and $Cs_5(NH_2)_4N_3$[254] contain both trigonal and tetragonal prisms connected into a three-dimensional network (Cs–N distances are 3.17–4.16 and 3.08–3.99 Å, respectively). The reaction of metallic caesium with barium in ammonia at 473 K (5 kbar) yielded $CsBa(NH_2)_3$,[255] which comprises strings of anionic, edge-sharing double octahedra containing barium ions and connected by caesium ions into a three-dimensional network. The Cs–N and Ba–N distances lie between 3.40 and 4.27 Å, and 2.69 and 2.94 Å.

Reaction of metallic caesium with $HN(SiMe_3)_2$ in an ammonia/$Fe(NO_3)_3{\cdot}9H_2O$ mixture yielded the tetrameric caesium amide $[\{CsNH(SiMe_3)\}_4]$.[256] Desilylation of the amide may be due to the presence of the hydrated iron salt. The complex has a heterocubane $(CsN)_4$ framework where the caesium ions have the unusual coordination number of three. The Cs–N distances are short 2.915(5) Å and are only 0.13 Å longer than an average K–N distance. In contrast, the reaction of metallic caesium with an excess of $HN(SiMe_3)_2$ and recrystallization from dioxane afforded $[\{CsN(SiMe_3)_2\}_2(1,4\text{-dioxane})_3]_\infty$.[216] The complex is isostructural with the rubidium analogue,[216] comprising an almost planar $(CsN)_2$ ring with Cs–N bond lengths of 3.067(1) and 3.388(2) Å. The direct reaction of $HN(SiMe_3)_2$ with caesium and crystallization from toluene afforded either the centrosymmetric dimer $[\{CsN(SiMe_3)_2\}_2]$ or its toluene solvate $[\{CsN(SiMe_3)_2\}_2{\cdot}PhMe]$.[248] In these, the Cs–N distances are in the range 3.016(3)–3.149(2) Å. In contrast to its K and Rb analogues the caesium carbazolate $[(CsNC_{12}H_8)_2]$ is dimeric with two Cs^+ ions sandwiched between the six-membered rings of the carbazolato ligands.[242] Reaction of $CsNC_{12}H_8$, prepared by metallation of $HNC_{12}H_8$ with $CsN(SiMe_3)_2$, with a crown ether yielded the dimer $[\{Cs(\mu\text{-}NC_{12}H_8)(18\text{-crown-}6)\}_2]$ or the salt $[Cs(15\text{-crown-}5)_2]$-$[NC_{12}H_8]$.[249]

Another method of making substituted caesium amides is by use of the parent amide. For example, $CsNH_2$ with P_3N_5 at 673 K gave $[Cs_5\{P(NH)_4\}(NH_2)_2]$.[257] The structure involves cubic close packing with caesium ions lying in apices with alternating layers of centrally located $[P(NH)_4]^{3-}$ (Cs–P distances lie between 3.69 and 4.77 Å) and $[NH_2]^-$ ions (Cs–N distances lie between 3.13 and 4.36 Å). The two crystallographically independent Cs ions are in a distorted square antiprism and a distorted octahedron corresponding to the structural formula $Cs_3[P(NH)_4]{\cdot}2CsNH_2$.

The reaction of $CsNH_2$ with other metal complexes yielded heterometallic complexes, such as $Cs_2Ni(NH_2)_4{\cdot}NH_3$.[258] Two nickel atoms and four caesium atoms form an octahedral unit (Cs_4 in the tetragonal plane with two apical Ni atoms). The bridging amido groups connect caesium and nickel atoms (Cs–N distances between 3.18 and 3.82 Å, Ni–N distances between 1.92 and 1.94 Å); the ammonia groups bridge caesium ions (Cs–N contacts 3.31 and 3.61 Å). The octahedral units are connected into a three-dimensional network through the ammonia groups (Cs–N contacts 3.41 and 3.76 Å). Similarly, the reaction of $CsNH_2$ with tetraphenyltin in liquid ammonia gave $Cs_2[Sn(NH_2)_6]$,[259] which contains isolated octahedral $[Sn(NH_2)_6]^{2-}$ units with the caesium ions coordinated anti-cuboctahedrally by 12 amide ions (Cs–N bond lengths between 3.19 and 3.74 Å). The anti-cuboctahedron are connected into a three-dimensional lattice by octahedral tin atoms.

Substituted caesium amides may be obtained from the reaction $CsCH_2Ph$ with an amine, as seen for $[\{CsN(SiMe_2Ph)(SiMe_3)\}_2]$,[209] which has two independent dimers in the unit cell, the Cs atom in each having a different coordination environment. Similarly, $CsCH_2Ph$, generated *in situ*, reacted with octamethylcyclotetrasilazane yielding the doubly metallated compound $[Cs_2\{NSi(Me)_2N(H)SiMe_2\}_2(thf)_3]_\infty$.[260]

The reaction of $[Li\{N(SiMe_3)C_5H_4N\text{-}2\}(12\text{-crown-4})]$ with $Cs(OC_8H_{17})$ at −78°C yielded the dimeric $[Cs\{\mu\text{-}\kappa^2{:}\kappa^2\text{-}N(SiMe_3)C_5H_4N\text{-}2\}(12\text{-crown-4})]_2$ with each of the four nitrogen atoms of the bridging 2-pyridylamido ligands coordinating to both caesium ions (Cs–N between 3.09 and 3.25 Å). Two 12-crown-4 macrocyclic polyether units cap the ends of the dimer with Cs–O bond lengths between 3.09 and 3.52 Å. The structure is similar to that found for the potassium and rubidium analogues.[261] An ambient temperature route led to the N–Si bond cleavage, yielding the polymeric complex $[\{Cs(PyNH)(12\text{-crown-4})\}_\infty]$[259] with Cs–N bond lengths between 3.09 and 3.25 Å and Cs–O distances between 3.06 and 3.25 Å. The complex $Cs[N(SO_2F)_2]$ contains heterotopic Cs^+ coordination by N, O and F atoms.[262] The anion is surrounded by seven cations and takes part in eleven Cs–X bonding interactions (Cs–N 3.21 Å, Cs–O between 3.08 and 3.75 Å, Cs–F between 3.37 and 3.69 Å).

References

1. M. F. Lappert, P. P. Power, A. R. Sanger and R. C. Srivastava, *Metal and Metalloid Amides: Synthesis, Structures, and Physical and Chemical Properties;* Horwood-Wiley: Chichester, 1980.
2. D. Mootz, A. Zinnius B. Böttcher, *Angew. Chem., Int. Ed.*, 1969, **8**, 378.
3. R. D. Rogers, J. L. Atwood and R. Grüning, *J. Organomet. Chem.*, 1978, **157**, 229.
4. R. Grüning and J. L. Atwood, *J. Organomet. Chem.*, 1977, **137**, 101.
5. W. N. Setzer and P. v. R. Schleyer, *Adv. Organomet. Chem.*, 1985, **24**, 353.
6. K. Gregory, P. v. R. Schleyer and R. Snaith, *Adv. Inorg. Chem.*, 1991, **37**, 47.
7. R. E. Mulvey, *Chem. Soc. Rev.*, 1991, **20**, 167.
8. F. Pauer and P. P. Power, Structures of lithium salts of heteroatom compounds in lithium chemistry; in *Lithium Chemistry: A Theoretical and Experimental Overview*, eds. A.-M. Sapse and P. v. R. Schleyer, Wiley: New York, 1995; Chapter 9, pp. 295–392.
9. D. S. Wright and M. A. Beswick, Alkali metals; in *Comprehensive Organometallic Chemistry II*, eds. E. W. Abel, F. G. A. Stone and G. Wilkinson, Pergamon, Oxford, 1995, Volume 1 (ed. C. E. Housecroft), Chapter 1.

10. R. E. Mulvey, *Chem. Soc. Rev.*, 1998, **27**, 339.
11. T. P. Hanusa, Group 1s and 2s metals; in *Comprehensive Coordination Chemistry II*, eds. J. A. McCleverty and T. J. Meyer, Elsevier, Amsterdam, 2004, Volume 3 (ed. G. F. R. Parkin), Chapter 3.1.
12. K. Ruhlandt-Senge, K. W. Henderson and P. C. Andrews, *Alkali Metal Organometallics – Structure and Bonding; in Comprehensive Organometallic Chemistry III*, eds; R. H. Crabtree and D. M. P. Mingos, Elsevier, Amsterdam, 2007, Volume 2 (ed. K. Meyer), Chapter 2.01.
13. D. B. Collum, *Acc. Chem. Res.*, 1993, **26**, 227.
14. B. L. Lucht and D. B. Collum, *Acc. Chem. Res.*, 1999, **32**, 1035.
15. D. B. Collum, A. J. McNeil and A. Ramirez, *Angew. Chem., Int. Ed.*, 2007, **46**, 3002.
16. J. Eames, *Science of Synthesis*, 2006, **8a**, 173.
17. A. Jonczyk and A. Kowalkowska, *Science of Synthesis*, 2006, **8b**, 1141.
18. P. Venturello and M. Barbero, *Science of Synthesis*, 2006, **8b**, 1399.
19. M. Majewski and D. M. Gleave, *J. Organomet. Chem.*, 1994, **470**, 1.
20. A. W. Titherley, *J. Chem. Soc.*, 1894, **65**, 504.
21. H. Jacobs and R. Juza, *Z. Anorg. Allg. Chem.*, 1972, **391**, 271.
22. J. B. Yang, X. D. Zhou, Q. Cai, W. J. James and W. B. Yelon, *Appl. Phys. Lett.*, 2006, **88**, 041914.
23. M. H. Sorby, Y. Nakamura, H. W. Brinks, T. Ichikawa, S. Hino, H. Fujii, and B. C. Hauback, *J. Alloys Comp.*, 2007, **428**, 297.
24. H. Barlage and H. Jacobs, *Z. Anorg. Allg. Chem.*, 1994, **620**, 479.
25. D. B. Grotjahn, P. M. Sheridan, I. Al Jihad and L. M. Ziurys, *J. Am. Chem. Soc.*, 2001, **123**, 5489.
26. A. M. Sapse, E. Kaufmann, P. v. R. Schleyer and R. Gleiter, *Inorg. Chem.*, 1984, **23**, 1569.
27. E.-U. Würthwein, K. D. Sen, J.A. Pople and P. v. R. Schleyer, *Inorg. Chem.*, 1983, **22**, 496.
28. D. R. Armstrong, P. G. Perkins and G. T. Walker, *J. Mol. Struct. (THEOCHEM)*, 1985, **122**, 189.
29. P. Burk and I. Koppel, *Int. J. Quantum Chem.*, 1994, **51**, 313.
30. E. Kaufmann, T. Clark and P. v. R. Schleyer, *J. Am. Chem. Soc.*, 1984, **106**, 1856.
31. A. M. Sapse, K. Raghavachari, P. v. R. Schleyer and E. Kaufmann, *J. Am. Chem. Soc.*, 1985, **107**, 6483.
32. K. Raghavachari, A. M. Sapse and D. C. Jain, *Inorg. Chem.*, 1987, **26**, 2585.
33. K. Raghavachari, A. M. Sapse and D. C. Jain, *Inorg. Chem.*, 1988, **27**, 3862.
34. A. J. Mountford, W. Clegg, S. J. Coles, R. W. Harrington, P. N. Norton, S. M. Humphrey, M. B. Hunsthorpe, J. A. Wright and S. J. Lancaster, *Chem. Eur. J.*, 2007, **13**, 4535.
35. T. Markmaitree, R. Ren and L. L. Shaw, *J. Phys. Chem. B*, 2006, **110**, 20710.
36. Y. Song and Z. X. Guo, *Phys. Rev. B: Condens. Matter*, 2006, **74**, 195120.
37. B. Magyari-Kope, V. Ozolins and C. Wolverton, *Phys. Rev. B: Condens. Matter*, 2006, **73**, 220101.
38. Y. H. Hu and E. Ruckenstein, *Ind. Eng. Chem. Res.*, 2006, **45**, 4993.
39. H. Y. Leng, T. Ichikawa, S. Hino, N. Hanada, S. Isobe and H. Fujii, *J. Power Sources*, 2006, **156**, 166.
40. P. Chen, Z. Xiong, J. Luo, J. Lin and K. L. Tan, *Nature*, 2002, **420**, 302.
41. P. Chen, Z. Xiong, J. Luo, J. Lin and K. L. Tan, *J. Phys. Chem. B*, 2003, **107**, 10967.
42. Y. Kojima and Y. Kawai, *Chem. Commun.*, 2004, **19**, 2210.
43. W. F. Luo, *J. Alloys Compd.*, 2004, **381**, 284.
44. Z. Xiong, G. Wu, J. Hu, P. Chen, W. Luo and J. Wang, *J. Alloys Compd.*, 2006, **417**, 190.
45. Y. Liu, Z. Xiong, J. Hu, G. Wu, P. Chen, K. Murata and K. Sakata, *J. Power Sources*, 2006, **159**, 135.
46. J. Lu, Z. Z. Fang and H. Y. Sohn, *J. Phys. Chem. B*, 2006, **110**, 14236.
47. Y. Chen, C. Z. Wu, P. Wang and H. M. Cheng, *Int. J. Hydrogen Energy*, 2006, **31**, 1236.
48. Y. Kojima, M. Matsumoto, Y. Kawai, T. Haga, N. Ohba, K. Miwa, S. I. Towata, Y. Nakamori and S. I. Orimo, *J. Phys. Chem. B*, 2006, **110**, 9632.

49. Y. Nakamori, A. Ninomiya, G. Kitahara, M. Aoki, T. Noritake, K. Miwa, Y. Kojima, S. Orimo, *J. Power Sources*, 2006, **155**, 447.
50. T. Noritake, M. Aoki, S. Towata, A. Ninomiya, Y. Nakamori and S. Orimo, *Appl. Phys. A*, 2006, **83**, 277.
51. C. Zhang and A. Alavi, *J. Phys. Chem. B*, 2006, **110**, 7139.
52. G. P. Meisner, M. L. Scullin, M. P. Balogh, F. E. Pinkerton and M. S. Meyer, *J. Phys. Chem. B*, 2006, **110**, 4186.
53. W. Luo and S. Sickafoose, *J. Alloys Compd.*, 2006, **407**, 274.
54. Y. Nakamori and S. Orimo, *Mater. Sci. Eng. B*, 2004, **108**, 48.
55. Y. Nakamori and S. Orimo, *J. Alloys Compd.*, 2004, **370**, 271.
56. S. Orimo, Y. Nakamori, G. Kitahara, K. Miwa, N. Ohba and T. Noritake, *Appl. Phys. A*, 2004, **79**, 1765.
57. P. A. Chater, W. I. F. Davis and P. A. Anderson, *Chem. Commun.*, 2007, 4770
58. P. A. Chater, P. A. Anderson, J. W. Prendergast, A. Walton, V. S. J. Mann, D. Book, W. I. F. David, S. R. Johnson and P. P. Edwards, *J. Alloys Compd.*, 2007, **446–447**, 350.
59. P. P. Power and X. J. Xu, *J. Chem. Soc., Chem. Commun.*, 1984, 358.
60. H. Chen, R. A. Bartlett, H. V. R. Dias, M. M. Olmstead and P. P. Power, *J. Am. Chem. Soc.*, 1989, **111**, 4338.
61. P. P. Power, *Acc. Chem. Res.*, 1988, **21**, 147.
62. R. A. Bartlett and P. P. Power, *J. Am. Chem. Soc.*, 1987, **109**, 6509.
63. U. Pieper, S. Walter, U. Klingebiel and D. Stalke, *Angew. Chem., Int. Ed.*, 1990, **29**, 209.
64. F. T. Edelmann, F. Knosel, F. Pauer, D. Stalke and W. Bauer, *J. Organomet. Chem.*, 1992, **438**, 1.
65. F. Pauer, J. Rocha and D. Stalke, *J. Chem. Soc., Chem. Commun.*, 1991, 1477.
66. K. W. Henderson, A. E. Dorigo, Q.-Y. Liu and P. G. Williard, *J. Am. Chem. Soc.*, 1997, **119**, 11855.
67. D. Stalke, U. Klingebiel and G. M. Sheldrick, *Chem. Ber.*, 1988, **121**, 1457.
68. R. Boese, U. Klingebiel and G. M. Sheldrick, *J. Organomet. Chem.*, 1986, **315**, C17.
69. F. Antolini, B. Gehrhus, P. B. Hitchcock, M. F. Lappert and J. C. Slootweg, *Dalton Trans.*, 2004, 3288.
70. R. A. Bartlett, X. Feng, M. M. Olmstead, P. P. Power and K. J. Wiese, *J. Am. Chem. Soc.*, 1988, **110**, 4858.
71. H. Chen, R. A. Bartlett, M. M. Olmstead, P. P. Power and S. C. Shoner, *J. Am. Chem. Soc.*, 1990, **112**, 1048.
72. P. Paetzold, C. Pelzer and R. Boese, *Chem. Ber.*, 1988, **121**, 51.
73. T. Fjeldberg, P. B. Hitchcock, M. F. Lappert and A. J. Thorne, *J. Chem. Soc., Chem. Commun.*, 1984, 822.
74. B. Çetinkaya, P. B. Hitchcock, M. F. Lappert, M. C. Misra and A. J. Thorne, *J. Chem. Soc., Chem. Commun.*, 1984, 148.
75. D. Barr, W. Clegg, R. E. Mulvey, R. Snaith and D. S. Wright, *J. Chem. Soc., Chem. Commun.*, 1987, 716
76. P. C. Andrews, P. J. Duggan, G. D. Fallon, T. D. McCarthy and A. C. Peatt, *J. Chem. Soc., Dalton Trans.*, 2000, 1937.
77. P. B. Hitchcock, M. F. Lappert, W. P. Leung, D. S. Liu, T. C. W. Mak and Z. X. Wang, *J. Chem. Soc., Dalton Trans.*, 1999, 1263.
78. P. I. Arvidsson, G. Hilmersson and P. Ahlberg, *J. Am. Chem. Soc.*, 1999, **121**, 1883.
79. R. Bulow, H. Gornitzka, T. Kottke and D. Stalke, *Chem. Commun.*, 1996, 1639
80. W. Clegg, L. Horsbaugh, S. T. Liddle, F. M. Mackenzie, R. E. Mulvey and A. Robinson, *J. Chem. Soc., Dalton. Trans.*, 2000, 1225
81. T. Kottke, U. Klingebiel, M. Noltemeyer, U. Pieper, S. Walter and D. Stalke, *Chem. Ber.*, 1991, **124**, 1941.

82. C. Reiche, S. Klein, U. Klingbiel, M. Noltemeyer, C. Voit, R. Herbst-Irmer and S. Schmatz, *J. Organomet. Chem.*, 2003, **667**, 24.
83. T. Fjeldberg, M. F. Lappert and A. J. Thorne, *J. Mol. Struct.*, 1984, **125**, 265.
84. F. Antolini, P. B. Hitchcock, M. F. Lappert and P. G. Merle, *Chem. Commun.*, 2000, 1301.
85. M. F. Lappert, M. J. Slade, A. Singh, J. L. Atwood, R. D. Rogers and R. Shakir, *J. Am. Chem. Soc.*, 1983, **105**, 302.
86. L. M. Engelhardt, A. S. May, C. L. Raston and A. H. White, *J. Chem. Soc., Dalton Trans.*, 1983, 1671.
87. L. M. Engelhardt, B. S. Jolly, P. C. Junk, C. L. Raston, B. W. Skelton and A. H. White, *Aust. J. Chem.*, 1986, **39**, 1337.
88. G. C. Forbes, A. R. Kennedy, R. E. Mulvey and P. J. A. Rodger, *Chem. Commun.*, 2001, 1400.
89. K. W. Henderson, A. E. Dorigo, P. W. Williard and P. R. Bernstein, *Angew. Chem., Int. Ed.*, 1996, **35**, 1322.
90. R. A. Bartlett, H. Chen, H. V. R. Dias, M. M. Olmstead and P. P. Power, *J. Am. Chem. Soc.*, 1988, **110**, 446.
91. D. Barr, D. J. Berrisford, L. Mendez, A. M. Z. Slawin, R. Snaith, J. Stoddard, D. J. Williams and D. S. Wright, *Angew. Chem., Int. Ed.*, 1991, **30**, 82.
92. D. Barr, W. Clegg, R. E. Mulvey and R. Snaith, *J. Chem. Soc., Chem. Commun.*, 1984, 285.
93. D. R. Armstrong, R. E. Mulvey, G. T. Walker, *et al.*, *J. Chem. Soc., Dalton Trans.*, 1988, 617.
94. G. Boche, M. Marsch and K. Harms, *Angew. Chem., Int. Ed.*, 1986, **25**, 373.
95. L. M. Engelhardt, A. S. May, C. L. Raston and A. H. White, *J. Chem. Soc., Dalton Trans.*, 1983, 1671.
96. M. Westerhausen and W. Schwarz, *Z. Anorg. Allg. Chem.*, 1993, **619**, 1053.
97. H. Dietrich, W. Mahdi and R. Knorr, *J. Am. Chem. Soc.*, 1986, **108**, 2462.
98. K. Gregory, M. Bremer, W. Bauer, *et al.*, *Organometallics*, 1990, **9**, 1485.
99. E. Egert, U. Kliebisch, U. Klingebiel and D. Schmidt, *Z. Anorg. Allg. Chem.*, 1987, **548**, 89.
100. M. Haase and G. M. Sheldrick, *Acta Crystallogr., Sect. C: Cryst. Struct. Commun.*, 1986, **42**, 1009.
101. D. Stalke, U. Klingebiel and G. M. Sheldrick, *J. Organomet. Chem.*, 1988, **344**, 37.
102. S. Daniele, C. Drost, B. Gehrhus, S. M. Hawkins, P. B. Hitchcock, M. F. Lappert, P. G. Merle and S. G. Bott, *J. Chem. Soc., Dalton Trans.*, 2001, 3179.
103. P. C. Andrews, M. Minopoulos and E. G. Roberston, *Eur. J. Inorg. Chem.*, 2006, 2865.
104. A. G. Avent, F. Antolini, P. B. Hitchcock, A. V. Khvostov, M. F. Lappert and A. V. Protchenko, *J. Chem. Soc., Dalton Trans.*, 2006, 919.
105. G. Boche, I. Langlotz, M. Marsch, K. Harms and G. Frenking, *Angew. Chem., Int. Ed.*, 1993, **32**, 1171.
106. G. Hilmersson, P. I. Arvidsson, O. Davidsson and M. Hakansson, *Organometallics*, 1997, **16**, 3352.
107. A. Johansson and O. Davidsson, *Organometallics*, 2001, **20**, 4185.
108. P. I. Arvidsson and O. Davidsson, *Angew. Chem., Int. Ed.*, 2000, **39**, 1467.
109. D. Seebach, W. Bauer, J. Hansen, T. Laube, W. B. Schweizer and J. D. Dunitz, *J. Chem. Soc., Chem. Commun.*, 1984, 853.
110. C. F. Caro, P. B. Hitchcock, M. F. Lappert and M. Layh, *Chem. Commun.*, 1998, 1297.
111. D. Doyle, Y. K. Gunko, P. B. Hitchcock and M. F. Lappert, *J. Chem. Soc., Dalton Trans.*, 2000, 4093.
112. M. Veith, J. Böhnlein and V. Huch, *Chem. Ber.*, 1989, **122**, 841.
113. L. A. Atagi, D. M. Hoffman and D. C. Smith, *Inorg. Chem.*, 1993, **32**, 5084.
114. G. L. J. van Vliet, F. J. J. de Kanter, M. Schakel, G. W. Klumpp, A. L. Spek and M. Lutz, *Chem. Eur. J.*, 1999, **5**, 1091.

115. A. Johansson, A. Pettersson and O. Davidsson, *J. Organomet. Chem.*, 2000, **608**, 153.
116. S. J. Trepanier and S. N. Wang, *Organometallics*, 1993, **12**, 4207.
117. D. Barr, D. J. Berrisford, R. V. H. Jones, A. M. Z. Slawin, R. Snaith, J. F. Stoddart and D. J. Williams, *Angew. Chem., Int. Ed.*, 1989, **28**, 1044.
118. M. Veith, M. Zimmer, K. Fries, J. Böhnlein-Maus and V. Huch, *Angew. Chem., Int. Ed.*, 1996, **35**, 1529.
119. S. Daniele, P. B. Hitchcock, M. F. Lappert, T. A. Nile and C. M. Zdanski, *J. Chem. Soc., Dalton Trans.*, 2002, 3980.
120. J.-F. Li, L.-H. Weng, X.-H. Wei and D.-S. Liu, *J. Chem. Soc., Dalton Trans.*, 2002, 1401
121. D. J. Brauer, H. Bürger and G. R. Liewald, *J. Organomet. Chem.*, 1986, **308**, 119.
122. K. Dippel, U. Klingebiel, M. Noltemeyer, F. Pauer and G. M. Sheldrick, *Angew. Chem., Int. Ed.*, 1989, **27**, 1074.
123. K. Dippel, U. Klingebiel, T. Kottke, F. Pauer, G. M. Sheldrick and D. Stalke, *Chem. Ber.*, 1990, **123**, 237.
124. H. Chen, R. A. Bartlett, H. V. R. Dias, M. M. Olmstead and P. P. Power, *Inorg. Chem.*, 1991, **30**, 2487.
125. B. Tecklenbring, U. Klingebiel and D. Schmid-Bäse, *J. Organomet. Chem.*, 1992, **426**, 287.
126. M. Veith, F. Goffing and C. Huch, *Chem. Ber.*, 1988, **121**, 943.
127. D. J. Brauer, H. Burger, G. R. Liewald and J. Wilke, *J. Organomet. Chem.*, 1985, **287**, 305.
128. D. R. Armstrong, D. Barr, S. T. Brooker, *et al.*, *Angew. Chem., Int. Ed.*, 1990, **29**, 410.
129. M. D. Fryzuk, G. R. Giesbrecht, S. A. Johnson, J. E. Kickham and J. B. Love, *Polyhedron*, 1998, **17**, 947.
130. W. N. Setzer, P. v. R. Schleyer, W. Mahdi and H. Dietrich, *Tetrahedron*, 1988, **44**, 3339.
131. C. Drost, P. B. Hitchcock and M. F. Lappert, *J. Chem. Soc., Dalton Trans.*, 1996, 3595.
132. D. Stalke, N. Keweloh, U. Klingebiel, M. Noltemeyer and G. M. Sheldrick, *Z. Naturforsch.*, 1987, **42b**, 1237.
133. S. Walter, U. Klingebiel and M. Noltemeyer, *Chem. Ber.*, 1992, **125**, 783.
134. K. Dippel, U. Klingebiel, G. M. Sheldrick and D. Stalke, *Chem. Ber.*, 1987, **120**, 611.
135. M. Rannenberg, H. D. Hausen and J. Weidlein, *J. Organomet. Chem.*, 1989, **376**, C27.
136. D. R. Armstrong, D. R. Baker, F. J. Craig, R. E. Mulvey, W. Clegg and L. Horsburgh, *Polyhedron*, 1996, **15**, 3533.
137. D. Barr, W. Clegg, L. Cowton, L. Horsburgh, F. M. Mackenzie and R. E. Mulvey, *J. Chem. Soc., Chem. Commun.*, 1995, 891
138. M. Veith, O. Schutt and V. Huch, *Z. Anorg. Allg. Chem.*, 1999, **625**, 1155.
139. G. Hilmersson and B. Malmros, *Chem. Eur. J.*, 2001, **7**, 337.
140. P. G. Williard and C. Sun, *J. Am. Chem. Soc.*, 1997, **119**, 11693.
141. D. Walther, F. Schramm, N. Theyssen, R. Beckert and H. Gorls, *Z. Anorg. Allg. Chem.*, 2002, **628**, 1938.
142. J. S. Bradley, F. Cheng, S. J. Archibald, R. Supplit, R. Rovai, C. W. Lehmann, C. Krüger and F. Lefebvre, *Dalton Trans.*, 2003, 1846
143. C. Jones, P. C. Junk and N. A. Smithers, *J. Organomet. Chem.*, 2000, **607**, 105.
144. H. Sachdev, C. Wagner, C. Preis, V. Huch and M. Veith, *J. Chem. Soc., Dalton Trans.*, 2002, 4709.
145. R. M. Garvin, N. Kyritsakas, J. Fischer and J. Kress, *Chem. Commun.*, 2000, 965
146. L. Ruwisch, U. Klingebiel, S. Rudolph, R. Herbst-Irmer and M. Noltemeyer, *Chem. Ber.*, 1996, **129**, 823.
147. D. R. Armstrong, D. Barr, W. Clegg, S. M. Hodgson, R. E. Mulvey, D. Reed, R. Snaith and D. S. Wright, *J. Am. Chem. Soc.*, 1989, **111**, 4719.
148. P. I. Arvidsson, P. Ahlberg and G. Hilmersson, *Chem. Eur. J.*, 1999, **5**, 1348.
149. K. W. Henderson, D. S. Walther and P. G. Williard, *J. Am. Chem. Soc.*, 1995, **117**, 8680.

150. J. F. Remenaur, B. L. Lucht, D. Kruglyak, F. E. Romesberg, J. H. Gilchrist and D. B. Collum, *J. Org. Chem.*, 1997, **62**, 5748.
151. K. B. Aubrecht, B. L. Lucht and D. B. Collum, *Organometallics*, 1999, **18**, 2981.
152. G. Boche, I. Langlotz, M. Marsch, K. Harris and N. E. S. Nudleman, *Angew. Chem., Int. Ed.*, 1992, **31**, 1205.
153. M. F. Lappert, M. J. Slade, A. Singh, J. L. Atwood, R. D. Rogers and R. Shakir, *J. Am. Chem. Soc.*, 1983, **105**, 302.
154. W. Clegg, S. T. Liddle, R. E. Mulvey and A. Robertson, *Chem. Commun.*, 2000, 223
155. D. R. Armstrong, D. Barr, R. Snaith, W. Clegg, R. E. Mulvey, K. Wade and D. Reed, *J. Chem. Soc., Dalton Trans.*, 1987, 1071.
156. W. Clegg, L. Horsburgh, F. M. Mackenzie and R. E. Mulvey, *J. Chem. Soc., Chem. Commun.*, 1995, 2011
157. D. Barr, W. Clegg, S. M. Hodgson, G. R. Lamming, R. E. Mulvey, A. J. Scott, R. Snaith and D. S. Wright, *Angew. Chem., Int. Ed.*, 1989, **28**, 1241.
158. K. W. Henderson and P. G. Willard, *Organometallics*, 1999, **18**, 5620.
159. N. D. R. Barnett, W. Clegg, L. Horsburgh, D. M. Lindsay, Q. Y. Liu, F. M. Mackenzie, R. E. Mulvey and P. G. Williard, *Chem. Commun.*, 1996, 2321.
160. M. Westerhausen, A. N. Kneifel and N. Makropoulos, *Inorg. Chem. Commun.*, 2004, **7**, 990.
161. W. Clegg, L. Horsburgh, P. R. Dennison, F. M. Mackenzie and R. E. Mulvey, *Chem. Commun.*, 1996, 1065.
162. N. D. R. Barnett, R. E. Mulvey, W. Clegg and P. A. O'Neil, *J. Am. Chem. Soc.*, 1991, **113**, 8187.
163. (a) M. P. Bernstein, F. E. Romesberg, D. J. Fuller, A. T. Harrison, D. B. Collum, Q.-Y. Liu and P. G. Williard, *J. Am. Chem. Soc.*, 1992, **114**, 5100; (b) W. Clegg, L. Horsburgh, S. A. Couper and R. E. Mulvey, *Acta. Crystallog. C*, 1999, **55**, 867.
164. J. P. Bezombes, P. B. Hitchcock, M. F. Lappert and P. G. Merle, *J. Chem. Soc., Dalton Trans.*, 2001, 816.
165. D. R. Armstrong, D. Barr, W. Clegg, *et al.*, *J. Chem. Soc., Chem. Commun.*, 1986, 869.
166. W. Clegg, S. T. Liddle, R. E. Mulvey and A. Robertson, *Chem. Commun.*, 1999, 511.
167. R. von Bulow, H. Gornitzka, T. Kottke and D. Stalke, *Chem. Commun.*, 1996, 1639.
168. A. R. Kennedy, R. E. Mulvey and A. Robertson, *Chem. Commun.*, 1998, 89.
169. J. L. Rutherford and D. B. Collum, *J. Am. Chem. Soc.*, 1999, **121**, 10198.
170. G. R. Kowach, C. J. Warren, R. S. Haushalter and F. J. DiSalvo, *Inorg. Chem.*, 1993, **37**, 156.
171. E. C. Franklin, *J. Phys. Chem.*, 1919, **23**, 36.
172. H. Jacobs and B. Harbrecht, *J. Less-Common Met.*, 1982, **85**, 87.
173. H. Jacobs and B. Harbrecht, *Z. Anorg. Allg. Chem.*, 1984, **518**, 87.
174. B. Harbrecht and H. Jacobs, *Z. Anorg. Allg. Chem.*, 1987, **546**, 48.
175. P. A. Chater, W. I. F. David, S. R. Johnson, P. P. Edwards and P. A. Anderson, *Chem. Commun.*, 2006, 2439.
176. P. G. Williard and M. A. Nichols, *J. Am. Chem. Soc.*, 1991, **113**, 9671.
177. J. J. Morris, B. C. Noll and K. W. Henderson, *Acta. Crystallogr.*, 2007, **E63**, m2477.
178. A. R. Kennedy, J. G. MacLellan and R. E. Mulvey, *Angew. Chem., Int. Ed.*, 2001, **40**, 3245.
179. A. R. Kennedy, R. E. Mulvey and R. B. Rowlings, *Angew. Chem., Int. Ed.*, 1998, **37**, 3180.
180. A. R. Kennedy, R. E. Mulvey and R. B. Rowlings, *J. Am. Chem. Soc.*, 1998, **120**, 7816.
181. W. Clegg, K. W. Henderson, R. E. Mulvey and P. A. O'Neil, *J. Chem. Soc., Chem. Commun.*, 1994, 769.
182. L. Barr, A. R. Kennedy, J. G. MacLellan, J. H. Moir, R. E. Mulvey and P. J. A. Rodger, *Chem. Commun.*, 2000, 1757.
183. D. Rutherford and D. A. Atwood, *J. Am. Chem. Soc.*, 1996, **118**, 11535.
184. S. Bock, H. Nöth and P. Z. Rahm, *Z. Naturforsch.*, 1988, **43b**, 53.
185. M. A. Petrie, K. Ruhlandt-Senge and P. P. Power, *Inorg. Chem.*, 1993, **32**, 1135.

186. W. E. Rhine, G. Stucky and S. W. Peterson, *J. Am. Chem. Soc.*, 1975, **97**, 6401.
187. J. García-Álvarez, D. V. Graham, A. R. Kennedy, R. E. Mulvey and S. Weatherstone, *Chem. Commun.*, 2006, 3208.
188. G. Linti, H. Nöth and P. Z. Rahm, *Z. Naturforsch.*, 1988, **43b**, 1101.
189. D. Atwood and D. Rutherford, *Chem. Commun.*, 1996, 1251.
190. W. Clegg, S. T. Liddle, K. W. Henderson, F. E. Keenan, A. R. Kennedy, A. E. McKeown and R. E. Mulvey, *J. Organomet. Chem.*, 1999, **572**, 283.
191. P. Böttcher, H. W. Roesky, M. G. Walawalkar and H. G. Schmidt, *Organometallics*, 2001, **20**, 790.
192. D. J. Brauer, H. Burger and G. R. Liewald, *J. Organomet. Chem.*, 1986, **307**, 177.
193. B. D. Murray and P. P. Power, *Inorg. Chem.*, 1984, **23**, 4584.
194. J. F. Eichler, O. Just and W. S. Rees, Jr., *Inorg. Chem.*, 2006, **45**, 6706.
195. J. L. Gay-Lussac and L. J. Thenard, *Physicochimiques (I)*, 1811, **337**, 354.
196. T. H. Vaughn, R. R. Vogt and J. A. Nieuwland, *J. Am. Chem. Soc.*, 1934, **56**, 2120.
197. J. M. Mcgee, *J. Am. Chem. Soc.*, 1921, **43**, 586.
198. L. Wöhler and F. Stang-Lund, *Z. Elektrochem.*, 1918, **24**, 261.
199. P. Caubere, *Acc. Chem. Res.*, 1974, **7**, 308.
200. M. G. Barker, A. Harper and P. Hubberstey, *J. Chem. Res.*, 1978, 432.
201. M. Becker, J. Nuss and M. Jansen, *Z. Anorg. Allg. Chem.*, 2000, **626**, 2505.
202. P. C. Andrews, P. J. Duggan, G. D. Fallon, T. D. McCarthy and A. C. Peatt, *Dalton Trans.*, 2000, 2505
203. J. Knizek, I. Krossing, H. Nöth, H. Schwenk and T. Seifert, *Chem. Ber. Recl.*, 1997, **130**, 1053.
204. I. Cragg-Hine, M. G. Davidson, A. J. Edwards, P. R. Raithby and R. Snaith, *J. Chem. Soc., Dalton Trans.*, 1994, 2901
205. M. Karl, G. Seybert, W. Massa, K. Harms, S. Agarwal, R. Maleika, W. Stelter, A. Greiner, W. Heitz, B. Neumüeller and K. Dehnicke, *Z. Anorg. Allg. Chem.*, 1999, **625**, 1301.
206. C. Knapp, E. Lork, T. Borrmann, W.-D. Stohrer and R. Mews, *Z. Anorg., Allg. Chem.*, 2005, **631**, 1885.
207. P. C. Andrews, P. J. Duggan, M. Maguire and P. J. Nichols, *Chem. Commun.*, 2001, 53.
208. P. C. Andrews, N. D. R. Barnett, R. E. Mulvey, W. Clegg, P. A. O'Neil, D. Barr, L. Cowton, A. J. Dawson and B. J. Wakefield, *J. Organomet. Chem.*, 1996, **518**, 85.
209. F. Antolini, P. B. Hitchcock, A. V. Khvostov and M. F. Lappert, *Eur. J. Inorg. Chem.*, 2003, 3391.
210. P. C. Andrews, D. R. Armstrong, W. Clegg, M. Macgregor and R. E. Mulvey, *J. Chem. Soc., Chem. Commun.*, 1991, 497.
211. P. G. Williard and M. J. Hintze, *J. Am. Chem. Soc.*, 1990, **112**, 8602.
212. P. C. Andrews, W. Clegg and R. E. Mulvey, *Angew. Chem. Int. Ed.*, 1990, **29**, 1440.
213. B. Gehrhus, P. H. Hitchcock, A. R. Kennedy, M. F. Lappert, R. E. Mulvey and P. J. A. Rodger, *J. Organomet. Chem.*, 1999, **587**, 88.
214. M. Driess, H. Pritzkow, M. Skipinski and U. Winkler, *Organometallics*, 1997, **16**, 5108.
215. N. P. Lorenzen, J. Kopf, F. Olbrich, U. Schümann and E. Weiss, *Angew. Chem., Int. Ed.*, 1990, **29**, 1441.
216. F. T. Edelmann, F. Pauer, M. Wedler and D. Stalke, *Inorg. Chem.*, 1992, **31**, 4143.
217. J. Eppinger, E. Herdtweck and R. Anwander, *Polyhedron*, 1998, **17**, 1195.
218. W. Clegg, K. W. Henderson, L. Horsburgh, F. M. Mackenzie and R. E. Mulvey, *Chem. Eur. J.*, 1998, **4**, 53.
219. J. J. Morris, B. C. Noll, G. W. Honeyman, C. T. O'Hara, A. R. Kennedy, R. E. Mulvey and K. W. Henderson, *Chem. Eur. J.*, 2007, **13**, 4418.
220. D. R. Baker, R. E. Mulvey, W. Clegg and P. A. O'Neil, *J. Am. Chem. Soc.*, 1993, **115**, 6472.
221. A. R. Kennedy, J. G. MacLellan, R. E. Mulvey and A. Robertson, *J. Chem. Soc., Dalton Trans.*, 2000, 4112.

222. D. R. Armstrong, A. R. Kennedy, R. E. Mulvey and R. B. Rowlings, *Angew. Chem., Int. Ed.*, 1999, **38**, 131.
223. G. C. Forbes, A. R. Kennedy, R. E. Mulvey, R. B. Rowlings, W. Clegg, S. T. Liddle and C. C. Wilson, *Chem. Commun.*, 2000, 1759.
224. K. F. Tesh, T. P. Hanusa, and J. C. Huffman, *Inorg. Chem.*, 1990, **29**, 1584.
225. J. B. Randazzo, J. J. Morris and K. W. Henderson, *Main Group Chem.*, 2006, **5**, 215.
226. R. Juza, H. Jacobs and W. Klose, *Z. Anorg. Allg. Chem.*, 1965, **338**, 171.
227. D. Peters, A. Tenten and H. Jacobs, *Z. Anorg. Allg. Chem.*, 2002, **628**, 1521.
228. F. Kraus and N. Korber, *Z. Anorg. Allg. Chem.*, 2005, **631**, 1032.
229. W. Schnick and H. Huppertz, *Z. Anorg. Allg. Chem.*, 1995, **621**, 1703.
230. M. Becker, M. Jansen, A. Lieb, W. Milius and W. Schnick, *Z. Anorg. Allg. Chem.*, 1998, **624**, 113.
231. M. Becker and M. Jansen, *Solid State Sciences*, 2000, **2**, 711.
232. F. Kraus and N. Korber, *J. Solid State Chem.*, 2005, **178**, 1241.
233. M. Monz and H. Jacobs, *Z. Anorg. Allg. Chem.*, 1995, **621**, 137.
234. N. Scotti and H. Jacobs, *Z. Anorg. Allg. Chem.*, 2000, **626**, 2275.
235. B. Gehrhus, P. B. Hitchcock, M. F. Lappert, and J. C. Slootweg, *Chem. Commun.*, 2000, 1427.
236. P. G. Williard, *Acta. Crystallog. C*, 1988, **C44**, 270.
237. J. Schneider, E. Popowski and H. Reinke, *Z. Anorg. Allg. Chem.*, 2003, **629**, 55.
238. R. Dinnebier, H. Esbak, F. Olbrich and U. Behrens, *Organometallics*, 2007, **26**, 2604.
239. F. M. Mackenzie, R. E. Mulvey, W. Clegg and L. Horsburgh, *J. Am. Chem. Soc.*, 1996, **118**, 4721.
240. A. R. Kennedy, R. E. Mulvey, C. L. Raston, B. A. Roberts and R. B. Rowlings, *Chem. Commun.*, 1999, 353.
241. P. C. Andrews, A. R. Kennedy, R. E. Mulvey, C. L. Raston, B. A. Roberts and R. B. Rowlings, *Angew. Chem., Int. Ed.*, 2000, **39**, 1960.
242. G. C. Forbes, F. R. Kenley, A. R. Kennedy, R. E. Mulvey, C. T. O'Hara and J. A. Parkinson, *Chem. Commun.*, 2003, 1140.
243. P. Bohger, T. Zeiske and H. Jacobs, *Z. Anorg. Allg. Chem.*, 1998, **624**, 364.
244. H. Jacobs and J. Kockelkorn, *Z. Anorg. Allg. Chem.*, 1979, **456**, 147.
245. H. Jacobs, P. Bohger and W. Kockelmann, *Z. Anorg. Allg. Chem.*, 2002, **628**, 1794.
246. H. Jacobs and T. J. Hennig, *Z. Anorg. Allg. Chem.*, 1998, **624**, 1823.
247. J. Bock and H. Jacobs, *J. Less-Common Met.*, 1988, **137**, 105.
248. S. Neander and U. Behrens, *Z. Anorg. Allg. Chem.*, 1999, **625**, 1429.
249. H. Esbak and U. Behrens, *Z. Anorg. Allg. Chem.*, 2005, **631**, 1581.
250. F. Golinski and H. Jacobs, *Z. Anorg. Allg. Chem.*, 1995, **621**, 29.
251. E. Renegade, *Compt. Rend.*, 1905, **140C**, 1183.
252. D. Peters and H. Jacobs, *Z. Anorg. Allg. Chem.*, 2003, **629**, 497.
253. B. Harbrecht and H. Jacobs, *Z. Anorg. Allg. Chem.*, 1983, **500**, 181.
254. H. Jacobs and D. Peters, *J. Less-Common Met.*, 1986, **118**, 261.
255. H. Jacobs, J. Birkenbeul and J. Kockelkorn, *J. Less-Common Met.*, 1982, **85**, 71.
256. K. F. Tesh, B. D. Jones, T. P. Hanusa and J. C. Huffman, *J. Am. Chem. Soc.*, 1992, **114**, 6590.
257. H. Jacobs and F. Golinski, *Z. Anorg. Allg. Chem.*, 1994, **620**, 531.
258. A. Tenten and H. Jacobs, *J. Alloys Compds.*, 1991, **177**, 193.
259. F. Flacke and H. Jacobs, *J. Alloys Compds.*, 1995, **227**, 109.
260. T. Kottke and D. Stalke, *Organometallics*, 1996, **15**, 4552.
261. S. T. Liddle and W. Clegg, *J. Chem. Soc., Dalton Trans.*, 2001, 402.
262. O. Hiemisch, D. Henschel, A. Blaschette and P. G. Jones, *Z. Anorg. Allg. Chem.*, 1997, **623**, 324.

3

Beryllium and the Alkaline Earth Metal Amides

3.1 Introduction

Even a cursory reading of Chapter 3 of the previous volume,[1] dealing with alkaline earth diorgano amides, shows that detailed information on their preparation, structure and reactivity was quite sparse. For example, there were no data available for strontium or barium amide derivatives and there was just one structure of a complex related to a calcium amide – that of $[(thf)_3Ca(AlH)_3(NBu^t)_4]$[2] – in which calcium was incorporated as part of an iminoalane cuboidal framework to afford a $CaAl_3N_4$ core. For magnesium, a related species $[(thf)Mg(AlH)_3(NBu^t)_4]$[2] had been structurally characterized, but only one other diorganoamide structure, the complex $[\{MeMgNMe(CH_2)_2NMe_2\}_2]$,[3] which featured four-coordinate magnesium had been described. In fact, more structural information was available for beryllium amides (the three complexes $[\{Be(NMe_2)_2\}_3]$,[4] $[Be\{N(SiMe_3)_2\}_2]$,[5–7] and $[K][Be(NH_2)_3]$[8]) than for the other group 2 metals. There had been much work in the early twentieth century on magnesium amides, particularly derivatives of the type '$XMgNR_2$,' which are amido analogues of Grignard reagents.[1] There were also reports of compounds that were assigned the formula '$(XMg)_2NR$' but, apart from some work on molecular weights, detailed knowledge of the composition and structures of even these magnesium compounds was unavailable.

The last two decades have witnessed a very large increase in our knowledge of alkaline earth amides and this is detailed in the following sections. It will be seen that the largest expansion has occurred for the alkaline earth metals Mg - Ba. Work on beryllium amides, although significant, has been limited by concerns about the health hazards associated with this element and its derivatives. In common with other chapters, we emphasize derivatives of monodentate amido ligands, and for the most part, we do not include coverage of compounds in which nitrogen is a part of a multidentate or delocalized array.

Metal Amide Chemistry Michael Lappert, Andrey Protchenko, Philip Power and Alexandra Seeber

3.2 Beryllium Amides

Because of its toxicity, the number of well-characterized beryllium compounds of all types is much lower than that of the other alkaline earth metals. The largest number of the published beryllium structures involve oxygen donor ligands,[9,10] and beryllium amides have been reviewed, at least in part, under the aegis of wider-ranging surveys[13–16] of group 2 metal silylamides,[11] group 2 heterometallic amides,[12] organoderivatives of the group 2 elements,[14,15] and alkaline earth element coordination chemistry.[16]

Common synthetic routes to beryllium amides, which were summarized in Ref. 1, involve the direct or indirect amination of beryllium hydride, beryllium chloride, a beryllium alkyl or amide precursor. Beryllium amides with bulky substituents are generally synthesized via the transmetallation of beryllium dichloride with the lithium amides.[5,17,18] The reaction of beryllium dichloride with secondary amines in the presence of an alkyllithium represents a less common synthetic route to beryllium amides.[19,20] The formation of beryllium amides via the reaction of an alkyl beryllium[21,22] as well as beryllium hydride species with amines is also known.[23,24]

Some structural and spectroscopic information for selected beryllium amides and other beryllium species complexed by nitrogen containing ligands is provided in Table 3.1.

In one of the earliest compounds,[5] the two-coordinate, monomeric species $[Be\{N(SiMe_3)_2\}_2]$ was reported to have a linear NBeN framework in the vapour phase with mutually perpendicular NSi_2 planes in an analogous manner to the allene structure.[6,7] The compounds $[Be(NPr^i_2)_2]$, $[Be(tmp)_2]$ (where tmp = 2,2,6,6-tetramethylpiperidido) and $[Be\{N(Bu^t)(SiMe_3)\}_2]$ were also reported to be two-coordinate monomers based on mass spectrometry and NMR data.[17,20] The only structurally characterized two-coordinate beryllium amide in the solid state, however, is $[(C_6H_3Mes_2\text{-}2,6)Be\{N(SiMe_3)_2\}]$ (Figure 3.1, where Mes = $C_6H_2Me_3$-2,4,6),[18] which was synthesized by reaction of the preformed $ArBeCl(OEt_2)$ (from LiAr + $BeCl_2$, Ar = $C_6H_3Mes_2$-2,6) with $LiN(SiMe_3)_2$ in toluene at room temperature, Equation (3.1).

$$LiAr + BeCl_2 \xrightarrow[-LiCl]{Et_2O} ArBeCl(OEt_2) \xrightarrow[toluene,-LiCl]{LiN(SiMe_3)_2} [ArBe\{N(SiMe_3)_2\}] \qquad (3.1)$$

In this complex the geometry at the beryllium is linear with a very short Be–N bond length (1.519(4) Å) and a Be–C distance of 1.700(4) Å. The short Be–N interaction is probably due to ionic resonance contributions with a minor contribution from π-bonding between the nitrogen lone pair and the beryllium 2p orbitals.

A number of three-coordinate beryllium amides have also been reported. For example, two of the three berylliums in the simple trimer $[(Me_2N)Be(\mu\text{-}NMe_2)_2Be(\mu\text{-}NMe_2)_2Be(NMe_2)]$ are three-coordinate. The bulky aryl and silylamido groups in the heteroleptic complex $[(C_6H_3Mes_2\text{-}2,6)Be\{N(H)SiPh_3\}(OEt_2)]$,[18] synthesized by salt elimination, stabilize its monomeric, three-coordinate structure. The Be–N bond length is short (1.577(4) Å), and the Be–C and Be–O distances are 1.780(4) and 1.655(4) Å, respectively. The dimeric complex $[\{\overline{BeN(Me)SiMe_2CH_2SiMe_2N}(Me)\}_2]$,[22] formed by the reaction of $BeMe_2$ with the aminosilane, also involves three-coordinate beryllium atoms, with the bidentate ligand acting both as a bridging (Be–N in $(BeN)_2$ ring = 1.683(3) and 1. 714(3) Å) as well as a terminating ligand (Be–N distance is 1.550(4) Å) to complete a distorted trigonal planar geometry at beryllium. Similarly, the structure of the ketimido complex

Table 3.1 *Selected structural and spectroscopic data for beryllium amides and related complexes*

Compound	Be–N (Å)	^{9}Be NMR(δ)	Reference
$[Be(NPr^i_2)_2]$	—	12.7	17,20
$[Be(tmp)_2]^a$	—	15.3	17
$[Be\{N(SiMe_3)_2\}_2]$	1.56(2)	12.5	6,7
$[Be\{N(Bu^t)SiMe_3\}_2]$	—	15.1	20
$[Be(C_6H_3Mes_2\text{-}2,6)\{N(SiMe_3)_2\}]^b$	1.519(4)	15.6	19
$[\{Be(N=CBu^t_2)_2\}_2]$	1.502(13) 1.674(15) 1.682(14)	—	2
$[Be\{piperidide)_2\}_2]$	—	9.4	17
$[(Me_2N)Be(\mu\text{-}NMe_2)_2Be(\mu\text{-}NMe_2)_2Be(NMe_2)]$	1.57(1) 1.653(7) 1.785(4)	9.24/1.09	4,17
$[Be(C_6H_3Mes_2\text{-}2,6)(NHSiPh_3)(Et_2O)]$	1.577(4)	12.6	18
$[Be(C_6H_3Mes_2\text{-}2,6)(\mu\text{-}NHPh)\}_2]$	1.711 avg.	—	22
$[\{BeN(Me)SiMe_2CH_2SiMe_2N(Me)\}_2]$	1.550(4) 1.683(3) 1.714(3)	—	22
$[\{PhC(NSiMe_3)_2\}(BeCl)_2\{N(SiMe_3)_2\}]$	1.649(6)- 1.686(5)	11.4	18
$[\{Be(\mu\text{-}NPr^i_2)(BH_4)\}_2]$	1.695(5) av.	7.76	20
$[K][Be(NH_2)_3]$	1.59	—	8
$[(HBe\{\mu\text{-}N(Me)(CH_2)_2NMe\})_2]$	1.747(3)	—	23,24
$[Be\{(NSiMe_3)_2CPh\}_2]$	1.72(3) av.	5.5	18
$[Be\{N(Ph)C(Ph)C(Ph)NPh\}]_2$	1.68(2) 1.73(2)	—	19
$[Be(\{N(SiMe_3)\}_2PPh_2)_2]$	1.765 av.	—	25
$[(Be(Cl)(thf)\{\mu\text{-}N(SnMe_3)_2\})_2]$	1.72(1)	—	26
$[\{Be(Cl)(\mu_3\text{-}NPEt_3)\}_4]$	1.752 av.	—	31

^{a} tmp = 2,2,6,6—tetramethylpiperidido, ^{b} Mes = $C_6H_2Me_3$-2,4,6

$[\{Be(N=CBu^t_2)_2\}_2]$,[21] formed by the reaction of $BeBu^t_2$ with the ketimine $Bu^t_2C=NH$, features imido groups as bridging and terminal ligands. The distances to the bridging nitrogen atoms (1.674(14) and 1.682(15) Å) are longer than those to the terminal nitrogen atoms, which feature the shortest Be–N bond length (1.502(13) Å) in a stable species although it is well within three standard deviations of the 1.519(4) Å in $[(C_6H_3Mes_2\text{-}2,6)Be\{N(SiMe_3)_3\}]$.[18] The $(BeN)_2$ ring in the heteroleptic dimer $[((C_6H_3Mes_2\text{-}2,6)Be\{\mu\text{-}N(H)\text{-}Ph\})_2]$[18] is planar and has slightly longer Be–N bond lengths than those observed in the $(BeN)_2$ framework species (between 1.69 and 1.73 Å) due to the steric crowding of the aryl substituents.

Four-coordinate beryllium structures related to amides are exemplified by $[Be(dad)_2]$[19] (dad = $N_2C_2Ph_4$) and $[Be\{(NSiMe_3)_2PPh_2\}_2]$.[26] These complexes feature a distorted tetrahedral geometry at beryllium (Be–N bond lengths range between 1.68–1.73 and 1.76–1.78 Å, respectively). The distorted tetrahedral geometry at the beryllium atom in the chelated dimeric complex $[(HBe\{\mu\text{-}N(Me)(CH_2)_2NMe\})_2]$[23,24] is due to the bidentate

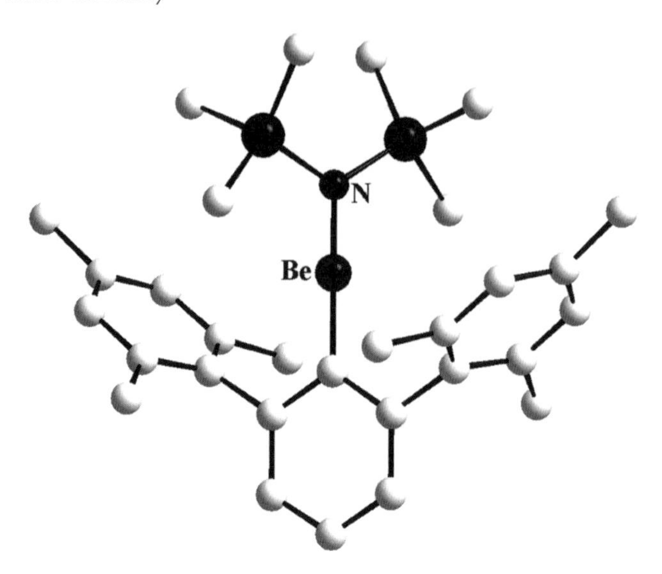

Figure 3.1 *Illustration of the linear coordination geometry in the monomeric complex* $[(C_6H_3Mes_2\text{-}2,6)Be\{N(SiMe_3)_2\}]$.[18] *The beryllium, nitrogen and silicon atoms are shown as black spheres and carbon atoms are white*

ligand, which coordinates via the amido nitrogen site in a bridging manner to form a planar $(BeN)_2$ core and the amino nitrogen in a terminal position (Be—N bond lengths are between 1.75–1.81 Å). The reaction of $(HB(3\text{-}Bu^tpz)_3)Tl$ ($3\text{-}Bu^tpz = 3\text{-}Bu^tC_3N_2H_2$) with $BeCl_2{\cdot}xH_2O$ and $BeBr_2(OEt_2)_2$ led to the four-coordinate complexes $[\{HB(3\text{-}Bu^tpz)_3\}\text{-}BeCl]$ and $[\{HB(3\text{-}Bu^tpz)_3\}BeBr]$, respectively.[27] Similar (BeN_3B) frameworks are observed in the complexes $[\{HB(3\text{-}Bu^tpz)_3\}BeH]$[27] and $[\{HB(3\text{-}Bu^tpz)_3\}BeCH_3]$,[28] which have different Be—N bond distances (1.78 and 1.86–1.88 Å, respectively), the lengthening in the Me derivative being due to the greater steric interactions between the Bu^t substituents of the ligand and BeMe groups.

The four-membered $(BeN)_2$ ring in the unusual species $[\{Be(NPr^i_2)(BH_4)\}_2]$[20] (Figure 3.2) has Be—N distances of *ca.* 1.70 Å, and the tetrahedral geometry at beryllium is completed by coordination of two hydridic hydrogen atoms from the borohydride groups. This complex was synthesized by the reaction of diisopropylamine with $BeCl_2$ to yield the intermediate $BeCl_2{\cdot}HNPr^i_2$, which was deprotonated with $LiBu^n$ at low temperature to afford $ClBe(NPr^i_2)$, which, upon reaction with $LiBH_4$, yielded $[\{Be(NPr^i_2)(BH_4)\}_2]$.

In reactivity studies of monomeric $Be(NPr^i_2)_2$, it has been shown that CO_2 and CS_2 insert into the Be—N bonds to give $[Be(O_2CNPr^i_2)_2]$ or $[Be(S_2CNPr^i_2)_2]$ whereas the dimeric beryllium dialkylamides can afford more complicated structures.[29,30] The cubanoid complex $[\{Be(Cl)(\mu_3\text{-}NPEt_3)\}_4]$ was obtained by the thermal Me_3SiCl elimination from $[BeCl_2(Me_3SiNPEt_3)]$.[31]

An important feature of beryllium amide studies since 1980 has been the acquisition of ^9Be NMR data. NMR signals for this nucleus, which has 100% abundance, $I = -3/2$ and a receptivity almost two orders of magnitude greater than the ^{13}C nucleus,[32] can be obtained without major difficulties. Some chemical shift data gathered from Refs. 17,18 and 20 are

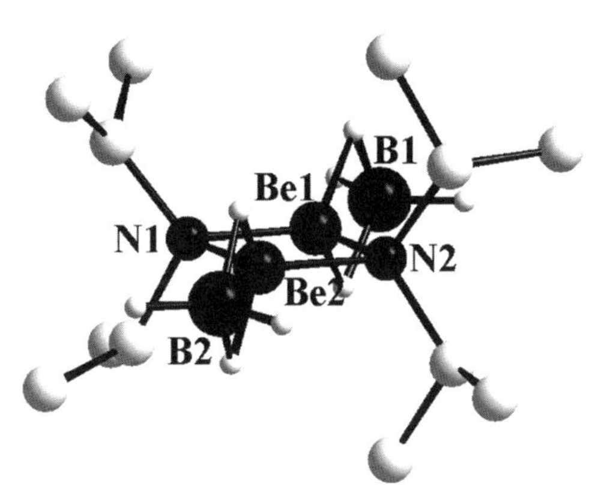

Figure 3.2 Representation of $[\{Be(NPr^i{}_2)(BH_4)\}_2]$[20] illustrating the $(BeN)_2$ core and a distorted tetrahedral beryllium coordination geometry. Beryllium, boron and nitrogen atoms are shown as black spheres and carbon atoms are white. Hydrogen atoms except B-H are not shown for clarity. Selected bond lengths: Be1–N1 1.690(4), Be1–N2 1.703(4), Be2–N1 1.690(4), Be2–N2 1.697(4), Be–H 1.06–1.53 Å

given in Table 3.1. Generally speaking, lower coordination produce signals at lower field; cf. signals at $\delta = 9.24$ and 1.09 for three- and four-coordinated beryllium in the trimer $[\{Be(NMe_2)_2\}_3]$,[17] whereas two-coordinate beryllium amides generally afford signals downfield of *ca.* 12 ppm. Data for several other structurally uncharacterized homo and heteroleptic beryllium amides are listed in Refs 17 and 20. A recent paper has demonstrated excellent agreement between calculated and observed ^{9}Be NMR chemical shifts in a wide range of beryllium compounds.[33]

Calculations[34] on simple beryllium amido derivatives such as $(HBe)_nNH_{3-n}$, $Be(NH_2)_2$, $MBe(NH_2)_3$ (M = Li or Na) and related species yielded short Be–N bonds for the neutral derivatives together with planar geometries for the terminal amido nitrogens. Although Be–N π-bonding was not discussed in detail, several of the compounds featured Be–N bond orders that were greater than one, which suggests the existence of Be–N π-bonding. The calculations also showed that the all-planar structures of $[Be(NH_2)_3]^-$ was 21.9 kcal mol^{-1} more stable than the structure where the amido groups were oriented perpendicularly to the BeN_3 plane. More recent computational work on bond dissociation energies has afforded a vapour phase Be–N bond dissociation energy of *ca.* 120 kcal mol^{-1} for the hypothetical species $HBeNH_2$.[35]

3.3 Magnesium Amides

3.3.1 Introduction

Magnesium amides are now the most numerous among the group 2 metals and they have been treated, either fully or in part, in a number of reviews.[11–16,36–39] The crystal structure of the simplest magnesium amide $Mg(NH_2)_2$ was investigated via X-ray crystallography in 1969.[40] The unit cell contains three crystallographically unique amido (NH_2^-) units with H–N–H angles between 104.1 and 129.6°. The crystal structure of $Mg(ND_2)_2$ was investigated by high-resolution powder neutron and synchrotron X-ray diffraction.[41] The neutron diffraction data indicate a tetragonal unit cell (space group $I4_1$/acd), although the powder synchrotron X-ray diffraction data suggest that the symmetry may be lower. The magnesium atom is tetrahedrally coordinated by four amido (ND_2^-) ligands (Mg–N bond lengths between 2.00 and 2.17 Å). The amido ions exist in three crystallographically unique units, with N–D distances and D–N–D angles between 0.95–1.07 Å and 101–107°, respectively. It was suggested that the $Mg(ND_2)_2$ structure can be viewed as a pseudo-fcc lattice of Mg_4 tetrahedra each surrounded by a slightly distorted octahedral arrangement of amido units with Mg·Mg distances > 3.4 Å. Like lithium amides, magnesium amides are of current interest as a component of hydrogen storage materials (e.g. $Mg(NH_2)_2/MgH_2$).[42–51]

Magnesium diorganoamides can be synthesized by a number of routes.[52–138] The oldest involves the displacement of an alkane from a diorganomagnesium upon reaction with an amine as shown in Equation (3.2).

$$MgR_2 + 2HNR'_2 \xrightarrow{-2RH} Mg(NR'_2)_2 \tag{3.2}$$

Grignard reagents also react with amines in a similar fashion, Equation (3.3).

$$RMgX + HNR'_2 \xrightarrow{-RH} Mg(X)NR'_2 \tag{3.3}$$

The salt elimination route, usually via reaction of a lithium amide with a magnesium halide, may be the most commonly used synthetic approach. It is exemplified by the reactions in Scheme 3.1 (DMAP = 4-$Me_2NC_5H_4N$).[52]

Both routes remain popular, however, and have been used to synthesize the dimeric three-coordinate $[(Bu^sMg\{\mu\text{-}N(SiMe_3)_2\})_2]$[53] or the monomeric $[Mg\{N(SiMePh_2)_2\}_2]$[54a] featuring two-coordinated magnesium; if a primary amine is used in the reaction described by Equation (3.2), a magnesium imide such as $[\{(thf)MgNPh\}_6]$ can be isolated (see Section 3.3.7). A wide variety of heteroleptic magnesium amides can also be obtained if only one equivalent of an amine HNR_2' is used.[54b] Several have been shown to be useful, regioselective deprotonating agents for weak CH acids.[55] The use of a Grignard reagent and an

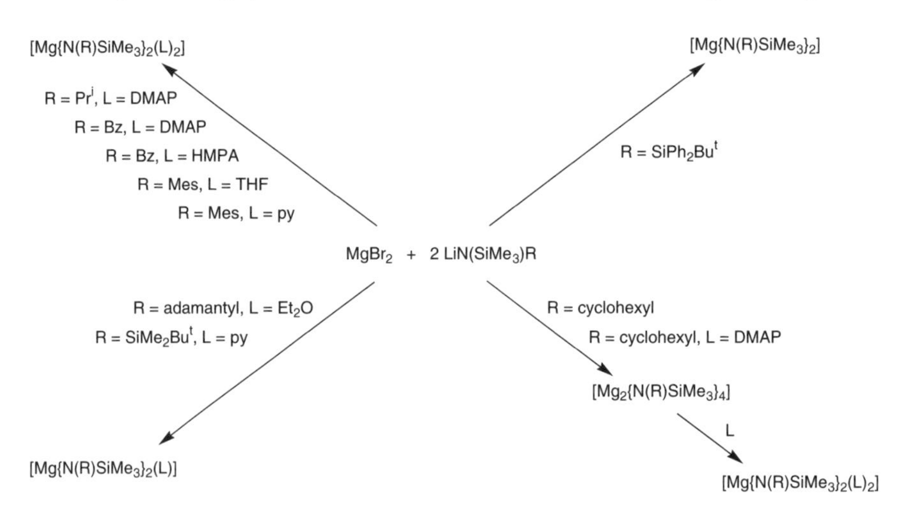

Scheme 3.1 *Transmetallation reactions of magnesium bromide with a range of lithium amides and neutral donors*[52]

amine, as in Equation (3.3), affords the so called Hauser base species[56] such as [BrMg(μ-NEt_2)(hmpa)] (hmpa = hexamethylphosphoramide)][54b] or [{$(Me_3Si)_2N$}Mg(μ-X)(OEt_2)] (X = Cl[54a] or Br[55a]). In combination with lithium halides, as in $R_2NMgCl{\cdot}LiCl$, they display enhanced reactivity and regiospecificity.[57] The direct reaction of magnesium hydrides with amines to yield magnesium amides is rare,[58,59] as is the direct reaction of amines with magnesium metal at elevated temperature and pressure.[60] Reduction of an α-diimine with Mg metal gave a magnesium diamide, as in [Mg{N(Ph)C(Ph) = C(Ph)N(Ph)}$(dme)_2$].[61]

3.3.2 Monomeric Magnesium Amides

Monomeric Bis-amides with magnesium coordination number ≥4. The simplest substituted magnesium amides are monomeric species, a large variety of which are now known. Tetrahedral geometry at Mg is the predominant coordination motif for most magnesium amido complexes (trigonal bipyramidal and octahedral complexes are also known), monomeric or otherwise, as exemplified in the primary amido derivative [(tmeda)Mg-{N(H)$SiPr^i_3$}$_2$] (tmeda = tetramethylethylenediamine).[63] This complex was synthesized by an unusual route involving ligand exchange between an amidozinc alkyl and a magnesium alkyl as shown in Equation (3.4).[63]

$$2RZn(NHSiPr^i_3) + MgR_2 \xrightarrow[+\,tmeda]{-2ZnR_2} [Mg(NHSiPr^i_3)_2(tmeda)] \tag{3.4}$$

The magnesium is coordinated in a distorted tetrahedral fashion by two amido ligands (Mg–N distance = 1.961(2) Å) and the bidentate donor tmeda (Mg–N = 2.239(2) Å). A similar structure is found in the primary amido species [Mg{N(H)Mes}$_2$$(hmpa)_2$],[100] which was obtained by alkane elmination and where the average Mg–N and Mg–O bond lengths are 2.01 and 1.97 Å, respectively).

The structures of numerous four-coordinate secondary amidomagnesium complexes have been determined. For example, in [$Mg(NPh_2)_2(hmpa)_2$][54b] or [$Mg(NPh_2)_2(thf)_2$],[64] synthesized by routine alkane or benzene elimination, the average Mg–N and Mg–O distances are 2.05 and 1.94 Å or 2.013 and 2.012 Å, respectively. The silylamido complex [Mg{$N(SiMe_3)_2$}$_2$$(thf)_2$] (average Mg–N and Mg–O distances are 2.02 and 2.09 Å, respectively),[65] was obtained by a transmetallation route as shown in Equation (3.5).

$$Hg\{N(SiMe_3)_2\}_2 + Mg \xrightarrow{thf} [Mg\{N(SiMe_3)_2\}_2(thf)_2] + Hg \tag{3.5}$$

Further reaction of [Mg{$N(SiMe_3)_3$}$_2$$(thf)_2$] with a Lewis base nitrogen donor ligand yielded a variety of complexes as shown in Scheme 3.2 and can be exemplified by

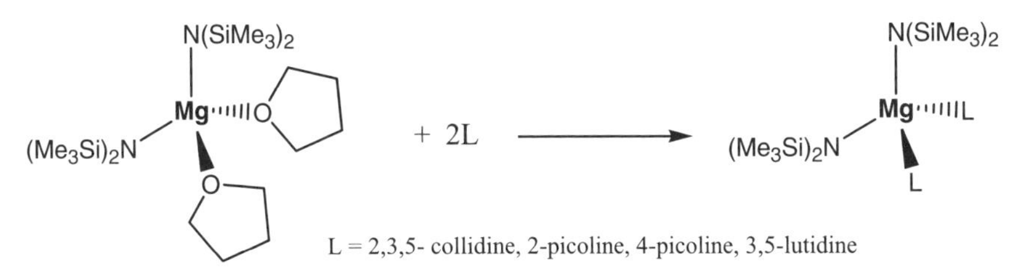

Scheme 3.2 *Formation of four-coordinate magnesium amides with distorted tetrahedral coordination geometry*[66]

the structure of $[Mg\{N(SiMe_3)_2\}_2(4\text{-picoline})_2]$[66] (av. $Mg{-}N_{(amide)} = 2.03$ Å, av. $Mg{-}N_{(picoline)} = 2.21$ Å). A similar $Mg{-}N_{amide}$ bond length of 2.023(1) Å is observed in the complex $[Mg\{N(Mes)(SiMe_3)\}_2(thf)_2]$[67] in which the distorted tetrahedral geometry is completed by two thf donor molecules (Mg–O distances are 2.055(2) Å).

Tetrahedral coordination at magnesium is also found in various monomeric chelated complexes such as $[Mg\{N(8\text{-quinolyl})(SiMe_3)\}_2]$,[69] $[Mg\{N(Bu^t)SiMe_2C_6H_4OMe\text{-}2\}_2]$,[70] $[(Mg\{N(Bu^t)SiHMe_2\}_2)_2]$[71](featuring agostic Si–H–Mg interactions) or $[Mg\{1,2\text{-}(NAr)_2C_{12}H_6\}(thf)_2]$,[72,73] ($Ar = C_6H_3Pr^i_2\text{-}2,6 = Dipp$ or $C_6H_3Bu^t_2\text{-}2,5$) and in other complexes related to amides (but outside our scope) such as $[Mg\{(NSiMe_3)_2SN(SiMe_3)_2\}_2]$,[74]$[Mg\{N(R)C(Bu^t)CHR\}_2]$ and $[Mg(\{N(R)C(Ph)\}_2CH)_2]$ ($R = SiMe_3$),[75] $[Mg\{(NBu^t)(NSiMe_3)P(NHBu^t)_2\}_2]$[76] and $[Mg\{(NSiMe_3)_2PPh_2\}_2]$,[25] while in $[Mg\{(NBu^t)_2(PNBu^t)_2\}(thf)_2]$[77] Mg atom has a trigonal bipyramidal coordination. Transamination of $Mg(NPr^i_2)_2$ with the β-diketimine $H[\{N(Dipp)C(Me)\}_2CH]$ yielded the compound $[Mg(NPr^i_2)(\{N(Dipp)C(Me)\}_2CH)(thf)]$,[78] in which the distorted tetrahedral environment at the metal is comprised of one short $Mg{-}N_{amide}$ bond to the diisopropylamido nitrogen atom (1.968 (2) Å) and longer interactions to the chelating nitrogen atoms (2.071(2) and 2.091(2) Å) and the thf oxygen (2.092(2) Å). Similar transamination of $Mg\{N(SiMe_3)_2\}_2$ with $H[\{N(C_6H_4OMe\text{-}2)C(Me)\}_2CH]$ gave the solvent-free compound $[Mg\{N(SiMe_3)_2\}(\{N(C_6H_4OMe\text{-}2)C(Me)\}_2CH)]$,[79] having five-coordinate Mg due to chelating OMe groups. Related derivatives include $[Mg(Me)(\{N(Dipp)C(Me)\}_2CH)(L)\}$ ($L = Et_2O$ or thf).[80,81] An alkane elimination reaction have afforded the triazenido derivative $[Mg\{(NDipp)_2N\}_2(L)]$ ($L = Et_2O$).[82]

Monomeric heteroleptic magnesium amides. Numerous heteroleptic magnesium amido complexes of the type $Mg(NR_2)X(L)$ (X = organo group, L = neutral donor) have been reported. For example, the reaction of $Mg(Bu^n)(Bu^s)$ with the preformed $Mg\{N(H)C_6H_3Pr^i_2\text{-}2,6\}_2$ gave the dodecamer $[\{Mg(Bu^n)(N(H)Dipp)\}_{12}]$, which afforded the monomeric complex $[Mg(Bu^n)\{N(H)Dipp\}(tmeda)]$ upon treatment with tmeda.[83] The tetrahedrally coordinated magnesium has Mg–N(H), Mg–C, and Mg–N(tmeda) bond lengths of 2.0042(17), 2.130(2) Å, and 2.1975(16) and 2.2443(17) Å, respectively. The heteroleptic product $[Mg(Bu^n)\{N(SiMe_3)(Dipp)\}(thf)_2]$ was obtained by the alkane elimination procedure.[84] The ligand redistribution reaction of $Mg\{N(SiMe_3)_2\}_2$ with $Mg\{SMes^*\}_2$ ($Mes^* = C_6H_2Bu^t_3\text{-}2,4,6$) gave $[Mg\{N(SiMe_3)_2\}(SMes^*)(thf)_2]$.[85] The magnesium has distorted tetrahedral coordination with a wide 117.3(1)° angle between the sterically demanding amido and thiolato ligands (Mg–N and Mg–S bond lengths of 1.998(3) and 2.431(2) Å, respectively) and a narrower 93.3(1)° angle between the smaller thf ligands (Mg–O distances = 2.042(3) and 2.051(3) Å). The use of less bulky thiols afforded the thiolato-bridged amido dimers $[\{Mg(\mu\text{-}SAr)\{N(SiMe_3)_2\}(thf)\}_2\}$ ($Ar = Ph$ or $C_6H_2Pr^i_3\text{-}2,4,6$).[85] Reaction of $Mg(Bu^n)Cl$ with a bidentate lithium amide unexpectedly proceeded with the elimination of both the Cl- and alkyl groups yielding the distorted octahedral magnesium complex $[Mg\{N(Ph)(2\text{-}Pyr)\}_2(thf)_2]$.[86] Two chelating ligands form the tetragonal plane (Mg–N bond lengths are 2.105(2) and 2.182(2) Å) with the thf molecules in axial positions (Mg–O distances are 2.212(2) Å). The reaction of $EtMgBH_4$ with $HN(SiMe_3)_2$ afforded the monomeric $Mg(BH_4)\{N(SiMe_3)_2\}(OEt_2)_2$ which, upon further reaction with diglyme, yielded the crystalline hexacoordinate complex $[Mg(\kappa^2\text{-}BH_4)\{N(SiMe_3)_2\}(diglyme)]$ (Mg–N distance 2.013(7) Å and Mg–H interactions near 2.1 Å).[68]

An unusual cationic magnesium amide $[(Et_2O)_3Mg\{N(SiMe_3)_2\}][B(C_6F_5)_4]$ was obtained from the reaction of $[H(OEt_2)_2][B(C_6F_5)_4]$ with $Mg\{N(SiMe_3)_2\}_2$.[87] The magnesium is four-coordinate with an Mg–N bond length of 1.986(3) Å and Mg–O distances near 2.05 Å. This compound and its zinc analogue were investigated as catalysts for the polymerization of cyclohexene oxide and ε-caprolactone. High molecular weight polymers were obtained although the magnesium compound was more sensitive and less stable than its zinc analogue.

Reaction of $[(Mg\{N(SiMe_3)_2\}_2)_2]$ or $[(Mg\{N(SiHMe_2)_2\}_2)_2]$ with surface SiOH groups of mesoporous silicas (SBA-1, SBA-2, SBA-16 or MCM-48) in hexane produced surface-bound amidomagnesium siloxides. Use of $Ti(NMe_2)_4$ together with a Mg amide yielded heterobimetallic hybrid materials.[88]

Three-coordinate monomeric amides. A smaller number of monomeric three-coordinate magnesium amides is known. They can be synthesized with use of a bulky ligand set by a variety of different routes (Scheme 3.3). For example, the trigonal complexes $[Mg\{N(SiMe_3)_2\}_2(L)]$[66] (L = 2,6-lutidine, 2-methylpyridine) were obtained by treatment of the thf solvate $[Mg\{N(SiMe_3)_2\}_2(thf)_2]$ with 2,6-lutidine or by sublimation of the four-coordinate 2-methylpyridine adduct; the magnesium atom in these compounds has an essentially planar geometry. The plane of the aromatic ring is approximately perpendicular to this plane; the bond to the lutidine (2.158(3) Å) is considerably longer than to the amido nitrogen (1.974(3) and 1.976(3) Å), but both of these distances are shorter than the corresponding ones in the related four-coordinate species discussed above. Trigonal planar geometry is also observed in complexes such as $[Mg\{N(SiMe_2Bu^t)(SiMe_3)\}_2(py)]$[52] and $[Mg\{N(Dipp)(SiMe_3)\}_2(Et_2O)]$.[89] Both complexes display considerable angular variation, which results from the size difference between the ligands. In the

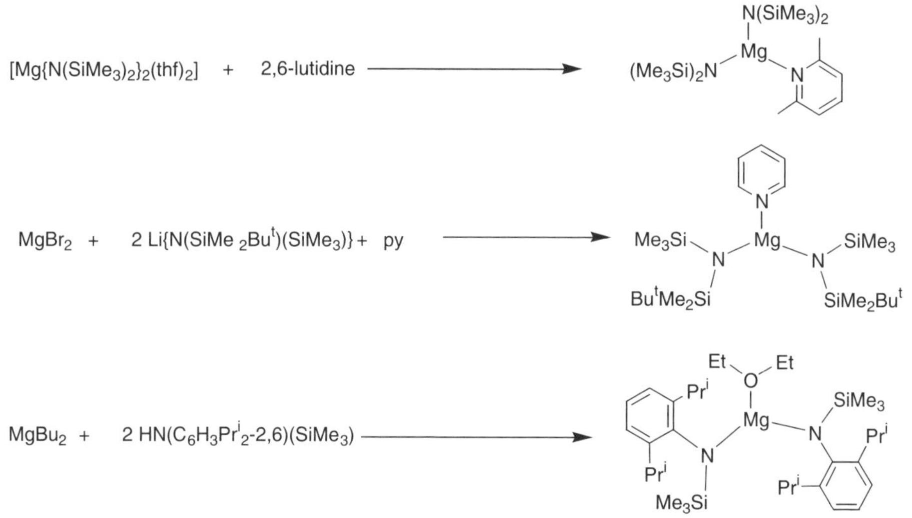

$[Mg\{N(SiMe_3)_2\}_2(thf)_2]$ + 2,6-lutidine ⟶ $[Mg\{N(SiMe_3)_2\}_2(2,6\text{-lutidine})]$

$MgBr_2$ + 2 $Li\{N(SiMe_2Bu^t)(SiMe_3)\}$ + py ⟶ $[Mg\{N(SiMe_2Bu^t)(SiMe_3)\}_2(py)]$

$MgBu_2$ + 2 $HN(C_6H_3Pr^i_2\text{-}2,6)(SiMe_3)$ ⟶ $[Mg\{N(C_6H_3Pr^i_2\text{-}2,6)(SiMe_3)\}_2(Et_2O)]$

Scheme 3.3 *Synthesis of selected three-coordinate magnesium amides with various donor ligands*[52,66,89]

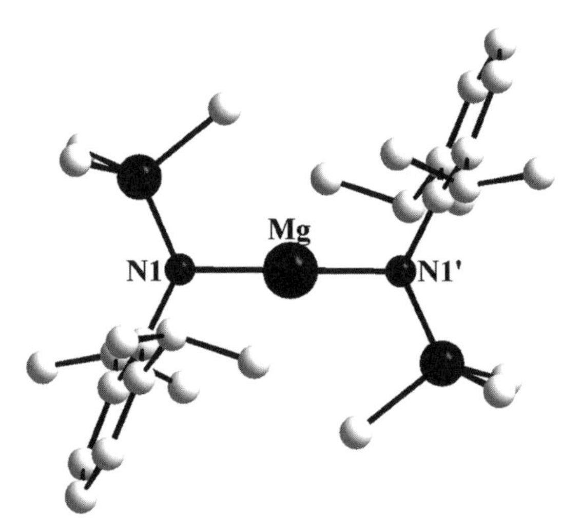

Figure 3.3 Representation of the linear complex $[Mg\{N(Dipp)(SiMe_3)\}_2]$[84] with magnesium and nitrogen atoms shown as black spheres and carbon atoms are white. Mg–N1 bond length: 1.919(2) Å

latter complex the N–Mg–N angle is very wide 140.69(6)°. The Mg–N distances average 1.97 Å and the Mg–O distance is 2.0361(13) Å, with an additional $Mg–C_{(ipso)}$ interaction of 2.799(2) Å.

Two-coordinate monomeric amides. The silylamido derivative $Mg\{N(SiMe_3)_2\}_2$ was first synthesized in 1972 by reaction of $MgEt_2$ with $HN(SiMe_3)_2$.[90] It has a linear monomeric N-M-N structure in the vapor (Mg–N = 1.91(3) Å) by gas electron diffraction.[91] It is a dimer in the solid state (see Section 3.3.3). The reaction of $MgBu^n_2$ with the more sterically demanding amine $HN(SiMePh_2)_2$ afforded $[Mg\{N(SiMePh_2)_2\}_2]$.[54] The average Mg–N distance is 1.97 Å in a non-linear geometry (N–Mg–N is 162°). The bending can be attributed to weak magnesium-aromatic group interactions. A further example of a low coordination environment at magnesium is seen in $[Mg\{N(SiPh_2Bu^t)(SiMe_3)\}_2]$,[52] in which the geometry is even more distorted from linearity (N–Mg–N is 140.2°) due to the '*syn*' positioning of the aromatic groups with respect to the magnesium centre. These groups have weak interactions to the magnesium atom (Mg·C between 2.56 and 2.79 Å) with short Mg–N distances (average 1.98 Å). In contrast, the two-coordinate magnesium atom in $[Mg\{N(Dipp)(SiMe_3)\}_2]$[84] displays an ideal linear geometry with short Mg–N bonds (Figure 3.3); this compound is not coordinatively saturated as is shown by its formation of a 1:1 complex with diethyl ether.[89]

3.3.3 Dimeric Magnesium Amides

Dimers with three-coordinate magnesium. The simplest dimeric magnesium amides are unsolvated and contain a planar $(MgN)_2$ core with trigonal planar coordinated magnesium atoms and bridging amides. The best known example is the crystalline silylamide $[(Mg\{N(SiMe_3)_2\})_2]$,[90] which has a planar $(MgN)_2$ core with terminal Mg–N bond lengths = 1.97 Å (avg.) and longer average bridging distances of 2.15 Å.[92] A similar structure is seen for the dibenzylamide $[(Mg\{N(CH_2Ph)_2\}_2)_2]$,[62] which also has a planar $(MgN)_2$ ring with mean terminal Mg–N and bridging distances of 1.935(2) and 2.09(1) Å, respectively. The amide $[(Mg\{N(Cy)(SiMe_3)\}_2)_2]$[52] also contains a planar central $(MgN)_2$ unit, which has a slightly distorted rectangular structure with longer bridging Mg–N bond lengths (2.1292(19) and 2.1033(18) Å) than terminal Mg–N distances (1.9733(18) Å). The heteroleptic dimers $[\{Bu^tMg(\mu\text{-}NR_2)_2\}_2]$ ($NR_2 = N(SiMe_3)_2$, NPr^i_2, $N(CH_2Ph)_2$ or tmp) featuring three-coordinate magnesiums (Mg–N = 2.08 – 2.13 Å) were obtained by the unusual reaction of $LiBu^t$ with Bu^nMgNR_2.[83] The trigonal planar geometry at the magnesium atom in the non-centrosymmetric $[\{Mg(Bu^n)(\mu\text{-tmp})\}_2]$[93] is highly distorted (mean C–Mg–N and N–Mg–N angles are 132.7 are 94.3°, respectively). Each μ-tmp ligand has average Mg–N and Mg···$C_{(2\text{- or }6\text{-Me})}$ distances of 2.12 and 2.82 Å, respectively. Reaction of CpMgMe-(OEt_2), prepared by dissolving $MgCp_2$ and $MgMe_2$ in Et_2O, with $HNPh_2$, $H_2NCHPr^i_2$, $H_2NC_6H_3Pr^i_2$-2,6, $HN(CH_2Ph)Pr^i$ or $HNPr^i_2$ yielded the bridged amido dimers of the general structure $[\{\eta^5\text{-}C_5H_5Mg(\mu\text{-}NR_2)\}_2]$ in which the Mg_2N_2 core is planar and almost perfectly square.[96]

Treatment of $[\{Bu^nMg(\mu\text{-OAr})\}_2]$ ($OAr = OC_6H_3Bu^t_2$-2,6) with $HNPr^i_2$ afforded $[\{ArOMg(\mu\text{-}NPr^i_2)\}_2]$ in which NPr^i_2 rather than aryloxide is the bridging ligand. The magnesiums have essentially planar three-coordination with bridging Mg–N distance in the range 2.079(1) to 2.114(1) Å.[94] In $[(\{(Me_3Si)_2N\}Mg(\mu\text{-tempo}))_2]$ (tempo = 2,2,6,6-tetramethylpiperidineoxide), obtained by treatment of $MgBu^n_2$ with $HN(SiMe_3)_2$ and tempo, the amido ligand is bound terminally (Mg–N = 1.952(3) and 1.965(3) Å) whereas the tempo ligands act as bridges through oxygen. There is also a long Mg–N (2.395(3) Å interaction between one of the magnesiums and a nitrogen from a bridging tempo ligand.[95]

Dimers with magnesium coordination number ≥4. A much larger number of dimeric magnesium amides with distorted tetrahedral metal coordination have been reported. In these, the degree of distortion depends on the coordinating ligand set.[62,65] The complex $[\{Mg(NPh_2)\{NPh(2\text{-}Py)\}\}_2]$[97] has bridging NPh_2 ligands that afford a planar $(MgN)_2$ core (Mg–N bond lengths between 2.08 and 2.11 Å) and the distorted tetrahedral geometry at Mg is completed by the chelating NPh(2-Py) ligand (Mg–N distances are between 2.03 and 2.11 Å). An interesting tricyclic structure is observed in the dimer [(Mg{*N*-trimethylsilylfurfurylamide}$_2$)$_2$],[98] in which the planar $(MgN)_2$ core is formed by coordination of two bridging amido ligands (mean Mg–N bond length is 2.13 Å) and further coordination of the oxygen atom from the same ligand results in a (MgOCCN) ring on each side of the $(MgN)_2$ core. The distorted tetrahedral geometry at the magnesium atom is completed by coordination of the second amido ligand via the nitrogen atom with an Mg–N distance of ca. 1.97 Å. Magnesium amides with $(MgN)_2$ cores can be formed from bridging diamido ligands and terminal donor molecules. For example, alkane elimination from $Mg(Bu^n)(Bu^s)$ by $C_6H_4\{N(H)SiMe_3\}_2$-1,2 in Et_2O yielded the centrosymmetric dimeric magnesium amide $[\{(Et_2O)Mg\{N(SiMe_3)\}_2C_6H_4\text{-}1,2\}_2]$, Equation (3.6), in which Mg–$N_{(bridging)}$, Mg–N and Mg–O distances are 2.083, 1.997 and 2.083 Å, respectively.[99] Protonolysis of $MgBu_2$ with a diamine $HN(R)CH_2CH_2N(R)H$ (R = Ph) in the presence of thf afforded the five-coordinate dimer $[\{(thf)_2Mg(PhNCH_2CH_2NPh)\}_2]$, while with $R = CH_2Ph$ and HMPA as a neutral donor the tetrahedral crystalline compound $[\{(HMPA)Mg(PhCH_2NCH_2CH_2NCH_2Ph)\}_2]$ was obtained.[100]

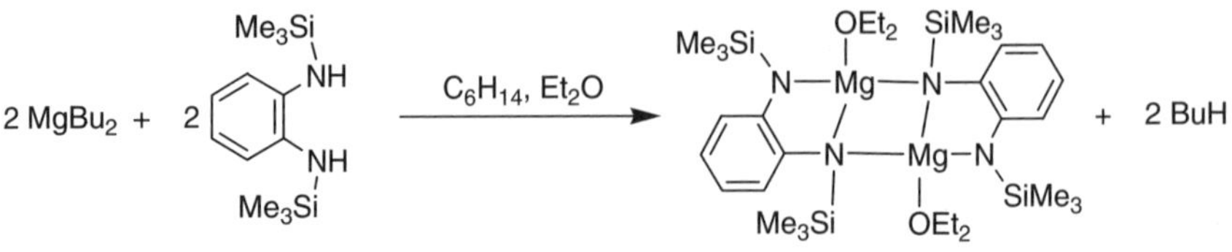

$$2\,MgBu_2 + 2\,C_6H_4(NHSiMe_3)_2 \xrightarrow{C_6H_{14},\ Et_2O} [Mg\{\mu\text{-}N(SiMe_3)C_6H_4N(SiMe_3)\}(OEt_2)]_2 + 2\,BuH \qquad (3.6)$$

The reaction of aniline with $Mg\{N(SiMe_3)_2\}_2$ in thf gave $[Mg\{\mu\text{-}N(H)Ph\}\{N(SiMe_3)_2\}(thf)]_2$; the average $Mg{-}N_{(bridging)}$, Mg–N and Mg–O distances are 2.12, 2.00 and 2.03 Å, respectively and the $(MgN)_2$ core lies perpendicular to the aromatic ring plane.[101]

Four-coordinate magnesium amido dimers based upon an (Mg_2NX) or $(MgX)_2$ (X non-amido) core are also known. Treatment of BuMgCl with a chelating secondary lithium amide yielded in the tricyclic dimeric complex $[(Mg(Bu^n)\{\mu\text{-}N(CH_2Ph)CH_2CH_2\text{-}NMe_2\})_2]$, Mg–C, 2.135(2) Å.[85] The magnesium atoms have a distorted tetrahedral geometry with chelation by NMe_2 group of the bridging ligand ($Mg{-}N_{(bridging)}$ and $Mg{-}NMe_2$ distances are *ca.* 2.11 and 2.20 Å, respectively). The reaction of the primary amine Bu^tNH_2 with $MgBu^t_2$ can result in compounds with three different donor atoms coordinating to the magnesium, as in $[\{Bu^tMg\{\mu\text{-}N(H)Bu^t\}(thf)\}_2]$,[102] which has a planar $(MgN)_2$ core (Mg–N, 2.098 Å); the Mg–C and Mg–O bond lengths are 2.194(4) and 2.085 (4) Å. As mentioned in Section 3.3.2.2, species with $(MgX)_2$ cores can be synthesized from the reaction of $Mg\{N(SiMe_3)_2\}_2$ with a magnesium thiolate to give complexes such as $[(Mg\{N(SiMe_3)_2\}(\mu\text{-}SPh)(thf))_2]$ and $[(Mg\{N(SiMe_3)_2\}(\mu\text{-}STrip)(thf))_2]$ (Trip = $C_6H_2Pr^i_3$-2,4,6).[85] The core unit of the compound $[(Et_2O)Mg(\mu\text{-}Cl)\{N(SiMe_3)_2\}]_2$[53] is the almost perfectly square $(MgCl)_2$. The magnesium atoms have a distorted tetrahedral geometry, with Mg–N and Mg–O bond lengths of 1.970(3) and 2.000(3) Å, respectively.

The deprotonation of ketones to give enolate synthons is a very important organic reaction. Although lithium reagents are often used for this purpose, organomagesium reagents can also be employed. Among the first investigations involving magnesium amides was the reaction of $MgBu_2$ with two equivalents of $HN(SiMe_3)_2$ followed by the addition of benzophenone. This afforded the product $[(\{(Me_3Si)_2N\}Mg\{\mu\text{-}OCHPh_2\}(OCPh_2))_2]$,[103] which is believed to form in accordance with Scheme 3.4 by β-H transfer.

The reaction of propiophenone, Ph(Et)CO, with $Mg\{N(SiMe_3)_2\}_2$ showed that in the presence of a donor solvent the ketone is deprotonated stereoselectively to afford the *Z*-isomer of $[(\{Me_3Si)_2N\}(thf)Mg\{\mu\text{-}OC(Ph)C{=}HMe\})_2]$, with $Mg{-}O_{(bridging)}$, Mg–N and $Mg{-}O_{(thf)}$ distances of 1.987, 2.019 and 2.046 Å, respectively.[104]

The reaction of $MgEt_2$ with an amine containing *O*-donor atoms, $HN\{CH_2CH_2OMe\}_2$ ($\equiv HL^1$) or aza-15-crown-5 ($\equiv HL^2$), gave 5-coordinate dimeric compounds

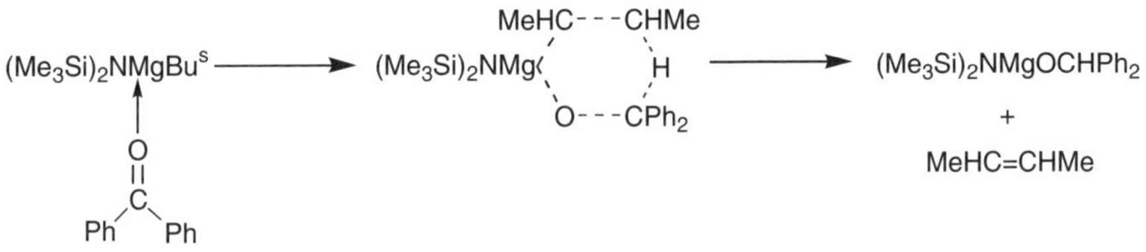

$$(Me_3Si)_2NMgBu^s + Ph_2C{=}O \longrightarrow (Me_3Si)_2NMg(\cdots MeHC{-}{-}{-}CHMe\cdots H\cdots O{-}{-}{-}CPh_2) \longrightarrow (Me_3Si)_2NMgOCHPh_2 + MeHC{=}CHMe$$

Scheme 3.4 *Proposed route for the formation of $[(\{Me_3Si)_2N\}Mg\{\mu\text{-}OCHPh_2\})_2]$*[103]

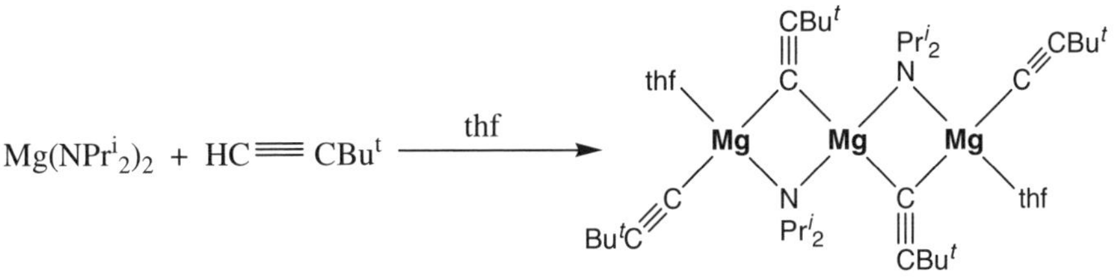

$$Mg(NPr^i_2)_2 + HC \equiv CBu^t \xrightarrow{thf}$$

Scheme 3.5 Synthesis of a trinuclear magnesium amide[55]

$[\{EtMg(\mu\text{-}L^1)\}_2]$ or $[\{EtMg(\mu\text{-}L^2)\}_2]$ with bridging amido nitrogen atoms (average Mg–N distances of 2.135 and 2.149 Å, respectively); only two of four O atoms of the aza-15-crown-5 anion are coordinated to magnesium.[105]

3.3.4 Higher Aggregates and Related Magnesium Amides

Several types of higher aggregate magnesium amides are known. The unusual trinuclear compound $[(Bu^tC{\equiv}C)(thf)Mg(C{\equiv}CBu^t)(NPr^i_2)Mg(C{\equiv}CBu^t)(NPr^i_2)Mg(thf)(C{\equiv}CBu^t)]$[55] is composed of two non-planar MgNMgC units connected through the central magnesium atom (Scheme 3.5). The complex was synthesized by the reaction of the magnesium amide $Mg(NPr^i_2)_2$ with the acetylene $HC \equiv CBu^t$ in thf at room temperature. Each of the magnesiums has distorted tetrahedral geometry, with endocyclic Mg–N and Mg–C bond lengths of 2.09–2.12 Å and 2.20–2.23 Å, respectively. The outer magnesiums are coordinated by a terminal ethynyl ligand (Mg–C, 2.092(3) Å) and a thf molecule (Mg–O, 2.057(2) Å).

The unusual compound $[\{MgBu^n(\mu_2\text{-}N(H)Dipp)_2Mg(\mu_3\text{-}OBu^n)\}_2]$,[93] which has been described by Mulvey *et al.* as the 'first homometallic inverse crown', has a tetrametallic magnesium amide structure with an octagonal cationic ring capped on either side by BuO groups (Scheme 3.6). There are two distorted tetrahedral magnesium environments within the chair-shaped ring: (i) two metal atoms are coordinated by two bridging amido ligands (Mg–N, 2.041(3) Å) and two Bu^nO groups (Mg–O = 2.039(3) and 2.068(2) Å), and (ii) two

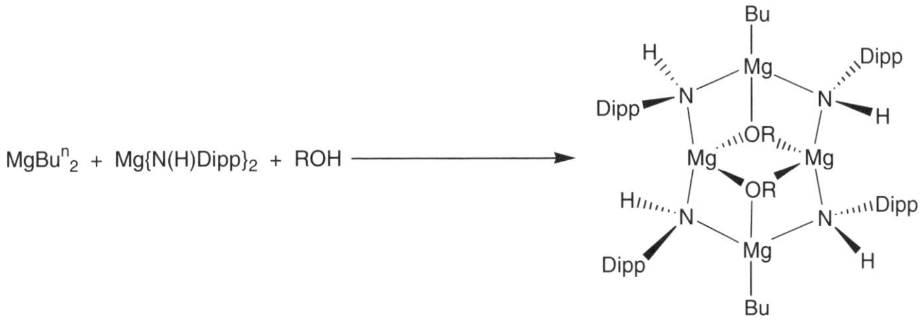

$$MgBu^n_2 + Mg\{N(H)Dipp\}_2 + ROH \longrightarrow$$

Scheme 3.6 Synthesis of the tetrametallic inverse crown $[\{MgBu^n(\mu_2\text{-}N(H)Dipp)_2Mg(\mu_3\text{-}OBu^n)\}_2]$ (R = Bu, Dipp = $C_6H_3Pr^i_2$-2,6)[93]

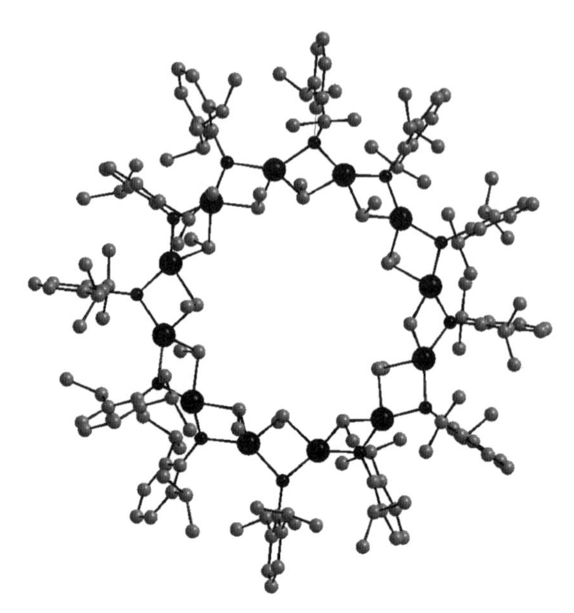

Figure 3.4 Representation of the dodecameric macrocycle $[Mg(\mu\text{-}Et)\{\mu\text{-}NH(Dipp)\}]_{12}$[102] with magnesium and nitrogen atoms shown as large and small black spheres, respectively, and carbon atoms in grey

metal ions are coordinated by two bridging amido ligands (Mg–N = 2.168(4) Å) and a triply bridging OBu^n (Mg–O = 2.121(8) Å) and a Bu group.

The highest degree of aggregation currently known for a magnesium amide molecule is observed in the dodecameric primary amide derivative $[(Mg(\mu\text{-}Et)\{\mu\text{-}NH(Dipp)\})_{12}]$,[102] (see also Ref. 83) which was obtained from the reaction of $MgEt_2$ with $2,6\text{-}Pr^i{}_2C_6H_3NH_2$. It has a cyclic structure composed of 12 (MgNMgC) rings in convex fusion that are linked through the common magnesium centres (Figure 3.4). The ethyl groups are located in the interior of the cycle and the more bulky $NHC_6H_3Pr^i{}_2$-2,6 groups project from the exterior. The magnesiums have distorted tetrahedral geometry, with average Mg–N and Mg–C distances of 2.09 and 2.21 Å, respectively.

3.3.5 Heterometallic Magnesium Amides

Magnesium amides can form a wide range of mixed metal amido complexes with alkali metal ions ($M_2[Mg(NH_2)_4]$) (M = K, Rb or Cs).[106,107] These contain tetrahedral Mg-centered $[Mg(NH_2)_4]^{2-}$ ions connected in three-dimensional networks by coordination of the amido groups to the group 1 metal ions. The most common 'hetero' metal is lithium and lithium amido magnesiates are readily accessible by the addition of a lithium amide to a magnesium amide.

For example, $[LiMg\{N(SiMe_3)_2\}_3]$[108] was obtained by the simple addition of $LiN(SiMe_3)_2$ to $Mg\{N(SiMe_3)_2\}_2$. It features a magnesium bound to three silylamides in a near-planar fashion (Mg–N bond lengths are between 1.998(4) and 2.125(4) Å). The lithium ion is bound to two of these amides and displays further interactions to methyl groups on the amido ligand (average Li–N and Li–C distances are 2.02 Å and 2.29 and 2.83 Å respectively). The lithium is coordinatively unsaturated and interacts with a Lewis base, as shown in [LiMg{N(SiMe$_3$)$_2$}$_3$(thf)], [LiMg{N(SiMe$_3$)$_2$}$_3$(py)] and [LiMg{NCy$_2$}$_3$(thf)].[109] The heteroleptic complex [LiMg{N(SiMe$_3$)$_2$}$_2$(Bu^t)][110] features a bimetallic, planar (LiNMgN) ring in which the lithium and magnesium atoms are bridged by two [N(SiMe$_3$)$_2$]$^-$ anions (mean Li–N and Mg–N bond lengths are 2.04 and 2.08 Å, respectively). The magnesium has distorted trigonal coordination geometry completed by a terminal alkyl Bu^t group. The lithium atom has quasi-two-coordinate non-linear geometry within the discrete bimetallic units with an additional short intermolecular contact to a methyl group of the Bu^t ligand of an adjacent unit – resulting in a linear polymeric structure.

Heterometallic diorganoamido species involving the heavier alkali metals are also known; for example, the polymeric compound [{KMg(NPr$^i_2$)$_3$}$_\infty$][111] has planar (MgNKN) rings which are linked to form an infinite spiral chain via bridging of the terminal [NPr$^i_2$]$^-$ ligand (Mg–N and K–N distances range between 2.03–2.05 and 2.91–2.95 Å, respectively). Addition of tmeda to the reaction mixture resulted in the formation of [KMg(NPr$^i_2$)$_3$(tmeda)][111] with distorted trigonal planar magnesium geometry (Mg–N bond lengths 1.97–2.03 Å) and distorted tetrahedral geometry at the potassium ions (K–N distances 2.86 - 2.91 Å). The sodium analogue, [NaMg(NPr$^i_2$)$_3$(tmeda)], is isostructural and features the four-membered (MgNNaN) cycle (Mg–N and Na–N bond lengths of 1.97–2.06 and 2.49–2.56 Å, respectively).[111a] A solvent separated ion pair [Na(pmdeta)$_2$][Mg(tmp)$_3$], which has a trigonal planar coordinated magnesium and Mg-N distances near 2.05 Å has also been reported.[111b] A (MgNNaC) ring is observed in the compound [(tmeda)Na(μ-Bu^n)(μ-tmp)Mg(tmp)].[112] The preference for Bu^n as one of the bridging ligands is presumably steric in origin. The four-element ring is slightly distorted with bridging butyl (Mg–C and Na–C distances are 2.200(2) and 2.669(2) Å, respectively) and tmp groups (Mg–N and Na–N distances are 2.0791(17) and 2.4523(18) Å, respectively). The magnesium atom has distorted trigonal planar geometry completed by a second tmp ligand, and the four-coordinate sodium ion is chelated by one tmeda ligand. The reaction of NH(SiMe$_3$)$_2$ with butylmagnesium and butylpotassium and an arene yielded a variety of heterometallic structures with the general formula [{K(arene)$_2$Mg{N(SiMe$_3$)$_2$}$_3$}].[113] For example, when arene = benzene, the structure is a linear, polymeric chain of [Mg{N(SiMe$_3$)$_2$}$_3$]$^-$ units connected via coordination of methyl substituents to potassium ions with mean K–C$_{(methyl)}$ distances of 3.26 Å. The magnesium atoms have distorted trigonal planar geometry, whereas the potassium ions form multiple interactions with two benzene molecules that lie approximately *trans* to each other across the potassium ion. One-dimensional polymeric structures are also observed where Ar = toluene or xylene, with additional coordinative interactions between methyl groups and the potassium ions to form successive six-membered (KCSiNSiC) rings along the chain. Alternatively, a second isomer can be formed when Ar = toluene, in which a tetrameric ring structure results from a variation in the approach angle of the methyl substituents to the potassium ions.

A remarkable aspect of the heterobimetallic magnesium salts is the discovery that compounds, derived from the pairing of two different metal atoms usually Li, Na, or K and Mg or Zn that are bound by amido ligands, can deprotonate arene rings;[114] *e.g.*, [(tmeda)Na(μ$_2$-Bu^n)(μ$_2$-tmp)Mg(tmp)][112] which quantitatively *meta*-deprotonated toluene, Equation (3.7).[115a]

$$[(\text{tmeda})\text{Na}(\mu_2\text{-Bu}^n)(\mu_2\text{-tmp})\text{Mg}(\text{tmp})] \xrightarrow[-\text{Bu}^n\text{H}]{+\text{PhMe}} [(\text{tmeda})\text{Na}(\mu_2:\eta^1\text{-C}_6\text{H}_4\text{Me-3})(\mu_2\text{-tmp})\text{Mg}(\text{tmp})] \quad (3.7)$$

More recent investigations have shown that the reaction of a Na(tmp)/Bu^tMg(tmp)/tmeda mixture with 0.5 equivalent of benzene leads to dideprotonation of the arene to afford [{(tmeda)Na(μ-tmp)Mg$(tmp)_2$}$_2C_6H_4$].[115b] The reaction of the magnesium amide $Mg(tmp)_2$ with $LiN(SiMe_3)_2$ resulted in a deprotonation/amine elimination process as shown in Scheme 3.7.[116] The product, a dimeric lithium/magnesium mixed amide has a conformationally locked structure with two stereogenic nitrogen centres.[113]

The trimetallic complex [Li_2Mg{N(CH_2Ph)($CH_2CH_2NMe_2$)}$_4$][117] is free of additional donor ligands and is composed of two heterometallic (MgNLiN) rings connected through the central magnesium atom in a spiro fashion. Each metal centre has a distorted tetrahedral coordination geometry to four nitrogen atoms with central Mg–N bond lengths between 2.07 and 2.12 Å. The lithium atoms are coordinated by two bridging amido ligands and two chelating amino nitrogens. A similar trimetallic framework is observed in [Li_2Mg{N-($CH_2Ph)_2$}$_4$],[118] with each lithium ion having a two-coordinate environment (Li–N bond lengths are in the range 1.92 to 2.04 Å). The central magnesium atom connects the two (MgNLiN) rings in a distorted tetrahedral geometry (Mg–N distances are between 2.07 and 2.12 Å). Addition of pyridine to this reaction mixture afforded [LiMg{N($CH_2Ph)_2$}$_3$(py)], which contains a planar (MgNLiN) ring with trigonal planar coordinated lithium and

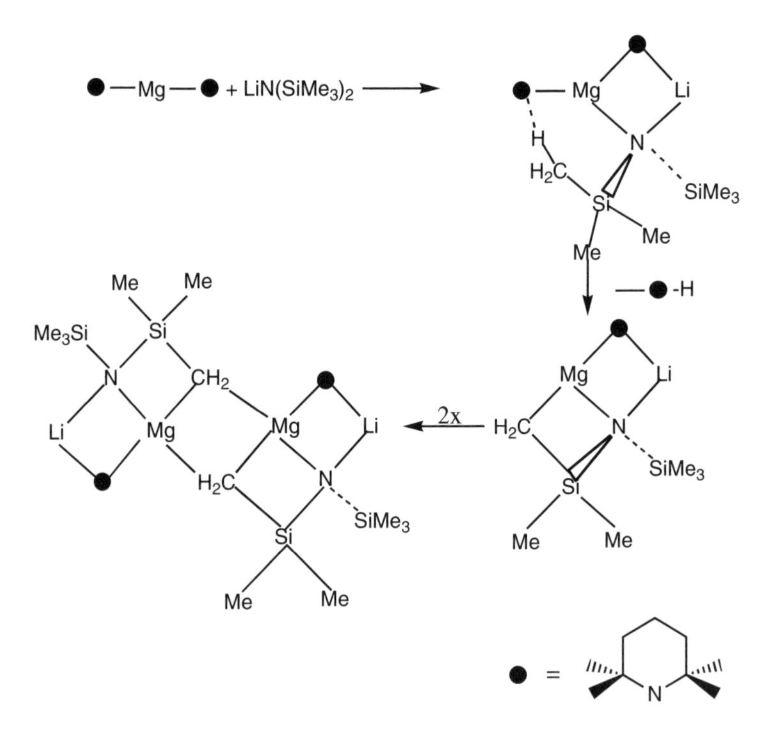

Scheme 3.7 *Deprotonation/amine elimination in a mixed lithium/magnesium amide*[116]

magnesium atoms (Mg–N and Li–N distances 1.94–2.05 and 2.01–2.06 Å, respectively).[118]

In addition to the heterometallic species involving group 1 and group 2 metals there is also an example featuring two alkaline earth elements in the heterobimetallic compound $[(Me_3Si)_2NMg\{\mu\text{-}N(SiMe_3)_2\}_2CaN(SiMe_3)_2]$[119] which has a puckered (MgNCaN) ring with two bridging amido ligands (average $Mg-N_{(bridging)}$ and $Ca-N_{(bridging)}$ distances are 2.14 and 2.51 Å, respectively) and two terminal amido ligands (average Mg–N and Ca–N bond lengths are 1.99 and 2.27 Å, respectively).

There have been several heterometallic species involving group 13 metals.[120–125] For example, the reaction of $AlMe_3$ with $Mg(NEt_2)_2$ or $Mg(NPr^i{}_2)_2$ afforded $[\{Me_2Al(\mu\text{-}NEt_2)_2MgMe\}_2]$ or $[\{Me_2Al(\mu\text{-}NPr^i{}_2)_2MgMe\}_4]$.[120] The former is tricyclic with a central $\{Mg(\mu\text{-}Me)\}_2$ ring and flanking $Mg(\mu\text{-}NEt_2)_2AlMe_2$ rings. The latter is weakly associated to form tetramers and involves four-coordinate aluminiums and quasi three-coordinate magnesiums; the $Mg-CH_3$ distance is 2.151(6) Å and the methyl is weakly associated with a magnesium from a neighbouring molecule, $Mg\cdots CH_3 = 2.493(6)$ Å. Another example is $[\{(R_2N)Mg(\mu\text{-}Me)\}_2\{Mg(NR_2)(\mu\text{-}NR_2)AlMe_3\}_2]$, obtained from $[\{Mg(NR_2)_2\}_2(1,4\text{-dioxane})]$ and $(AlMe_3)_2$ ($R = SiMe_3$), which involves the central $MeMg\{\mu\text{-}N(SiMe_3)_2\}_2MgMe$ unit linked via the terminal MgMe groups to the magnesiums in two $Mg\{N(SiMe_3)_2\}\{\mu\text{-}N(SiMe_3)_2\}AlMe_3$ moieties.[121] Treatment of $[\{Me_2Al(\mu\text{-}NPr^i{}_2)_2MgMe\}_4]$ (i) with $CHCl_3$ gave $[\{Me_2Al(\mu\text{-}NPr^i{}_2)_2MgCl\}_2]$, which with LiR ($R = Bu^t$ or Ph) yielded $[\{Me_2Al(\mu\text{-}NPr^i{}_2)_2MgR\}_n]$ ($n = 1$ or 2);[122] and (ii) with HNR_2 ($R = Et$, Pr^i) or I_2 afforded $[\{Me_2Al(\mu\text{-}NPr^i)_2MgNR_2\}_2]$ or $[\{Me_2Al(\mu\text{-}NPr^i{}_2)_2Mg(\mu\text{-}I)\}_2]$.[123] Degradation reactions on the same starting material with a phenol gave a variety of products, *e. g.*, when $HOC_6H_2Bu^t{}_2$-2,6-Me-4 (HOAr) was used the monomeric product $[Me_2Al(\mu\text{-}NPr^i)_2MgOAr]$ with three-coordinate magnesium was obtained.[124] A series of ethynyl-bridged polynuclear Al/Mg complexes $[\{Me_2Al(\mu\text{-}NR_2)_2MgC{\equiv}CR'\}_2]$ ($R = Et$ or Pr^i; $R' = C_6H_4CH_3$-4, Bu^t or Ph) was obtained by treatment of $[(MeAl(\mu\text{-}NR_2)_2MgMe)_n]$ with the alkyne.[125] A number of reactions with heterocumulenes to afford chelated products have also been reported,[126–128] but these are outside our scope (see Ref. 35 for details).

Heterometallic amides featuring nitrido ligands and transition metals have also been described, including the single-cubane structures $[\{(Me_3Si)_2N\}Mg\{(\mu_3\text{-}N)(\mu_3\text{-}NH)_2Ti_3(\eta^5\text{-}C_5Me_5)_3(\mu_3\text{-}N)\}]$[129,130] and $[\{4\text{-}MeC_6H_4(H)N\}Mg\{(\mu_3\text{-}N)(\mu_3\text{-}NH)_2Ti_3(\eta^5\text{-}C_5Me_5)_3(\mu_3\text{-}N)\}]$.[131] The latter was synthesized by the addition of the former to a toluene solution of $4\text{-}MeC_6H_4NH_2$; its structure features a (Ti_3MgN_4) cubane framework with the magnesium atom occupying a distorted tetrahedral coordination site.

Another example of nitride incorporation is shown in Equation (3.8). The product features a nitride surrounded by six Mg atoms ($Mg-N = 2.148(3)$ Å avg.) that describe a trigonal prism. Each metal is also bound to three $\mu_2\text{-}NHBu^t$ groups ($Mg-N = 2.093(2)$ Å avg.) to afford four-coordinate magnesium centers.[60]

$$4Mg(NHBu^t)_2 + 2Mg + 2Bu^tNH_2 \longrightarrow [(\mu_6\text{-}N)Mg_6(NHBu^t)_9] + Bu^tH + H_2 \quad (3.8)$$

A variety of alkali metal and metallocene-containing magnesium amides are known. Thus, attempted metallation of ferrocene using potassium tris(amido)magnesiate $KMg\{N(SiMe_3)_2\}_3$ unexpectedly led to π-coordination of neutral ferrocene moieties to a potassium cation in the crystalline $[K\{Fe(C_5H_5)_2\}_2(PhMe)_2][Mg\{N(SiMe_3)_2\}_3]$.[132] Selective

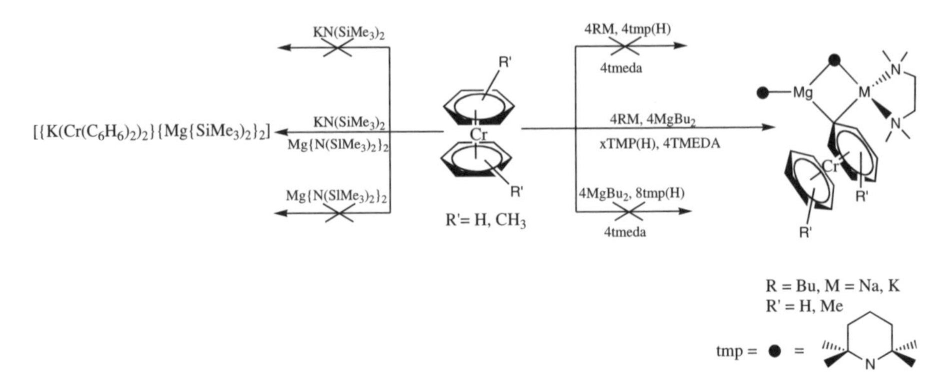

Scheme 3.8 *Metallation reactions of bis(benzene)chromium and bis(toluene)chromium with alkali metal alkyls and magnesium amides*[134,135]

1,1′-double deprotonation of ferrocene was achieved with a synergic base mixture M(tmp)/$Mg(tmp)_2$ to yield $[\{Fe(C_5H_4)_2\}_3M_2Mg_3(tmp)_2(L)_2]$ (M = Li or Na, L = tmpH or py).[133] A different outcome was observed in the reaction of bis(benzene)chromium or bis(toluene) chromium with sodium, potassium and magnesium amides.[134,135] No metallation was observed in the reaction between bis(benzene)chromium with $KN(SiMe_3)_2$ and $Mg\{N(SiMe_3)_2\}_2$. The resulting compound was the unusual ion-separated complex $\{[K\{Cr(C_6H_6)_2\}_2][Mg\{N(SiMe_3)_2\}_3]\}_\infty$,[134] in which the bis(benzene) chromium units form π-interactions with the potassium cation, which in turn forms $K\cdots C_{(methyl)}$ contacts with the anion (K–C distances 3.16–3.69 Å). Use of the stronger base tetramethylpiperidido (tmp^-) facilitates the metallation of bis(benzene)chromium with Mg/(Na or K)[134] and of bis (toluene)chromium with Mg/Na amides.[135] The compounds $[NaMg\{Cr(C_6H_5)(C_6H_6)\}(tmp)_2(tmeda)]$[134] and $[KMg\{Cr(C_6H_5)(C_6H_6)\}(tmp)_2(tmeda)]$[134] both contain a central (MgNMC) ring (M = Na, K), where the N atom is from a bridging tmp ligand and the C atom is a coordinating aromatic carbon from the monodeprotonated bis(benzene)chromium unit. In each case the magnesium atom has distorted trigonal planar geometry (Mg–N and Mg–C are 2.00–2.09 and 2.20 Å, respectively) and the group 1 metal ion has distorted tetrahedral geometry due to chelation by the tmeda ligand (Na–N and Na–C are 2.46–2.52 and 2.71 Å, respectively, and K–N and K–C are 2.83–2.92 and 2.92–3.22 Å, respectively). The monometallation reactions of bis(benzene)chromium and bis(toluene)chromium are summarized in Scheme 3.8.

3.3.6 Magnesium Inverse Crown Complexes

An important class of alkali and alkaline earth metal amides are Mulvey's inverse crown complexes (also discussed in Chapter 2, dealing with sodium and potassium amides), in which cationic homo- or heterometallic macrocycles are hosts to anionic guest moieties.[38,111] The term 'inverse crown' indicates that the Lewis acidic/Lewis basic sites are reversed or exchanged in comparison to conventional crown ether complexes. Scheme 3.9 illustrates the range of recently published alkali and alkaline earth metal amide inverse crown complexes (for related Zn species see Chapter 7 on group 12 amides).

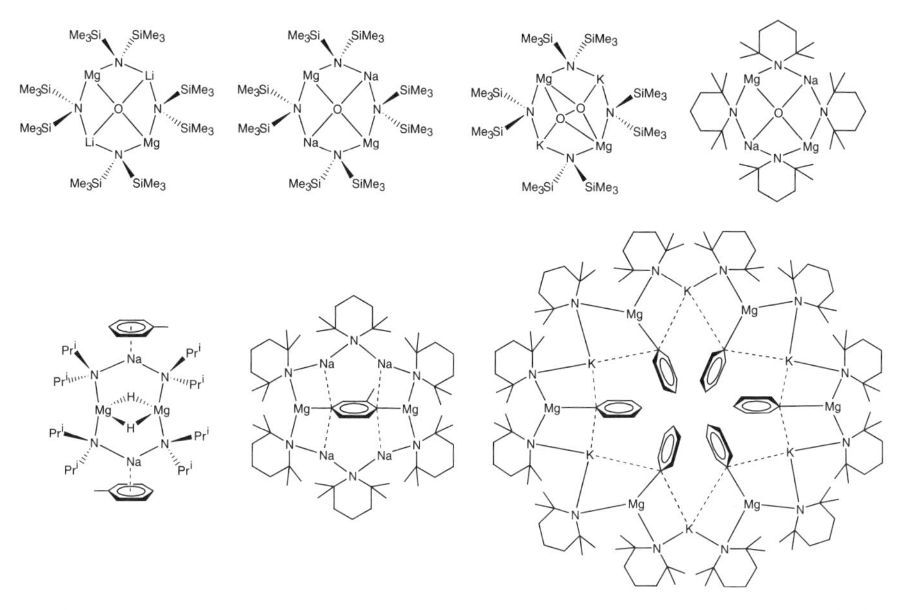

Scheme 3.9 *A selection of magnesium amide-based inverse crown complexes*[139]

The first alkaline earth metal inverse crowns were discovered in the late 1990s via the fortuitous reaction of traces of O_2 with a mixture of an alkali metal and a magnesium amide to afford complexes of the formulae (see Scheme 3.9) $M_2Mg_2\{N(SiMe_3)_2\}_4(O_2)_x(O)_y$ ($M = Li$[108] or Na[136]) and $[Li_2Mg_2(tmp)_4(O)]$,[136] or $[\{K_2Mg_2\{N(SiMe_3)_2\}_4(O_2)\}_\infty]$[137] which is associated via $K{\cdot}CH_3$ interactions.

Structures related to inverse crowns are also formed from the reaction of $MgBu^n{}_2$, MBu^n ($M = Li$, Na) and $HNPr^i{}_2$ and subsequent addition of HOR ($R = Oct^n$, Bu^n).[138] The three compounds $[LiMg\{NPr^i{}_2\}_2(OOct^n)]_2$, $[NaMg\{NPr^i{}_2\}_2(OOct^n)]_2$ and $[NaMg\{NPr^i{}_2\}_2(O\text{-}Bu^n)]_2$ are isostructural and contain eight-membered $(MgNMN)_2$ rings ($M = Li$, Na) with two internal OR groups. For example, the metal centres in $[NaMg\{NPr^i{}_2\}_2(OBu^n)]_2$ (Fig. 3.5) have distorted tetrahedral geometry with coordination by two bridging nitrogen atoms (Mg–N and Na–N distances both average 2.05 Å) and two central alkoxo oxygen atoms (mean Mg–O and Na–O distances are 2.02 and 2.05 Å, respectively).

Inverse crown compounds hosting metallocenyl moieties have also been described. For example, a series of complexes was synthesized by the multiple deprotonation of ferrocene,[139] ruthenocene or osmocene[140] using sodium-magnesium tris(diisopropylamide) as base. The inverse crown structures comprise a 16-membered $[(NaNMgN)_4]^{4+}$ ring with a metallocenyl $[M(C_5H_3)_2]^{4-}$ 'guest' ion. For example, the magnesium atoms in the compound $[\{Ru(C_5H_3)_2\}Na_4Mg_4\{NPr^i{}_2\}_8]$ (Figure 3.6) have a distorted trigonal planar environment with two bridging amido contacts and one $Mg{-}C_{(Cp)}$ interaction, whereas there are two distinctly different types of coordination for sodium atoms – distorted tetrahedral geometry for Na1 and pseudo-trigonal for Na2.

Hydride encapsulation in inverse crown molecules has also been observed. The reaction of NaBu, $MgBu_2$ and three equivalents of $HNPr^i{}_2$ led to $[Na_2Mg_2(NPr^i{}_2)_4(\mu\text{-}H)_2]$[141] which

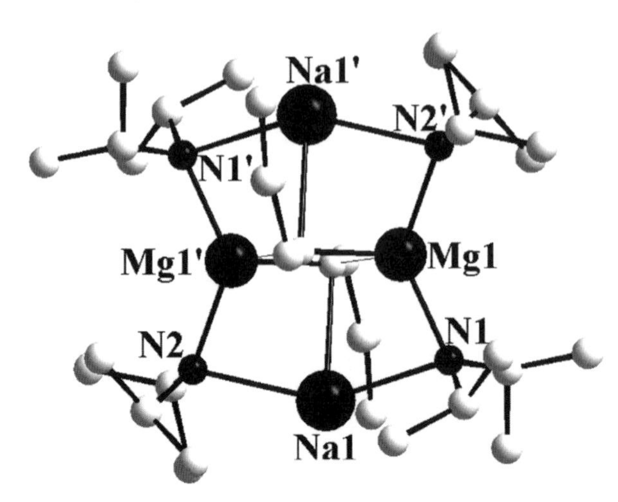

Figure 3.5 *The heterometallic cluster* $[NaMg\{NPr^i_2\}_2(OBu^n)]_2$[138] *with sodium, magnesium and nitrogen atoms shown as black spheres and carbon and oxygen atoms are white*

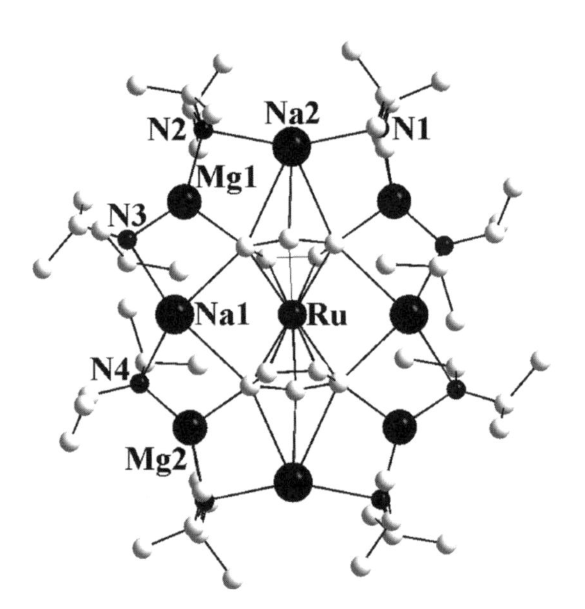

Figure 3.6 *Representation of the* $[\{Ru(C_5H_3)_2\}Na_4Mg_4\{NPr^i_2\}_8]$,[140] *showing the* $(MgNNaN)_4$ *ring as host to a deprotonated ruthenocene guest molecule. Metal atoms are shown as large black spheres, nitrogen atoms as small black spheres and carbon atoms are white. Selected bond lengths: Mg1–N1 2.044(3), Mg1–N2 2.014(3), Mg2–N3 1.997 (3), Mg2–N4 2.047(3), Mg1–C27 2.136(3), Mg1–C29 2.150(3), Na1–N2 2.542(3), Na1–N3′ 2.737(3), Na2–N1 2.511(3), Na2–N4 2.475(3), Na1–C27 2.818(3), Na1–C29 2.661(3), Na2···C25 3.286(3) Å*

was believed to arise by cleavage of an Mg–NPr^i_2 bond in $NaMg(NPr^i_2)_3$ with elimination of $Pr^iN=CMe_2$ and formation of an Mg–H bond leading to formation of the dimerized product.

3.3.7 Magnesium Imides

Simple magnesium imides in which nitrogen carries a hydrogen have been known for several decades.[40,142] They are attracting current interest as components of chemical hydrogen storage materials.[143–145] However, well-characterized magnesium imides in which the nitrogen carries an organic group are of more recent origin. The structure of $[\{(thf)Mg\}(AlH)_3(NBu^t)_4]_2$ with a cubane $MgAl_3N_4$ core has already been mentioned.[2] It was prepared either by the reaction of a mixture of $Mg(AlH_4)_2$ and $AlH_3{\cdot}NMe_3$ with H_2NBu^t or by the direct reaction of the metals with H_2NBu^t in the presence of thf under H_2 pressure.[146]

In earlier sections it has been shown that the reaction of a magnesium dialkyl with one equivalent of a primary amine usually affords the primary amido derivative. However, it is also possible to effect elimination of the second equivalent of alkane to give magnesium imides. The reaction of $MgEt_2$ with aniline in thf proceeded via an amido intermediate which eliminated ethane to yield the hexameric imido compound $[\{(thf)MgNPh\}_6]$ which was the first homometallic magnesium imide to be structurally characterized, Equation (3.9).[147]

$$6\,MgEt_2 + 6\,PhNH_2 \xrightarrow{thf} [\{(thf)Mg(NPh)\}_6] + 12C_2H_6 \tag{3.9}$$

The structure has a slightly distorted hexagonal-prismatic arrangement with alternating magnesium and nitrogen atoms located at the apices. The mean Mg–N bond length within the parallel six-membered Mg_3N_3 rings is 2.08 Å. These two rings are linked through *inter*-cyclic Mg–N bonds that have a slightly shorter average distance of 2.05 Å. Each magnesium is further coordinated by a thf molecule (average Mg–O is 2.04 Å). Likewise, the reaction of $H_2NC_{10}H_7$ ($C_{10}H_7$ = 1-naphthyl) with $MgBu_2$ afforded the naphthyl analogue $[\{(thf)MgNC_{10}H_7\}_6]$ which had an almost identical structure.[148] Several reactions of $[(thf)MgNPh\}_6]$ were explored, Scheme 3.10.

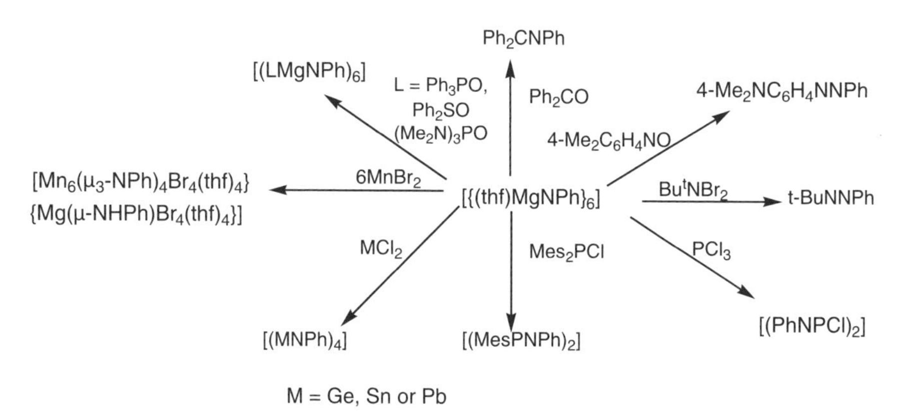

Scheme 3.10 *Selected ligand transfer and replacement reactions of $[\{(thf)MgNPh\}_6]$*[148,149]

Further reactions between $[\{(thf)MgNPh\}_6]$ and $[(\eta^5\text{-}C_5H_5)MCl_2]$ (M, Ti or Zr) led to the imide $[(\eta^5\text{-}C_5H_5)(Cl)Ti(\mu\text{-}NPh)_2Ti(\eta^5\text{-}C_5H_5)_2]$ or $[\{(\eta^5\text{-}C_5H_5)_2(Cl)M(\mu\text{-}NPh)\}_2]$.[150] Treatment of $[\{(thf)MgNPh\}_6]$ with SO_2 led to cleavage of the MgN bonds and formation of the cage species $[((thf)Mg\{O_2S(\mu\text{-}NPh)SO_2\})_6]$.[151] From aniline and two equivalents of EtMgBr in ether the related magnesium imide aggregate $[\{(Et_2O)Mg\}_6(NPh)_4Br_4]$ was obtained, which features an adamantanyl Mg_6N_4 framework composed of six ether-solvated magnesium atoms and four bridgehead imido nitrogens.[152] Each hexagonal face of the framework is capped by a bromide. In recent work the reaction of $MgBu_2$ with $H_2NC_6H_2Cl_3$-2,4,6 in suitable solvents led to the first tetramers $[\{(L)MgNC_6H_2Cl_3\text{-}2,4,6\}_4]$ (L = thf or 1,4-dioxane). These complexes possess an Mg_4N_4 core structure with average Mg–N and Mg–O distances of 2.094(2) and 2.085(2) Å.[153]

3.4 Calcium Amides

3.4.1 Introduction

At the time of publication of our previous edition,[1] only six substituted calcium amide structures had been reported. No structural data were available and only the related imide $[(thf)_3Ca(AlH)_3(NBu^t)_4]$, where a calcium atom was included as part of a $CaAl_3N_4$ cubane array, had been structurally characterized.[2] Since then, a large number of well-characterized calcium amides have appeared, including rare low-coordinate (coordination number less than four, see below) calcium amides. The simplest calcium amide is $Ca(NH_2)_2$, which has a defect NaCl type lattice with $[NH_2]^-$ ions occupying the atomic positions of the anions and calcium ions and lattice defects occupying the atomic positions of the cations.[154] The metal ions have octahedral environments, with Ca–N bond lengths of 2.57 Å. Calcium amides are of current interest owing to their possible applications in hydrogen storage materials.[47,155,156]

Other than $[(thf)_3Ca(AlH)_3(NBu^t)_4]$,[2] no molecular calcium imides related to those described for magnesium in Section 3.3.7, are currently known. Crystals of the parent imide CaNH have been shown to have a structure similar to that of NaCl.[157]

3.4.2 Monomeric Calcium Amides

The first well characterized alkaline earth metal amides were of the general formula $[M\{N(SiMe_3)_2\}_2L]$ (L = thf, $(thf)_2$, or dme, M = Ca, Sr, Ba)[65,158–160] and were synthesized in 1990/1991 from $M'\{N(SiMe_2)_2\}_2$ (M′ = Hg[65] or Sn[158]) or from the activated metal powder (prepared by co-condensation of metal vapour and toluene) and $HN(SiMe_3)_2$ (for M = Ca).[160] The Ca amide $[Ca\{N(SiMe_3)_2\}\{\mu\text{-}N(SiMe_3)_2\}(thf)]_2$ was the first to be made from $[Ca(OC_6H_2Bu^t{}_2\text{-}2,6\text{-}Me\text{-}4)_2(thf)_3]$ and $2Li\{N(SiMe_3)_2\}$.[160]

The synthesis of substituted calcium amides is generally achieved either via salt elimination between a halide (often CaI_2) and an alkali metal amide or via transamination using bis(trimethylsilylamido)calcium derivatives, most commonly $[Ca\{N(SiMe_3)_2\}_2(thf)_2]$. The latter was also conveniently prepared from $Ca(OSO_2CF_3)_2$ and $2Na\{N(SiMe_3)_2\}$ in thf.[161]

Most monomeric calcium amides involve the use of neutral donor molecules to prevent association. The only example of a three-coordinate mononuclear Ca amide, $[Ca\{N(SiMe_3)(SiBu^tPh_2)\}_2(thf)]$,[162] was prepared from CaI_2 and $KN(SiMe_3)(SiBu^tPh_2)$; similar reaction with less bulky amides $KN(SiMe_3)(SiBu^tMe_2)$ or $KN(SiMe_3)(SiPh_3)$ resulted in bis-thf adducts.[162] Several four-coordinate complexes have been characterized, including $[Ca\{N(SiMe_3)_2\}_2(dme)]$,[159] $[Ca\{N(SiMe_3)_2\}_2(thf)_2]$,[161] and $[Ca\{(N(Dipp)C(Me))_2CH\}\{N(SiMe_3)_2\}(thf)]$,[163] which have distorted tetrahedral coordination environments with average Ca–N bond lengths of 2.27, 2.30 and 2.34 Å, respectively and average Ca–O distances of 2.40, 2.38 and 2.35 Å, respectively. Structural parameters for the complexes $[Ca\{N(SiMe_3)_2\}_2(py)_2]$ and $[Ca\{N(SiMe_3)_2\}_2(tmeda)]$ were listed in Ref. 65 although full details of the structures were not given. The complexes $[Ca\{N(Ar)(SiMe_3)\}_2(thf)_2]$ (Ar = Mes,[67] Dipp[84]) were formed by salt elimination, e.g. from CaI_2 and $[KN(C_6H_3Pr^i{}_2\text{-}2,6)(SiMe_3)]$ in thf. The bulkier $[Ca\{N(Dipp)(SiMe_3)\}_2(thf)_2]$ has a distorted four-coordinate geometry with Ca–N and Ca–O of 2.326(8) and 2.536(8) Å, respectively. The distorted trigonal bipyramidal metal environment in $[Ca\{N(Mes)(SiMe_3)\}_2(thf)_2]$[67] comprises two amido ligands (mean Ca–N is 2.30 Å) and two thf solvent molecules (mean Ca–O is 2.36 Å) as well as one agostic interaction with a hydrogen atom from one methyl group (Ca···H, 2.89 Å). The large amido groups and one of the thf ligands are positioned in the equatorial plane with an N–Ca–N angle of 137.55(1)°.

Monomeric calcium amides with formal metal coordination numbers of five and higher have been reported. For example, trigonal bipyramidal $[\{HB(3\text{-}Pr^ipz)_3\}Ca\{N(SiMe_3)_2\}(thf)]$, was obtained from of CaI_2 with $[KN(SiMe_3)_2]$ and $[K\{HB(3\text{-}Pr^ipz)_3\}]$ in thf.[163] The oxygen atom occupies an axial position (Ca–O, 2.378(3) Å) whereas the amido nitrogen atom is in an equatorial position (Ca–N, 2.339(3) Å) and the tridentate, tris(pyrazolyl)borate ligand occupies the remaining axial and two equatorial positions (Ca–N, 2.464(3) and 2.494(3) Å). The diphenylamido compound $[Ca(NPh_2)_2(dme)_2]$ was obtained by protonolysis of the heavier Grignard analogue $[Ca(I)Ph(thf)_4]$ with $HNPh_2$ followed by ligand redistribution.[64] The reaction of $Ca\{N(SiMe_3)_2\}_2$ with propiophenone and pmdeta in thf gave $[\{MeCH{=}C(Ph)O\}Ca\{N(SiMe_3)_2\}(pmdeta)]$[104] (see Section 3.4.3 for reaction without pmdeta); the $Ca{-}N_{(amido)}$ bond length is 2.3283 (12) Å. The short Ca–O distance of 2.1701(10) Å is noteworthy. The Ca–N and Ca–O bond lengths in the five-coordinate calcium bis(amide) $[Ca\{N(SiMe_2CH_2)_2\}_2(thf)_3]$[164] and the heteroleptic β-diketiminatocalcium amide $[Ca(\{N(Dipp)C(Me)\}_2CH)\{N(H)Dipp\}(thf)]$[165] are unexceptional (Ca–N, 2.34 and 2.28–2.38 Å, and Ca–O, 2.39–2.41 and 2.31 Å). The complex $[Ca\{(NSiMe_3)_2PPh_2\}_2(thf)]$[25] was prepared as shown in Equation (3.10).

$$Me_3SiN{=}PPh_2(NHSiMe_3) + [Ca\{N(SiMe_3)_2\}_2(thf)_2] \xrightarrow{\text{thf / hexane}} [Ca\{(Me_3SiN)_2PPh_2\}_2(thf)] \qquad (3.10)$$

The calcium ion has distorted trigonal bipyramidal coordination geometry with the equatorial positions occupied by two nitrogen atoms (Ca–N, *ca.* 2.39 Å) and the oxygen atom (Ca–O, 2.353(1) Å). The axial Ca–N bond lengths are *ca.* 2.55 Å.

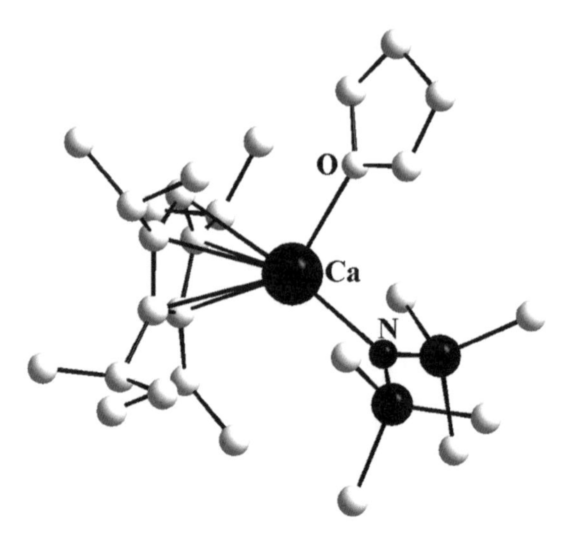

Figure 3.7 Molecular structure of $[(Cp')Ca\{N(SiMe_3)_2\}(thf)]$;[166] calcium, nitrogen and silicon atoms are shown as black spheres, oxygen and carbon atoms are white. Selected bond lengths: Ca–N 2.30(1), Ca–O 2.35(1), Ca–M 2.397 Å

The reaction of the tetra(isopropyl)cyclopentadienyl (Cp′) calcium iodide $(Cp')CaI(thf)_2$ with $KN(SiMe_3)_2$ yielded $[(Cp')Ca\{N(SiMe_3)_2\}(thf)]$[166] (Figure 3.7), in which the calcium ion lies in the centre of the pseudo-trigonal NO*M* (*M* = centroid of Cp′) plane.

The preparation of the six-coordinate calcium complex $[Ca(tmhd)\{N(SiMe_3)_2\}(thf)_3]$ (Htmhd = 2,2,6,6-tetramethylheptane-3,5-dione) is shown in Equation (3.11).[167] This derivative and related species were investigated in connection with the catalysis of ring-opening polymerization of lactones. The calcium ion is in a distorted octahedral environment; the pseudo-tetragonal plane is comprised of O atoms of the three coordinating thf molecules (Ca–O, 2.40 – 2.47 Å) and one oxygen atom from the β-diketonate, whose second oxygen atom binds in an axial position. The remaining axial position is occupied by the amido nitrogen atom, Ca–N 2.377(6) Å.

$$[Ca\{N(SiMe_3)_2\}_2(thf)_2] \quad + \quad \text{Htmhd} \xrightarrow{\text{thf}} [Ca\{N(SiMe_3)_2\}(\text{tmhd})(thf)_3] \qquad (3.11)$$

A similar coordination environment was observed in $[Ca\{\kappa^2\text{-}(NSiMe_3)_2SPh\}_2(thf)_2]$,[74] (Ca–N, 2.42 and 2.44 Å) and (mean Ca–O, 2.46 Å). Reaction of CaI_2 with the corresponding potassium amide followed by the addition of tmeda yielded $[Ca\{N(C_6H_2Me_3\text{-}2,4,6)(SiMe_3)\}_2(tmeda)]$ (average $Ca{-}N_{(amido)}$ and $Ca{-}N_{(tmeda)}$ are 2.31 and 2.53 Å). The distorted octahedral coordination environment at the calcium atom is completed by two agostic Ca···H interactions (mean 2.77 Å).[67] The reaction of the aminotroponimine $C_7H_5(NPr^i)_2$-1,2-H with $Ca\{N(SiMe_3)_2\}_2$ in the presence of thf afforded the five-coordinate Ca complex $[Ca\{N(SiMe_3)_2\}\{(NPr^i)_2C_7H_5\}(thf)_2]$, which showed a high reactivity as a catalyst for the intramolecular hydroamination/cyclization reaction of non-activated amino-alkenes.[168]

Several monomeric amides were obtained by reduction of an α-diimine with metallic calcium in a donor solvent, including $[Ca\{N(Ar)C(Ph)=C(Ph)N(Ar)\}(dme)_2]$ (Ar = C_6H_4Me-4),[169] $[Ca\{1,2\text{-}(NAr)_2C_{12}H_6\}(thf)_3]$ (Ar = C_6H_4Ph-2),[73] and $[Ca\{1,2\text{-}(NAr)_2\text{-}C_{12}H_6\}(thf)_4]$ (Ar = $C_6H_3Pr^i{}_2$-2,6).[72]

The calcium silylamide $Ca\{N(SiMe_3)_2\}_2$, or its ether adducts, has been used as a synthon to generate numerous other related derivatives. These include phosphanides,[170–172] arsanides,[173] thiolates,[174] selenolates and tellurolates,[175–177] as well as a variety of diaza-substituted complexes, some of which were discussed above (see also Ref. 11).

3.4.3 Dimeric Calcium Amides and Higher Aggregates

The formation of alkaline earth metal bis[bis(trialkylsilyl)amides] has been discussed in detail elsewhere.[11] Like all heavier group 2 metal bis[bis(trisalkylsilyl)amides], the complex $[(Ca\{N(SiMe_3)_2\}_2)_2]$ has a dimeric structure both in solution[158] and the solid state (Figure 3.8),[159] in which the calcium atoms are in a distorted trigonal planar environment.

The compound $[\{Ca\{N(SiMe_3)_2\}\{\mu\text{-}N(SiMe_3)_2\}(thf)\}_2]$, prepared from Ca and $2HN(SiMe_3)_2$ or $[Ca(OC_6H_2Bu^t{}_2\text{-}2,6\text{-}Me\text{-}4)_2(thf)_3]$ and $2Li\{N(SiMe_3)_2\}$, was characterized spectroscopically.[160] Each Ca atom in the fluorenyl complex $[\{Ca\{\eta^5\text{-}C_{13}H_8(SiMe_3)\text{-}9\}(\mu\text{-}NPr^i{}_2)\}_2]$ is formally three-coordinate (av. Ca–N, 2.39 Å) but has three short agostic Ca···Me contacts (2.72 and 2.81 Å).[178] Treatment of this compound with $PhSiH_3$ gave an unstable heteroleptic calcium hydride; similar reaction of $[Ca\{(N(Dipp)C(Me))_2CH\}$-

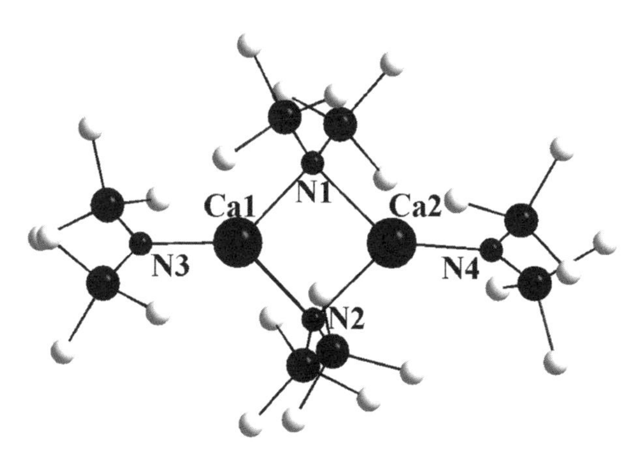

Figure 3.8 *Molecular structure of $[(Ca\{N(SiMe_3)_2\}_2)_2]$.[159] Calcium, nitrogen and silicon atoms are shown as black spheres and carbon atoms are white. Selected bond lengths: Ca1–N1 2.482(6), Ca1–N2 2.520(6), Ca1–N3 2.282(6), Ca2–N1 2.430(6), Ca2–N2 2.466(6), Ca2–N4 2.267(6) Å*

$\{N(SiMe_3)_2\}$(thf)] yielded the first well-defined hydrocarbon-soluble Ca hydride [{Ca{(N-(Dipp)C(Me))$_2$CH}(μ-H)(thf)}$_2$].[178]

If the 1:1 reaction stoichiometry in Equation (3.11) was changed to 2:3, the complex [{(thf)Ca}$_2$(tmhd)$_3${N(SiMe$_3$)$_2$}][168] was obtained. The structure comprises two calcium centred octahedra connected via a common face. One of the two diketonato ligands and the amido ligand form the bridging units between the calcium atoms, with notably long average Ca–N distances of 2.47 Å. The terminal groups involve the remaining diketonato ligand and a thf. An unusual coordination environment is seen at the two calcium atoms in the dimeric structure of [{(η^5-C_5Me_4Et)Ca(μ-2NSiMe$_2$CH$_2$CH$_2$SiMe$_2$)}^{2}],[179] in which the central (CaN)$_2$ ring (Ca–N, 2.41 and 2.49 Å) is flanked by two η^5-cyclopentadienyl groups (mean Ca–C, 2.93 Å) that are rotated by 162° with respect to each other. This complex was formed by the reaction of CaI$_2$ with successively the potassium peralkylcyclopentadienyl and the potassium amide.

In contrast to its magnesium analogue, reaction of Ca{N(SiMe$_3$)$_2$}$_2$ with propiophenone afforded the ion-separated salt [Ca$_2${μ-OC(Ph)=CHMe}$_3$(thf)$_6$][Ca{N(SiMe$_3$)$_2$}$_3$].[104] The cation is dinuclear with six-coordinate calciums in a slightly distorted octahedral geometry. The anion has the trigonal planar calcium bound to three silylamido ligands with Ca–N distances between 2.30 and 2.32 Å. Deprotonation of 4-t-butylcyclohexanone with various bases, including Ca{N(SiMe$_3$)$_2$}$_2$ or its 1:1 and 1:2 mixtures with K{N(SiMe$_3$)$_2$}, was studied by trapping the resulting enolate with Me$_3$SiCl to afford the silyl-enol ether.[180] When 2′,4′,6′-trimethylacetophenone was treated with Ca{N(SiMe$_3$)$_2$}$_2$ the crystalline amidocalcium enolate [{Ca{N(SiMe$_3$)$_2$}(μ-OC(Mes)=CH$_2$)(OEt$_2$)}$_2$] was isolated.[180]

The tetranuclear primary amide [{Ca(NHPh)$_2$(thf)$_2$}$_4$] was prepared by various routes including metallation of aniline with diphenylcalcium or Ca(Ph){N(SiMe$_3$)$_2$}, or metathetical exchange between CaI$_2$ and 2KNHPh in thf.[181] The calcium amidotrihydroborate [{Ca(thf)$_2$(NH$_2$BH$_3$)$_2$}]$_\infty$ has an unusual polymeric structure in which the Ca(thf)$_2$-(NH$_2$BH$_3$)$_2$ units are linked by Ca···H$_3$B interactions.[156]

3.4.4 Heterometallic Calcium Amide Derivatives

Several calcium amide structures that incorporate alkali metal ions are known. A simple example is [NaCa(NH$_2$)$_3$],[182] which comprises distorted cubic close-packed [Ca(NH$_2$)$_3$]$^-$ anions with calcium-nitrogen distances between 2.48 and 2.60 Å. The alkali metal salts [KCa(NH$_2$)$_3$][183] and [RbCa(NH$_2$)$_3$][184] have one-dimensional, infinite chains of face-sharing octahedral anions (mean Ca–N bond length is 2.50 Å) that are connected via coordination of the amido groups to the second metal (K–N and Rb–N distances are 2.98–4.00 and 3.12 Å, respectively). On the other hand, a monomeric structure based on a planar (LiN$_2$Ca) core is observed for [Li{μ-N(SiMe$_3$)$_2$}$_2$Ca{N(SiMe$_3$)$_2$}], obtained from the two metal bis(trimethylsilyl) amides,[185] which has a Li–N bond length of 2.005(3) Å and bridging Ca–N distances of 2.414(1), 2.426(1) with a Ca–N terminal distance of 2.288 (1) Å. The coordination geometries at the lithium and calcium ions are distorted tetrahedral and distorted trigonal bipyramidal, respectively, due to a M···CH$_3$ interaction (average Li–C and Ca–C distances are 2.39 and 2.85 Å, respectively). A related compound, [K(thf)-{μ-N(SiMe$_3$)$_2$}$_2$Ca{N(SiMe$_3$)$_2$}]$_\infty$ is a polymer having intermolecular K···CH$_3$ interactions.[180] The heterobimetallic species [(Me$_3$Si)NCa{μ-N(SiMe$_3$)$_2$}$_2$MgN(SiMe$_3$)$_2$] has already been discussed in Section 3.3.5.[119]

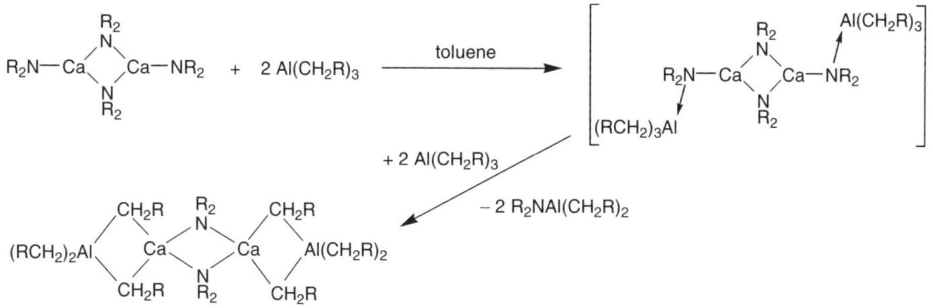

Scheme 3.11 Formation of calcium-carbon bonds by reaction of a calcium amide with an aluminium trialkyl ($R = SiMe_3$)[186]

The aluminium-calcium tetranuclear amide was prepared as shown in Scheme 3.11.[186] It comprises a slightly distorted planar central $(CaN)_2$ ring (Ca–N bond lengths are 2.407(2) and 2.487(2) Å) that links two (AlCCaC) rings into a tricyclic structure (Ca–C and Al–C distances are 2.64–2.68 and 2.06–2.08 Å, respectively).

Similar to the titanium-containing magnesium amide structures mentioned in Section 3.3.5, the addition of $[(thf)\{(Me_3Si)_2N\}Ca\{(\mu_3\text{-}N)(\mu_3\text{-}NH)_2Ti_3(\eta^5\text{-}C_5Me_5)_3(\mu_3\text{-}N)\}]$ to a solution of $2,4,6\text{-}Me_3C_6H_2NH_2$ in toluene yielded the double cubane calcium amido complex $[\{\mu\text{-}2,4,6\text{-}Me_3C_6H_2(H)N\}Ca\{(\mu_3\text{-}N)(\mu_3\text{-}NH)_2Ti_3(\eta^5\text{-}C_5Me_5)_3(\mu_3\text{-}N)\}]_2$[131] in which two azaheterometallocubane moieties (Ca–N distances 2.44 - 2.54 Å) are held together by a four-membered $(CaN)_2$ ring comprising bridging 2,4,6-trimethylanilido ligands (Ca–N bond lengths, 2.438(3) and 2.447(3) Å).[131]

The crystalline lithium-calcium diamide $[Li(thf)(\mu\text{-}I)(\mu\text{-}\{N(CH_2Bu^t)\}_2C_6H_4\text{-}1,2)Ca(thf)_3]$, featuring a seven-membered CCNCaILiN cage, was obtained as shown in Equation (3.12).[187]

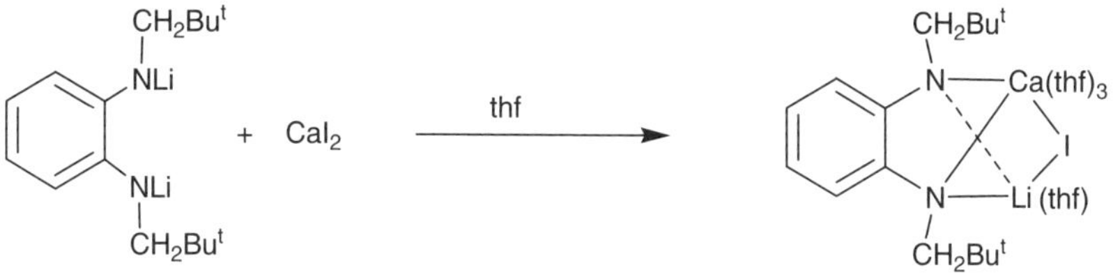

$$1,2\text{-}C_6H_4\{N(Li)CH_2Bu^t\}_2 + CaI_2 \xrightarrow{thf} [Li(thf)(\mu\text{-}I)(\mu\text{-}\{N(CH_2Bu^t)\}_2C_6H_4\text{-}1,2)Ca(thf)_3] \quad (3.12)$$

3.5 Strontium Amides

3.5.1 Introduction

Prior to the previous edition of this book, no examples of substituted strontium amides had been well characterized. Even now, only a few strontium amide crystal structures are known and have been briefly reviewed in the context of more general aspects of the chemistry of group 2 metal compounds.[11,12,14]

The crystal structure of $Sr(NH_2)_2$, synthesized by the reaction of strontium metal with ammonia at room temperature, was first reported in 1963, and was shown to have a cubic face-centred network of $[NH_2]^-$ anions held together by coordination to six-coordinate strontium ions.[154] Because of their potential in synthetic, polymer and solid-state applications, attention is now focussed on the heavier alkaline earth metal derivatives as suitable reagents and source materials. The synthesis of substituted heavier alkaline earth metal amides is often challenging due to their large metal ionic radius, which often leads to their deterioration during storage. To prevent this, and to reduce aggregation during synthesis, it is necessary to coordinatively saturate the metal centre via the use of bulky substituents on the nitrogen and also often the neutral co-donor. Even if the heavy alkaline earth metals are highly reactive, the direct reaction of strontium metal with a secondary amine occurred only for the activated metal powder prepared by cocondensation of metal vapour with toluene.[188] The direct metallation can also be achieved by dissolving strontium in liquid ammonia[189] or by using ammonia-saturated thf as a reaction medium.[67,84] Other common synthetic routes to strontium amides proceed via the salt elimination from SrI_2 with an alkali metal amide[67,84,165,190] or by further reaction of $Sr\{N(SiMe_3)_2\}_2$.[25,130,191–193] The latter, or its neutral donor adducts, can be synthesized from strontium metal by reaction with $HN(SiMe_3)_2$[188] or $M[N(SiMe_3)_2]_2$ (M = Hg [65] or Sn [158]). The choice of solvent system is a deciding factor in the structure of the resulting complex (Scheme 3.12). Carrying out the reaction in a donor solvent such as thf or dme yields the monomeric strontium amides $[Sr\{N(SiMe_3)_2\}_2(thf)_2]$[158] (also obtained from $Sr(OSO_2CF_3)_2$ and $2NaN(SiMe_3)_2$ in thf[161]) or $[Sr\{N(SiMe_3)_2\}_2(dme)_2]$.[194] The polymer $[\{Sr\{N(SiMe_3)_2\}_2(1,4\text{-dioxane})\}_\infty]$, with a square planar Sr environment, was obtained with 1,4-dioxane.[188] In the thf complex the Sr–N and Sr–O distances average 2.46 Å and 2.53 Å and there is a wide N–Sr–N angle of

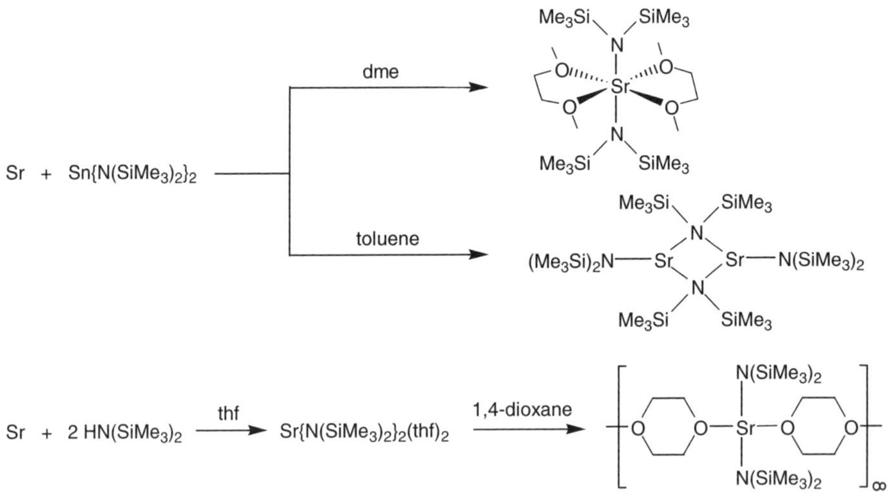

Sr + $Sn\{N(SiMe_3)_2\}_2$ → (dme) / (toluene)

Sr + 2 $HN(SiMe_3)_2$ →(thf) $Sr\{N(SiMe_3)_2\}_2(thf)_2$ →(1,4-dioxane)

Scheme 3.12 *Formation of $Sr\{N(SiMe_3)_2\}_2$ and its ether solvates*[158,188,194]

120.6° and a narrow O–Sr–O angle near 84.7° reflecting the size disparity of the ligands. In the dme compound, oxygen donors form a tetragonal plane (Sr–O bond lengths are 2.596(6) and 2.713(6) Å) in an octahedral geometry with coordination of the monodenate silylamido ligands above and below (Sr–N distance is 2.538(7) Å). In contrast, the use of a non-coordinating medium such as toluene yielded the dimeric strontium amide $[\{Sr(N(SiMe_3)_2)_2\}_2]$,[158,194] which is structurally similar to the magnesium and calcium analogues. It features a planar $(SrN)_2$ core with distorted trigonal planar coordination at the strontium ions. The Sr–N bond lengths are longer to the bridging (2.61–2.67 Å) than to the terminal nitrogen atoms (2.43–2.44 Å).

3.5.2 Monomeric Strontium Amides

In addition to the complexes mentioned above, there exist several other monomeric strontium amides. In $[Sr\{N(Dipp)(SiMe_3)\}_2(thf)_2]$, synthesized by salt elmination, the Sr–N and Sr–O bond lengths are 2.480(6) and 2.511(4) Å, respectively.[84] The monomeric compound $[Sr\{CH(Ph_2P{=}NC_6H_2Me_3\text{-}2,4,6)_2\}\{N(SiMe_3)_2\}(thf)]$[190] contains a five-coordinate strontium atom with one tridentate ligand coordinating via two nitrogens (Sr–N, 2.557(17) and 2.5951(17) Å) and a carbon atom (Sr–C, 2.861(2) Å). The distorted tetragonal pyramidal geometry is completed by one amido (Sr–N, 2.4740(18) Å) and a thf ligand (Sr–O, 2.5674(18) Å); the isoleptic Ca compound was also reported. An unusual metal coordination environment is observed in $[Sr\{N(C_6H_2Me_3\text{-}2,4,6)(SiMe_3)\}_2(pmdeta)]$, in which two amido ligands (Sr–N is 2.501(4) Å) and one tridentate pmdeta ligand (Sr–N is 2.857(6) Å) coordinate to the strontium ion with four additional agostic interactions involving hydrogen atoms from the methyl groups (Sr·H distances are between 2.90 and 3.19 Å).[67]

As exemplified by the complex $[Sr\{N(SiMe_3)_2\}_2(dme)_2]$, a coordination number of six rather than four is often observed in monomeric strontium amides because of the relatively large size of the Sr^{2+} ion. In $[Sr\{(N(SiMe_3))_2CPh\}_2(thf)_2]$,[193] there are two chelating benzaminidate ligands in a tetragonal plane (mean Sr–N bond length is 2.58 Å) and the slightly distorted octahedral geometry is completed by *trans* coordination of two thf molecules (mean Sr–O distance is 2.52 Å). The two structurally related complexes $[Sr\{(NSiMe_3)_2SPh\}_2(thf)_2]$[74] and $[Sr\{(NSiMe_3)_2PPh_2\}_2(thf)_2]$[25] were formed by the reaction of $Sr\{N(SiMe_3)_2\}_2$ with the corresponding aminoimino-phenylsulfine or -diphenylphosphorane in thf (cf. Equation (3.10)). Both complexes show highly distorted octahedral metal geometries (average Sr–N or Sr–O bond lengths are 2.57 and 2.65 Å or 2.59 and 2.62 Å).

The six-coordinate diphenylamido compound $[Sr(NPh_2)_2(thf)_4]$ (and its Ba congener) was obtained by a salt elimination method.[64] The compound $[Sr\{1,2\text{-}(NDipp)_2C_{12}H_6\}(thf)_4]$[195] was prepared similarly to its Ca analogue (see Section 3.4.2); with acetonitrile, it formed a dinuclear complex containing the bridging primary amido ligand $[NC{-}C(H){=}C(Me){-}NH]^-$.[195]

Monomeric Sr–N compounds with coordination numbers higher than six have been reported, as in the benzamidinate shown in Figure 3.9.[193]

A seven-coordinate Sr compound, having chelating imino-bis(2-pyridyl)phosphoraneamido ligands, was prepared as shown in Equation (3.13);[192] the Ba analogue was similarly prepared.

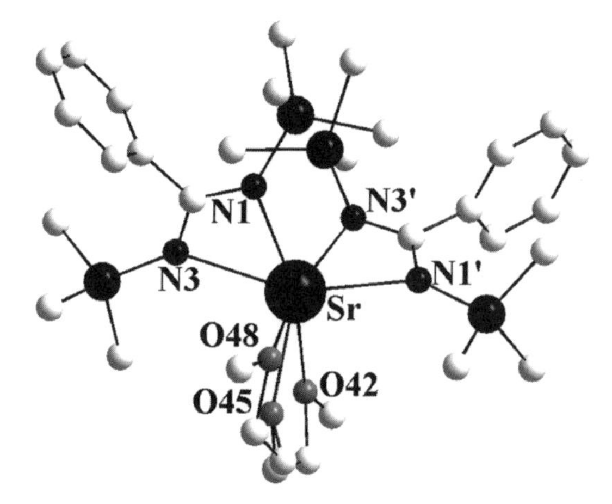

Figure 3.9 *Representation of the structure of [Sr{(N(SiMe$_3$))$_2$CPh)}(diglyme)].*[193] *Strontium, nitrogen and silicon atoms are shown as black spheres, oxygen atoms are grey and carbon atoms are white. Selected bond lengths: Sr–N1 2.594(4), Sr–N3 2.598(3), Sr–O42 2.848(8), Sr–O45 2.612(6), Sr–O48 2.681(8) Å*

[Sr{N(SiMe$_3$)$_2$}$_2$(thf)$_2$] + Py$_2$P(NHSiMe$_3$)(NSiMe$_3$) $\xrightarrow[-\,2\ HN(SiMe_3)_2]{thf\ /\ pentane}$ [Sr{Py$_2$P(NSiMe$_3$)$_2$}$_2$(thf)]

(3.13)

3.5.3 Higher Aggregate Strontium Amides

Attempted metallation of triphenylmethane with Sr{N(SiMe$_3$)$_2$}$_2$ and 18-crown-6 unexpectedly produced a dinuclear amidostrontium alkoxide as a result of crown ether deprotonation and C–O bond cleavage, Equation (3.14).[196] The Sr atoms are seven-coordinate with bridging alkoxide and terminal amido ligands (average Sr–O$_{(bridging)}$ and Sr–N bond lengths are 2.441 and 2.574 Å, respectively). Metallation of 2,2,4,4,6,6,8,8-octamethylcyclotetrasilazane with Sr{N(SiMe$_3$)$_2$}$_2$ gave the dinuclear six-coordinate compound [{Sr(μ-NSiMe$_2$N(H)SiMe$_2$)$_2$(thf)}$_2$]; the Ca compound was isostructural, while Mg formed the monomer [Mg(NSiMe$_2$N(H)SiMe$_2$)$_2$(thf)$_2$].[197]

[Sr{N(SiMe$_3$)$_2$}$_2$(thf)$_2$] + 18-crown-6 $\xrightarrow[-\,HN(SiMe_3)_2]{thf,\ Ph_3CH}$ 1/2 [Sr{N(SiMe$_3$)$_2$}(alkoxide)]$_2$

(3.14)

As mentioned in Section 3.5.1 (see Scheme 3.12), a one-dimensional polymeric structure is obtained by the coordination of 1,4-dioxane molecules in the compound $[Sr\{N(SiMe_3)_2\}_2(1,4\text{-dioxane})]_\infty$.[188] A polymeric structure in the solid state was also found in the six-coordinate primary amide $[Sr\{\mu\text{-}N(H)Ph\}_2(thf)_2]_\infty$, obtained by a salt elimination route.[181]

3.6 Barium Amides

3.6.1 Introduction

In our 1980 book, no examples of substituted barium amides had been well-characterized. The synthesis of $Ba(NH_2)_2$ by the streaming of gaseous ammonia at 300 °C over a nickel vessel containing barium metal had been published in 1963,[154] but the crystal structure was not solved until 1975.[198] Single crystals were grown from the reaction of barium and ammonia at −70 °C over a period of 3–4 months. The crystal data showed two barium ion environments: a seven- and an eight-coordinate metal centre with Ba−N bond lengths ranging 2.75–3.34 or 2.78–3.17 Å, respectively. The compound $Ba(NH_2)_2$ has subsequently found use as a convenient synthetic precursor for amido-free barium coordination complexes.[199,200]

3.6.2 Monomeric Barium Amides

Since 1980, several of barium amides have been synthesized and structurally characterized.[11–16,38] These have coordination numbers at barium that range from three to eleven. Monomeric barium amido complexes can be obtained if sterically demanding amido ligands and neutral donor co-ligands are employed as in $[Ba\{N(SiMe_3)_2\}_2(thf)_2]$[201] (Fig. 3.10).

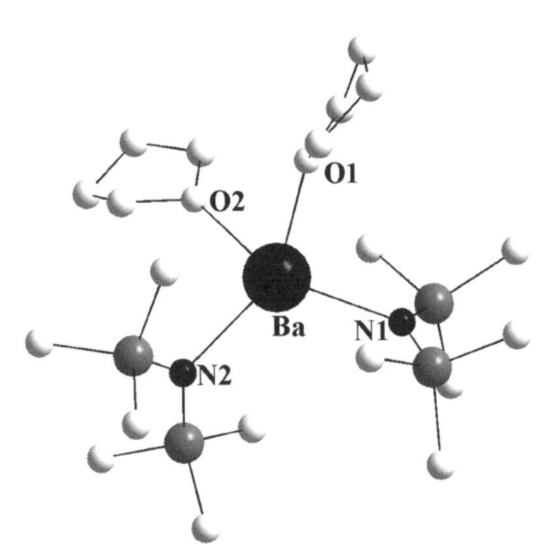

Figure 3.10 *The structure of $[Ba\{N(SiMe_3)_2\}_2(thf)_2]$.[201] Barium and nitrogen atoms are shown as black spheres, silicon atoms are grey and carbon and oxygen atoms are white. Selected bond lengths: Ba−N1 2.587(6), Ba−N2 2.596(6), Ba−O1 2.745(6), Ba−O2 2.717(6) Å*

The latter was obtained from Ba and $HN(SiMe_3)_2$ in the presence of NH_3,[201] Hg{N-$(SiMe_3)_2$},[65] or, for the dme analogue, Sn{N$(SiMe_3)_2$}$_2$.[158]

The complexes [Ba{N($C_6H_3Pr^i_2$-2,6)($SiMe_3$)}$_2$(thf)$_2$][67] and [Ba{N($C_6H_2Me_3$-2,4,6)(Si-Me_3)}$_2$(L)][84] (L = $(thf)_2$, tmeda or pmdta) were made from BaI_2 and the appropriate potassium amide and L or Ba, HN(Ar)($SiMe_3$), NH_3 and thf. X-ray data for these complexes were compared with those of the corresponding Mg, Ca and Sr amides. The complex [Ba-{(κ^2-N($C_6H_3Pr^i_2$-2,6)C(Me))$_2$CH}{N$(SiMe_3)_2$}{thf}] was prepared from BaI_2, the β-diketimine and K{N$(SiMe_3)_2$} in thf.[165] The four-coordinate barium atom has distorted tetrahedral geometry comprising the chelating ligand (mean Ba–N bond length, 2.67 Å), the amide (Ba–N, 2.593(2) Å) and thf (Ba–O, 2.766(2) Å).

Barium silylamides have also been used as precursors for closely related complexes. For example the reaction of Ba{N$(SiMe_3)_2$}$_2$ with bis(trimethylsilyl)-aminoiminophenylsulfine or bis(trimethylsilyl)aminoiniminodiphenylphosphorane yielded the distorted octahedral barium complexes [Ba{$(NSiMe_3)_2$SPh}$_2$(thf)$_2$][74] or [Ba{$(NSiMe_3)_2PPh_2$}$_2$(thf)$_2$],[25] respectively, having mean Ba–N or Ba–O bond lengths of 2.72 and 2.78 Å, or 2.76 and 2.79 Å, respectively. For a related Ca compound, see Equation (3.10). Ba{N$(SiMe_3)_2$}$_2$ served also as a starting material for the benzyl derivative {Ba$(CH_2Ph)_2$}$_n$ (via metathesis with [Li-(CH_2Ph)(tmeda)]),[202] as well as phosphanides,[172,203] selenolates and tellurolates.[177]

3.6.3 Dimeric Barium Amides

Desolvation,[201] or treatment with a β-diketimine,[204] crown ether,[196,205] primary[172] or secondary[206] phosphane, or $H_2(NBu^t)_4S$,[207] of [Ba{N$(SiMe_3)_2$}$_2$(thf)$_2$] gave the dinuclear barium amides of Scheme 3.13. The oxygen-bridged siloxobarium amide [(Ba{μ-OSi-$(Bu^t)_2(CH_2)_3NMe_2$}{N$(SiMe_3)_2$})$_2$] and its thf solvate were synthesized by the protonolysis of [Ba{N$(SiMe_3)_2$}$_2$(thf)$_2$] with the amino-functionalized silanol; in thf solution the compounds exist in equilibrium with their homoleptic ligand redistribution products.[208] The compounds [Ba{N(Ph)C(Ph) = C(Ph)N(Ph)}(dme)$_2$][169] and [(Ba{1,2-$(NAr)_2$-$C_{12}H_6$}{dme})$_2$(μ-dme)] (Ar = $C_6H_3Pr^i_2$-2,6),[195] were prepared by reduction of the appropriate α-diimine with metallic Ba in dme.

Like its lighter alkali earth metal congeners, [{Ba$(N(SiMe_3)_2)_2$}$_2$] has a dimeric structure in which each barium atom is coordinated by one terminal (Ba–N, 2.576(3) Å) and two bridging (Ba–N bond lengths = 2.798(3) and 2.846(4) Å) ligands which form a planar $(BaN)_2$ core. The thf complex [{Ba{N$(SiMe_3)_2$}$_2$(thf)}$_2$] maintains a planar $(BaN)_2$ ring structure[201] wherein the bariums have distorted tetrahedral geometry. The structure of the mixed-ligand [Ba(L)(μ-L)$_2$Ba{N$(SiMe_3)_2$}] (L = {N(Cy)C(Me)}$_2$CH) is shown in Fig. 3.11.[204]

The structure of Ba anilide, obtained by the salt elimination route from BaI_2 and 2KN(H)-Ph in thf, was described as a one-dimentional polymer composed of dinuclear moieties [{Ba-{μ-N(H)Ph}$_2$(thf)$_2$}{Ba{μ-N(H)Ph}$_2$(thf)}]$_\infty$, where the Ba atom with one thf ligand has an additional Ba···Ph π-arene interaction.[181]

3.6.4 Heterometallic Barium Amides

The mixed metal complexes KBa$(NH_2)_3$,[182] RbBa$(NH_2)_3$[182] and CsBa$(NH_2)_3$[209] contain isostructural distorted [Ba$(NH_2)_3$]$^-$ anions (Ba–N, 2.84–2.90, 2.69–2.94, and 2.78–3.20 Å,

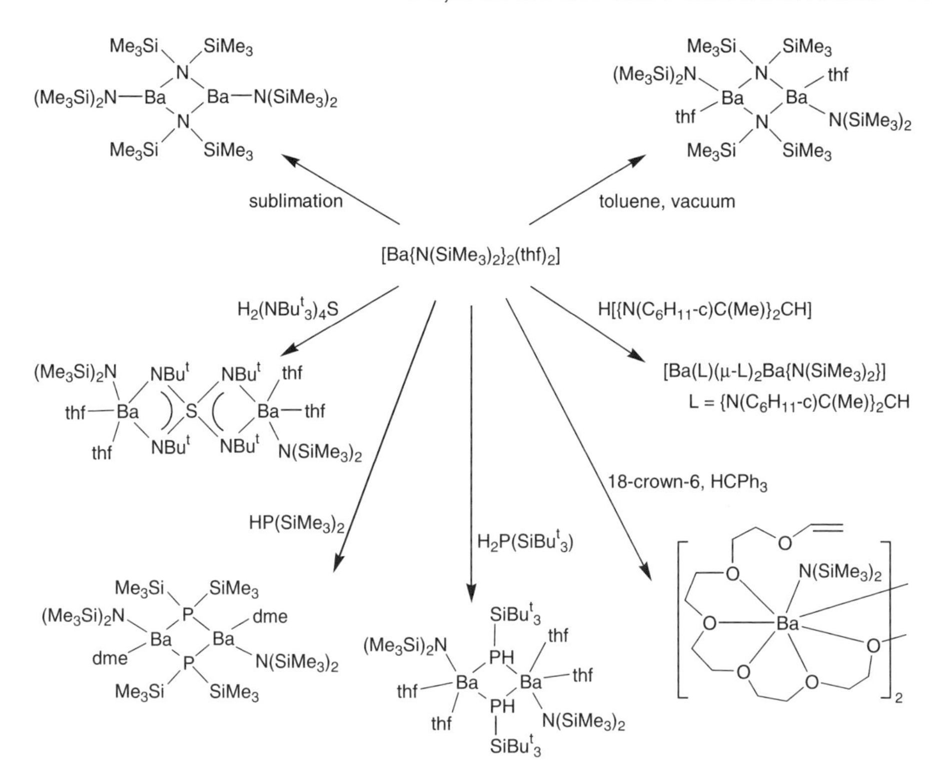

Scheme 3.13 *Routes to various dimeric barium amides*[172,196,201,204–207]

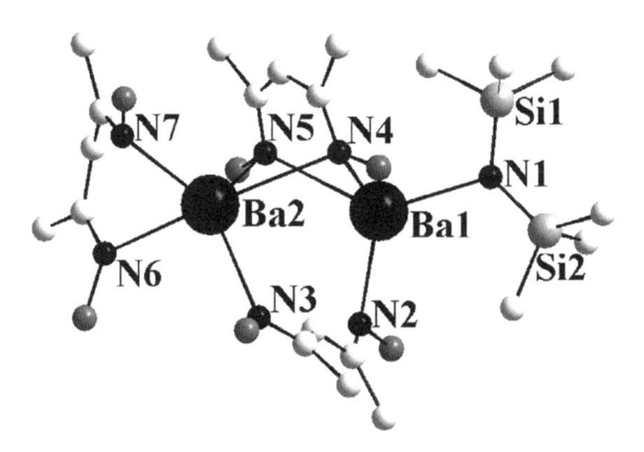

Figure 3.11 *The structure of* $[Ba(L)(\mu\text{-}L)_2Ba\{N(SiMe_3)_2\}]$ *(L =* $\{N(Cy)C(Me)\}_2CH$*). Barium and nitrogen atoms are shown as black spheres and silicon atoms are grey. Cyclohexyl groups are represented by the α-C (small grey spheres). Selected bond lengths: Ba1–N1 2.613(4), Ba1–N2 2.703(4), Ba1–N4 2.829(4), Ba1–N5 2.841(4), Ba2–N3 2.873(4), Ba2–N4 2.860(4), Ba2–N5 3.108(4), Ba2–N6 2.689(4), Ba2–N7 2.635(4) Å*

respectively) connected into strands by coordination of the amido nitrogen atoms to the second metal centres (M–N interactions have average distances of 2.87, 3.83 and 3.01 Å, respectively). In the compound $[(thf)_2Ba_2Li\{N(SiMe_3)_2\}\{(NBu^t)_3S\}_2]$, prepared as shown in Equation (3.15), the $N(SiMe_3)_2$ ligand is terminally bound (Ba–N, 2.711(6) Å).[210]

$$[Li_4\{(NBu^t)_3S\}_2] + 2[Ba\{N(SiMe_3)_2\}_2(thf)_2] \xrightarrow{thf} [(thf)_2Ba_2Li\{N(SiMe_3)_2\}\{(NBu^t)_3S\}_2] + 3LiN(SiMe_3)_2 \quad (3.15)$$

Treatment of $Ba[OS(O)_2CF_3]_2$ with $2NaN(SiMe_3)_2$ in thf afforded the benzene-soluble complex $Ba\{N(SiMe_3)_2\}\{\mu\text{-}N(SiMe_3)_2\}_2Na(thf)_2$.[161] Reaction of $[Ba(C_5Me_5)_2(thf)_2]$ and $LiN(SiMe_3)_2$ in thf gave the hydrocarbon-insoluble compound $Li[Ba(C_5Me_5)_2\{N(SiMe_3)_2\}](thf)_2$.[179] Treatment of $Ba\{N(SiMe_3)_2\}_2$ with $LiN(SiMe_3)_2$ in thf produced the co-crystals $[Ba\{N(SiMe_3)_2\}_2(thf)_3]\cdot[\{Li(N(SiMe_3)_2)(thf)\}_2]$, while the Ca analogue gave the heterobimetallic amide $[Li\{\mu\text{-}N(SiMe_3)_2\}_2Ca\{N(SiMe_3)_2\}(thf)]$ (*cf.* Section 3.4.4).[211]

The barium-zinc cage compound $[(Ba\{N(SiMe_3)_2\}(thf)(\mu\text{-}Et)(\mu\text{-}PSiBu^t{}_3)_2Zn_2)_2]$, featuring terminal $N(SiMe_3)_2$ ligands (Ba–N, 2.55(1) Å), was obtained by deprotonation of $H_2PSiBu^t{}_3$ with successively $ZnEt_2$ and $[\{Ba(N(SiMe_3)_2)_2(thf)\}_2]$ in toluene.[212]

References

1. M. F. Lappert, P. P. Power, A. R. Sanger and R. C. Srivastava, *Metal and Metalloid Amides*, Horword-Wiley, Chichester, 1980.
2. G. Del Piero, M. Cesari and A. Mazzei, *J. Organomet. Chem.*, 1977, **137**, 265.
3. V. R. Magnuson and G. D. Stucky, *Inorg. Chem.*, 1969, **8**, 1427.
4. J. L. Atwood and G. D. Stucky, *J. Am. Chem. Soc.*, 1969, **91**, 4426.
5. H. Bürger, C. Forker and J. Goubeau, *Monatsh. Chem.*, 1965, **96**, 5976.
6. A. H. Clark and A. Haaland, *J. Chem. Soc. D, Chem. Commun.*, 1969, 912.
7. A. H. Clark and A. Haaland, *Acta. Chem. Scand.*, 1970, **24**, 3024.
8. L. Guemas-Brisseau, M. G. B. Drew and J. E. Goulter, *J. Chem. Soc., Chem. Commun.*, 1972, 916.
9. C. Y. Wong and J. D. Woollins, *Coord. Chem. Rev.*, 1994, **130**, 243.
10. H. Schmidbaur, *Coord. Chem. Rev.*, 2001, **215**, 223.
11. M. Westerhausen, *Coord. Chem. Rev.*, 1998, **176**, 157.
12. M. Westerhausen, *Dalton Trans.* 2006, 4755.
13. T. P. Hanusa, *Chem. Rev.*, 1993, **93**, 1023.
14. N. A. Bell (Beryllium) and W. E. Lindsell (Magnesium, Calcium, Strontium and Barium); in *Comprehensive Organometallic Chemistry II* (eds. E. W. Abel, F. G. A. Stone and G. Wilkinson), Vol., 1 (ed. C. E. Housecroft), Ch. 2, Elsevier, Oxford, 1995.
15. T. P. Hanusa, Alkaline Earth Organometallics; in *Comprehensive Organometallic Chemistry III* (eds. R. H. Crabtree and D. M. P. Mingos). Vol., **2** (ed. K. Meyer), Ch. 2.02, Elsevier, Amsterdam, 2007.
16. T. P. Hanusa, Group 1s and 2s metals, *in Comprehensive Coordination Chemistry II* (eds. J. McCleverty and T. J. Meyer,), Vol., 3 (ed. G. F. R. Parkin), Ch. 3.1, Elsevier, Oxford, 2004.
17. H. Nöth and D. Schlosser, *Inorg. Chem.*, 1983, **22**, 2700.
18. M. Niemeyer and P. P. Power, *Inorg. Chem.*, 1997, **36**, 4688.

19. K. H. Thiele, V. Lorenz, G. Thiele, P. Zönnchen and J. Scholz, *Angew. Chem., Int. Ed.*, 1994, **33**, 1372.
20. H. Nöth and D. Schlosser, *Eur. J. Inorg. Chem.,* 2003, 2245.
21. B. Hall, J. B. Farmer, H. M. M. Shearer, J. D. Sowerby and K. Wade, *J. Chem. Soc., Dalton Trans.*, 1979, 102.
22. D. J. Brauer, H. Bürger, H. H. Moretto, U. Wannagat and K. Wiegel, *J. Organomet. Chem.*, 1979, **170**, 161.
23. N. A. Bell, G. E. Coates, M. L. Schneider and H. M. M. Shearer, *J. Chem. Soc., Chem. Commun.*, 1983, 828.
24. N. A. Bell, G. E. Coates, M. L. Schneider and H. M. M. Shearer, *Acta Crystallogr., Sect. C: Cryst. Struct. Commun.*, 1984, **40**, 608.
25. R. Fleischer and D. Stalke, *Inorg. Chem.*, 1997, **36**, 2413.
26. J. F. Eichler, O. Just and W. S. Rees, *Inorg. Chem. Comm.*, 2005, **8**, 936.
27. R. Y. Han and G. Parkin, *Inorg. Chem.*, 1992, **31**, 983.
28. R. Han and G. Parkin, *Inorg. Chem.*, 1993, **32**, 4968.
29. H. Nöth and D. Schlosser, *Chem. Ber.*, 1988, **121**, 1711.
30. H. Nöth and D. Schlosser, *Chem. Ber.*, 1988, **121**, 1715.
31. B. Neumüller and K. Dehnicke, *Z. Anorg. Allg. Chem.*, 2004, **630**, 799.
32. J. Emsley, *The Elements*, Clarendon Press, Oxford 2nd edn., 1991, p. 30.
33. P. G. Plieger, K. D. John, T. S. Keizer, T. M. McCleskey, A. K. Burrell and R. L. Martin, *J. Am. Chem. Soc.*, 2004, **126**, 14651.
34. D. R. Armstrong, H. L. Benham and G. T. Walker, *J. Mol. Struct. (Theochem)*, 1988, **165**, 65.
35. O. Mo, M. Yanez, M. Eckert-Maksic, Z. B. Maksic, I. Alkorta and J. Elguero, *J. Phys. Chem. A*, 2005, **109**, 4359.
36. C.-C. Chang and M. A. Ameerunisha, *Coord. Chem. Rev.*, 1999, **189**, 199.
37. A. G. Pinkus, *Coord. Chem. Rev.*, 1978, **25**, 173.
38. F. T. Edelmann, *Coord. Chem. Rev.*, 1994, **137**, 403.
39. R. E. Mulvey, *Chem. Soc. Rev.*, 1998, **27**, 339.
40. H. Jacobs and R. Juza, *Z. Anorg. Allg. Chem.*, 1969, **370**, 254.
41. M. H. Sørby, Y. Nakamura, H. W. Brinks, T. Ichikawa, S. Hino, H. Fujii and B. C. Hauback, *J. Alloys Comp.*, 2007, **428**, 297.
42. J. Hu, Z. Xiong, G. Wu, P. Chen, K. Murata and K. Sakata, *J. Power Sources*, 2006, **159**, 120.
43. J. Hu, G. Wu, Y. Liu, Z. Xiong, P. Chen, K. Murata, K. Sakata and G. Wolf, *J. Phys. Chem. B*, 2006, **110**, 14688.
44. P. Chen, Z. Xiong, J. Luo, J. Lin and K. L. Tan, *J. Phys. Chem. B*, 2003, **107**, 10967.
45. H. Leng, T. Ichikawa and H. Fujii, *J. Phys. Chem. B*, 2006, **110**, 12964.
46. Z. Xiong, G. Wu, J. Hu, P. Chen, W. Luo and J. Wang, *J. Alloys Comp.*, 2006, **417**, 190.
47. H. Y. Leng, T. Ichikawa, S. Hino, N. Hanada, S. Isobe and H. Fujii, *J. Power Sources*, 2006, **156**, 166.
48. J. Hu, Z. Xiong, G. Wu, P. Chen, K. Murata and K. Sakata, *J. Power Sources*, 2006, **159**, 116.
49. Y. Chen, C. Z. Wu, P. Wang and H. M. Cheng, *Int. J. Hyd. Energy*, 2006, **31**, 1236.
50. C. Zhang and A. Alavi, *J. Phys. Chem. B*, 2006, **110**, 7139.
51. W. Luo and S. Sickafoose, *J. Alloys Comp.*, 2006, **407**, 274.
52. Y.-J. Tang, L. N. Zakharov, A. L. Rheingold and R. A. Kemp, *Organometallics*, 2005, **24**, 836.
53. L. M. Engelhardt, B. S. Jolly, P. C. Junk, C. L. Raston, B. W. Skelton and A. H. White, *Aust. J. Chem.*, 1986, **39**, 1337.
54. (a) R. A. Bartlett, M. M. Olmstead and P. P. Power, *Inorg. Chem.*, 1994, **33**, 4800; (b) K.-C. Yang, C.-C. Chang, J.-Y. Huang, C.-C. Lin, G.-H. Lee, Y. Wang and M. Y. Chiang, *J. Organomet. Chem.*, 2002, **648**, 176.
55. M.-X. Zhang and P. E. Eaton, *Angew. Chem. Int. Ed.*, 2002, **41**, 2169.

56. (a) C. R. Hauser and H. G. Walker, *J. Am. Chem. Soc.*, 1947, **69**, 295; (b) P. E. Eaton, C.-H. Lee and Y. Xiong, *J. Am. Chem. Soc.*, 1989, **111**, 8016.
57. A. Krasoviskiy, V. Krasovskaya and P. Knochel, *Angew. Chem. Int. Ed.*, 2006, **45**, 2958.
58. K. Angermund, B. Bogdanovic, G. Koppetsch, C. Krüger, R. Mynott, M. Schwickardi and Y. H. Tsay, *Z. Naturforsch.*, 1986, **41b**, 455.
59. E. C. Ashby and A. B. Goel, *J. Organomet. Chem.*, 1981, **204**, 139.
60. G. Dozzi, G. Del Piero, M. Cesari and S. Cucinella, *J. Organomet. Chem.*, 1980, **190**, 228.
61. V. Lorenz, K.-H. Tiele and B. Neumüller, *Z. Anorg. Allg. Chem.*, 1994, **620**, 691.
62. W. Clegg, F. J. Craig, K. W. Henderson, A. R. Kennedy, R. E. Mulvey, P. A. O'Neil and D. Reed, *Inorg. Chem.*, 1997, **36**, 6238.
63. M. Westerhausen, T. Bollwein, N. Makropoulos and H. Piotrowski, *Inorg. Chem.*, 2005, **44**, 6439.
64. M. Gartner, R. Fischer, J. Langer, H. Görls, D. Walther and M. Westerhausen, *Inorg. Chem.*, 2007, **46**, 5118.
65. D. C. Bradley, M. B. Hursthouse, A. A. Ibrahim, K. M. A. Malik, M. Motevalli, R. Möseler, H. Powell, J. D. Runnacles and A. C. Sullivan, *Polyhedron*, 1990, **9**, 2959.
66. J. L. Sebestl, T. T. Nadasdi, M. J. Heeg and C. H. Winter, *Inorg. Chem.*, 1998, **37**, 1289.
67. M. Gillett-Kunnath, W.-J. Teng, W. Vargas, K. Ruhlandt-Senge, *Inorg. Chem.*, 2005, **44**, 4862.
68. M. Bremer, G. Linti, H. Nöth, M. Thomann-Albach and G. E. W. J. Wagner, *Z. Anorg. Allg. Chem.*, 2005, **631**, 683.
69. L. M. Englehardt, P. C. Junk, W. C. Patalinghug, R. E. Sue, C. L. Raston, B. W. Skelton and A. H. White, *J. Chem. Soc., Chem. Commun.*, 1991, 930.
70. B. Goldfuss, Pv. R Schleyer, S. Handschuh and F. Hampel, *J. Organomet. Chem.*, 1998, **552**, 285.
71. B. Goldfuss, Pv. R Schleyer, S. Handschuh, F. Hampel and W. Bauer, *Organometallics*, 1997, **16**, 5999.
72. I. L. Fedushkin, A. A. Skatova, V. A. Chudakova, G. K. Fukin, S. Dechert and H. Schumann, *Eur. J. Inorg. Chem.*, 2003, 3336.
73. I. L. Fedushkin, V. A. Chudakova, A. A. Skatova, N. M. Khvoinova, Y. A. Kurskii, T. A. Glukhova, G. K. Fukin, S. Dechert, M. Hummert and H. Schumann, *Z. Anorg. Allg. Chem.*, 2004, **630**, 501.
74. R. Fleischer and D. Stalke, *J. Organomet. Chem.*, 1998, **550**, 173.
75. C. F. Caro, P. B. Hitchcock and M. F. Lappert, *Chem. Commun.*, 1999, 1433.
76. S. D. Robertson, T. Chivers and J. Konu, *J. Organomet. Chem.*, 2007, **692**, 4327.
77. I. Schranz, L. Stahl and R. J. Staples, *Inorg. Chem.*, 1998, **37**, 1493.
78. M. H. Chisholm, J. Gallucci and K. Phomphrai, *Inorg. Chem.*, 2002, **41**, 2785.
79. M. H. Chisholm, J. Gallucci and K. Phomphrai, *Inorg. Chem.*, 2005, **44**, 8004.
80. P. J. Bailey, C. M. E. Dick, S. Fabre and S. Parsons, *J. Chem. Soc., Dalton Trans.* 2000, 1655.
81. V. C. Gibson, J. A. Segal, A. J. P. White and D. J. Williams, *J. Am. Chem. Soc.*, 2000, **122**, 7120.
82. N. Nimitsiriwat, V. C. Gibson, E. L. Marshall, P. Takolpuckdee, A. K. Tomov, A. J. P. White, D. J. Williams, M. R. J. Elsegood and S. H. Dale, *Inorg. Chem.*, 2007, **46**, 9988.
83. B. Conway, E. Hevia, A. R. Kennedy, R. E. Mulvey and S. Weatherstone, *Dalton. Trans.*, 2005, 1532.
84. W. Vargas, U. Englich and K. Ruhlandt-Senge, *Inorg. Chem.*, 2002, **41**, 5602.
85. W. J. Teng, U. Englich and K. Ruhlandt-Senge, *Inorg. Chem.*, 2000, **39**, 3875.
86. K. W. Henderson, R. E. Mulvey, W. Clegg, P. A. O'Neil, *J. Organomet. Chem.*, 1992, **439**, 237.
87. Y. Sarazin, M. Schormann and M. Bochmann, *Organometallics*, 2004, **23**, 3296.
88. C. Zapilko, Y.-C. Liang and R. Anwander, *Chem. Mater.*, 2007, **19**, 3171.

89. A. R. Kennedy, R. E. Mulvey and J. H. Schulte, *Acta Crystallogr., Sect. C: Cryst. Struct. Commun.*, 2001, **57**, 1288.
90. U. Wannagat, H. Autzen and H. Kuckertz, *Z. Anorg. Allg. Chem.*, 1972, **394**, 254.
91. T. Fjeldberg and R. A. Andersen, *J. Mol. Struct.*, 1984, **125**, 287.
92. M. Westerhausen and W. Schwarz, *Z. Anorg. Allg. Chem.*, 1992, **609**, 39.
93. E. Hevia, A. R. Kennedy, R. E. Mulvey and S. Weatherstone, *Angew. Chem., Int. Ed.*, 2004, **43**, 1709.
94. K. W. Henderson, G. W. Honeyman, A. R. Kennedy, R. E. Mulvey, J. A. Parkinson and D. C. Sherrington, *Dalton. Trans.*, 2003, 1365.
95. G. C. Forbes, A. R. Kennedy, R. E. Mulvey and P. J. A. Rodger, *Chem. Commun.*, 2001, 1400.
96. A. Xia, M. J. Heeg and C. H. Winter, *Organometallics*, 2002, **21**, 4718.
97. K. W. Henderson, R. E. Mulvey, W. Clegg, P. A. O'Neil, *Polyhedron*, 1993, **12**, 2535.
98. H. Sachdev, C. Wagner, C. Preis, V. Huch and M. Veith, *J. Chem. Soc., Dalton Trans.*, 2002, 4709.
99. A. W. Duff, P. B. Hitchcock, M. F. Lappert, R. G. Taylor and J. A. Segal, *J. Organomet. Chem.*, 1985, **293**, 271.
100. W. Clegg, L. Horsburgh, R. E. Mulvey, M. J. Ross, R. B. Rowlings and V. Wilson, *Polyhedron*, 1998, **17**, 1923.
101. D. R. Armstrong, W. Clegg, R. E. Mulvey and R. B. Rowlings, *J. Chem. Soc., Dalton Trans.*, 2001, 409.
102. M. M. Olmstead, W. J. Grigsby, D. R. Chacon, T. Hascall and P. P. Power, *Inorg. Chim. Acta*, 1996, **251**, 273.
103. K. W. Henderson, J. F. Allan and A. R. Kennedy, *Chem. Commun.*, 1997, 1149.
104. X. Y. He, J. F. Allan, B. C. Noll, A. R. Kennedy and K. W. Henderson, *J. Am. Chem. Soc.*, 2005, **127**, 6920.
105. E. P. Squiller, A. D. Pajerski, R. R. Whittle and H. G. Richey, Jr., *Organometallics*, 2006, **25**, 2465.
106. H. Jacobs, J. Birkenbeul and J. Kockelkorn, *J. Less-Common Met.*, 1984, **97**, 205.
107. H. Jacobs, J. Birkenbeul and D. Schmitz, *J. Less-Common Met.*, 1982, **85**, 79.
108. A. R. Kennedy, R. E. Mulvey and R. B. Rowlings, *J. Am. Chem. Soc.*, 1998, **120**, 7816.
109. G. C. Forbes, A. R. Kennedy, R. E. Mulvey, P. J. A. Rodger and R. B. Rowlings, *J. Chem. Soc., Dalton Trans.*, 2001, 1477.
110. P. C. Andrikopoulos, D. R. Armstrong, A. R. Kennedy, R. E. Mulvey, C. T. O'Hara, R. B. Rowlings and S. Weatherstone, *Inorg. Chim. Acta*, 2007, **360**, 1370.
111. (a) E. Hevia, F. R. Kenley, A. R. Kennedy, R. E. Mulvey and R. B. Rowlings, *Eur. J. Inorg. Chem.,* 2003, 3347; (b) D. V. Graham, E. Hevia, A. R. Kennedy, R. E. Mulvey, C. T. O'Hara and C. Talmard, *Chem. Commun.,* 2006, 417.
112. E. Hevia, D. J. Gallagher, A. R. Kennedy, R. E. Mulvey, C. T. O'Hara and C. Talmard, *Chem. Commun.,* 2004, 2422.
113. G. C. Forbes, A. R. Kennedy, R. E. Mulvey, B. A. Roberts and R. B. Rowlings, *Organometallics*, 2002, **21**, 5115.
114. R. E. Mulvey, *Chem. Commun.*, 2001, 1049.
115. (a) P. C. Andrikopoulos, D. R. Armstrong, D. V. Graham, E. Hevia, A. R. Kennedy, R. E. Mulvey, C. T. O'Hara and C. Talmard, *Angew. Chem. Int. Ed.*, 2005, **44**, 3459; (b) D. R. Armstrong, W. Clegg, S. H. Dale, D. V. Graham, E. Hevia, L. M. Hogg, G. W. Honeyman, A. R. Kennedy and R. E. Mulvey, *Chem. Commun.,* 2007, 598.
116. L. Barr, A. R. Kennedy, J. G. MacLellan, J. H. Moir, R. E. Mulvey and P. J. A. Roger, *Chem. Commun.*, 2000, 1757.
117. W. Clegg, K. W. Henderson, R. E. Mulvey, P. A. O'Neil, *J. Chem. Soc., Chem. Commun.*, 1993, 969.
118. W. Clegg, K. W. Henderson, R. E. Mulvey, P. A. O'Neil, *J. Chem. Soc., Chem. Commun.*, 1994, 769.

119. L. T. Wendell, J. Bender, X. He, B. C. Noll and K. W. Henderson, *Organometallics*, 2006, **25**, 4953.
120. T.-Y. Her, C.-C. Chang and L.-K. Liu, *Inorg. Chem.*, 1992, **31**, 2291.
121. T.-Y. Her, C.-C. Chang, G.-H. Lee, S.-M. Peng and Y. Wang, *Inorg. Chem.*, 1994, **33**, 99.
122. C.-C. Chang, T.-Y. Her, F.-Y. Hsieh, C.-Y. Yang, M.-Y. Chiang, G.-H. Lee, Y. Wang and S.-M. Peng, *J. Chin. Chem. Soc.*, 1994, **41**, 783.
123. C.-C. Chang, W.-H. Lee, T.-Y. Her, G.-H. Lee, S.-M. Peng and Y. Wang, *J. Chem. Soc., Dalton Trans.*, 1994, 315.
124. C.-C. Chang, T.-Y. Her, M.-D. Li, R. Williamson, G.-H. Lee, S.-M. Peng and Y. Wang, *Inorg. Chem.*, 1995, **34**, 4296.
125. C.-C. Chang, B. Srinivas, M.-L. Wu, W.-H. Chiang, M.-Y. Chiang and C.-S. Hsiung, *Organometallics*, 1995, **14**, 5150.
126. M.-D. Li, C.-C. Chang, Y. Wang and G.-H. Lee, *Organometallics*, 1996, **15**, 2571.
127. C.-C. Chang, J.-H. Chen, B. Srinivas, M. Y. Chiang, G.-H. Lee and S.-M. Peng, *Organometallics*, 1997, **16**, 4980.
128. B. Srinivas, C.-C. Chang, C.-H. Chen, M. Y. Chiang, I.-T. Chen, Y. Wang and G.-H. Lee, *J. Chem. Soc., Dalton Trans.*, 1997, 957.
129. A. Martín, M. Mena, A. Perez-Redondo and C. Yelamos, *Organometallics*, 2002, **21**, 3308.
130. A. Martín, M. Mena, A. Perez-Redondo and C. Yelamos, *Inorg. Chem.*, 2004, **43**, 2491.
131. A. Martín, M. Mena, A. Perez-Redondo and C. Yelamos, *Dalton. Trans.*, 2005, 2116.
132. G. W. Honeyman, A. R. Kennedy, R. E. Mulvey and D. C. Sherrington, *Organometallics*, 2004, **23**, 1197.
133. K. W. Henderson, A. R. Kennedy, R. E. Mulvey, C. T. O'Hara and R. B. Rowlings, *Chem. Commun.*, 2001, 1678.
134. E. Hevia, G. W. Honeyman, A. R. Kennedy, R. E. Mulvey and D. C. Sherrington, *Angew. Chem., Int. Ed.*, 2005, **44**, 68.
135. P. C. Andrikopoulos, D. R. Armstrong, E. Hevia, A. R. Kennedy and R. E. Mulvey, *Organometallics*, 2006, **25**, 2415.
136. A. R. Kennedy, R. E. Mulvey and R. B. Rowlings, *Angew. Chem., Int. Ed.*, 1998, **37**, 3180.
137. A. R. Kennedy, R. E. Mulvey, C. L. Raston, B. A. Roberts and R. B. Rowlings, *Chem. Commun.*, 1999, 353.
138. K. J. Drewette, K. W. Henderson, A. R. Kennedy, R. E. Mulvey, C. T. O'Hara and R. B. Rowlings, *Chem. Commun.*, 2002, 1176.
139. W. Clegg, K. W. Henderson, A. R. Kennedy, R. E. Mulvey, C. T. O'Hara, R. B. Rowlings and D. M. Tooke, *Angew. Chem., Int. Ed.*, 2001, **40**, 3902.
140. P. C. Andrikopoulos, D. R. Armstrong, W. Clegg, C. J. Gilfillan, E. Hevia, A. R. Kennedy, R. E. Mulvey, C. T. O'Hara, J. A. Parkinson and D. M. Tooke, *J. Am. Chem. Soc.*, 2004, **126**, 11612.
141. D. J. Gallagher, K. W. Henderson, A. R. Kennedy, C. T. O'Hara, R. E. Mulvey and R. B. Rowlings, *Chem. Commun.*, 2002, 376.
142. R. Juza and E. Eberius, *Naturwiss.*, 1962, **49**, 104.
143. L. Luo, *J. Alloys Comp.*, 2004, **381**, 284.
144. Y. Liu, T. Liu, Z. Xiong, *et al.*, *Eur. J. Inorg. Chem.*, 2006, 4368.
145. C. M. Araujo, R. H. Scheicher, P. Jena and R. Ahuja, *App. Phys. Lett.*, 2007, **91**, 091924.
146. S. Cucinella, G. Dozzi, G. Perego and A. Mazzei, *J. Organomet. Chem.*, 1977, **137**, 257.
147. T. Hascall, Ruhlandt-Senge K. and P. P. Power, *Angew. Chem., Int. Ed.*, 1994, **33**, 356.
148. W. J. Grigsby, T. Hascall, J. J. Ellison, M. M. Olmstead and P. P. Power, *Inorg. Chem.*, 1996, **35**, 3254.
149. W. J. Grigsby and P. P. Power, *J. Chem. Soc., Dalton Trans.*, 1996, 4613.
150. W. J. Grigsby, M. M. Olmstead and P. P. Power, *J. Organomet. Chem.*, 1996, **513**, 173.

151. J. K. Brask, T. Chivers and M. Parvez, *Angew. Chem., Int. Ed.*, 2000, **39**, 958.
152. T. Hascall, M. M. Olmstead and P. P. Power, *Angew. Chem., Int. Ed.*, 1994, **33**, 1000.
153. J. A. Rood, B. C. Noll and K. W. Henderson, *Inorg. Chem.*, 2007, **46**, 7259.
154. R. Juza and H. Schumacher, *Z. Anorg. Allg. Chem.*, 1963, **324**, 278.
155. Y. Liu, Z. Xiong, J. Hu, G. Wu, P. Chen, K. Murata and K. Sakata, *J. Power Sources*, 2006, **159**, 135.
156. H. V. K. Diyabalanage, R. P. Shrestha, T. A. Semelsberger, B. L. Scott, M. E. Bowden and A. K. Burrell, *Angew. Chem., Int. Ed.*, 2007, **46**, 8995.
157. T. Sichla and H. Jacobs, *Z. Anorg. Allg. Chem.*, 1996, **622** 2079.
158. M. Westerhausen, *Inorg. Chem.*, 1991, **30**, 96.
159. M. Westerhausen and W. Schwarz, *Z. Anorg. Allg. Chem.*, 1991, **604**, 127.
160. P. B. Hitchcock, M. F. Lappert, G. A. Lawless and B. Royo, *J. Chem. Soc., Chem. Commun.*, 1990, 1141.
161. A. D. Frankland, P. B. Hitchcock, M. F. Lappert and G. A. Lawless, *J. Chem. Soc., Chem. Commun.*, 1994, 2435.
162. Y.-J. Tang, L. N. Zakharov, W. S. Kassel, A. L. Rheingold and R. A. Kemp, *Inorg. Chim. Acta*, 2005, **358** 2014.
163. M. H. Chisholm, J. C. Gallucci and K. Phomphrai, *Inorg. Chem.*, 2004, **43**, 6717.
164. M. Westerhausen, J. Greul, H.-D. Hausen and W. Schwarz, *Z. Anorg. Allg. Chem.*, 1996, **622**, 1295.
165. A. G. Avent, M. R. Crimmin, M. S. Hill and P. B. Hitchcock, *Dalton Trans.*, 2005, 278.
166. D. J. Burkey, E. K. Alexander and T. P. Hanusa, *Organometallics*, 1994, **13**, 2773.
167. M. Westerhausen, S. Schneiderbauer, A. N. Kneifel, Y. Söltl, P. Mayer, H. Nöth, Z.-Y. Zhong, P. J. Dijkstra and J. Feijen, *Eur. J. Inorg. Chem.*, 2003, 3432.
168. S. Datta, P. W. Roesky and S. Blechert, *Organometallics*, 2007, **26**, 4392.
169. V. Lorenz, B. Neumüller and K.-H. Tiele, *Z. Naturforsch.*, 1995, **50b** 71.
170. M. Westerhausen and W. Schwarz, *Z. Anorg. Allg. Chem.*, 1996, **622**, 903.
171. M. Westerhausen, R. Löw and W. Schwarz, *J. Organomet. Chem.*, 1996, **513**, 213.
172. M. Westerhausen, M. H. Digeser, M. Krofta, N. Wiberg, H. Nöth, J. Knizek, W. Ponikwar and T. Seifert, *Eur. J. Inorg. Chem.* 1999, 743.
173. M. Westerhausen and W. Schwarz, *Z. Naturforsch.*, 1995, **50b**, 106.
174. S. Chadwick, U. Englich, B. Noll, K. Ruhlandt-Senge, *Inorg. Chem.*, 1998, **37**, 4718.
175. G. Becker, K. W. Klinkhammer, W. Schwarz, M. Westerhausen and T. Hildenbrand, *Z. Naturforsch.*, 1992, **47b**, 1225.
176. D. E. Gingelberger and J. Arnold, *J. Am. Chem. Soc.*, 1992, **114**, 6242.
177. D. E. Gindelberger and J. Arnold, *Inorg. Chem.*, 1994, **33**, 6293.
178. S. Harder and J. Brettar, *Angew. Chem., Int. Ed.*, 2006, **45**, 3474.
179. S. C. Sockwell, T. P. Hanusa and J. C. Huffman, *J. Am. Chem. Soc.*, 1992, **114**, 3393.
180. X. He, B. C. Noll, A. Beatty, R. E. Mulvey and K. W. Henderson, *J. Am. Chem. Soc.*, 2004, **126**, 7444.
181. M. Gartner, H. Görls and M. Westerhausen, *Inorg. Chem.*, 2007, **46**, 7678.
182. H. Jacobs, J. Kockelkorn and J. Birkenbeul, *J. Less-Common Met.*, 1982, **87**, 215.
183. H. Jacobs and U. Fink, *Z. Anorg. Allg. Chem.*, 1977, **435**, 137.
184. H. Jacobs and J. Kockelkorn, *Z. Anorg. Allg. Chem.*, 1979, **456**, 147.
185. A. R. Kennedy, R. E. Mulvey and R. B. Rowlings, *J. Organomet. Chem.*, 2002, **648**, 288.
186. M. Westerhausen, C. Birg, H. Nöth, J. Knizek and T. Seifert, *Eur. J. Inorg. Chem.*, 1999, 2209.
187. P. B. Hitchcock, M. F. Lappert and X.-H. Wei, *Dalton Trans.* 2006, 1181.
188. F. G. N. Cloke, P. B. Hitchcock, M. F. Lappert, G. A. Lawless and B. Royo, *J. Chem. Soc., Chem. Commun.*, 1991, 724.

189. G. Mösges, F. Hampel, M. Kaupp, P. v. R. Schleyer, *J. Am. Chem. Soc.*, 1992, **114**, 10880.
190. M. S. Hill and P. B. Hitchcock, *Chem. Commun.*, 2003, 1758.
191. H. M. El Kaderi, M. J. Heeg and C. H. Winter, *Organometallics*, 2004, **23**, 4995.
192. S. Wingerter, M. Pfeiffer, A. Murso, C. Lustig, T. Stey, V. Chandrasekhar and D. Stalke, *J. Am. Chem. Soc.*, 2001, **123**, 1381.
193. M. Westerhausen, H. D. Hausen and W. Schwarz, *Z. Anorg. Allg. Chem.*, 1992, **618**, 121.
194. M. Westerhausen and W. Schwarz, *Z. Anorg. Allg. Chem.*, 1991, **606**, 177.
195. I. L. Fedushkin, A. G. Morozov, O. V. Rassadin and G. K. Fukin, *Chem. Eur. J.*, 2005, **11**, 5749.
196. J. S. Alexander, K. Ruhlandt-Senge and H. Hope, *Organometallics*, 2003, **22**, 4933.
197. L. Lameyer, O. Abu Salah, S. Deuerlein, T. Stey and D. Stalke, *Z. Anorg. Allg. Chem.*, 2004, **630**. 1801.
198. H. Jacobs and C. Hadenfeldt, *Z. Anorg. Allg. Chem.*, 1975, **418**, 132.
199. S. L. Castro, O. Just, W. S. Rees, Jr. *Angew. Chem., Int. Ed.*, 2000, **39**, 933.
200. A. P. Purdy, A. D. Berry and C. F. George, *Inorg. Chem.*, 1997, **36**, 3370.
201. B. A. Vaartstra, J. C. Huffman, W. E. Streib and K. G. Caulton, *Inorg. Chem.*, 1991, **30**, 121.
202. A. Weeber, S. Harder and H. H. Brintzinger, *Organometallics*, 2000, **19**, 1325.
203. M. R. Crimmin, A. G. M. Barrett, M. S. Hill, P. B. Hitchcock and P. A. Procopiou, *Inorg. Chem.*, 2007, **46**, 10410.
204. W. Clegg, S. J. Coles, E. K. Cope and F. S. Mair, *Angew. Chem., Int. Ed.*, 1998, **37**, 796.
205. J. S. Alexander and K. Ruhlandt-Senge, *Angew. Chem., Int. Ed.*, 2001, **40**, 2658.
206. M. Westerhausen, H.-D. Huasen and W. Schwarz, *Z. Anorg. Allg. Chem.*, 1995, **621**, 877.
207. R. Fleischer, B. Walfort, A. Gbureck, P. Scholz, W. Kiefer and D. Stalke, *Chem. Eur. J.*, 1998, **4**, 2266.
208. P.-C. Shao, D. J. Berg and G. W. Bushnell, *Can. J. Chem.*, 1995, **73**, 797.
209. H. Jacobs, J. Birkenbeul and J. Kockelkorn, *J. Less-Common Met.*, 1982, **85**, 71.
210. R. Fleischer and D. Stalke, *Organometallics*, 1998, **17**, 832.
211. R. P. Davies, *Inorg. Chem. Commun.*, 2000, **3**, 13.
212. M. Westerhausen, G. Sapelza and P. Mayer, *Angew. Chem., Int. Ed.*, 2005, **44**, 6234.

4

Amides of the Group 3 and Lanthanide Metals

4.1 Introduction

In our 1980 book, amides of Sc, Y and the lanthanide metals (hereafter 'Ln') were discussed with those of thorium and uranium in $2^{1}/_{2}$ pages of text, 4 pages devoted to X-ray structures (4 for Ln amides) and one page of tabulated data.[1] The Ln amides described were $Ln(NPr^{i}_{2})_{3}$ (Ln = Y, Nd, Yb), $[Ln\{N(SiMe_{3})_{2}\}_{3}]$ (Ln = Y, Sc,* La, Ce, Pr, Nd,* Sm, Eu,* Gd, Ho, Yb,* Lu), $[Ln\{N(SiMe_{3})_{2}\}_{3}(OPPh_{3})]$ (Ln = La,* Eu, Lu) and $[\{Ln(N(SiMe_{3})_{2})_{2}(OPPh_{3})\}_{2}\{(\mu\text{-}O\text{-}O)^{2-}\}]$ (Ln = La,* Pr, Sm, Eu, Lu); X-ray data were available for the six compounds with an asterisk. The most far-reaching discovery was of the three-coordinate homoleptic Ln bis(trimethylsilyl)amides by Bradley, *et al.* Part II of our book dealing with chemical properties of metal amides (reactions with protic compounds, metal hydrides, metal halides, Lewis acids or bases; their role as dehydrochlorinating agents or polymerisation initiators; or as substrates for oxidative addition, or reductive elimination, or homolysis) and their applications had no entries for Ln amides.

This review deals with Ln compounds having the amido ligand $[NRR']^{-}$, in which R and R′ may be the same or different and each is a hydrocarbyl or a silyl group or one of them is hydrogen. Thus, excluded are Ln derivatives of hydrazines such as pyrazolates (some compounds involving pyrazolato ligands are discussed in Section 4.3.4).[2a] Also omitted are 1-azaallyls even if bonded as the enamido tautomer,[2b] and Ln complexes of *N,N′*-centred monoanionic ligands such as β-diketiminates,[2c] amidinates,[2d] and guanidinates.[2e] Amido-based macrocycles such as porphirinates and Schiff bases are also not discussed.

The most complete subsequent review concerned with the Ln amides is the 1996 chapter by Anwander,[3] described in outline in Section 4.2. Post-1996 surveys have dealt with selected areas. Cotton's contribution on Ln coordination chemistry up to 2001 (and in part 2002) has 95 pages and 1135 references;[4a] it has a substantial section on Ln amides (ca. 50 refs.), which inevitably were dealt with briefly.

Metal Amide Chemistry Michael Lappert, Andrey Protchenko, Philip Power and Alexandra Seeber

A review of the organometallic chemistry of lanthanides included several references to Ln amides.[4b] A paper dealing with selected highlights of amidometal chemistry cited 12 papers on Ln amide catalysts.[5a] Chiral lanthanide complexes and their applications as catalysts included complexes such as $[M(thf)_2]_3[Ln'(binol)_3]$ (M = Li or Na; Ln′ = Y, Eu or Yb), prepared from $[Ln'\{N(SiMe_3)_2\}_3]$ and 3M(Hbinol); H_2binol = (*R*)- or (*S*)-2,2′-dihydroxy-1, 1′-binaphthyl.[5b] Reviews on the synthesis and structural chemistry of non-cyclopentadienyl-Ln complexes,[6] or their role as olefin polymerisation catalysts,[7] have small sections dealing with Ln amides. Monocyclopentadienyl-Ln compounds have been surveyed.[8a] This topic was taken up in more detail in the context of Ln complexes containing linked amido-cyclopentadienyl dianionic ligands as *ansa*-metallocene mimics and constrained geometry catalysts for hydrosilylation, hydroamination or polymerisation of an α-olefin, acrylonitrile or *t*-butyl acrylate;[9a] cationic alkyl-Ln complexes featured in another publication, but amides were mentioned only briefly.[9b] Amido-Ln-containing complexes as catalysts for the hydroamination of aminoalkenes have been surveyed.[9c,9d]

TetraphenylboratoYb^{II} bis(trimethylsilyl)amides derived from $[Yb\{N(SiMe_3)_2\}_2(thf)_2]$ and $[HNMe_3][BPh_4]$ have been reviewed.[8b] Aspects of Ln^{III} chemistry, derived largely from $[Ln\{N(SiMe_3)_2\}_3]$ as precursors, were discussed in the context of their low coordination numbers[10a] and were illustrated with examples taken from Dehnicke's group, including $[ScCl_2\{N(SiMe_3)_2\}(thf)_2]$,[10b] $[Na(thf)_2\{\mu\text{-}\eta^6\text{-}PhN(Ph)\}_2Sm\{N(SiMe_3)_2\}_2]$,[10c,10d] $[Na(thf)_3\{CH_2Si(Me)_2N(SiMe_3)Ln(N(SiMe_3)_2)_2\}]$,[10e] $[Na(thf)_6][\{Ln(N(SiMe_3)_2)(\mu\text{-}NH_2)(\mu\text{-}NSiMe_3)\}_2]$[10f] and $[Na(thf)_3(\mu\text{-}C{\equiv}CPh)Sm\{N(SiMe_3)_2\}_3]$ (a catalyst for ring-opening polymerisation of ε-caprolactone or δ-valerolactone).[10f,10g]

Scandium compounds containing a dianionic bis(amido) ligand have been reviewed: $[NC_5H_4(2\text{-}CH_2\{N(CH_2)_2NSiMe_3\}_2]^{2-}$, $[NC_5H_4\{2\text{-}C(Me)(CH_2NSiMe_3)_2\}]^{2-}$, $[MeN\{(CH_2)_nNSiMe_3\}_2]^{2-}$ (n = 2 or 3) and $[Me_3SiN\{(CH_2)_2NSiMe_3\}_2]^{2-}$.[11] A survey of pyrazolylborate-Ln complexes included the Ln^{II} compounds $[Ln(HB\{NC(Me)C(H)C(Bu^t)N\}_3)\{N(SiMe_3)_2\}]$;[12a] carbene-alkylamido ligands $CN(R)CH{=}CHN(CH_2)_2\bar{N}Bu^t$ in Ln^{III} chemistry have been explored by P. L. Arnold, *et al.*[12b] A review by Trifonov, entitled 'Reactions of ytterbocenes with diimines: steric manipulation of reductive reactivity', dealt with reduction of α-diimines and related compounds, which led to Ln^{III} amides via electron transfer and/or C-H bond activation.[12c]

Non-cyclopentadienyl ancillaries in group 3 metal (M) chemistry have been described in detail by Piers and Emslie.[13] Included were (i) the use of $[M\{N(SiMe_3)_2\}_3]$ and the less hindered $[M\{N(SiHMe_2)_2\}_3]$ as precursors for reactions with protic compounds, (ii) complexes containing amides as co-ligands as in $[Y\{(N(C_6H_3Et_2\text{-}2,6))_2C(C_6H_4Me\text{-}4)\}_2\{N(SiMe_3)_2\}]$ or $[M(chel)\{N(SiHMe_2)_2\}_2]$ (chel = an aminotroponiminato or a bisoxazolinato ligand), and (iii) complexes containing bifunctional ligands. Item (iii) dealt with ligands such as $[N(Bu^t)Si(Me)_2OBu^t]^-$, salicylaldiminates, Fryzuk's diphosphine-amides $[N\{Si(Me)_2CH_2PMe_2\}_2]^-$, the chiral diamide $[6,6'\text{-}Me_2\text{-}2,2'\text{-}\{N(SiBu^tMe_2)\}_2\text{-}biphenyl]^{2-}$, the diamide $[\{N(Bu^t)C_6H_4\text{-}2\}_2O]^{2-}$, bis(amido)pyridines, bis(amido)troponiminates, and the bis(*N*-aryliminomethyl)pyrrolyl $[NC_4H_2\{(CH{=}NC_6H_3Me_2\text{-}2',6')_2\text{-}2,5\}]^-$.

4.2 The Pre-1996 Literature: Anwander's Review[3]

4.2.1 Introduction

Anwander's comprehensive survey of Ln amides covered in 78 pages, with X-ray data for more than 200 compounds, and 330 references. This clearly demonstrates that the 1979–1996 period had shown a massive interest in the topic, even if we exclude a number of themes dealt therein which we regard to be outside the remit of the present chapter. These include compounds involving the ligands $[NH_2]^-$, $[N(H)NR^1]^{2-}$, $[N(R^1)NR^2]^{2-}$, $[N_3]^-$, macrocycles containing $\bar{N}$-centres and polypyrazolylborates.

In this Section 4.2, the focus is on the individual relevant parts of Anwander's review. For each, a limited choice of references is made, thus reflecting our view as to their particular significance. Additionally, data from a few papers absent from the review are included.

4.2.2 Ln^{III} Complexes with *N*-Hydrocarbyl-Amido Ligands

The first well characterised Ln alkylamides either had lipophilic co-ligands as in $[LaCp^*_2\{N(H)Me\}\{N(H)_2Me\}]$,[14] or were adducts with $AlMe_3$, $GaMe_3$ or a lithium phosphide, exemplified by $[Nd\{N(Me)_2{\cdot}MMe_3\}_3]$ (**1**)[15] or $[La(NPr^i_2)_2\{\mu\text{-}P(C_6H_4OMe\text{-}4)_2\}_2Li(thf)]$;[16] each was X-ray-characterised. Compound **1** (M = Al) with an excess of $AlMe_3$ afforded $[Nd\{(\mu\text{-}Me)_2AlMe_2\}_3]$.[17] Bis(diisopropylamido)Ln complexes included $[Ln(NPr^i_2)_3(thf)]$ (Ln = Y,[18a] La,[18a] Nd,[18b] Yb,[18a]), $[Ln(NPr^i_2)_2(\mu\text{-}NPr^i_2)_2Li(thf)]$ (Ln = Y,[18a] La,[18a] Nd,[18b] Yb,[18a]) and $[Nd(NPr^i_2)\{(\mu\text{-}NPr^i_2)(\mu\text{-}Me)AlMe_2\}\{(\mu\text{-}Me)_2AlMe_2\}]$.[18b]

X-ray-characterised Ln arylamides were those containing the ligand $[N(H)Ph]^-$, $[N(H)\text{-}C_6H_3Pr^i_2\text{-}2,6]^-$ and $[NPh_2]^-$, such as $[SmCp^*_2\{N(H)Ph\}(thf)]$,[19] $[Y\{N(H)C_6H_3Pr^i_2\text{-}2,6\}_2\{\mu\text{-}N(H)C_6H_3Pr^i_2\text{-}2,6\}]_2$ (**2**),[20] $[Ln\{N(H)C_6H_3Pr^i_2\text{-}2,6\}_3(thf)_n]$ (Ln = Nd and n = 3, Ln = Yb and n = 2),[20] $[Li(dme)_3][La(\eta^5\text{-}C_5H_4Me)(NPh_2)_3]$,[21a] $[Li(thf)_4][Yb(\eta^5\text{-}C_5H_4Bu^t)(NPh_2)_3]$,[21b] and $[Li(thf)_4][Yb(NPh_2)_4]$.[21c]

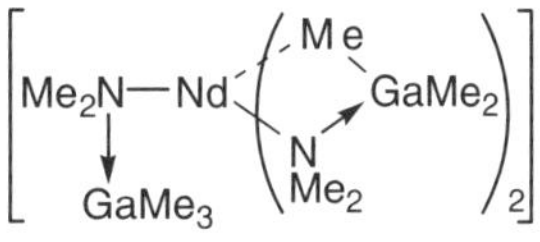

1 (M = Ga)

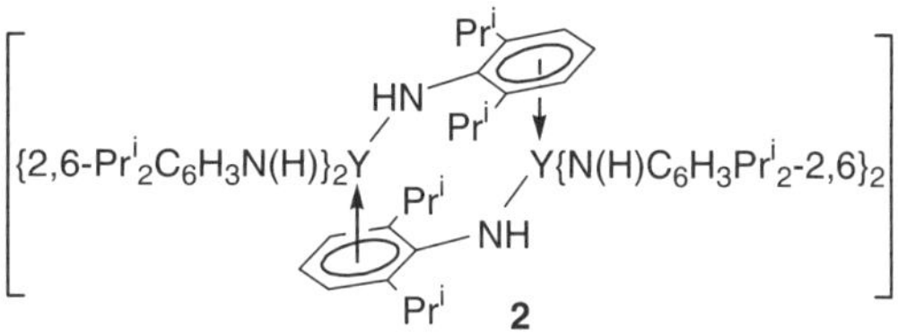

2

4.2.3 Ln^{III} Complexes having Silylamido Ligands

Homoleptic bis(trimethylsilyl)amido-Ln^{III} complexes were established for the complete (not the radioactive Pm) Ln series. The original syntheses were based on $LnCl_3/3Li[N(SiMe_3)_2]$ systems, but complications could involve LiCl incorporation, as in $[Nd\{N(SiMe_3)_2\}_3\text{-}(\mu\text{-}Cl)Li\text{-}(thf)_3]$[22] or $[Li(thf)_4][Y\{N(SiMe_3)_2\}_3Cl]$.[23] The use of the Na or the K rather than the Li amide was a favoured option (Na > K), preferably with a slight excess of $LnCl_3$. A useful alternative synthesis used the anhydrous triflate $Ln(OTf)_3$ in place of $LnCl_3$;[24a] $[Sm\{N(SiMe_3)_2\}_3]$ was cleanly obtained from $[Sm\{CH(SiMe_3)_2\}_3]$ and $3HN(SiMe_3)_2$.[24b]

There has been much discussion as to the pyramidal, rather than the planar LnN_3 skeleton, which was shown to persist even in the gas phase (Ln = La).[25] The favoured explanation took account of the short Ln···Si contacts; thus the solid state structure of $[Dy\{N(SiMe_3)_2\}_3]$ could be expressed in terms of a distorted trigonal prismatic arrangement at Dy, each (chelating) ligand having Dy–N and Dy···Si contacts.[26]

Heteroleptic Ln^{III} bis(trimethylsilyl)amido complexes reported included neutral adducts such as $[Y\{N(SiMe_3)_2\}_3(NCPh)_2]$,[23] complexes involving Cl^-[27] (see also above[22,23]), I^-, $\bar{O}R'$, $\bar{S}Bu^t$, COT^{2-},[28] or a cyclopentadienyl.[28] Examples included $[Y\{N(SiMe_3)_2\}_2(OR')]$ $[R' = Si(Bu^t)\{C_6H_4(CH_2NMe_2)\text{-}2\}_2]$,[29] $[Yb\{N(SiMe_3)_2\}\{OC(Bu^t)(2\text{-}CH_2NC_5H_3Me\text{-}6)_2\}_2]$ (**3**),[30] $[Gd\{N(SiMe_3)_2\}_2(\mu\text{-}SBu^t)]_2$,[31] $[Ln(COT)\{N(SiMe_3)_2\}(thf)]$,[32] and (*R*)- or (*S*)- $[Ln\{Me_2Si(C_5Me_4)(C_5H_3R^*)\}\{N(SiMe_3)_2\}]$ (Ln = Y, La, Sm, Lu; R^* = (−)-menthyl or (+)-neomenthyl).[33]

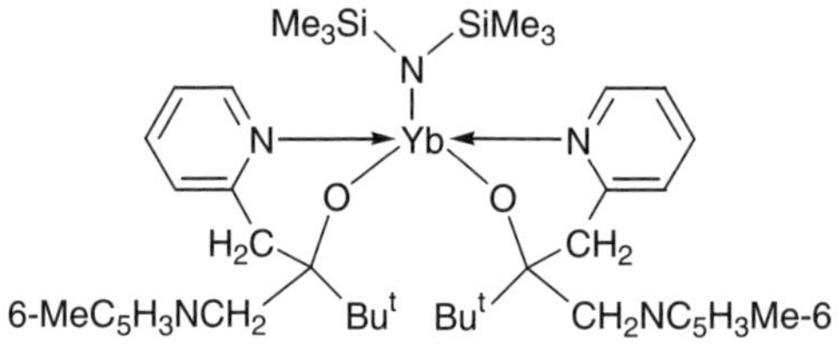

3

The less sterically demanding $[N(SiHMe_2)_2]^-$ ligand was first used in Ln chemistry in 1994; treatment of $LnCl_3$ with $3Li[N(SiHMe_2)_2]$ in thf afforded the trigonal bipyramidal [*trans*-$Ln\{N(SiHMe_2)_2\}_3(thf)_2$] (Ln = Y, Nd).[34] Their synthetic utility was demonstrated by showing that treatment with $HOCBu^t_3$ yielded $[Ln(OCBut_3)_3(thf)]$, whereas $[Ln\{N(SiMe_3)_2\}_3]$ did not react with this bulky alcohol under the same (ambient) conditions.[34] Three $[N(aryl)SiMe_3]^-$ ligands featured in $[Ln\{N(Ph)SiMe_3\}_3(thf)_n]$ (Ln = La and n = 2, or Nd - Lu and n = 1), $[Nd\{N(C_6H_3Me_2\text{-}2,6)SiMe_3\}_2(thf)(\mu\text{-}Cl)_2Li(thf)_2]$ and $[Ln\{N(C_6H_3Pr^i_2\text{-}2,6)SiMe_3\}_2Cl(thf)]$ (Ln = Nd, Lu), each obtained from the appropriate $LnCl_3$ + $3Li[N(aryl)SiMe_3]$ in thf.[35]

4.2.4 Bis(Trimethylsilyl)Amido-Ln^{II} Complexes and a Ce^{IV} Analogue

This area of chemistry was pioneered by Andersen's group. Crystalline complexes $[Yb\{N(SiMe_3)_2\}_2(OEt_2)_2]$[36a] [and its dmpe[36b] or $NC_5H_4\{C(H)(SiMe_3)_2\}\text{-}2$ or $NC_5H_4\{CH_2(SiMe_3)\}\text{-}2$ analogues[36c]] and the remarkable compound $[Yb\{N(SiMe_3)_2\}\{\mu\text{-}N(SiMe_3)_2\}_2Na]_\infty$ (**4**) were obtained from YbI_2 + $2Na\{N(SiMe_3)_2\}$ in Et_2O, or

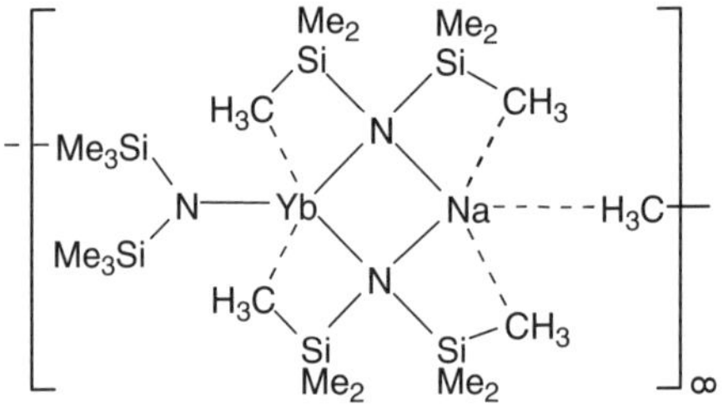

4

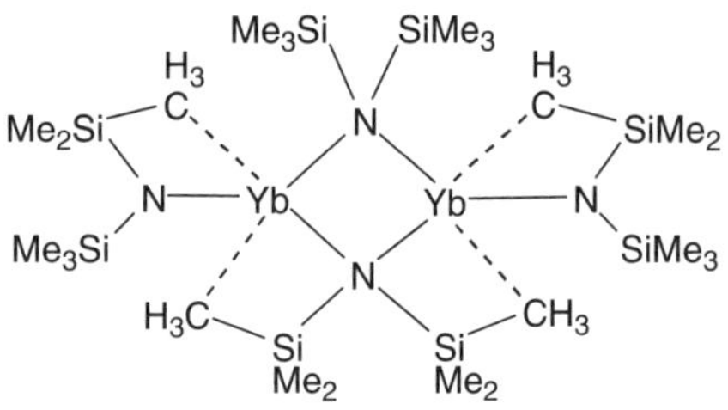

5

successively in Et_2O and PhMe, respectively.[36a] These compounds (as well as other Yb^{II} complexes) were used for the first high resolution direct NMR spectral observations of an *f*-block element in solution, namely of ^{171}Yb ($I = ^1/_2$, natural abundance 14.27%).[37] The homoleptic crystalline Yb^{II} amide showed strong intramolecular $CH_3\cdots Yb$ contacts (**5**) and formed an adduct with $AlMe_3$, $[Yb\{N(SiMe_3)_2 \cdot AlMe_3\}_2]$;[38] the Sm^{II}/Ga analogue has also been structurally charaterised.[30] Europium[36] and samarium amides, isoleptic with **4**, as well as $KSm\{N(SiMe_3)_2\}_3$ have been prepared and structurally characterised. The compounds Ln-$\{N(SiMe_3)_2\}_2(dme)$ (Ln = Sm, Eu, Yb) were obtained from $Hg\{N(SiMe_3)_2\}_2$ + Ln in dme Equation (4.1).[39] Similarly, treatment of Yb with $Sn[N(SiMe_3)_2]_2$ in refluxing thf and crystallisation from dme-C_6H_{14} yielded $[Yb\{N(SiMe_3)\}_2(dme)_2]$.[40a] Reacting $[Yb\{N(SiMe_3)_2\}_2(OEt_2)_2]$ with $Li[N(SiMe_3)_2]$ and tmeda yielded $[Li(tmeda)][Yb\{N(SiMe_3)_2\}_3]$.[40b]

$$Ln + Hg\{N(SiMe_3)_2\}_2 \xrightarrow[Ln = Sm,\ Eu,\ Yb]{dme} Ln\{N(SiMe_3)_2\}_2(dme) + Hg \qquad (4.1)$$

Heteroleptic Ln^{II} amides containing a mono-anionic co-ligand included $[Sm\{N(SiMe_3)_2\}(\mu\text{-}I)(dme)(thf)]_2$,[41] $[Yb\{N(SiMe_3)_2\}(\mu\text{-}OR')]_2$ ($R' = C_6H_2Bu^t_2$-2,6-Me-4, or $OCBu^t_3$)[40,42] and $[YbCp^*\{\mu\text{-}N(SiMe_3)_2\}]_2$.[43a] Each of the OR′ compounds was accessible from the homoleptic Ln^{II} amide and the appropriate protic acid HOR′.[40,42]

The sole simple Ce^{IV} amide reported at this time was $[Ce\{N(SiMe_3)_2\}_2(OBu^t)_2]$, in a conference abstract,[43b] although Ce^{IV} amides having a porphyrin or phthalocyanine were known.[3]

4.2.5 Ln^{III} Complexes having Donor-Functionalised Amido Ligands

The title ligands are the monoanionic **6–10** and the dianionic **11** and **12**. Doubly deprotonated diazacrowns have also featured in Ln^{III} chemistry, as well as *N,N′*-dianionic macrocycles. Exceptionally, an $[N(H)aryl]^-$ ligand figured as a chelate, implicating the aryl π-electrons, as in **2**.[20] Structurally characterised complexes using these ligands included $[Nd(\mathbf{6})_3]$,[44] $Li[Ho(\mathbf{7})_4]$ (the homoleptic compound was not accessible from $HoCl_3$ + $Li[N(Me)CH_2CH_2NMe_2]$),[45] $[Sm(\mathbf{8})(Cl)(\mu\text{-}\mathbf{8})(\mu\text{-}Cl)Li]$,[46] $[Sc(\mathbf{9})Cp^*_2]$,[47] $[\{Y(\mathbf{10})(\eta^3\text{-}C_3H_5)(\mu\text{-}Cl)\}_2]$,[48a] $[Y(\mathbf{11})(\mathbf{10})]$,[48b] $[Y(\mathbf{12})\{N(SiMe_3)_2\}]$[49] and $[\{Sc(\mathbf{12})(PMe_3)\}_2(\mu\text{-}C_2H_4)]$.[50] The conversion of the coordinated ligand **10** into **11** is illustrated in Equation (4.2) (Ln = Y, La, Lu and MR = LiPh or KCH_2Ph).[49]

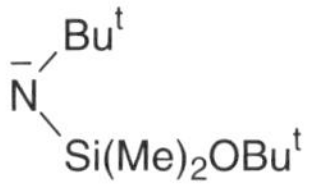

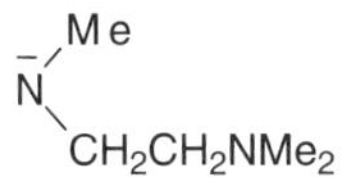

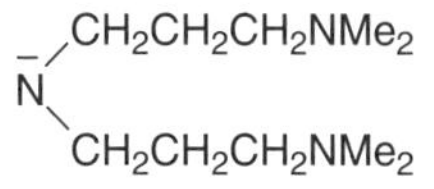

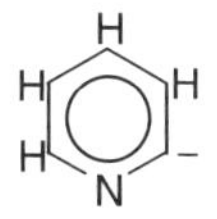

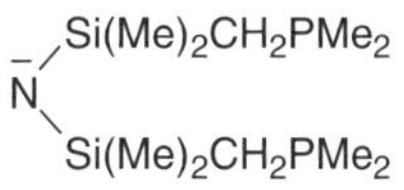

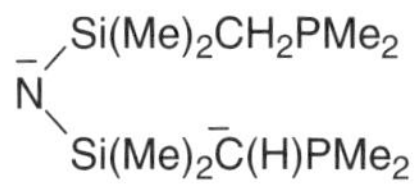

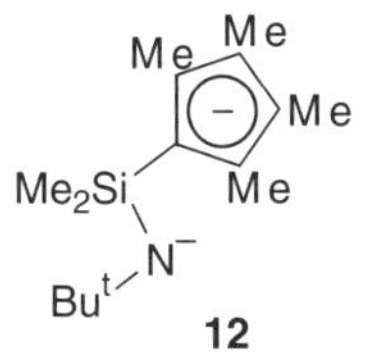

$$[Ln(\mathbf{10})_2Cl] + MR \xrightarrow{THF} [Ln(\mathbf{10})_2R] \xrightarrow[-RH]{Ln = Y,\ heat} [Y(\mathbf{10})(\mathbf{11})] \qquad (4.2)$$

Ligands **8** and **10** are tridentate and **11** is tetradentate (**13**).

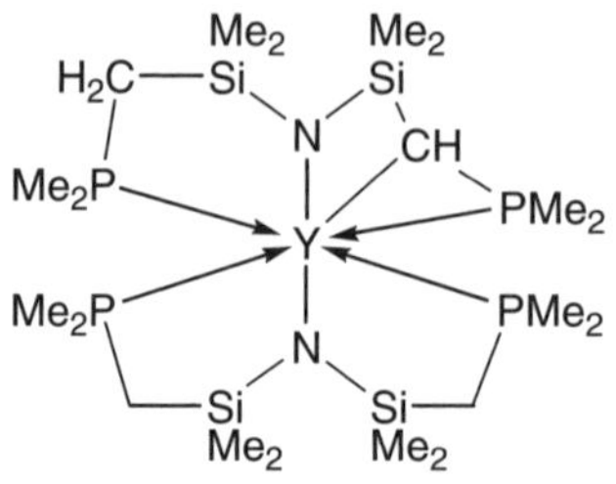

13

The 1,4-diazabuta-1,3-diene molecule may act as a radical anionic or a dianionic chelating ligand. The former, as in [Yb(Bu^t_2DAB)$_3$], the latter as in [{Lu{(μ-Bu^t_2DAB)-Li(thf)}{(μ-Bu^t_2DAB)Li(OEt_2)}Cl], can be regarded as Ln amides, see Section 4.3.8. On the other hand, a related ligand shown in **14** and Equation (4.3),[51] is an eneamide.

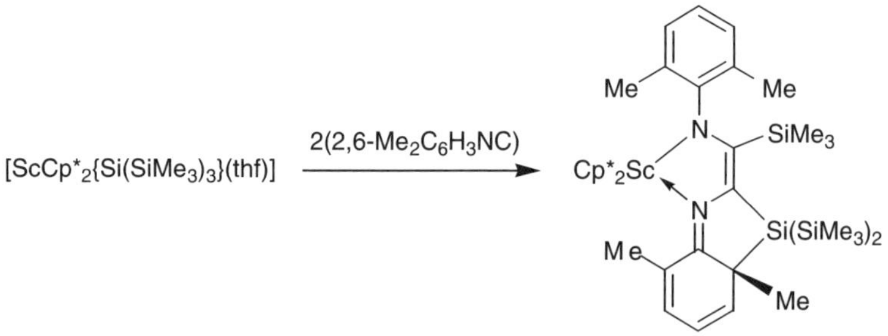

$$[ScCp^*_2\{Si(SiMe_3)_3\}(thf)] \xrightarrow{2(2,6\text{-}Me_2C_6H_3NC)} \mathbf{14} \qquad (4.3)$$

14

The numerous Ln^{III} and Ln^{II} polypyrazolylborates include [Yb(HB{NC(Me)C(H)-C(Bu^t)N}$_3$){N($SiMe_3$)$_2$}],[52] prepared from [Yb^{II}($Tp^{Bu^t,Me}$)I] and Na[N($SiMe_3$)$_2$].

Both pyrrolyl- and carbazolyl-Ln complexes can be found in *N*-σ-bonded forms, provided that steric demands were minimised. Thus, [Lu($NC_4H_2Me_2$-2,5)Cp_2(thf)],[53] unlike [{Nd(η^5-$NC_4H_2Bu^t_2$-2,5)$_2$(μ-Cl)$_2$Na(thf)}$_2$],[54] has a Ln–N σ-bond as do [Ln^{II}(cbz)$_2$(thf)$_4$] (Ln = Sm, Eu, Yb).[55]

N,N′-Dianionic fused six-membered heterocyclic ligands derived from phenazine and quinoxaline have been found in Ln complexes including [($SmCp^*$)$_2$(μ-η^3:η^3-$C_{12}H_8N_2$)] (**15**)[56] and [($LaCp^*_2$)$_2$(μ-η^3:η^3-$C_{12}H_8N_2$)].[57]

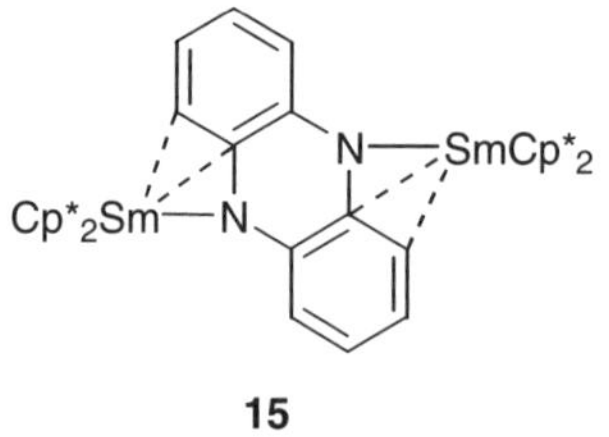

15

4.2.6 Ln Amides as Precursors for Ln Coordination or Organometallic Compounds

Protic compounds HA which are more acidic than $HN(SiMe_3)_2$, if steric effects are not excessive, readily displaced the amide from compounds containing Ln-$N(SiMe_3)_2$ bonds, replacing them by Ln-A bonds. Such reactions have been reported for HA = HC≡CR,[58] HCp,[59] $HSnR_3$,[60] HNR_2,[20] HPR_2,[61] HOR,[40b,62] HOAr,[63] $HOSiPh_3$,[64] HSR,[39,40a] HOS-$(O)_2CF_3$,[40b] HSeR,[39] HTeR,[65] and HCl.[66] Illustrative examples are in Equations (4.4),[61](4.5),[63](4.6)[40a] and (4.7).[65]

$$[La\{N(SiMe_3)_2\}_2(OPPh_3)] + HPPh_2 + OPPh_3 \longrightarrow [La\{N(SiMe_3)_2\}_2(PPh_2)(OPPh_3)] + HN(SiMe_3)_2 \tag{4.4}$$

$$[La\{N(SiMe_3)_2\}_3 + 3HOC_6H_2Bu^t{}_2\text{-}2,6\text{-}Me\text{-}4 \longrightarrow [La(OC_6H_2Bu^t{}_2\text{-}2,6\text{-}Me\text{-}4)_3] + 3HN(SiMe_3)_2 \tag{4.5}$$

$$[Yb\{N(SiMe_3)_2\}_2(dme)_2] + 2HSC_6H_2Bu^t{}_3\text{-}2,4,6 \longrightarrow [Yb(SC_6H_2Bu^t{}_3\text{-}2,4,6)_2(dme)_2] + 2HN(SiMe_3)_2 \tag{4.6}$$

$$[La\{N(SiMe_3)_2\}_3 + 3HTeSi(SiMe_3)_3 \xrightarrow{dmpe} [La\{TeSi(SiMe_3)_3\}_3(dmpe)_2] + 3HN(SiMe_3)_2 \tag{4.7}$$

Such reactions for Ln^{III} amides can be susceptible to steric effects. For example, [Ln{N-$(SiMe_3)_2\}_3] + 3HOCBu^t{}_3$ in thf readily yielded $[Ln\{OCBu^t{}_3\}_3]$ for Ln = Nd, whereas the smaller Ln = Y or Tm amides were unreactive under the same conditions.[67] As noted in Section 4.2.3, using the less bulky $[N(SiHMe_2)_2]^-$ ligand, such homoleptic alkoxides were accessible even with the smallest Ln^{III} amides. For Ln^{II} amides, such steric effects were shown to be minimal since even the bulky tritoxides $[Yb\{N(SiMe_3)_2\}(\mu\text{-}OCBu^t{}_3)]_2$[42] and $[\{Yb(OCBu^t{}_3)(\mu\text{-}OCBu^t{}_3)\}_2]$[42,68] were obtained from $[Yb\{N(SiMe_3)_2\}\{\mu\text{-}N(SiMe_3)_2\}]_2$ and the appropriate portion of $HOCBu^t{}_3$. The first Ln^{II} aryloxide, $[Yb(OC_6H_2Bu^t{}_2\text{-}2,6\text{-}Me\text{-}4)_2(OEt_2)_2]$, was prepared from $[Yb\{N(SiMe_3)_2\}_2(OEt_2)_2] + 2HOC_6H_2Bu^t{}_2\text{-}2,6\text{-}Me\text{-}4$.[69]

Alkylation reactions, as in $[Nd\{N(Me)_2{\cdot}AlMe_3\}_3] \xrightarrow{AlMe_3} [Nd\{(\mu\text{-}Me)_2AlMe_2\}_3]$,[17] were mentioned in Section 4.2.2.

The compound $[Ln\{N(SiMe_3)_2\}_3]$ (Ln = Pr, Nd) or $[Pr\{N(SiMe_3)_2\}_2\{Sn(CH_2SiMe_3)_3\}]$ in dme absorbed two equivalents of carbon dioxide under ambient conditions;[70] the product may well have been carbamates, such as $[Ln\{OC(O)NMe_2\}_2\{N(SiMe_3)_2\}]$.

4.2.7 Applications as Materials or Catalysts

The homoleptic Ln^{III} bis(trimethylsilyl)amides are among the most volatile Ln compounds, and have featured in the doping of Ge/Ln semiconductors,[71] or in the formation of lanthanide nitrides.[72]

The Ln^{III} catalysed hydroamination/cyclisation of *N*-unprotected alkenes to form heterocyclic nitrogen compounds has been intensively studied by Marks' group.[14,73] An example of enantioselective transformation is shown in Equation (4.8), the catalyst being (*R*)- or (*S*)-[Sm-$\{(\eta^5\text{-}C_5Me_4)Si(Me)_2(\eta^5\text{-}C_5H_3R^*)\}\{N(SiMe_3)_2\}]$ [R^* = (+)-neomenthyl, (−)-menthyl, or (−)-phenylmenthyl].[73c] Similar Ln amides were even more active as catalysts for the corresponding regiospecific hydroamination/cyclisation of aminoalkynes $RC{\equiv}C(CH_2)_nNH_2$.[74]

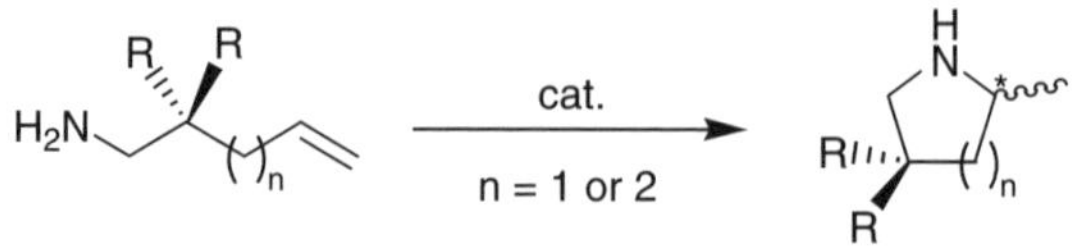

(4.8)

The scandium compound, containing the constrained geometry dianionic ligand **12**, $[Sc\{(\eta^5\text{-}C_5Me_4)Si(Me)_2NBu^t\}(PMe_3)(\mu\text{-}H)]_2$ was a catalyst for the oligomerisation of propene, but-1-ene or pent-1-ene and polymerisation of C_2H_4.[50] Using stoichiometric amounts of $RCH{=}CH_2$ and the Sc amide, the μ-hydrides were replaced by $\mu\text{-}\eta^2{:}\eta^2\text{-}C_2H_4$ or, for C_3H_6 the product was $[\{Sc(\mathbf{12})(\mu\text{-}Pr^n)\}_2]$. Other ethylene polymerisation catalysts to be reported were $[Yb\{N(SiMe_3)_2\}_2(AlEt_3)_2]$[36a] and $[Y(\eta^3\text{-}C_3H_5)_2\{N(SiMe_2CH_2\text{-}PMe_2)_2\}]$,[48] the latter involving the ligand **10**. Samarium(II) amides showed high reactivity in catalytic ε-caprolactone polymerisation.[75] Soluble organolanthanide reagents,[11] prepared from $[Ln\{N(SiMe_3)_2\}_3]$ (Ln = Ce, Sm), were used for ring-opening alkylation of epoxides, as in Equation (4.9).[76]

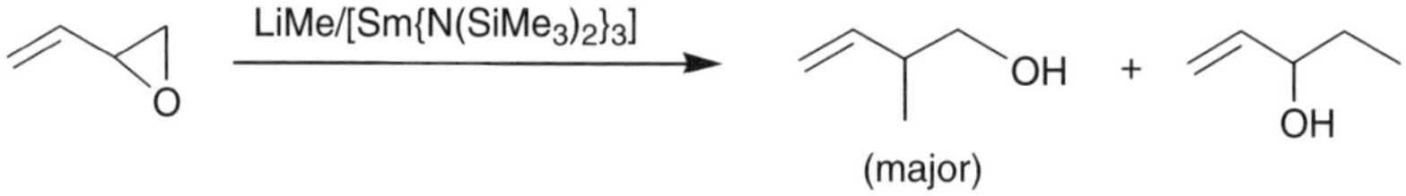

(4.9)

4.3 The Recent (Post-1995) Literature

4.3.1 Introduction

The layout follows that adopted for Section 4.2.

4.3.2 Ln^{III} Complexes with *N*-Hydrocarbyl Substituted Ligands

Compounds containing $[N(H)R]^-$ as the sole amido ligand have included $[Ln_2Br_4\{\mu\text{-}N(H)\text{-}Ph\}_2(thf)_5]$ (Ln = Sm, Yb)[77a] and $[Yb^{III}\{N(H)C_6H_3Pr^i{}_2\text{-}2,6\}_2\{\mu\text{-}N(H)C_6H_3Pr^i{}_2\text{-}2,6\}_2Na\text{-}(thf)]$,[77b] each obtained from the appropriate Na[N(H)R] and $LnBr_3$ or $YbCl_3$; the latter with $LiBu^t$ gave $[\{Li(thf)\}\{Na(thf)\}\{\mu\text{-}N(H)C_6H_3Pr^i{}_2\text{-}2,6\}Yb\{NC_6H_3Pr^i{}_2\text{-}2,6\}\{\mu\text{-}NC_6H_3Pr^i{}_2\text{-}2,6\}]_2$.[77b] An oxidative route to the compounds $[LnI_2\{N(H)Pr^i\}(NH_2Pr^i)_n]$ (Ln = Nd, n = 4; Ln = Dy, n = 3) involved decomposition of the thermally unstable solvates $[LnI_2(NH_2\text{-}Pr^i)_m]$ (from LnI_2 and Pr^iNH_2), H_2 being eliminated.[77c]

The dinuclear compound **16** was prepared from $LnCp_3$ and 2-amino-4,6-dimethylpyrimidine.[78]

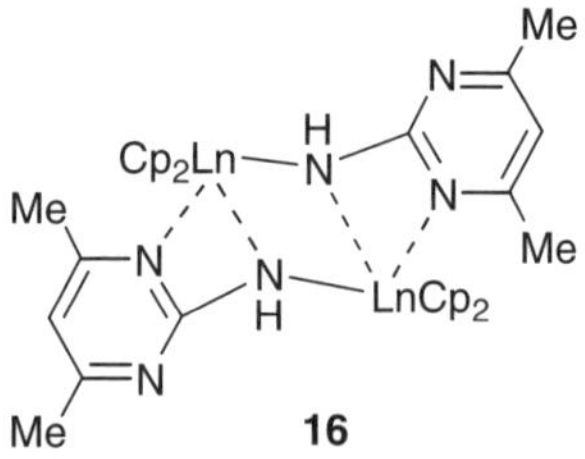

16

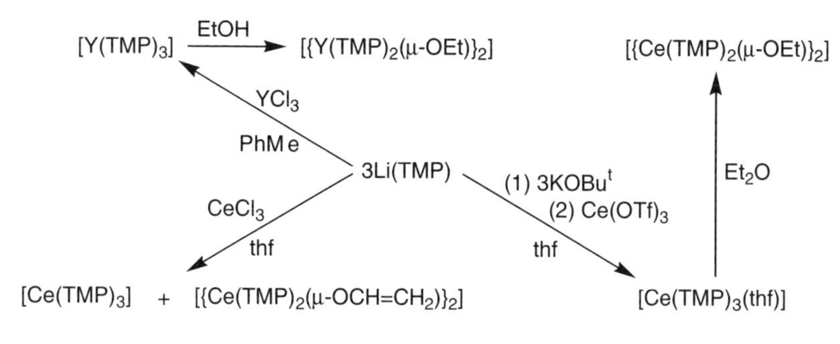

Scheme 4.1

The Ln^{III} dialkylamides $[Nd(NPr^i_2)_3(thf)]$ and $[Ln(NPr^i_2)(\mu\text{-}NPr^i_2)_2Li(thf)]$ (Ln = Nd,[10e] Sm[80]) (for Y, La, Yb, see Ref. 18a; the Sc, Y and Lu analogues were obtained from $[Ln(CH_2SiMe_3)_3(thf)_2]$ and $3HNPr^i_2$[123b]) were prepared from $LnCl_3/Li[NPr^i_2]$. The ligand $[NCy_2]^-$ has featured in $[Ln(NCy_2)_3(thf)]$ (Ce,[79] Sm[80]), $[Ce(NCy_2)_2(\mu\text{-}NCy_2)_2Li(thf)]$,[79] $[\{Sm(NCy_2)_2(\mu\text{-}Cl)\}_2]$[80] and $[Sm(NCy_2)\{C(PPh_2{=}NSiMe_3)_2\}(thf)]$.[81]

The 2,2,6,6-tetramethylpiperidinato ligand $[TMP]^-$ has appeared in the experiments illustrated in Scheme 4.1. Each of the compounds $[Ln(TMP)_3]$ (Ln = Y,[82] Ce[83]), $[\{Ln(TMP)_2(\mu\text{-}OEt)\}_2]$ (Y, Ce),[82] $[Ce(TMP)_3(thf)]$,[82] and $[\{Ce(TMP)_2(\mu\text{-}OCH{=}CH_2)\}_2]$,[83] has been crystallographically characterised. The Ln^{III}-TMP complexes are the first *N*-hydrocarbylamides to have a three-coordinate metal environment. Moreover, unlike the trigonal pyramidal LnN_3 for $[Ln\{N(SiMe_3)_2\}_3]$, the $[Ln(TMP)_3]$ compounds are trigonal planar;[82,83] and $[TMP]^-$ is sterically more demanding than the ubiquitous $[N(SiMe_3)_2]^-$,[82] as evident *inter alia* by the observation that in the $YCl_3/3Li[TMP]$ system the product is free from Cl^- or LiCl incorporation.[82] The formation of the Ce^{III} compound containing a bridging ethoxide[82] or vinyloxide[83] was attributed to Et_2O cleavage for the former[82] (or deprotonation of Et_2O and C_2H_4 extrusion[83]); and for the latter, deprotonation of thf and C_2H_4 extrusion.[83] Another example of C–O cleavage is described in Section 4.3.4.

Diarylamido ligands have featured in $[Li(thf)_4][Er(NPh_2)_4]$[84] (for the Yb analogue, see Ref. 21c) and $[Nd\{N(C_6F_5)_2\}_3(\eta^6\text{-}PhMe)]$.[85a] The crystalline homoleptic complexes $[\{Ln(L)_2(\mu\text{-}L)\}_2]$ [Ln = Sc, Ce, Nd, Sm, Ho, Er, Tm, Yb; L = $N(C_5H_4N\text{-}2)_2$], having eight-coordinate Ln centres, were obtained by the oxidation of Ln by the molten amine HL.[85b] Piperidinato, $[NPr^i_2]^-$ and $[NPh_2]^-$ ligands have appeared in the complexes $[Yb(L)(NPr^i_2)(\mu\text{-}Cl)_2Li(thf)_2]$, $[\{Yb(L)(NPr^i_2)(\mu\text{-}Cl)\}_2]$, $[Yb(L)(NPh_2)Cl(thf)]$ and $[\{Yb(L)(NC_5H_{10})(\mu\text{-}Cl)\}_2]$ [L = $\{N(C_6H_3Pr^i_2\text{-}2,6)C(Me)\}_2CH$].[152a]

4.3.3 Ln^{III} Complexes having Silylamido Ligands

Crystalline $[Ln\{N(SiMe_3)_2\}_3]$ (Ln = Ce,[86] Tb,[172] Yb,[91a] Lu[91b]) were shown to have the typical trigonal pyramidal LnN_3 core; the Yb[91a] and Lu[91b] complexes had short Ln···Si and Ln···C β- and γ-agostic contacts, respectively. Unpolarised absorption and luminescence as well as the σ and π absorption spectra of a large single crystal of $[Yb\{N(SiMe_3)_2\}_3]$ were recorded and the derived crystal field splitting patterns could be simulated; the Raman spectra, as well as those of the Y analogue were analysed.[90b]

The Ce amide had been obtained in poor yield from $CeCl_3/3Li[N(SiMe_3)_2]$,[86] but in high yield from $Ce(OTf)_3/3Na[N(SiMe_3)_2]$.[179] The $Ln(OTf)_3$ route (see also Ref. 24a) was also used for the La, Nd, Sm and Er homoleptic analogues.[87] Treatment of $[Ce\{N(SiMe_3)_2\}_3]$ with two equivalents of $CeCp_3$ in toluene yielded $[CeCp_2\{N(SiMe_3)_2\}]$.[88] DFT calculations on $[Ln\{N(SiH_3)_2\}_3]$ as a model for $[Ln\{N(SiMe_3)_2\}_3]$ reproduced the experimental Ln–N bond lengths, but showed β-Si-H agostic (to Ln) interactions which clearly is inappropriate for the bis(trimethylsilyl)amides and consequently did not reproduce the experimentally observed conformations.[89]

Heteroleptic Ln^{III} bis(trimethylsilyl)amides have included $[Ln\{N(SiMe_3)_2\}_3(CNCy)_2]$ (Ln = Y, La, Ce, Pr, Nd, Sm, Eu, Tb, Dy, Ho, Tm, Yb); the isonitriles were shown to be axial in the trigonal bipyramidal Nd compound,[90a] as in the Y-benzonitrile analogue.[23] A number of $(\mu\text{-Hal})_2$-bridged binuclear bis(amido)Ln complexes have been reported: $[Yb\{N(SiMe_3)_2\}_2(\mu\text{-Cl})(thf)_n]_2$ [n = 0 (with agostic $Yb\cdots CH_3$ contacts) or 1],[91a] $[La\{N(SiMe_3)_2\}_2(\mu\text{-I})(thf)]_2$,[92] $[Ln\{N(SiMe_3)_2\}_2(\mu\text{-Cl})(thf)]_2$ (Ln = Ce,[93] Nd,[94a] Sm,[95a] Y;[96a] see also Eu, Gd, Yb[27]) and $[Ln\{N(SiMe_3)_2\}_2(\mu\text{-Br})(thf)]_2$ (Ln = Ce,[94a] Sm[95a]). LiCl-containing complexes $[Ln\{N(SiMe_3)_2\}_2(\mu\text{-Cl})Li(thf)_n(\mu\text{-Cl})]_2$ [Ln = Nd (n = 3), Eu and Ho (n = 2)] and $[Ln\{N(SiMe_3)_2\}(\mu\text{-Cl})_2Li(thf)_2(\mu\text{-Cl})]_2$ (Ln = Nd, Sm, Eu, Ho, Yb) were isolated when $LiN(SiMe_3)_2$ was used as a starting material.[94b] Lanthanate(III) complexes $[Na(thf)_3][Sm\{N(SiMe_3)_2\}_3(C{\equiv}CPh)]$,[10f] $[Na(thf)_6][\{Lu\{N(SiMe_3)_2\}_2(\mu\text{-NH}_2)(\mu\text{-NSiMe}_3)\}_2]$[10f] and $[Na(thf)_2][Ln\{N(SiMe_3)_2\}_2(NPh_2)_2]$ (Ln = Sc, Gd, Yb) have been reported; the latter had close $Na\cdots\eta^6$-Ph contacts.[10d] The crystalline complexes $[Y\{N(SiMe_3)_2\}_2\{(NCy)_2CN(SiMe_3)_2\}]$ and $[Y\{N(SiMe_3)_2\}_2(\mu\text{-Cl})(thf)]_2$ were obtained from $[Y\{N(SiMe_3)_2\}_3]$ with $C(NCy)_2$ (via insertion of the carbodiimide into an Y–N bond) and YCl_3 with $[Li\{N(SiMe_3)_2\}(OEt_2)]_2$ in presence of the evidently unreactive $C(NSiMe_3)_2$, respectively.[96a]

Three amidosamarium(III) amidinates were obtained from a Sm^{II} precursor and a carbodiimide by oxidative routes (*cf.*, insertion for $[Y\{N(SiMe_3)_2\}_3]$) shown in Scheme 4.2.[96b]

Mononuclear amidometal dihalides (obtained from the appropriate LnX_3) were the distorted pentagonal bipyramidal $[CeBr_2\{N(SiMe_3)_2\}(thf)_3]$ having agostic $CH_3\cdots Ce$

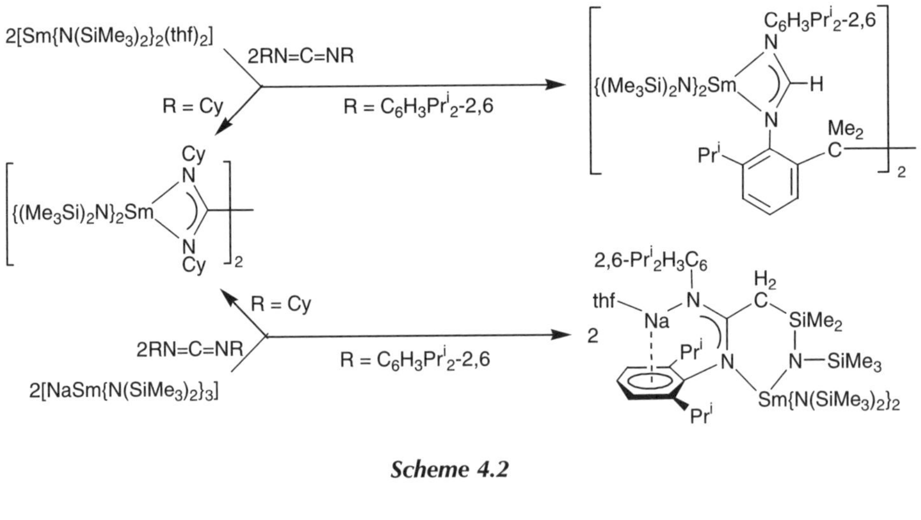

Scheme 4.2

with N and Br axial,[93] and the trigonal bipyramidal $[ScCl_2\{N(SiMe_3)_2\}(thf)_2]$ with axial $(thf)_2$.[10b] A co-product of the $YbCl_3/2Na[N(SiMe_3)_2]$ reaction in thf was the trinuclear complex **17**;[91a] the bromo analogue had earlier been reported.[95b] Related guanidinates $[Ln\{N(SiMe_3)_2\}L_2]$ $[L = N(Cy)C\{N(SiMe_3)_2\}NCy]$ (Ln = Sm, Yb) were obtained from $[LnL_2(\mu\text{-}Cl)_2Li(tmeda)]$ and $Li[N(SiMe_3)_2]$;[97a] $[Yb(C_6H_3Mes_2\text{-}2,6)\{N(SiMe_3)_2\}(\mu\text{-}Cl)_2Li(thf)_2]$ was prepared from $YbCl_3$, LiAr and $K[N(SiMe_3)_2]$ in thf,[97b] $[La(I)(L)\{N(SiMe_3)_2\}(thf)]$ from the diiodide $[L = C(SiMe_3)_2\{SiMe_2(OMe)\}]$,[97c] and $[La(L')_2\{N(SiMe_3)_2\}]$ from $[LaI_3(thf)_4]$, $K[N(SiMe_3)_2]$ and KL′ $[L' = P(C_6H_3Pr^i{}_2\text{-}2,6)(C_6H_4CH_2NMe_2\text{-}2)]$.[97d]

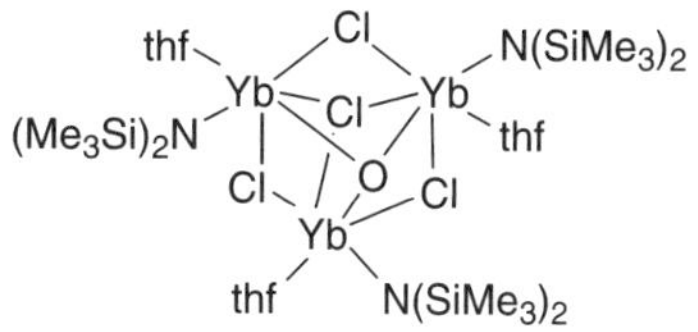

Although μ-peroxoLn^{III} amides $[\{Ln(N(SiMe_3)_2)_2(L)\}_2(\mu\text{-}\eta^2\text{:}\eta^2\text{-}O_2)]$ (Ln = La, Pr, Sm, Eu, Lu and $L = OPPh_3$) had been reported (from Ln^{III} precursors) in 1977,[98] a more recent example (Ln = Yb, L = thf) (**18**) was made from $[Yb\{N(SiMe_3)_2\}_2(thf)_2]$ in C_5H_{12} and O_2.[91a] The peroxo bridge was supported by a more definitive X-ray structure, *l* (O–O) 1.543(4) Å, and also the single crystal Raman spectrum, [ν(O–O) 755 cm^{-1}] and DFT calculations.[91a]

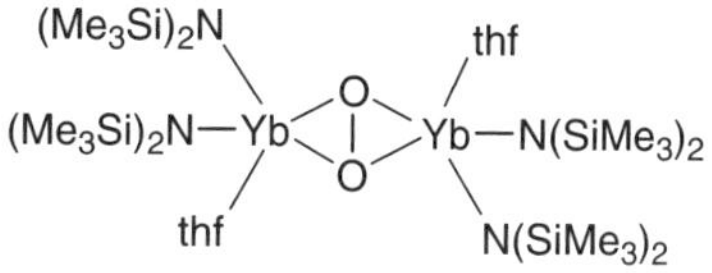

A particularly noteworthy discovery, by Evans, *et al.*, was of the bridging dinitrogen compounds $[\{Ln(N(SiMe_3)_2)_2(thf)\}_2(\mu\text{-}\eta^2\text{:}\eta^2\text{-}N_2)]$ (Ln = Y,[100a] Dy,[99a] Ho,[100a] Tm,[99a,100a] Lu[100a]) (**19**). The Dy and Tm compounds were obtained from the appropriate LnI_2,[99a] but remarkably not only the Tm but also the Y, Ho, Tm and Lu compounds **19** were derived from $[Ln\{N(SiMe_3)_2\}_3]$,[100a] as summarised in Scheme 4.3. These adducts **19** are clearly Ln^{III} amides, since the Y and Lu compounds were shown to be diamagnetic. DFT calculations on the model compounds $[\{Ln(N(SiH_3)_2)_2(solv)\}_2(\mu\text{-}\eta^2\text{:}\eta^2\text{-}N_2)]$ (Ln = Y, Gd) and $[\{Gd(N(SiMe_3)_2)_2(solv)\}_2(\mu\text{-}\eta^2\text{:}\eta^2\text{-}N_2)]$ (solv = thf, OMe_2 and OH_2) have been reported.[99b]

The hetereoleptic Sm^{III} amides **20–22** were derived from $[Sm^{II}\{N(SiMe_3)_2\}_2(thf)_2]$;[101] thus, treatment with an equimolar portion of PhNO or $(\overline{OCH_2CH}-)_2$ yielded **20**; **21** was obtained with 4-phenylpyridine *N*-oxide and **22** by reaction with azobenzene (0.5 molar).

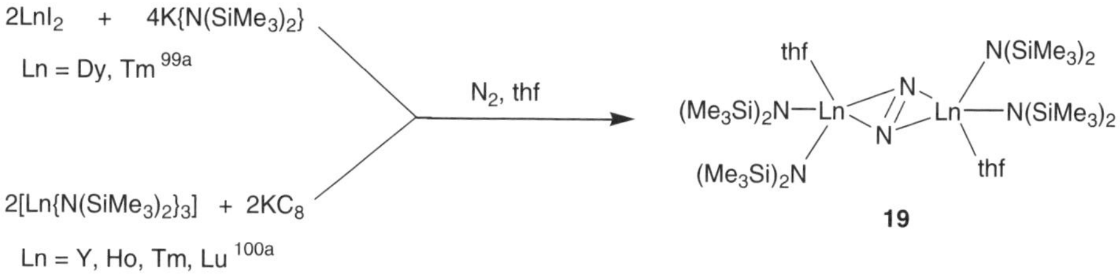

Scheme 4.3

$[\{Sm(N(SiMe_3)_2)_2(thf)\}_2(\mu\text{-}O)]$
20

$[\{Sm(N(SiMe_3)_2)_2(NC_5H_4Ph\text{-}4)\}_2(\mu\text{-}O)]$
21

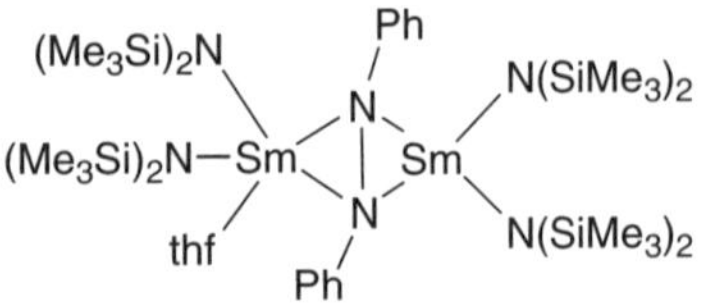

Ph
$(Me_3Si)_2N$ N $N(SiMe_3)_2$
$(Me_3Si)_2N-Sm$ Sm
thf N $N(SiMe_3)_2$
Ph
22

Further studies on the isostructural trigonal bipyramidal compounds $[Ln\{N(SiHMe_2)_2\}_3(thf)_2]$ (Ln = Y, La, Nd, Er, Lu) (*cf.*,[34] Y, Nd), with axial thf groups, have been reported.[102a] The solvent-free, binuclear amide $[Y\{N(SiHMe_2)_2\}_2\{\mu\text{-}N(SiHMe_2)_2\}]_2$ was prepared via protonolysis of the polymeric trimethylyttrium with $HN(SiHMe_2)_2$.[102b] Some have been grafted by Anwander's group to generate Ln-supported mesoporous silicates or treated with an appropriate compound LH_2 to furnish $[LnL\{N(SiHMe_2)_2\}]$ (see Sections 4.3.6 and 4.3.7). The ligand $[N(SiMe_2Ph)_2]^-$ featured in $[LaCp^*_2\{N(SiMe_2Ph)_2\}]$, prepared from $[LaCp^*_2(Ph_2BPh_2)]$ and the potassium amide.[100b]

The trigonal planar $[Ce\{N(Bu^t)SiMe_3\}_3]$, contaminated with $[Ce\{(N(Bu^t)SiMe_3)_2(\mu\text{-}OEt)\}_2]$, was obtained from $CeCl_3/3Li[N(Bu^t)SiMe_3]$.[83] The compounds $[Nd\{N(Ph)SiMe_3\}_3(thf)]$,[103a] $[Yb\{N(C_6H_5)SiMe_3\}_2(\mu\text{-}Cl)(thf)]_2$,[103b] $[Yb\{N(C_6H_3Pr^i_2\text{-}2,6)SiMe_3\}Cl_2(thf)_3]$,[103b] $[Ln\{N(C_6H_3Pr^i_2\text{-}2,6)SiMe_3\}_2X(thf)]$ (Ln = Nd and X = Cl;[103a] La and X = Br[82]) were prepared from LnX_3 and the alkali metal amide, as was $[Nd\{N(C_6H_3Pr^i_2\text{-}2,6)SiMe_3\}_2(\mu\text{-}Cl)_2Li(thf)_2]$.[103a] The *p*-phenylenediamido ligand featured in the macrocyclic compounds $[Yb\{\mu\text{-}(NSiMe_3)_2C_6H_4\text{-}1,4\}(Cl)(thf)_2]_2$ and $[Nd\{\mu\text{-}(NSiMe_3)_2C_6H_4\text{-}1,4\}(\mu\text{-}Cl)(thf)]_4$.[103c] $[Sm\{N(C_6F_5)SiMe_3\}_3]$ has *o*-F···Sm and Me···Sm close contacts,[85a] as has the thf adduct.[104] The compound $[Er\{N(Bu^t)SiHMe_2\}_3]$ has three agostic Er···H-Si contacts;[105a] the compounds $[Ln(L)_3(\mu\text{-}Cl)Li(L')_3]$ [Ln = Sm, Eu, Gd-Tm, Lu; L′ = thf, OEt_2; or $L'_3 = (thf)_2(OEt_2)$] have featured in the cyclic bis(silyl)-amido ligand $[\overline{N(SiMe_2CH_2CH_2Si}Me_2)]^-$ (≡ L).[105b]

4.3.4 Ln^{II} and Ce^{IV} Amides

The synthesis of Yb^{II} amides by the method of Equation (4.10) [see also Equation (4.1)[39]] was pioneered by Deacon, *et al.*, originally for the 3,5-disubstituted pyrazolates as L^- yielding $[Yb(R_2pz)_2(dme)_2]$ (R = Ph,[106a] Bu^t [106b]). This method was also used for the preparation of Sm^{II}, Eu^{II} and Yb^{II} amides containing the bulky ligand $[N(C_6H_3Pr^i_2\text{-}2,6)SiMe_3]^-$.[107] Surprisingly, the same procedure but with LH = $HN(C_6H_4OR'\text{-}2)(SiMe_3)$ (R′ = Me, Ph), gave in low yield the Yb^{III} amides $[\{Yb(N(C_6H_4OMe\text{-}2)SiMe_3)_2(\mu\text{-}OMe)\}_2]$ or $[Yb\{N(C_6H_4OPh\text{-}2)SiMe_3\}_2(OPh)(thf)]$, there having been C–O cleavage of the ligand.[108] The Yb^{II} amide $[Yb\{N(C_6H_4OMe\text{-}2)SiMe_3\}_2(thf)_2]$, obtained by transamination from $[Yb\{N(SiMe_3)\}_2(thf)_2]$ (the Sm analogue was prepared similarly[109a]) was converted in low yield into the above bridging Yb^{III} methoxide by heating with $HN(C_6H_4OMe\text{-}2)SiMe_3$. The procedure of Equation (4.10) and

$$Yb + HgPh_2 + 2LH \longrightarrow YbL_2 + Hg + 2C_6H_6 \qquad (4.10)$$

crystallisation from dme yielded not only $[Yb(Ph_2pz)_2(dme)_2]$[106a,109a] [also available[109a] from $2Yb(Ph_2pz)_3 + Yb$, $Yb(C_6F_5)_2 + 2LH$, or $YbI_2 + 2K(Ph_2pz)$] but also $[Eu(Bu^t_2pz)_2(dme)_2]$.[109a] A redox transmetallation of Ln metal and 2 TlL in thf and crystallisation from dme yielded $[YbL_2(dme)_2]$ (L = MePhpz, bind, or azin) or $[Eu(Ph_2pz)_2(dme)_2]$ [bindH = 4,5-dihydro-2*H*-benz[g]indazole; azinH = 7-azaindole].[109a] Transamination of $[Sm\{N(SiMe_3)_2\}_2(thf)_2]$ with $HN(SiHMe_2)_2$ gave the crystalline trinuclear $[Sm(\{\mu\text{-}N(SiHMe_2)_2\}_2Sm\{N(SiHMe_2)_2\}thf)_2]$ featuring a bent (Sm···Sm···Sm, 130°) array.[109b]

The reaction of YbI_2 with the appropriate 2K[N(Ph)R] in thf furnished $[Yb(NPh_2)_2(thf)_4]$, $[Yb\{N(Ph)SiMe_3\}_2(thf)_3]$, or $[\{Yb(N(Ph)SiMe_3)(\mu\text{-}N(Ph)SiMe_3)(thf)\}_2]$.[110] From $[Yb(NPh)_2(thf)_4]$ or $[Yb\{N(SiMe_3)_2\}_2(thf)_2]$ and [18]crown-6 the product was $[Yb(NPh)_2([18]crown\text{-}6)]$ or $[Yb\{N(SiMe_3)_2\}([18]crown\text{-}6)][Yb\{N(SiMe_3)_2\}_3]$. The former has the eight-coordinate Yb atom at the centre of a distorted hexagonal bipyramid having six equatorial oxygen atoms and the two axial N atoms. The latter complex is noteworthy for having both an amido-Yb^{II} cation and a tris(amido)Yb^{II} anion.[110] A series of Ln^{II} arylamides having the $[N(Ar)SiMe_3]^-$ (Ar = Ph, $C_6H_3Pr^i_2$-2,6) ligand was prepared by the metathesis of LnI_2 (Ln = Sm) with a potassium amide or by the Na/K alloy reduction of the respective Ln^{III} chloride (Ln = Sm, Yb).[103b] Treatment of $[Yb(L)\{N(SiMe_3)_2\}(thf)]$ [obtained from YbI_2, $KN(SiMe_3)_2$ and the β-diketimine HL (L = $\{N(C_6H_3Pr^i_2\text{-}2,6)C(Me)\}_2CH$)] with SiH_3Ph or H_2O yielded the crystalline $[\{Yb(L)(\mu\text{-}H)(thf)\}_2]$ or $[\{Yb(L)(\mu\text{-}OH)(thf)\}_2]$, respectively.[111]

The heteroleptic amide $[Yb\{N(SiMe_3)_2\}(thf)(BPh_4)]$ (two crystalline forms with $\eta^6:\eta^4$- and $\eta^6:\eta^1$-bound BPh_4 ligand),[112a] obtained from $[Yb\{N(SiMe_3)_2\}_2(thf)_2]$ and $[HNMe_3][BPh_4]$, yielded the crystalline, thf-free compound $[Yb\{N(SiMe_3)_2\}(BPh_4)]$ (by desolvation in PhMe)[112b] and $[Yb(Bu^t_2pz)(thf)(BPh_4)]$ with 3,5-Bu^t_2-pyrazole;[112b] both complexes contained the ligand $(\eta^6\text{-}Ph)_2BPh_2$. Complex $[YbL(thf)_2]$ having an unusual amidoborate ligand $[N(SiMe_3)(SiMe_2CH_2BPh_3)]^{2-}$ (≡L) was isolated as a minor product in the synthesis of $[Yb\{N(SiMe_3)_2\}(thf)(BPh_4)]$.[112a]

The first Ce^{IV} amide to be structurally characterised, prepared by Scott and co-workers, was the five-coordinate, diamagnetic Ce^{IV} compound $[Ce(I)(NN'_3)]$, $[NN'_3]^{3-}$ being $[N(CH_2CH_2NSiMe_2Bu^t)_3]^{3-}$.[113] It was obtained by iodination of the trigonal pyramidal $[Ce^{III}(NN'_3)]$ precursor, prepared from $[CeCl_3(thf)_4]$ and $[Li_3(NN'_3)(thf)_3]$. However, the corresponding reaction of $[Ce(NN'_3)]$ with Cl_2 or Br_2 gave the Ce^{III}/Ce^{IV} mixed valence compound $[\{Ce(NN'_3)\}_2(\mu\text{-}X)]$ (X = Cl or Br) having a Ce^{IV}–X→Ce^{III} core.[113]

Although $[Ce\{N(SiMe_3)_2\}_3]$ proved to be unreactive towards Cl_2,[113] Br_2 or I_2,[93] the C_3-symmetric, diamagnetic Ce^{IV} amide $[CeCl\{N(SiMe_3)_2\}_3]$ was obtained from the homoleptic Ce^{III} amide and $TeCl_4$ in toluene[114] or thf;[93] in thf, the paramagnetic Ce^{III} compound $[Ce\{N(SiMe_3)_2\}_2(\mu\text{-}Cl)(thf)]_2$ was a co-product (see Section 4.3.3). The isomorphous, crystalline Ce^{IV} bromide **23** was prepared from $[Ce\{N(SiMe_3)_2\}_3]$ and PBr_2Ph_3 in diethyl ether, together with $CeBr_3(OEt_2)_n$ which with thf gave $[CeBr_3(thf)_4]$.[93] That the supposedly weak oxidising agents $TeCl_4$ and PBr_2Ph_3, unlike the stronger Hal_2, were able to affect these $Ce^{III} \rightarrow Ce^{IV}$ transformations was suggested to be related to their dissociation in solution to form the halogenonium cations $[TeCl_3]^+$ and $[PBrPh_3]^+$.[93]

Computational studies on $[CeCl\{N(SiH_3)_2\}_3]$ were carried out;[114] the optimised model showed close agreement with the experimentally observed $CeClN_3$ core geometric parameters for $[CeCl\{N(SiMe_3)_2\}_3]$, including the elongated (compared with the Ce^{III} precursor) N–Si bond lengths.

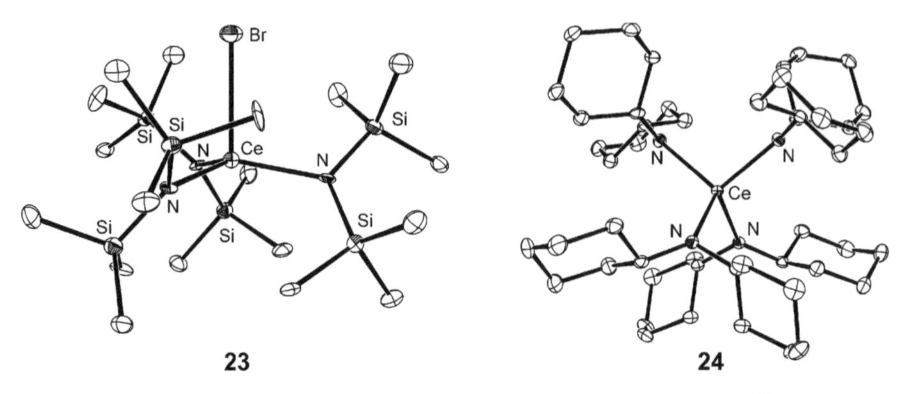

The diamagnetic, crystalline, four-coordinate, distorted tetrahedral Ce^{IV} amide **24** was obtained by aerial oxidation of $[Ce(NCy_2)_3(thf)]$ or $[Ce(NCy_2)_2(\mu\text{-}NCy_2)_2Li(thf)]$ (for these Ce^{III} compounds, see Section 4.3.3).[79] The average Ce^{IV}–N bond length in **24** is 2.243 Å[79] and 2.219 Å in **23**,[93] while the Ce^{III}–N bond length in $[Ce(NCy_2)_3(thf)]$ is 2.317 Å[79] and 2.320 Å in $[Ce\{N(SiMe_3)_2\}_3]$.[86] Possible intermediates to **24** may have included Ce^{IV} superoxo-, peroxo- and oxo-complexes which may have disproportionated into **24** and a polymeric amido-oxo compound such as $\{Ce(NCy_2)_2(\mu\text{-}O)\}_n$.[79]

The Sussex group has also shown that oxidation of $[Ce\{N(SiMe_3)_2\}_3]$ by O_2 at ambient pressure and −27 °C, in the absence or presence of $M[N(SiMe_3)_2]$ (M = Na, K) and H_2 or H_2O, provides a range of X-ray-characterised crystalline products as summarised in Scheme 4.4 (the cited yields refer to isolated X-ray quality crystals).[115] The compound $[\{Ce(N(SiMe_3)_2)_2(\mu\text{-}O)\}_3]$ was the hydrolysate of $[\{(Me_3Si)_2N\}_3Ce(OC_6H_4O)Ce\{N(SiMe_3)_2\}_3]$ formed from $[Ce\{N(SiMe_3)_2\}_3]$ and *p*-benzoquinone.

The compound $[\{Ce(N(SiMe_3)_2)_3\}_2(\mu\text{-}\eta^2\text{:}\eta^2\text{-}O_2)]$ (**25**) had limited solubility and stability in a hydrocarbon solution and hence reliable NMR or Raman solution spectra were not obtained.

$[Ce(NR_2)_3]$ — $p\text{-}C_6H_4O_2$, trace H_2O → $[\{(R_2N)_2Ce(\mu\text{-}O)\}_3]$ Purple, 10%

O_2, hexane, −27 °C → $[(R_2N)_3Ce(\mu\text{-}\eta^2\text{:}\eta^2\text{-}O_2)Ce(NR_2)_3]$ **25** Dark brown, 21%

trace H_2O → $[(R_2N)_2Ce(\mu\text{-}O)_2Ce(NR_2)_2]$ Red, 38%

MNR_2 → $[(R_2N)_3Ce\text{—}O(\mu\text{-}M)_2O\text{—}Ce(NR_2)_3]$ M = Na, 17% or K, 23% Dark brown

KNR_2, H_2 → $[(R_2N)_3Ce\text{—}O\text{—}Ce(NR_2)_2\text{—}O(\mu\text{-}K)_2O\text{—}Ce(NR_2)_2\text{—}O\text{—}Ce(NR_2)_3]$ Red-brown, 10%

Scheme 4.4 *(R = SiMe₃)*

The X-ray structure showed a mean Ce–N bond length of 2.259 Å; the O–O distance of 1.328(6) Å is indicative of a superoxide $[O_2]^-$ (rather than a peroxide $[O_2]^{2-}$), which implies that the cerium oxidation state is intermediate between +3 and +4.[115]

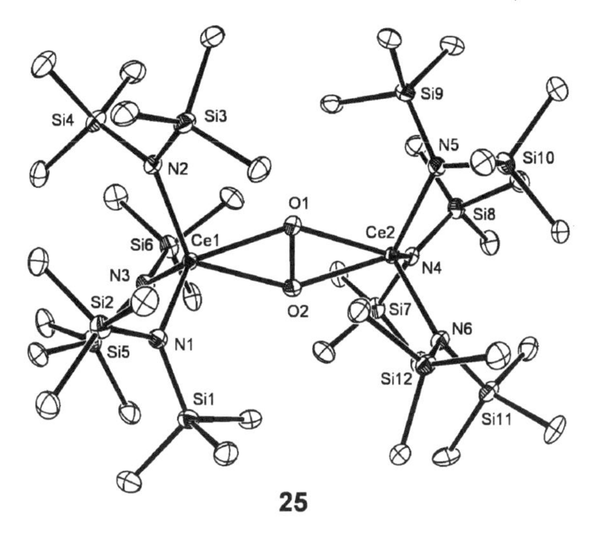

4.3.5 Ln Complexes having Donor-Functionalised Amido Ligands

The title ligands are the monoanionic **10** (see Section 4.2.5), **26–29** and **54** (see also **79**, **80** and **83**), the dianionic **30–52** (see also **77**) and the trianionic **53** (see also **84**). These are chelating ligands, hence firmly bound to the Ln^{3+} core. Their function has largely been that of supporting ligands. Some of them have had a role not only in *f*-block but also early *d*-block metal chemistry; certain of the derived compounds have been examined as catalysts for a variety of organic chemical transformations.

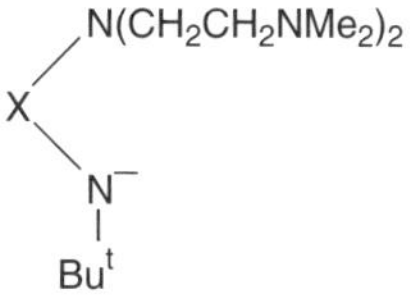

$N(CH_2CH_2NMe_2)_2$ / X / N^- / Bu^t

26 **a** $X = (CH_2)_2$
b $X = SiMe_2$

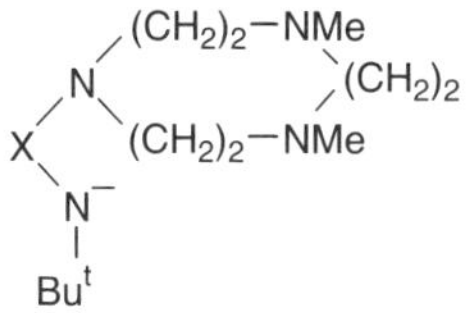

$(CH_2)_2$–NMe / N / $(CH_2)_2$ / $(CH_2)_2$–NMe / X / N^- / Bu^t

27 **a** $X = (CH_2)_2$
b $X = SiMe_2$

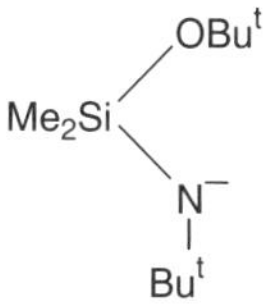

OBu^t / Me_2Si / N^- / Bu^t

28

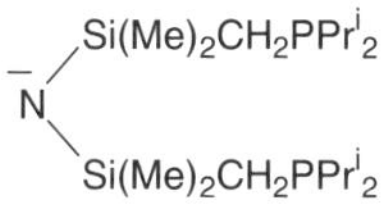

$Si(Me)_2CH_2PPr^i_2$ / $\bar{N}$ / $Si(Me)_2CH_2PPr^i_2$

29

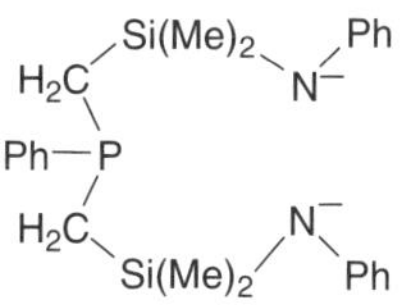

$Si(Me)_2$ / Ph / H_2C / N^- / Ph–P / H_2C / N^- / $Si(Me)_2$ / Ph

30

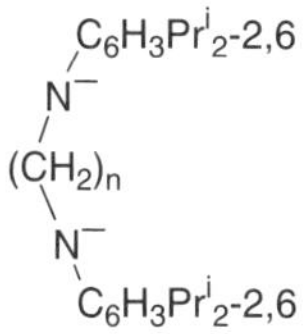

$C_6H_3Pr^i_2$-2,6 / N^- / $(CH_2)_n$ / N^- / $C_6H_3Pr^i_2$-2,6

31 **a** n = 2
b n = 3

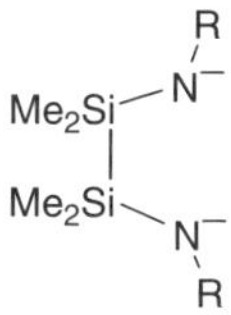

R / Me_2Si–N^- / Me_2Si / N^- / R

32 **a** $R = Bu^t$
b R = Ph

$SiMe_3$ / N / $(CH_2)_2$–N^- / N / CH_2 / $(CH_2)_2$–N^- / $SiMe_3$

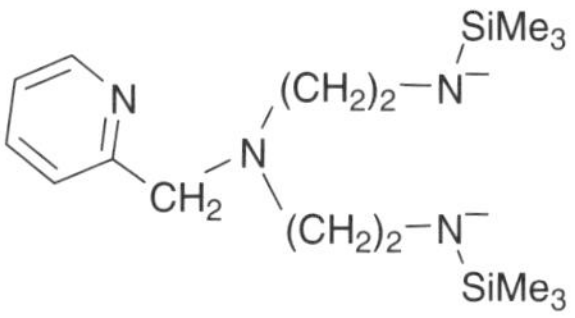

33

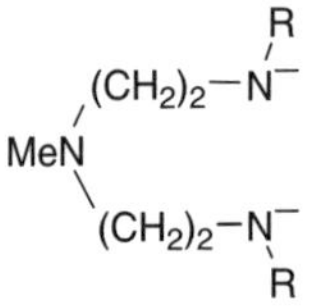

34 **a** R = $C_6H_2Me_3$-2,4,6
b R = $C_6H_3Et_2$-2,6
c R = $C_6H_3Cl_2$-2,6

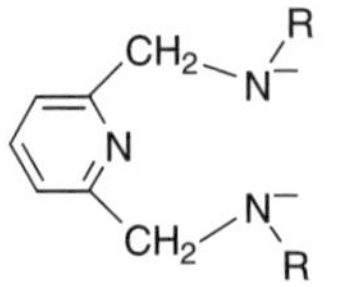

35 **a** R = $C_6H_3Pr^i_2$-2,6
b R = $C_6H_2Me_3$-2,4,6

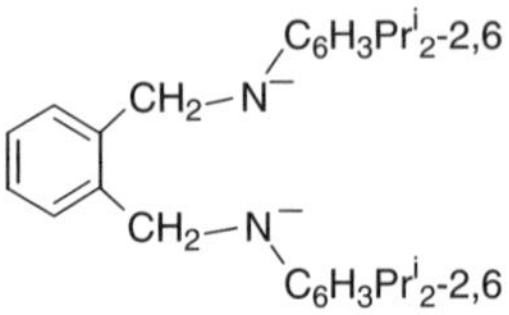

36

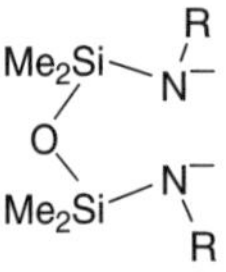

37 **a** R = Bu^t
b R = 2-C_5H_3NMe-4

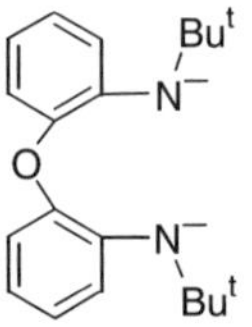

38

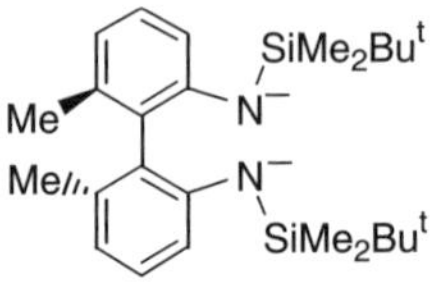

39

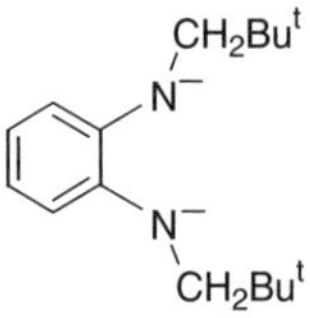

40

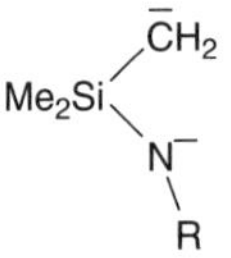

41 **a** R = $SiMe_3$
b R = $C_6H_3Pr^i_2$-2,6

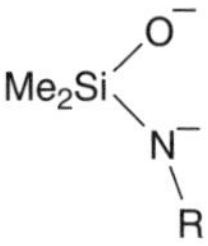

42 **a** R = Ph
b R = Bu^t

43

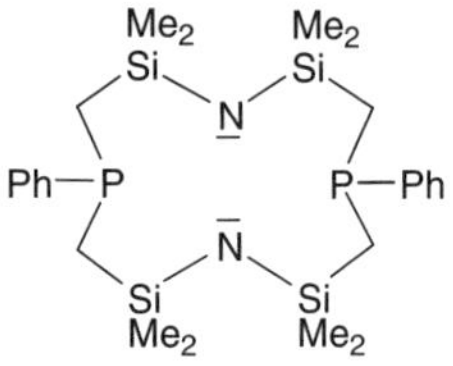

44

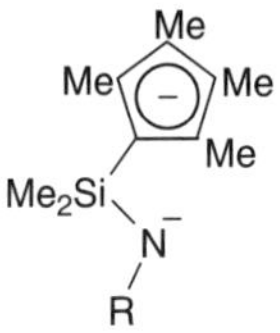

45

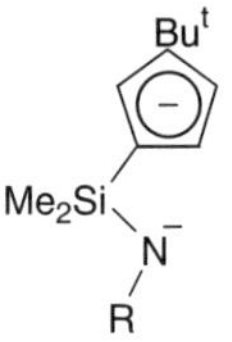

46

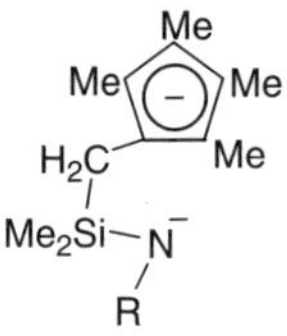

47

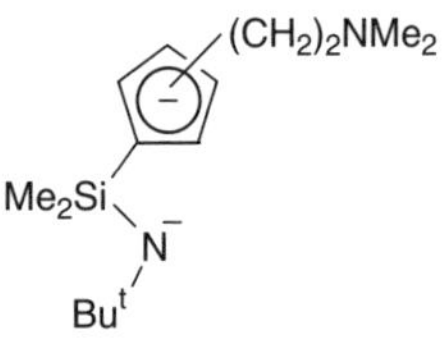

48

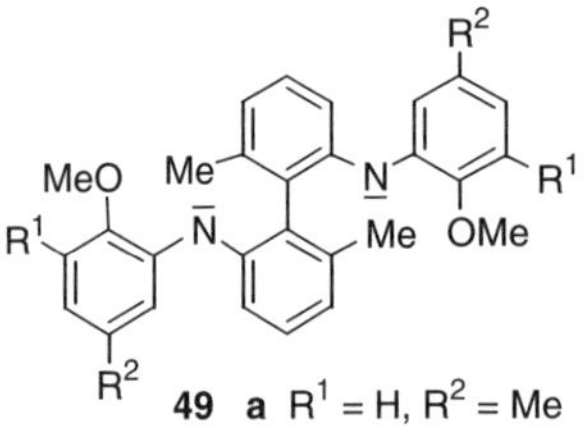

49 **a** R^1 = H, R^2 = Me
b R^1 = Bu^t = R^2

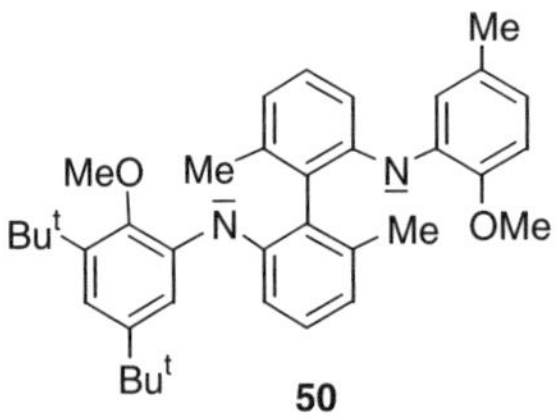

50

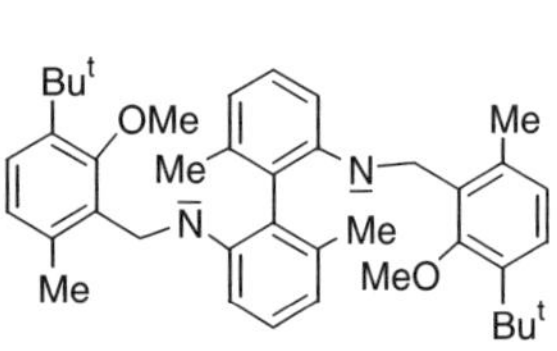

51

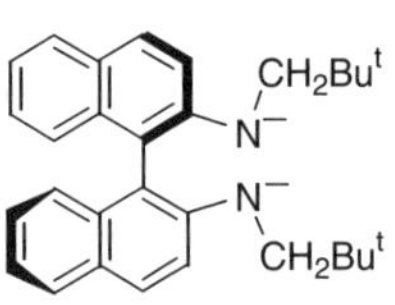

52

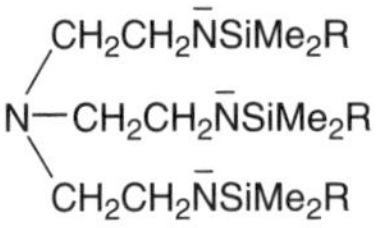

53 **a** R = Bu^t
b R = Me

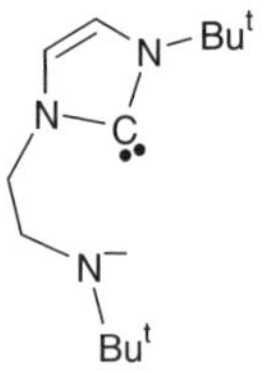

54

From H(**26a**) or H(**26b**) and [$Y(CH_2SiMe_3)_3(thf)_2$] in pentane, Hessen and co-workers obtained [Y(**26a**)($CH_2SiMe_3)_2$] or **55**, respectively.[116] Each with [$HNMe_2Ph$][$B(C_6F_5)_4$] in thf yielded the labile (stable in thf) [Y(**26a**)(CH_2SiMe_3)][$B(C_6F_5)_4$], a poor catalyst for polymerisation of ethylene. By contrast, the isoleptic salt derived from the conformationally more rigid **27a**, obtained from [Y(**27a**)($CH_2SiMe_3)_2$] and [$HNMe_2Ph$][$B(C_6F_5)_4$], was an effective C_2H_4-polymerisation catalyst.[117a,117b] Treatment of *in situ*-prepared $La(CH_2SiMe_3)_3(thf)_x$ with **27a** or **27b** gave [La(**27a**)($CH_2SiMe_3)_2$] or **56**, respectively.[117b] The former with [$HNMe_2Ph$][$B(C_6F_5)_4$] was a catalyst for the *cis*-reductive linear dimerisation of PhC≡CH. The reactions of compound **57** (the **26b**-analogue of **55**) with C_2H_4 or pyridine are shown in Equation (4.11).[116]

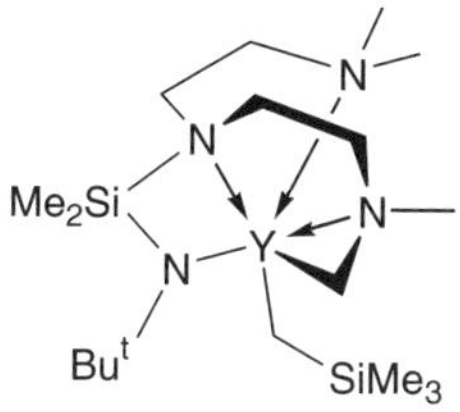

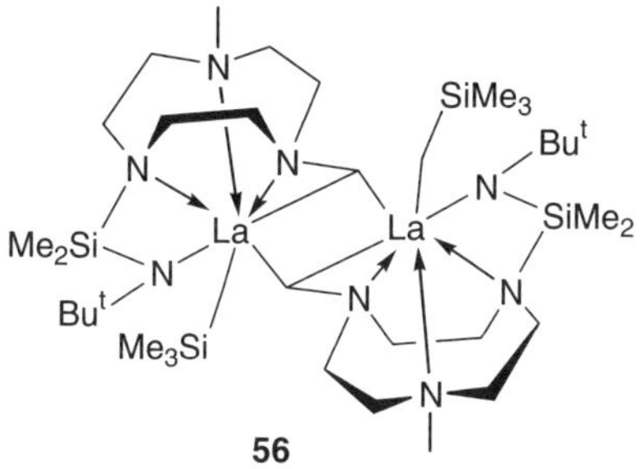

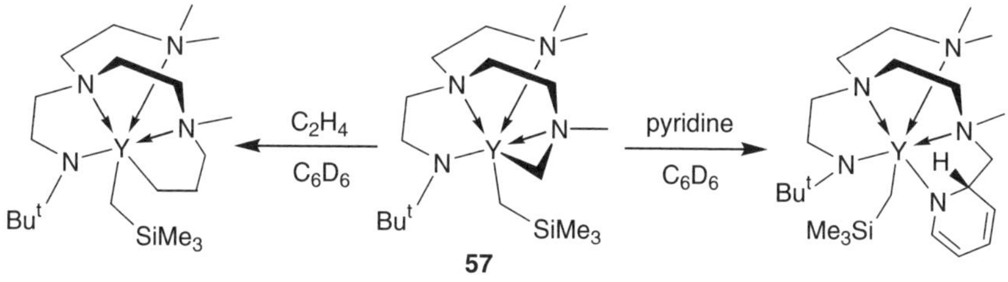

(4.11)

Treatment of $YCl_3(thf)_{3.5}$ with 2Li(**28**) gave [Y(**28**)$_2$Cl(thf)], from which the following four Y(**28**)$_2$X compounds were derived: X = $CH(SiMe_3)_2$, $N(SiMe_3)_2$, $OC_6H_3Bu^t{}_2$-2,6 and BH_4(thf).[118a] Reactions of the alkyl are illustrated in Scheme 4.5,[118a] including the reactions with pyridine or 2-picoline.[118b]

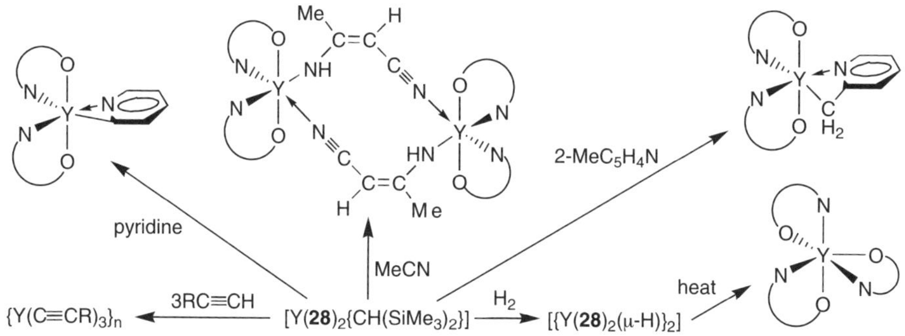

Scheme 4.5

The ligand **29** (a variant of **10**) has featured in the distorted octahedral compound [$ScCl_2$(**29**)(thf)] (*trans*-chlorides), prepared from $ScCl_3$(thf) and Li(**29**).[119a] It was converted into [Sc(**29**)R_2] (R = Me, Et, CH_2SiMe_3)[119a] or [ScCp(Cl)(**29**)R] (R = Me, Ph, N(H)Bu^t, N(H)Ph, BH_4).[119b] The compound [{ScCp(Cl)(μ-H)}$_2$] was obtained from the tetrahydridoborate upon addition of PMe_3,[119b] and the trigonal bipyramidal compound [Y(**30**)-(CH_2SiMe_3)(thf)] (thf/P axial) from [Y(CH_2SiMe_3)$_3$(thf)$_2$] and H_2(**30**).[119c]

Treatment of YI_3 with K_2(**31a**) or K_2(**31b**) gave [{YI(**31a**)(thf)$_2$}$_2$] or, in lower yield, [YI(**31b**)(thf)$_2$]. Each, with KCH_2Ph, was converted into [Y(CH_2Ph)(**31a**)(thf)$_2$] or [Y(CH_2Ph)(**31b**)(thf)$_2$].[120] Whereas [Y($OC_6H_2Bu^t_2$-2,6-Me-4)(**31a**)(thf)] was obtained from the iodide and K[$OC_6H_2Bu^t_2$-2,6-Me-4], the corresponding reaction involving the **31b** ligand was less effective. Other yttrium derivatives available from these iodides were the bis(trimethylsilyl)-methyls and -amides. Whereas [Y{CH($SiMe_3$)$_2$}(**31a**)(thf)] underwent hydrogenolysis to give [Y_3(**31a**){H(**31a**)}$_2$(μ-H)$_3$($μ_3$-H)$_2$(thf)] (**58**) [the atoms Y2 and Y3 are each bonded to H(**31a**)], the isoleptic alkyl from **31b** was unreactive. However, each with $PhSiH_3$ gave Si(**31a** or **31b**)H(Ph) and H_2Si{CH($SiMe_3$)$_2$}Ph.[120]

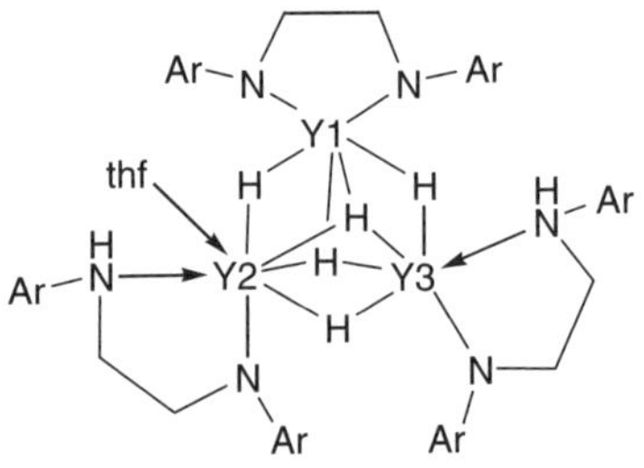

58 Ar = $C_6H_3Pr^i_2$-2,6

The compounds [{Ln(**32a**)(μ-Cl)}$_2$] and [{Ln(**32b**)(μ-Cl)}$_2$] were isolated from in situ-prepared Li_2(**32a**) and Li_2(**32b**) and $LnCl_3$ in thf (Ln = Nd, Gd, Yb); from the appropriate chloride and Na[OC(O)CF_3] the product was [{Ln(**32a**)(μ-OC(O)CF_3)}$_2$] (Ln = Nd, Gd).[121]

Treatment of $LnCl_3$ with Li_2(**33**) gave **59** (Ln = Sc[122a] and Y[122b]). The chloride ligand was readily displaced by [CH_2SiMe_3]$^-$ yielding [Ln(**33**)(CH_2SiMe_3)];[122a,122b] the Sc compound was also accessible from [Sc(CH_2SiMe_3)$_3$(thf)$_2$] and H_2(**33**).[122a] Other Cl^-/X^- exchanges in the yttrium series were those with X = Me, N(H)Bu^t, N(H)$C_6H_3Pr^i_2$-2,6, N(H)C_6F_5, N($SiMe_3$)$_2$ and {N($SiMe_3$)}$_2$CPh.[122b] The ligand **33** promotes trigonal bipyramidal coordination in its complexes.[122]

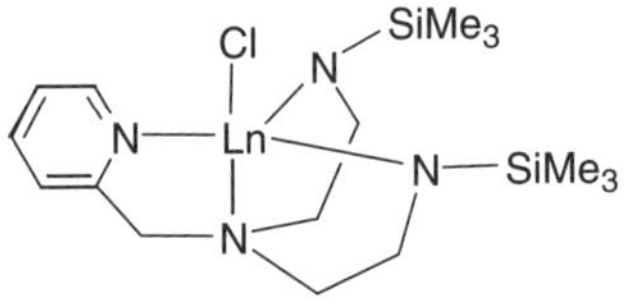

59

Treatment of the appropriate triamido-Ln^{III} compound with the triamine H_2(**34a**), H_2(**34b**) or H_2(**34c**) yielded the corresponding Ln(**34**) amide, Scheme 4.6.[123a] Each of

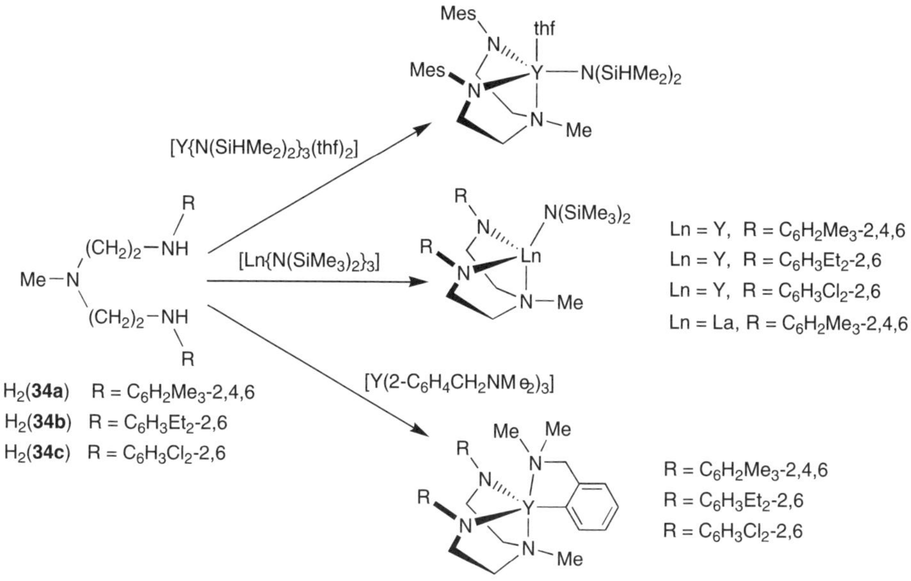

[$Y\{N(SiHMe_2)_2\}_3(thf)_2$]

[$Ln\{N(SiMe_3)_2\}_3$]

[$Y(2\text{-}C_6H_4CH_2NMe_2)_3$]

H_2(**34a**) R = $C_6H_2Me_3$-2,4,6
H_2(**34b**) R = $C_6H_3Et_2$-2,6
H_2(**34c**) R = $C_6H_3Cl_2$-2,6

Ln = Y, R = $C_6H_2Me_3$-2,4,6
Ln = Y, R = $C_6H_3Et_2$-2,6
Ln = Y, R = $C_6H_3Cl_2$-2,6
Ln = La, R = $C_6H_2Me_3$-2,4,6

R = $C_6H_2Me_3$-2,4,6
R = $C_6H_3Et_2$-2,6
R = $C_6H_3Cl_2$-2,6

Scheme 4.6

these complexes catalysed the intramolecular hydroamination/cyclisation of pent-4-enylamine, 1-methylpent-4-enylamine (high *trans* selectivity) and 5-phenylpent-4-ynylamine, and effected the bicyclisation of 2-allyl-2-methylpent-4-enylamine.[123a]

The ligands **35** and **36** featured in complexes **60** (n = 1 for Ln = Sc, or n = 2 for Ln = Y or Lu) and **61**, respectively.[123b] The alkyls (**60**, X = CH_2SiMe_3) and **61** were obtained from

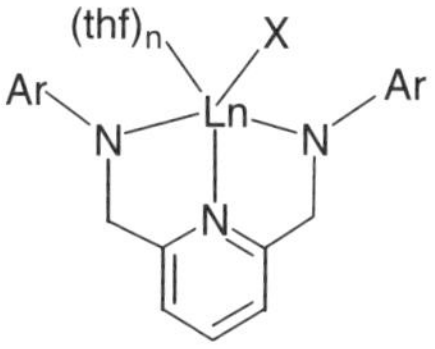

60 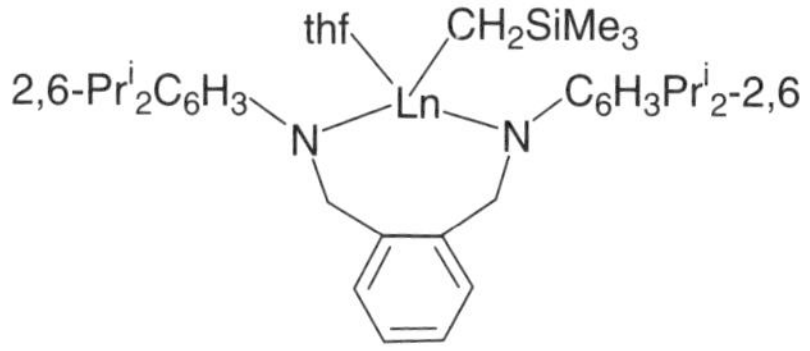 **61**

[$Ln(CH_2SiMe_3)_3(thf)_2$] and H_2(**35a**) or H_2(**35b**) and H_2(**36**), respectively. The Sc or Lu amides **60** (X = NR_2) were prepared from the alkyl by addition of HNR_2 (R = $SiHMe_2$ or Et). The related Sc or Lu amides **60** (X = NPr^i_2 and $n = 1$) were formed by addition of H_2(**35a**) or H_2(**35b**) to [$Ln(NPr^i_2)_3(thf)$]. The mono-thf adducts were kinetically inert in solution whereas the bis-thf adducts were fluxional. The mesityl complexes **60** (Ar = $C_6H_2Me_3$-2,4,6) were less stable than the 2,6-diisopropylphenyl analogues. The 4- and 5-coordinate [Sc(**36**)(CH_2SiMe_3)(thf)] and [Sc(**35a**)(CH_2SiMe_3)(thf)] complexes were X-ray-characterised. Each of the 5-coordinate [Sc(**35a**)X(thf)] complexes effectively polymerised methyl methacrylate in a living manner affording mainly atactic P(MMA) at ambient temperature.[123b]

The asymmetrically bridged diamido compound [$\{YbCp_2(thf)\}_2\{\mu\text{-}\kappa^2{:}\kappa^1\text{-}C_5H_3N(NH)_2\text{-}2,6\}$], obtained from $YbCp_3$ and 2,6-diaminopyridine in thf, with $Pr^iN{=}C{=}NPr^i$ gave bi- or tetranuclear Yb^{III} guanidinates.[124] The compounds [Lu(**62**)(CH_2SiMe_3)$_2$] and [Lu-(**62**)(Cp*)(CH_2SiMe_3)] (in **62**, R = H or Bu^t) were obtained by the remarkable dearomatisation and functionalisation of the terpyridine ligand by its reaction with [Lu(CH_2Si-Me_3)$_3$(thf)$_2$] and [Lu(Cp*)(CH_2SiMe_3)$_2$(thf)], respectively.[125]

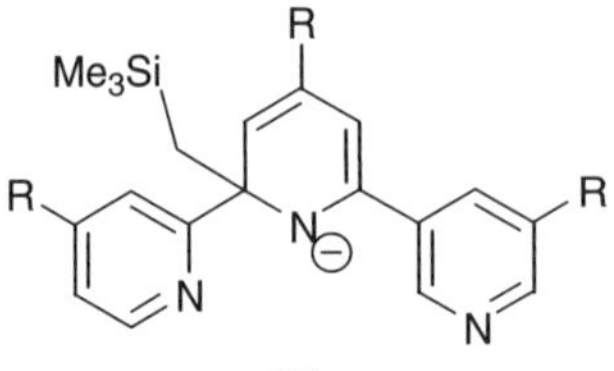

62 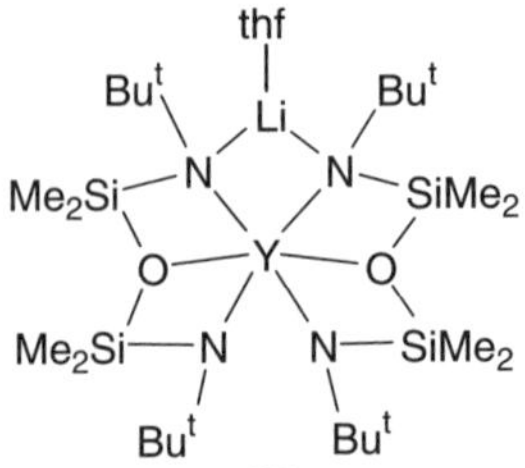 **63**

The compound [{Y(**37a**)(μ-Cl)(thf)}$_2$] was isolated from YCl_3 and Li_2(**37a**) in thf. It was converted into [Y(**37a**){CH($SiMe_3$)$_2$}(thf)], [Y(**37a**)Cp(thf)] and **63** (reversibly) by treatment with Li[CH($SiMe_3$)$_2$], NaCp and Li_2(**37a**), respectively.[126a] The salt Li_2(**37b**) with $LnCl_3$ or $LaBr_3$ in thf gave [Ln(**37b**)Cl(thf)$_n$] (Ln = Y and n = 2; Ln = Sm and n = 3) or [Li-(thf)$_4$][La(**37b**)$_2$], respectively.[126b] The latter was a highly efficient initiator for the ring-opening polymerisation of lactones.[126b] Upon reaction with [{RhCl(cod)}$_2$] or [{RhCl-(C_2H_4)}$_2$] (LiCl elimination) [Li(thf)$_4$][La(**37b**)$_2$] produced a series of bimetallic complexes, featuring short La···Rh contacts.[126c] YCl_3 had been the source of [Y(**38**)Cl(thf)$_2$] and therefrom [Y(**38**)(CH_2SiMe_3)(thf)$_2$] and [Y(**38**){CH($SiMe_3$)$_2$}(thf)}] which were relatively unreactive towards H_2 or C_2H_4.[127]

The compounds **64** [X = Cl, Me, CH($SiMe_3$)$_2$, Et, C_6H_{13}-*n*] have been prepared.[128] The chloride [from YCl_3(thf)$_3$ and Li_2(**39**)] with the appropriate LiR gave **64** [with X = Me or CH($SiMe_3$)$_2$]. The latter with H_2 or $PhSiH_3$ gave [{Y(**39**)(μ-H)(thf)}$_2$], which with RCH=CH_2 or pyridine gave **64** (with X = Et or C_6H_{13}-*n*) or the labile [Y(**39**)(N)(py)$_2$], which was isomerised yielding [Y(**39**)(N)(py)$_2$].[128]

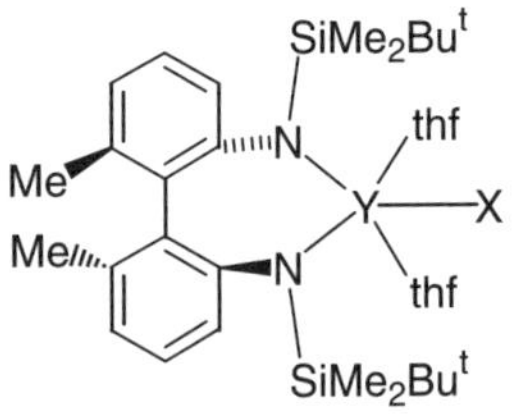

64

The heterodimetallic, crystalline complexes **65** or **66** were obtained by reaction in diethyl ether of YCl_3 with Li_2(**40**) or $CeCl_3$ with Na_2(**40**), respectively.[129] The centrosymmetric molecule **65** has two Y(**40**)(μ-Cl)$_2$Li(OEt_2)$_2$ moieties associated by Y···M1′ and Y′···M1 close contacts; each Y···M vector is essentially orthogonal to the aromatic ring of which M is the centroid. The six-coordinate Ce atom of **66** is at the centre of a distorted trigonal prism Ce(NN′)$_3$ and each of the sodium ions is not only attached to the N and N′ atoms of its ligand **40** but also has close η^4-contacts to a neighbouring aromatic ring.[129]

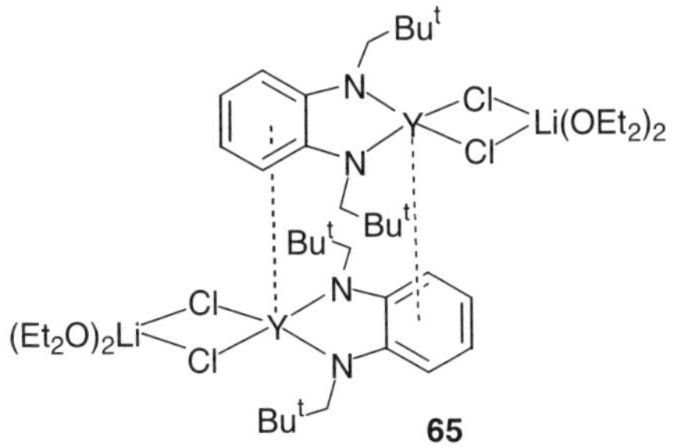

65

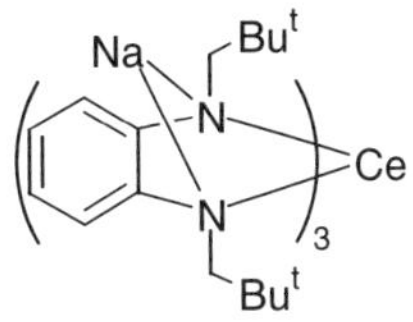

66

The Ln-coordinated ligand **41a** was generated by an interesting deprotonation reaction, (i) in Scheme 4.7; the product underwent an insertion with 2,6-$Me_2C_6H_3NC$.[10e] Treatment of [$Ln(CH_2SiMe_3)_3(thf)_2$] (Ln = Gd) with $HN(C_6H_3Pr^i{}_2$-2,6)$SiMe_3$ yielded [Gd_2{μ-(**41b**)}$_3$(thf)$_3$], while for Ln = Sc, Y, Ho, Lu the product was [$Ln(CH_2SiMe_3)_2${N-($C_6H_3Pr^i{}_2$-2,6)$SiMe_3$}(thf)]; each of the latter showed high activity for the living polymerisation of isoprene and the Sc complex also for the polymerisation of hex-1-ene.[130c]

The ligand **42a** has featured in a number of oligonuclear Ln^{III} complexes.[77a] From the system $LnBr_3$/Na[N(H)Ph]/{Me_2Si(μ-O)}$_3$ in thf the crystalline tetranuclear complexes [$Ln_4(\mu_4$-O){N(H)Ph}(**42a**)$_6Na_5$(thf)$_7$] were obtained (Ln = Gd, Yb) with, for Ln = Yb, as a minor product [Na_4(thf)$_6Yb_2${OSi(Me)$_2$N(Ph)Si(Me)$_2$O}$_2$(**42a**)$_2${N(H)Ph}$_2$]. The former has a tetrahedral array of Ln atoms surrounding a central μ_4-O atom, each tetrahedral edge has a bridging ligand **42a**; the $[N(H)Ph]^-$ ligands are bound terminally to three of the Ln atoms. Similarly, the $LnCl_3$/Na[N(H)Bu^t]/{Me_2Si(μ-O)}$_3$ system yielded the complex [Li_2Ln{N(H)Bu^t}(**42b**)$_2$(thf)]$_2$ (Ln = Sm, Gd, Yb) containing a central *O*- and *N*- bridged Ln-Li polyhedral $Ln_2Li_2O_4N_2$ core.[77a]

The crystalline, diamagnetic, five-coordinate Yb^{II} complex [Yb(**43**)(hmpa)$_3$] was prepared from Ph_2C=NPh and Yb metal in thf/hmpa, which in thf with Me_2CO, Pr^iNCO or air gave $Ph_2C(H)N(H)Ph$, Ph_2C{N(H)Ph}C(O)N(H)Pr^i or Ph_2C=NPh, respectively.[130a] The dianionic *N,C*-centred ligand **43** featured also in the Sm^{III} complexes [Sn(**43**)($OC_6H_2Bu^t{}_2$-

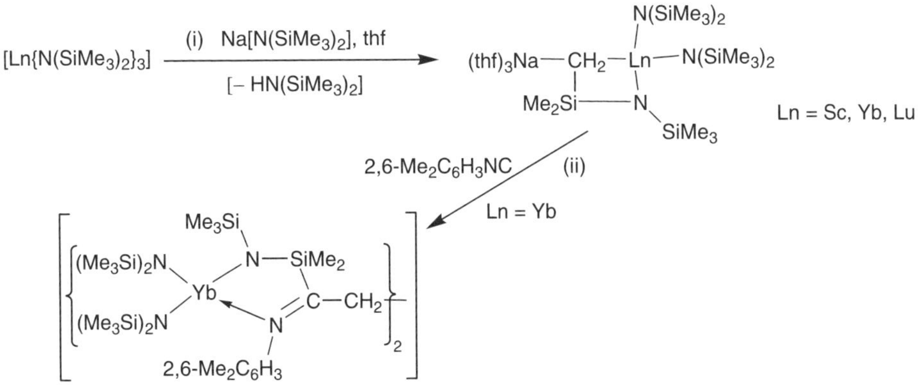

Scheme 4.7

2, 6-Me-4)(thf)$_3$] and [Sn(**43**){N(SiMe$_3$)$_2$}(thf)$_3$] prepared from Ph$_2$C=NPh and the respective Sm^{II} aryloxide and amide.[130b]

Reaction of [Li$_2$(**44**)(thf)] with LnCl$_3$(thf)$_3$ in thf yielded [{Ln(**44**)(μ-Cl)}$_2$] (Ln = Y,[131a] Lu[131b]). The yttrium compound with LiR gave the crystalline [Y(**44**){CH(SiMe$_3$)$_2$}] or the oily CH$_2$SiMe$_3$ analogue.[131a] From the latter and benzene, or the chloride and LiPh, the product was **67**.[131a] Reduction of the Y or Lu chloride with KC$_8$ in toluene/diethyl ether in the presence of naphthalene, 2-MeC$_{10}$H$_7$ or anthracene gave crystalline [{Ln(**44**)}$_2$-(μ-η^4:η^4-C$_{10}$H$_7$R)] (R = H or Me) or [{Ln(**44**)}$_2$(μ-η^4:η^4-C$_{14}$H$_{10}$)], which were fluxional in solution.[131b]

67

The ligands **45**–**48**, being of general formula [C$_5$R$_4$ZNR′]$^{2-}$, are extensions of **12** (R = Me, R′ = Bu^t, Z = SiMe$_2$, Section 4.2.5). Their role as *ansa*-metallocene mimics and constrained geometry Ln catalysts has been surveyed in detail by Okuda,[9a] and is here described only in outline. In **45**: R = Bu^t (**45a**), Am^t (**45b**), Ph (**45c**), (CH$_2$)$_2$NMe$_2$ (**45d**), (CH$_2$)$_2$OMe (**45e**), or 2-C$_5$H$_4$N (**45f**); in **46**: R = Bu^t (**46a**), (CH$_2$)$_2$NMe$_2$ (**46b**) or (CH$_2$)$_2$OMe (**46c**). Both mono- and binuclear Ln^{III}-X complexes of these ligands are known, in which X represents an alkyl, aryl, hydrido, amido, alkoxo or halide. The hydrido complexes are almost invariably binuclear, having an Ln(μ-H)$_2$Ln core. The cyclopentadienyl ligands in binuclear Ln^{III} complexes may be *cis* or *trans*, and if each Ln atom has an additional neutral ligand L, then each may also have *homo* or *hetero* chirality, as exemplified in the *homo-trans*-**68** and *hetero-cis*-**69**.

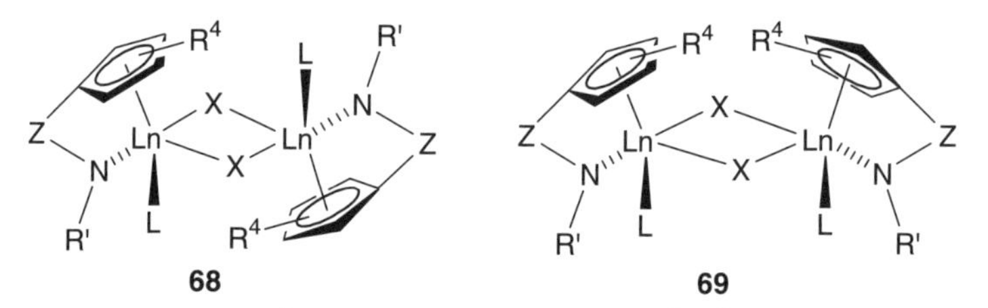

68 **69**

Hydrocarbyl (or amido) (R^2) complexes Ln(C$_5$R$_4$ZNR^1)R^2 have been prepared by salt elimination, alkane (or amine) elimination, ligand exchange, an α-olefin insertion into an Ln–H bond, or an ortho-metallation, as exemplified for the synthesis of the compounds **70**,[132] **71**,[133] **72**,[134] **73**[135] and **74**.[136] The reagents for these preparations were: YCl$_3$/Li(**45e**)/LiCH$_2$C$_6$H$_4$NMe$_2$-2 (**70**),[132] [Ln{E(SiMe$_3$)$_2$}$_3$]/H$_2$(**45a**) (**71**),[133] [{Y(**45b**)(μ-H)(thf)}$_2$]/EtCH=CH$_2$/dme (**72**),[134] [Sc(CH$_2$SiMe$_3$)$_3$(thf)$_2$]/H$_2$(**48**) (**73**)[135] and [Y(**47**)(CH$_2$SiMe$_3$)-(thf)]/C$_6$H$_5$OMe (**74**);[136] for [Ln(C$_5$Me$_4$ZNR′(CH$_2$SiMe$_3$)(thf)$_n$] (Ln = Y, Lu; Z = SiMe$_2$, CH$_2$SiMe$_2$; R′ = Bu^t, Ph, C$_6$H$_4$Bu^t-4, C$_6$H$_4$Bu^n-4), catalysts for hydrosilylation of olefins such as dec-1-ene, see Ref. 137c. Compounds **71** were active catalysts (superior to metallocene analogues) for amino-alkene hydroamination/cyclisation reactions [cf. Equation (4.8)].[133] Complex **72** was inactive for styrene polymerisation but the thf analogue (instead of dme) was active when a large excess of PhCH=CH$_2$ was used;[134] from [{Y(**45a**)-

(μ-H)(thf)}$_2$][137a] and styrene the mono-insertion product [Y(**45a**){CH(Me)Ph}(thf)] was obtained.[137b] The formation of **73** was diastereoselective.[135] If instead of anisole (for **74**), 2-methylanisole was used, then [Y(**47**)(CH$_2$C$_6$H$_4$OMe-2)(thf)] was isolated.[136]

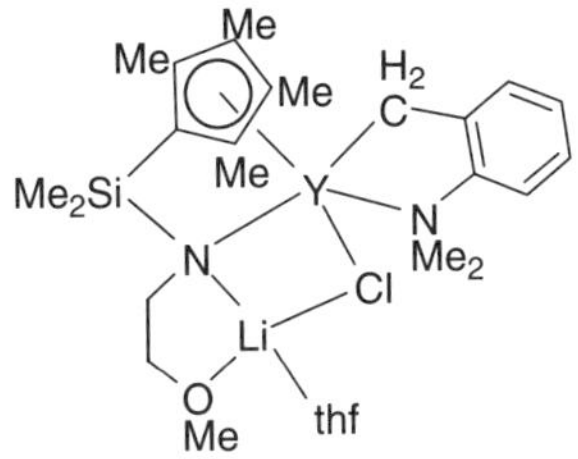

70

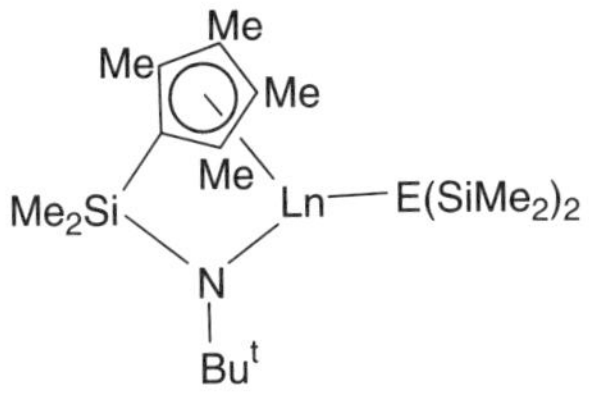

71 (E = CH, Ln = Yb, Lu)
(E = N, Ln = Nd, Sm, Lu)

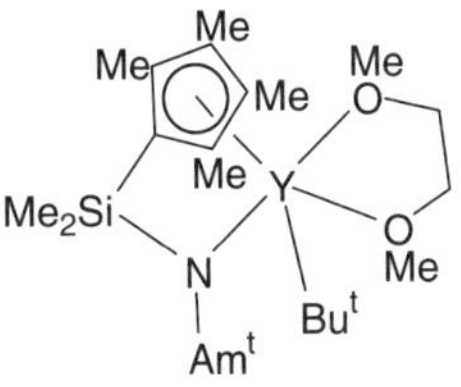

72

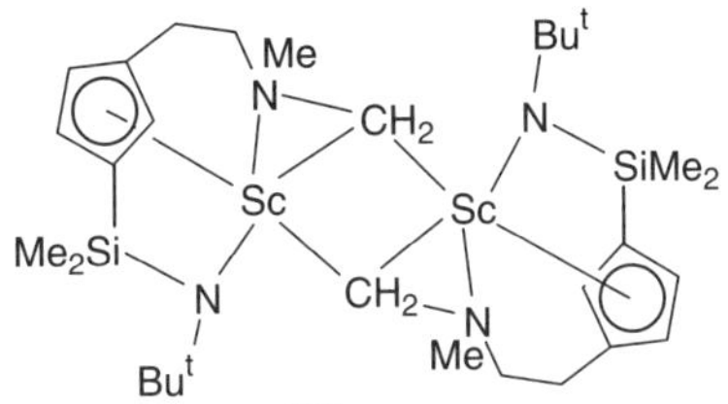

73

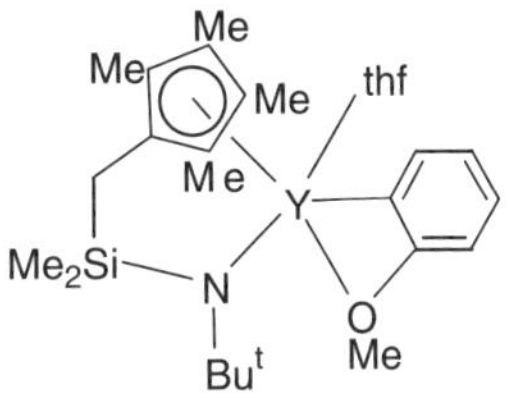

74

A derivative of the ligand **45** is the fluorenyl analogue **45g**, as featured in **75**, obtained from [Y(CH$_2$SiMe$_3$)$_3$(thf)$_2$] and H$_2$(**45g**), which upon hydrogenolysis gave [{Y(**45g**)(μ-H)-(thf)}$_2$].[138] From LnCl$_3$ (Ln = Y, La, Nd) and Li$_2$(**45g**) compounds such as [{Nd(**45g**)(μ-Cl)-(thf)}$_2$] and [Li(OEt$_2$)$_2$][La(**45g**)$_2$] were isolated. Their performance as catalysts for C$_2$H$_4$ or MMA polymerisation was evaluated; only the Nd complex with Li[CH(SiMe$_3$)$_2$] showed modest activity for PE, but [Li(thf)$_4$][La(**45g**)$_2$] was effective for P(MMA).[138]

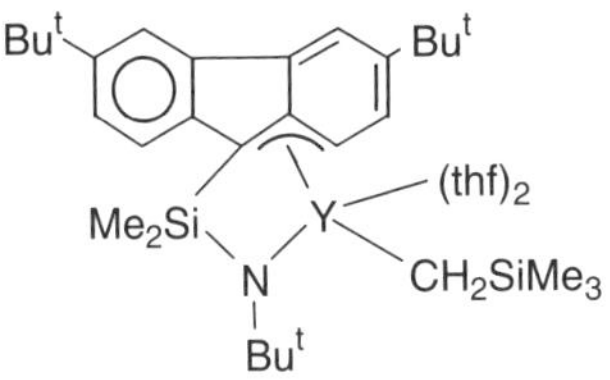

75

Bridged hydrido- or chloro-Ln^{III} amides include [{Lu(**45b**)(μ-H)(thf)}$_2$] and [{Y(**45b**)-(thf)}$_2$(μ-H)(μ-Cl)]; they were made by hydrogenolysis, or ligand exchange using PhSiH$_3$, from the appropriate alkyl (*e.g.*, CH$_2$SiMe$_3$) precursor.[139]

Ln^{II} amides include Yb(**45a**)(thf)[140] or [Yb(**45c**)(thf)$_3$],[141] prepared from [Yb(C$_{10}$H$_8$)-(thf)$_2$] and H$_2$(**45a**) or [Yb{N(SiMe$_3$)$_2$}$_2$(thf)$_2$] and H$_2$(**45c**), respectively. Recrystallisation of the latter from toluene-hexane gave [{Yb(**45c**)(thf)}$_2$], in which each NPh group had η^6-contact with the neighbouring Yb atom, while treatment with azobenzene gave [{Yb(**45c**)-(thf)}(μ-η^2:η^3-PhN$_2$Ph){Yb(**45c**)}].[141] Sm(**45c**)(thf) was a moderately active catalyst for PE formation.[141]

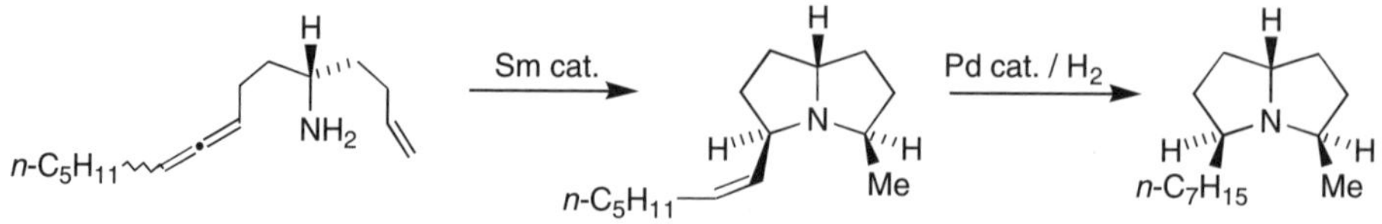

Scheme 4.8

The compound [Y(**47**)(CH_2SiMe_3)(thf)], obtained from H_2(**47**) and [Y(CH_2Si-$Me_3)_3(thf)_2$], yielded [{Y(**47**)(μ-H)(thf)}$_2$].[142] The latter with styrene gave [Y(**47**){C(H)-(Me)Ph}(thf)], and was also a catalyst for the hydrosilylation of dec-1-ene or styrene with $PhSiH_3$, giving $PhSi(H)_2C_{10}H_{21}$-*n* (exclusively) or $PhSi(H)_2(CH_2)_2Ph$ (1 part) and PhSi-$(H)_2${C(H)(Me)Ph} (2 parts).[142] Likewise, [{Y(**45a**)(μ-H)(thf)}$_2$] with hex-1-ene or styrene gave the corresponding 1:1-insertion product; each was catalytically active for the polymerisation of C_2H_4 or a polar monomer such as *t*-butyl acrylate or acrylonitrile.[143]

The amide [Sm(**45a**){N($SiMe_3)_2$}] was a catalyst for an allene-based hydroamination/cyclisation. As an illustration, one such product upon hydrogenation yielded a naturally occurring alkaloid, Scheme 4.8.[144] The same samarium(III) amide was also active for the intramolecular hydrophosphination/cyclisation of phosphino-alkenes or -alkynes; *e.g.*, $H_2P(CH_2)_3C{\equiv}CPh$ was transformed into **76**.[145]

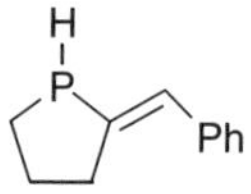

76

The diamido ligands **49–51** are potentially tetradentate by virtue of the two methoxy substituents. The racemic crystalline yttrium complexes [Y(**49a**){H(**49a**)}] and [Y(**49a**)-{N($SiHMe_2)_2$}(thf)] were X-ray-characterised.[146] These and the racemic [Ln(**50**){N-($SiHMe_2)_2$}(thf)] (Ln = Y, La) were obtained from the appropriate [Ln{N(SiH-$Me_2)_2\}_3(thf)_2$] and H_2(**49a**), H_2(**49b**) or H_2(**50**). Using the in situ-prepared chiral Y or La amide Ln(**50**){N($SiHMe_2)_2$}, only modest enantioselectivity and low reactivity for the catalytic hydroamination/cyclisation as in Equation (4.8) (n = 2) was achieved.[146]

The optically active salts [Li(thf)$_4$][Ln(**52**)$_2$],[147a] prepared from Li_2(**52**)[147b] and $LnCl_3$ (Ln = Sm, Yb) in thf, were highly effective for the enantioselective transformation of Equation (4.12).[147a]

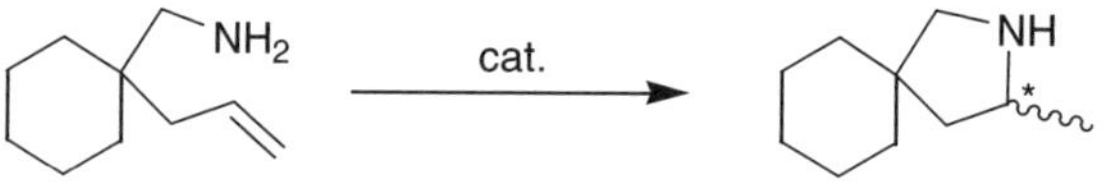

(4.12)

The ligands **45d**, **45e**, **45f**, **46a**, **46b** and **46c** have also appeared in lanthanate(III) anions (or their equivalents), as in Li[Y(**45d**)$_2$], Li[Y(**45e**)$_2$], Li[Y(**46b**)$_2$], Li[Y(**46c**)$_2$],[148] Li[Y-(**45f**)$_2$],[149] Li[Lu(**45e**)$_2$],[148] [K(dme)][Yb(**46a**)$_2$][150] and Cu[Y(**45f**)$_2$] (**77**).[149] Several of these are efficient ring-opening polymerisation catalysts for ε-caprolactone and lactide,[151] as are the compounds [Yb{(N($C_6H_3Pr^i_2$-2,6)C(Me))$_2$CH}(NPh_2)Cl(thf)] and [Yb-{(N($C_6H_3Pr^i_2$-2,6)C(Me))$_2$CH}L(μ-Cl)]$_2$ (L = NPr^i_2 or NC_5H_{10}) (also for PMMA).[152a]

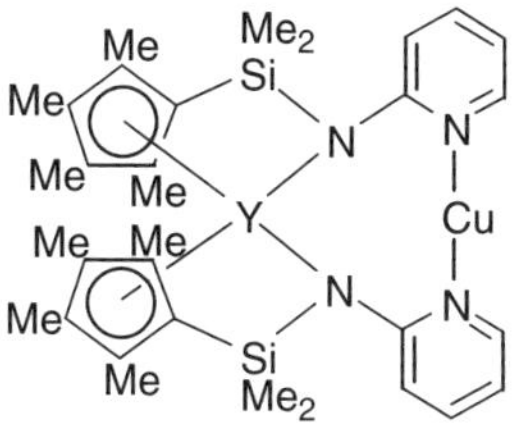

77

The trianionic ligand **53a** has been discussed in Section 4.3.4 in the context of Ce^{III}, Ce^{IV} and mixed valence Ce^{III}/Ce^{IV} amides and halogeno-amides.[113] Reactions of the less bulky ligand **53b** (as a Li salt) with $LnCl_3$ (Ln = Nd or Y) led to polynuclear clusters, which included μ_5-O[153a] or μ_5-O and μ_2-OCH=CH_2 ligands.[153b] The dinuclear tripodal triamido compound $[Sm\{(N(C_6H_4Me\text{-}4)SiMe_2)_3SiMe\}]_2$ was prepared from $[Sm\{N(SiMe_3)_2\}_3]$ via the amine elimination route.[154]

The ligand **78** has recently featured in the complexes $\{Yb(\mathbf{78})Cl(tmeda)\}_2$, $[Li(dme)_3]$-$[Ln(\eta^5\text{-}C_5H_4R)_2(\mathbf{78})]$ (R = H and Ln = Sm, Yb; R = Me and Ln = Yb) and **79**, each showing high activity for the polymerisation of MMA yielding syndiotactic-rich polymers of high molecular weight and relatively low molecular weight distribution.[152b]

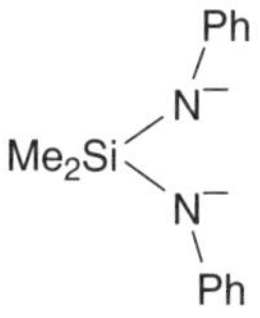

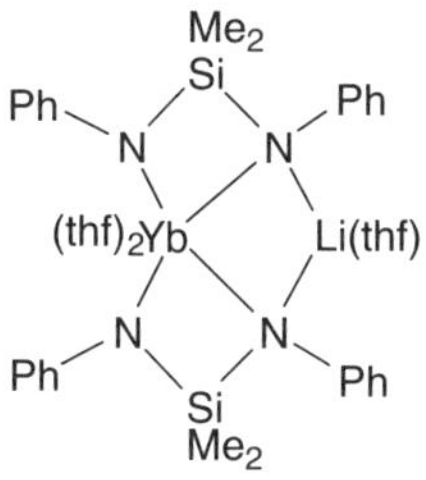

78 **79**

The ligands **80** and **81** have appeared in the *fac*-octahedral complexes $[Ln\{\kappa^2\text{-}(\mathbf{80})\}_3]$ (Ln = La, Sm, Eu, Er, Lu), prepared from 3Li(**80**) and $[(LaBr_3)(thf)_3]$, $LnCl_3$(Ln = Sm, Eu, Er) or $[LuCl_3(thf)_2]$ in thf; the crystalline complex $[\{Er(\mathbf{80})_2(\mu\text{-}Cl)\}_2]$ resulted from a 2:1 $ErCl_3$ reaction.[155] From 4Li(**81a**) and $2[Yb(\eta^5\text{-}C_5H_4Me)Cl_2(thf)]$, $[\{Yb(\mathbf{81a})_2(\mu\text{-}Cl)\}_2]$ was obtained;[156a] using 2Li(**81a** or **81b**), the product was $[\{Yb(\eta^5\text{-}C_5H_4Me)(\mathbf{81a}$ or $\mathbf{81b})(\mu\text{-}Cl)\}_2]$.[156a] The compound $[Yb(\mathbf{81b})_2(OC_6H_2Bu^t_3\text{-}2,4,6)]$ was isolated from $[Yb(OC_6H_2Bu^t_3\text{-}2,4,6)_3]$ and 2Li(**81b**).[156a] The homoleptic, distorted octahedral, crystalline compounds $[Ln(\mathbf{81a})_3]$ (Ln = Pr, Nd, Sm, Er, Yb) and $[Ln(\mathbf{81b})_3]$ were prepared from $[Li(\mathbf{81a})(OEt_2)]_2$ and $YbCl_3$ in thf, or *in situ*-prepared Li(**81a** or **81b**) and the appropriate $LnCl_3$ in thf.[156b]

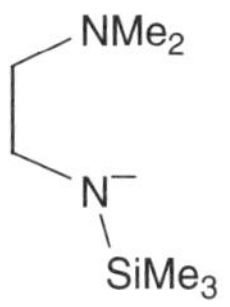

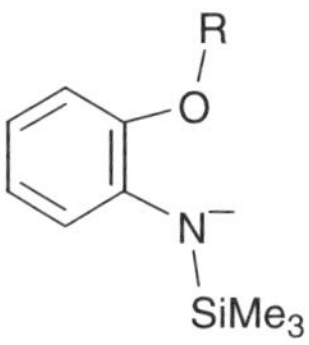

80 **81** **a** R = Me
b R = Ph

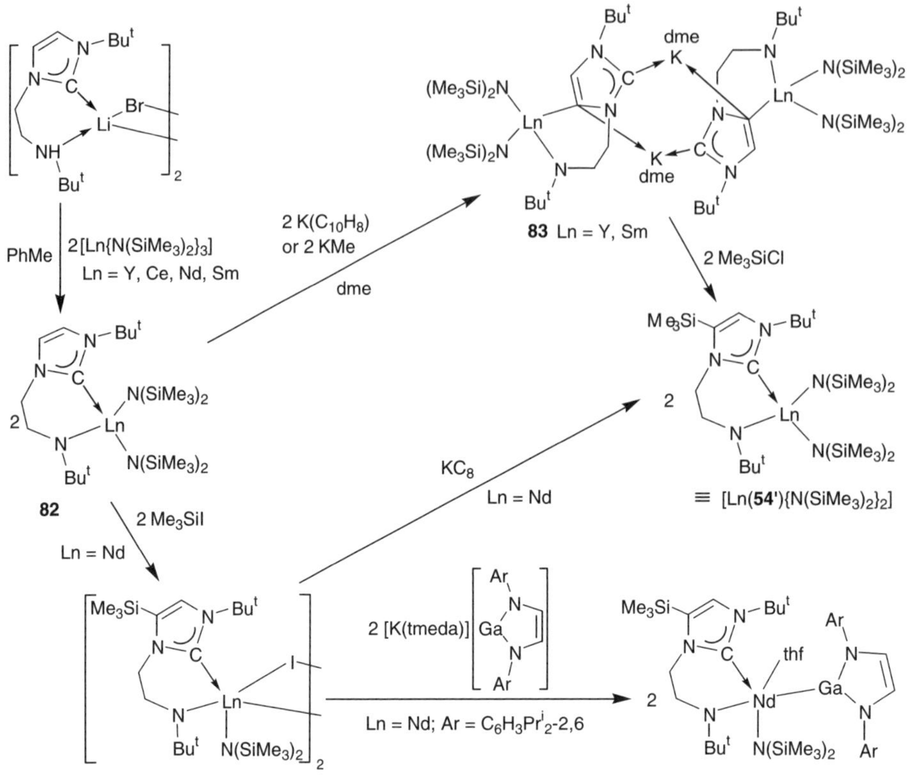

Scheme 4.9

The *N*-heterocyclic carbene-tethered amido ligand **54** has been used for the preparation of heteroleptic amides [Ln(**54**){$N(SiMe_3)_2$}$_2$] (**82**) (Ln = Y, Ce, Nd, Sm)[157] by the reaction of the corresponding amide [Ln{$N(SiMe_3)_2$}$_3$] with H(**54**)[157a] or LiBr·H(**54**).[157a,b] Reaction of cerium triiodide with K(**54**) failed to produce amido-Ce iodides, but treatment of [Ce(**54**){N-($SiMe_3)_2$}$_2$] with LiI gave the dimeric [{Ce(**54**)(μ-I)($N(SiMe_3)_2$)}$_2$].[157b] Deprotonation of the carbene moiety in complexes [Ln(**54**){$N(SiMe_3)_2$}$_2$] (Ln = Y, Sm) with K-naphthalene or KMe yielded hydrocarbyl-amido compounds **83** (Scheme 4.9), which with Me_3SiCl gave the silylated carbene-amido complex [Ln(**54′**){$N(SiMe_3)_2$}$_2$] (Ln = Y).[157c] Lanthanide complexes containing **54′** were readily available by the direct silylation of **82** with Me_3SiI;[157d] the resulting compound [{Nd(**54′**){$N(SiMe_3)_2$}(μ-I)}$_2$] yielded the first stable complex with an *f*-element-gallium bond.[157e] Complex [Y(**54**){$N(SiMe_3)_2$}$_2$] has been employed as a bifunctional catalyst for *D,L*-lactide ring-opening polymerisation.[157f]

4.3.6 Ln Amides as Precursors for Ln Coordination or Organometallic Compounds

Homoleptic silylamido-Ln^{III} compounds [Ln{$N(SiMe_3)_2$}$_3$] or the more readily displaceable [Ln{$N(SiHMe_2)_2$}$_3$(thf)$_n$] have been widely used as substrates for reaction with protic reagents H_nL. Some examples have already been described in Sections 4.3.3 and 4.3.5; others are summarised in Table 4.1[158–167] and Table 4.2.[160,168–174,194] From these, it is

Table 4.1 *Products of reactions of [Ln{N(SiMe$_3$)$_2$}$_3$] with protic compounds*

Ln in [Ln{N(SiMe$_3$)$_2$}$_3$]	Protic reagent	Product	Reference
Gd, La	HOSi(SiMe$_3$)$_3$	[Ln{OSi(SiMe$_3$)$_3$}$_3$(thf)$_n$] Ln = Gd, n = 2; Ln = La, n = 4	158
Y, La, Yb	(c-C$_6$H$_{11}$)$_7$Si$_7$O$_9$(OH)$_3$	[Ln(silasesquioxanate)(thf)$_2$] (Ln = Y, La) [Ln(silasesquioxanate)(pmdeta)] (Ln = Y, La, Yb)	159
Sc, Y, La	(C$_3$F$_7$)C(O)CH$_2$C(O)CMe$_3$ (≡fodH)	[Ln(fod)$_3$]	161
La, Ce, Pr, Nd	HSC$_6$H$_2$Bu^t$_3$-2,4,6	[Ln(SMes*)$_3$]	162a
Y	HN(EPPh$_2$)$_2$ (E = S, Se)	[Y{N(EPPh$_2$)$_2$}$_3$]	162b
Sm, Nd	N{(CH$_2$)$_2$N=CH(2-OH-3,5-Bu^t$_2$C$_6$H$_2$)}$_3$ (a Schiff base)	[LnN{(CH$_2$)$_2$N=CH(2-O-3,5-Bu^t$_2$C$_6$H$_2$)}$_3$]	163
Sm	MeSi{Si(Me)$_2$N(H)C$_6$H$_4$Me-4}$_3$	[Sm{(N(C$_6$H$_4$Me-4)SiMe$_2$)$_3$SiMe}]$_2$	154
Nd	Mesoporous MCM-41 (a Na aluminosilicate)	Nd{N(SiMe$_3$)$_2$}$_2$ … OSiMe$_3$ Na$^+$ … Nd{N(SiMe$_3$)$_2$} (schematic representation of surface)	164
La	(R)- [binaphthol with Si(C$_6$H$_3$Me$_2$-3,5)$_3$, OH, OH, Si(C$_6$H$_3$Me$_2$-3,5)$_3$]	(R)- [Si(C$_6$H$_3$Me$_2$-3,5)$_3$, O, La(thf)$_2$—N(SiMe$_3$)$_2$, O, Si(C$_6$H$_3$Me$_2$-3,5)$_3$] (catalyst for asym. hydroamination/cyclisation of aminoalkenes)	166
Ce	H[{N(SiMe$_3$)C(C$_6$H$_4$Bu^t-4)}$_2$CH] (≡LH)	[Ce{N(SiMe$_3$)$_2$}$_2$L]	165

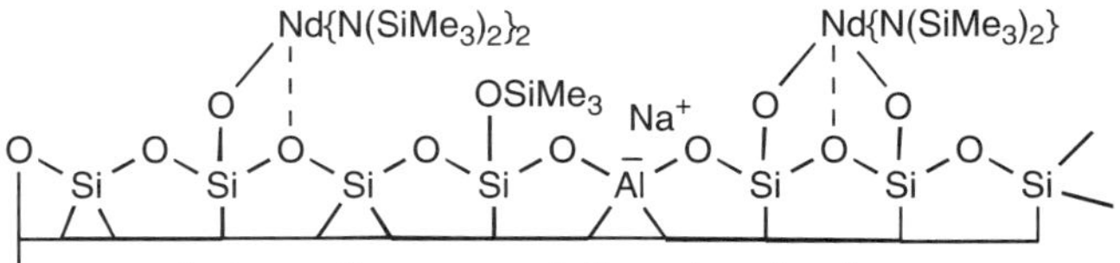

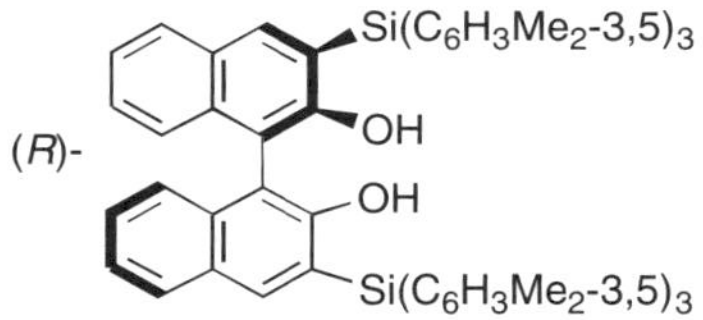

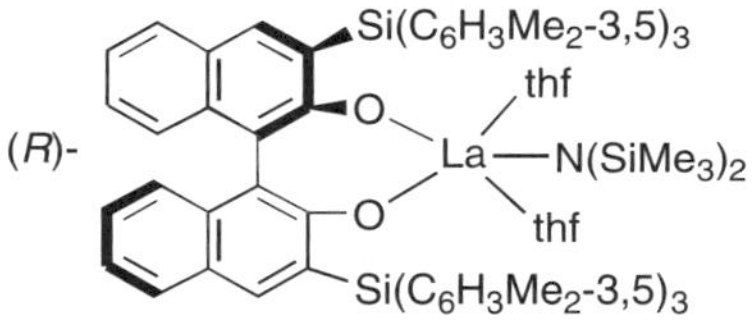

(continued)

Table 4.1 *(Continued)*

Ln in $[Ln\{N(SiMe_3)_2\}_3]$	Protic reagent	Product	Reference
La, Pr, Er, Lu	$C_5H_5(CH_2CH(OH)CH_2OBu^n)$ (≡LH_2)	$[Ln(L)\{N(SiMe_3)_2\}]$	167
La	$C_9H_7(CH_2CH(OH)CH_2OBu^n)$ (≡$L'H_2$)	$[La(L')\{N(SiMe_3)_2\}]$	167
Ce	$(c\text{-}C_6H_{11})_8Si_8O_{11}(OH)_2$	$[Ce^{IV}(silasesquioxanate)_2(py)_3]$	168
La, Nd, Sm, Yb	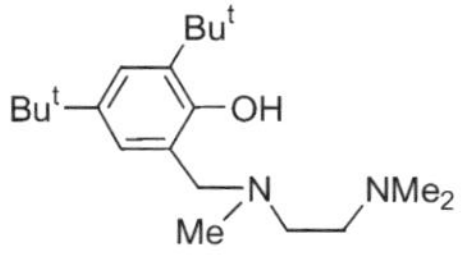 (≡ ArOH)	$[Ln(OAr)_2\{N(SiMe_3)_2\}]$ (catalysts for polymerisation of lactide and ε-caprolactone)	178

Table 4.2 *Products of reactions of $[Ln\{N(SiHMe_2)_2\}_3(thf)_n]$ with protic compounds*

Ln in $[Ln\{N(SiHMe_2)_2\}_3(thf)_n]$	Protic reagent	Product	Application	Ref.
Sc, Y, La	Mesoporous MCM-41	$[MCM\text{-}41]Ln\{N(SiHMe_2)_2\}_x(thf)_y$ (an immobilised catalyst)	Catalyst for hetero-Diels-Alder Danishefsky reactions)	161
Y, La	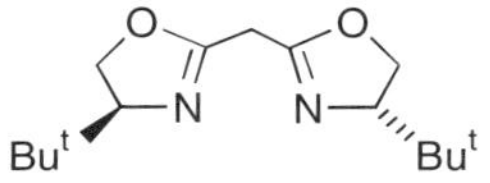 [a bis(oxazoline) ≡LH]	$[Y(L)\{N(SiHMe_2)_2\}_2]$ or $[Y(L)_2\{N(SiHMe_2)_2\}]$	Potential enantioselective catalyst	169
Y	$[p\text{-Bu}^t\text{-calix[4]arene}]H_4$	$[Y\{p\text{-Bu}^t\text{-calix[4]arene}(SiHMe_2)\}(thf)]_2$	A model oxo-surface	170
Y	$N\{CH_2C_6H_6Bu^t_2\text{-}3,5\text{-}OH\text{-}6\}_2CH_2CH_2OMe$ (≡ $L'H_2$)	$[Y(L')\{N(SiHMe_2)_2\}(thf)]$	Catalyst for ring-opening polymerisation of ε-caprolactone; sluggish for PMMA (isotactic)	171
Y	$-[CH_2N{=}CHC_6H_2Bu^t_2\text{-}3,5\text{-}OH\text{-}2]_2$ (a Schiff base) (≡ L^2H_2)	$[Y(L^2)\{N(SiHMe_2)_2\}(thf)]$	Immobilised on MCM-41; reacted with $HOSiPh_3$	172
Y	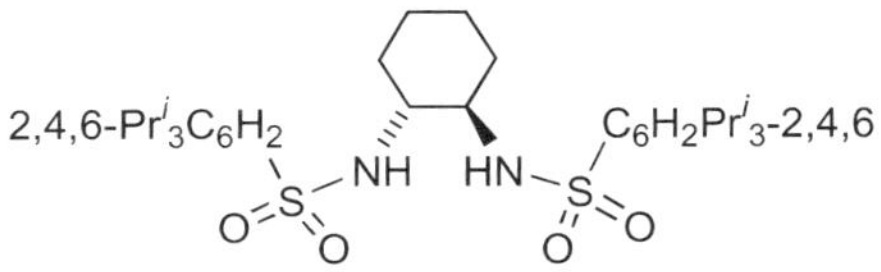 $2,4,6\text{-}Pr^i_3C_6H_2$ / $C_6H_2Pr^i_3\text{-}2,4,6$ (≡ L^3H_2)	$[Y(L^3)\{N(SiHMe_2)_2\}(thf)]$	Immobilised on MCM-41; reacted with $HOSiPh_3$	172
Y, La, Sm	(S)- 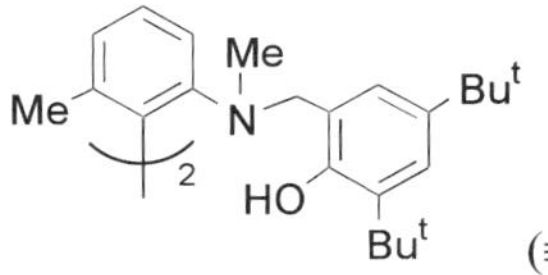 (≡ LH_2)	(S)-Λ-*cis*-$[Sm(L)\{N(SiHMe_2)_2\}]$ also Y, La	Catalysts for enantioselective hydroamination/ cyclisation	173

(continued)

Table 4.2 *(Continued)*

Ln in $[Ln\{N(SiHMe_2)_2\}_3(thf)_n]$	Protic reagent	Product	Application	Ref.
Y	PMO[MCM-41] PMO[KIT-5] PMO[SBA-1]	$PMO[MCM\text{-}41]Y[N(SiHMe_2)_2]_x(THF)_y$ $PMO[KIT\text{-}5]Y[N(SiHMe_2)_2]_x(THF)_y$ $PMO[SBA\text{-}1]Y[N(SiHMe_2)_2]_x(THF)_y$		174
Y	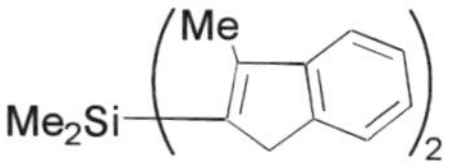 Me, Me_2Si, 2	$[Y\{(\eta^5\text{-}C_9H_5Me\text{-}2)_2SiMe_2\}\{N(SiHMe_2)_2\}]$		175
Y	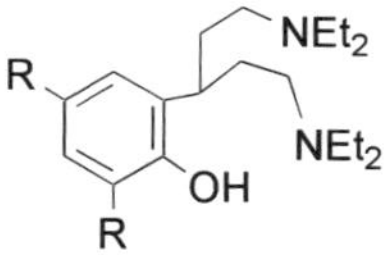 R, R, OH, NEt_2, NEt_2 $(\equiv L^{Me}H, R = Me$ $L^{Bu^t}H, R = Bu^t)$	$[Y(L^{Me})\{N(SiHMe_2)_2\}_2]$ $[Y(L^{Bu^t})\{N(SiHMe_2)_2\}_2]$	Catalysts for polymerisation of lactide and ε-caprolactone	194

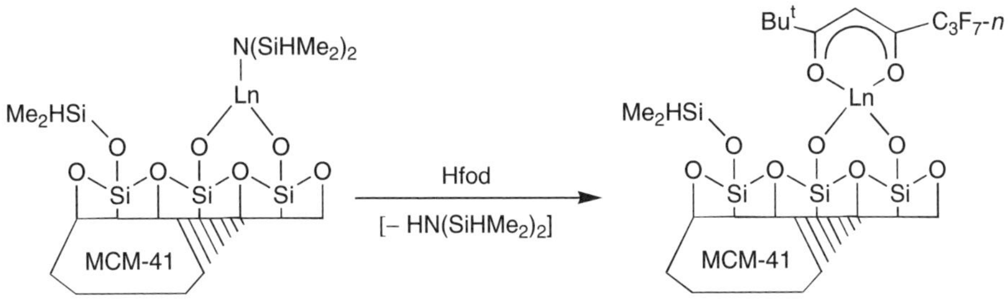

$$[Ln\{N(SiHMe_2)_2\}]\text{-}MCM\text{-}41 \xrightarrow[\text{[}- HN(SiHMe_2)_2\text{]}]{\text{Hfod}} [Ln(fod)]\text{-}MCM\text{-}41$$

Scheme 4.10

evident that in most cases the products have been molecular Ln^{III} compounds. But for others, as shown by Anwander, *et al.*, these Ln^{III} tris(amides) are useful precursors for organolanthanide-modified mesoporous silicates (derived from MCM-41, MCM-48, KIT-5 or SBA-1), which have the potential to act as silica-supported heterogeneous catalysts for various organic reactions. As an illustration, Scheme 4.10 is a representation showing such a surface-supported Ln amide and its further transformation into the β-diketonate 'fod'; while Equation (4.13) shows a catalytic hetero-Diels-Alder Danishefsky transformation.[161]

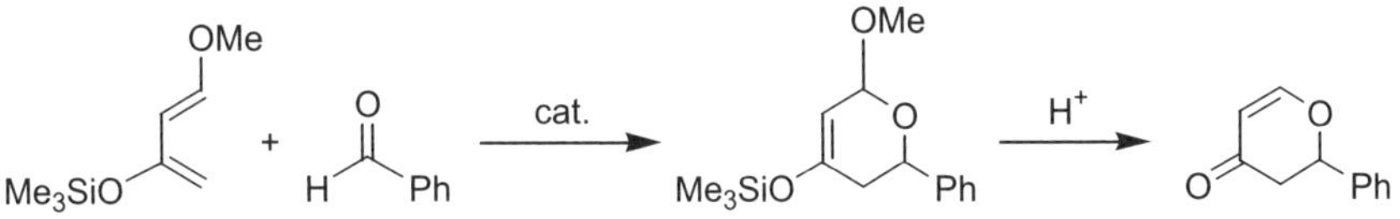

$$\text{Me}_3\text{SiO-diene(OMe)} + \text{PhCHO} \xrightarrow{\text{cat.}} \text{cycloadduct} \xrightarrow{H^+} \text{dihydropyranone} \quad (4.13)$$

Similar surface-supported amides have been derived from the Sm^{II} amide $Sm\{N(SiHMe)_2\}_2(thf)_x$ by grafting on MCM-41, MCM-48 or AS-200; further elaboration led to the formation of the corresponding Sm-fluorenone ketyl, which was shown to contain surface-confined ketyl radicals.[176] Treatment of $Sm\{N(SiHMe_2)_2\}(thf)_x$@MCM-41 with MeOH, $AlHBu^i_2$ or $Si(H)Me_2$-substituted indene gave surface-supported catalysts for methyl methacrylate polymerisation.[177]

Treatment of $[Ln\{N(SiMe_3)_2\}_3]$ with dialkylthiuram disulfide displaced the amido ligands and replaced them by dialkyldithiocarbamates $[Ln(\kappa^2\text{-}S_2CNR_2)_3]$ (Ln = Ce, Nd, Tm; R = Me, Et), as in Equation (4.14).[179] The cerium experiment had been designed as a route to a Ce^{IV} product; the displacement reaction was unexpected, but $[Ce(\kappa^2\text{-}S_2CNEt_2)_3]$ with oxygen gas gave $[Ce(\kappa^2\text{-}S_2CNEt_2)_4]$ and CeO_2.[179] From $[Ln\{N(SiMe_3)_2\}_3]$ (Ln = Dy, Ho, Er, Lu) and Li(bipy) in thf, the homoleptic neutral complexes $[Ln(bipy)_4]$ or the salts $[Li(thf)_4][Ln(bipy)_4]$ (depending on stoichiometry) were obtained.[180]

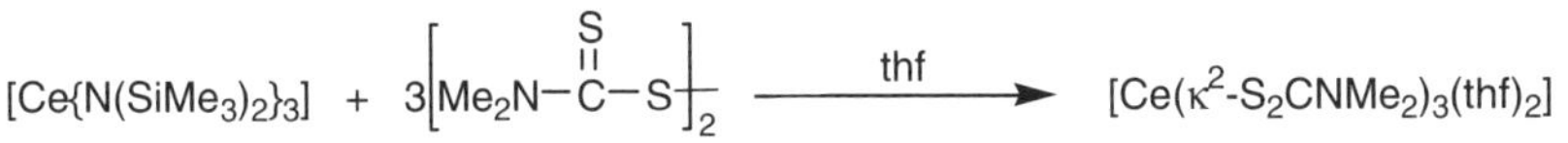

$$[Ce\{N(SiMe_3)_2\}_3] + 3[Me_2N\text{-}C(=S)\text{-}S]_2 \xrightarrow{\text{thf}} [Ce(\kappa^2\text{-}S_2CNMe_2)_3(thf)_2] \quad (4.14)$$

Reaction of $[Yb\{N(SiMe_3)_2\}_3(\mu\text{-}Cl)Li(thf)_3]$ or $[Yb\{N(SiMe_3)_2\}_3]$ and an amino-fuctionalised indene Ind'H (Ind'H = XC_9H_7, $XC_9H_6SiMe_3$, $XC_9H_6SiMe_2NBu^tH$, X = $CH_2CH_2NMe_2$) unexpectedly produced Yb^{II} bis(indenyl) compounds via homolysis of the Yb–N bond, while the Yb^{III} amides $[Yb(\eta^5\text{-}C_9H_7)_2\{N(SiMe_3)_2\}]$ and $[Yb\{(\eta^5\text{-}C_9H_6CH_2)_2\}\{N(SiMe_3)_2\}]$ were obtained with a non-fuctionalised indene.[183] Treatment of the appropriate [Ln{N(SiMe$_3$)$_2$}$_2$(thf)$_2$] (Ln = Sm, Eu, Yb) with the amino bis(phenol) (**84**)H$_2$ or (**85**)H$_2$ yielded [Sm^{II}Sm^{III}(**84**)$_4$], [Yb(**84**)(thf)$_3$], [{Eu(**84**)}$_2$], [Yb(**85**)(thf)$_2$] and [{Eu(**85**)}$_2$].[184]

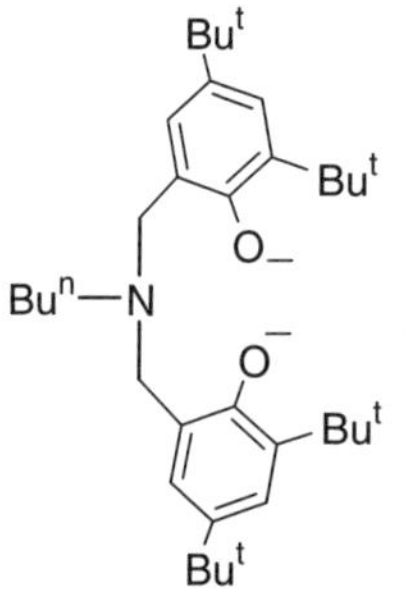

4.3.7 Applications as Materials or Catalysts

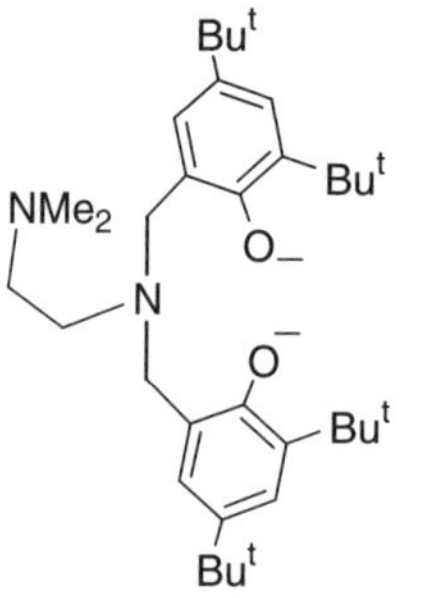

Various Ln amides have already been described in Sections 4.3.5 and 4.3.6 as catalysts for polymerisation of ethylene,[117a,138,141,143] hex-1-ene,[130c] isoprene,[130c] styrene,[134] methyl methacrylate or other polar monomers such as t-butyl acrylate or acrylonitrile,[123b,138,143,152,177] or ring-opening polymerisation catalysts for ε-caprolactone or lactide.[151,171,178] Catalysts for the following transformations have also been mentioned: hydrosilylation of α-olefins,[142] hydroamination/cyclisation of appropriate amino-alkenes or amino-alkynes,[117c,123a,133,144,147a,166,172,182] hydrophosphination/cyclisation of corresponding phosphino-alkenes or -alkynes,[145,182] and dimerisation of phenylacetylene.[116] Mesoporous silica-supported heterogeneous catalysts for various oganic reactions have been prepared and their surface properties studied.[160,164,172,174,176,177]

The cationic Y complex [Y(**86**)(CH$_2$SiMe$_3$)][B(C$_6$F$_5$)$_4$], derived from [Y(**86**)(CH$_2$SiMe$_3$)$_2$] and [HN(Me)$_2$Ph][B(C$_6$F$_5$)$_4$], was a competent catalyst for ethylene polymerisation.[181a] The trianionic ligand **87** has featured in the Ln^{III} complexes [Ln(**87**)] (Ln = Y, Eu, Yb) and [La(**87**)(thf)].[181b]

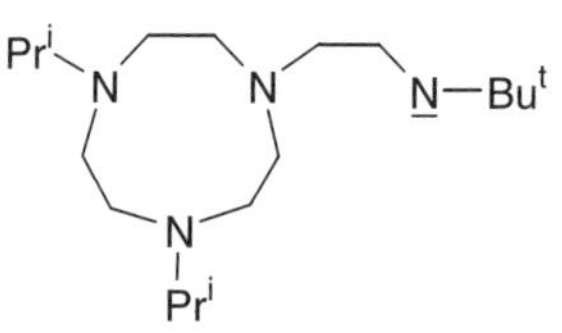

The chiral *ansa*-lanthanocene amides (*S*)-[Ln(L){N(SiMe$_3$)$_2$}] were prepared by the metathetical exchange reaction between (*S*)-[Ln(L)(μ-Cl)$_2$Li(dme)$_2$] and K[N(SiMe$_3$)$_2$] [Ln = Y, Sm, Lu; L = [(η^5-C$_5$H$_4$R*)Si(Me)$_2${η^5-(OHF)}]$^{2-}$ (R* = (−)-menthyl, OHF = octahydrofluorenyl)].[182] The crystallographically characterised yttrium amide was a precatalyst (activated by addition of Pr^{n}NH$_2$) for the highly enantioselective hydroamination/cyclisation or highly diasteroselective hydrophosphination/cyclisation of an appropriate amino- or phosphino-alkene H$_2$NCH$_2$CR$_2$(CH$_2$)$_n$CH=CH$_2$ (R = H or Me and n = 1 or 2)

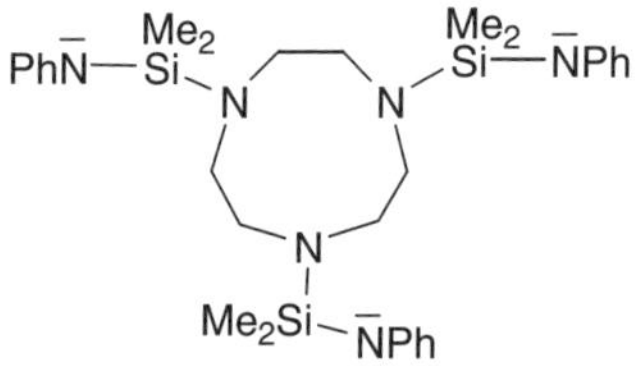

or $H_2PC(H)R(CH_2)_2CH{=}CH_2$ (R = H or Me).[182] Compounds **88** (Ln = Sc, Y, La, Nd, Sm, Lu), generated in situ from the corresponding (*R*)-H_2BINOL-{P(E)R_2}$_2$ (E = O and R = Et, Bu^t, CH_2Ph, Ph, $C_6H_3Me_2$-3,5, OEt; E = S and R = Ph) and [Ln{N(SiMe$_3$)$_2$}$_3$], have shown modest-to-good (Ln = Nd, E = O, R = Ph) enantioselectivity for the hydroamination/cyclisation of $CH_2{=}CHCH_2CMe_2CH_2NH_2$.[185a]

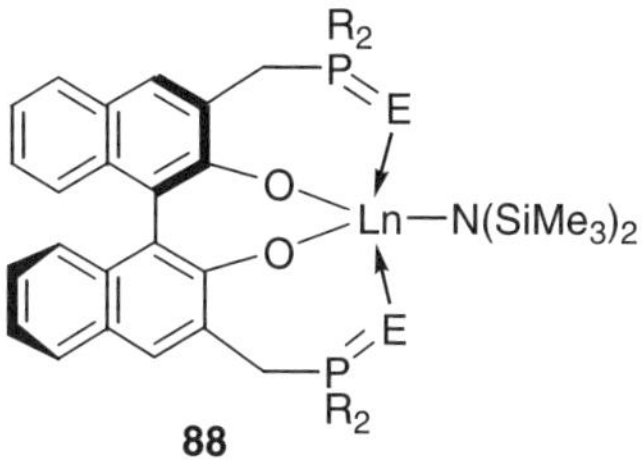

88

The regulation of ring-substituent diastereoselectivity in the intramolecular hydroamination/cyclisation of α-substituted aminodienes by constrained geometry Ln catalysts has been studied by DFT; the variation in the Ln^{3+} radius and the introduction of extra steric pressure at the substrate's α-position or the *N*-centre in $[Me_2Si(\eta^5\text{-}C_5Me_4)NBu^t]^{2-}$ (≡ **45a**, see Section 4.3.5) as in [Sm^{III}{N(SiMe$_3$)$_2$}(**45a**)] were identified as effective handles for tuning the selectivity.[185b]

The Ce^{III} amide [Ce{N(SiMe$_3$)$_2$}$_3$] has been used as a source for Sr:S:Ce thin film generation by atomic layer epitaxy.[86] The Lu analogue (with H_2O or O_3) has been employed for deposition of Lu silicate films.[91b]

4.3.8 Ln Complexes having 1,4-Disubstituted-1,4-Diazabutadiene Ligands, R_2-DAD

The redox behaviour of R_2-DAD arises from its energetically low-lying π^* orbitals making it available in appropriate metal (M) complexes (M having readily accessible oxidation states) to function as the neutral, monoanionic or dianionic ligand. Although only the last of these can properly be discussed as an amido ligand, the other options in Ln chemistry are also here considered. The R_2-DAD studies in Ln chemistry are largely due to the groups of Trifonov, Bochkarev and Schumann.[12c]

The first R_2-DAD-Ln complexes to be described were [Ln(Bu^t_2-DAD)$_3$] (Ln = Nd, Sm, Yb), obtained by Cloke, *et al.*, by metal vapour synthesis from the appropriate Ln and Bu^t_2-DAD;[186a] the crystalline samarium compound was shown to be the octahedral tris-chelate and formulated as [Sm^{III}(Bu^t_2-DAD$^{\bullet-}$)$_3$].[186b] The Yb compound was also prepared from 3K[Bu^t_2-DAD] and $YbCl_3$ in thf or [Yb($C_{10}H_8$)(thf)$_3$] + 3Bu^t_2-DAD,[187] and on the basis of magnetic measurements, was assigned as [Yb^{III}(Bu^t_2-DAD$^{\bullet-}$)$_3$] above 100 K but [Yb^{II}-(Bu^t_2-DAD$^{\bullet-}$)$_2$(Bu^t_2-DAD)] below that temperature.[187] Likewise this dichotomy was proposed for [Sm^{III}(Bu^t_2-DAB)$_2$(bipy)], synthesised from [Sm(bipy)$_4$] and Bu^t_2-DAD.[188]

Further studies focused on the bis(cyclopentadienyl)- or bis(indenyl)Yb^{II} as substrate or $YbCp_2Cl$. The ytterbocene(III) complex [$YbCp_2$(Bu^t_2-DAD$^{\bullet-}$)], characterised by magnetochemical and X-ray crystallographic techniques, was obtained from [$YbCp_2$(thf)$_2$] and Bu^t_2-DAD, or $YbCp_2Cl$ with K[Bu^t_2-DAD] or ½Na_2[Bu^t_2-DAD].[189a] Upon treatment with 9-fluorenone it yielded the pinacol-dimerisation product [{YbCp(μ-$OC_{13}H_8$-$C_{13}H_8O$)-

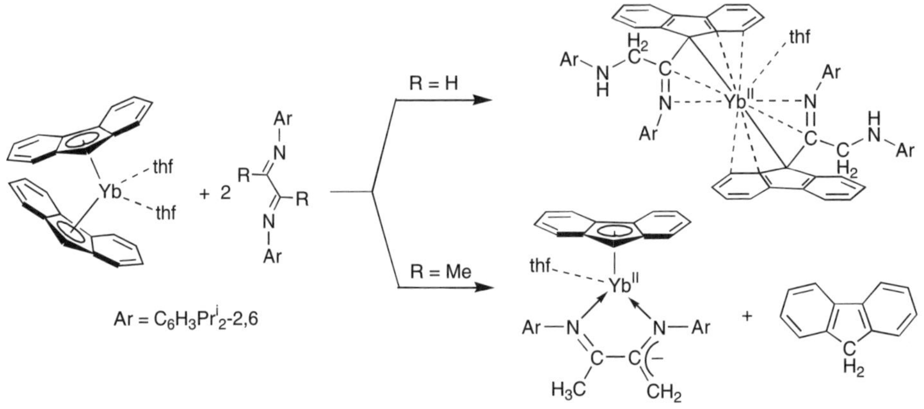

thf
thf
+ 2
Ar
N
R
R
N
Ar
$Ar = C_6H_3Pr^i{}_2$-2,6
R = H
R = Me
Ar
N
H
H_2
C
C
N
Ar
Yb^{II}
thf
Ar
N
C
C
H_2
H
N
Ar
thf
Yb^{II}
Ar—N
N—Ar
C—C
H_3C
CH_2
+
C
H_2

Scheme 4.11

$(thf)\}_2]$.[189b] A related redox reaction involved a bis(indenyl)Yb^{II} substrate, Equation (4.15);[190] attempted recrystallisation resulted in reversion to the starting materials.[190] Similarly, $[Yb^{III}(\eta^5\text{-}C_9H_7)_2\{(2,6\text{-}Pr^i{}_2C_6H_3)_2\text{-}DAD^{\bullet -}\}]$ was isolated from $[Yb(C_9H_7)_2(thf)_2]$

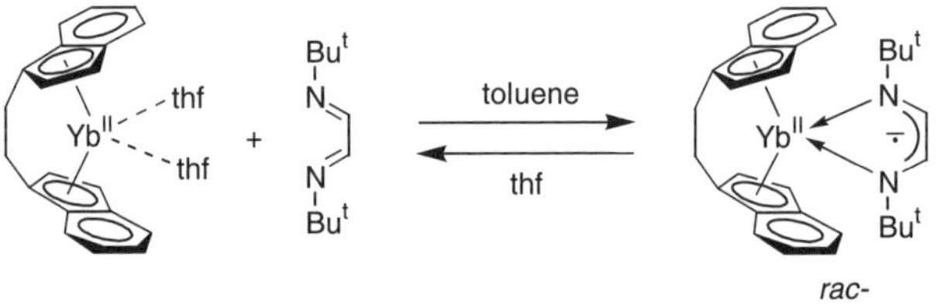

Yb^{II}
thf
thf
+
Bu^t
N
N
Bu^t
toluene
thf
Yb^{II}
Bu^t
N
N
Bu^t
rac-

(4.15)

and $(2,6\text{-}Pr^i{}_2C_6H_3)_2$-DAD.[191] The corresponding fluorenyl Yb^{II} substrate and two equivalents of $(2,6\text{-}Pr^i{}_2C_6H_3)_2$-DAD, or the 2,3-dimethyl derivative led to C–C-coupled or deprotonated products, Scheme 4.11.[192] The bis(indenyl)Yb^{II} complex was also the source of two binuclear, mixed valence (Yb^{II}/Yb^{III}) compounds, Scheme 4.12.[193a] The

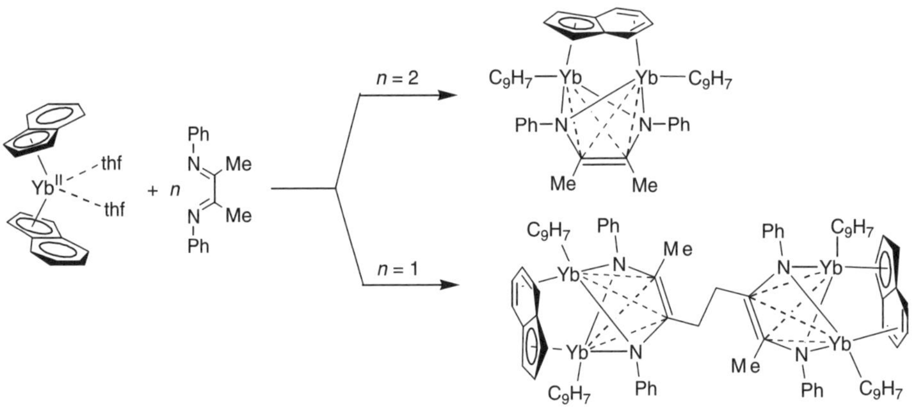

Yb^{II}
thf
thf
+ n
Ph
N
Me
N
Me
Ph
n = 2
n = 1
C_9H_7—Yb
Yb—C_9H_7
Ph—N
N—Ph
Me
Me
C_9H_7
Ph
Me
Yb
N
Yb
N
C_9H_7
Ph
Ph
C_9H_7
N
Yb
Me
N
Yb
Ph
C_9H_7

Scheme 4.12

ytterbocenes [Yb-$(Cp^x)_2(thf)_2$] ($Cp^x = \eta^5$-C_5H_4Me or η^5-C_5Me_4R, R = Me, H) behaved as a 1- or 2-electron reductant upon treatment with (2,6-$Pr^i{}_2C_6H_3)_2$-DAD giving [$Yb^{III}(\eta^5$-$C_5H_4Me)_2${(2,6-$Pr^i{}_2C_6H_3)_2$-$DAD^{\bullet -}$}] or [$Yb^{III}(\eta^5$-$C_5Me_4R)${(2,6-$Pr^i{}_2C_6H_3)_2$-DAD^{2-}}], respectively.[193b]

References

1. M. F. Lappert, P. P. Power, A. R. Sanger and R. C. Srivastava, *Metal and Metalloid Amides*, Horwood-Wiley, Chichester, 1980.
2. Recent papers: (a) S. Beaini, G. B. Deacon, M. Hilder, P. C. Junk and D. R. Turner, *Eur. J. Inorg. Chem.*, 2006, 3434; (b) C. F. Caro, M. F. Lappert and P. G. Merle, *Coord. Chem. Rev.*, 2001, **219–221**, 605; (c) S. K. Ibrahim, A. V. Khvostov, M. F. Lappert, L. Maron, L. Perrin, C. J. Pickett and A. V. Protchenko, *Dalton Trans.*, 2006, 2591; (d) S. Bambirra, A. Meetsma and B. Hessen, *Organometallics*, 2006, **25**, 3454; (e) A. A. Trifonov, D. M. Lyubov, E. A. Fedorova, G. K. Fukin, H. Schumann, S. Mühle, M. Hummert and M.N. Bochkarev, *Eur. J. Inorg. Chem.*, 2006, 351.
3. R. Anwander, *Lanthanide Amides,* in *Topics Current Chem*, 1996, **179**, 33.
4. (a) S. A. Cotton, Scandium, yttrium and the lanthanides, in *Comprehensive Coordination Chemistry II*, (eds J. A. McCleverty and T. J. Meyer). Vol. **3** (ed. G. F. R. Parkin), Elsevier, Oxford, 2004.(b) F. T. Edelmann, Scandium, yttrium and the lanthanide and actinide elements, excluding their zero oxidation state complexes, in *Comprehensive Organometallic Chemistry II* (eds E. W. Abel, F. G. A. Stone and G. Wilkinson,). Vol. **4** (ed. M. F. Lappert), Elsevier, Oxford, 1995.
5. (a) R. Kempe, *Angew. Chem., Int. Ed.*, 2000, **39**, 468; (b) H. C. Aspinall, *Chem. Rev.,* 2002, **102,** 1807.
6. F. T. Edelmann, D. M. M. Freckmann and H. Schumann, *Chem. Rev.*, 2002, **102** 1851.
7. (a) V. C. Gibson and S. K. Spitzmesser, *Chem. Rev.*, 2003, **103**, 283; (b) J. Gromada, J.-F. Carpentier and A. Montreux, *Coord. Chem. Rev.*, 2004, **248**, 397.
8. (a) S. Arndt and J. Okuda, *Chem. Rev.*, 2002, **102**, 1953; (b) G. B. Deacon, D. J. Evans, C.M. Forsyth and P. C. Junk, *Coord. Chem. Rev.*, 2007, **251**, 1699.
9. (a) J. Okuda, *Dalton Trans.*, 2003, 2367; (b) S. Arndt and J. Okuda, *Adv. Synth. Cat.*, 2005, **347**, 339; (c) K. C. Hultzsch, D. V. Gribkov and F. Hampel, *J. Organomet. Chem.*, 2005, **690**, 4441; (d) S. Hong and T. J. Marks, *Acc. Chem. Res.*, 2004, **37**, 673.
10. (a) K. Dehnicke and A. Greiner, *Angew. Chem., Int. Ed.*, 2003, **42**, 1340; (b) M. Karl, G. Seybert, W. Massa and K. Dehnicke, *Z. Anorg. Allg. Chem.*, 1999, **625**, 375; (c) M. Karl, A. Dashti-Mommertz, B. Neumüller and K. Dehnicke, *Z. Anorg. Allg. Chem.*, 1998, **624**, 355; (d) M. Karl, K. Harms and K. Dehnicke, *Z. Anorg. Allg. Chem.*, 1999, **625**, 1774; (e) M. Karl, K. Harms, G. Seybert, W. Massa, S. Fau, G. Frenking and K. Dehnicke, *Z. Anorg. Allg. Chem.*, 1999, **625**, 2055; (f) M. Karl, G. Seybert, W. Massa, K. Harms, S. Agarwal, R. Maleika, W. Stelter, A. Greiner, W. Heitz, B. Neumüller and K. Dehnicke, *Z. Anorg. Allg. Chem.*, 1999, **625**, 1301; (g) S. Agarwal, C. Mast, K. Dehnicke and A. Greiner, *Makromol. Rapid Commun.*, 2000, **21**, 195.
11. P. Mountford and B. D. Ward, *Chem. Commun.*, 2003, 1797.
12. (a) M. Marques, A. Sella and J. Takats, *Chem. Rev.*, 2002, **102**, 2137; (b) P. L. Arnold and S. T. Liddle, *Chem. Commun.*, 2006, 3959; (c) A. A. Trifonov, *Eur. J. Inorg. Chem.*, 2007, 3151.
13. W. E. Piers and D. J. H. Emslie, *Coord. Chem. Rev.*, 2003, **233–234**, 131.
14. M. R. Gagné, C. L. Stern and T. J. Marks, *J. Am. Chem. Soc.*, 1992, **114**, 275.
15. W. J. Evans, R. Anwander, R. J. Doedens and J. W. Ziller, *Angew. Chem., Int. Ed.*, 1994, **33**, 1641.

16. H. C. Aspinall, S. R. Moore and A. K. Smith, *J. Chem. Soc., Dalton Trans.*, 1993, 993.
17. W. J. Evans, R. Anwander and J. W. Ziller, *Oranometallics*, 1995, **14**, 1107.
18. (a) H. C. Aspinall and M. R. Tillotson, *Polyhedron*, 1994, **13**, 3229; (b) W. J. Evans, R. Anwander, J. W. Ziller and S. I. Khan, *Inorg. Chem.*, 1995, **34**, 5927.
19. W. J. Evans, G. Kociok-Köhn, V. S. Leong and J. W. Ziller, *Inorg. Chem.*, 1992, **31**, 3592.
20. W. J. Evans, M. A. Ansari, J. W. Ziller and S. I. Khan, *Inorg. Chem.*, 1996, **35**, 5435.
21. (a) J. Guan, S. Jin, Y. Lin and Q. Shen, *Organometallics*, 1992, **11**, 2483; (b) L. Mao, Q. Shen and S. Jin, *Polyhedron*, 1994, **13**, 1023; (c) W.-K. Wong, L. Zhang, F. Xue and T. C. W. Mak, *Polyhedron*, 1996, **15**, 345.
22. F. T. Edelmann, A. Steiner, D. Stalke, J. W. Gilje, S. Kagner and H. Håkansson, *Polyhedron*, 1994, **13**, 539; and refs. therein.
23. M. Westerhausen, M. Hartmann, A. Pfitzner and W. Schwarz, *Z. Anorg. Allg. Chem.*, 1995, **621**, 837.
24. (a) A. D. Frankland, P. B. Hitchcock, M. F. Lappert and G. A. Lawless, *J. Chem. Soc., Chem. Commun.*, 1994, 2435; (b) P. B. Hitchcock, M. F. Lappert, R. G. Smith, R. A. Bartlett and P. P. Power, *J. Chem. Soc., Chem Commun.*, 1988, 1007.
25. T. Fjeldberg and R. A. Andersen, *J. Mol. Struct.*, 1985, **128**, 49; **129**, 93.
26. W. Scherer, Thesis, Technical University, Munich, 1994.
27. H. C. Aspinall, D. C. Bradley, M. B. Hursthouse, K. D. Sales, N. P. C. Walker and B. Hussain, *J. Chem. Soc., Dalton Trans.*, 1989, 623.
28. H. J. Heeres, A. Meetsma, J. H. Teuben and R. D. Rogers, *Organometallics*, 1989, **8**, 2637.
29. P. Shao, D. J. Berg and G. W. Bushnell, *Inorg. Chem.*, 1994, **33**, 6334.
30. W. J. Evans, R. Anwander, U. H. Berlekamp and J. W. Ziller, *Inorg. Chem.*, 1995, **34**, 3583.
31. H. C. Aspinall, D. C. Bradley, M. B. Hursthouse, K. D. Sales and N. P. C. Walker, *J. Chem. Soc., Chem. Commun.*, 1985, 1585.
32. H. Schumann, J. Winterfeld, L. Esser and G. Kociok-Köhn, *Angew. Chem. Int. Ed.*, 1993, **32**, 1208.
33. (a) M. R. Gagné, L. Brard, V. P. Conticello, M. A. Giardello, C. L. Stern and T. J. Marks, *Organometallics*, 1992, **11**, 2003; (b) M. A. Giardello, V. P. Conticello, L. Brard, M. Sabat, A. L. Rheingold, C. L. Stern and T. J. Marks *J. Am. Chem. Soc.*, 1994, **116**, 10212.
34. W. A. Herrmann, R. Anwander, F. C. Munck, W. Scherer, V. Dufaud, N. W. Huber and G. R. J. Artus., *Z. Naturforsch. Teil B*, 1994, **49**, 1789.
35. H. Schumann, J. Winterfeld, E. C. E. Rosenthal, H. Hemling and L. Esser, *Z. Anorg. Allg. Chem.*, 1995, **621**, 122.
36. (a) T. D. Tilley, R. A. Andersen and A. Zalkin, *Inorg. Chem.*, 1984, **23**, 2271; (see, also T. D. Tilley, J. M. Boncella, D. J. Berg, C. Burns and R. A. Andersen, *Inorg. Synth.*, 1990, **27**, 146); (b) T. D. Tilley, R. A. Andersen and A. Zalkin, *J. Am. Chem. Soc.* 1982, **104**, 3725; (c) J. R. van den Hende, P. B. Hitchcock, S. A. Holmes, M. F. Lappert and S. Tian, *J. Chem. Soc., Dalton Trans.*, 1995, 3933.
37. A. G. Avent, M. A. Edelman, M. F. Lappert and G. A. Lawless, *J. Am. Chem. Soc.*, 1989, **111**, 3423.
38. (a) T. D. Tilley, PhD Thesis, University of California, Berkeley, 1982; (b) J. M. Boncella and R. A. Andersen, *Organometallics*, 1985, **4**, 205.
39. Yu. F. Rad'kov, E. A. Fedorova, S. Ya. Khorshev, G. S. Kalinina, M. N. Bochkarev and G. A. Razuvaev, *J. Gen. Chem. USSR*, 1985, **55**, 1911.
40. (a) B. Çetinkaya, P. B. Hitchcock, M. F. Lappert and R. A. Smith, *J. Chem. Soc., Chem. Commun.*, 1992, 932; (b) J. R. van den Hende, P. B. Hitchcock and M. F. Lappert, *J. Chem. Soc., Dalton Trans.*, 1995, 2251.
41. W. J. Evans, D. K. Drummond, H. Zhang and J. L. Atwood, *Inorg. Chem.*, 1988, **27**, 575.
42. J. R. van den Hende, P. B. Hitchcock, S. A. Holmes and M. F. Lappert, *J. Chem. Soc., Dalton Trans.*, 1995, 1435.

43. (a) J. M. Boncella, PhD Thesis University of California, Berkeley, 1982, cited in ref. 3; (b) A. F. England, T. M. Frankcom, C. J. Burns and S. L. Buchwald, unpublished work reported at the 207th ACS National Meeting, San Diego, 1994.
44. A. Recknagel, A. Steiner, S. Brooker, D. Stalke and F. T. Edelmann, *J. Organomet. Chem.*, 1991, **415**, 315.
45. H. Schumann, P. R. Lee and J. Loebel, *Chem. Ber.*, 1989, **122**, 1897.
46. W. J. Evans, S. E. Foster and J. W. Ziller, cited in ref. 3.
47. M. E. Thompson, S. M. Baxter, A. R. Bulls, B. J. Burger, M. C. Nolan, B. D. Santarsiero, W. P. Schaefer and J. E. Bercaw, *J. Am. Chem. Soc.*, 1987, **109**, 203.
48. (a) M. D. Fryzuk, T. S. Haddad and S. J. Rettig, *Organometallics*, 1992, **11**, 2967; (b) M. D. Fryzuk, T. S. Haddad and S. J. Rettig, *Organometallics*, 1991, **10**, 2026.
49. Y. Mu, W. E. Piers, M. A. MacDonald and M. J. Zaworotko, *Can. J. Chem.*, 1995, **73**, 2233.
50. P. J. Shapiro, W. P. Schaefer, J. A. Labinger, J. E. Bercaw and W. D. Cotter, *J. Am. Chem. Soc.*, 1994, **116**, 4623, and refs. therein.
51. B. K. Campion, R. H. Heyn and T. D. Tilley, *J. Am. Chem. Soc.*, 1990, **112**, 2011.
52. L. Hasinoff, J. Takats, X. W. Zhang, A. H. Bond and R. D. Rogers, *J. Am. Chem. Soc.*, 1994, **116**, 8833.
53. H. Schumann, P. R. Lee and A. Dietrich, *Chem. Ber.*, 1990, **123**, 1331.
54. H. Schumann, J. Winterfeld, H. Hemling and N. Kuhn, *Chem. Ber.*, 1993, **126**, 2657.
55. (a) W. J. Evans, G. W. Rabe and J. W. Ziller, *Organometallics*, 1994, **13**, 1641; (b) G. B. Deacon, C. M. Forsyth and B. M. Gatehouse, *Aust. J. Chem.*, 1990, **43**, 795.
56. W. J. Evans, S. L. Gonzales and J. W. Ziller, *J. Am. Chem. Soc.*, 1994, **116**, 2600.
57. J. Scholz, A. Scholz, R. Weimann, C. Janiak and H. Schumann, *Angew. Chem., Int. Ed.*, 1994, **33**, 1171.
58. L. N. Bochkarev, S. B. Shustov, T. B. Guseva and S. F. Zhil'tsov, *J. Gen. Chem. USSR*, 1988, **58**, 819.
59. (a) M. J. Booij, N. H. Kiers, H. J. Heeres and J. H. Teuben, *J. Organomet. Chem.*, 1989, **364**, 79; (b) S. D. Stults, R. A. Andersen and A. Zalkin, *Organometallics*, 1990, **9**, 115 and 1623.
60. G. A. Razuvaev, G. S. Kalinina and E. A. Fedorova, *J. Organomet. Chem.*, 1980, **190**, 157.
61. H. C. Aspinall, S. R. Moore and A. K. Smith, *J. Chem. Soc., Dalton Trans.*, 1992, 153.
62. H. A. Stecher, A. Sen and A. L. Rheingold, *Inorg. Chem.*, 1989, **28**, 3280.
63. (a) P. B. Hitchcock, M. F. Lappert and A. Singh, *J. Chem. Soc., Chem Commun.*, 1983, 1499; (b) M. F. Lappert and A. Singh, *Inorg. Synth.*, 1990, **27**, 164.
64. M. J. McGeary, P. S. Coan, K. Folting, W. E. Streib and K. G. Caulton, *Inorg. Chem.*, 1989, **28**, 3283.
65. D. R. Cary and J. Arnold, *J. Am. Chem. Soc.*, 1993, **115**, 2520.
66. E. A. Fedorova, G. S. Kalinina, M. N. Bochkarev and G. A. Razuvaev, *J. Gen. Chem. USSR*, 1982, **52**, 1041,
67. W. A. Herrmann, R. Anwander, M. Kleine and W. Scherer, *Chem. Ber.*, 1992, **125**, 1971.
68. J. R. van den Hende, P. B. Hitchcock and M. F. Lappert, *J. Chem. Soc., Chem. Commun.*, 1994, 1413.
69. G. B. Deacon, P. B. Hitchcock, S. A. Holmes, M. F. Lappert, P. MacKinnon and R. H. Newnham, *J. Chem. Soc., Chem. Commun.*, 1989, 935.
70. M. N. Bochkarev, E. A. Fedorova, Yu. F. Rad'kov, S. Ya. Khorshev, G. S. Kalinina and G. A. Razuvaev, *J. Organomet. Chem.*, 1983, **258**, C29.
71. A. C. Greenwald, W. S. Rees, Jr. and U. W. Lay, in *Rare Earth Doped Semiconductors* (eds. G. S. Pomrenke, D. B. Klein, D. W. Langers). MRS, Pittsburgh, 1993.
72. R. L. LaDuca and P. T. Wolczanski, *Inorg. Chem.*, 1992, **31**, 1311.
73. (a) M. R. Gagné and T. J. Marks, *J. Am. Chem. Soc.*, 1989, **111**, 4108; (b) M. R. Gagné, S. P. Nolan and T. J. Marks, *Organometallics*, 1990, **9**, 1716; (c) M. A. Giardello, V. P. Conticello, L. Brard, M. R. Gagné and T. J. Marks, *J. Am. Chem. Soc.*, 1994, **116**, 10241.

74. Y. Li, P.-F. Fu and T. J. Marks, *Organometallics*, 1994, **13**, 439.
75. W. J. Evans and H. Katsumata, *Macromolecules*, 1994, **27**, 2330.
76. (a) I. Mukerji, A. L. Wayda, G. Dabbagh and S. H. Bertz, *Angew. Chem., Int. Ed.*, 1986, **25**, 760; (b) M. T. Reetz, H. Haning and S. Stanchev, *Tetrahedron Lett.*, 1992, **33**, 6963.
77. (a) S. Kraut, J. Magull, U. Schaller, M. Karl, K. Harms and K. Dehnicke, *Z. Anorg. Allg. Chem.*, 1998, **624**, 1193; (b) H.-S. Chan, H.-W. Li and Z. Xie, *Chem. Commun.*, 2002, 652; (c) M. N. Bochkarev, G. V. Khoroshenkov, D. M. Kuzyaev, A. A. Fagin, M. E. Burin, G. K. Fukin, E. V. Baranov and A. A. Maleev, *Inorg. Chim. Acta*, 2006, **359**, 3315.
78. R. A. Layfield, A. Bashall, M. McPartlin, J. M. Rawson and D. S. Wright, *Dalton Trans.*, 2006, 1660.
79. P. B. Hitchcock, M. F. Lappert and A. V. Protchenko, *Chem. Commun.*, 2006, 3546.
80. R. K. Minhas, Y. Ma, J.-I. Song and S. Gambarotta, *Inorg. Chem.*, 1996, **35**, 1866.
81. K. Aparna, M. Ferguson and R. G. Cavell, *J. Am. Chem. Soc.*, 2000, **122**, 726.
82. P. B. Hitchcock, Q.-G. Huang, M. F. Lappert and X.-H. Wei, *J. Mater. Chem.*, 2004, **14**, 3266.
83. S. D. Daniel, J.-S. Lehn, J. D. Korp and D. M. Hoffman, *Polyhedron*, 2006, **25**, 205.
84. W.-K. Wong, L. Zhang, F. Xue and T. C. W. Mak, *Polyhedron*, 1997, **16**, 2013.
85. (a) D. R. Click, B. L. Scott and J. G. Watkin, *Chem. Commun.*, 1999, 633; (b) K. Müller-Buschbaum and C. C. Quitmann, *Inorg. Chem.*, 2006, **45**, 2678.
86. W. S. Rees, Jr., O. Just and D. S. Van Derveer, *J. Mater. Chem.*, 1999, **9**, 249.
87. S. A. Schuetz, V. W. Day, R. D. Sommer, A. L. Rheingold and J. A. Belot, *Inorg. Chem.*, 2001, **40**, 5292.
88. U. Baisch, S. Pagano, M. Zeuner, J. Schmedt auf der Günne, O. Oeckler and W. Schnick, *Organometallics*, 2006, **25**, 3027.
89. L. Maron and O. Eisenstein, *New J. Chem.*, 2000, **25**, 255.
90. (a) S. Jank, J. Hanss, H. Reddmann, H.-D. Amberger and N. M. Edelstein, *Z. Anorg. Allg. Chem.*, 2002, **628**, 1355; (b) S. Jank, H. Reddmann, C. Apostolidis and H.-D. Amberger, *Z. Anorg. Allg. Chem.*, 2007, **633**, 398.
91. (a) M. Niemeyer, *Z. Anorg. Allg. Chem.*, 2002, **628**, 647; (b) G. Scarel, C. Wiemer, M. Fanciulli, I. L. Fedushkin, G. K. Fukin, G. A. Domrachev, Y. Lebedinskii, A. Zenkevich and G. Pavia, *Z. Anorg. Allg. Chem.* 2007, **633**, 2097.
92. J. Collin, N. Giuseppone, N. Jaber, A. Domingos, L. Maria and I. Santos, *J. Organomet. Chem.*, 2001, **628**, 271.
93. P. B. Hitchcock, A. G. Hulkes and M. F. Lappert, *Inorg. Chem.*, 2004, **43**, 1031.
94. (a) D. J. Berg and R. A. L. Gendron, *Can. J. Chem.*, 2000, **78**, 454; (b) H.-X. Li, Q.-F. Xu, J.-X. Chen, M.-L. Cheng, Y. Zhang, W.-H. Zhang, J.-P. Lang and Q. Shen, *J. Organomet. Chem.*, 2004, **689**, 3438.
95. (a) M. Karl, G. Seybert, W. Massa, S. Agarwal, A. Greiner and K. Dehnicke, *Z. Anorg. Allg. Chem.*, 1999, **625**, 1405; (b) M. Karl, B. Neumüller, G. Seybert, W. Massa and K. Dehnicke, *Z. Anorg. Allg. Chem.*, 1997, **623**, 1203.
96. (a) A. A. Trifonov, D. M. Lyubov, E. A. Fedorova, G. G. Skvortsov, G. K. Fukin, Yu. A. Kurskii and M. N. Bochkarev, *Russ. Chem. Bull.*, 2006, **55**, 435; (b) G. B. Deacon, C. M. Forsyth, P. C. Junk and J. Wang, *Inorg. Chem.*, 2007, **46**, 10022.
97. (a) Y. Zhou, G. P. A. Yap and D. S. Richeson, *Organometallics*, 1998, **17**, 4387; (b) G. W. Rabe, C. S. Strissel, L. M. Liable-Sands, T. E. Concolino and A. L. Rheingold, *Inorg. Chem.*, 1999, **38**, 3446; (c) L. J. Bowman, K. Izod, W. Clegg and R. W. Harrington, *Organometallics,* 2006, **25**, 2999; (d) K. Izod, S. T. Liddle, W. Clegg and R. W. Harrington, *Dalton Trans.*, 2006, 3431.
98. D. C. Bradley, J. S. Ghotra, F. A. Hart, M. B. Hursthouse and P. R. Raithby, *J. Chem. Soc., Dalton Trans.*, 1977, 1166.

99. (a) W. J. Evans, G. Zucchi and J. W. Ziller, *J. Am. Chem. Soc.*, 2003, **125**, 10; (b) L. E. Roy and T. Hughbanks, *J. Am. Chem. Soc.*, 2006, **128**, 568.
100. (a) W. J. Evans, D. S. Lee and J. W. Ziller, *J. Am. Chem. Soc.*, 2004, **126**, 454; (b) W. J. Evans, D. B. Rego and J. W. Ziller, *Inorg. Chem.*, 2006, **45**, 3437.
101. E. D. Brady, D. L. Clark, D. W. Keogh, B. L. Scott and J. G. Watkin, *J. Am. Chem. Soc.*, 2002, **124**, 7007.
102. (a) R. Anwander, O. Runte, J. Eppinger, G. Gerstberger, E. Herdtweck and M. Spiegler, *J. Chem. Soc., Dalton Trans.*, 1998, 847; (b) H. M. Dietrich, C. Meermann, K. W. Törnroos and R. Anwander, *Organometallics*, 2006, **25**, 4316.
103. (a) H. Schumann, J. Winterfeld, E. C. E. Rosenthal, H. Hemling and L. Esser, *Z. Anorg. Allg. Chem.*, 1995, **621**, 122; (b) L. Zhou, J. Wang, Y. Zhang, Y. M. Yao and Q. Shen, *Inorg. Chem.*, 2007, **46**, 5763; (c) B. Zhao, H. H. Li, Q. Shen, Y. Zhang, Y. M. Yao and C. R. Lu, *Organometallics*, 2006, **25**, 1824.
104. D. R. Click, B. L. Scott and J. G. Watkin, *Acta. Crystallogr. Sect. C.*, 2000, **56**, 1095.
105. (a) W. S. Rees, Jr., O. Just and H. Schumann and R. Weimann, *Angew. Chem., Int. Ed.*, 1996, **35**, 419; (b) O. Just and W. S. Rees, Jr., *Inorg. Chem.*, 2001, **40**, 1751.
106. (a) G. B. Deacon, E. E. Delbridge, B. W. Skelton and A. H. White, *Eur. J. Inorg. Chem.*, 1998, 543; (b) *Idem, Angew. Chem., Int. Ed.*, 1998, **37**, 543.
107. G. B. Deacon, G. D. Fallon, C. M. Forsyth, H. Schumann and R. Weimann, *Chem. Ber. Recl.*, 1997, **130**, 409.
108. G. B. Deacon, C. M. Forsyth and N. M. Scott, *Eur. J. Inorg. Chem.*, 2000, 2501.
109. (a) G. B. Deacon, E. E. Delbridge, B. W. Skelton and A. H. White, *Eur. J. Inorg. Chem.*, 1999, 751; (b) I. Nagl, W. Scherer, M. Tafipolsky and R. Anwander, *Eur. J. Inorg. Chem.*, 1999, 1405.
110. P. B. Hitchcock, A. V. Khvostov, M. F. Lappert and A. V. Protchenko, *J. Organomet. Chem.*, 2002, **647**, 198.
111. C. Ruspic, J. Spielmann and S. Harder, *Inorg. Chem.*, 2007, **46**, 5320.
112. (a) G. B. Deacon and C. M. Forsyth, *Chem. Commun.*, 2002, 2522; (b) G. B. Deacon, C. M. Forsyth and P. C. Junk, *Eur. J. Inorg. Chem.*, 2005, 817.
113. C. Morton, N. W. Alcock, M. R. Lees, I. J. Munslow, C. J. Sanders and P. Scott, *J. Am. Chem. Soc.*, 1999, **121**, 11255.
114. O. Eisenstein, P. B. Hitchcock, A. G. Hulkes, M. F. Lappert and L. Maron, *Chem. Commun.*, 2001, 1560.
115. A. G. Avent, P. B. Hitchcock, A. V. Khvostov, M. F. Lappert, Z. Li and A. V. Protchenko, unpublished work.
116. S. Bambirra, S. J. Boot, D. van Leusen, A. Meetsma and B. Hessen, *Organometallics*, 2004, **23**, 1891.
117. (a) S. Bambirra, D. van Leusen, A. Meetsma, B. Hessen and J. H. Teuben, *Chem. Commun.*, 2001, 637; (b) G. C. J. Tazelaar, S. Bambirra, D. van Leusen, A. Meetsma, B. Hessen and J. H. Teuben, *Organometallics*, 2004, **23**, 936; (c) S. Bambirra, H. Tsurugi, D. van Leusen and B. Hessen, *Dalton Trans.*, 2006, 1157.
118. (a) R. Duchateau, T. Tuinstra, E. A. C. Brussee, A. Meetsma, P. T. van Duijnen and J. H. Teuben, *Organometallics*, 1997, **16**, 3511; (b) R. Duchateau, E. A. C. Brussee, A. Meetsma and J. H. Teuben, *Organometallics*, 1997, **16**, 5506.
119. (a) M. D. Fryzuk, G. Giesbrecht and S. J. Rettig, *Organometallics*, 1996, **15**, 3329; (b) M. D. Fryzuk, G. Giesbrecht and S. J. Rettig, *Can. J. Chem.*, 2000, **78**, 1003; (c) M. D. Fryzuk, P. H. Yu and B. O. Patrick, *Can. J. Chem.*, 2001, **79**, 1194.
120. A. G. Avent, F. G. N. Cloke, B. R. Elvidge and P. B. Hitchcock, *Dalton Trans.*, 2004, 1083.
121. A. A. S. Shah, H. Dorn, H. W. Roesky, P. Lubini and H.-G. Schmidt, *Inorg. Chem.*, 1997, **36**, 1102.
122. (a) M. E. G. Skinner, B. R. Tyrrell, B. D. Ward and P. Mountford, *J. Organomet. Chem.*, 2002, **647**, 145; (b) M. E. G. Skinner and P. Mountford, *J. Chem. Soc., Dalton Trans.*, 2002, 1694.

123. (a) K. C. Hultzsch, F. Hampel and T. Wagner, *Organometallics*, 2004, **23**, 2601; (b) F. Estler, G. Eickerling, E. Herdtweck and R. Anwander, *Organometallics*, 2003, **22**, 1212.
124. C. F. Pi, Z. Y. Zhu, L. H. Weng, Z. X. Chen and X. G. Zhou, *Chem. Commun.*, 2007, 2190.
125. K. C. Jantunen, B. L. Scott, J. Hay, J. C. Gordon and J. L. Kiplinger, *J. Am. Chem. Soc.*, 2006, **128**, 6322.
126. (a) E. W. Y. Wong, A. K. Das, M. J. Katz, Y. Nishimura, R. J. Batchelor, M. Onishi and D. B. Leznoff, *Inorg. Chim. Acta*, 2006, **359**, 2826; (b) H. Noss, M. Oberthür, C. Fischer, W. P. Kretschmer and R. Kempe, *Eur. J. Inorg. Chem.*, 1999, 2283; (c) H. Noss, W. Baumann, R. Kempe, T. Irrgang and A. Schulz, *Inorg. Chim. Acta*, 2003, **345**, 130.
127. D. D. Graf, W. M. Davis and R. R. Schrock, *Organometallics*, 1998, **17**, 5820.
128. T. I. Gountchev and T. D. Tilley, *Organometallics*, 1999, **18**, 2896.
129. P. B. Hitchcock, Q. Huang, M. F. Lappert, X.-H. Wei and M. Zhou, *Dalton Trans.*, 2006, 2991.
130. (a) Y. Makioka, Y. Taniguchi, Y. Fujiwara, K. Takaki, Z. Hou and Y. Wakatsuki, *Organometallics*, 1996, **15**, 5476; (b) Z. Hou, C. Yoda, T. Koizumi, M. Nishimura, Y. Wakatsuki, S. Fukuzawa and J. Takats, *Organometallics*, 2003, **22**, 3586; (c) Y. Luo, M. Nishimura and Z. Hou, *J. Organomet. Chem.*, 2007, **692**, 534.
131. (a) M. D. Fryzuk, J. B. Love and S. J. Rettig, *J. Am. Chem. Soc.*, 1997, **119**, 9071; (b) M. D. Fryzuk, L. Jafarpour, F. M. Kerton, J. B. Love and S. J. Rettig, *Angew. Chem., Int. Ed.*, 2000, **39**, 767.
132. K. C. Hultzsch, T. P. Spaniol and J. Okuda, *Organometallics*, 1998, **17**, 485.
133. S. Tian, V. M. Arredondo, C. L. Stern and T. J. Marks, *Organometallics*, 1999, **18**, 2568.
134. P. Voth, S. Arndt, T. P. Spaniol, J. Okuda, L. J. Ackerman and M. L. H. Green, *Organometallics*, 2003, **22**, 65.
135. Y. Mu, W. E. Piers, D. C. MacQuarrie, M. J. Zaworotko and V. G. Young, Jr., *Organometallics*, 1996, **15**, 2720.
136. T. P. Spaniol, J. Okuda, M. Kitamura and T. Takahashi, *J. Organomet. Chem.*, 2003, **684**, 194.
137. (a) K. C. Hultzsch, T. P. Spaniol and J. Okuda, *Angew. Chem., Int. Ed.*, 1999, **38**, 227; (b) K. C. Hultzsch, P. Voth, K. Beckerle, T. P. Spaniol and J. Okuda, *Organometallics*, 2000, **19**, 228; (c) D. Robert, A. A. Trifonov, P. Voth and J. Okuda, *J. Organomet. Chem.*, 2006, **691**, 4393.
138. E. Kirillov, L. Toupet, C. W. Lehmann, A. Razavi and J.-F. Carpentier, *Organometallics*, 2003, **22**, 4467.
139. S. Arndt, P. Voth, T. P. Spaniol and J. Okuda, *Organometallics*, 2000, **19**, 4690.
140. A. A. Trifonov, E. N. Kirillov, Y. A. Kurskii and M. N. Bochkarev, *Russ. Chem. Bull.*, 2000, **49**, 744.
141. Z. Hou, T. Koizumi, M. Nishiura and Y. Wakatsuki, *Organometallics*, 2001, **20**, 3323.
142. A. A. Trifonov, T. P. Spaniol and J. Okuda, *Organometallics*, 2001, **20**, 4869.
143. J. Okuda, S. Arndt, K. Beckerle, P.-J. Sinnema, K. C. Hultzsch, P. Voth and T. P. Spaniol, *J. Mol. Catalysis*, 2002, **190**, 215.
144. V. M. Arredondo, S. Tian, F. E. McDonald and T. J. Marks, *J. Am. Chem. Soc.*, 1999, **121**, 3633.
145. M. R. Douglas and T. J. Marks, *J. Am. Chem. Soc.*, 2000, **122**, 1824.
146. P. N. O'Shaughnessy, K. M. Gillespie, P. D. Knight, I. J. Munslow and P. Scott, *Dalton Trans.*, 2004, 2251.
147. (a) J. Collin, J.-C. Daran, E. Schulz and A. A. Trifonov, *Chem. Commun.*, 2003, 3048; (b) C. Drost, P. B. Hitchcock and M. F. Lappert, *J. Chem. Soc., Dalton Trans.*, 1996, 2323.
148. K. C. Hultzsch, T. P. Spaniol and J. Okuda, *Organometallics*, 1997, **16**, 4845.
149. M. S. Hill and P. B. Hitchcock, *Angew. Chem., Int. Ed.*, 2001, **40**, 4089.
150. A. A. Trifonov, T. P. Spaniol and J. Okuda, *Eur. J. Inorg. Chem.*, 2003, 926.
151. K. Beckerle, K. C. Hultzsch and J. Okuda, *Macromol. Chem. Phys.*, 1999, **200**, 1702.
152. (a) Y. Yao, Z. Zhang, H. Peng, Y. Zhang, Q. Shen and J. Lin, *Inorg. Chem.*, 2006, **45**, 2175; (b) L. Zhou, Y. Yao, C. Li, Y. Zhang and Q. Shen, *Organometallics*, 2006, **25**, 2880.

153. (a) T. G. Wetzel and P. W. Roesky, *Z. Anorg. Allg. Chem.*, 1999, **625**, 1953. (b) H. C. Aspinall and M. R. Tillotson, *Inorg. Chem.*, 1996, **35**, 2163.
154. H. Zhu and E.Y.-X. Chen, *Organometallics*, 2007, **21**, 5395.
155. G. B. Deacon, C. M. Forsyth, P. C. Junk, B. W. Skelton and A. H. White, *J. Chem. Soc., Dalton Trans.*, 1998, 1381.
156. (a) G. B. Deacon, C. M. Forsyth and N. M. Scott, *Dalton Trans.*, 2003, 3216; (b) G. B. Deacon, C. M. Forsyth and N. M. Scott, *Eur. J. Inorg. Chem.*, 2002, 1425; (c) G. B. Deacon, C. M. Forsyth, N. M. Scott, B. W. Skelton and A. H. White, *Austr. J. Chem.*, 2001, **54**, 439.
157. (a) P. L. Arnold, S. A. Mungur, A. J. Blake and C. Wilson, *Angew. Chem., Int. Ed.*, 2003, **42**, 5981; (b) S. T. Liddle and P. L. Arnold, *Organometallics*, 2005, **24**, 2597; (c) P. L. Arnold and S. T. Liddle, *Organometallics*, 2006, **25**, 1485; (d) P. L. Arnold and S. T. Liddle, *Chem. Commun.*, 2005, 5638; (e) P. L. Arnold, S. T. Liddle, J. McMaster, C. Jones and D. P. Mills, *J. Am. Chem. Soc.*, 2007, **129**, 5360; (f) D. Patel, S. T. Liddle, S. A. Mungur, M. Rodden, A. J. Blake and P. L. Arnold, *Chem. Commun.*, 2006, 1124.
158. A. N. Kornev, T. A. Chesnokova, E. V. Zhezlova, L. N. Zakharov, G. K. Fukin, Y. A. Kursky, G. A. Domrachev and P. D. Lickiss, *J. Organomet. Chem.*, 1999, **587**, 113.
159. J. Annand and H. C. Aspinall, *J. Chem. Soc., Dalton Trans.*, 2000, 1867.
160. J. Annand, H. C. Aspinall and A. Steiner, *Inorg. Chem.*, 1999, **38**, 3941.
161. G. Gerstberger, C. Palm and R. Anwander, *Chem. Eur. J.*, 1999, **5**, 997.
162. (a) M. Roger, N. Barros, T. Arliguie, P. Thuéry, L. Maron and M. Ephritikhine, *J. Am. Chem. Soc.*, 2006, **128**, 8790; (b) C. G. Pernin and J. A. Ibers, *Inorg. Chem.*, 2000, **39**, 1222.
163. M. W. Essig, D. W. Keogh, B. L. Scott and J. G. Watkin, *Polyhedron*, 2001, **20**, 373.
164. R. Anwander and R. Roesky, *J. Chem. Soc., Dalton Trans.*, 1997, 137.
165. A. G. Avent, C. F. Caro, P. B. Hitchcock, M. F. Lappert, Z. Li and X.-H. Wei, *Dalton Trans.*, 2004, 1567.
166. D. V. Gribkov, K. C. Hultzsch and F. Hampel, *J. Am. Chem. Soc.*, 2006, **128**, 3748.
167. A. A. Trifonov, E. A. Fedorova, E. N. Kirillov, S. E. Nefedov, I. L. Eremenko, Y. A. Kurskii, A. S. Shavyrin and M. N. Bochkarev, *Russ. Chem. Bull.*, 2002, **51**, 684.
168. Y. K. Gun'ko, R. Reilly, F. T. Edelmann and H.-G. Schmidt, *Angew. Chem., Int. Ed.*, 2001, **40**, 1279.
169. H.-W. Görlitzer, M. Spiegler and R. Anwander, *J. Chem. Soc., Dalton Trans.*, 1999, 4287.
170. R. Anwander, J. Eppinger, I. Nagl, W. Scherer, M. Tafipolsky and P. Sirsch, *Inorg. Chem.*, 2000, **39**, 4713.
171. C.-X. Cai, L. Toupet, C. W. Lehmann and J.-F. Carpentier *J. Organomet. Chem.*, 2003, **683**, 131.
172. R. Anwander, H.-W. Görlitzer, G. Gerstberger, C. Palm, O. Runte and M. Spiegler, *J. Chem. Soc., Dalton Trans.*, 1999, 3611.
173. P. N. O'Shaughnessy, P. D. Knight, C. Morton, K. M. Gillespie and P. Scott, *Chem. Commun.*, 2003, 1770.
174. Y. Liang and R. Anwander, *Dalton Trans.*, 2006, 1909.
175. W. A. Herrmann, J. Eppinger, M. Spiegler, O. Runte and R. Anwander, *Organometallics*, 1997, **16**, 1813.
176. I. Nagl, M. Widenmeyer, S. Grasser, K. Köhler and R. Anwander, *J. Am. Chem. Soc.*, 2000, **122**, 1544.
177. R. Anwander, I. Nagl, C. Zapilko and M. Widenmeyer, *Tetrahedron*, 2003, **59**, 10567.
178. P. I. Binda and E. E. Delbridge, *Dalton Trans.*, 2007, 4685.
179. P. B. Hitchcock, A. G. Hulkes, M. F. Lappert and Z. Li, *Dalton Trans.*, 2004, 129.
180. I. L. Fedushkin, T. V. Petrovskaya, F. Girgsdies, V. I. Nevodchikov, R. Weimann, H. Schumann and M. N. Bochkarev *Russ. Chem. Bull.*, 2000, **49**, 1869.
181. (a) S. Bambirra, A. Meetsma, B. Hessen and A. Bruins, *Organometallics*, 2006, **25**, 3486; (b) B. Monteiro, D. Roitershtein, H. Ferreira, J. R. Ascenso, A. M. Martins, A. Domingos and N. Marques, *Inorg. Chem.*, 2003, **42**, 4223.

182. M. R. Douglas, M. Ogasawara, S. Hong, M. V. Metz and T. J. Marks, *Organometallics*, 2002, **21**, 283.
183. E. Sheng, S. Wang, G. Yang, S. Zhou, L. Cheng, K. Zhang and Z. Huang, *Organometallics*, 2003, **22**, 684.
184. H. D. Guo, H. Zhou, Y. M. Yao, Y. Zhang and Q. Shen, *Dalton Trans.*, 2007, 3555.
185. (a) X. Yu and T. J. Marks, *Organometallics*, 2007, **26**, 365; (b) S. Tobisch, *Chem. Eur. J.*, 2007, **13**, 9127.
186. (a) F. G. N. Cloke, H. C. de Lemos and A. A. Sameh, *J. Chem. Soc., Chem. Commun.*, 1986, 1344; (b) F. G. N. Cloke, *Chem. Soc. Rev.*, 1993, **22**, 17.
187. M. N. Bochkarev, A. A. Trifonov, F. G. N. Cloke, C. I. Dalby, P. T. Matsunaga, R. A. Andersen, H. Schumann, J. Loebel and H. Hemling, *J. Organomet. Chem.*, 1995, **486**, 177.
188. T. V. Petrovskaya, I. L. Fedushkin, M. N. Bochkarev, H. Schumann and R. Weimann, *Russ. Chem. Bull.*, 1997, **46**, 1766.
189. (a) A. A. Trifonov, E. N. Kirillov, M. N. Bochkarev, H. Schumann and S. Mühle, *Russ. Chem. Bull.*, 1999, **48**, 382; (b) A. A. Trifonov, Y. A. Kurskii, M. N. Bochkarev, S. Mühle, S. Dechert and H. Schumann, *Russ. Chem. Bull.*, 2003, **52**, 601.
190. A. A. Trifonov, E. A. Fedorova, V. N. Ikorskii, S. Dechert, H. Schumann and M. N. Bochkarev, *Eur. J. Inorg. Chem.*, 2005, 2812.
191. A. A. Trifonov, E. A. Fedorova, G. K. Fukin, V. N. Ikorskii, Y. A. Kurskii, S. Dechert, H. Schumann and M. N. Bochkarev, *Russ. Chem. Bull.*, 2004, **53**, 2736.
192. A. A. Trifonov, E. A. Fedorova, G. K. Fukin, N. O. Druzhkov and M. N. Bochkarev, *Angew. Chem., Int. Ed.*, 2004, **43**, 5045.
193. (a) A. A. Trifonov, E. A. Fedorova, G. K. Fukin, E. V. Baranov, N. O. Druzhkov and M. N. Bochkarev, *Chem. Eur. J.*, 2006, **12**, 2752. (b) A. A. Trifonov, I. A. Borovkov, E. A. Fedorova, G. K. Fukin, J. Larionova, N. O. Druzhkov and V. K. Cherkasov, *Chem. Eur. J.*, 2007, **13**, 4981.
194. I. Westmoreland and J. Arnold, *Dalton Trans.*, 2006, 4155.

5

Amides of the Actinide Metals

5.1 Introduction

In our 1980 book (references to mid-1978), 14 papers on the diamagnetic thorium(IV) and the f^2-uranium(IV) amides were cited.[1] The earliest was Gilman's (1956) on $U(NEt_2)_4$ [which on alcoholysis gave $U(OR)_4$ (R = Me, Et)] and Bradley's (1969) on $Th(NEt_2)_4$, each obtained from the appropriate $AnCl_4$ and $LiNR_2$. Crystallographic characterisation of the former as $[\{U(NEt_2)_3(\mu\text{-}NEt_2)\}_2]$, as well as of four others, was described by Edelstein and co-workers about 20 years later: $[U(NPh_2)_4]$, $[U_3L_6]$, $[U_4L_8]$ and $[\{LiU(\mu\text{-}O)(NPh_2)_3(OEt_2)\}_2]$ [L = N(Me)(CH_2)$_2$NMe]. The synthesis and characterisation of the homoleptic amides $U(NR_2)_4$ (R = Me, Pr, Bu, Bu^i, Ph, $NC_4H_2Me_2$-2,5) and $Th(NR_2)_4$ (R = Me, Pr, Bu, Bu^i) were reported, as well as of the heteroleptic compounds $UCp_2(NR_2)_2$ and $ThCl\{N(SiMe_3)_2\}_3$.

The only reactions of An^{IV} (henceforth, An = Th, U) amides then recorded were insertions (Ch. 10) and protonolyses (Ch. 11).[1] Thus, from $An(NR_2)_4$ and an excess of CO_2, C(O)S, CS_2 or CSe_2 there were obtained $An\{EC(E)NR_2\}_4$ (E = O, S, Se) or $An\{OC(S)NR_2\}_4$. From $U(NEt_2)_4$ and an excess of cyclopentadiene, $UCp_2(NEt_2)_2$ was isolated; the latter and the appropriate thiol, dithiol, hydroxythiol or catechol gave UCp_2X_2 [X_2 = $(SEt)_2$, $(SBu^t)_2$, $S_2C_6H_3$-1,2-Me-4, SCH_2CH_2S, $O(S)C_6H_4$-1,2, $O_2C_6H_4$-1,2] or $UCp_3(SEt)$.

Early (1987–1995) reviews of organoAn chemistry, including bibliographies on amides, are in Refs. 2–6. The greatest and most up-to-date information is to be found in the survey by Burns *et al.*,[7] which includes material in the 2002 literature.

The bulk of the researches on the amidoAn compounds have been on U and, to a lesser extent, Th. The only examples in the open literature of transuranic amides are on the homoleptic Np^{III} and Pu^{III} bis(trimethylsilyl)amides, prepared from the appropriate $[AnI_3(thf)_4]$ and $3Na\{N(SiMe_3)_2\}$ in thf;[8a] this was also the preferred procedure for $[U\{N(SiMe_3)_2\}_3]$. $Pu\{N(SiMe_3)_2\}_3$ was cited as a catalyst precursor;[8b] its molecular structure has been reported recently.[8c]

Metal Amide Chemistry Michael Lappert, Andrey Protchenko, Philip Power and Alexandra Seeber

The literature is most prolific on neutral An^{IV} amides, but neutral U^{III} and U^{V} amides are also well documented. There are only few publications on U^{VI} amides,[9,10] one being on three alkali metal amidouranate(VI)s.[10] There is a paper on a compound (**18**) which *could* be classified as of U^{II},[11] and two on mixed valence U^{III}/U^{IV} amides[12,13] (see Section 5.4 for these types). There are a number of articles on monocationic amidoU^{IV} tetraphenylborates, three on dicationic analogues, three on a cationic U^{V} salt; and four on alkali metal amidoactinate(IV)s and three on corresponding U^{III} salts. For crystalline uranium amides containing only monodentate σ-bonded ligands, metal coordination numbers of 3–7 are established, exemplified by the pyramidal $[U\{N(SiMe_3)_2\}_3]$,[14] tetrahedral $[U(NPh_2)_4]$,[15] distorted trigonal bipyramidal $[\{U(NEt_2)_3(\mu\text{-}NEt_2)\}_2]$,[16] pseudooctahedral $[U(NEt_2)_3(thf)_3][BPh_4]$,[17] and $[U(NEt_2)_2(py)_5][BPh_4]_2$,[17] respectively. Using the potentially tridentate ligand $[N(CH_2CH_2PEt_2)_2]^-$, the uranium atoms of $[\{U\{\kappa^3\text{-}N(CH_2CH_2PEt_2)_2\}\{\kappa^2\text{-}N(CH_2CH_2PEt_2)_2\}Cl(\mu\text{-}Cl)\}_2]$ are eight-coordinate.[18]

5.2 Neutral Amidouranium(IV) and Thorium(IV) Complexes

5.2.1 Introduction

Studies of amidoAn(IV) complexes dominate aminoactinide chemistry. Such neutral compounds are either free of π-centred coligands, or they contain one or more cyclopentadienyl, cycloheptatrienyl or cyclooctatetraene. In the former category, silylamido complexes have been prominent, but hydrocarbylamides have also featured. Most of the researches have focused on monodentate amides. There is one publication on the monoanionic $[DAD]^{\dot{-}}$ ligand **1**;[19a] the monoanionic ligand **2** is bi- or tridentate,[18] but can be monodentate.[40] The monoanionic carbene-tethered ligand **3** is bidentate.[19b] Much used has been the dianionic *C,N*-centred bidentate ligand **4**.[20] The dianionic cyclopentadienyl-amido ligand **5** was used for the preparation of constrained geometry organoactinide complexes.[66] The pyrrole-based ligands are the dianionic tridentate **6**, the trianionic tridentate **7**, the dianionic bidentate **9** and the monoanionic bidentate **8**;[21a] the macrocyclic ligand **10** has two tetradentate dianionic moieties allowing synthesis of bimetallic uranium/transition metal complexes.[94] The trianionic tetradentate ligands **11a**,[22,58,59] **11b**,[12,13,23,57,59,60] and the tetraanionic derivative **11c**[23] have proved to be of interest, as has the trianionic, potentially hexacoordinate **12**.[24] Diamido ligands **13** and **14** have been reported recently.[41]

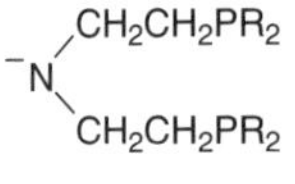

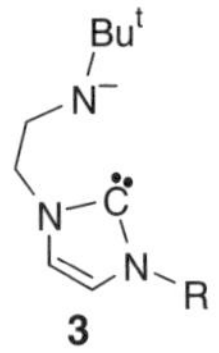

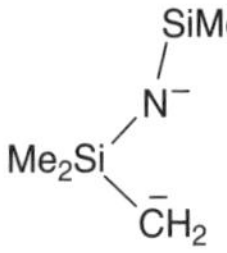

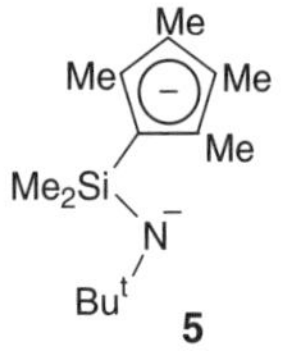

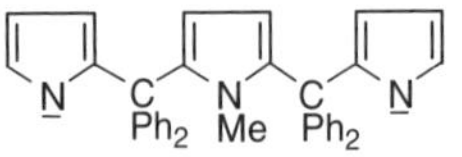

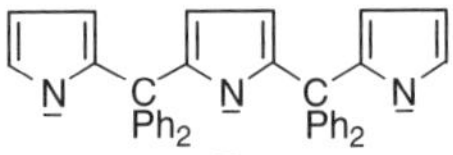

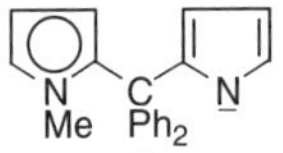

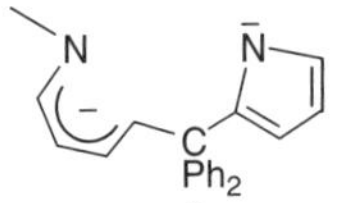

10 (R = H or Me)

11a (R = Me)
11b (R = Bu^t)

11c

12

13

14

5.2.2 Hydrocarbylamido-An^{IV} Compounds Free of π-Centred Ligands

The compound $U(NEt_2)_4$ has been shown to be a useful precursor for a range of organic uranium complexes. The majority of its reactions have been with protic compounds HA, leading, at least in the first place, to $U(NEt_2)_{4-n}(A)_n$ ($n = 1, 2, 3$). Factors favouring the displacement include the nucleophilicity of HA, its protic character, steric effects if A is particularly bulky; and the favourable entropy, diethylamine being volatile. The first such reaction (1983) was that with 2,6-$Bu^t{}_2C_6H_3OH$; even with an excess of this phenol, the homoleptic U^{IV} aryloxide was not accessible.[25] Reactions with Group 16 and 15 protic compounds are shown in Schemes 5.1 and 5.2, respectively, and with cyclopentadienes also in Scheme 5.2. The compounds $[Th(NEt_2)_n(Cp)_{4-n}]$ ($n = 1, 2, 3, 4$) were obtained[33] from $Th(NEt_2)_4$[35] and CpH.

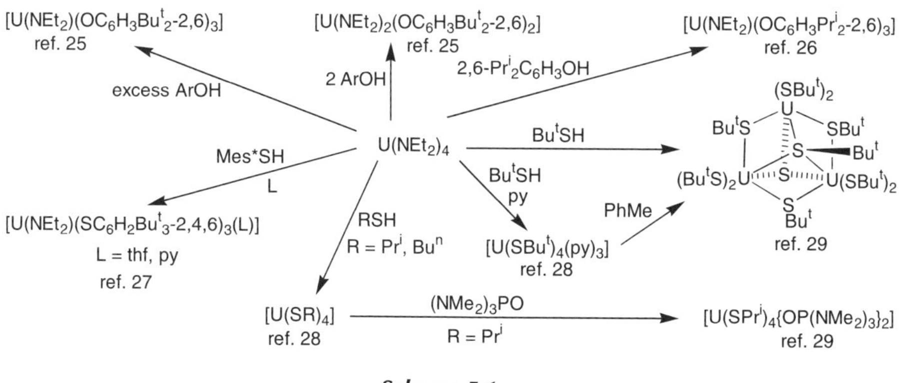

Scheme 5.1

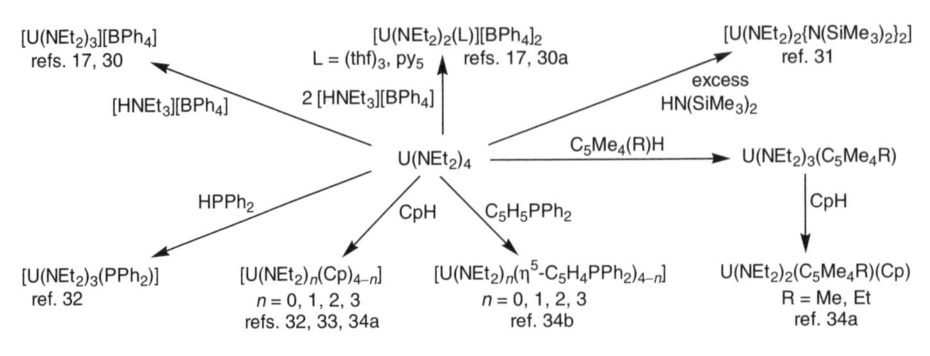

Scheme 5.2

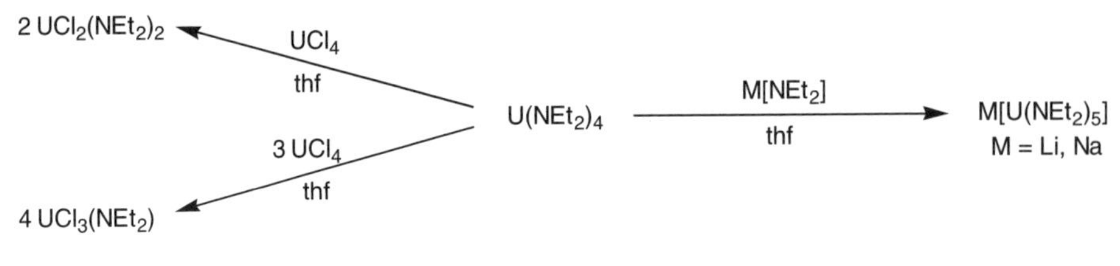

Scheme 5.3

Two metathetical NEt_2/Cl exchange reactions[17,30c] and a nucleophilic addition reaction[30a] of $U(NEt_2)_4$ have been reported, Scheme 5.3.

The complex $U(NEt_2)_3PPh_2$ with cyclopentadiene afforded $UCp_2(NEt_2)_2$ via dismutation of the unstable mixed-ligand species $UCp(NEt_2)_2PPh_2$.[32] Dimethylamido compounds $An(NMe_2)_4$ (An = U, Th), prepared from $AnCl_4$ and $4LiNMe_2$, were contaminated with *ate*-complexes $Li[An(NMe_2)_5(thf)_n]$, but they were used successfully for the preparation via amine elimination with H_2(**5**) of constrained geometry compounds [An(**5**)$(NMe_2)_2$], effective catalysts for the intramolecular hydroamination/cyclisation of amino-alkenes or alkynes.[66]

Cummins and coworkers have carried out interesting experiments on compounds of the general formula $U\{N(Bu^t)C_6H_3Me_2\text{-}3,5\}_3X$, as summarised in Scheme 5.4. Particularly noteworthy are: (i) the formation of the U^{IV} iodide **15** (X = I) from a U^{III} precursor;[36] (ii) the reduction of **15** to the U^{III} complex **16** (X = thf);[36] (iii) the oxidative reactions of **16** with N_2 in the presence of a homoleptic Mo^{III} amide under ambient conditions furnishing the N_2-bridged binuclear (U^{IV}/Mo^{IV}) complexes **17** and **18** [X = N=N–Mo{N(R)Ar}$_3$] (there was no reaction in the absence of the Mo amide);[36] and (iv) the formation of the carbodiimido- (≡ cyanoimido-) bridged binuclear U^{IV} complex **19** [X = N=C=N–U$\{N(Bu^t)C_6H_3Me_2\text{-}3,5\}_3$] from the U^{III} amide **16** and the redox ligand transfer agent **20** (anthracene being eliminated).[37]

20

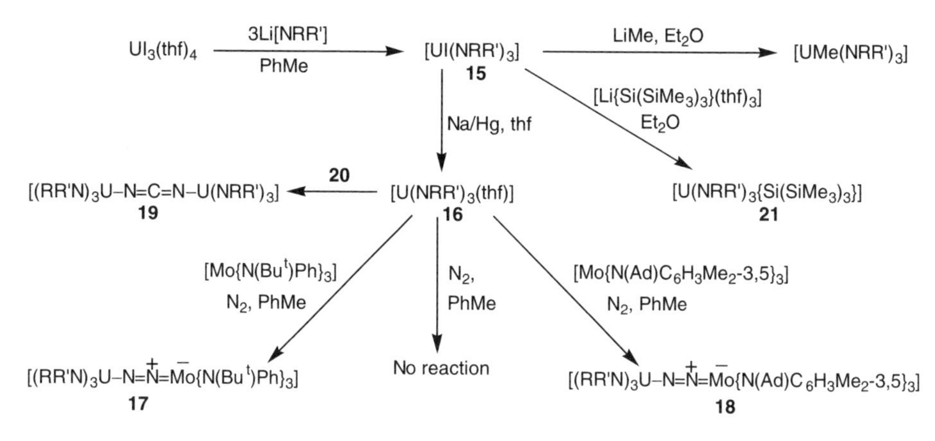

Scheme 5.4 *(R = Bu^t; R′ = $C_6H_3Me_2$-3,5; Ad = 1-adamantyl)*

The supersilyl U^{IV} complex **21** [X = $Si(SiMe_3)_3$] proved to be unreactive towards CO, H_2, nitriles, isocyanates or isonitriles (except Pr^iNC), attributed to sterically imposed inhibition.[38] The crystalline complexes **15**,[36] **16**,[36] **17**[36] (see Section 5.3), **18**,[36] **19**,[37] **21**[38] and [U(Me){N(Bu^t)$C_6H_3Me_2$-3,5}$_3$][38] were X-ray-characterised. The geometric parameters of **17** and **18** are consistent with their valence bond representations in Scheme 5.4.[36] The N–N bond length of 1.232(11) Å in **17** is slightly longer than in N_2; the U–N–N and Mo–N–N bond angles are 173.8(7) and 179.1(7)°, respectively; and the N_2 ligand is inside a 'cage' of six Bu^t groups.[36] The U–N–C angle in **19** is 162.6(15)°.[37] DFT calculations on the model compound [{$NH_3(NH_2)_3U$}$_2$(μ-$\eta^2:\eta^2$-N_2)] (*cf.*, **17**) provided evidence for U→N_2 π-back-bonding.[39]

Several heteroleptic bis(hydrocarbylamido)uranium(IV) complexes have already been mentioned: $U(NEt_2)_2X_2$ [X = $OC_6H_3Bu^t_2$-2,6 (Scheme 5.1),[25] $N(SiMe_3)_2$[31] or $C_5H_4PPh_2$[34b] (Scheme 5.2), Cl or PPh_2 (Scheme 5.3)[17]] and the eight-coordinate binuclear U^{IV} compound [{U{κ^3-$N(CH_2CH_2PEt_2)_2$}{κ^2-$N(CH_2CH_2PEt_2)_2$}Cl(μ-Cl)}$_2$] (Section 5.1).[18,40] The latter, as well as some analogues, were prepared from the appropriate $AnCl_4$ and 2LiNR$_2$ (R = $CH_2CH_2PEt_2$ or $CH_2CH_2PPr^i_2$, abbreviated as L^{Et} and L^{Pr^i}, respectively): [{UCl(L^{Et})$_2$(μ-Cl)}$_2$] (one of the L^{Et} ligands was *N,P,P′*-tridentate and the other *N,P*-bidentate), [{UCl(L^{Pr^i})$_2$(μ-Cl)}$_2$], [{ThCl(L^{Et})$_2$(μ-Cl)}$_2$], [{ThCl(L^{Pr^i})$_2$(μ-Cl)}$_2$] (each of the L^{Pr^i} ligands was *N,P*-bidentate).[18] The bis(cyclopentadienyl)bis(diethylamido)U^{IV} compound [{(η^5-$C_5H_4PPh_2$)$_2$Mo(CO)$_4$}U(NEt_2)$_2$] was prepared from [U(NEt_2)$_2$(η^5-$C_5H_4PPh_2$)$_2$] and [Mo(CO)$_3$(C_7H_8)].[34b] Further (cf. Scheme 5.4) remarkable redox experiments involving the moiety U[N(Bu^t)$C_6H_3Me_2$-3,5]$_n$ have been reported by the Cummins group, as summarised in Scheme 5.5;[11] some discussion on the subvalent U compound **22** is in Section 5.4. Analogues of **22** [NRR′ = N(Ad)$C_6H_3Me_2$-3,5] and **24** [(μ-NC_6D_5)$_2$, using perdeuterated azobenzene as the reagent] were also characterised.[11]

Gambarotta and coworkers have reported some unusual crystalline X-ray-characterised complexes based on monoanionic (**8**), dianionic (**6**, **9**) and trianionic (**7**) pyrrolide ligands, as summarised in Scheme 5.6.[21a] Although the formation of compounds **26** or **27** used the strong reducing agent K or K/naphthalene, it was the ligand rather than the metal which underwent electron-transfer. Thus, for **26**: 2dme + 2$\bar{e}$ → 2$\bar{O}(CH_2)_2OMe$ + CH_4 + C_2H_6

$[UI(NRR')_3]$ **15** —(KC_8, PhMe)→ $[(RR'N)_2U(\mu\text{-}\eta^6{:}\eta^6\text{-PhMe})U(NRR')_2]$ **22**

22 —(PhS-SPh)→ $[\{U(NRR')_2(SPh)(\mu\text{-SPh})\}_2]$ **23**

22 —(PhN=NPh)→ **24** ($(RR'N)_2U$, $U(NRR')_2$)

Scheme 5.5 *(R = Bu^t; R′ = $C_6H_3Me_2$-3,5)*

and dme + 2ē → $2\bar{O}Me + CH_2{=}CH_2$; while for **27**, the monoanionic ligand **8** underwent pyrrolide ring-opening with the formation of the dianionic **9**.[21a] Transient formation of a low-valent Th species (or its synthetic equivalent) was observed in the K/naphthalene reduction of a bis(diamido)thorium complex, which led to the formation of a metallacycle resulting from C–H bond activation of the ligand, Scheme 5.7.[21b]

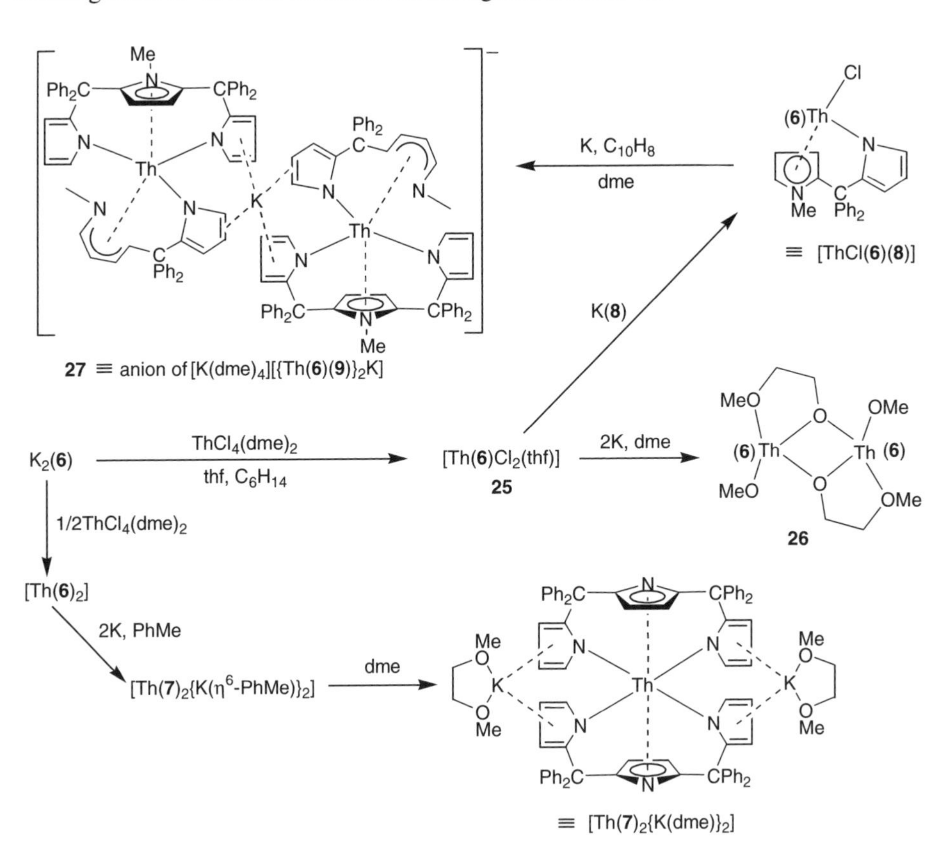

Scheme 5.6

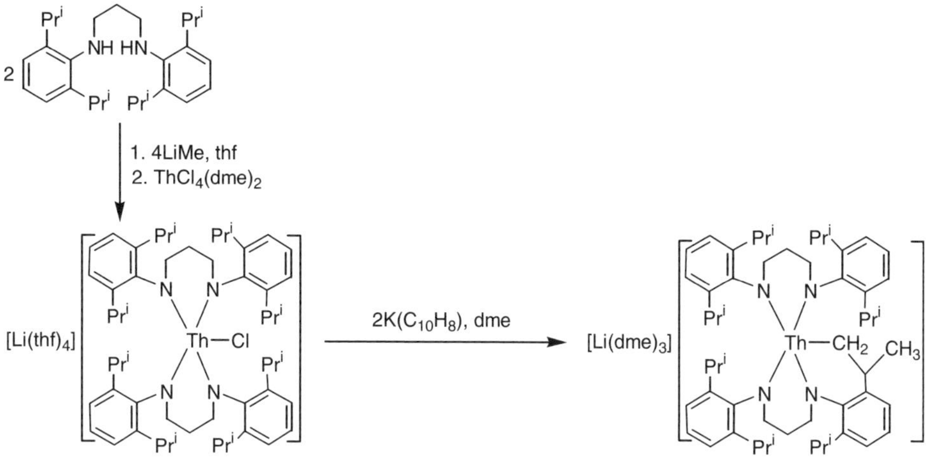

Scheme 5.7

Remarkably robust dialkylthorium complexes $[Th(L)(CH_2SiMe_3)_2]$ (L = **13** or **14**), supported by bulky bis(anilido) ligands based on a xanthene or 2,6-dimethylpyridine framework, were prepared by salt elimination from $[ThCl_2(L)(dme)]$ and $2LiCH_2SiMe_3$ or by alkane elimination from $[Th(CH_2SiMe_3)_4]$ and H_2(**13**) or H_2(**14**).[41]

5.2.3 Silylamido-An^{IV} Compounds Free of π-Centred Ligands

Some of Andersen's pioneering studies on silylamidoAn chemistry began with the tris(amido)An^{IV} chloride **28**, obtained from $AnCl_4$ and $3Na[N(SiMe_3)_2]$ in thf.[42] The chloride ligand of **28** was displaced by various nucleophiles, yielding the borohydrides **29**,[42] methyls **30**,[43] or hydrides **31**,[44] as summarised in Scheme 5.8 (see also Ref. 43). The tridentate borohydrides **29** were also obtained from **31** and $BH_3{\cdot}thf$.[44] Treatment of the hydrides **31** with D_2 in C_5H_{12} yielded the perdeuterated analogues.[20] The metallacycles **32** were isolated by pyrolysis of **30** or **31**, or from **28** (An = U) with LiR (R = Et, CH_2SiMe_3);[20] while **32** with H_2 or D_2 furnished **31** (X-ray structure of the U hydride[45]) or its ^2D equivalent.[44] $[\{U(N(SiMe_3)_2)_2(\mu\text{-}NC_6H_4Me\text{-}4)\}_2]$ was obtained from **28** and $Li[N(H)C_6H_4Me\text{-}4]$.[46] The metallacycle **32** underwent a 1 : 1-addition reaction with Me_3SiN_3 or the insertion reactions shown in Scheme 5.9.[47]

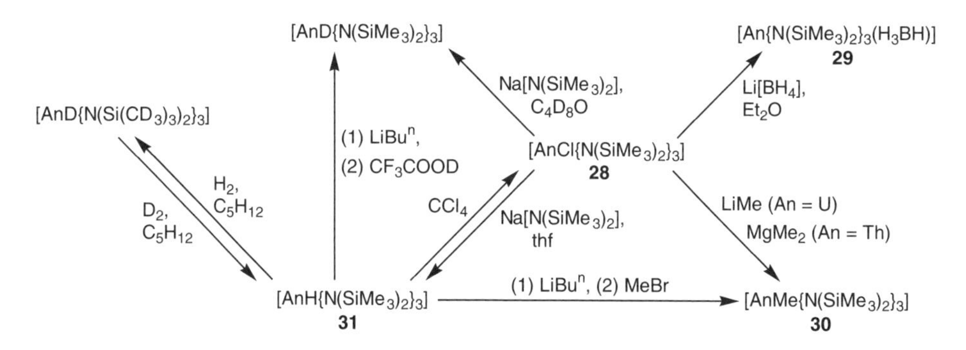

Scheme 5.8

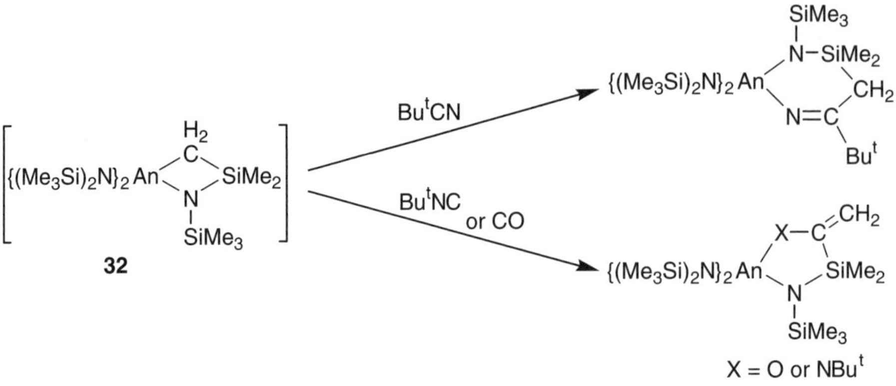

Scheme 5.9

Dormond *et al.* have exploited compounds **30** and particularly **32** as synthetically useful reagents. The uranium-mediated methylenation of carbonyl compounds (cf., the Wittig, Peterson or Tebbe reactions) is illustrated in Equation (5.1). The instantaneous formation under ambient conditions of the insertion product **33** was essentially quantitative (e.g.,[48c] $R^1R^2 =$ Me_2, Ph_2 or [Me, H]), as was its hydrolysis (preferably in d. HCl);[48a] the insertion reaction

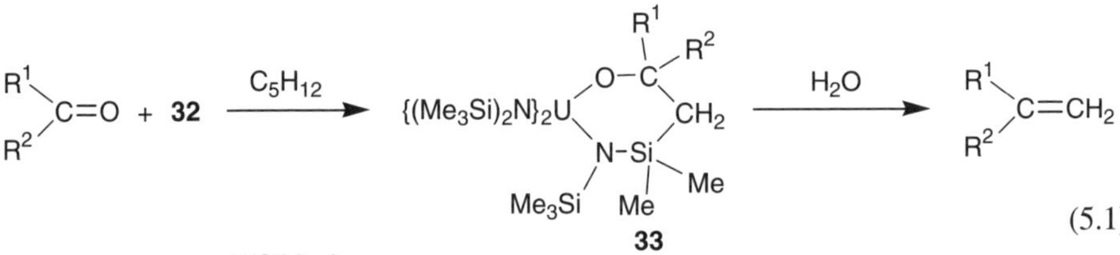

(5.1)

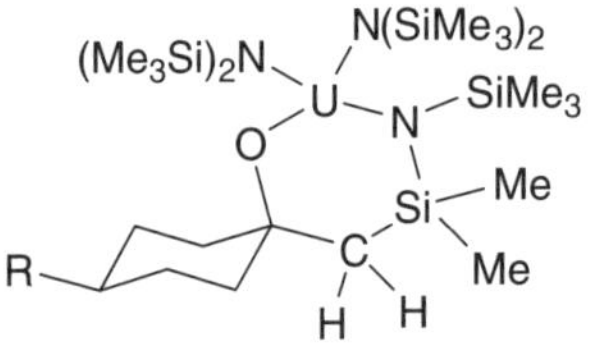

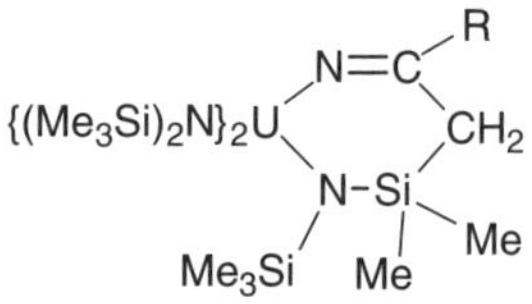

was stereoselective, *exo*-**33** being the major product (**33a**) from equatorial attack of **32** on a substituted cyclohexanone.[48b,48d] Polycyclic ketones, such as camphor, gave *exo*-adducts (analogues of **33a**) which showed sharp ^{1}H NMR signals and thus may be useful shift-reagents.[49] A ketone synthesis from a nitrile RCN was developed;[50b] thus, the azomethine complex **34** was obtained from **32** and RCN (R = Me, Pr^n, Ph;[50a] $CH{=}CH_2$, $CHPh_2$[50b]); **34** upon hydrolysis (d. HCl) yielded the methyl ketone RC(O)Me.[50b] Protonolysis of **32** with ROH, PhC≡CH or 2- (or 4-) picoline yielded the appropriate compound $[U\{N(SiMe_3)_2\}_3X]$.[51,52] The methyl complex **30** was useful as a methyl transfer agent for aldehydes or ketones, generating alcohols R(R′)MeCOH (R′ = H or hydrocarbyl).[48c,48d,52] For example, treatment of a deficiency of **30** with a mixture of PhC(O)Me and PhC(O)H selectively afforded $[U\{N(SiMe_3)_2\}_3\{OCH(Me)Ph\}]$ which upon hydrolysis gave Ph(Me)CHOH; from 4-Bu^t-cyclohexanone, this sequence provided *exo*-1-Me-4-Bu^t-cyclohexanol as the major product.[48c] Further reactions of **30** with RNC (R = Bu^t, *c*-C_6H_{11}, $C_6H_3Me_2$-2,6), RCN, HNR_2 ($HNEt_2$, indole or carbazole) or $[MH(CO)_3Cp]$ (M = Mo, W) furnished $[U\{N(SiMe_3)_2\}_3X]$ [X = C(Me)NR, N=C(Me)R, NR_2 or $(OC)M(CO)_2Cp$, respectively.[50a]

$\{(Me_3Si)_2N\}_2U^+$—N—$SiMe_3$ | $SiMe_2$—H_2C—$^-B(C_6F_5)_3$

35

The chloroU^{IV} complex **28** with $Li[N(H)C_6H_4Me$-4] gave $[\{U(N(SiMe_3)_2)_2$-$(\mu$-NC_6H_4Me-4$)\}_2]$.[45] Treatment of the hydride **31** (An = U) with $B(C_6F_5)_3$ gave the X-ray- and neutron diffraction-characterised zwitterionic adduct **35**.[53] The U^{IV} hydride **31** and the corresponding metallacycle **32** were shown to be in dynamic equilibrium with H_2, possibly implicating a κ^2-H_2 adduct of **32**, as H_2 evolution was not observed.[53] Various further protonolyses of **32** have been reported with: (i) $5Bu^tOH \rightarrow [U_2(OBu^t)_8(HOBu^t)]$,[54] (ii) $HOC_6H_3R'_2$-2,6 $\rightarrow [An\{N(SiMe_3)_2\}_3(OC_6H_3R'_2$-2,6$)]$ (R′ = Pr^i, Bu^t),[55b] (iii) $nHOC_6H_3$-Bu^t_2-2,6 $\rightarrow [Th\{N(SiMe_3)_2\}_{4-n}(OC_6H_3Bu^t_2$-2,6$)_n]$ ($n = 2$, 3),[55b] (iv) $4HOC_6H_3R'_2$-2,6 $\rightarrow$ $[U(OC_6H_3R'_2$-2,6$)_4]$ (R′ = Bu^t,[55a] Pr^i[55b]), (v) $4Pr^iOH$/py $\rightarrow [\{Th_4(OPr^i)_{16}(py)_2]$,[56a] (vi) $4Et_2CHOH$/py $\rightarrow [\{Th_2(OCHEt_2)_8(py)_2]$,[56a] (vii) $4Pr^i_2CHOH \rightarrow [\{Th(OCHPr^i_2)_3$-$(\mu$-$OCHPr^i_2)\}_2]$,[56b] (viii) $HSC_6H_3Me_2$-2,6 $\rightarrow [U\{N(SiMe_3)_2\}_3(SC_6H_3Me_2$-2,6$)]$,[57] and (ix) $HSC_6H_2Bu^t_3$-2,4,6 $\rightarrow [U\{N(SiMe_3)_2\}_3(SC_6H_2Bu^t_3$-2,4,6$)]$.[27]

Scott and coworkers have developed some interesting chemistry using the trianionic tetradentate ligands **11a** and **11b** and the tetraanionic **11c**, see also Sections 5.3 and 5.4. Treatment of $AnCl_4$ with 3Li(**11a**) or 3Li(**11b**) yielded [{An(**11a**)(μ-Cl)}$_2$] [An = Th or U (**36**)],[22] [An(**11b**)Cl] [An = Th[58] or U (**38**)[12,58]]; compound $U\{N(CH_2CH_2NSiMePh_2)_3\}Cl$ was also prepared.[12] Crystalline **36** is a dimer of distorted octahedra having edge-sharing $UClN'_3N$ units with an inversion centre as the mid-point of the UCl_2U plane.[22] The chloride ligands of **36** or **37** were readily displaceable, as shown in Schemes 5.10[59] and 5.11,[23,58,60,61] respectively. As for the former, noteworthy are the volatility of the *tert*-butoxides **38**, and the

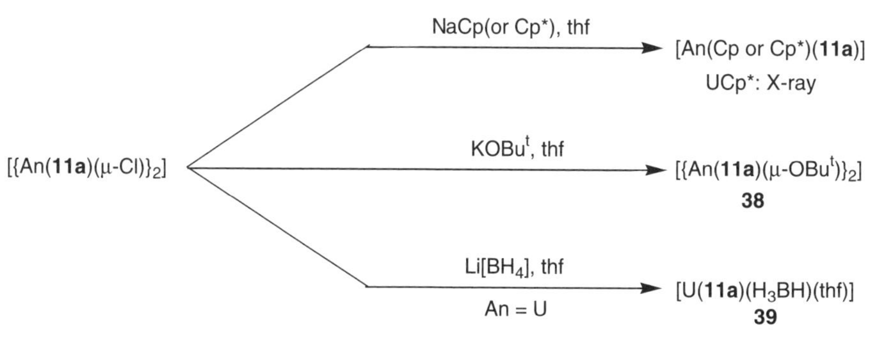

Scheme 5.10

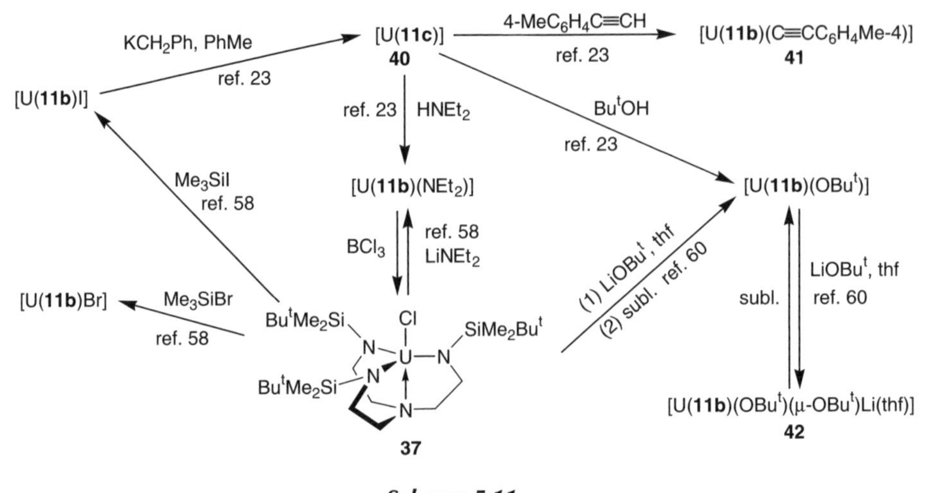

Scheme 5.11

triple bridging (assigned on the basis of the U-B distance) in the crystalline distorted octahedral complex **39**.[59]

The thorium complexes [Th(**11b**)X] were obtained from [Th(**11b**)Cl] and Me_3SiX (X = Br, I).[58] Each of the crystalline complexes [An(**11b**)X′] showed three-fold symmetry, illustrated schematically for **37** in Scheme 5.11, as established crystallographically for the U compounds [U(**11b**)Y] [Y = Cl (**37**), Br, I, NEt_2,[58] $C{\equiv}CC_6H_4Me$-4 (**41**)[23]]; the acetylide **41** had the angle U–C–C 156°.[23] The crystalline [U(**11c**)] (**40**) (structure in Figure 5.1, having a long U–C bond of ~2.75 Å) readily underwent protonolyses; as well as the three examples shown in Scheme 5.10, it reacted with $[HNMe_3]Cl$ yielding **37**.[23] A further U^{IV}(**11b**)-containing compound is the distorted octahedral crystalline uranium hydroxide [U(**11b**)(OH)(CH_2PMe_3)], obtained adventitiously as a minor product from the reaction of [U^{III}(**11b**)] (see Section 5.3) and Me_3PCH_2.[61]

Uranium(IV) halides derived from 1,4,7-tris(dimethylsilylphenylamine)-1,4,7-triazacyclononane H_3(**12**) have been prepared by Marques and coworkers as shown in Scheme 5.12.[24,62] Crystalline [U(**12**)Cl] (**43**) is a bicapped trigonal (N2, N5, N6) bipyramidal

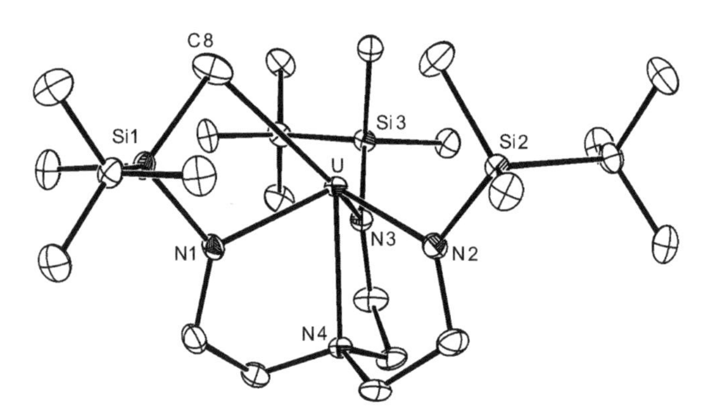

Figure 5.1 *Molecular structure of* ***40***[23]

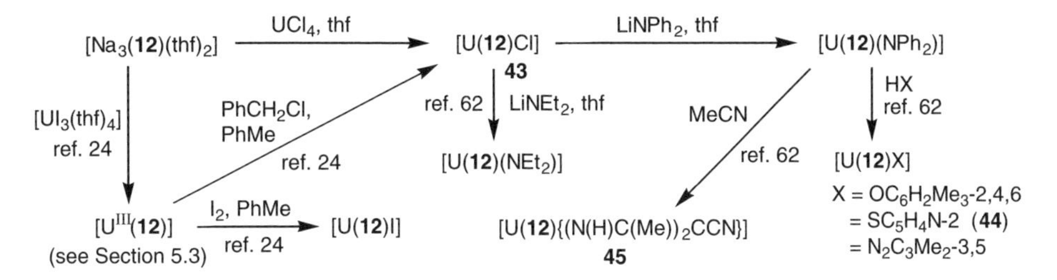

Scheme 5.12

polyhedron, sketched schematically in Figure 5.2 and as an ORTEP representation in Figure 5.3.[24] It proved to be highly fluxional in CD_2Cl_2 solution; such dynamic processes were less facile for the isoleptic iodide.[24] Treatment of **43** with $LiNR_2$ gave the crystalline trigonal prismatic diethylamide or the NPh_2 analogue.[62] The latter readily underwent protonolysis; complex **44** [also obtained from [U(**12**)] and $(C_5H_4N)SS(C_5H_4N)$] is dodecahedral, the thiopyridyl ligand being an *S,N*-chelate.[62] The reaction of the diphenylamide with acetonitrile surprisingly gave the X-ray-characterised crystalline dodecahedral β-diketiminate **45**, shown schematically in Figure 5.4.[62]

The amides $[UCl_2\{N(SiEt_3)_2\}_2]$[63] and the crystalline C_2-symmetric octahedral $[UCl_2\{N(SiMe_3)_2\}_2(dme)]$[63] were obtained from UCl_4 and two equivalents of the appropriate alkali metal amide.

The tetraamide $[U\{Fc(NSiBu^tMe_2)_2\}_2]$ was prepared from the potassium salt of 1,1′-ferrocenylenediamine $Fc\{N(H)SiBu^tMe_2\}_2$ and $UI_3(thf)_4$. Its oxidation by I_2 followed by treatment with $NaBPh_4$ produced the mixed-valence (Fe^{II}/Fe^{III}) compound $[U\{Fc(NSiBu^tMe_2)_2\}_2][BPh_4]$, in which the U^{IV} centre mediates the electronic communication. The model of electron transfer between Fe^{II} and Fe^{III} via a direct U-Fe interaction was confirmed by magnetic measurements, EPR, NIR and IR spectroscopy and DFT calculations on model systems.[65]

5.2.4 An^{IV} Amides Containing π-Centred Co-Ligands

Presently known amidouranium(IV) compounds containing a single cyclopentadienyl ligand are $[U(\eta^5\text{-}C_5Me_4R)(NEt_2)_3]$ (R = Me, Et),[34a] $[AnCp(NEt_2)_3]$ (An = U,[32,33] Th[33]), $\{UCp(NEt_2)_2(PPh_2)\}$,[32] $[U(\eta^5\text{-}C_5H_4PPh_2)(NEt_2)_3]$;[34a,34b] and the X-ray-characterised

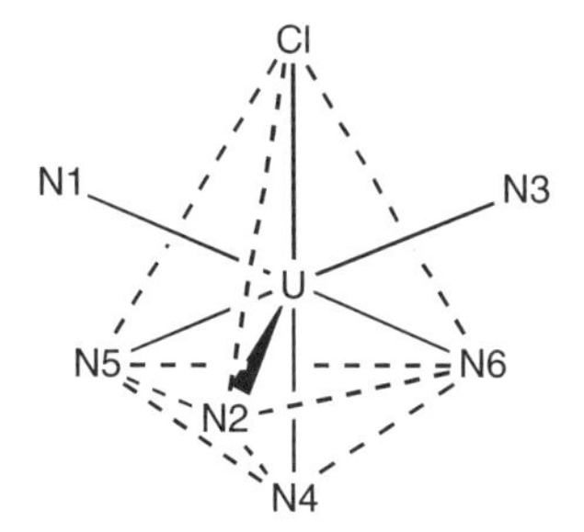

Figure 5.2[24]

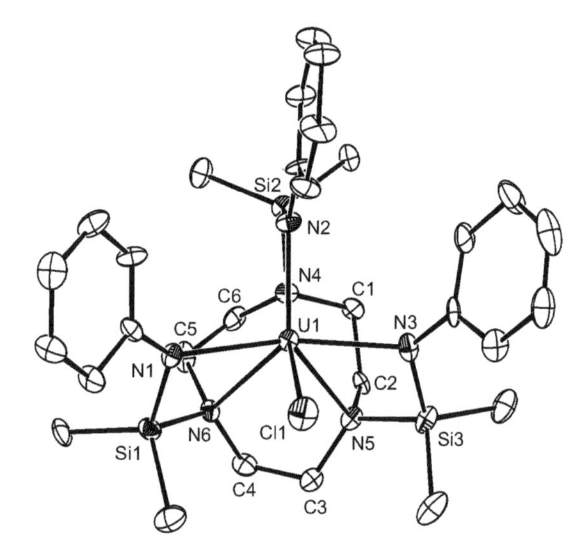

Figure 5.3[24]

crystalline [Th(Cp*)(cot){N($SiMe_3$)$_2$}], obtained from Th(Cp*)(cot)Cl(thf)$_x$ and Na[N-($SiMe_3$)$_2$].[64] The other compounds were prepared by protonolysis from the appropriate An(NEt_2)$_4$ or [UCp(NEt_2)$_3$] with $HPPh_2$.[32] [UCp(NEt_2)$_3$] was stable to disproportionation,[33] unlike the postulated {UCp(NEt_2)$_2$(PPh_2)}.[32] Compounds [An(**5**)(NMe_2)$_2$] (An = U, Th) were used as hydroamination/cyclisation catalysts.[66]

The reported bis(cyclopentadienyl)amido-An^{IV} compounds are rather more numerous. As mentioned in Ref. 1, the first such compound [UCp_2(NEt_2)$_2$] was obtained from U(NEt_2)$_4$ and 2HCp[67] (later also from UCl_4 and 2TlCp and $2LiNEt_2$, or for an analogue: $2KNPh_2$; [$ThCp_2$(NEt_2)$_2$] was prepared similarly[33]). Some protolytic cleavages [UCp_2-(NEt_2)$_2$]/1 or 2HX are long known;[67] post-1984 examples are in Scheme 5.13. Noteworthy features are that using (i) the bulky 2,6-$Bu^t_2C_6H_3OH$ only one NEt_2 group was displaceable;[70] and (ii) 2HX, with X = tropolonato[68] or diphenylphosphido,[32] the appropriate disproportionation product [UCp_3(PPh_2)] or [UCp(trop)$_3$] was isolated. [UCp_3(PPh_2)] was also obtained from [UCp(NEt_2)$_3$] and $HPPh_2$ or [U(NEt_2)$_3$(PPh_2)] and HCp.[32]

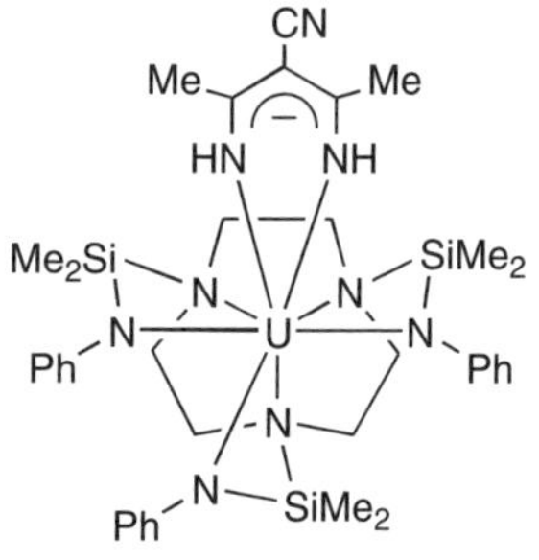

Figure 5.4[62]

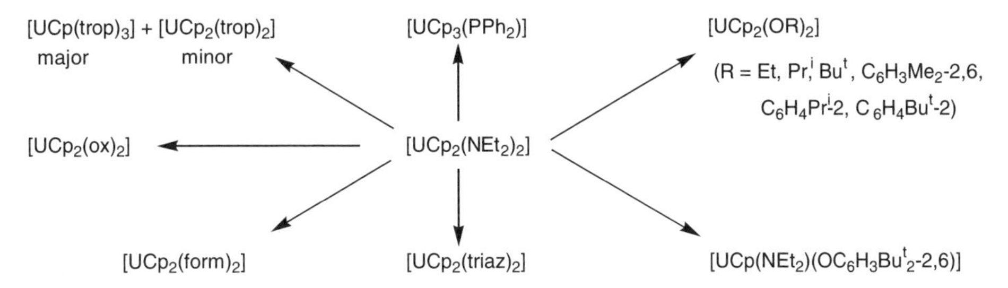

Scheme 5.13 *Reactions of [UCp$_2$(NEt$_2$)$_2$] with 2HX [X = trop,[68] PPh$_2$,[32] OR,[70] N{N(C$_6$H$_4$Me-4)}$_2$ (≡ triaz),[69] {N(C$_6$H$_4$Me-4)}$_2$CH (≡ form),[69] ox[71]]*

The compound [UCp″$_2$(NMe$_2$)$_2$], [Cp″ = η^5-C$_5$H$_3$(SiMe$_3$)$_2$-1,3] (obtained from [UCp″$_2$Cl$_2$] + 2LiNMe$_2$)[26] when treated with H$_2$O or BF$_3$·OEt$_2$ yielded [{UCp″$_2$(μ-O)}$_2$][26] or [UCp″$_2$F$_2$];[72a] [UCp″$_2$(Cl){N(SiMe$_3$)$_2$}] was prepared by the redox reaction of Equation (5.2).[26]

$$2[\mathrm{UCp''_2(Cl)(thf)}] + [\mathrm{Sn\{N(SiMe_3)_2\}_2}] \longrightarrow 2[\mathrm{UCp''_2(Cl)\{N(SiMe_3)_2\}}] + \mathrm{Sn} \qquad (5.2)$$

Several primary (alkyl or aryl)amidouranocene(IV) complexes have been prepared from the appropriate dimethyluranocene and RNH$_2$: the thermally robust [U(η^5-C$_5$H$_3$R$_2$-1,3)$_2${N(H)C$_6$H$_4$Me-4}$_2$] (R = Bu^t, SiMe$_3$) and [U(η^5-C$_5$H$_2$Bu$^t_3$-1,2,4)$_2${N(H)R}$_2$] (R = Me, CH$_2$Ph, C$_6$H$_4$Me-4). In solution the latter compounds were in equilibrium with the imides [U(η^5-C$_5$H$_2$Bu$^t_3$-1,2,4)$_2$(=NR)] + RNH$_2$; and the compound with R = Me reacted with R′C≡CR′ to give [(η^5-C$_5$H$_2$Bu$^t_3$-1,2,4)$_2$U{N(Me)C(R′)=CR′}] which with excess of MeNH$_2$ produced [U(η^5-C$_5$H$_2$Bu$^t_3$-1,2,4)$_2${N(H)Me}$_2$] and the hydroamination product MeN=C(R′)CH$_2$R′ (R′ = Me, Ph).[72b]

The He(I) and He(II) photoelectron spectra of [UCp$_2$(NEt$_2$)$_2$] have been recorded; the first and second ionisation potentials were assumed as arising from electron loss from the 5f and a nitrogen-centred orbital, respectively.[73]

The first 1 : 1- insertions into An^{IV}NR$_2$ bonds, Equation (5.3), were reported in 1981 by Marks' group in the context of [An(Cp*)$_2$Cl$_{2-n}$(NEt$_2$)$_n$] chemistry (n = 1 or 2). The latter compounds were prepared from [AnCp$^*_2$Cl$_2$] and LiNR$_2$ (R = Me, Et) or for [UCp$^*_2$(NEt$_2$)$_2$] from [UCp$^*_2$Cl$_2$] and an excess of HNEt$_2$;[74] treatment of [Th(Cp*)$_2$Cl(NEt$_2$)] with LiMe-LiBr yielded [Th(Cp*)$_2$Me(NEt$_2$)], which with CO furnished [Th(Cp*)$_2${η^2-C(O)Me}(NEt$_2$)]. The X-ray structures of two of the crystalline carbamoylAn^{IV} compounds are shown schematically in **46** and **47**.[74] The An–O bonds are significantly shorter than the An–C bonds, reflecting the high oxophilicity of An^{4+}; as also evident from the abnormally low IR ν(CO) stretching frequencies for the various carbamoylAn^{IV} complexes, indicative also of appreciable N–C π-dative-bonding.

$$[\mathrm{An(Cp^*)_2(NR_2)X}] \xrightarrow{\mathrm{CO,\ PhMe}} [(\eta^5\text{-}\mathrm{Cp^*})_2\mathrm{An(X)}\{\eta^2\text{-}\mathrm{C(O)NR_2}\}] \qquad (5.3)$$

X = Cl, NR$_2$, η^2-C(O)NR$_2$

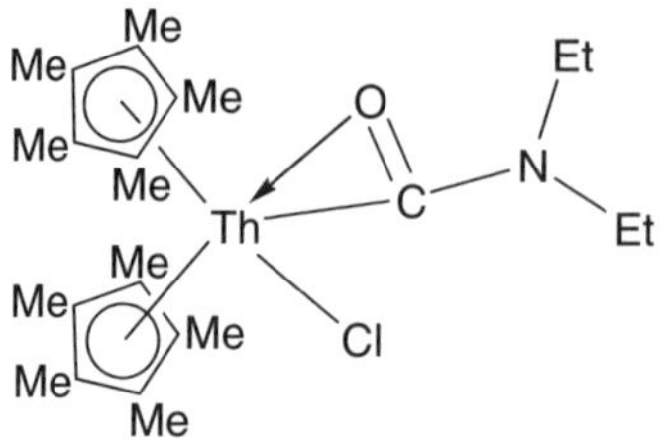

46 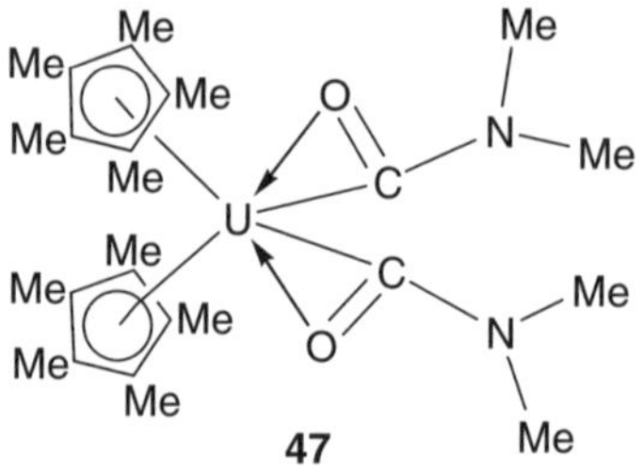 **47**

Insertion reactions of RNC ($R = 2,6\text{-}Me_2C_6H_3$, $c\text{-}C_6H_{11}$) into U–N bonds of $[UCp_2(NEt_2)_2]$ and $[U(Cp^*)_2Cl(NEt_2)]$ or into the U–C bond of $[U(Cp^*)_2Me(NEt_2)]$ have yielded $[UCp_2\{\eta^2\text{-}C(NR)NEt_2\}_n(NEt_2)_{2\text{-}n}]$ ($n = 1$ or 2), $[U(Cp^*)_2Cl\{\eta^2\text{-}C(NC_6H_{11})NEt_2\}]$ or $[U(Cp^*)_2\{\eta^2\text{-}C(NC_6H_{11})Me\}(NEt_2)]$, respectively.[75] Contrary to an earlier report,[75] $[UCp_3(NEt_2)]$ (see Scheme 5.2) and $2,6\text{-}Me_2C_6H_3NC$ in Et_2O afforded the X-ray-characterised crystalline insertion product **48**,[76] and with CO gave the carbamoyl complex $[UCp_3\{\eta^2\text{-}C(O)NEt_2\}]$.[77] Each of the compounds $[UCp_4]$,[32,36] **49**,[33] $[UCp_3(OR)]$,[69] $[UCp_2(NEt_2)_2]$,[33] $[UCp_2(form)_2]$[69] or $[UCp_2(triaz)_2]$[69] was obtained from $[UCp_3(NEt_2)]$ and C_5H_6, pyrazole, ROH, $LiNEt_2$, $HN(C_6H_4Me\text{-}4)C(H){=}NC_6H_4Me\text{-}4$ or $HN(C_6H_4Me\text{-}4)\text{-}N{=}NC_6H_4Me\text{-}4$, respectively; but this amidotri(cyclopentadienyl)uranium compound failed to react with $HPPh_2$.[32] The compound $[U(\eta^5\text{-}C_5H_4PPh_2)_4]$ was isolated from $[U(NEt_2)_4]$ and 4 equivalents of $C_5H_5PPh_2$.[34b] The complexes $[An(Cp^x)_2(CH_2Si(Me)_2NSiMe_3)]$ and RNC gave **50** ($Cp^x = Cp$, $R = Bu^t$ or Cy; $Cp^x = C_5H_4Me$, $R = Bu^t$) (see also Scheme 5.9).[51]

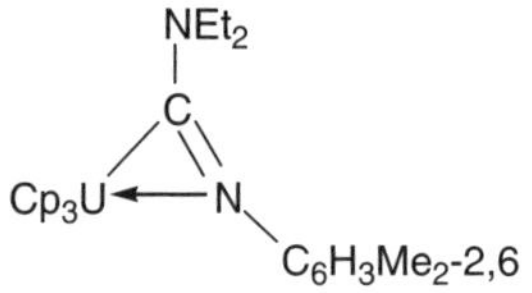

48 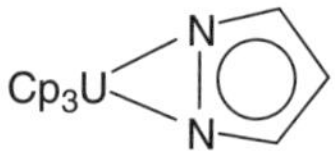 **49** 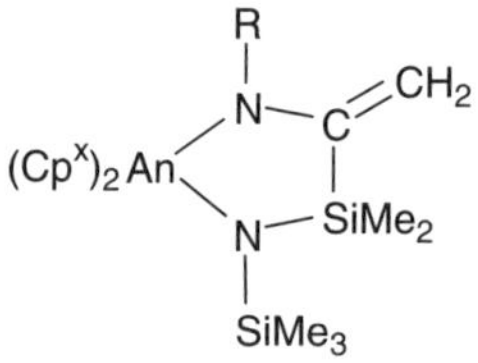 **50**

Treatment of $[UCp''_2Cl_2]$ with $LiNMe_2$ or $Li[N(SiMe_3)_2]$ afforded $[U(Cp'')_2Cl(NR_2)]$ [$R = Me$[25] or $SiMe_3$;[26] $Cp'' = \eta^5\text{-}C_5H_3(SiMe_3)_2\text{-}1,3$]. $[U(cot)(NEt_2)(thf)_2][BPh_4]$ and KCp^x in thf furnished $[U(Cp^x)(cot)(NR_2)]$ ($Cp^x = Cp$ or Cp^*);[30b] while the crystalline, X-ray-characterised $[Th(Cp^*)(cot)\{N(SiMe_3)_2\}]$ was isolated from $[Th(Cp^*)(cot)Cl(thf)_n]$ and $Na[N(SiMe_3)_2]$ in toluene.[64]

A number of cyclopentadienyl-free amidoAn^{IV}-cyclooctatetraene complexes have been reported. Sattelberger's group found that $[An(cot)Cl_2(thf)_2]$ with $2Na[N(SiMe_3)_2]$ in toluene yielded the crystalline $[An(cot)\{N(SiMe_3)_2\}_2]$, which for An = Th was X-ray-characterised as a 'four-legged piano stool', with the Th···Centroid distance of 2.04 Å, each amido ligand having one short Th···C contact (3.15 or 3.04 Å).[78] The compounds $[U(cot)(NEt_2)_2]$[30b] and its thf adduct[30c] were obtained from $[U(NEt_2)_2(thf)_3][BPh_4]_2$ and $K_2[cot]$; $[U(cot)(NEt_2)\{N(SiMe_3)_2\}]$ was prepared from $[U(cot)(NEt_2)(thf)_2][BPh_4]$ and $Na[N(SiMe_3)_2]$.[79a] This tetraphenylborate salt with $Li[CH(SiMe_3)_2]$ gave $[U(cot)\{CH(SiMe_3)_2\}(NEt_2)]$, and with LiCl or $K[BH_4]$ in thf furnished $[U(cot)(NEt_2)X(thf)_n]$ ($X = Cl$ or BH_4).[79a] Protolytic reaction of $[U(cot)\{N(SiMe_3)_2\}_2]$ with a tetraamine produced the dinuclear $[\{U(cot)\}_2\{\mu\text{-}\eta^4:\eta^4\text{-}N(H)(CH_2)_3N(CH_2)_2N(CH_2)_3NH\}]$.[79b]

5.3 Neutral U^{III} Amides

Andersen prepared the volatile [U{N(SiMe$_3$)$_2$}$_3$] (**51**), the first U^{III} amide in 1979, from UCl$_3$ and 3Na[N(SiMe$_3$)$_2$] in thf.[80] An improved synthesis used [UI$_3$(thf)$_4$] as precursor;[81a] the isoleptic Np^{III} and Pu^{III} amides were obtained similarly.[8] The crystalline **51** is pyramidal;[14] its He(I) and He(II) photoelectron spectra have been recorded.[82] Reactions of **51** are summarised in Scheme 5.14; each of the crystalline products, except the U^{V} compound [U{N(SiMe$_3$)$_2$}$_3$(=O)],[80] was X-ray-characterised. The tetra(amido)uranate-(III) (**52**) has a distorted tetrahedral arrangement of the four nitrogen atoms around the central U.[84] Treatment of [UI{N(Bu^t)C$_6$H$_3$Me$_2$-3,5}$_3$] (**15**) with sodium-amalgam in thf gave the crystalline distorted tetrahedral amide [U{N(Bu^t)C$_6$H$_3$Me$_2$-3,5}$_3$(thf)] (**16**) having close U$\cdots$C$_{ipso}$ contacts of *ca.* 2.9 Å; some of its oxidative reactions are shown in Scheme 5.4.[36] The hexadentate ligand **53** is a triphenolate, based on an *N*,*N*′,*N*″-triazacyclononane skeleton.[86]

53

Using the lignd N(CH$_2$CH$_2\bar{N}$SiMe$_2$Bu^t)$_3$ (**11b**), Scott and coworkers have carried out some interesting chemistry based on [U(**11b**)] (**54**)[13] [prepared as shown in Equation (5.4)]; this is summarised in Scheme 5.15 (all reactions in C$_5$H$_{12}$). The crystalline dinitrogen

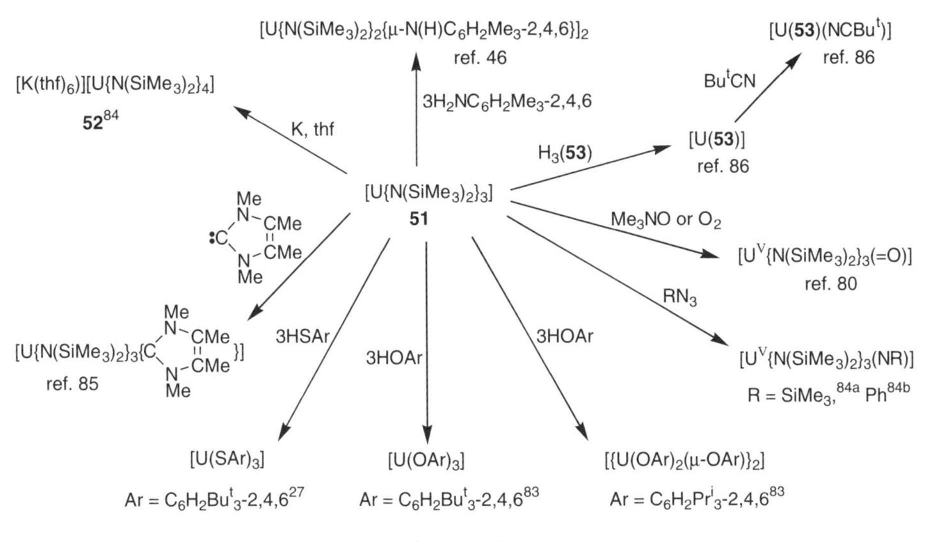

Scheme 5.14

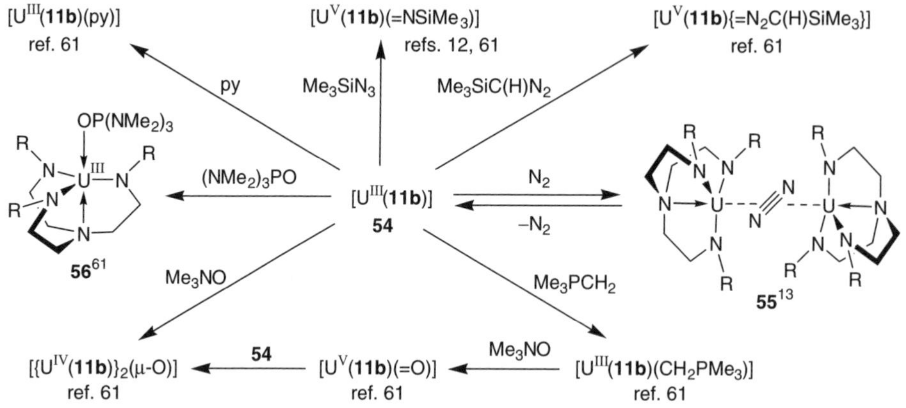

Scheme 5.15 ($R = SiMe_2Bu^t$)

complex **55** has the two trigonal monopyramidal mutually staggered U(**11b**) fragments with a side-on bridging N_2 moiety; the U^{III} assignment was supported by UV-visible spectroscopy and magnetochemistry.[13] In contrast, DFT calculations on a model compound (see Section 5.2.2) were consistent with a $U^{IV}(N_2)^{2-}U^{IV}$ assignment.[39] The other crystalline U^{III} compounds of Scheme 5.15 were formulated as having the three-fold symmetrical structure, as established crystallographically for the hmpa adduct **56**,[61] but were shown to be fluxional in solution.[61]

$$[U^{IV}(\mathbf{11b})Cl] \xrightarrow{K,\ C_5H_{12}} [\{U(\mathbf{11b})\}_2(\mu\text{-}Cl)] \xrightarrow{120\ ^{\circ}C/10^{-6}\,mbar} [U^{III}(\mathbf{11b})] \qquad (5.4)$$

37 (Scheme 5.11) → **58**[12] [see eqn. (5.5), Section 5.4] → **54**[13]

Scheme 5.12 summarised some U^{IV} amido chemistry based on the trianionic ligand **12**; three structures were illustrated in Figures 5.2–5.4.[24,62] Reacting $[UI_3(thf)_4]$ with Na_3(**12**) in thf afforded the six-coordinate crystalline complex $[U^{III}$(**12**)]·0.5PhMe, described as a slightly distorted trigonal prism with the two trigonal planes defined by the two sets (amine and amido) of nitrogen donor atoms being nearly parallel.[24]

Evans *et al.* obtained $[U(Cp^*)_2\{N(SiMe_3)_2\}]$ (this compound was earlier prepared from $[\{U(Cp^*)_2(\mu\text{-}Cl)\}_3]$ and $Na[N(SiMe_3)_2]$[88]) by replacing one of the Cp^* ligands in the highly crowded U^{III} compound $[UCp^*_3]$ using $K[N(SiMe_3)_2]$.[89a] Starting from $[\{U(Cp^*)_2\}_2(\mu\text{-}\eta^6:\eta^6\text{-}C_6H_6)]$ (which was regarded as a six-electron reducing agent) the same approach gave the crystalline mixed ligand derivative $[(U(Cp^*)\{N(SiMe_3)_2\})_2(\mu\text{-}\eta^6:\eta^6\text{-}C_6H_6)]$ (**57**, Figure 5.5).[89a] The application of U^{III} compounds including [U(**12**)], **16**, **51**, [U(**53**)] and **54** as reducing agents has been reviewed.[89b]

$[U\{N(SiMe_3)_2\}Cl_2]$ and 2NaCp in thf gave $[UCp_3(thf)]$.[90]

5.4 Neutral Mixed Valence (U^{III}/U^{IV}), U^{II}, U^{V} and U^{VI} Amides

As shown in Equation (5.4) (Section 5.3), treatment of $[U^{IV}$(**11b**)Cl] with a potassium film in pentane yielded the mixed valence compound $[\{U^{III/IV}$(**11b**)$\}_2(\mu\text{-}Cl)]$ (**58**).[12] The

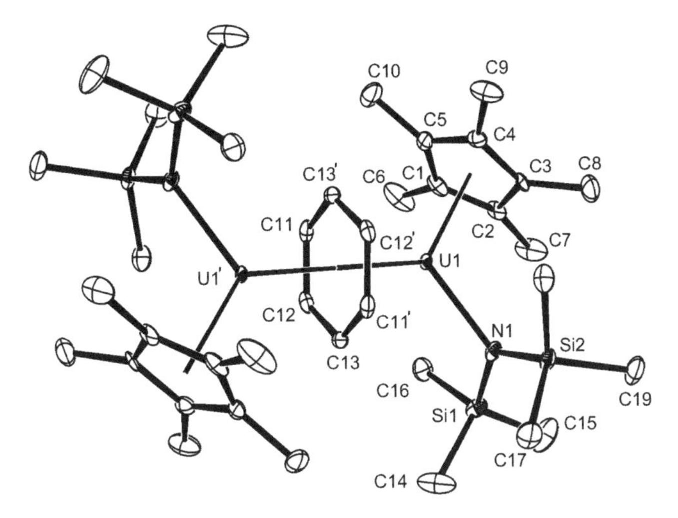

Figure 5.5 *Molecular structure of **57**[89a]*

crystalline compound has D_{3d} symmetry with a C_3 axis (N–U–Cl–U–N) and an inversion centre at the Cl atom. The $U^{III/IV}$ assignment was partly based on the reaction with trimethylsilyl azide, Equation (5.5).[12]

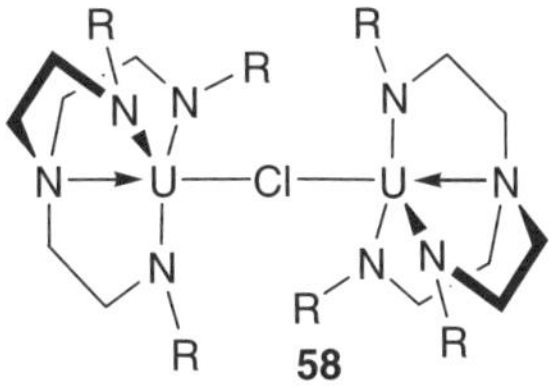

$$\mathbf{58} + Me_3SiN_3 \longrightarrow [U^{V}(\mathbf{11b})(=NSiMe_3)] + [U^{IV}(\mathbf{11b})Cl] \qquad (5.5)$$

As shown in Scheme 5.5 (Section 5.2.2), Cummins and coworkers have reported that reduction of $[U^{IV}I(NRR')_3]$ ($R = Bu^t$, $R' = C_6H_3Me_2$-3,5) (**15**) with KC_8 in toluene yielded $[\{U(NRR')_2\}_2(\mu\text{-}\eta^6:\eta^6\text{-}PhMe)]$ (**22**).[11] An analogue (**22′**) with R = 1-adamantyl was prepared similarly [as was the $\mu\text{-}\eta^6:\eta^6\text{-}C_6H_6$ ($R = Bu^t$) compound]. The X-ray structure of crystalline **22′** is sketched in Figure 5.6. The U–C distances range from the shortest 2.503 Å (U2–C3) to the longest (U1–C3), 2.660 Å; the C3 atom is displaced slightly from the mean plane of the toluene ligand. The preferred assignments for **22** and **22′** were $(U^{II})_2$/[PhMe], but $(U^{III})_2/[PhMe]^{2-}$ was not excluded.[11] The calculated structure of

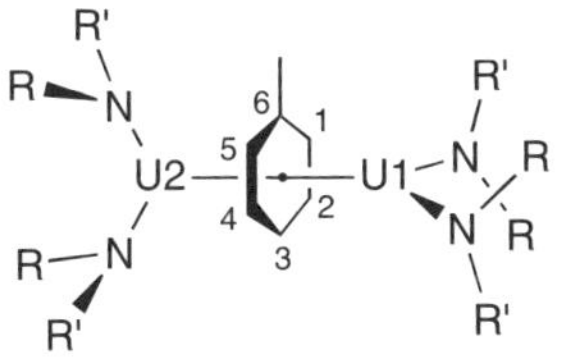

Figure 5.6 *Structure of **22′** (R = 1-Ad, R′ = $C_6H_3Me_2$-3,5)*

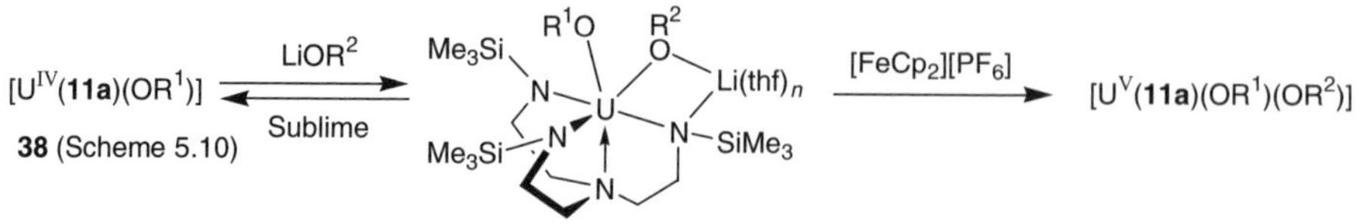

$[U^{IV}(\mathbf{11a})(OR^1)]$ **38** (Scheme 5.10) $\underset{\text{Sublime}}{\overset{LiOR^2}{\rightleftarrows}}$ **42'** $\xrightarrow{[FeCp_2][PF_6]}$ $[U^{V}(\mathbf{11a})(OR^1)(OR^2)]$

42' (see Scheme 5.11 for the **11b** analogue)

Scheme 5.16

$[\{U(NH_2)_2\}_2(\mu\text{-}\eta^6:\eta^6\text{-}C_6H_6)]$, constrained to D_2 symmetry, reproduced the key features of the structure of **22′** quite closely; it was suggested that δ symmetry back-bonding represents a vehicle for gaining access to a U^{II} synthon in the context of arene binding.[11] Similar bridging arenes or a cycloheptatrienyl (C_7H_7) have been observed *inter alia* in **57** {assigned as $(U^{III})_2/[PhMe]^{2-}$},[89a] $[K(18\text{-crown-6})(\eta^2\text{-}C_6H_6)_2][\{La(\eta^5\text{-}C_5H_3Bu^t{}_2\text{-}1,3)_2\}_2(\mu\text{-}\eta^6:\eta^6\text{-}C_6H_6)]$ {assigned as $(La^{II})_2/[C_6H_6]^-$},[91] and $[U(BH_4)_2(thf)_5][\{U(BH_4)_3\}_2(\mu\text{-}\eta^7:\eta^7\text{-}C_7H_7)]$ and $[K(18\text{-crown-6})][\{U(NEt_2)_3\}_2(\mu\text{-}\eta^7:\eta^7\text{-}C_7H_7)]$ {assigned as $(U^{IV})_2/[C_7H_7]^{3-}$}.[92]

The first amidouranium(V) compounds $[U\{N(SiMe_3)_2\}_3(=O)]$[80] and $[U\{N(SiMe_3)_2\}_3(=NR)]$ ($R = SiMe_3$,[87a] Ph[87b]) were prepared about 20 years ago by oxidation of $[U\{N(SiMe_3)_2\}_3]$ with Me_3NO or O_2[80] and the azide RN_3,[87] respectively (Scheme 5.14, Section 5.2.5); the latter were crystallographically characterised.[84] Likewise (Scheme 5.15, Section 5.2.5) the oxidation of $[U^{III}(\mathbf{11b})]$ with Me_3SiN_3 yielded $[U^{V}(\mathbf{11b})(=NSiMe_3)]$,[12,61] and $[U^{III}(\mathbf{11b})(CH_2PMe_3)]$ with Me_3NO gave $[U^{V}(\mathbf{11b})(=O)]$.[61] The compounds $[U(\mathbf{11a})(OR^1)(OR^2)]$ ($R^1 = Ph = R^2$; $R^1 = Bu^t = R^2$; $R^1 = Ph$, $R^2 = Bu^t$) were obtained as shown in Scheme 5.16.[60] Another oxidative route is illustrated in Equation (5.6).[30a,30c]

$$Na[U^{IV}(NR_2)_3X_2] \underset{Na/Hg}{\overset{Tl[BPh_4]}{\rightleftarrows}} [U^{V}(NR_2)_3X_2] + Tl + Na[BPh_4] \qquad (5.6)$$

$R = Me, Et; X = NR_2$

$R = Me, X_2 = cot$

Treatment of UCl_4 with $3Li[N\{(CH_2)_2PPr^i{}_2\}_2]$ ($\equiv 3Li[\mathbf{2}]$) in thf adventitiously afforded the crystalline *trans*-octahedral U^V compound **59**, which was isolated in a better yield when the reaction was conducted in the presence of the stoichiometric amount of O_2.[40]

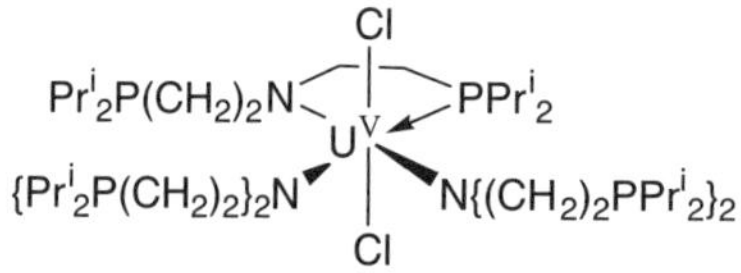

Cl; $Pr^i{}_2P(CH_2)_2N$; $PPr^i{}_2$; U^V; $\{Pr^i{}_2P(CH_2)_2\}_2N$; $N\{(CH_2)_2PPr^i{}_2\}_2$; Cl

59

Andersen prepared the first U^{VI} amide *trans*-$[UO_2\{N(SiMe_3)_2\}_2(thf)_2]$ (**60**) in 1979, from UO_2Cl_2 and $2Na[N(SiMe_3)_2]$ in thf.[9] The structure of the crystalline material was confirmed by X-ray data, albeit of poor quality;[93a] recently a high-quality structure was obtained.[93b] The amido groups in **60** were readily displaced by treatment with ArOH and pyridine, yielding $[UO_2(OAr)_2(py)_3]$ ($Ar = C_6H_3Pr^i{}_2\text{-}2,6$).[93a] Reaction of **60** with the macrocyclic compound $H_4(\mathbf{10})$ ($R = H$ or Me) resulted in the exclusive formation of the mono-uranyl complex $[UO_2\{H_2(\mathbf{10})\}(thf)]$ (**61**).[94a] The latter upon addition of

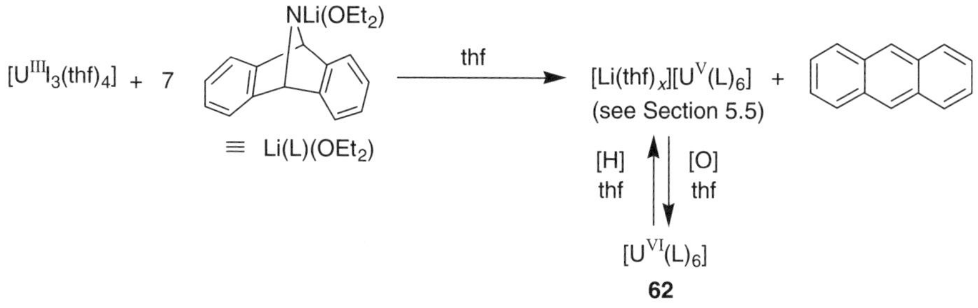

$[U^{III}I_3(thf)_4]$ + 7 NLi(OEt$_2$) ≡ Li(L)(OEt$_2$) —thf→ $[Li(thf)_x][U^V(L)_6]$ + anthracene (see Section 5.5); [H] thf ↑ ↓ [O] thf; $[U^{VI}(L)_6]$ **62**

Scheme 5.17

$[(M\{N(SiMe_3)_2\}_2)_2]$ gave the heterobimetallic $[(thf)UO_2(\mathbf{10})M(thf)]$ (M = Mn, Fe, Co; R = H),[94b] in which one O atom of the uranyl cation acts as a donor to the transition metal. Unprecedented monosilylation of the uranyl group (accompanied by U^{VI} to U^V reduction) occurred when **61** was treated with successively $K\{N(SiMe_3)_2\}_2$ (or KH and the silylating reagent $N(SiMe_3)_3$ or $PhCH_2SiMe_3$) and FeI_2 yielding $[UO(OSiMe_3)(thf)(\mathbf{10})Fe_2I_2]$ (R = Me); similar trinuclear compounds were obtained with ZnI_2 and $ZnCl_2$.[94c]

Whereas $U(NMe_2)_6$ {prepared by AgI oxidation of $[Li(thf)]_2[U(NMe_2)_6]$[95]} was labile,[95] the homoleptic crystalline, octahedral U^{VI} amide **62** was robust; its synthesis is illustrated in Scheme 5.17.[96] The oxidant was $[FeCp_2][OTf]$, AgOTf, I_2 or air; and the reducing agent was $[CoCp_2]$ or Na/Hg. Compound **62** underwent reversible one-electron reduction at −1.01 V relative to $[FeCp_2]/[FeCp_2]^+$.[96] The crystalline trigonal bipyramidal (F and NR axial) U^{VI} compounds **63a** and **63b** were prepared as shown in Equation (5.7);[97]

$$[U^V\{N(SiMe_3)_2\}_3(=NR)] + Ag[PF_6] \longrightarrow [U^{VI}(F)\{N(SiMe_3)_2\}_3(=NR)] + Ag + PF_5 \quad (5.7)$$

63a R = $SiMe_3$
63b R = Ph

each of the U^V precursors underwent reversible one-electron oxidation at *ca.* –0.40 V relative to ferrocene.[97] The uranyl(VI) amides **64a** and **64b**, based on the ligand **3** were obtained as shown in Scheme 5.18;[19b,98] **64a** showed significant distortion around the U-carbene bonds, while **64b** was virtually undistorted.[98]

5.5 Amidouranates

Two homoleptic amidouranate(III)s have been reported. The crystalline complex $[K(thf)_2]_2[U\{N(H)C_6H_3Pr^i_2\text{-}2,6\}_5]$ was isolated from the reaction of $[UI_3(thf)_4]$ and a large excess of $K[N(H)C_6H_3Pr^i_2\text{-}2,6]$ in thf; the anion is trigonal bipyramidal having close $(K\cdots\eta^6\text{-aryl})_2$ contacts.[99] The crystalline tetraamidouranate(III) salt $[K(thf)_6][U\{N(SiMe_3)_2\}_4]$ (**52**) was obtained from $[U\{N(SiMe_3)_2\}_3]$ and potassium in thf under N_2 (Scheme 5.14);[84] the formation of **52** was unexpected as Ln dinitrogen complexes $[\{Ln(N(SiMe_3)_2)_2(thf)\}_2(\mu\text{-}\eta^2:\eta^2\text{-}N_2)]$ (Ln = Nd, Gd, Tb, Dy) were isolated under the same reaction conditions.[84]

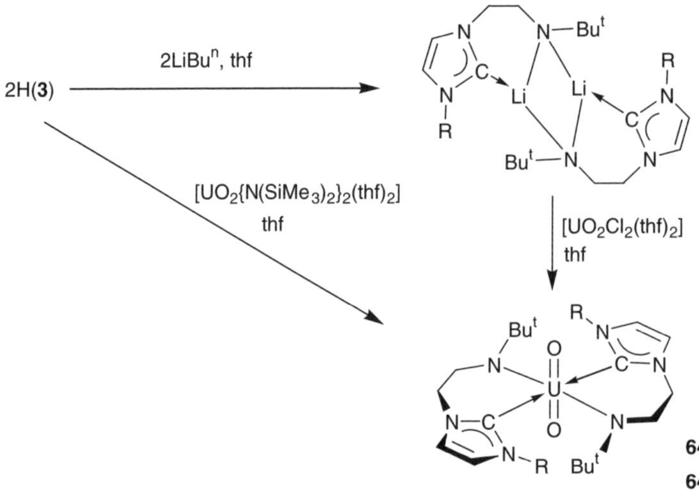

2H(**3**) — $2LiBu^n$, thf; $[UO_2\{N(SiMe_3)_2\}_2(thf)_2]$, thf; $[UO_2Cl_2(thf)_2]$, thf

64a R = Bu^t
64b R = $C_6H_2Me_3$-2,4,6

Scheme 5.18

The amidouranate(IV)s are more numerous. The homoleptic compounds $M[U(NEt_2)_5]$ [M = Li(thf) or Na] [see also Equation (5.6), Section 5.4] were prepared either from $[\{U(NEt_2)_3(\mu\text{-}NEt_2)\}_2]$ and $MNEt_2$ in thf or from UCl_4 and an excess of $LiNEt_2$ in thf.[30a] The compounds $[U(\mathbf{11a})(OR)_2Li(thf)_x]$ (R = Bu^t and $x = 1$, R = Ph and $x = 2$) and the X-ray-characterised capped trigonal bipyramidal $[U(\mathbf{11a})(OBu^t)(OPh)Li(thf)]$ (**42′**) were synthesised [Scheme 5.16 (Section 5.4); see also Scheme 5.11 (Section 5.2.3) for the **11b** analogue **42**] from $[\{U(\mathbf{11a})(\mu\text{-}Cl)\}_2]$ and excess LiOR or $[U(\mathbf{11a})(OPh)]$ and $LiOBu^t$.[60] Some chemistry involving $M[U(cot)(NEt_2)_3]$ is outlined in Scheme 5.19.[30a,30c] A similar $[U^{IV}]^-$/U^V interconversion involved $Na[UCp(cot)(NEt_2)_2]$ and $[UCp(cot)(NEt_2)_2]$.[30c]

Treatment of UO_2Cl_2 with $4Na[N(SiMe_3)_2]$ gave a compound tentatively formulated as $[Na(thf)_2]_2[U^{VI}O_2\{N(SiMe_3)_2\}_4]$,[93a] which with C_5Me_5H yielded the crystalline salt $[Na(thf)_2][U^{VI}O_2\{N(SiMe_3)_2\}_3]$ containing the trigonal bipyramidal U^{VI} atom with transoid O^{2-} ligands.[10] Although $[UO_2Cl_2(thf)_2]$ did not furnish this salt with $3Na[N(SiMe_3)_2]$ in thf, the potassium amide gave $[K(thf)_2][UO_2\{N(SiMe_3)_2\}_3]$.[10] The homoleptic lithium amidouranate(V) $[Li(thf)_x][U(L)_6]$ (see Scheme 5.17) upon crystallisation from a solution containing $[PPh_4]$Br afforded the crystalline X-ray-characterised salt $[PPh_4][U(L)_6]$.[96]

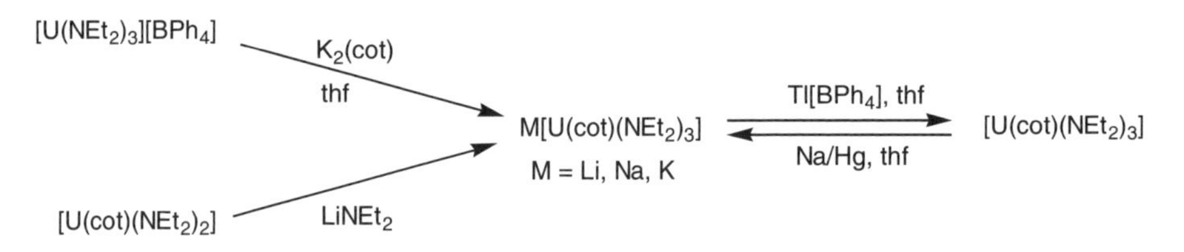

Scheme 5.19

The crystalline potassium U^{V}/U^{VI} mixed valent diuranate salt **65** was prepared as shown in Equation (5.8) (for **11b** see Section 5.2.1); each uranium atom is in a capped trigonal bipyramidal environment.[100]

$$[K(18\text{-cr-}6)]_2[UO_2Cl_4] + Li_3(\mathbf{11b}) \xrightarrow{Et_2O} [K(18\text{-cr-}6)(OEt_2)][\ldots] \quad \mathbf{65} \tag{5.8}$$

5.6 Amidouranium Tetraphenylborates

This is an area developed by Ephritikhine and coworkers; the majority (**66–71**) of the salts are monoanionic U^{IV} amides.[101] The synthetic strategy was based on the amine elimination reactions of Equation (5.9), as summarised in Table 5.1. From Table 5.1 it is evident that NEt_2 is more readily displaced from U than $N(SiMe_3)_2$ (**67**), Cp* (**68**), cot (**69**, **70**, **73**), or Cl (**71**) upon protolysis.

$$L_nU^{IV}(NEt_2)_2 \xrightarrow[-HNEt_2,-NEt_3]{[HNEt_3][BPh_4]} [L_nU^{IV}(NEt_2)][BPh_4] \xrightarrow[-HNEt_2,-NEt_3]{[HNEt_3][BPh_4]} [L_nU^{IV}][BPh_4]_2 \tag{5.9}$$

Treatment of $[UCl_2(NEt_2)_2]$, $[UCl_3(NEt_2)(thf)]$, or $[U(cot)(C_5R_5)(NEt_2)]$ with $n[NHEt_3]$-$[BPh_4]$ in thf afforded $[UCl_2(thf)_4][BPh_4]_2$ ($n = 2$),[17] $[UCl_3(thf)_2][BPh_4]$ ($n = 1$),[17] or $[U$-$(cot)(C_5R_5)(thf)_2][BPh_4]$ ($n = 1$, R = H or Me),[30b] respectively. $[UCp^*_2\{N(SiMe_3)_2\}]$ was unreactive towards $[NHEt_3][BPh_4]$ in thf, but with $[NH_4][BPh_4]$ it furnished $[UCp^*_2(thf)_2]$-$[BPh_4]$ or, in C_6H_6 in presence of dmpe, $[UCp^*_2(dmpe)][BPh_4]$.[31]

The uranium environment in each of the well separated cations of **66a**, **68**, **69**, **72a**, and **73** is distorted *fac*-octahedral (**66a**),[17] distorted trigonal bipyramidal (O-U-O′, 157°) (**68**),[30b] five-coordinate with angles at U ranging from 73° (O1-U-O2) to 133° (centroid of cot-U-N) (**69a**),[79] pentagonal bipyramidal (5N atoms ± 0.33 Å out of the mean plane) (**72a**), or distorted tetrahedral (**73**);[30] each of Cp* or cot is regarded as occupying a single coordination site.

Two reactions of $[U(NEt_2)_3][BPh_4]$ (**66**) in thf-d_8 have been reported, Scheme 5.20,[30a,30c,31] but the most extensive studies of reactivity of amidouranium tetraphenylborates have been of $[U(cot)(NEt_2)(thf)_2][BPh_4]$ (**69**), Scheme 5.21.[30b,31,79] Studies on the bis-(tetraphenylborate) (**72**) and the U^{V} (**73**) compounds are shown in Equation (5.10)[30a] and Scheme 5.22,[30c] respectively.

$$\underset{\mathbf{72}}{[U(NEt_2)_2(thf)_3][BPh_4]_2} \xrightarrow{K_2(cot)} [U(cot)(NEt_2)_2] \tag{5.10}$$

The amidouranium tetraphenylborate $[U(NEt_2)_3][BPh_4]$ (**66**) was an effective catalyst for the selective dimerisation of terminal alkynes, as shown in Scheme 5.23; a suggested key intermediate was $[U(C{\equiv}CR)(NEt_2)_2(\eta^2\text{-}CH{\equiv}CR)][BPh_4]$.[102,103]

Table 5.1 Amidouranium tetraphenylborates from reactions of an amidoU compound with $[NHEt_3][BPh_4]$ in thf

AmidoU tetraphenylborate	Substrate	X-ray structure/Comments	References
$[U(NEt_2)_3][BPh_4]$ (**66**)	$[U(NEt_2)_4]_2$	Reactions (Scheme 5.20);[30a,30c,31] Catalyst[102,103,104]	17, 30b
$[U(NEt_2)_3(thf)_3][BPh_4]$ (**66a**)	$[U(NEt_2)_4]_2$	X-ray[17]	17, 30a
$[U(NEt_2)\{N(SiMe_3)_2\}_2(thf)_x][BPh_4]$ (**67**)	$[U(NEt_2)_2\{N(SiMe_3)_2\}_2]$		31
$[UCp^*(NEt_2)_2(thf)_2][BPh_4]$ (**68**)	$[UCp^*(NEt_2)_3]$	X-ray[30b]	30b
$[U(cot)(NEt_2)(thf)_2][BPh_4]$ (**69**)	$[U(cot)(NEt_2)_2]$[30a]	X-ray of $(thf)_3$ cation **69a**;[79] Reactions (Scheme 5.21;[30a,30c,31] see also Section 5.2.4)[79]	30a, 79
$[U(cot)\{N(SiMe_3)_2\}(thf)_x][BPh_4]$ (**70**)	$[U(cot)(NEt_2)\{N(SiMe_3)_2\}]$		31, 76
$[UCl_2(NEt_2)(thf)_2][BPh_4]$ (**71**)	$[UCl_2(NEt_2)_2]$		17
$[U(NEt_2)_2(thf)_3][BPh_4]_2$ (**72**)	$[U(NEt_2)_3][BPh_4]$	Reaction [Equation (5.10)][30a]	17, 30a
$[U(NEt_2)_2(py)_5][BPh_4]_2$ (**72a**)	$[U(NEt_2)_3][BPh_4]$/py	X-ray[17]	17
$[U^V(cot)(NEt_2)_2(thf)][BPh_4]$ (**73**)	$[U(cot)(NEt_2)_3]$	X-ray;[30c] Reactions (Scheme 5.22)[30c]	30a, 30b, 30c

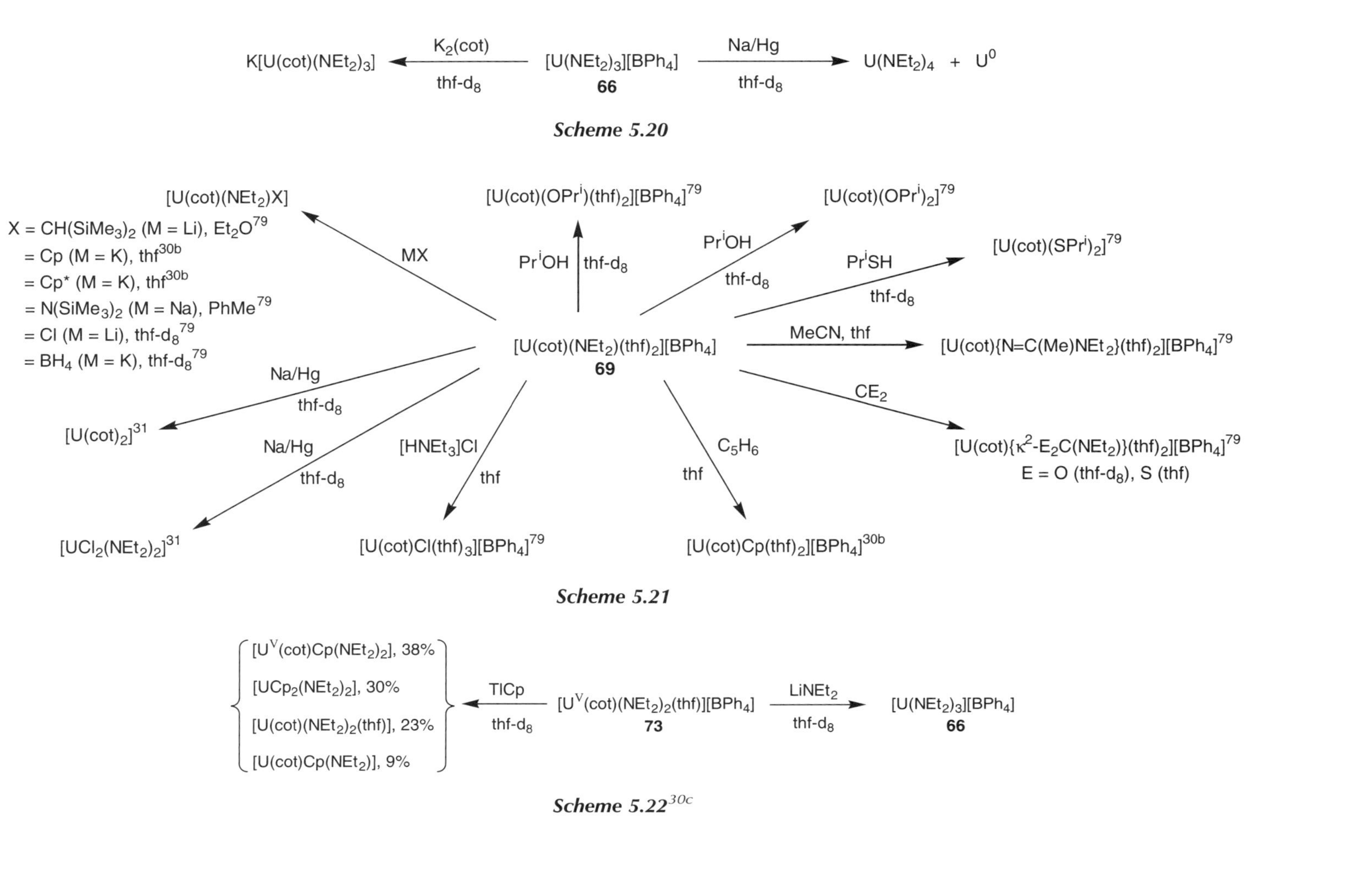

Scheme 5.20

Scheme 5.21

Scheme 5.22[30c]

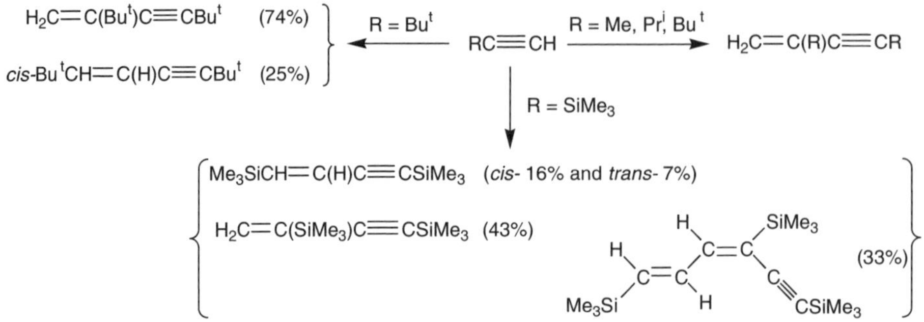

$H_2C{=}C(Bu^t)C{\equiv}CBu^t$ (74%)
cis-$Bu^tCH{=}C(H)C{\equiv}CBu^t$ (25%)
← R = Bu^t — $RC{\equiv}CH$ — R = Me, Pr^i, Bu^t → $H_2C{=}C(R)C{\equiv}CR$
↓ R = $SiMe_3$

$Me_3SiCH{=}C(H)C{\equiv}CSiMe_3$ (*cis*- 16% and *trans*- 7%)
$H_2C{=}C(SiMe_3)C{\equiv}CSiMe_3$ (43%)
$Me_3Si(H)C{=}C(H)-C(H){=}C(SiMe_3)C{\equiv}CSiMe_3$ (33%)

Scheme 5.23[103]

Using the same catalyst **66**, the hydrosilylation of a terminal alkyne $RC{\equiv}CH$ with $PhSiH_3$ at ambient temperature afforded a mixture of *cis*- and *trans*-$RCH{=}C(H)Si(H)_2Ph$, $CH_2{=}C(R)Si(H)_2Ph$, $RCH{=}CH_2$ and $RC{\equiv}CSi(H)_2Ph$; at higher temperatures, the corresponding doubly hydrosilylated compounds $RCH{=}C\{Si(H)_2Ph\}_2$ were also formed.[103] The salt $[UH(NEt_2)_2][BPh_4]$ (**74**) was implicated as a crucial intermediate.[103]

The catalytic activity of **66** was also explored in the context of the dehydrocoupling of RNH_2 ($RR'NH$ less reactive) with $PhSiH_3$.[104] For the case of $R = Pr^n$, Pr^i or Bu^t, the aminosilanes $PhSi(H)_{3-n}\{N(H)R\}_n$ ($n = 1$, 2, or 3) were obtained; **74** was again postulated to be a significant intermediate.[104] Oligomers resulted when 1,2-diaminoethane was the substrate.

References

1. M. F. Lappert, P. P. Power, A. R. Sanger and R. C. Srivastava, *Metal and Metalloid Amides*, Horwood-Wiley, Chichester, 1980.
2. K. W. Bagnall, The actinides; in *Comprehensive Coordination Chemistry* (eds. G. Wilkinson, R. D. Gillard and J. A. McCleverty), Pergamon Press, Oxford, 1987, vol. 3, ch. 40.
3. T. J. Marks and A. Streitwieser, Jr., in *The Chemistry of the Actinide Elements*, (eds. J. J. Katz, G. T. Seaborg and L. R. Morss), Chapman and Hall, London, 2nd edn., 1986, vol. 2, ch. 22.
4. T. J. Marks, in *The Chemistry of the Actinide Elements*, (eds. J. J. Katz, G. T. Seaborg and L. R. Morss), Chapman and Hall, London, 2nd edn., 1986, vol. 2, ch. 23.
5. P. J. Fagan, E. A. Maatta, J. M. Manriquez, K. G. Moloy, A. M. Seyam and T. J. Marks, in *Actinides in Perspective* (ed. N. M. Edelstein), Pergamon Press, Oxford, 1982.
6. F. T. Edelmann, Scandium, yttrium and the lanthanide and actinide elements, excluding their zero oxidation state complexes; in *Comprehensive Organometallic Chemistry II* (eds. E. W. Abel, F. G. A. Stone and G. Wilkinson), Elsevier, Oxford, 1995, vol. 4 (ed. M. F. Lappert), ch. 2.
7. C. J. Burns, M. P. Neu, H. Boukhalfa, K. E. Gutowski, N. J. Bridges and R. D. Rogers, The actinides; in *Comprehensive Coordination Chemistry II* (eds. J. A. McCleverty and T. J. Meyer), Elsevier, Oxford, 2004, vol. 3 (ed. G. F. R. Parkin), ch. 3.3.
8. (a) L. R. Avens, S. G. Bott, D. L. Clark, A. P. Sattelberger, J. G. Watkin and B. D. Zwick, *Inorg. Chem.*, 1994, **33**, 2248; (b) B. P. Warner, J. A. D'Alessio, A. N. Morgan, III, C. J. Burns, A. R. Shake and J. C. Watkin, *Inorg. Chim. Acta*, 2000, **309**, 45; (c) A. J. Gaunt, A. E. Enriquez, S. D. Reilly, B. L. Scott and M. P. Neu, *Inorg. Chem.*, 2008, **47**, 26.

9. R. A. Andersen, *Inorg. Chem.*, 1979, **18**, 209.
10. C. J. Burns, D. L. Clark, R. J. Donohoe, P. B. Duval, B. L. Scott and C. D. Tait, *Inorg. Chem.*, 2000, **39**, 5464.
11. P. L. Diaconescu, P. L. Arnold, T. A. Baker, D. J. Mindiola and C. C. Cummins, *J. Am. Chem. Soc.*, 2000, **122**, 6108.
12. P. Roussel, P. B. Hitchcock, N. D. Tinker and P. Scott, *Chem. Commun.*, 1996, 2053.
13. P. Roussel and P. Scott, *J. Am. Chem. Soc.*, 1998, **120**, 1070.
14. J. L. Stewart and R. A. Andersen, *Polyhedron,* 1998, **17**, 953.
15. J. G. Reynolds, A. Zalkin, D. H. Templeton and N. M. Edelstein, *Inorg. Chem.*, 1977, **16**, 1090.
16. J. G. Reynolds, A. Zalkin, D. H. Templeton, N. M. Edelstein and K. L. Templeton, *Inorg. Chem.*, 1976, **15**, 2498.
17. J.-C. Berthet, C. Boisson, M. Lance, J. Vigner, M. Nierlich and M. Ephritikhine, *J. Chem. Soc., Dalton Trans.*, 1995, 3019.
18. S. J. Coles, A. D. Danopoulos, P. G. Edwards, M. B. Hursthouse and P. W. Read, *J. Chem. Soc., Dalton Trans.*, 1995, 3401.
19. (a) P. Scott and P. B. Hitchcock, *J. Chem. Soc., Chem. Commun.*, 1995, 579; (b) P. L. Arnold and S. T. Liddle, *Chem. Commun.*, 2006, 3959 (a review).
20. S. J. Simpson, H. W. Turner and R. A. Andersen, *J. Am. Chem. Soc.*, 1979, **101**, 7728.
21. (a) A. Arunachalampillai, P. Crewdson, I. Korobkov and S. Gambarotta, *Organometallics*, 2006, **25**, 3856; (b) A. Arunachalampillai, S. Gambarotta and I. Korobkov, *Organometallics*, 2005, **24**, 1996.
22. P. Scott and P. B. Hitchcock, *Polyhedron*, 1994, **13**, 1651.
23. R. Boaretto, P. Roussel, A. J. Kingsley, I. J. Munslow, C. J. Sanders, N. W. Alcock and P. Scott, *Chem. Commun.*, 1999, 1701.
24. B. Monteiro, D. Roitershtein, H. Ferreira, J. R. Ascenso, A. M. Martins, Â Domingos and N. Marques, *Inorg. Chem.*, 2003, **42**, 4223.
25. P. B. Hitchcock, M. F. Lappert, A. Singh, R. G. Taylor and D. Brown, *J. Chem. Soc., Chem. Commun.*, 1983, 561.
26. P. C. Blake, M. F. Lappert, R. G. Taylor, J. L. Atwood and H. Zhang, *Inorg. Chim. Acta*, 1987, **139**, 13.
27. M. Roger, N. Barros, T. Arliquie, P. Thuéry, L. Maron and M. Ephritikhine, *J. Am. Chem. Soc.*, 2006, **128**, 8790.
28. P. C. Leverd, M. Lance, J. Vigner, M. Nierlich and M. Ephritikhine, *J. Chem. Soc., Dalton Trans.*, 1995, 237.
29. P. C. Leverd, T. Arliquie, M. Ephritikhine, M. Nierlich and J. Vigner, *New J. Chem.*, 1993, **17**, 769.
30. (a) J.-C. Berthet, and M. Ephritikhine, *J. Chem. Soc., Chem. Commun.*, 1993, 1566; (b) J.-C. Berthet, C. Boisson, M. Lance, J. Vigner, M. Nierlich and M. Ephritikhine, *J. Chem. Soc., Dalton Trans.*, 1995, 3027; (c) C. Boisson, J.-C. Berthet, M. Lance, J. Vigner, M. Nierlich and M. Ephritikhine, *J. Chem. Soc., Dalton Trans.*, 1996, 947.
31. C. Boisson, J.-C. Berthet, M. Ephritikhine, M. Lance and M. Nierlich, *J. Organomet. Chem.*, 1997, **533**, 7.
32. G. Paolucci, G. Rossetto, P. Zanella and R. D. Fischer, *J. Organomet. Chem.*, 1985, **284**, 213.
33. F. Ossola, G. Rossetto, P. Zanella, G. Paolucci and R. D. Fischer, *J. Organomet. Chem.*, 1986, **309**, 55.
34. (a) A. Dormond, *J. Organomet. Chem.*, 1983, **256**, 47; (b) A. Dormond, P. Hepiégne, A. Hafid and C. Moise, *J. Organomet. Chem.*, 1990, **398**, C1.
35. D. C. Bradley and M. H. Gitlitz, *J. Chem. Soc. (A)*, 1969, 980.
36. A. L. Odom, P. L. Arnold and C. C. Cummins, *J. Am. Chem. Soc.*, 1998, **120**, 5836.
37. D. J. Mindiola, Y.-C. Tsai, R. Hara, Q. Chen, K. Meyer and C. C. Cummins, *Chem. Commun.*, 2001, 125.

38. P. L. Diaconescu, A. L. Odom, T. Agapie and C. C. Cummins, *Organometallics*, 2001, **20**, 4993.
39. N. Kaltsoyannis and P. Scott, *Chem. Commun.*, 1998, 1665.
40. S. J. Coles, P. G. Edwards, M. B. Hursthouse and P. W. Read, *J. Chem. Soc., Chem. Commun.*, 1994, 1967.
41. C. A. Cruz, D. J. H. Emslie, L. E. Harrington, J. F. Britten and C. M. Robertson, *Organometallics*, 2007, **26**, 692.
42. H. W. Turner, R. A. Andersen, A. Zalkin and D. H. Templeton, *Inorg. Chem.*, 1979, **18**, 1221.
43. H. W. Turner, S. J. Simpson and R. A. Andersen, *J. Am. Chem. Soc.*, 1979, **101**, 2782.
44. S. J. Simpson, H. W. Turner and R. A. Andersen, *Inorg. Chem.*, 1981, **20**, 2991.
45. R. A. Andersen, A. Zalkin and D. H. Templeton, *Inorg. Chem.*, 1981, **20**, 622.
46. J. L. Stewart and R. A. Andersen, *New J. Chem.*, 1995, **19**, 587.
47. S. J. Simpson and R. A. Andersen, *J. Am. Chem. Soc.*, 1981, **103**, 4063.
48. (a) A. Dormond, A. El Bouadili and C. Moise, *J. Org. Chem.*, 1987, **52**, 688; (b) A. Dormond, A. El Bouadili and C. Moise, *J. Organomet. Chem.*, 1989, **369**, 171; (c) A. Dormond, A. El Bouadili, A. Aaliti and C. Moise, *J. Organomet. Chem.*, 1985, **288**, C1; (d) A. Dormond, A. Aaliti and C. Moise, *J. Org. Chem.*, 1988, **53**, 1034.
49. A. Dormond, C. Moise, A. El Bouadili and H. Bitar, *J. Organomet. Chem.*, 1989, **371**, 175.
50. (a) A. Dormond, A. Aaliti, A. El Bouadili and C. Moise, *J. Organomet. Chem.*, 1987, **329**, 187; (b) A. Dormond, A. El Bouadili and C. Moise, *J. Org. Chem.*, 1989, **54**, 3747.
51. A. Dormond, A. El Bouadili and C. Moise, *J. Chem. Soc., Chem. Commun.*, 1985, 914.
52. A. Dormond, A. Aaliti and C. Moise, *Tetrahedron Lett.*, 1986, **27**, 1497.
53. M. Müller, V. C. Williams and L. H. Doerrer, M. A. Leech, S. A. Mason, M. L. H. Green and K. Prout, *Inorg. Chem.*, 1998, **37**, 1315.
54. W. G. Van der Sluys, A. P. Sattelberger and M. W. McElfresh, *Polyhedron*, 1990, **9**, 1843.
55. (a) W. G. Van der Sluys, A. P. Sattelberger, W. E. Streib and J. C. Huffman, *Polyhedron*, 1989, **8**, 1247; (b) J. M. Berg, D. L. Clark, J. C. Huffman, D. E. Morris, A. P. Sattelberger, W. W. Streib, W. G. Van der Sluys and J. G. Watkin, *J. Am. Chem. Soc.*, 1992, **114**, 10811.
56. (a) D. M. Barnhart, D. L. Clark, J. C. Gordon, J. C. Huffman and J. G. Watkin, *Inorg. Chem.*, 1994, **33**, 3939; (b) D. L. Clark, J. C. Huffman and J. G. Watkin, *J. Chem. Soc., Chem. Commun.*, 1992, 266.
57. D. L. Clark, M. M. Miller and J. G. Watkin, *Inorg. Chem.*, 1993, **32**, 772.
58. P. Roussel, N. W. Alcock, R. Boaretto, A. J. Kingsley, I. J. Munslow, C. J. Sanders and P. Scott, *Inorg. Chem.*, 1999, **38**, 3651.
59. P. Scott and P. B. Hitchcock, *J. Chem. Soc., Dalton Trans.*, 1995, 603.
60. P. Roussel, P. B. Hitchcock, N. G. Tinker and P. Scott, *Inorg. Chem.*, 1997, **36**, 5716.
61. P. Roussel, R. Boaretto, A. J. Kingsley, N. W. Alcock and P. Scott, *J. Chem. Soc., Dalton Trans.*, 2002, 1423.
62. M. A. Antunes, M. Dias, B. Monteiro, Â. Domingos, I. C. Santos and N. Marques, *Dalton Trans.*, 2006, 3368.
63. L. G. McCullough, H. W. Turner, R. A. Andersen, A. Zalkin and D. H. Templeton, *Inorg. Chem.*, 1981, **20**, 2869.
64. T. M. Gilbert, R. R. Ryan and A. P. Sattelberger, *Organometallics*, 1989, **8**, 857.
65. M. J. Monreal, C. T. Carver and P. L. Diaconescu, *Inorg. Chem.*, 2007, **46**, 7226.
66. B. D. Stubbert and T. J. Marks, *J. Am. Chem. Soc.*, 2007, **129**, 4253.
67. J. D. Jamerson and J. Takats, *J. Organomet. Chem.*, 1974, **78**, C23.
68. P. Zanella, G. Rossetto, A. Berton and G. Paolucci, *Inorg. Chim. Acta*, 1984, **95**, 263.
69. G. Paolucci, P. Zanella and A. Berton, *J. Organomet. Chem.*, 1985, **295**, 317.
70. A. Berton, M. Porchia, G. Rossetto and P. Zanella, *J. Organomet. Chem.*, 1986, **302**, 351.
71. P. Zanella, G. Rossetto and G. Paolucci, *Inorg. Chim. Acta*, 1984, **82**, 227.

72. (a) W. W. Lukens, Jr., S. M. Beshouri, L. L. Blosch, A. L. Stuart and R. A. Andersen, *Organometallics*, 1999, **18**, 1235; (b) G. Zi, L. L. Blosch, L. Jia and R. A. Andersen, *Organometallics*, 2005, **24**, 4602.
73. A. L. Arduini, J. Malito, J. Takats, E. Ciliberto, I. Fragala and P. Zanella, *J. Organomet. Chem.*, 1987, **326**, 49.
74. P. J. Fagan, J. M. Manriquez, S. H. Vollmer, C. S. Day, V. W. Day and T. J. Marks, *J. Am. Chem. Soc.*, 1981, **103**, 2206.
75. A. Dormond, A. Aaliti and C. Moise, *J. Chem. Soc., Chem. Commun.*, 1985, 1231.
76. P. Zanella, N. Brianese, U. Casellato, F. Ossola, M. Porchia, G. Rossetto and R. Graziani, *J. Chem. Soc., Dalton Trans.*, 1987, 2039.
77. G. Paolucci, G. Rossetto, P. Zanella, K. Yünlü and R. D. Fischer, *J. Organomet. Chem.*, 1984, **272**, 363.
78. T. M. Gilbert, R. R. Ryan and A. P. Sattelberger, *Organometallics*, 1988, **7**, 2514.
79. (a) C. Boisson, J.-C. Berthet, M. Ephritikhine, M. Lance and M. Nierlich, *J. Organomet. Chem.*, 1996, **522**, 249; (b) T. Le Borgne, M. Lance, M. Nierlich and M. Ephritikhine, *J. Organomet. Chem.*, 2000, **598**, 313.
80. R. A. Andersen, *Inorg. Chem.*, 1979, **18**, 1507.
81. (a) D. L. Clark, A. P. Sattelberger, S. G. Bott and R. N. Vrtis, *Inorg. Chem.*, 1989, **28**, 1771; (b) D. L. Clark, A. P. Sattelberger and R. A. Andersen, *Inorg. Synth.*, 1997, **31**, 307.
82. J. C. Green, M. Payne, E. A. Seddon and R. A. Andersen, *J. Chem. Soc., Dalton Trans.*, 1982, 887.
83. W. G. Van der Sluys, C. J. Burns, J. C. Huffman and A. P. Sattelberger, *J. Am. Chem. Soc.*, 1988, **110**, 5924.
84. W. J. Evans, D. S. Lee, D. B. Rego, J. M. Perotti, S. A. Kozimor, E. K. Moore and J. W. Ziller, *J. Am. Chem. Soc.*, 2004, **126**, 14574.
85. H. Nakai, X. Hu, L. N. Zakharov, A. L. Rheingold and K. Meyer, *Inorg. Chem.*, 2004, **43**, 855.
86. (a) I. Castro-Rodriguez, K. Olsen, P. Gantzel and K. Meyer, *Chem. Commun.*, 2002, 2764; (b) I. Castro-Rodriguez, K. Olsen, P. Gantzel and K. Meyer, *J. Am. Chem. Soc.*, 2003, **125**, 4565.
87. (a) A. Zalkin, J. G. Brennan and R. A. Andersen, *Acta. Crystallogr. Sect. C.*, 1988, **44**, 1553; (b) J. G. Brennan (1985), J. L. Stewart (1988), Ph.D. Theses, University of California, Berkeley.
88. P. J. Fagan, J. M. Manriquez, T. J. Marks, C. S. Day, S. H. Vollmer and V. W. Day, *Organometallics*, 1982, **1**, 170.
89. (a) J. W. Ziller and N. Kaltsoyannis, *J. Am. Chem. Soc.*, 2004, **126**, 14533; (b) W. J. Evans and S. A. Kozimor, *Coord. Chem. Rev.*, 2006, **250**, 911.
90. H. J. Wasserman, A. J. Zozulin, D. C. Moody, R. R. Ryan and K. V. Salazar, *J. Organomet. Chem.*, 1983, **254**, 305.
91. M. C. Cassani, D. J. Duncalf and M. F. Lappert, *J. Am. Chem. Soc.*, 1998, **120**, 12958.
92. (a) T. Arliguie, M. Lance, M. Nierlich, J. Vigner and M. Ephritikhine, *J. Chem. Soc., Chem. Commun.*, 1994, 847; (b) T. Arliguie, M. Lance, M. Nierlich and M. Ephritikhine, *J. Chem. Soc., Dalton Trans.*, 1997, 2501.
93. (a) D. M. Barnhart, C. J. Burns, N. N. Sauer and J. G. Watkin, *Inorg. Chem.*, 1995, **34**, 4079; (b) A. E. Vaughn, C. L. Barnes and P. B. Duval, *J. Chem. Crystallogr.*, 2007, **37**, 779.
94. (a) P. L. Arnold, A. J. Blake, C. Wilson and J. B. Love, *Inorg. Chem.*, 2004, **43**, 8206; (b) P. L. Arnold, D. Patel, A. J. Blake, C. Wilson and J. B. Love, *J. Am. Chem. Soc.*, 2006, **128**, 9610; (c) P. L. Arnold, D. Patel, C. Wilson and J. B. Love, *Nature*, 2008, **451**, 315.
95. C. Boisson, Dissertation, University of Orsay 1996, (cited in ref. 7).
96. K. Meyer, D. J. Mindiola, T. A. Baker, W. M. Davis and C. C. Cummins, *Angew. Chem., Int. Ed.*, 2000, **39**, 3063.

97. C. J. Burns, W. H. Smith, J. C. Huffman and A. P. Sattelberger, *J. Am. Chem. Soc.*, 1990, **112**, 3237.
98. S. A. Mungur, S. T. Liddle, C. Wilson, M. J. Sarsfield and P. L. Arnold, *Chem. Commun.*, 2004, 2738.
99. J. E. Nelson, D. L. Clark, C. J. Burns and A. P. Sattelberger, *Inorg. Chem.*, 1992, **31**, 1973.
100. P. B. Duval, C. J. Burns, W. E. Buschmann, D. L. Clark, D. E. Morris and B. L. Scott, *Inorg. Chem.*, 2001, **40**, 5491.
101. Reviews: (a) J.-C. Berthet and M. Ephritikhine, *Coord. Chem. Rev.*, 1998, **178–180**, 83; (b) E. Barnea and M. S. Eisen, *Coord. Chem. Rev.*, 2006, **250**, 855; (c) M. Ephritikhine, *Dalton Trans.*, 2006, 2501.
102. J. Q. Wang, J.-C. Berthet, M. Ephritikhine and M. S. Eisen, *Organometallics*, 1999, **18**, 2407.
103. A. K. Dash, J. X. Wang, J.-C. Berthet, M. Ephritikhine and M. S. Eisen, *J. Organomet. Chem.*, 2000, **604**, 83.
104. J. X. Wang, A. K. Dash, J.-C. Berthet, M. Ephritikhine and M. S. Eisen, *J. Organomet. Chem.*, 2000, **610**, 49.

6

Amides of the Transition Metals

6.1 Introduction

Since 1980, interest in the transition metal derivatives of amide ligands has led to the creation of a very large and diversified research field. It has attracted workers from many areas whose interests include olefin polymerization, nitrogen fixation, hydroamination, carboamination, chemical vapor deposition, exploratory inorganic synthesis as well as fundamental bonding questions involving transition metal complexes. The currently available data for the synthesis, physical properties and reactions of these compounds are far beyond the capacity of a single chapter to cover in detail and are more than sufficient to justify a separate volume. A conservative estimate of the increase in the number of publications since the original survey was published in 1980 is about a tenfold one. As a result of the prodigious increase in the information available, we have had to make some difficult decisions on the material selected for presentation.

A key objective on our part was to complement material that had already been reviewed as well as to provide an overview of the key developments. Several reviews and commentaries[1–54] have appeared since the 1980 book[55] and almost half of these have been published since 2000. These have dealt with, either fully or in part, derivatives of specific types of amido and related ligands, the applications of amido substituted complexes in chemical transformations, and the use of amido complexes as precursors for electronic materials or catalysis. The increasing interest in the use of multidentate amido and similar ligands of various types, which had been a notable development of mainstream amide chemistry since 1980, has resulted in the largest numbers of reviews. These cover ligands such as amidinate[11,40] amidophosphine,[5,7,41,45,52,53] azaatranes,[9,17,48] amidoamine,[26,30,31,35] amidopyridine,[31,33,40,44] β-diketiminate,[26,34,42,43] azaallyl,[32] benzamidinate,[13] phosphoraneiminato,[21] and various macrocyclic ligands.[15,16,22,41] Some of these ligands have been studied for the imposition of regio- and stereospecific environments at the transition metal centre. Approximately two-thirds of the reviews deal with such derivatives. The remaining reviews, which in several cases can also include ligands in the above categories, concern a

Metal Amide Chemistry Michael Lappert, Andrey Protchenko, Philip Power and Alexandra Seeber

variety of topics that include sterically hindered amides,[4,10,15,17,18,20,28,45,51] *ansa*-cyclopentadienyl/amido ligands,[23,42] the involvement of late transition metal amides in C–H activation or hydroamination reactions,[3,8,27,39,49] olefin polymerization[23,42] the catalysis of C–N bond formation,[24,25,49,50,52,54] nitrogen fixation,[19,43,46,48,53] and chemical vapor deposition.[12,29] Clearly, many aspects of transition metal amide chemistry, in particular those pertaining to the use of multidentate amido ligands, have been very well reviewed. However, with the exception of one early overview[1] and a small number reviews that have focused on complexes of bulky ligands,[4,10,19,20,45,51] and their reactions, there has been no general treatment that deals specifically with the monodentate diorganoamido and closely related derivatives of transition metals. Yet, since 1980, there has been a continuing interest in these simple ligands for a variety of reasons that include their use to expand the known range of transition metal coordination geometries or bonding configurations, the stabilization of metal-metal multiple bonds, the exploration of the reactivity of the M–N unit toward organic substrates, the catalysis of C–N bond formation, the activation of C–H bonds or dinitrogen, synthons for other transition metal species and their employment in the chemical vapor deposition of transition metal nitrides and oxides. As a result of this widespread and continued interest, the number of well-characterized transition metal derivatives of monodentate amido ligands continues to exceed that of other types of amido ligands by a significant margin. For these reasons the current chapter focuses primarily on complexes of the monodentate amido ligands although due attention is given to the parallel development of multidentate and other related amido derivatives in Section C.

6.2 Transition Metal Derivatives of Monodentate Amides

6.2.1 Overview

Prior to 1980, well-characterized, homoleptic and heteroleptic diorganoamido derivatives of almost all the earlier (i.e. groups 3–6) transition metals had been reported. These continue to attract much attention and new examples are a constant feature of the literature. Tables 6.1[56–237] and 6.2[238–506] provide a listing of the structurally characterized (X-ray or gas electron diffraction (ged)) transition metal amido complexes that have been published since 1980 and that do not involve chelating amido ligands in their metal coordination spheres. These tables provide the reader with a rough guide to the level of activity for each metal. Over 600 compounds are listed but it should be borne in mind that there are several hundred structurally characterized derivatives that incorporate multidentate amido ligands of various kinds that are not included in these tables. These numbers may be compared to the total of ca. 40 transition metal amide crystal structures of all types available at the time the previous volume was published.[55] Thus, the growth in the available structural information has been enormous and this is due to the greatly heightened interest in these compounds as well as the widespread availability and efficiency of X-ray crystallographic data collection facilities.

It will be immediately apparent from Tables 6.1 and 6.2 that amides of the groups 4–6 elements are by far the most numerous. Furthermore, the number of second and third row element amido derivatives of these groups approaches or exceeds those of the first row. It can also be seen that for the first row elements homoleptic amides in the highest oxidation

Table 6.1 *Structurally characterized first row transition metal derivatives of non-chelating diorganoamido ligands*[a]

	Scandium[b]
Sc(III):	$[Sc\{N(SiMe_3)_2\}_3]$ (ged),[56] $[Sc\{N(SiHMe_2)_2\}_3(thf)]$,[57] $[Na(THF)_3\{CH_2SiMe_2N(SiMe_3)\}Sc\{N(SiMe_3)_2\}_2]$,[58] $[(salen)Sc(NR_2)]$, ($R = Pr^i$, $SiHMe_2$),[59] $[Sc(HC\{C(Bu^t)NC_6H_3Pr^i_2\text{-}2,6\}_2)\{N(H)Bu^t\}][MeB(C_6F_5)_3]$,[60] $[ScCl_2\{N(SiMe_3)_2\}(thf)_2]$,[61] $[Sc(NHR')(HC\{C(R)NC_6H_3Pr^i_2\text{-}2,6)\}_2)X]$, ($R = R' = Bu^t$, $X = Cl$; $R = Bu^t$, $R' = C_6H_3Pr^i_2$-2,6, $X = Cl$ or Me).[62]
	Titanium
Ti(IV):	$[Ti(NMe_2)_4]$ (ged),[63a] $[Ti(NMe_2)_4]$,[63b] $[Ti(NPh_2)_4]$,[64] $[Ti(NMe_2)_3\{N(SiMe_3)_2\}]$,[65] $[Ti(NR_2)_3\{NH_2B(C_6F_5)_3\}]$ (R = Me or Et),[66] $[Ti(NMe_2)_2(2\text{-mesitylpyrrolide})_2]$.[67]
	$[Ti(Cl)(NR_2)_3]$ (R = Me or Et),[68] $[Ti(\eta^5\text{-}C_2B_9H_{11})(NMe_2)_3]$,[69] $[Ti(tpb)(NMe_2)_3]$,[70] $[Ti(NHMe_2)(NMe_2)_3][B(C_6F_5)_4]$,[71a] $[Ti(NHMe_2)_2(NMe_2)_3]$ $[TiCl(NMe_2H)_2\{NH_2B(C_6F_5)_3\}\{NB(C_6F_5)_3\}]$[71b] $[Ti(py)_2(NMe_2)_3][BPh_4]$,[72] $[Ti(\eta^5\text{-}C_5H_5)(NMe_2)_2\{NH_2B(C_6F_5)_3\}]$,[66] $[Ti(\eta^5\text{-}C_5Me_5)(NMe_2)_3]$,[73] $[Ti\{M(\eta^5\text{-}C_5H_5)(CO)_2\}(NMe_2)_3]$ (M = Fe[74] or Ru[75]), $[Ti(Cl)\{N(SiMe_3)_2\}_3]$,[76] $[Ti(Me)\{N(SiMe_3)_2\}_3]$,[77] $[Ti(Cl)(NCy_2)_3]$,[78] $[\{Ti(NPh_2)_3\}_2(\mu\text{-}O)]$,[65] $[Ti(Cl)\{N(R)Ar\}_3]$,[79] $[\{Ar(R)N\}_3Ti(\mu\text{-}O)Mo(N)(OBu^t)_2]$,[80] $[O_2MoOTi\{N(Bu^t)Ph\}_3]$,[80] $[Ti\{N(H)C_6H_3Pr^i_2\text{-}2,6\}_3(NPBu^t_3)]$,[81] $[(4\text{-}Me\text{-}H_4C_6)(H_2C)COTi\{N(Bu^t)Ar\}_3]$,[82] $[PhN(Me)C(CH_2)OTi\{N(Bu^t)Ar\}_3]$,[82] $[H(O)COTi\{N(Bu^t)Ar\}_3]$,[83] $[(Et_2O)_2LiOTi\{N(Bu^t)Ar\}_3]$,[83] $[\{Ph(Bu^t)N\}_3TiOC(O)C(Ph)NMo\{N(R)Ar\}_3]$,[84] $[\{Ph(Bu^t)N\}_3Ti(\mu\text{-}N_2)Mo\{N(R)Ar\}_3]$.[85]
	$[Ti(\eta^5\text{-}C_5Me_5)(NBu^t)(NHBu^t)_2\}LiNH_2Bu^t\}]$,[86] $[\{Ti(NMe_2)_2F_2\}_4]$,[87] $[Ti(NMe_2)_2(OC_6H_3Bu^t_2\text{-}2,6)_2]$,[88] $[Ti(OC_6H_2\{\eta^5\text{-indenyl}\}\text{-}2\text{-}Bu^t_2\text{-}4,6)(NMe_2)_2]$,[89] $[\{Ti(NMe_2)_2(\mu\text{-}NBu^t)\}_2]$,[90a] $[\{Ti(NMe_2)_2(\mu\text{-}NSO_2C_6H_4Me\text{-}4)\}_2]$,[90b] $[Ti(NMe_2)_2\{N(Pr^i)Tos\}_2]$,[90c] $Ti(Cl)(NMe_2)_2\{(NPr^i)_2CN(SiMe_3)_2\}$,[91] $[Ti(NMe_2)_2L_2]$,[92] (L = amino-oxazolinate, aminopyrrolinate), $[Ti(NMe_2)_2(\sigma{:}\sigma\text{-}Pr^i_2NPOC_9H_6C_2B_{10}H_{10})]$,[93a] $[Ti(NR_2)_2\{\sigma{:}\eta^1{:}\eta^5\text{-}(OCH_2)(Me_2NCH_2)C_2B_9H_9\}]$ (R = Me or Et),[93b] $[Ti(NMe_2)_2(OAl(Me)\{N(C_6H_3Pr^i_2\text{-}2,6)C(Me)\}_2CH)_2]$,[93c] $[Ti(\eta^5\text{-}C_2B_9H_{11})(NMe_2)_2(HNMe_2)_2]$,[94] $[TiCl_2(NMe_2)_2\{CN(Mes)CH=CHN(Mes)\}]$,[95] $[Ti(\eta^5\text{-}C_5Me_5)(NMe_2)\{\mu\text{-}OC(O)NMe_2\}W(CO)_5]$,[96] $[Ti(\eta^5\text{-}C_5H_5)_2(NC_4H_4)_2]$,[97] $[Ti(NEt_2)_2\{S(4\text{-}Me\text{-}6\text{-}Bu^tC_6H_2O)_2\}]$,[98] $[Ti(NEt_2)_2\{CH_2(OC_6H_2Me\text{-}4\text{-}Bu^t\text{-}6)_2\}]$,[98] $[Ti(NEt_2)_2(OC(Ph)NSiMe_3)_2]$,[99] $[TiCH_2CMe_2C_6H_4(NCy_2)_2]$,[100] $[\{TiCl(NCy_2)_2\}_2O]$,[100] $[TiMe_2(NCy_2)_2]$[101] $[\{Ti(NCy_2)_2(\mu\text{-}CH_2)\}_2]$,[101] $[Ti(CH_2Ph)_2\{N(SiMe_3)_2\}_2]$,[101] $[TiCl_2\{N(SiCl_3)_2\}_2]$.[102,103]
	$[Ti(\eta^5\text{-}C_5H_4R)Cl_2(NHBu^t)]$,[104] (R=H or Me) $[Ti(\eta^5\text{-}C_5Me_5)Cl_2(NHBu^t)]$,[105] $[Ti(\eta^5\text{-}C_5H_5)(NHAr)(OC_6H_3Pr^i_2\text{-}2,6)_2]$,[106] $[Ti(CH_2CH_2Bu^t)\{N(H)SiBu^t_3\}(OSiBu^t_3)_2]$,[107] $[Ti(\eta^5\text{-}C_5Me_5)(CH_2Ph)_2(NMe_2)]$,[108] $[Ti_4(NMe_2)_6(\mu\text{-}NPh)_5]$,[109] $[\{Ti(NMe_2)(N_3)(\mu\text{-}NMe_2)\}_3(\mu^3\text{-}N_3)(\mu^3\text{-}NH)]$,[110] $[\{Ti(Cl)(NMe_2)(NHMe_2)(\mu\text{-}O)\}_4]$,[90b] $[Ti(\eta^5\text{-indenyl})(NMe_2)Cl_2]$,[111] $[Ti(\eta^5\text{-}C_5H_3Me_2\text{-}1,3)Cl_2(NMeCy)]$,[112] $[Ti(\eta^5\text{-}C_5Me_5)Cl_2(NMeCy)]$,[112] $[Ti(\eta^5\text{-}C_5H_3Me_2\text{-}1,3)Cl_2(NCy_2)]$,[112] $[Ti\{NPr^i(C_6H_3\text{-}2,6\text{-}Me_2)\}(OC_6H_3\text{-}2,6\text{-}Ph_2)(OC_6H_3Ph\text{-}\eta^1\text{-}C_6H_4)]$,[113] $[(TiCl_3\{N(Bu^t)SiMe_3\})_2]$,[114] $[(TiCl_3\{N(SiMe_2Cl)(SiMe_2NH_2)\})_2]$,[103] $[Ti(\eta^5\text{-}C_5H_3Me_2\text{-}1,3)Cl_2\{N(SiMe_3)(C_6H_3Me_2\text{-}2,6)\}]$,[115] $[\{TiCl\{N(SiMe_3)_2\}(\mu\text{-}NSiMe_3)\}_2]$,[116] $[TiCl\{N(SiMe_3)_2\}(\mu\text{-}NSiMe_3)(\mu\text{-}NSi(Me_2)NSiMe_3)Ti\{N(SiMe_3)\}]$,[116] $[Ti(\eta^5\text{-}C_5H_5)Cl_2\{N(SiMe_3)_2\}]$,[117] $[Ti(NC_6H_3Pr^i_2\text{-}2,6)(NHMe_2)_3(NMe_2)][B(C_6F_5)_4]$.[71]

(continued)

Table 6.1 *(Continued)*

Ti(III):	$[Ti\{N(C_6F_5)_2\}_4\{Na(thf)_2\}]$,[118] $[Ti(NCy_2)_2(\mu\text{-}Me)_2Li(tmeda)]$,[101] $[Ti(N\{C(CD_3)_2Me\}(C_6H_5Me_2\text{-}3,5))_3]$,[119] $[Ti(thf)X\{N(SiMe_3)_2\}_2]$[120] (X = Cl or BH_4), $[Ti(tmeda)_2][TiCl_2\{N(SiMe_3)_2\}_2]$,[121] $[Ti\{N\{C(CD_3)_2Me\}(C_6H_5Me_2\text{-}3,5)\}_2\{CH(SiMe_3)_2\}]$,[122] $[(Ti\{N(Bu^t)Ar\}_2\{\mu\text{-}O_2CN(Bu^t)Ph\})_2]$,[123] $[Ti\{N(SiMe_3)_2\}(py)_2(BH_4)_2]$,[120] $[Ti(\eta^5\text{-}C_5Me_5)_2(NH_2)]$,[124] $[Ti(\eta^5\text{-}C_5Me_5)_2(NHMe)]$,[125] $[Ti(\eta^5\text{-}C_5Me_5)_2(NMePh)]$,[126] $[\{Ti(N(H)SiBu^t_3)(\mu\text{-}NSiBu^t_3)\}_3]$.[127]
Ti(II):	$[(\mu_2{:}\eta^1{:}\eta^1\text{-}N_2)(Ti(Cl)\{N(SiMe_3)_2\}(py)_2)_2]$,[121] $[(\mu_2{:}\eta^1{:}\eta^1\text{-}N_2)\{Ti(Cl)\{N(SiMe_3)_2\}(tmeda)\}_2]$.[128]
Ti(I)/(III):	$[Li(tmeda)_2][(\mu_2{:}\eta^2\text{-}N_2)_2(Ti\{N(SiMe_3)_2\}_2)_2]$.[128]
	Vanadium
V(V):	$[V(NEt_2)_4][B(C_6F_5)_4]$,[129] $[VI(N_3Mes)\{N(C(CD_3)_2Me)(C_6H_3F\text{-}2\text{-}Me\text{-}5)\}_2]$,[130] $[V(Se)\{SeSi(SiMe_3)_3\}\{N(SiMe_3)_2\}_2]$,[131] $[V(Cl)\{N(SiMe_3)_2\}_2N_2CPh_2]$,[132] $[V(NBu^t)(NPr^i_2)Cl_2]$,[133] $[V(NBu^t)(\eta^5\text{-}C_5H_5)(NPr^i_2)Cl]$,[133] $[V(NC_6H_3Pr^i_2\text{-}2,6)(NMePh)(CH_2Ph)_2]$.[134]
V(V)/V(IV):	$[\{(Me_3P)_3V\}_2(\mu\text{-}H)_3][(\{(Me_3Si)_2N\}_2V)_2(\mu\text{-}N)_2]$.[135]
V(IV):	$[V(NMe_2)_4]$ (ged),[63a] $[V(NMe_2)_4]$ (X-ray),[136] $[V(NMe_2)_2(NC_{12}H_8)_2]$,[137a] $[V\{N(SiMe_3)_2\}_3][CN]$,[138] $[VCl_2(NR_2)_2]$ (R = Pr^i or Cy),[139] $[\{V(NMe_2)_2(\mu\text{-}NR)\}_2]$ (R = Ph,[137a] $C_6H_3Me_2$-2,6,[137b] $NSO_2C_6H_4Me$-4[90b]) $[(Ar_FBu^tN)_3VNCMo\{N(1\text{-}Ad)Ar\}_3]$ ($Ar_F = C_6H_3F$-2-Me-5),[140] $[(\mu\text{-}E)_2V_2\{N(SiMe_3)_2\}_4]$ (E = O[141], S[142], Se[143]), $[(VMe\{N(SiMe_3)_2\}_2(\mu\text{-}O)]$,[143] $[VCl_2\{N(SiMe_3)_2\}_2]$,[132] $[V(NEt_2)_2(thf)_4][B(C_6F_5)_4]_2$.[129]
V(III):	$[V\{N(1\text{-}Ad)(C_6H_5Me_2\text{-}3,5)\}_3]$,[144] $[V(thf)(NPh_2)_3]$,[145] $[V(thf)\{N(Ph)SiMe_3\}_3]$,[146] $[V(Cl)\{N(C_6F_5)_2\}_3K]$.[118] $[V(Cl)(thf)\{N(SiMe_3)_2\}_2]$,[147,148] $[V(Me)(thf)\{N(SiMe_3)_2\}_2]$,[143] $[(V\{N(SiMe_3)_2\}_2\{\mu\text{-}CH_2SiMe_2NSiMe_2\})_2]$,[147a] $[(V\{N(SiMe_3)_2\}_2\{\mu\text{-}O(CH_2)CH_2SiMe_2NSiMe_3\})_2]$,[147a] $[V\{N(SiMe_3)_2\}\{CH_2SiMe_2NSiMe_3\}py]$,[147b] $[V\{N(SiMe_3)_2\}\{C(Ph)CPhCH_2SiMe_2NSiMe_3\}]$,[147b] $[V(BH_4)(thf)\{N(SiMe_3)_2\}_2]$,[148] $[V\{N(SiMe_3)_2\}_2ER]$ (E = Se, R = $Si(SiMe_3)_3$; E = Te, R = $Si(SiMe_3)_3$, $SiPh_3$],[131] $[\{Li(tmeda)_2\}(VX_2\{N(SiMe_3)_2\}_2)]$ (X = Cl or Me),[132] $[V\{N(SiMe_3)_2\}\{N(SiMe_3)SiMe_2CH_2C(Ph)C(Ph)C(Ph)C(Ph)\}]$,[149] $[\{V\{N(SiMe)_3\}_2N(SiMe_3)SiMe_2CH_2\}_2\{K(THF)_2\}_2]$,[150] $[\{LiV\{N(Pr^i)C(CH_2)CH_2\}(NPr^i_2)_2\}_2]$,[145] $[V(NCy_2)_2(\mu\text{-}Cl)]_2$,[148] $[V(NCy_2)_2(\mu\text{-}Cl)_2Li(thf)_2]$.[148]
	Chromium
Cr(VI):	$[NCr(NPr^i_2)_3]$,[151] $[NCr(NPr^i_2)_2(CH_2SiMe_2Ph)]$,[152] $[NCr(NPr^i_2)(SBu^t)_2]$,[153] $[NCr(NPr^i_2)(\mu\text{-}SBu^t)_2CuS_2C(NEt_2)_2]$,[153] $[NCr(NPr^i_2)(Bu^tNCCH_2SiMe_3)(CH_2SiMe_3)]$,[153] $[NCr(I)(NRAr_F)_2]$ (R = $C(CD_3)_2Me$; $Ar_F = C_6H_3F$-2-Me-5), $[NCr(dbabh)(NRAr_F)_2]$[154] (dbabH = 2,3:5,6-dibenzo-7-azabicyclo[2.2.1]hepta-2,5-diene), $[Cr\{N(Bu^tAr^F\}_2O_2]$,[155] $[Cr(NCy_2)_2O_2]$,[156] $[Cr\{N(1\text{-}Ad)(C_6H_3Me_2\text{-}3,5)_2\}_2O_2]$,[156] $[Cr(Cl)(NHBu^t)(NC_6H_3Pr^i_2\text{-}2,6)_2]$,[157] $[Cr(NMes)_3(NHMes)\{\mu\text{-}Li(OEt_2)_2\}]$,[158] $[Cr(Cl)(NHAr)(NAr)_2]$ (Ar = $C_6H_3Pr^i_2$-2,6),[159] $[Cr(NPr^i_2)(NAr)(O)Ar]$.[160]
Cr(V):	$[\{Cr(\mu\text{-}N)(NPr^i_2)_2\}_2]$,[152] $[(OCr\{N(1\text{-}Ad)(C_6H_3Me_2\text{-}3,5)_2\}_2)_2(\mu\text{-}O)]$,[156] $[OCr(OSiPh_3)\{N(Bu^t)Ar_F\}_2]$,[155] $[OCr\{OC(O)Ph\}\{N(Bu^t)Ar_F\}_2]$,[155] ($Ar_F = C_6H_3F$-2-Me-5).

Table 6.1 *(Continued)*

Cr(IV):	$[Cr(I)_2\{N(SiMe_3)_2\}_2]$,[161a] $[(Cr\{N(1\text{-}Ad)(C_6H_3Me_2\text{-}3,5)\}_2\{\mu\text{-}O\})_2]$,[156] $[\{Cr(NCy_2)_2(\mu\text{-}S)\}_2]$[161b]
Cr (III/IV):	$[(\eta^5\text{-}C_5Me_5)(Br)Cr(\mu\text{-}NHBu^t)_2Cr(NHBu^t)(NBu^t)]$.[162]
Cr(III):	$[Cr\{N(CHMePh)_2\}_3]$,[161a] $[Cr(NCy_2)_3]$,[161b] $[Cr(\overline{NCMe_2(CH_2)_3CMe_2})_3]$[156] $[Cr\{N(SiMe_3)_2\}_3](Si_2Me_6)_{0.5}]$,[164] $[Cr\{N(1\text{-}Ad)(C_6H_3Me_2\text{-}3,5)\}_3]$,[156] $[Cr\{N(SiMe_3)_2\}\{OCMe_2Ph\}_2]$,[161a] $[\{Cr(\eta^5\text{-}C_5H_5)(\mu\text{-}CH_2SiMe_2NSiMe_3)\}_2]$.[165]
Cr(II):	$[\{Cr(NPr^i{}_2)(\mu\text{-}NPr^i{}_2)\}_2]$,[166] $[\{Cr(NCy_2)(\mu\text{-}NCy_2)\}_2]$,[167] $[(Cr\{N(1\text{-}Ad)(C_6H_3Me_2\text{-}3,5)\}\{\mu\text{-}N(1\text{-}Ad)(C_6H_3Me_2\text{-}3,5)\})_2]$,[156] $[\{Cr(NPh_2)(\mu\text{-}NPh_2)(thf)\}_2]$,[167] $[Cr(NPh_2)_2(py)_2]$,[167] $[\{(py)Li\}_2\{Cr(NEt_2)_4\}]$,[167] $[\{(thf)_2Li\}_2Cr\{N(H)Dipp\}_4]$,[158] $[Cr(NPhBMes_2)_2]$,[168] $[Cr(NMesBMes_2)_2]$.[169] $[Cr(\eta^5\text{-}C_5H_5)(I)(NO)(NPh_2)]$,[170] $[Cr(\eta^5\text{-}C_5H_5)\{OC(O)Ph\}(NO)(NPr^i{}_2)]$,[171] $[Cr(\eta^5\text{-}C_5H_5)(\eta^1\text{-}C_5H_5)(NO)(NPr^i{}_2)]$,[171] $[Cr(\eta^5\text{-}C_5H_5)(OSiMe_3)(NO)(NPr^i{}_2)]$,[171] $[Cr(CH_2Ph)_2(NO)(NPr^i{}_2)]$,[171] $[Cr(CH_2SiMe_3)_2(NO)(NPr^i{}_2)]$.[171]
	Manganese
Mn(III):	$[Mn\{N(SiMe_3)_2\}_3]$.[172]
Mn(II):	$[Na_2Mn(NPh_2)_4]$,[173] $[Li(thf)_4]_2[Mn(NC_{12}H_8)_4]$,[174] $[Na(12\text{-}crown\text{-}4)_2][Mn\{N(SiMe_3)_2\}_3]$,[175] $[Mn\{N(SiMe_3)_2\}_3Li(thf)]$,[176] $[Mn\{N(SiMe_3)_2\}_2]$ (ged),[177] $[(Mn\{N(SiMe_3)_2\}_2)_2]$,[176,178] $[\{Mn(NPr^i{}_2)_2\}_2]$,[179,184] $[Mn\{N(SiMePh_2)_2\}_2]$,[180a] $[Mn\{\mu\text{-}NHC_6H_3Pr^i{}_2\text{-}2,6\}_4\{MnN(SiMe_3)_2\}_2]$,[181] $[Mn(NPhBMes_2)_2]$,[168] $[Mn(NMesBMes_2)_2]$,[168] $[Mn\{N(SiMe_3)_2\}_2(pyrazine)_2]$,[180b] $[Mn\{N(SiMe_3)(C_6H_3Pr^i{}_2\text{-}2,6)\}_2(thf)]$,[181] $[Mn(NPh_2)_2(thf)_2]$,[173] $[Mn\{N(SiMe_3)_2\}_2(thf)_2]$,[182] $[Mn\{N(SiMe_3)_2\}_2(py)_2]$,[180b] $[Mn\{N(SiMe_3)_2\}_2(1,10\text{-}phenanthroline)]$,[183] $[\{Mn\{N(SiMe_3)_2\}_2(4,4'\text{-}bipyridyl)\cdot thf\}_n]$,[180b,183] $[Mn(NC_{12}H_8)_2(thf)_3]$,[174] $[Mn_3(NEt_2)_6(\mu\text{-}Cl)_2\{Li(thf)_2\}_2]$.[179,184] $[(tmp)Mn(\mu\text{-}R)(\mu\text{-}tmp)Na(tmeda)]$ ($R = CH_2SiMe_3$ or Ph),[185a] $[Na_4Mn_2(tmp)_6(C_6H_4)]$,[185a] $[Mn(CH_2SiMe_3)_2(\mu\text{-}tmp)Li(tmeda)]$.[185b] $[Mn\{N(SiMe_3)_2(AlMe_3)\}(\mu\text{-}Me)]_2$,[186] $[LiMn\{N(SiMe_3)_2\}(OCBu^t{}_3)_2]$,[187] $[\{Mn\{N(SiMe_3)_2\}(\mu\text{-}EMes_2)\}_2]$ ($E = P$ or As),[188] $[\{Mn\{N(SiMe_3)_2\}(\mu\text{-}SeC_6H_3Pr^i{}_3\text{-}2,4,6)(thf)\}_2]$,[189] $[(Mn\{N(SiMe_3)_2\}(Bu^tNCHC_6H_4O))_2]$,[190] $[(Mn\{N(SiMe_3)_2\}(Bu^tNCHC_6H_4NH))_2]$,[190] $[Mn\{N(SiMe_3)_2\}\{HC(C(Me)N(C_6H_3Pr^i{}_2\text{-}2,6))_2\}]$,[191] $[Mn\{NH(C_6H_3Pr^i{}_2\text{-}2,6)\}\{PhB(CH_2PPr^i{}_2)_3\}]$,[192] $[Mn(L)\{PhB(CH_2PPr^i{}_2)_3\}]$ ($L = NC_{14}H_8$, 1-phenylisoindolate),[192] $[\{Mn(\eta^5\text{-}C_5H_5)(NHR)\}_4\{MnNR\}_4]$, ($R = 4,6\text{-}Me_2$-pyrimidinyl, 4-MeO-6-Me-pyrimidinyl, 4,6-$(OMe)_2$-pyrimidinyl).[193]
	Iron
Fe(III):	$[Fe(Cl)\{N(SiMe_3)_2\}_2(thf)]$,[194] $[Li(tmeda)_2][FeCl_2\{N(SiMe_3)_2\}_2]$,[194] $[Fe(I)\{N(Bu^t)Ar_F\}_2(py\text{-}d_5)]$,[195] ($Ar_F = C_6H_3F\text{-}2\text{-}Me\text{-}5$), $[Fe_4S_4\{N(SiMe_3)_2\}_4]$,[196] $[(Fe\{N(SiMe_3)_2\}(\mu\text{-}S)\{SC(NMe_2)_2\})_2]$,[196] $[Fe(NHC_6H_3\text{-}2,6\text{-}Pr^i{}_2)(HC\{C(Me)NC_6H_3Pr^i{}_2\text{-}2,6\}_2)(OTf)]$.[197]
Fe(III/FeII):	$[(\eta^5\text{-}C_5Me_5)Fe(\mu\text{-}NHPh)(\mu\text{-}NPh)Fe(\eta^5\text{-}C_5Me_5)]$.[198]
Fe(II):	$[Na(12\text{-}crown\text{-}4)_2][Fe\{N(SiMe_3)_2\}_3]$,[175] $[Fe(NBu^t{}_2)_2]$,[199] $[Fe\{N(SiMe_3)_2\}_2]$ (ged),[177] $[(Fe\{N(SiMe_3)_2\}_2)_2]$,[200] $[Fe\{N(SiMe_2Ph)_2\}_2]$,[180] $[Fe\{N(SiMePh_2)_2\}_2]$,[201a,180] $[Fe(NMesBMes_2)_2]$,[168] $[\{Fe(NPh_2)_2\}_2]$,[200] $[Fe\{N(CH_2Bu^t)(C_6H_3Pr^i{}_2\text{-}2,6)\}_2]$,[201b] $[Fe\{N(SiMe_3)_2\}_2(thf)]$,[200] $[Fe\{N(SiMe_3)_2\}_2(py)]$,[182] $[\{Fe\{N(SiMe_3)_2\}(4,4'\text{-}bipyridyl)\}_n]$,[182,184] $[Fe\{N(SiMe_3)_2\}_2(py)_2]$,[182] $[Fe\{N(C_6F_5)_2\}_2$

(continued)

Table 6.1 *(Continued)*

	$(thf)_2]$,[118] $[Fe\{N(Bu^t)Ar_F\}_2(NO)(L)]$,[195] (L = py-$d_5$, PEt_3; $Ar_F = C_6H_3F$-2-Me-5), $[(thf)_3Li(\mu\text{-}Cl)Fe\{N(SiMe_3)_2\}_2]$,[202]
	$[Fe(dmpe)_2(H)(NH_2)]$,[203] $[Fe(NHR)\{HC(CR'NC_6H_3Pr^i_2\text{-}2,6)_2\}(L)]$,[197] (R = $C_6H_3Pr^i_2$-2,6, R′ = Me, L = py-4-Bu^t, MeCN, THF; R = 1-Ad, R′ = Me, L = py-4-Bu^t), $[Fe\{N(SiMe_3)_2\}(HC\{C(Me)N(C_6H_3Pr^i_2\text{-}2,5)_2\}_2)]$,[191] $[Fe\{N(SiMe_3)_2\}(HC\{C(Me)N(C_6F_5)\}_2)]$,[191] $[\{Fe(\eta^5\text{-}C_5Me_5)(\mu\text{-}NHPh)\}_2]$,[204] $[Fe(\eta^5\text{-}C_5Me_5)\{N(SiMe_3)_2\}]$,[205] $[Fe\{N(SiMe_3)_2\}(SC_6H_3Mes_2\text{-}2,6)]$,[206] $[\{Fe\{N(SiMe_3)_2\}(\mu\text{-}OC_6H_2Bu^t_3\text{-}2,4,6)\}_2]$,[207] $[\{Fe\{N(SiMe_3)_2\}(\mu\text{-}PMes_2)\}_2]$,[208] $[\{Fe\{N(SiMe_3)_2\}(\mu\text{-}SSiPh_3)\}_2]$,[209] $[Fe_3\{N(SiMe_3)_2\}_2(\mu\text{-}SC_6H_2Pr^i_3\text{-}2,4,6)_4]$,[210] $[Fe\{N(SiMe_3)_2\}\{Ph_2PCH(C_5H_4N\text{-}2)\}]_2$,[211] $[Fe\{N(SiMe_3)_2\}(\mu\text{-}SMes)(thf)]_2$,[212] $[Fe\{N(SiMe_3)_2\}\{CH(PPh_2NC_6H_3Pr^i_2\text{-}2,6)_2\}]$,[213] $[Fe(Cl)\{N(Bu^t)Ar_F\}(tmeda)]$,[195] $[Fe(tmeda)(Me)\{N(CH_2Bu^t)C_6H_3Pr^i_2\text{-}2,6\}]$,[201b] $[Fe\{C_5H_3N\text{-}2,6\text{-}(C(=CH_2)NC_6H_3Pr^i_2\text{-}2,6)_2\}(\mu\text{-}NMe_2)K(Et_2O)_2]$.[214]
	Cobalt
Co(III):	$[Co\{N(SiMe_3)_2\}_3]$.[172]
Co(II):	$[Na(12\text{-}crown\text{-}4)_2][Co\{N(SiMe_3)_2\}_3]$,[175] $[Co\{N(SiMe_3)_2\}_2]$ (ged),[177] $[(Co\{N(SiMe_3)_2\}_2)_2]$,[176] $[Co\{N(SiMePh_2)_2\}_2]$,[201a,180] $[\{Co(NPh_2)_2\}_2]$,[215] $[Co\{N(SiMe_3)_2\}\{HC(CMeNC_6H_3Pr^i_2\text{-}2,6)_2\}]$,[191] $[Co\{NMesBMes_2\}_2]$,[168] $[Co\{N(SiMe_3)_2\}_2(py)]$,[182] $[Co\{N(SiMe_3)_2\}_2(py)_2]$,[182] $[Co\{N(SiMe_3)_2\}_2(4,4'\text{-}bipyridyl)]$,[182] $[Co\{N(C_6F_5)_2\}_2(py)_2]$.[118]
	$[Co(Me)(tmeda)\{N(CH_2Bu^t)(C_6H_2Pr^i_2\text{-}2,6)\}]$,[201b] $[Li(THF)_{4.5}][Co\{N(SiMe_3)_2\}(OCBu^t_3)_2]$,[216] $[Li(\mu\text{-}OCBu^t_3)_2Co\{N(SiMe_3)_2\}]$,[216] $[(Et_2O)_2(thf)Li(\mu\text{-}Cl)Co(NPhBMes_2)_2]$,[217a] $[(\eta^5\text{-}C_5Me_5)Co(\mu\text{-}Cl)(\mu\text{-}NMe_2)Co(\eta^5\text{-}C_5Me_5)]$,[217b] $[\{Li(thf)_2\}_3\{Co_3(\mu_2\text{-}NHMes)_3Cl_6\}]$,[218,219] $[Li(dme)_3][Li(dme)][Co_6(\mu_4\text{-}NPh)_3(\mu_3\text{-}NPh)_2(NHPh)_6\}]$,[219] $[Li_2Cl(thf)_6][Co_{18}(\mu_4\text{-}NPh)_3(\mu_3\text{-}NPh)_{12}(NHPh)_3]$.[219]
	Nickel
Ni(II):	$[Li(thf)_4][Ni(NPh_2)_3]$,[215] $[Ni(NPhBMes_2)_2]$,[168] $[Ni(NMesBMes_2)_2]$,[169] $[\{Ni(NPh_2)_2\}_2]$,[215] $[Ni\{C_6H_3(CH_2PPr^i_2)_2\text{-}2,6\}(NH_2)]$,[220a] $[Ni(Pr^i_2PCHCH_2PPr^i_2)(Me)\{N(H)C(H)MePh\}]$,[220b] [*trans*-$Ni(PMe_3)_2(Mes)(NHPh)$],[221] $[Ni(\eta^5\text{-}C_5Me_5)(PEt_3)(NHC_6H_4Me\text{-}4)]$,[222] [*cis*-$\{Ni(\eta^5\text{-}C_5Me_4Et)(\mu\text{-}NHR)\}_2$],[223] (R = Bu^t, C_6H_4Me-4, $C_6H_3Me_2$-2,6) $[Ni\{HC(CMeNC_6H_3Pr^i_2\text{-}2,6)_2\}\{N(SiMe_3)_2\}]$,[224] $[Ni\{HC(CMeNMes)_2\}\ \{NH(1\text{-}Ad)\}]$.[225]
Ni(I):	$[Ni(Bu^t_2PCH_2CH_2PBu^t_2)NHC_6H_3Pr^i_2\text{-}2,6]$,[226] $[Ni_3(\mu\text{-}NHPh)_3(PEt_3)_3][Br]$.[219]
	Copper
Cu(I):	$[\{Li(OEt_2)_2\}_2Cu(NPh_2)_3]$,[227] $[Li(thf)_4][Cu\{N(SiMePh_2)_2\}_2]$,[228] $[\{CuN(SiMePh_2)_2\}_3]$,[229] $[(CuNEt_2)_4]$,[230] $[(CuNRR')_4]$,[231] (NRR′ = NMe_2, $N(CH_2)_3CH_2$, $N(Me)CH_2CH_2NEt_2$), $[\{CuN(SiMe_3)_2\}_4]$,[232a,232b] $[\{CuN(SiMe_2Ph)_2\}_4]$,[229] $[\{CuN(SnMe_3)_2\}_4]$,[227] $[(CuNHBu^t)_8]$,[227] $[(LiCu(Mes)\{N(CH_2Ph)_2\})_2]$,[233] $[(2,6\text{-}Pr^i_2H_3C_6)NCH_2CH_2N(C_6H_3Pr^i_2\text{-}2,6)CCu(NHPh)]$,[234] $[(Bu^t_2PCH_2CH_2PBu^t_2)CuNHPh]$,[235] $[Cu(C_6H_4PPh_3)(NHMes)]$,[227] $[(Cu_2\{N(SiMe_3)_2\}Mes)_2]$,[236] $[\{Li(dme)Cu(NHMes)(NHPh)\}_2]$,[227] $[Li(dme)_3][Cu_6(NHMes)_3(NMes)_2]$,[227] $[W_2Cu_5(NBu^t)_2(\mu\text{-}NBu^t)_6(NHBu^t)_2][BF_4]$.[237]

[a] 1980–2007. Complexes are arranged by element in the order of descending oxidation state and degree of amide substitution, Ad = adamantanyl, Cy = cyclohexyl, dme = 1,2-dimethoxyethane, dmpe = 1,2-bis(dimethylphosphino)-ethane, dmpm = bis(dimethylphosphine)methane, dipp = 1,2-bis(diphenylphosphino)methane, dppm = bis-(diphenylphosphino)methane, dppp = 1,3-bis(diphenylphosphino)propane, Mes* = $C_6H_2Bu^t_3$-2,4,6, tpb = tris(3,5-dimethylpyrazolyl) borate, Tf = $CF_3SO_3^-$, thf = tetrahydrofuran, Ts = 4-$MeC_6H_4SO_3^-$, R = $C(CD_3)_2Me$, Ar = $C_6H_3Me_2$-3,5 unless otherwise stated.

[b] Scandium amides are described in more detail in Chapter 4.

Table 6.2 *Structurally characterized second and third row transition metal derivatives of non-chelating diorganoamido ligands*[a]

	Yttrium[b]
Y(III):	$[Y\{N(SiMe_3)_2\}_3]$,[238] $[Y\{NCMe_2(CH_2)_3CMe_2\}_3]$,[239]
	$[Y\{N(SiHMe_2)_2\}_3(CN(Me)CCH_2CH_2NMe)_{1\text{ or }2}]$,[240] $[Y\{N(SiMe_3)_2\}_3(NCPh)_2]$,[238] $[Y_2\{N(SiMe_3)_2\}_4(thf)_2(\mu\text{-}\eta^2{:}\eta^2\text{-}N_2)]$,[241] $[Y\{N(SiHMe_2)_2\}$ $\{3,3'\text{-}Bu^t_2\text{-}5,5',6,6'\text{-}Me_4\text{-}1,1'\text{-biphenyl-}2,2'\text{-}O_2\}(thf)_2]$.[242a]
	$[Y\{N(SiMe_3)_2\}(Cl)(N\{CH_2CH_2(1\text{-}CNCHCHNMes)\}_2)]$.[242b]
	Zirconium
Zr(IV):	$[\{(thf)Li\}_2Zr(NMe_2)_6]$,[243] $[\{(Et_2O)Li\}_2[Zr(NMe_2)_6]$,[244] $[Na(thf)_6]_2[Zr(\eta^1\text{-}NC_4H_4)_6]$,[97] $[Zr(NMe_2)_4]$ (ged),[245] $[\{Zr(NMe_2)_4\}_2]$,[243] $[Zr(NPh_2)_4]$,[246] $[Zr(NH_2)\{N(SiMe_3)_2\}_3]$,[247] $[Zr(NMe_2)_3(NHMe_2)\{H_2NB(C_6F_5)_3\}]$.[66]
	$[Zr(\eta^5\text{-}C_5Me_5)(NHBu^t)_3]$,[248] $[Zr(\eta^5\text{-}C_5H_5)(NMe_2)_2\{NH_2(B(C_6F_5)_3\}]$,[66] $[Zr(\eta^5\text{-}C_5H_2(SiMe_3)_3\text{-}1,2,4)(NMe_2)_3]$,[249] $[\{Zr(NMe_2)_3\}_2\{Zr(\mu\text{-}H)_2$ $(\mu\text{-}NMe_2)_4\}]$,[250,251] $[\{Zr(NMe_2)_2(\mu\text{-}NMe_2)I\}_2]$,[252] $[Zr(NMe_2)_3(tpb)]$,[253] $[Zr(NMe_2)_3\{Si(SiMe_3)_3\}]$,[254] $[Zr(NMe_2)_3\{SiPhBu^t_2\}\cdot 0.5thf]$,[254] $[\{Zr(Cl)(NMe_2)_2$ $(\mu\text{-}NMe_2)\}_2]$,[255] $[(Me_2N)_3Zr(\mu\text{-}Cl)_2(\mu\text{-}NMe_2)Zr(NMe_2)_2(thf)]$,[256] $[Zr(\eta^5\text{-}C_5Ph_5)$ $(NMe_2)_3]$,[257] $[\{Zr(NMe_2)_3\}_2\{\mu\text{-}C_6H_3(CH_2NC_6H_3Pr^i_2\text{-}2,6)_2\text{-}1,3\}]$,[258]
	$[K(18\text{-crown-}6)_{1.5}[Zr(NMe_2)_3\{(Me_3Si)_2Si(CH_2)_2Si(SiMe_3)_2\}]$, $[Li(thf)_4]$
	$[Zr(NMe_2)_3(SiBu^tPh_2)_2]$,[259,265] $[Zr\{N(SiMe_3)_2\}_3Me]$,[77] $[Zr\{N(SiMe_2)_2\}_3]$ $[MeB(C_6F_5)_3]$,[260] $[Zr(Cl)(NMe_2)\{N(SiMe_3)_2\}_2]$,[261] $[Zr(L)\{N(1\text{-}Ad)(C_6H_3$ $Me_2\text{-}3,5)\}_3]$,[262] ($L = Me, BH_4$), $[Zr(NMe_2)_2\{N(SiMe_3)_2\}(SiPh_2Bu^t)]$,[263] $[ZrCl\{N(SiMe_3)NMe_2\}_3]$,[264] $[Zr\{N(SiMe_3)_2\}_2(\mu\text{-}NHSiMe_3)(\mu\text{-}NSiMe_3)$ $Li(thf)_2]$,[247] $[K(\eta^6\text{-}C_6H_5CH_3)_2][ZrCl_2\{N(C_6F_5)_2\}_3]$.[118]
	$[Zr(\eta^5\text{-}C_5H_5)_2\{N(H)C_6H_4SMe\text{-}2\}_2]$,[265] $[Zr\{\text{ethylene-}1,2\text{-bis}(\eta^5\text{-}4,5,6,7\text{-tetrahydro-}1\text{-indenyl})\}(NHC_6H_3Me_2\text{-}2,6\}_2]$,[267] $[Zr(\eta^5\text{-}C_5H_5)(NPBu^t_3)$ $(NHC_6H_3Pr^i_2\text{-}2,6)_2]$,[81] $[Zr\{N(H)SiBu^t_3\}_2(NSiBu^t_3)(thf)]$,[268] $[\{Zr(NMe_2)_2$ $(\mu\text{-}NBu^t)\}_2]$,[269] $[ZrCl_2(NR_2)_2(thf)_2]$ ($R = Me$ or Et),[270] $[Zr(\eta^5{:}\eta^1\text{-}C_5Me_4SiMe_2$ $C_2B_{10}H_{10})(NMe_2)_2]$,[271] $[Zr(\eta^5{:}\eta^1\text{-}Pr^i_2BNB(C_9H_7)(C_2B_{10}H_{11})(NMe_2)_2]$,[93a] $[Zr(NMe_2)_2(OAl(Me)\{N(C_6H_3Pr^i_2\text{-}2,6)CMe\}_2CH)_2]$,[93c] $[Zr(\text{salicylaldiminato})_2$ $(NMe_2)_2]$,[272] $[Zr(\eta^5{:}\eta^5\text{-}C_5H_4CMe_2C_9H_6)(NMe_2)_2]$,[273] $Zr(\text{bis(amido) cyclodiphosphazene})(NMe_2)_2]$,[274] $[Zr\{Me_2Si(\eta^5\text{-}C_9H_6)_2\}(NMe_2)_2]$,[274] $[Zr\{MeN(CH_2CH_2NMes)_2\}Me_2]$,[276] $[Zr(\eta^1{:}\eta^1\text{-}Pr^i_2NP(O)C_9H_6C_2B_{10}H_{10})$ $(NMe_2)_2(NHMe_2)]$,[277] $[Zr\{Me_2Si(\eta^5\text{-}C_5H_2Pr^i_2\text{-}2,4)_2\}(NMe_2)_2]$,[278] $[Zr(NEt_2)_2I_2]$,[253] $[Zr(NEt_2)_2(t\text{-}Bu_2\text{malonato})_2]$,[279] $[Zr\{Mo(\eta^5\text{-}C_5H_5)(CO)_3\}_2$ $(NEt_2)_2(NHEt_2)]$,[280] $[Zr(2,2'\text{-di}(3\text{-methylindolyl})\text{methane})(NEt_2)_2(thf)]$,[281] $[Zr(\eta^5\text{-}C_5H_5)_2(\eta^1\text{-}NC_4H_4)_2]$,[97] $[Zr\{Me_2Si(\eta^5\text{-}C_5H_2Me\text{-}2\text{-}Bu^t\text{-}6)_2\}(NC_4H_4)_2]$,[282] $[(Zr(Cl)(\mu\text{-}Cl)\{N(SiHMe_2)_2\}_2)_2]$,[283] $[Zr\{N(SiMe_3)C_6H_3Pr^i_2\text{-}2,6\}_2X_2]$ ($X = F$,[284] Cl,[284] or Me[285]).
	$[K][\{(Zr(\eta^5\text{-}C_5Me_5))_3(\mu_3\text{-}N)(\mu_3\text{-}NH)(\mu\text{-}NH_2)_3\}_4(NH_2)_5(NH_3)_7]$,[286] $[\{Zr(\eta^5\text{-}C_5H_4Me)\}_5(\mu_5\text{-}N)(\mu^3\text{-}NH)_4(\mu\text{-}NH_2)_4]$,[287] $[\{Zr(OCBu^t_3)\}_5(\mu^5\text{-}N)(\mu^3\text{-}NH)_4$ $(\mu^2\text{-}NH_2)_4]$,[286] $[Zr(\eta^5\text{-}C_5Me_5)(NHBu^t)(NBu^t)(py)]$,[248] $[Zr(\eta^5\text{-}C_5Me_5)\{(NPr^i)_2$ $C{=}CH_2\}(NHBu^t)]$,[289] $[Zr(\eta^5\text{-}C_5Me_5)\{(NPr^i_2)_2CMe\}(NHBu^t)][B(C_6F_5)_4]$,[289] $[\{Zr(NEt_2)(I)_2(\mu\text{-}I)\}_2]$,[253] $[Zr(NEt_2)Cl_3(Et_2O)_2]$,[290] $[Zr(\eta^5\text{-}C_5H_5)_2\{NBu^t(SiMe_2H)\}$ $(X)]$,[291] ($X = H$ or Cl), $[Zr(\eta^5\text{-}C_5H_5)_2(Cl)\{N(C_6H_3Pr^i_2\text{-}2,6)(C(CH_2)Ph)\}]$,[292] $[Zr(\eta^5\text{-}C_5Me_5)\{N(SiMe_3)C_6H_3Pr^i_2\text{-}2,6\}(Cl)(F)]$,[285] $[Zr(\eta^5\text{-}C_5Me_5)\{N(SiMe_3)$ $C_6H_3Pr^i_2\text{-}2,6\}Me_2]$,[285] $[(\eta^5\text{-}C_5H_5)_2Zr(\mu\text{-}P\{Si(CMe_2Pr^i)Me_2\})_2Zr(\eta^5\text{-}C_5H_5)$ $(NEt_2)]$.[293]

(*continued*)

Table 6.2 *(Continued)*

	Hafnium
Hf(IV):	$[Hf(NPh_2)_4]$[294] $[Hf(\eta^5-C_5Me_5)(NHC_6H_3Pr^i_2-2,6)_3]$,[249] $[Li(thf)_4][Hf(NMe_2)_3(SiBu^tPh_2)_2]$,[259] $[K(18-crown-6)_{1.5}][Hf(NMe_2)_3\{(Me_3Si)_2Si(CH_2)_2Si(SiMe_3)_2\}]$,[259] $[Hf(Cl)\{N(SiMe_3)_2\}_3]$,[76] $[Hf\{N(SiMe_3)_2\}_3][MeB(C_6F_5)_3]$,[260] $[Hf\{N(SiMe_3)_2\}_2(\mu-NHSiMe_3)(\mu-NSiMe_3)Li(thf)_2]$.[247] $[Hf\{\eta^5:\eta^5(CH_2C_9H_6)_2\}(NMe_2)_2]$,[295] $[Hf\{(NPr^i)_2CNRR'\}_2(NRR')_2]$,[296] (R/R′ = Me/Me, Me/Et, Et/Et), $[Hf(HC\{C(OBu^t)O\}_2)_2(NEt_2)_2]$,[297a] $[Hf\{N(SiMe_3)C_6H_3Pr^i_2-2,6\}_2Cl_2]$.[284] $[Hf(\eta^5-C_5H_5)_2(Me)_2\{NH_2B(C_6F_5)_3\}]$,[66] $[\{Hf(\eta^5-C_5Me_5)(NHPh)(\mu-NPh)\}_2]$,[248] $[Hf(\eta^5-C_5Me_5)(Cl)\{N(SiMe_3)C_6H_3Pr^i_2-2,6\}_2]$.[284]
	Niobium
Nb(V):	$[Nb(NMe_2)_4\{HC(C(O)Bu^t)_2\}]$,[297b] $[Nb(NC_6H_3Pr^i_2-2,6)(NMe_2)_3]$,[298] $[Nb(3,1,2-C_2B_9H_{11})(NMe_2)_3]$,[299a] $[Nb(NPh_2)_3Br_2(thf)]$,[300] $[\{Nb(NCy_2)_3\}_2(\mu-\eta^1:\eta^1-N_2)]$,[301] $[Nb(\eta^2-OCPh_2)\{N(Pr^i)Ar\}_3]$,[302] $[Ar(Bu^t)N\}_3Mo(\mu-N_2)Nb\{N(Pr^i)Ar\}_3]$,[303] $[Nb(NX)\{N(R)Ar\}_3]$ (R = $C(CD_3)_2CH_3$; X = Na, $P(NMe_2)_2$, $SO_2(CF_3)$),[304] $[\{Ar(R)N\}_3NbNP(\mu-NBu^t)PNNb\{N(R)Ar\}_3]$,[305] $[Na(12-crown-4)_2][Nb(P)\{N(CH_2Bu^t)Ar\}_3]$,[306] $[Nb(\eta^2-PPR_2)\{N(CH_2Bu^t)Ar\}_3]$ (R = Bu^t or Ph),[304] $[Nb(O)\{N(CH_2Bu^t)Ar\}_3]$,[307] $[Nb(PX)\{N(CH_2Bu^t)Ar\}_3]$ (X = Na,[307] $C(O)Bu^t$,[307] C(O)1-Ad,[307] PNMes*,[308]] $[Nb(NX)\{N(CH_2Bu^t)Ar\}_3]$ (X = $Na(thf)_3$,[309] $NC(O)Bu^t$,[309] Mes*[308]), $[Nb(SO_2CF_3)_2\{N(CH_2Bu^t)Ar\}_3]$,[307] $[Nb\{N(CH_2Bu^t)Ar\}_3(\eta^2-NCMes)]$,[310] $[(\mu^2:\eta^3:\eta^3-cycloEP_2)(Nb\{N(CH_2Bu^t)Ar\}_3)_2]$ (E = Ge or Sn),[311] $[(\mu^2:\eta^4-P_2)(Nb\{N(CH_2Bu^t)Ar\}_3)_2]$.[310] $[Nb(tpb)(O)\{NH(SiMe_3)\}_2]$,[312] $[Nb(NCy_2)_2(C_6H_{10}NCy)Cl]$,[301] $[Nb(\eta^5-C_5H_5)(NR)\{NHR\}_2]$,[313] (R = $C_8H_3-2,6-Me_2$), $[Nb(X)(\eta^2-CMe_2NAr)\{N(Pr^i)Ar\}_2]$,[302] (X = $OCHPh_2$ or BH_4), $Nb(H)(\eta^2-Bu^t(H)CNAr)\{N(CH_2Bu^t)Ar\}_2]$,[299] $[(NNb\{N(Pr^i)Ar\}_2)_3]$.[303] $[Nb(\eta^5-C_5H_5)(NC_6H_3-2,6-Me_2)\{NH(C_6H_3Me_2-2,6)\}X]$[314] (X = Cl or Me), $[\{Nb(Cl)(\mu-OMe)\{N(SiMe_3)_2\}(NSiMe_3)\}_2]$.[315]
Nb(V)/ Nb(IV):	$[\{Nb(NCy_2)_2\}_2(\mu-N)\{\mu-NLi(tmeda)\}]$.[316]
Nb(IV):	$[Nb(NPh_2)_4]$,[317] $[NbBr_2\{N(SiMe_3)_2\}_2]$[317], $[Nb(Cl)(Ph)\{N(SiMe_3)_2\}_2]$,[318] $[(Nb(C_6H_3Me_2-2,6)\{N(1-Ad)Ar\})_2\{\mu-N(1-Ad)\}_2]$.[316]
	Tantalum
Ta(V):	$[Ta(NMe_2)_5]$ (ged),[319] $[Ta(NMe_2)_5]$,[299a] $[Ta(NEt_2)_5]$,[299b] $[Ta\{\overline{N(CH_2)_4CH_2}\}_5]$.[299c] $[Ta(NMe_2)_4Bu^t]$,[320] $[Ta(NMe_2)_4\{(NCy)_2CNMe_2\}]$,[321] $[Ta(NMe_2)_4(OC_6H_2Bu^t-2,4-R-6)]$,[322] (R = 1-naphthyl, 2,3-dihydro-1-naphthyl or inden-3-yl), $[Ta(NMe_2)_4\{HC(C(O)OBu^t)\}]$,[297b] $[TaCl(NMe_2)_3\{N(SiMe_3)_2\}]$,[323] $[Ta(NMe_3)_3\{N(SiMe_3)_2\}(SiPh_2Bu^t)]$.[261] $[Ta(NMe_2)_3Cl_2]$,[324] $[Ta(NMe_2)_3Cl_2(HNMe_2)]$,[324] $[Ta(NMe_2)_3Br(C_6H_4Me-4)]$,[320] $[Ta(NMe_3)_3(Cl)\{Si(SiMe_3)_3\}]$,[265] $[Ta(NMe_3)_3(NC_6H_3Pr^i_2-2,6)]$,[298] $[Ta(NMe_2)_3(N_2C_3H_2Me_2-3,5)]$,[325] $[Ta(NMe_2)_3\{(NCy)_2C{=}NCy\}]$,[326] $[TaCl(NMe_2)_3\{(NPr^i)_2CNHPr^i\}]$,[327] $[Ta(NMe_2)_3\{(NPr^i)_2C{=}NPr^i\}]$,[328] $[Ta(NMe_2)_3(CH_2Ph)\{(Pr^iN)_2CNPr^iSiMe_3\}]$,[328] $[Ta(NMe_2)_3\{(NBu^t)_2SO_2\}]$,[329] $[Ta(NMe_2)_3L_2]$ (L_2 = closo-1,2-$C_2B_9H_{10}Me$,[330] closo-1,2-[299a] or closo-1,7-$C_2B_9H_{11}$[331]), $[Ta(NMe_2)_3\{OC_6H_2(\eta^2-indenyl)-2-Bu^t_2-4,6\}]$,[322] $[Ta(NBu^t)(NMe_2)_2\{N(SiMe_3)NMe_2\}]$.[332] $[TaCl_3\{N(SiMe_3)_2\}_2]$,[333] $[TaCl_3(NC_7H_{10})_2]$,[334] $[(Ta(Cl)(NBu^t)\{N(SiMe_3)_2\}_2)_2]$,[335] $[TaCl_3(NMe_2)_2(NC_5H_4Ph-4)]$,[336] $[\{TaCl_2(NMe_2)_2(HNMe_2)_3\}_2O]$,[324] $[NH_2Me_2][cis-TaCl_4(NMe_2)_2]$.[336a] $[Ta(Cl)(NCy_2)_2\{N(Cy)C_6H_{10}\}]$.[336b]

Table 6.2 *(Continued)*

	$[Ta(Cl)_2(NHPr^i)(NPr^i)(NH_2Pr^i)_2]$,[337] $[\{Ta(Cl)(\mu\text{-}Cl)(NBu^t)(NHBu^t)(NH_2Bu^t)\}_2]$,[338] $[TaCl_2(NBu^t)(NHBu^t)(py)_2]$,[339] $[TaX_2(NAr)(NHAr)]$,[340] (Ar = $C_6H_3Mes_2$-2,6; X = Me or CH_2Bu^t), $[TaCl_4(NMe_2)(Et_2O)]$,[341] $[Ta(NMe_2)(NBu^t)\{(NPr^i)_2CNMe_2\}_2]$,[342] $[TaCl_2(NMe_2)\{(OC_6H_3Ph)_2CH_2)_2\}]$,[336] $[Ta(\eta^5\text{-}C_5Me_5)_2(NHBu^t)X][B(C_6F_5)_4]$,[343] (X = H, Cl, CCPh), $[\{Ta(H)(\mu\text{-}H)_2(PMe_2Ph)_2(NC_{12}H_8)\}_2]$,[344] $[Ta(X)(\mu\text{-}X)(NSiMe_3)\{N(SiMe_3)_2\}]_2$,[345] (X = Cl or OMe), $[Ta(NR^1R^2)(NR^4)\{(NR^3)_2CNR^1R^2\}_2]$,[346] ($R^1$, R^2, R^3, R^4 = Me, Et, Pr^i, Bu^t; Et, Et, Pr^i, Bu^t; Me, Me, Cy, Bu^t; Et, Et, Cy, Bu^t).
Ta(IV):	$[Ta(NEt_2)_2\{N(SiMe_3)_2\}_2]$,[347a] $[Ta\{N(SiMe_3)_2\}_2X_2]$,[347a] (X = Cl, Ph), $[\{Ta(NCy_2)_2Cl(\mu\text{-}H)\}_2]$.[347b]
	Molybdenum
Mo(VI):	$[Mo(NMe_2)_6]$,[348] $[Mo(N)(NPh_2)_3]$,[349] $[Mo(N)\{N(Bu^t)Ph\}_3]$,[350] $[Mo(N)\{N(CH_2Bu^t)Ar\}_3]$,[309] $[Mo(CH)\{N(Bu^t)Ar\}_3]$,[351] $[(=C(Ph)C\equiv Mo\{N(Pr^i)Ar\}_3)_2]$,[352] $[c\text{-}C_7H_{10}\text{-}1,2\text{-}(C\equiv Mo\{N(Pr^i)Ar\}_3)_2]$,[352] [K(benzo-15-crown-5)$_2$]$[CMo\{N(C(CD_3)_2Me)Ar\}_3]$,[353] $[(Et_2O)(thf)_2Na(Ph)PC\equiv Mo\{N(Pr^i)Ar\}_3]$,[354] $[\{(Ar_F)(Bu^t)N\}_3VNCMo\{N(1\text{-}Ad)Ar\}_3]$,[140] $[PMo\{N(C(CD_3)_2Me)Ar\}_3]$,[355] $[PMo\{N(2\text{-}Ad)Ar\}_3]$,[356] $[MesNPMo\{N(C(CD_3)_2Me)Ar\}_3]$,[355] $[OPMo\{N(C(CD_3)_2Me)Ar\}_3]$,[357] $[\{(1\text{-}Ad)S\}_3Mo\leftarrow NMo\{N(Bu^t)Ph\}_3]$,[358] [*trans*-$Mo\{NH(1\text{-}Ad)\}_2(OSiMe_3)_4]$,[359] $[Mo(NC_6H_3Pr^i_2\text{-}2,6)_2(NMe_2)_2]$,[360] $[Mo(NC_6H_3Pr^i_2\text{-}2,6)_2(NMe_2)\{N(SiMe_3)_2\}]$,[360] $[Mo(NAr)\{N(H)Ar\}(CH_2EMe_3)_2][OTf]$ (E = Si or Ge; Ar = $C_6H_3Pr^i_2$-2,6),[361] $[Mo(NC_6H_3Pr^i_2\text{-}2,6)_2(NMe_2)\{Si(SiMe_3)_3\}]$,[360] $[PMo\{N(Pr^i)Ar\}(OC_6H_3Me_2\text{-}2,6)_2]$,[362] $[PPh_4][MoNCl_3\{N(SiMe_3)_2\}]$,[363] $[Mo(NC_6H_3Me_2\text{-}2,6)\{N(SiMe_2H)C_6H_3Me_2\text{-}2,6\}Cl(PMe_3)_2]$.[364]
Mo(V):	$[EMo\{N(C(CD_3)_2Me)Ar\}_3]$,[365] (E = O, S, Se or Te), $[(\eta^2\text{-}OCPh_2)Mo\{N(Pr^i)Ar\}_3]$,[366] $[(\eta^2\text{-}Me_2CNAr)Mo(H)\{N(Pr^i)Ar\}_2]$,[366] $[Mo(NC_6H_3Pr^i_2\text{-}2,6)(NPh_2)_2(CHCMe_2Ph)]$.[367]
Mo(V/IV):	$[(Me_2N)_3MoNMo(NMe_2)_3]$,[368] $[(Me_2N)_3MoNMo\{N(C(CD_3)_2Me)(C_6H_4F\text{-}4)\}_3]$,[368] $[\{Bu^t(Ph)N\}_3MoPMo\{N(Ph)Bu^t\}_3]$.[368]
Mo(IV):	$[\{(thf)Li\}_2Mo(NMe_2)_6]$,[369] $[Mo(\eta^5\text{-}C_5H_5)(NMe_2)_3]$,[370] $[Ph_2PNNMo\{N(Bu^t)Ar\}_3]$,[371] $[Ph_2(Me)PNNMo\{N(Bu^t)Ar\}_3][OTf]$,[372] $[\{Ar(Me(CD_3)_2C)N\}_3UN_2Mo\{N(Bu^t)Ph\}_3]$,[373] $[PhS(Ph)CNMo\{N(Bu^t)Ar\}_3]$,[374] $[\{\mu\text{-}NC(Ph)C(Ph)N\}(Mo\{N(Pr^i)Ar\}_3)_2]$,[375] $[(\mu\text{-}NCN)(Mo\{N(Pr^i)Ar\}_3)_2]$,[376] $[\{Ph(Bu^t)N\}_3TiN_2Mo\{N(Bu^t)Ar\}_3]$,[85] $[Me_3SiNNMo\{N(C(CD_3)_2Me)Ar\}_3]$,[85] [Na(12-crown-4)$_2$]$[NNMo\{N(C(CD_3)Me)Ar\}]$,[85] $[(thf)_3NaN_2Mo\{N(1\text{-}Ad)Ar\}_3]$,[85] $[Mo(NMe_2)_2(OR)_2]$,[377] (R = 4-MeCy, $CHPr^i_2$), $[Mo(NMe_2)_2(PPh_2)_2]$,[378] $[Mo(NMe_2)L_3]$,[379] (L = 3,5-Bu^t_2pyrazolyl), $[Mo(NH_2)(OH)(dppe)_2][OTf]_2$.[380]
Mo(III):	$[Mo\{N(C(CD_3)_2Me)Ar\}_3]$,[381] $[Mo\{N(1\text{-}Ad)Ar\}_3]$,[85] $[Mo\{N(2\text{-}Ad)Ar\}_3]$,[356] $[Bu^tNCMo\{N(Bu^t)Ar\}_3]$,[382] $[(1\text{-}AdNC)_2Mo\{N(Pr^i)Ar\}_3]$,[382] $[Mo_2(NMe_2)_4L_2]$ (L = Pr^i,[383] CH_2Ph,[384,385] C_6H_4Me-2,[385] C_6H_4Me-4,[385] $C_6H_3Me_2$-2,6,[386] $OC_6H_3Bu^t$-2-Me-6,[387] C_9H_7,[388] CH_2-2(py-6-Me),[389] 4-$MeH_4C_6N_3C_6H_4Me$-4,[390] O-2(pyMe-6),[391] $OCPh_3$,[393] $Si(SiMe_3)_3$,[394] $Sn(SnMe_3)_3$,[394,395] PBu^t_2,[396] $AsBu^t_2$[397]), $[Mo_2(OBu^t)_4(NHPh)_2(NH_2Ph)_2]$,[398] $[Mo_2(NMe_2)_2(OC_6H_3Me_2\text{-}2,6)_4]$,[384] $[Mo_2(NMe_2)_2(HNMe_2)_2(OC_6F_5)_4]$,[392] $[Mo_2(NMe_2)_2L_2]$ (L = $CH_2(C_6H_3Bu^t\text{-}6\text{-}Me\text{-}4\text{-}O\text{-})_2$,[399] $MeCH(C_6H_3\text{-}4,6\text{-}Bu^t\text{-}O\text{-})_2$[400]), $[Mo_2Me_2(NMe_2)_2L_2]$ (L = 4-$MeH_4C_6NNNC_6H_4Me$-4[401]), $[Mo_2(\eta^5\text{-}C_5H_5)_2(\mu\text{-}NH_2)(\mu\text{-}SMe)_3]$.[402]
Mo(II):	$[Mo(tpb)(NO)(NHPh)(OAc)]$.[403]
Mo(I):	[*trans*-$Mo(dmpe)_2(NO)\{N(R^1)CH_2R\}]$,[404] (R = R′ = Ph; R = $(\eta^5\text{-}C_5H_5)(\eta^5\text{-}C_5H_4)Fe$, R′ = Ph); R = R′ = $(\eta^5\text{-}C_5H_5)(\eta^5\text{-}C_5H_4)Fe$).

(continued)

Table 6.2 *(Continued)*

	Tungsten
W(VI):	$[W(NMe_2)_6]$ (ged),[405] $[W(NPh)(NMe_2)_4]$,[406] $[W(Cy_7Si_7O_{12})(NMe_2)_3]$,[407] [*fac*-$WCl_3(NEt_2)_3$],[408] $[NW\{N(Pr^i)Ar\}_3]$,[409a] $[PW\{N(Pr^i)Ar\}_3]$,[409b] $[W\{N(Pr^i)Ar\}_3\{NC(O)CF_3\}\{OC(O)CF_3\}]$,[409] $[W(tpb)(CBu^t)(NHPh)_2]$,[410] $[\{W(\mu\text{-}NPr^i)(NPr^i)(NHPr^i)_2\}_2]$,[411] $[W(tpb)(CBu^t)(Cl)(NHPh)]$,[410] $[NH_3(C_6H_3Pr^i_2\text{-}2,6)][WCl_4(NC_6H_3Pr^i_2\text{-}2,6)(NHC_6H_3Pr^i_2\text{-}2,6)]$,[412] $[W(H)(NSiBu^t_3)_2(NHSiBu^t_3)]$,[413] $[W(NAr)\{N(H)Ar\}(CH_2EMe_3)_2\}Cl]$ (E = Si or Ge; Ar = $C_6H_3Pr^i_2$-2,6),[361] $[WCl_3(NSiCl_3)\{N(SiCl_3)_2\}]$.[414]
W(V):	$[OW\{N(Pr^i)Ar\}_3]$,[409a] $[W_4N_4(NPh_2)_6(OBu^n)_2]$.[349]
W(V/IV):	$[W_2Cl_3(NMe_2)_2(CH_2NMe)(CHCH_2)(PMe_2Ph)_2]$,[415] $[W_2Cl_2(NMe_2)_2(\mu\text{-}Cl)(\mu\text{-}NMe_2)(PMe_3)_2(\mu\text{-}C_2H_2)]$,[415] $[W_2(NMe_2)_2(\mu\text{-}C_2Me_2)Cl_4(py)_2]$,[416] $[(dppp)(Cl)W(\mu\text{-}Cl_3)WCl_2(NR_2)]$ (dppp = 1,3-bis(diphenylphospino)propane; R = Et or Bu^n).[417]
W(III):	$[W_2(NMe_2)_4L_2]$ (L = C_3H_5,[418,419] C_4H_7,[418,419] C_5H_4Me,[420] $L_2 = Me_2Si(C_5H_4)_2$,[420] L = $CH(SiMe_3)_2$,[421] $Si(SiMe_3)_3$,[394] $GePh_3$,[422,423] $Sn(SnMe_3)_3$,[394] PBu^t_2,[396,424] $P(SiMe_3)_2$,[424] PCy_2,[425] $PCy_2/P(SiMe_3)_2$,[424] $AsBu^t_2$,[397] $OCMe(CF_3)_2$,[425] $OCNMe_2(CF_3)_2$,[427] $OBMes_2$[428] $OSiPh_3$,[393] SBu^t,[429] 2-thienyl,[430a] OTf,[430b] $OCMe_2(py)$[431]), $[W_2(NMe_2)_3(OTf)_3(PMe_3)_2)$,[430b] $[W_2(\mu\text{-}C_3Ph_3)(\mu\text{-}CPh)(NMe_2)_4]$,[432] ($[W_2(NMe_2)_3Cl_3(PMe_2Ph)_2]$,[433] $[W_2(NHBu^t)_2Cl_4(PR_3)_2]$,[434] ($PR_3 = PMe_3$, PEt_3or PMe_2Ph), $[W_2(NHBu^t)_2Cl_4(P\text{-}P)]$ (P-P = dmpm, dmpe, dppm, or dppe),[435] $[W_2(NHSiBu^t_3)_2Cl_4]$,[413] $[W(NMe_2)_2\{OCMe(CF_3)_2\}_4]$,[436] $[W(NMe_2)_2Et_2(Bu^tacac)_2]$,[437] $[W_2(NMe_2)_2(O_2CNMe_2)_2(O_2CPBu^t_2)_2]$,[438] $[W_2(NBu^n_2)_2Cl_4(dppp)]$,[417] $[C(W_2\{N(SiMe_3)SiMe_2CH_2\}_2\{N(SiMe_3)SiMe_2\})_2]$.[439]
W(II):	$[W(tpb)(CO)_2\{N(H)Ts\}]$,[440] $[W(tpb)(CO)_2(NMe_2)]$.[441]
	Rhenium
Re(V):	$[Re(O)\{N(SiMe_3)_2\}_3]$,[442] $[Re(Cl)(O)(tpb)(NEt_2)]$.[443]
Re(II):	$[Re(\eta^5\text{-}C_5Me_5)(NO)(PPh_3)(NHPh)]$.[444,445]
Re(I):	$[Re(bipy)(CO)_3NR_2]$,[446] (R_2 = H, Ph; H; C_6H_4Me-4; Ph, Ph), $[Re(bipy)(CO)_3(N(C_6H_4Me\text{-}4)\{C(O)CHPh_2\})]$.[447]
	Ruthenium
Ru(II):	$[\{Ru(\eta^6\text{-}C_6Me_6)\}_2(\mu^2{:}\eta^1,\eta^1\text{-}NH_2NH_2)(\mu^2\text{-}NH_2)(\mu^2\text{-}H)][OTf]_2$,[448] *trans*-$[Ru(dmpe)_2(H)(NH_2)]$,[449] $[\{Ru(\eta^6\text{-}C_6Me_6)\}\}_2(\mu\text{-}NHPh)_3][BF_4]$,[450] $[Ru(H)(NHPh)(PMe_3)_4]$,[451] $[\{Ru(\eta^5\text{-}C_5Me_5)(\mu\text{-}NHPh)\}_2]$,[452] $[Ru(\eta^6\text{-}C_6Me_6)(Cl)\{NH(C_6H_3\text{-}2,6\text{-}Me_2)\}]$,[453] $[Ru(\eta^6\text{-}C_6Me_6)(PMe_3)(Ph)(NHPh)]$,[454] $[Ru(tpb)(CO)(PPh_3)(NHPh)]$,[455] $[Ru(tpb)\{P(OMe)_3\}_2(NHPh)]$,[456] $[\{Ru(tpb)\{P(OMe)_3\}_2(NHC_6H_4\text{-})\}_2]\ [OTf]_2$.[457]
Ru(I):	$[Ru(PMe_3)_4\{NHC(NPh)(NHPh)\}]$.[458]
	Osmium
Os(IV):	$[Os(Cl)_2(NHPh)(tbp)]$,[459] $[Os(Cl)_2(NC_6H_8)(tpb)]$.[460]
Os(II):	$[Os(\eta^6\text{-}1,3,5\text{-}C_6H_3Me_3)(PMe_3)(NHPh)][PF_6]$.[461]
Os(I):	$[Os_3(CO)_{10}\{\mu\text{-}N(H)(C_5H_3NBr)\}(\mu\text{-}NC_{10}H_3NO)]$,[462] $[Os_3(\mu\text{-}H)(\mu\text{-}NH_2)(CO)_{10}(PPh_3)]$.[463]

Table 6.2 *(Continued)*

	Rhodium
Rh(III):	$[(\eta^5\text{-C}_5\text{Me}_5)\text{Rh}(\mu\text{-Br})(\mu\text{-NHPh})_2\text{Rh}(\eta^5\text{-C}_5\text{Me}_5)][\text{Br}]$.[464]
Rh(I):	$[\{\text{Rh}(\mu\text{-NHPh})(\text{PPh}_3)_2\}_2]$,[465] $[\{\text{Rh(COD)}(\mu\text{-NHAr})\}_2]$ (Ar = Ph, $\text{C}_6\text{H}_4\text{Me-4}$),[466] $[\{\text{Rh}(\text{CNBu}^t)_2(\mu\text{-NHPh}\}_2]$,[466] $[\{\text{Rh}(\mu\text{-NHPh})(\text{PPhEt}_2)_2\}_2]$,[467] (*R,S*)-$[\{\text{Rh}_2([\mu\text{-NHC}_6\text{H}_4\text{Me-4}](\eta^1\text{-CH(Me)CO}_2\text{Me})(\text{CNBu}^t)_4(\mu\text{-X})[\text{X}]$ (X = Cl or I),[466b] $\{\text{Rh}(\text{CNBu}^t)_2(\mu\text{-NPh}_2)\}_2]$,[463] $[\text{Rh}_2(\mu\text{-NHMe})(\text{CO})_2(\mu\text{-dppm})(\mu\text{-dppm-H})]$,[468] $[\{\text{Rh(COD)}\}_2\{\mu\text{-N(H)C}_6\text{H}_4\text{NMe}_2\text{-4}\}(\mu\text{-OMe})]$,[469] $[\{\text{Rh(COD)}\}_2\{\mu\text{-N(H)}$(1-naphthyl)$\}(\mu\text{-OH})]$,[469] $[\text{Rh}(\text{PEt}_3)_2\{\text{N}(\text{SiMePh}_2)_2\}]$.[467]
Rh(I)/(III):	(*R* + *S*)-$[(\text{cod})\text{Rh}(\mu\text{-NHC}_6\text{H}_4\text{Me-4})_2\text{Rh(Cl)}(\eta^1\text{-CH(Me)CO}_2\text{Me})(\text{CNBu}^t)_2]$,[466b] $[\{\text{Rh}_2(\mu\text{-NHC}_6\text{H}_4\text{Me-4})_2(\eta^1\text{-CH(Me)CO}_2\text{Me})(\text{CNBu}^t)_4(\mu\text{-Cl})\}[\text{OTf}]$.[466b]
	Iridium
Ir(III):	$[\text{Ir}(\eta^5\text{-C}_5\text{Me}_5)\{\text{N(H)Ts}\}_2]$,[469b] $[\text{Ir}(\eta^5\text{-C}_5\text{Me}_5)(\text{Ph})(\text{PMe}_3)(\text{NH}_2)]$,[470] $[\text{Ir}\{\text{CH}(\text{CH}_2\text{CH}_2\text{PBu}^t{}_2)_2\}(\text{H})(\text{NH}_2)]$,[471] $[\text{Ir(H)(NHPh)}\{\text{C}_6\text{H}_3\text{-2,6-}(\text{CH}_2\text{PBu}^t{}_2)_2\}(\text{L})]$ (L = CO or CNBu^t),[472] $[\{\text{Ir(H)}(\mu\text{-NH}_2)(\text{NH}_3)(\text{PEt}_3)_2\}_2][\text{BPh}_4]_2$,[473] [*mer*-$\text{Ir(indolide)(H)(Cl)}(\text{PMe}_3)_3]$,[474] $[\{\text{Ir}(\eta^5\text{-C}_5\text{Me}_5)(\mu\text{-NH}_2)(\text{Ph})(\text{PMe}_3)\}_2$ NaI(thf)],[475] $[\{\text{Ir(I)(CO)}_2(\mu\text{-NHC}_6\text{H}_4\text{Me-4})_3\text{Ir}(\eta^5\text{-C}_5\text{Me}_5)][\text{Cl}]$,[476] $[(\eta^5\text{-C}_5\text{Me}_5)\text{Ir}(\mu\text{-NHC}_6\text{H}_4\text{Me-4})_3\text{Ir}(\eta^5\text{-C}_5\text{Me}_5)][\text{Cl}]$.[477]
Ir(II):	$[\text{Ir}_2\text{(Me)(I)}(\mu\text{-NHC}_6\text{H}_4\text{Me-4})_2(\text{CO})_4]$,[478] $[\{\text{Ir}(\eta^5\text{-C}_5\text{Me}_5)(\mu\text{-NHPh})\}_2]$,[479] $[\{\text{Ir(CO)}_2(\mu\text{-NHC}_6\text{H}_4\text{Me-4})\}_2]$.[480]
	Palladium
Pd(II):	$[\{\text{Pd(Cl)}_2(\mu\text{-NH}_2)\}_2][\text{PPh}_4]_2$,[481] [*trans*-$\text{Pd(Ph)(NHPh)}(\text{PMe}_3)_2]$,[482] $[\{\text{Pd(Ph)}(\text{PMe}_3)(\mu\text{-NHPh})\}_2]$,[482] $[\{\text{Pd}(\text{C}_6\text{F}_5)(\text{CNBu}^t)(\mu\text{-NHPh})\}_2]$,[483] $[\text{Pd}(\eta^3\text{-terpy})(\text{NHTs})][\text{BF}_4]$,[484] $[\text{NBu}^n{}_4]_2[\{\text{Pd}(\text{C}_6\text{F}_5)_2(\mu\text{-NHC}_6\text{F}_5)\}_2]$,[485] $[\text{Pd}_2(\text{tmeda})_2(\mu\text{-OH})(\mu\text{-NHC}_6\text{H}_4\text{Cl-4})][\text{BPh}_4]_2$,[486] $[\text{Pd}_2(\mu\text{-OH})(\mu\text{-NHC}_6\text{H}_4\text{Me-4})(\text{PPh}_3)_4][\text{BPh}_4]_2$,[487] $[\text{Pd}_2(\mu\text{-NHMes})_2(\text{PPh}_3)_2\text{Cl}_2]$,[219] $[\{\text{Pd}\{\text{CMeN}(\text{C}_6\text{H}_4\text{-OMe})\}_2(\mu\text{-NHC}_6\text{H}_4\text{Me-4}\}_2][\text{BF}_4]$,[488] $[\text{Pd}\{\text{Fe}(\eta^5\text{-C}_5\text{H}_4\text{PPh}_2)_2\}(\text{C}_6\text{H}_4\text{Bu}^t\text{-4})\{\text{N}(\text{C}_6\text{H}_4\text{Me-4})_2\}]$,[489] $[\text{Pd}\{\text{Fe}(\eta^5\text{-C}_5\text{H}_4\text{PPh}_2)_2\}\{\text{N}(\text{C}_6\text{H}_4\text{Me-4})_2\}(\text{L})]$,[490] (L = 2 or 3-furyl, thiofuryl), $[\text{Pd}(\text{C}_6\text{H}_4\text{-4-OMe})\{\text{N}(\text{C}_6\text{H}_3\text{Bu}^t{}_2\text{-3,5})_2\}(\text{PBu}^t{}_3)]$,[491] $[\text{Pd(Cl)(tmeda)}\{\text{N}(\text{SiMe}_3)_2\}]$.[492]
	Platinum
Pt(IV):	$[(\text{H}_3\text{N})_4\text{Pt}(\mu\text{-NH}_2)_2\text{Pt}(\text{NH}_3)_4][\text{Cl}]_6$,[493] $[(\text{H}_3\text{N})_3\text{Pt}(\mu\text{-NH}_2)(\mu\text{-N}_2\text{H}_5)\text{Pt}(\text{NH}_3)][\text{ClO}_4]_5$,[494]
Pt(II):	$[\{\text{Pt}(\mu\text{-NH}_2)(\text{P(O)Ph}_2)(\text{PMePh}_2)\}_2]$,[495] $[\{\text{Pt}(\mu\text{-NH}_2)(\text{PMePh}_2)_2\}_2][\text{BF}_4]_2$,[496] [*trans*-$\text{Pt(Cl)PEt}_3)_2(\text{NHC}_6\text{H}_4\text{I-4})]$,[497] [*trans*-$\text{Pt(Cl)}(\text{PEt}_3)_2(\text{NHC}_6\text{H}_4\text{NMe}_2\text{-2-Cl-4})]$,[497] $[\{\text{Pt}\{\text{C(Me)NC}_6\text{H}_4\text{Me-4}\}_2(\mu\text{-NHC}_6\text{H}_4\text{Me-4})\}_2]$,[498] [*cis*-$\text{Pt(Cl)}(\text{NPh}_2)(\text{PEt}_3)_2]$.[499]
	Silver
Ag(I):	$[(\text{Ag}\{\text{N}(\text{SiMe}_3)_2\})_4]$,[500] $[(\text{AgNCMe}_2(\text{CH}_2)_3\text{CMe}_2)_4]$,[500] $[\text{Ag}_6(\text{NHPh})_4(\text{PPr}^n{}_3)_6\text{Cl}_2]$.[219,501]
	Gold
Au(III):	$[(\text{Me}_2\text{AuNH}_2)_3]$,[502] $[(\text{Me}_2\text{AuNMe}_2)_2]$,[503] $[\text{Au}\{\text{N}_2\text{C}_{10}\text{H}_7(\text{CMe}_2\text{C}_6\text{H}_4)\text{-6}\}\{\text{N(H)C}_6\text{H}_3\text{Me}_2\text{-2,6}\}][\text{PF}_6]$.[504]
Au(I):	$[\{\text{AuN}(\text{SiMe}_3)_2\}_4]$,[505] $[\text{Au}\{\text{N}(\text{SiMe}_3)_2\}\text{PPh}_3]$.[506]

[a] 1980–2007. Complexes are arranged by element in the order of descending oxidation state and degree of amide substitution, for abbreviations see footnote[a] in Table 6.1.
[b] Yttrium amides are described in more detail in Chapter 4.

state (i.e. homoleptic d^0 derivatives) are available only for scandium and titanium whereas several such derivatives have been reported for all of the second and third row group 3–6 elements. Group 7–11 amido complexes of the first row elements (i.e. Mn–Cu) are less numerous than those of the earlier groups and no homoleptic derivatives of the 4d and 5d elements of group 7–10 are known as well-characterized stable species. Thus, the overall pattern described in the 1980 book – lower numbers of amido derivatives of the later transition elements in comparison to those of the earlier groups – continues to the present day. Nonetheless, this should not obscure the fact that interest in late transition element amides of the second and third rows is expanding rapidly (see below). This is primarily due to interest in their catalytic properties. Partly as a result of this attention, the imbalance in the numbers between the early and late element complexes is becoming less pronounced. For example Table 6.2 shows that over eighty structures of monodentate amido derivatives of the second and third row metals of groups 7–11 have been determined, whereas in 1980 no detailed structures of amido complexes of these elements were available.

The listings in Tables 6.1 and 6.2 also illustrate that a surprisingly large number of structures of simple homoleptic dialkyl- or diaryl-amido derivatives were not known in 1980. These include $[Ti(NMe_2)_4]$ (ged,[63] and X-Ray[64]), $[Ti(NPh_2)_4]$[65] (the first transition metal amide, which was originally reported in 1934[507]), $[V(NMe_2)_4]$ (ged),[63] $[V(NMe_2)_4]$ (X-ray),[136a] $[\{Cr(\mu\text{-}NPr^i_2)(NPr^i_2)\}_2]$,[166] $[\{Mn(NPr^i_2)_2\}_2]$,[184] $[\{M(NPh_2)_2\}_2]$ (M = Fe,[201a] Co,[215] or Ni,[215]), $[(CuNEt_2)_4]$,[230] $[Zr(NMe_2)_4]$ (ged),[245] $[\{Zr(NMe_2)_4\}_2]$,[243] $[M(NPh_2)_4]$ (M = Zr,[246] Hf[296] or Nb[317]), $[Ta(NMe_2)_5]$ (ged),[319] $[Ta(NMe_2)_5]$,[299a] $[Mo(NMe_2)_6]$,[348] $[W(NMe_2)_6]$ (ged).[399]

An illustrative example of a simple amide is provided by the structure of the zirconium dimethylamide in Figure 6.1, which shows that it is dimerized through two bridging-NMe_2 ligands whereas in the vapor phase it has a monomeric structure with tetrahedrally coordinated zirconium. The dimeric structure may also be contrasted with that of $[Ti(NMe_2)_4]$ which is a monomer in the crystal phase.[64]

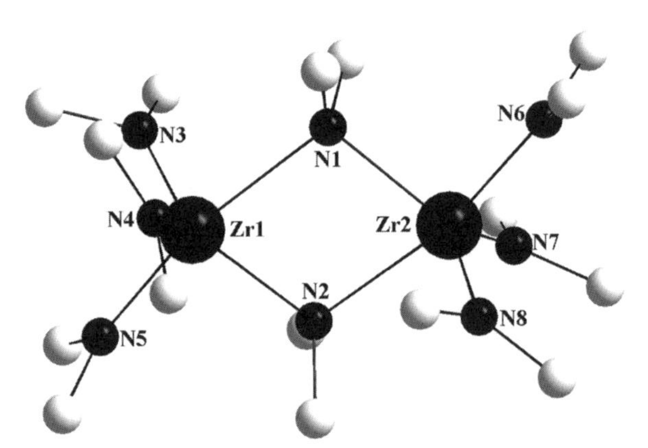

Figure 6.1 *View of the structure of $[\{Zr(NMe_2)_4\}_2]$[244] showing the dimeric $(ZrN)_2$ core. Zirconium and nitrogen atoms are shown as black spheres and carbon atoms are depicted in white shadow. Selected bond lengths: Zr1-N1 2.22, Zr1-N2 2.45, Zr1-N3 2.11, Zr1-N4 2.05, Zr1-N5 2.05, Zr2-N6 2.05, Zr2-N7 2.05, Zr2-N8 2.10 Å*

An additional aspect of the literature since 1980 has been an increasing interest in the characterization of charged amido complexes. In 1980 none had been structurally characterized. Several anionic complexes, including, for example, the homoleptic species $[Na(thf)_2Ti\{N(C_6F_5)_2\}_4]$,[118] $[Na(thf)_6]_2[Zr(\eta^1\text{-}NC_4H_4)_6]$,[97] $[\{Li(thf)\}_2Zr(NMe_2)_6]$,[244] $[\{Li(py)\}_2Cr(NEt_2)_4]$,[167] $[\{Li(thf)\}_2(Cr\{NH(C_6H_3Pr^i_2\text{-}2,6)\}_4)]$,[158] $[\{Li(thf)\}_2Mo(NMe_2)_6]$,[369] $[Mn_3(\mu\text{-}NEt_2)_6(\mu\text{-}Cl)_2\{Li(thf)_2\}_2]$,[179] $[Li(thf)_4][Ni(NPh_2)_3]$,[215] and the quasi-higher order cuprate salt $[\{Li(Et_2O)_2\}_2Cu(NPh_2)_3]$,[227] have been isolated and characterized. In both d^4 chromium species the square planar coordination of the CrN_4 array is noteworthy as is the trigonal planar geometry observed for the d^8 tris(amido)nickelate and d^{10} cuprate anions $[Ni(NPh_2)_3]^-$ [215] and $[Cu(NPh_2)_3]^{2-}$.[227] The reaction mixture 4 $NaBu^n$/6 tmp/2 $MnPh_2$/C_6H_6 afforded the transition metal 'inverse crown' species $[Na_4Mn_2(tmp)_6(C_6H_4)]$[185a] in which a 12-membered $Na_4Mn_4tmp_6$ ring encapsulates a doubly deprotonated benzene moiety C_6H_4.[185a]

Other interesting anionic transition metal amido derivatives have come from the reactions of primary amido ligands with metal halides by Fenske and coworkers. This has allowed the synthesis of a range of cobalt, nickel and copper amido cluster species with unusual structures which also incorporate NPh^{2-} as coligand (cf. Table 6.1).[218,219,227] These display a fascinating variety of structures. For example, the cobalt imido/amido cluster anion $[Co_6(\mu_4\text{-}NPh)_3(\mu_3\text{-}NPh)_2(NHPh)_6]^{4-}$ has a trigonal prismatic Co_6 arrangement with three and two imido ligands capping the rectangular and trigonal faces and a terminal −NHPh amido ligand attached to each metal. The 18 metal cluster $[Co_{18}(\mu_4\text{-}NPh)_3(\mu_3\text{-}NPh)_{12}(NHPh)_3]^-$ features a hexagonal prism of twelve cobalts. The hexagonal faces are each capped by a cobalt and there is also a 'metalloid' cobalt at the centre of the structure. Three cobalt amido moieties and three μ_4-NPh groups cap alternating tetragonal faces of the hexagonal prism and there are also twelve triply bridging NPh moieties on the periphery of the cluster. Another interesting feature of the Co_{18} cluster is that the average metal oxidation state is less than two.

Cationic transition metal amide complexes have been investigated in part because of their potential in catalysis particularly for olefin polymerization. Much of this work has concerned polydentate amido, linked cyclopentadienyl-amido or delocalized nitrogen centred bidentate ligands (see later). However, the structures of a small number of cationic complexes containing monodentate amido ligands have been determined. These include $[Ti(NHMe_2)(NMe_2)_3][B(C_6F_5)_4]$,[71] $[Ti(NC_6H_3Pr^i_2\text{-}2,6)(NHMe_2)_3(NMe_2)][B(C_6F_5)_4]$,[71] $[Ti(py)_2(NMe_2)_3][BPh_4]$,[72] $[M\{N(SiMe_3)_2\}_3][MeB(C_6F_5)_3]$ (M = Zr or Hf)[260] and $[V\{N(SiMe_3)_2\}_3][CN]$.[138]

The data in Tables 6.1 and 6.2 also show that the range of structurally characterized transition metal oxidation states and coordination numbers has been greatly expanded to include derivatives of Ti(II) (e.g. $[(\mu\text{:}\eta^1\text{:}\eta^1\text{-}N_2)\{Ti(Cl)N(SiMe_3)_2(tmeda)\}_2]$),[128] V(V) (e.g. $[V(NBu^t)(NPr^i_2)Cl_2]$),[133] V(IV, 4-coordinate) (e.g. $[V(NMe_2)_4]$),[136] V(IV, 3-coordinate) ($[V\{N(SiMe_3)_2\}_3][CN]$),[138] Nb(IV) (e.g. $[Nb(NPh_2)_4]$),[317] Ta(IV) (e.g. $[Ta(NEt_2)_2\{N(SiMe_3)_2\}_2]$,[347] Cr(II) (e.g. 2 or 3-coordinate) (e.g. $[\{Cr(NMesBMes_2)_2\}_2]$[169] or $[\{Cr(\mu\text{-}NPr^i)(NPr^i_2)\}_2]$),[166] Mo(VI) (e.g. $[Mo(NMe_2)_6]$),[348] Mn(III) (e.g. $[Mn\{N(SiMe_3)_2\}_3]$),[172] Fe (II, 2 or 3 − coordinate) (e.g. $[Fe\{N(SiMePh_2)_2\}_2]$,[201a,180a] or $[(Fe\{\mu\text{-}N(SiMe_3)_2\}\{N(SiMe_3)_2\})_2]$),[201a] Co(III) (e.g. $[Co\{N(SiMe_3)_2\}_3]$),[172] as well as a stable homoleptic Ni(II) amide (e.g. $[\{Ni(NPh_2)_2\}_2]$.[215]

6.2.2 Synthesis

The majority of the complexes listed in Tables 6.1 and 6.2 were synthesized by the eleven methods originally outlined in Chapter 3, page 58 of the 1980 volume.[55] The most common synthetic route remains transmetallation in which a metal halide is treated with an alkali metal amide, as shown in Equation (6.1).

$$L_mMX_n + nLiNRR' \rightarrow L_mM(NRR')_n + nLiX \quad (6.1)$$

Other important synthetic approaches also have maintained their popularity. Those most commonly employed are transamination, alkane or hydrogen elimination or, especially in the earlier groups, disproportionation and redistribution routes. However, in this section we focus on three further synthetic pathways that were either unknown or poorly investigated in 1980.

These three emerging synthetic approaches appear to be of growing importance in the current literature. The first route concerns the deceptively simple oxidative addition of N–H bonds to metal centres which is considered to be an essential step in the development of catalytic cycles for the addition of ammonia or amines to hydrocarbon substrates. A generalized cycle for the catalytic addition is illustrated in Scheme 6.1.

It can be seen that a key step in this process is the oxidative addition of the amine to the metal centre. The above catalytic scheme was shown to be feasible for the addition of aniline to norbornene with use of the Ir(I) complex $[Ir(PEt_3)_2(C_2H_4)Cl]$ as the catalyst.[508]

The addition of amines to transition metal centres is exemplified by the reaction shown in Equation (6.2).[474]

$$[Ir(COD)(PMe_3)_3][Cl] \xrightarrow[\text{HNR}_2=\text{pyrrole, indole, 3-methylindole 7-azaindole}]{\text{HNR}_2} [Ir(Cl)(H)(PMe_3)_3(NR_2)] \quad (6.2)$$

Known examples of N–H oxidative addition prior to 1980 usually involved the addition of relatively acidic N–H bonds, for example that in phthalimide, to Pd(0) or Pt(0) centres. Work by Milstein in the 1980s showed that the parent amine NH_3 can oxidatively add to Ir(I) centres (see Section 6.2.4) under mild conditions to afford parent amido hydrides in good

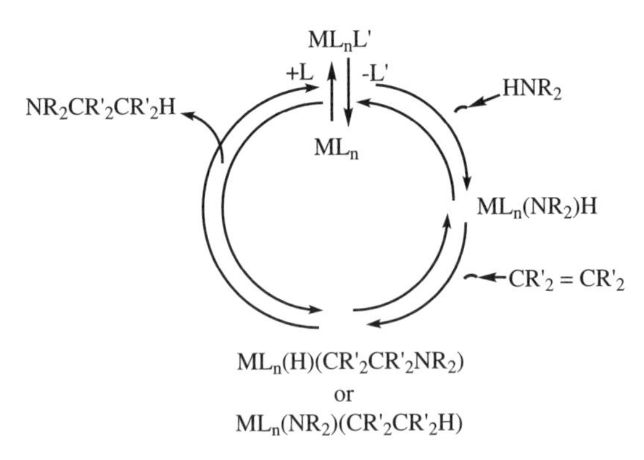

Scheme 6.1 *Proposed catalytic cycle for the addition of an amine to an olefin*[508]

yield.[473] These results have heralded the preparation of other late transition metal examples from ammonia or primary or secondary amino complexes of metals.

Two other routes to transition metal amides were not generally discussed in the 1980 book. The first of these involves the deprotonation of aminometal complexes as shown in Equation (6.3), which has afforded several new amido complexes.

$$\begin{gathered} L_nMNHR_2 \xrightarrow{[Base]^-} [L_nMNR_2]^- + [BaseH] \\ (L = \text{ligand, } R = \text{H or alkyl or aryl}) \end{gathered} \tag{6.3}$$

A number of examples of this reaction type have been reported, as in Equation (6.4),[509] in which $NaNH_2$ (or $NaOC_6H_2Bu^t{}_2$-2,6-Me-4) deprotonates a complexed NH_3.

$$trans\text{-}[Pt(H)(NH_3)(PPh_3)_2][ClO_4] \xrightarrow{NaNH_2} 1/2[\{Pt(H)(\mu\text{-}NH_2)PPh_3\}_2] + NH_3 + [Na][ClO_4] \tag{6.4}$$

This reaction is believed to proceed through a hydrido amido intermediate [*trans*-$Pt(H)(NH_2)(PPh_3)_2$] which eliminates PPh_3 to form the dimeric $[\{Pt(H)(\mu\text{-}NH_2)(PPh_3)\}_2]$. The use of NEt_3 is sufficient to deprotonate a rhenium complex, as in Equation (6.5).[510]

$$\begin{gathered} [Re(\eta^5\text{-}C_5Me_5)Me_3][OTf] \xrightarrow{2NH_3} [Re(\eta^5\text{-}C_5Me_5)Me_3(NH_3)_2][OTf] \\ [Re(\eta^5\text{-}C_5Me_5)Me_3(NH_2)] \xleftarrow[-NH_3,\ -[NEt_3H][OTf]]{2NEt_3} \end{gathered} \tag{6.5}$$

In addition ruthenium ammine complexes may be deprotonated in accordance with Equation (6.6).[511]

$$\begin{gathered} [(tbp)Ru(L)(L')NH_3][OTf] \xrightarrow[thf,\ -78^\circ C]{LiMe} [(tbp)Ru(L)(L')NH_2] \\ (L/L' = PMe_3/PMe_3; P(OMe)_3/P(OMe)_3; CO/PPh_3; NH_3/PPh_3) \end{gathered} \tag{6.6}$$

Similarly $Na\{N(SiMe_3)_2\}$ can be used to deprotonate the ruthenium(I) ammine of Equation (6.7).[512]

$$[\{C_6H_3(CH_2PBu^t_2\text{-}2,6)_2\}Ru(CO)(NH_3)] \xrightarrow[-HN(SiMe_3)_2,NaCl]{Na\{N(SiMe_3)_2\}} [C_6H_3(CH_2PBu^t_2\text{-}2,6)\}Ru(CO)(NH_2)] \tag{6.7}$$

An analogous reaction in which complexed aniline was deprotonated is shown in Equation (6.8).[456]

$$[(tbp)RuL_2(NH_2Ph)][OTf] \xrightarrow[-HN(SiMe_3)_2,-NaOTf]{NaN(SiMe_3)_2} (tbp)RuL_2(NHPh) \tag{6.8}$$

The third new synthetic route to transition metal amides has its origins in the recognition that the usual source of amido ligands in transition metal chemistry is an alkali metal amide transfer agent. Because of the anionic nature of the nitrogen centre in these salts it can act as a reducing agent. However, if a halogeno-amide is used instead of an alkali metal amide the polarity of the bond to nitrogen is changed such that the nitrogen retains little of its reducing power:

$\overset{+}{M}—\overset{-}{N}<$ $\qquad$ $\overset{-}{X}—\overset{+}{N}<$ or $\overset{\bullet}{X}$----$\overset{\bullet}{N}<$

alkali metal amide $\qquad$ halogeno amide $\quad$ (X = halogen)

The halogeno amide becomes in effect an oxidizing agent. This behaviour is exemplified by the treatment of the divalent metal amides $[(M\{N(SiMe_3)_2\}_2)_2]$ (M = Mn, Fe or Co) with $BrN(SiMe_3)_2$ to yield the trisamides $[M\{N(SiMe_3)_2\}_3]$ as shown in Equation (6.9).[172]

$$[M\{N(SiMe_3)_2\}_2] + BrN(SiMe_3)_2 \rightarrow [M\{N(SiMe_3)_2\}_3] + 1/2Br_2 \quad (M = Mn,\ Fe\ or\ Co) \tag{6.9}$$

However the existence of the bromine co-product is speculative and has not been experimentally verified. The reaction may be regarded as a one electron oxidation by the $N(SiMe_3)_2$ aminyl radical, and it is interesting to note that aminyl radical complexes of transition metals such as $[Rh(Ntrop_2)(bipy)][OTf]$[513] (trop = 5-*H*-dibenzo[a,d]cycloheptene) have been isolated. The manganese and cobalt compounds were the first examples of three-coordinate Mn(III) and Co(III) complexes to have been structurally characterized.

6.2.3 Structure and Bonding

6.2.3.1 Early transition metal amides

As Tables 6.1 and 6.2 clearly show, early transition metal amides consitute the largest body of metal amido derivatives. Those currently known span the largest range of coordination numbers and oxidation states of any group of compounds in the Periodic Table. In addition, amido ligands have been used (mainly through the work of Chisholm and coworkers, cf. Table 6.2) to support multiple bonding between transition metals. Although no structural parameters are provided for the compounds listed in Tables 6.1 and 6.2, a detailed examination would show that the vast majority the terminal amido ligands display planar, three-coordination at nitrogen. This fact, together with the observation that the M–N bond lengths are generally shorter than those predicted from the sum of the covalent radii of M and N, has often been interpreted in terms of the existence of M–N π-bonding as shown in **1** and **2**.

The π-overlap between a p-orbital lone pair is facilitated by the planar coordination geometry at nitrogen (**2**) whereas a pyramidal geometry at nitrogen (**1**) may diminish orbital overlap. For the π-bonding case the amido ligand thus behaves as both a σ and a π-donor as shown in **2**. In the early transition elements, i.e. metals of groups 3, 4, 5 and 6, their valence orbitals are not yet filled and the π-electron density from the ligand may be accommodated by the empty or partly occupied d-orbitals. For example, in the case of the hexavalent, six-coordinate group 6 metal derivatives $[M(NMe_2)_6]$ (M = Mo[348] or W;[514] T_h symmetry for the $M(NC_2)_6$ skeleton), half the π-electron density from six potentially 2π electron donor amido ligands can be accommodated in the t_g metal orbital set.

The availability of empty valence orbitals and the planarity of the coordination geometry at nitrogen can, however, be unreliable predictors of M–N π-bonding. This is because other factors such as the polarity of the M–N bond[515,517] and the reduced inversion barrier at nitrogen[513,514] could also cause the observed structures. For instance, the low inversion barrier in amines is further diminished by the electropositive metal substituent in the amide so that the inversion barrier becomes essentially zero and a planar geometry results. Nonetheless there can be little doubt that some π-bonding occurs in the early transition metal amides. Photoelectron spectroscopy (PES),[370,517–519] the measurement of core electron binding energies and theoretical calculations have provided considerable insight, although the extent of π-bonding is difficult to quantify.[519] Thermochemically measured M–N bond strengths (Table 6.3)[523–527] for early transition metal amides show that the bonds are quite strong (ca. 320–420 k J mol^{-1}) which tends to support M–N π-bonding. Examination of the core electron binding energies of the dimethylamido derivatives $[M(NMe_2)_4]$ (M = Ti, Zr or Hf), $[M(NMe_2)_5]$ (M = Nb or Ta), $[Mo(NMe_2)_4]$ and $[W(NMe_2)_6]$ led to the conclusion that the metal amide lone pairs are significantly stabilized by p–d π bonding. It was concluded that the t_g lone pairs in $[W(NMe_2)_6]$ were stabilized by 0.8–1.2 eV as a result of W–N π-bonding.[522] PES data for $[Cr(NPr^i{}_2)_3]$ have indicated a small but significant amount of nitrogen lone pair electron density in a Cr–N π-bonding orbital.[520] Electron diffraction, PES, NMR and theoretical calculations for $[Mo(\eta^5{-}C_5H_5)(NMe_2)_3]$[370] show that one of the NMe_2 groups lies axial to the Mo–η^5–C_5H_5 vector. Moreover all three–NMe_2 groups display Mo-N π-bonding although that from the axial –NMe_2 group is greatest as indicated by a shorter Mo-N bond. The theoretical data and split PE bands are consistent with the slippage of the cyclopentadienyl ring to an η^3-coordination mode. This is consistent with increased electron density at molybdenum as a result of Mo-N π-bonding. A theoretical study of tricoordinate amido complexes of transition metals led to the conclusion that there was a significant interaction between the metal d-orbitals and the π-lone pairs of the amido ligands on the basis of the variation of d-orbital energies with ligand rotation.[529] On the other hand a population analysis obtained from SCF MO calculations on the model species $Mn(NH_2)_2$ suggests that the Mn bonds are very polar and that $p\pi \rightarrow d\pi$ bonding is negligible, although Fenske-Hall MD Calculations and Mulliken population analysis for $Mo(NH_2)_4$ concluded that Mo–N π-bonding was present.[2,177] Extended Hückel calculations on the borylamido derivatives $M(NRBR'_2)_2$ (M = Cr or Ni, R/R′ = H, Ph, Mes) suggested that

Table 6.3 *Thermochemical bond energies (kJ mol^{-1}) of the M–N bonds in some early transition metal amides*

$Ti(NMe_2)_4{}^a$	340	$Ti(NMe_2)_4{}^b$	332				
$Ti(NEt_2)_4{}^a$	340	$Ti(NEt_2)_4{}^b$	317				
$Ti(NEt_2)_4{}^e$	305						
$Zr(NMe_2)_4{}^a$	381	$Zr(NMe_2)_4{}^b$	339	$Ta(NMe_2)_5{}^b$	319	$Mo_2(NMe_2)_6{}^{b,d}$	363
$Zr(NEt_2)_4{}^a$	372	$Zr(NEt_2)_4{}^b$	361	$Ta(NMe_2)_5{}^c$	328	$Mo(NMe_2)_4{}^c$	255
		$Zr(\eta^5{-}C_5H_5)_2(NH_2)(H)$	421[f]				
		$Hf(NMe_2)_4{}^b$	383			$W_2(NMe_2)_6{}^b$	429
$Hf(NEt_2)_4{}^a$	397	$Hf(NEt_2)_4{}^b$	381			$W(NMe_2)_6{}^{c,d}$	222

[a]Ref 523, [b]Ref 524, [c]Ref 525, [d]Ref 526, [e]Ref 527, [f]Ref 528.

metal d interactions with the BN π-orbital play an important role in determining the NMN bending angle.[530] In spite of the extensive evidence for M–N π-bonding, it is noteworthy that currently available thermochemical data for amides and related alkoxides suggest that the bonds are dominated by polar σ-effects since bond strengths decrease in the sequence M–O > M–N > M–C.

6.2.3.2 Late transition metal amides

Historically, the number of well-characterized amido derivatives of the late transition elements has been much lower than that of the earlier metals. Nonetheless, the imbalance in numbers is changing rapidly because of growing interest in the potential catalytic applications of the later metal amido complexes, especially of the second and third row metals (Table 6.2). This increase is also observed in the group 11 metal amides where, in 1980, a partial structure of just one derivative $[\{CuN(SiMe_3)_2\}_4]$ was described in a review (for a complete structure, see Ref. 232). Currently more than two dozen such structures are known (Tables 6.1 and 6.2). Over the years several reasons had been advanced to account for the relatively low numbers of late metal amides. Perhaps the simplest is that, since an $-NR_2^-$ ligand may act as both a two-electron σ- and a two-electron π-donor, strong bonding to electron-rich later transition metals, is not as favored as it is in the early metals for electrostatic and electronic reasons.[531] It can be argued also that hard Lewis bases (such as the amides) do not readily bind to late transition elements because the latter are considered soft Lewis acids. These arguments appear plausible but data to support the idea that late transition metal nitrogen bonds are inherently weaker than those to the early metals are scarce. Not the least of the problems in attempting to make such comparisons is the general absence of isoleptic metal amido complexes (i.e. metal amides with the same ligand set) for both the earlier and late elements. There are no experimental thermochemical data similar to those listed in Table 6.3 available for later element amides.

Most of the information on bond strengths of the later metal amides concerns relative bond energies of M–X and H–X (X = amide or alkoxide) moieties.[39,451] These studies do not support inherently weaker late transition metal-amide bonding. They have provided a different view of the M–X bond strength based on the M–X bond polarity in which the greater electrostatic character of the $M-NR_2$ (or M–OR) bonds relative to the N–H (or O–H) bond can account for observed thermochemical trends in ligand exchange reactions. The data show that late transition metal bonds to heteroatoms such as N or O are surprisingly strong despite the traditionally assumed weakening because of p-d π electron repulsions. There has been much discussion over the past two decades or so[3,15,27,39,54] on the effects of repulsion of late metal d-electrons in filled orbitals by the lone pair electrons of the nitrogen or other ligand heteroatom. The heteroatom (e.g. nitrogen or oxygen) orbitals of π-symmetry can stabilize the complex if there is an empty d-orbital of suitable energy available, as there generally is in 16-electron metal complexes with low coordination numbers (<6). Calculations on unsaturated (16-electron) d^8 and d^{10} metal complexes show that M–N π-bonding has a large effect on the differential in M–H and $M-NH_2$ bond energies.[532] For Ir species especially the Ir-N bond is strengthened sufficiently to reduce the differential to zero. In contrast, if the metal d-orbitals are full, as they are in six-coordinate complexes, which have eighteen valence electrons, p-d π-delocalization cannot take place and a nucleophilic nitrogen centre with pyramidal coordination can result. For example, in the amides $[\{M(NPh_2)_2\}_2]$ (M = Fe,[200] Co,[215] or Ni[215]), there are three-coordinate metals with both bridging and terminal amido ligands. In these the valence orbitals are not fully occupied and both association and N(p)–M(d) π-bonding may occur and planar geometries are observed for the terminal nitrogens. In contrast, there are several structures of six-coordinate amido species that display pyramidal coordination and nucleophilic character at amido nitrogens. Examples include $[Re(\eta^5\text{-}C_5H_5)(NO)(PPh_3)(NHPh)]$,[444,445] $[M(dmpe)(H)(NH_2)]$ (M = Fe,[203] or Ru,[449]), $[Ru(tpb)(CO)(PPh_3)(NHPh)]$,[455] and $[\{Ir(\eta^5\text{-}C_5Me_5)(PMe_3)(Ph)(NH_2)\}_2NaI(THF)]$.[471] Nonetheless, there are complexes where pyramidal nitrogen coordination might be expected on the basis of the considerations outlined above but where planar coordination is experimentally observed as in $[Re(bipy)(CO)_3(NRR')]$,[446] (R/R′ = H, Ph; Ph, Ph; H/C_6H_4Me-4), $[Ru(\eta^6\text{-}C_6Me_6)(P(OMe)_3)(Ph)(NHPh)]$,[454] and $[OsCl_2(tbp)(NHPh)]$.[455] In several of these there is thought to be delocalization of the nitrogen lone pair onto the aryl substituent rather than the metal.

Structural and spectroscopic data for several octahedral ruthenium(II) NHPh complexes have provided supporting evidence for the existence of varying degrees of N–C multiple bonding which is strongly influenced by the donor/acceptor properties of the co-ligands.[54] For example, there is an inverse correlation of the Ru–N and N–C bond lengths in $[(tbp)Ru(CO)PPh_3(NHPh)]$,[455] $[(tbp)Ru\{P(OMe)_3\}_2(NHPh)]$,[456] $[(\eta^6\text{-}C_6Me_6)Ru(Ph)(PMe_3)NHPh]$,[454] and $[(Me_3P)_4Ru(H)(NHPh)]$.[451] In addition, variable temperature ^1H NMR spectroscopy[456] on complexes of the type [(tbp)Ru(L)(L′)NHPh] shows that for L/L′ = CO/PPh_3 there is a rotational barrier of ca. 12 kcal mol^{-1} around the Ru-N bond with no rotational barrier observed for the C–N bond. For the strong σ-donor set L/L′ = PMe_3/PMe_3 there is no barrier observed for the Ru–N bond, whereas there is a rotational barrier of ca. 12.7 kcal mol^{-1} observed for the NC bond. The complex with L/L′ = $P(OMe)_3/P(OMe)_3$ displays an intermediate situation with a C–N rotation barrier near 10 kcal mol^{-1} and an upper limit of ca. 9 kcal mol^{-1} for Ru–N bond rotation. Thus, increasing electron density at the metal corresponds to lower Ru–N barrier and an increased N–C barrier. These considerations and the use of these complexes for the activation of H–H bonds or species with non-polar C–H bonds have been discussed by Gunnoe.[54]

6.2.4 Parent Amido ($-NH_2$) Derivatives

Transition metal derivatives of the parent amido ligand, i.e. $-NH_2$, were not treated in the original 1980 survey. However, in the ensuing period there have been numerous reports of such compounds as well as a rapid growth of interest in their chemistry. They may be divided into 'inorganic' derivatives that have the binary or ternary formulae of the type $M(NH_2)_n$ or $M'_xM_y(NH_2)_n$ where M = transition metal and M′ = alkali metal. These complexes are not listed in Tables 6.1 and 6.2. They usually form three-dimensional polymeric ionic structures and are exemplified by the structures of $[MY(NH_2)_4]$ (M = K or Rb),[533] $[Cs_3Y_2(NH_2)_9]$,[534] $[Rb_4Cr_2(NH_2)_{10}]$,[535] $[Mn(NH_2)_2]$,[536] $[Na_2Mn_2(NH_2)_4]$,[536] $[Na_4Ni_2(NH_2)_8\cdot 4NH_3]$,[537] and $[Cs_2Ni(NH_2)_4]\cdot NH_3$.[538] In contrast, $Ni(NH_2)_2$ possesses the molecular, hexameric formula $[Ni_6(NH_2)_{12}]$.[539] As illustrated in Figure 6.2, it has a structure that is composed of an octahedral array of six nickel atoms (Ni---Ni = 2.83–2.87 Å) which are linked by twelve doubly bridging NH_2 ligands located along the edges of the Ni_6 octahedron. Other $-NH_2$ derivatives include the salts $[Na(tacn)Cr(\mu\text{-}NH_2)(\mu\text{-}OH)Cr(tacn)][ClO_4]_4$,[540] $[Cr_2(\mu\text{-}NH_2)_3(NH_3)_6][I]_3$,[541] $[Mo(NH_2)(OH)(dppe)_2][OTf]_2$,[380] $[(NH_3)_4Ru(\mu\text{-}NH_2)_2Ru(NH_3)_4][Cl]_4$,[541] and $[\{(\eta^6\text{-}C_6Me_6)Ru\}_2(\mu_2{:}\eta^1,\eta^1\text{-}H_2NNH_2)(\mu_2\text{-}NH_2)(\mu\text{-}H)]$

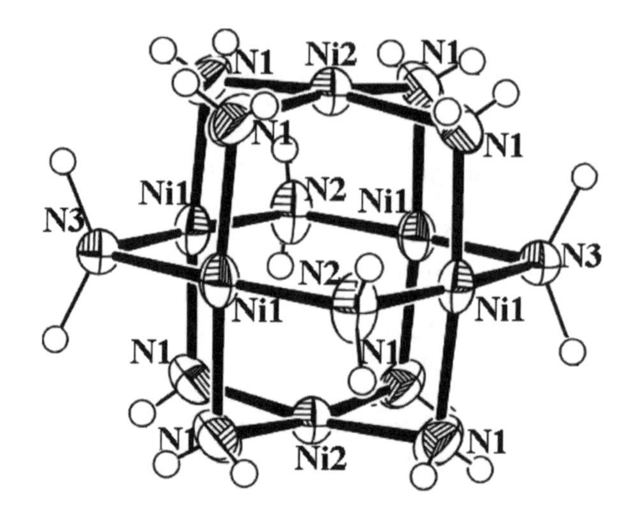

Figure 6.2 *Thermal ellipsoid drawing of $[Ni_6(NH_2)_{12}]$ highlighting the octahedral nickel framework. Selected bond lengths: Ni1-N1 1.89, Ni1-N2 1.89, Ni1-N3 1.93, Ni2-N1 1.92 Å*[539]

$[OTf]_2$[448] as well as the organometallic cluster species $[\{Zr(OCBu^t_3)\}_5(\mu_5\text{-}N)(\mu_3\text{-}NH)_4(\mu_2\text{-}NH_2)_4]$,[288] $[\{Zr(\eta^5\text{-}C_5H_4Me)\}_5(\mu_5\text{-}N)(\mu_3\text{-}NH)_4(\mu\text{-}NH_2)_4]$,[287] and $[K][\{(Zr(\eta^5\text{-}C_5Me_5))_3(\mu_3\text{-}N)(\mu_3\text{-}NH)(\mu\text{-}NH_2)_3\}_4(NH_2)_5(NH_3)_7]$.[286]

In addition to these simple derivatives, there are the neutral, 'organometallic' parent amido molecular complexes, which have been synthesized by a variety of routes and are usually stabilized by a saturated metal coordination sphere and/or bulky organic or closely related coligands or phosphines. Interest in neutral molecular species containing an $-NH_2$ ligand has rapidly increased over the last decade. Although examples of earlier transition element complexes with parent amido ligands are known, e.g. $[Ti(\eta^5\text{-}C_5Me_5)_2(NH_2)]$,[124] $[M(\eta^5\text{-}C_5Me_5)_2(H)(NH_2)]$ (M = Zr or Hf),[543] or $[M(NH_2)\{N(SiMe_3)_2\}_3]$ (M = Zr or Hf),[247] most of the recent work has focused on derivatives of the late transition metals of the second and third rows, for example $[Re(\eta^5\text{-}C_5Me_5)Me_3(NH_2)]$,[510] $[Os_3(\mu\text{-}H)(\mu\text{-}NH_2)(CO)_{10}(PPh_3)]$,[463] $[Pd_2Cl_4(NH_2)_2][PPh_4]_2$,[481] $[Pt(NH_3)_4(NH_2)AgCl][NO_3]_3$,[544] $[(H_3N)_4Pt(\mu\text{-}NH_2)_2Pt(NH_2)_4][Cl]_6$,[493] $[(H_3N)_3Pt(\mu\text{-}NH_2)(\mu\text{-}N_2H_5)Pt(NH_3)_3][ClO_4]_5$,[494] $[\{Pt(\mu\text{-}NH_2)(PMe_2Ph)_2\}_2][BF_4]_2$,[496] and $[(Me_2AuNH_2)_n]$.[503,504] Ruthenium and iridium complexes, as exemplified by *trans*-$[Ru(dmpe)_2(H)(NH_2)]$,[449] $[\{Ru(\eta^6\text{-}C_6Me_6)\}_2(\mu_2\text{-}\eta^1,\eta^1\text{-}H_2NNH_2)(\mu^2\text{-}NH_2)(\mu\text{-}H)][OTf]_2$,[448] $[Ru(tpb)LL'(NH_2)]$ (L/L' = PMe_3, $P(OMe)_3$, CO/PPh_3),[511] $[Ru(C_6H_3\text{-}2,6\text{-}CH_2PBu^t_2)(CO)(NH_2)]$,[512] $[\{Ir(H)(\mu\text{-}NH_2)(NH_3)(PEt_3)_2\}_2][BPh_4]_2$,[473] $[\{Ir(\eta^5\text{-}C_5Me_5)(Ph)(PMe_3)(NH_2)\}_2NaI(thf)]$[475] and $[Ir\{CH(CH_2CH_2PBu^t_2)_2\}(H)(NH_2)]$,[471] have received the most attention with comprehensive chemical studies being carried out on some complexes such as [*trans*-$Ru(dmpe)_2(H)(NH_2)$] and its iron congener.[545] The main reason for the interest in these species concerns the probable involvement of the $M-NH_2$ moiety in catalytic cycles similar to those seen in hydrocarbon based systems, particularly those involving the formation of carbon-nitrogen bonds to primary or secondary amines. The catalytic addition of amines to olefins, as illustrated in Scheme 6.1, or the coupling of ammonia with arenes under mild conditions have been listed among the top ten challenges for the catalytic field.[546] The facile oxidative addition of ammonia to form an amido-hydrido complex represents a key step in such a catalytic process. For such a process to be useful, ammonia is required to displace a weakly bound ligand from the metal coordination sphere and then form an amido-hydride as shown in Equation (6.10).

$$L_{n+1}M + NH_3 \xrightarrow[-L]{} \underset{\text{ammine}}{L_nM \leftarrow NH_3} \longrightarrow \underset{\text{amido hydride}}{L_nM(H)(NH_2)} \quad (6.10)$$

The thermodynamics of the oxidative addition process tends to be favored by increased electron density at the metal centre, hence the focus on later transition metal derivatives. Furthermore, as discussed above, it is believed that M–N π-bonds to the later transition metals are of significance only if the transition metal complex is unsaturated. Saturated late transition metal amides (parent or substituted) often exhibit the so-called π-conflict[15] (see above) so that the nitrogen centre displays no π-bonding to the metal and retains its lone pair character and basicity.

Theoretical studies[532] on unsaturated systems such as the d^8 moieties [M(η^5-C_5Me_5)(CO)] and [*trans*-M$(PH_3)_2$X] (M = Rh or Ir) show that the thermodynamics of addition of NH_3 to form M(H)NH_2 are more favorable for the third row element due to stronger M–H and M–NH_2 bonding. Experimental work available to date is in harmony with these calculations and has shown that it is possible to add NH_3 to displace a weakly bound ligand and form an amido-hydride. The first such reactions under mild conditions involved an Ir(I) complex, as shown in Equation (6.11).[473]

$$[Ir(PEt_3)_2(C_2H_4)Cl] \xrightarrow[\text{2. NaBPh}_4]{\text{1. NH}_3\ (l),\ 25\ ^\circ C} [(Et_3P)_2(NH_3)(H)Ir(\mu\text{-}NH_2)_2Ir(H)(NH_3)(PEt_3)_2][BPh_4]_2$$

$$[Ir(PEt_3)_2(C_2H_4)Cl] \xrightarrow[\text{2. pyridine, 110 } ^\circ C]{\text{1. NH}_3\ (l),\ 25\ ^\circ C} [(Et_3P)_2(H)(Cl)Ir(\mu\text{-}NH_2)_2Ir(H)(Cl)(PEt_3)_2] \quad (6.11)$$

The Ir(III) metal centres in the products, which are bound to a terminal hydride and a bridging –NH_2 group, represented the first X-ray structural authentication of a transition metal species with both amide and hydride bound to a metal centre. The reactivity of the complexes is low, however, and appears to be dominated by the stability of the Ir(μ-NH_2)$_2$Ir bridging unit. More recent work has shown that olefin-iridium(I) complexes, such as the propene species [{HC($CH_2CH_2PBu^t_2$)$_2$}Ir(CH_2CHMe)], react directly with ammonia at room temperature as shown in Equation (6.12).[471]

$$[\{HC(CH_2CH_2PBu^t_2)_2\}Ir(CH_2CHMe)] \xrightarrow[<5\text{ min}]{NH_3, 25^\circ C} [\{HC(CH_2CH_2PBu^t_2)_2\}Ir(H)(NH_2)] + CH_2CHMe \quad (6.12)$$

The product amido-hydride is the first structurally characterized transition metal complex that features both a terminal amide and hydride. The X-ray crystal structure shows that the nitrogen centre has planar coordination. This geometry results from π-bonding between the nitrogen electron pair and the lowest unoccupied metal orbital which lies in the equatorial coordination plane of the atoms Ir($C_{methine}$)(H)(N). This is consistent with earlier calculations on related species.[532] The possibility that the addition of

NH_3 to the iridium centre could have occurred by interaction with an Ir–H moiety derived from insertion of Ir(I) into a CH bond of the Bu^t groups or the methanide centre of the pincer ligand was eliminated by labelling studies.[471] All the available experimental data point to the simple insertion of the Ir(I) centre into an N–H bond of ammonia.

6.2.5 Low-coordinate Transition Metal Amides

6.2.5.1 Introduction

Low coordinate metal complexes (i.e. having coordination number three or lower) constitute a very important class of transition metal amides. They have played a key role in the development of transition metal amide chemistry and many of the first well-characterized amido derivatives of several transition metals were derivatives of bulky amido ligands. There has been continued interest in their synthesis and characterization. In addition, beginning around 1990, the use of new types of sterically crowded amido ligands such as tris(amido) amines (Section 6.3.3) and 'two-sided' amido ligands (Section 6.2.6) has led to spectacular developments in small molecule activation chemistry particularly of N_2. The stabilization of low-coordinate transition metal amides originated with the work of Wannagat and Bradley and their coworkers in the 1960s who introduced the $-N(SiMe_3)_2$ and $-NPr^i_2$ ligands to stabilize several three-coordinate, first row transition metal and 4f metal complexes.[531,547,548] Many of these were the first examples of three-fold coordination for these elements. For example, their work resulted in the isolation of the three-coordinate M(III) complexes $[M\{N(SiMe_3)_2\}_3]$ (M = Sc, Ti, V, Cr, and Fe) as well as $[Cr\{NPr^i_2\}_3]$.[549] In addition, the M(II) derivatives $[M\{N(SiMe_3)_2\}_2]$ (M = Mn,[550] Co[551] or Ni (unstable)[550]) were described. The manganese derivative was shown to have the dimeric $[(Me_3Si)_2NMn\{\mu\text{-}N(SiMe_3)_2\}_2MnN(SiMe_3)_2]$ structure with three-coordinate metals bound to two bridging and a terminal ligand in the crystalline phase.[178] In all, the structures of eleven three-coordinate amido derivatives, mainly as a result of the work of Bradley and Hursthouse, were known by 1980.[55] In addition, the structure of the tetramer $[\{CuN(SiMe_3)_2\}_4]$ had been described in a review,[548] although the complete structure was not published until later.[232a,232b] Currently, over eighty low (two- or three-) coordinate transition metal amides (including d^{10} species) have been structurally characterized, and these complexes have played a leading role in the study of transition metal complexes with low-coordination. Most low coordinate complexes have involved the use of secondary amido ($^-NR_2$) ligands. However, it seems probable that sufficiently large primary ($^-N(H)R$) amido ligands[107,286] can also be developed to support low coordination and new elimination reactions.[286]

6.2.5.2 Trivalent metal amides

Work in the 1980s and 1990s supplied numerous new examples of low-coordinate geometries for amido complexes of metals in various oxidation states then previously unknown. For example the 'missing' M(III) complexes $[M\{N(SiMe_3)_2\}_3]$ (M = Mn or Co),[172] featuring extremely rare examples of three-coordinate Mn(III) and Co(III) (Figure 6.3), were obtained *via* the unusual reaction of $[M\{N(SiMe_3)_2\}_2]$ with $BrN(SiMe_3)_2$, thereby completing the series of three-coordinate M(III) amides for M = Sc–Co.

The range of transition metal silylamides was also expanded to include the synthesis and structure of the yttrium species $[Y\{N(SiMe_3)_2\}_3]$,[239] which, like its previously reported scandium counterpart, has a pyramidal MN_3 array. In the vapor phase, the

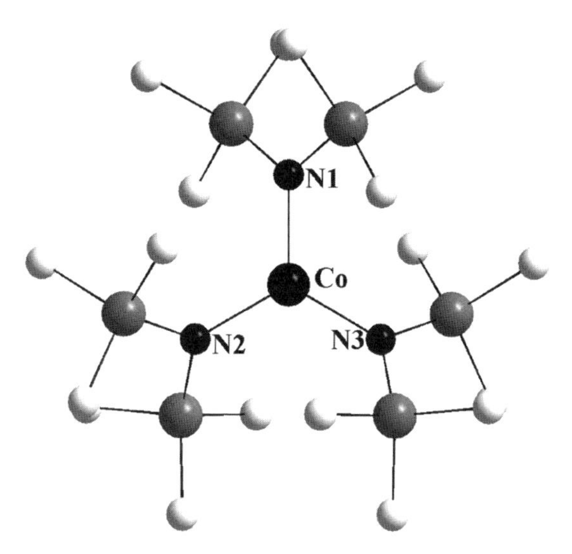

Figure 6.3 *Illustration of the three-coordinate complex [Co{N(SiMe$_3$)$_2$}$_3$].[172] Cobalt and nitrogen atoms are shown as black spheres, silicon atoms are grey and carbon atoms are given in white shadow. The Co–N bond length is 1.87 Å*

scandium species has a trigonal planar MN_3 arrangement and the difference between the vapor and crystalline phase structures was attributed to crystal packing effects.[56] In addition, the number of trigonal Cr(III) amides has been increased to include [Cr(tmp)$_3$] (tmp = 2,2,6,6-tetramethylpiperidinato),[156] [Cr(NRR′)$_3$] (R = R′ = CHMePh;[163] R = R′ = Cy;[161b] R = 1-adamantyl, R′ = C$_6$H$_3$Me$_2$-3,5[156]). The vanadium analogue of the latter [V{N(1-Ad)(C$_6$H$_3$Me$_2$-3,5)}$_3$][144] has also been characterized. Furthermore, a more efficient route to [Cr{N(SiMe$_3$)$_2$}$_3$] and a new crystal structure determination has been described.[164] Three-coordinate metal amides have been treated in a general review that covers three-coordinate transition metal species with hard ligands.[20] The electronic structure and bonding in tricoordinate amido complexes of transition metals have also been detailed and reviewed.[28]

6.2.5.3 Divalent metal amides

The synthesis of Fe{N(SiMe$_3$)$_2$}$_2$[177] completed the divalent M(II) series M = Mn–Ni. It was shown to have a two-coordinate monomeric structure in the vapor phase by gas electron diffraction.[177] Its crystal structure[200] showed that it was a symmetrically bridged dimer like its manganese and cobalt counterparts. Variable temperature ^{1}H NMR spectroscopy indicated an association energy of only 3 kcal mol^{-1}. The divalent series was also expanded to Cr(II) with use of the –NPr$^i_2$ and –NCy$_2$ (Cy = cyclohexyl) groups to afford the bridged dimers [R$_2$NCr(μ-NR$_2$)$_2$CrNR$_2$] (R = Pr^i or Cy).[166,167] In addition, it was shown that the dimers [Ph$_2$NM(μ-NPh$_2$)$_2$MNPh$_2$] (M = Fe,[200] Co,[215] or Ni,[215] see also above) could be obtained with the less bulky –NPh$_2$ ligand. The –NPh$_2$ ligand was earlier used by Fröhlich and coworkers to isolate several iron and cobalt derivatives.[552–556] These included [KFe(NPh$_2$)$_4$],[552] [Fe(NPh$_2$)$_2$(bipy)$_2$],[549] [Fe(NPh$_2$)$_2$(dioxane)$_2$)],[553] [Fe(NPh$_2$)$_2$(PEt$_3$)$_2$],[552] and [Fe(NPh$_2$)$_2$(NO)$_2$][553] which were not structurally characterized. In addition, the cobalt derivatives [Co(NPh$_2$)$_2$(NO)$_2$][551] and [{Co(NPh$_2$)$_2$}$_2$][556] were reported. The latter was described as having a very interesting non-bridged [(Ph$_2$N)$_2$CoCo(NPh$_2$)$_2$] structure with an unsupported Co–Co bond.[556] However this structural assignment was based on relatively few crystallographic data and, in view of the doubly amido bridged structure published later,[215] may be in error. It is also worthy of note that the use of the related anilido –NHPh ligand by Fenske and coworkers has allowed the synthesis of a range of cobalt, nickel and copper cluster species that have unusual structures which often incorporate the doubly deprotonated inido NPh^{2-} imido coligand (cf. Table 6.1).[218,219,227]

It is interesting to observe that the use of the $-NPh_2$ group affords a stable Ni(II) derivative whereas the $-N(SiMe_3)_2$ ligand does not. The reasons for the differences in stability are not well understood but it is possible that the greater steric repulsion in the $-N(SiMe_3)_2$ may induce instability. There is also a possibility that the $-NPh_2$ ligand may permit some delocalization of the nitrogen lone pair onto the aryl substituents thereby reducing greater d-π electron repulsion in the later more electron-rich elements.[54] It is noteworthy that the only other stable homoleptic nickel(II) amides involve boryl amido ligands of the type $-N(R)BR_2$ (see below) where nitrogen lone pair delocalization involving N–B, p–p π-bonding in the ligand is a prominent feature.

Numerous physical and chemical investigations of the dimeric M(II) silylamides showed that they were monomeric in hydrocarbon solution.[200,201] This suggested that the use of more crowded amido ligands would afford monomeric metal bisamides with formally two-coordinate metals in the solid state. This proved to be the case with use of the bulkier $-N(SiMe_2Ph)_2$ and $-N(SiMePh_2)_2$ ligands which yielded monomeric complexes such as $M\{N(SiMePh_2)_2\}_2$ (M = Mn, Fe (Figure 6.4), or Co),[180,201a] the neutral copper trimer $[(Cu\{N(SiMePh_2)\})_3]$[229] and the cuprate species $[Cu\{N(SiMePh_2)_2\}_2]^{-}$[228] which had two-coordinate metal geometries with NMN arrays that displayed deviations from strict linearity as well as secondary interactions between the metal and aryl groups.

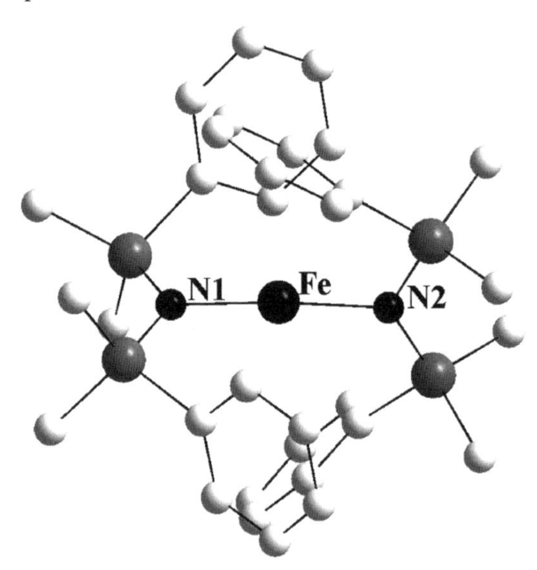

Figure 6.4 *Example of a two-coordinate iron complex, $[Fe\{N(SiMePh_2)_2\}_2]$.[201a] Iron and nitrogen atoms are shown as black spheres, silicon atoms are grey and carbon atoms are white. Selected bond lengths and an angle: Fe-N1 1.90, Fe-N2 1.91 Å; N-Fe-N 172.1°*

This theme was pursued further with use of ligands modified by boryl substituents as in $[N(BMes_2)R]^-$ (R = Ph or Mes), where the nitrogen lone pair is partially delocalized into the vacant p-orbital on boron with the intention of reducing its bridging tendency. Use of these ligands permitted a series of compounds of the type $M[N(BMes_2)R]_2$ (M = Cr–Ni; R = Ph or Mes) and related species to be characterized.[168,169,217] These also displayed various degrees of bending within the NMN unit which depended on the size of the nitrogen substituent and the metal electron configuration. Extended Hückel calculations on $M(NRBR^1R^2)_2$ model species did not provide an explanation of the experimentally observed geometries although they did provide useful insight into the variation of the d-p π interactions with the bending of the metal coordination.[530] Bis(amido)borane ligands, *e.g.*, $[PhB(NBu^t)_2]^{2-}$, have allowed derivatives of formula $[M\{(NBu^t)_2BPh\}_2]$ (M = Zr, Hf, V)[557] and $TiX_2\{(NBu^t)_2BPh\}$ (X = Cl, CH_2Ph, NMe_2)[558] to be prepared. Other examples include $M(R_2BN(CH_2)_2NBR_2)X_2$ (M = Zr, R = $C_6H_3Pr^i{}_2$-2,6, X = Cl, Me, Et, Bu; M = Ti, Zr, Hf, R = Cy, X = Cl, I, Me CH_2SiMe_3)[557] as well as $X_2\overline{M\{N(C_6H_3Pr^i_2\text{-}2,6)\ B(NMe_2)B(NMe_2)N}(C_6H_3Pr^i_2\text{-}2,6)\}$ (M = Ti; X = Cl or Me; Zr, X = CH_2Ph) which were isolated and structurally characterized.[560]

6.2.5.4 Reactions

Neutral homoleptic transition amide molecules are for the most part soluble in commonly used aprotic solvents. They are, in effect, hydrocarbon-soluble sources of M^{n+} ions, and because of the polarity of the M–N bond they can readily be used for the synthesis of a wide variety of reagents. The most common reactions involve the insertion of unsaturated species into the M–N bond or M–N bond cleavage reactions with protic molecules. Their utility can be illustrated by the reactions of some divalent Co(II) amides, shown in Scheme 6.2.

The coordinatively unsaturated divalent transition metal silylamides behave as Lewis acids and display a facility to be complexed either by one or two equivalents of Lewis base ligands to form three- or four- coordinate adducts.[173,181,180b] In addition, three-coordinate M(II) anionic species, for example $[M\{N(SiMe_3)_2\}_3]^-$, could be isolated by the addition of the alkali metal salt $M'N(SiMe_3)_2$ (M′ = alkali metal) to the neutral M(II) amide.[175,176] Adducts are also formed by interaction with small molecules, e.g. $[(NO)_2Co(NPh_2)_2]$,[555] and reaction of $Co\{N(SiMe_3)_2\}_2$ with $BrN(SiMe_3)$ produces the unique $[Co\{N(SiMe_3)_2\}_3]$. Several hetero-substituted three-coordinate transition metal complexes with alkoxide or halogenide coligands can also be isolated.[187,202,216,217] A range of low-coordinate species was obtained by treatment of the dimeric silylamides with β-diketimine,[191] bulky phosphines,[188] arsines,[188] alcohols,[207,216] phenols,[207] silanols,[559,560] thiols,[206,209,210,212] selenols,[131,189] and tellurols.[131]

6.2.6 'Two-sided' Amido Ligands

An important recent development in the area of the monodentate amido metal chemistry has come from the group of Cummins who introduced[79] ligands of the type –N(R)Ar (for example, R = Bu^t, 1-adamantyl (1-Ad), 2-adamantyl (2-Ad), $-C(CD_3)_2(CH_3)$; Ar = Ph, $C_6H_3Me_2$-3,5, C_6H_4F-4).[19] These allowed the isolation of an $Mo\{N(R)Ar\}_3$ species having three-coordinate molybdenum as well as three-coordinate derivatives of several other metals. Apart from $[M\{N(SiMe_3)_2\}_3]$ (M = Y,[239] La), the molybdenum compounds are the only examples of structurally characterized open shell, three-coordinate homoleptic amides

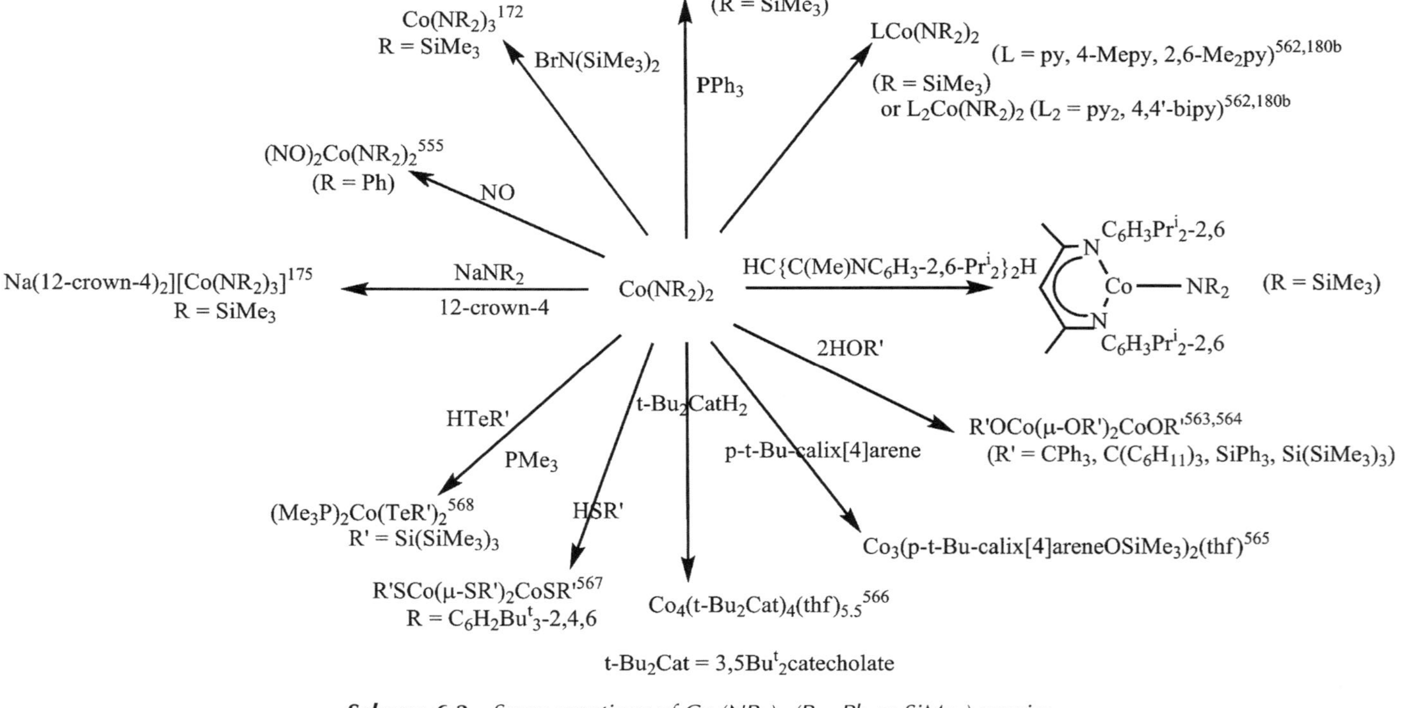

Scheme 6.2 *Some reactions of Co $(NR_2)_2$ (R = Ph or $SiMe_3$) species*

from the heavier transition elements,[85,358,381] although some three coordinate heteroleptic heavier derivatives, e.g. $[Rh(PEt_3)_2\{N(SiMePh_2)_2\}]$[467] are known. ^1H NMR data indicate that the three-coordinate molybdenum species adopts a structure in which the three aryl substituents are located on one side of the MoN_3 coordination plane and the three alkyl groups are on the opposite side. In this way, the two sides of the trigonal plane of the molecule have different steric properties. For example, the more open, alkyl-substituted side permits access of small molecules to the three metal valence orbitals (of *a* and *e* symmetry) of the Mo(III), d^3 centre which are isolobal to three singly occupied atomic orbitals of a nitrogen atom. Partly as a result of this, the metal centre can react directly with N_2[371] (or N_2O)[381] under ambient conditions to afford, inter alia, nitrido complexes such as $[NMo\{(N(R)Ar)\}_3]$ (which contain a four-coordinate molybdenum atom in complexes similar to that shown in Figure 6.5) thereby affording the remarkable room temperature cleavage of the NN triple bond.

In a similar vein, it was shown that molybdenum amido derivatives of the isolobal terminal phosphide[355] and carbide,[353,351] and related chalcogenide atoms,[365] could be obtained. Detailed mechanistic studies of the $Mo\{N(R)Ar\}_3/N_2$ system involving X-ray, EXAFS, magnetic, Raman, and isotopically labelled NMR spectroscopy showed that the reaction proceeded through an intermediate involving an end-on bound N_2 bridging the molybdenum centres (Scheme 6.3).[569,351]

Further investigation showed that the salt $[(thf)_xNa][N_2Mo(N(R)Ar)_3]$ is a key intermediate in the acceleration of N_2 reduction in the presence of Na amalgam.[85] Mechanistic studies on the remarkable cleavage of the NN bond in N_2O by $[Mo\{(N(R)Ar\}_3]$ showed that the nitride $[NMo\{N(R)Ar\}_3]$ and the nitrosyl $[ONMo\{N(R)Ar\}_3]$ were obtained in the first instance (Scheme 6.4).[357]

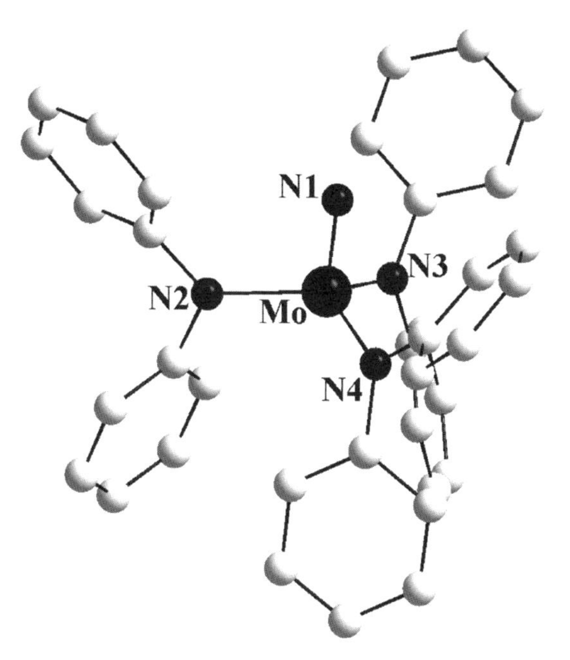

Figure 6.5 *The structure of a four-coordinate molybdenum amido/nitride $[MoN(NPh_2)_3]$.*[349] *Molybdenum and nitrogen atoms are shown as black spheres and carbon are given in white shadow. Selected bond lengths: Mo-N1 1.63, Mo-N2 2.00, Mo-N3 1.99, Mo-N4 2.00 Å*

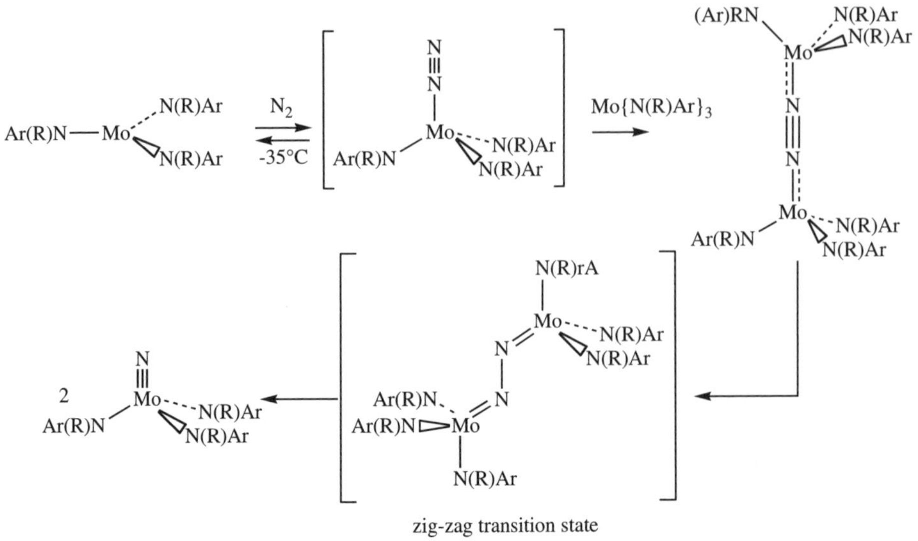

(Ar)RN N(R)Ar N(R)Ar Mo
N ≡ N
N(R)Ar
Ar(R)N—Mo N(R)Ar
N_2
-35°C
Ar(R)N Mo N(R)Ar N(R)Ar
Mo{N(R)Ar}$_3$
Ar(R)N Mo N(R)Ar N(R)Ar
N(R)rA
Mo N(R)Ar N(R)Ar
N=N
2 N≡Mo N(R)Ar N(R)Ar Ar(R)N
Ar(R)N Ar(R)N Mo
N(R)Ar
zig-zag transition state

Scheme 6.3 Depiction of the stepwise cleavage of N_2 by a three-coordinate Mo(III) amide[351]

Numerous studies involving similar amido derivatives of metals such as Ti (e.g. heterometallic cross-coupling of benzonitrile with carbon dioxide, pyridine, and benzophenone with mixed [Ti{N(Bu^t)Ar}$_3$] and [Mo{N(Bu^t)Ar}$_3$] systems),[84] V (e.g. denitrogenation of a vanadium nitride by carbon disulfide and dioxide),[570] Nb (e.g. the

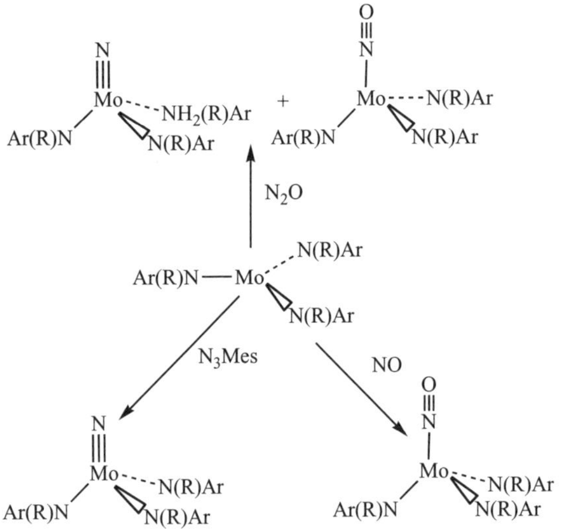

N≡Mo NH$_2$(R)Ar N(R)Ar Ar(R)N
+
O=N Mo N(R)Ar N(R)Ar Ar(R)N
N_2O
Ar(R)N—Mo N(R)Ar N(R)Ar
N_3Mes
NO
N≡Mo N(R)Ar N(R)Ar Ar(R)N
O=N Mo N(R)Ar N(R)Ar Ar(R)N

Scheme 6.4 Reactions of [Mo{N(R)Ar}$_3$] (R = $C(CD_3)_2CH_3$, Ar = $C_6H_3Me_2$-2,5) with N_2O and $MesN_3$[381]

stabilization of EP_2 triangles (E = Ge, Sn or Pb) supported by [Nb{N(Np)Ar}$_3$] (Np = neopentyl),[311] have been undertaken in investigations that continue to the present day.[309] An interesting feature of both niobium and molybdenum trisamides is that in certain cases they can exhibit instability with respect to a cyclometallated (β-H eliminated) species as illustrated by the molybdenum species in Equation (6.13).[366]

$$MoCl_3(thf)_3 + 3LiN(Pr^i)Ar \rightarrow [Mo(H)\{N(Pr^i)Ar\}_2(\eta^2\text{-}CMe_2 = NAr)] + 3LiCl \quad (6.13)$$

This transformation is in contrast to the corresponding more sterically crowded [Mo{N(Bu^t)Ar}$_3$] which does not exhibit this type of behaviour. Initially it was assumed that the cyclometallated structure would inhibit the type of reactions (e.g. with N_2 or N_2O) shown by the t-Bu substituted tris-amido complex. Investigation of the chemistry, however, showed that this is not the case and the imine-hydrido product behaves as an effective source of '[Mo{N(Pr^i)Ar}$_3$]' which, because of its less sterically encumbered nature, exhibits reactivity not observed for its more hindered [Mo{N(Bu^t)Ar}$_3$] counterpart. A simple example is provided by complex formation with benzophenone to give [Mo{N(Pr^i)Ar}$_3$(η^2-OCPh$_2$)], whereas the Bu^t analogue did not react with this reagent.[366]

The corresponding chemistry of analogous niobium complexes[51] was inhibited by the requirement of a more complicated synthetic approach for the isolation of the niobaziridine hydride. The use of the isopropyl substitued –N(Pr^i)Ar amido ligand proved unsuitable for the stabilization of [Nb(H){N(Pr^i)Ar}$_2$(η^2-Me$_2$CNAr)] because of insertion into the Nb–H bond.[302] These difficulties were overcome with use of the –N(CH$_2$Bu^t)Ar substituent and a synthetic approach based upon [Nb(O){N(CH$_2$Bu^t)Ar}$_3$]$_3$ which enabled the isolation of [Nb(H){N(CH$_2$Bu^t)Ar}$_2$(η^2-CH(Bu^t)=NAr)] via reduction.[310] The synthesis of this species has opened routes to some unusual chemistry as shown in Scheme 6.5.[51]

A more recent application, involving the nobium system, concerns the reaction of NaPNbN'$_3$ with [Cl(O)W{N(Pr^i)Ar}$_3$] to afford [P≡W{N(Pr^i)Ar}$_3$] with elimination of [ONbN'$_3$] and NaCl.[409b]

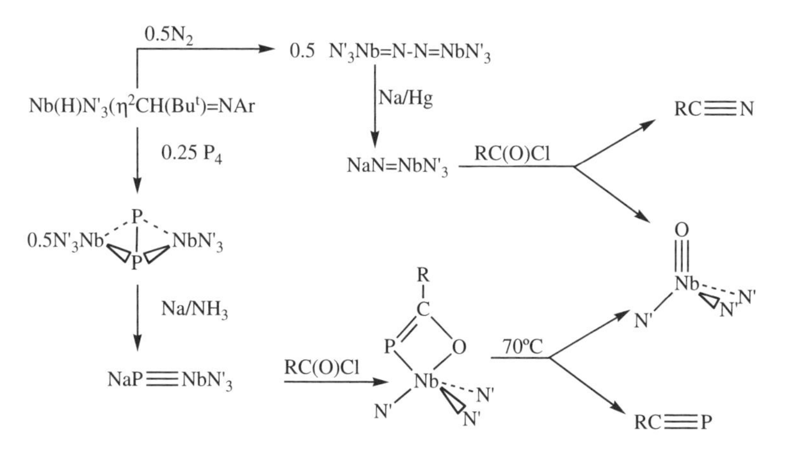

Scheme 6.5 *Reactions of [Nb(H)N'$_2$(η^2-CH(Bu^t) = NAr)] (N' = –N(CH$_2$Bu^t)Ar) to yield Nb(O)N'$_3$ and a nitrile or phosphalkyne*[51,306–311]

6.3 Transition Metal Complexes of Polydentate Amido Ligands

6.3.1 Introduction

The use of several new types of amido ligands that are based on a multiplicity of anionic and/or neutral donor sites has been a major feature of transition metal amide chemistry since around 1980.[5,9,17,18,22,23,26,30,31,33,36,37,40,44,46–48,52,53] One of the reasons for the development of multidentate amido ligands was a perceived need in organometallic chemistry to explore the use of ancillary ligands other than the ubiquitous cyclopentadienyl and related ligands. These had been widely investigated as supporting ligands in organometallic chemistry since the 1950s but there was a scarcity of systems in which they could be compared with other ligands. As a result it was not known if they were the best ligands for the promotion of catalytic efficiency. This motivation, combined with their synthetic accessibility, led to the development of a very wide range of multidentate amido ligands. As will become clear in Sections 6.3.2–6.3.4 the new ligands were designed, for the most part, to provide specific electronic or stereochemical environments at the metal that would promote catalytic activity at the metal centre – most often the catalysis of olefin polymerization or nitrogen fixation. However, there also exist a number of catalytic systems in which a multidentate amido ligand is directly involved in an intimate fashion in the catalytic process. Perhaps the foremost example is the hydroamination reaction in which an amine is added to an unsaturated alkene or alkyne to afford an amine/imine product. Although the catalysis of such reactions has been studied for a long time,[571,572] the intramolecular catalytic hydroamination was first observed with an electron-rich iridium complex (cf. Scheme 6.1[504]). In a series of classical investigations beginning in the 1990s,[573–579] Bergman and his group elucidated a general mechanism for hydroamination reactions [576] catalysed by early transition metals as shown in Scheme 6.6. This implicates a metal imide **A**, which with an alkyne affords the adduct **B** which carries a bidentate enamido ligand. Mechanistic investigations were carried out using the cyclopentadienyl supported zirconium amide $[Zr(\eta^5\text{-}C_5H_5)_2(NHAr)_2]$[573,574] which can eliminate amine to afford the

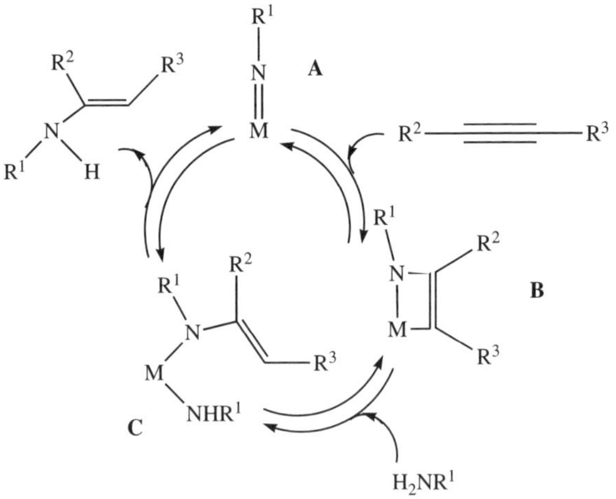

Scheme 6.6 *The Bergman mechanism for hydroamination*[373]

imide 'Zr(η^5-C_5H_5)$_2$NAr' (**A** in Scheme 6.6) which adds the alkyne to form the chelated amido azazirconacyclobutene complex (**B**). An example of a chelated amido complex, $[\overline{Zr(\eta^5\text{-}C_5H_5)_2N(C_6H_3Me_2\text{-}2,6)C(Ph)C}(Ph)]$, was isolated and structurally characterized for $R' = C_6H_3Me_2$-2,6, $R^2 = R^3 = C_6H_5$.[573]

Catalysis of the addition of amines to alkynes with use of titanium catalysts generated from titanium amides has been the subject of a number of investigations.[49,580] It is also noteworthy that carboamination reactions catalysed by [Zr(η^5-C_5H_5)$_2$(NHAr)$_2$] proceed by a route very similar to that described in Scheme 6.6 with the difference that the azazirconacyclobutene **B** reacts with an imine to form a 1,3-diazazirconacyclohexene which may then eliminate an α, β-unsaturated imine with regeneration of the metal imide **A**.[581] Examples of catalysis involving chelating titanium amides are also known.[582,583] The remainder of Section 6.3 is devoted to an overview of the wide variety of multidenate amides in use as supporting ligands for catalytic transition metal complexes.

6.3.2 Amido Phosphine Ligands

Among the earliest multidenate undelocalized amido ligands to be investigated in a systematic way were hybrid ligands involving amido and phosphine donor sites such as **D–F** that were

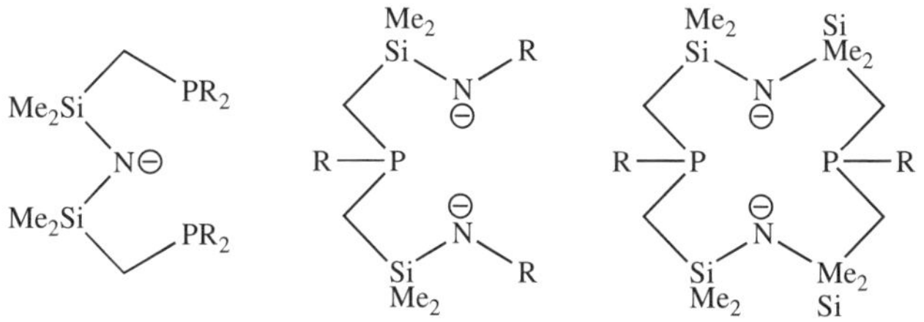

pioneered by Fryzuk and coworkers and which have been regularly reviewed.[4,6,38,48,53] These ligands which combine a 'hard' amido anionic centre(s) with the 'soft' neutral, phosphine donor site(s) were based on the proposition that the phosphine donor would contribute to the stability of the complex. This proved to be the case for amido derivatives of the 2nd and 3rd row elements of groups 8, 9 and 10 where the use of these ligands permitted the isolation of the first well-characterized amides of many of these elements. The reactivity of the M–N bond in such complexes, which includes protonation, β-elimination, migratory insertion and hydrogenolysis reactions, has been summarized in an early review.[4] However, later work showed that such ligands also supported a rich chemistry in the earlier transition elements of groups 4 and 5.[46] These complexes have played a prominent role in the activation of N_2 and various other reactions that include the formation of N–C, N–B and N–Si bonds. For example, the reduction of zirconium derivatives of a phenyl substituted ligand of type **F** (R = Ph) as in the complexes[584–587] shown in Scheme 6.7, afforded a dimeric complex in which N_2 bridges two zirconium centres in a side-on bound fashion in which the N–N distances (1.46 Å) are indicative of a single N–N bond. This complex also reacted with primary silanes or dihydrogen under mild conditions. NMR data in toluene solution indicated that a monohydrogen and NNH bridged complex had been formed. However, crystals grown from hexane showed that a μ-η-H_2 and μ-η-N_2 bridge species forms in this solvent. Ligands of type **D** and **E** also stabilize zirconium[587] or tantalum[587a,588a] dinitrogen complexes in

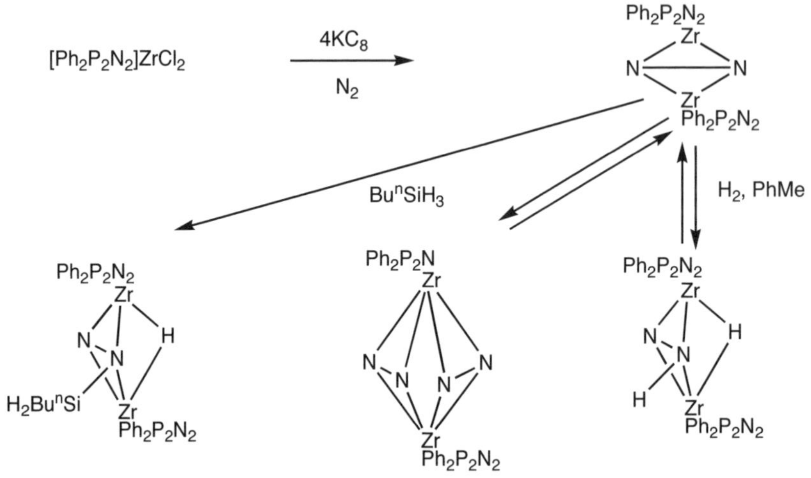

Scheme 6.7 Some reactions of the zirconium complex $[Zr(\mathbf{F})Cl_2]$[584–587a]

which dinitrogen can behave as an end-on and side-on bound ligand.[149] This and related chemistry have also been detailed in recent reviews.[46,53] Ligand **D** (R = Bu^t) also forms a stable T-shaped cobalt(I) amido complex which interacts weakly with N_2(end-on). However, with CO a Co(I) complex with square planar geometry can be isolated.[588b] A notable recent addition to the PNP pincer-type framework class of ligands similar to **D** has been the $[N(C_6H_3Me\text{-}4\text{-}PPr^i{}_2\text{-}2)_2]^-$ and related species.[588–592] These have been used as supporting ligands in transition metal complexes that can stabilize unsaturated moieties[53,591] and catalyse alkyne dimerization[593] or coupling of acetonitrile and aldehydes.[592]

6.3.3 Multidentate Podand Ligands

Several other classes of polydentate amide ligands in which the ligating atoms are various types of nitrogen (and occasionally oxygen) centres have been investigated. Prominent examples of these are the podand type of ligand as shown in **G–J**, introduced by Verkade, Schrock, Gade and coworkers.

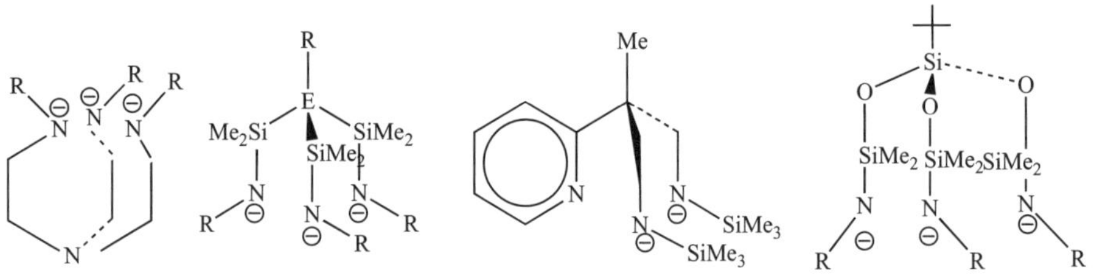

G **H** (E = C or Si) **I** **J**

These and related ligands (see below) carry a 3- (or 2-) charge and were developed in the early 1990s to stabilize metal complexes with high charges and to induce various coordination geometries at a metal centre.[594–636] They have been used almost exclusively with the early transition elements of groups 4–6. Ligands of type **H** and **J**, in which the bridgehead, linking and nitrogen substituents can be varied, favor tetrahedral geometries at the metal to which they are ligated. For example, the use of such ligands with a group 4 element permits complexes of the type (triamido)MCl to be isolated. These can be used to form bonds to other transition metals.[603,604] Such bimetallic complexes (involving Zr bonds to Fe or Ru) permit activation of molecules with relatively low reactivity such as CO_2.[604,605] Cationic zirconium and hafnium complexes of a ligand analogous to **I** with mesityl substituents at nitrogen have been shown to be effective living polymerization catalysts for 1-hexene.[607–609] A related class of podand ligands are the tripyrrolylmethane trianion, and 1,1-dipyrrolylcyclohexane dianion which have been shown to form derivatives of niobium, manganese, iron and cobalt with unusual structures.[610,612]

Of all the podand ligands investigated, the trisamidoamines have proven to be the most fruitful in generating new chemistry. Although methyl substituted trisamidoamine ligands of type **G** (azatranes) had been previously employed in titanium chemistry by Verkade,[9] Schrock and coworkers[17,18] devised the more crowded trimethylsilyl substituted tranionic $[N\{CH_2CH_2(Me_3Si)N\}_3]^{3-}$ ligand, a tris (trimethylsilyl) derivative, of G in order to take advantage of the relatively rigid four-coordinate core with a 'pocket' at the fifth coordination position, to synthesize high oxidation state inorganic or organometallic complexes with unusual structures and reactivities. These ligands were indeed found to bind to transition metals in the predicted tetradentate manner to create a sterically protected pocket in which three metal orbitals (of *a* and *e* symmetry corresponding to d_z^2 and d_{xy} and d_{yz} orbitals) are available for bonding.[17,18] The metal environment in these complexes thus resembles that of the $[Mo\{N(R)Ar\}_3]$ species described in Section B.6, although they preceded the latter in their development. Unlike the triamido-methanide and related ligands **H** and **J**, the trisamidoamine $[N\{CH_2CH_2(Me_3Si)N\}_3]^{3-}$ (abbreviated as $[(Me_3SiN)_3N]^{3-}$) favors structures that are based upon a trigonal bipyramidal geometry when complexed to the trivalent first row transition metal Ti-Fe. Complexes that are 'missing' an axial ligand, i.e. having trigonal monopyramidal geometry (the first stable complexes with such coordination), can thus be obtained.[609,610]

The complexes of this ligand proved to have a remarkable chemistry. For example the vanadium derivative $[V\{(Me_3SiN)_3N\}]$ forms complexes of the type $[\{(Me_3SiN)_3N\}V{=}E]$ (E = O, S, Se, Te, or NR) when reacted with various atom transfer reagents.[618] Tantalum analogues $[\{(Me_3SiN)_3N\}Ta{=}E]$ (E = Se or Te) were obtained by reaction of $[\{(Me_3SiN)_3N\}TaCl_2]$ with $LiESi(SiMe_3)_3$.[616] The iron derivative $[Fe\{(Me_3SiN)_3N\}]$ can be derivatized to yield $[Fe(CN)\{(Me_3SiN)_3N\}]$,[637] which is a relatively rare example of a well-characterized Fe(IV) complex. Reaction of $[(Me_3SiN)_3N]^{3-}$ with $TaCl_5$ afforded $[\{(Me_3SiN)_3N\}TaCl_2]$ which upon reaction with two equivalents of $LiCH_2CH_2R$ afforded $[\{(Me_3SiN)_3N\}Ta(CH_2CH_2R)_2]$ that subsequently underwent competitive α-hydrogen abstraction[18,617,618] to give an alkylidene complex or β-hydrogen elimination to give an unstable olefin complex as shown in Equation (6.14).

$$\{(Me_3SiN)_3N\}Ta(CH_2CH_2R)_2 \begin{cases} \xrightarrow{\alpha} \{(Me_3SiN)_3N\}Ta{=}CHCH_2R + CH_3CH_2R \\ \xrightarrow{\beta} \{(Me_3SiN)_3N\}Ta\text{----}(CHR{=}CH_2) + CH_3CH_2R \end{cases} \quad (6.14)$$

In a similar manner, reaction of $[\{(Me_3SiN)_3N\}TaCl_2]$ with two equivalents of LiEHR (E = N or P) afforded $\{(Me_3SiN)_3N\}TaER$ (Figure 6.6) with elimination of H_2ER.[619]

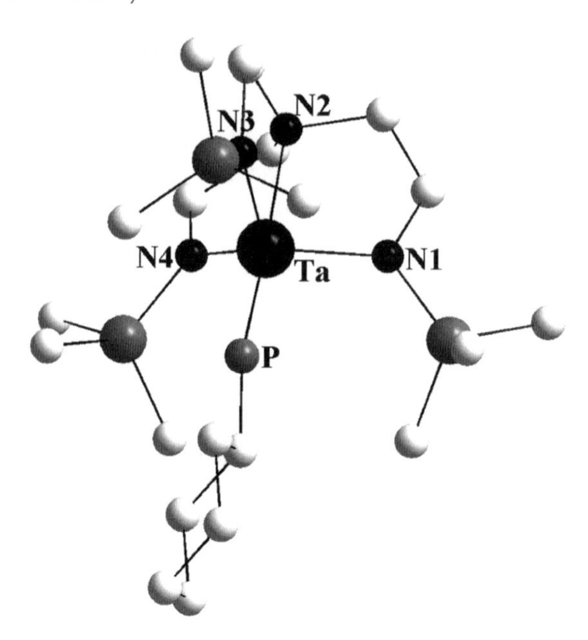

Figure 6.6 *Illustration of the phosphinidene complex [{(Me$_3$SiN)$_3$N}Ta = PCy].*[619] *Tantalum and nitrogen atoms are shown as black spheres, silicon and phosphorus atoms are grey and carbon atoms are indicated in white shadow. The Ta=P bond length is 2.15 Å*

Treatment of these phosphinidene complexes with R′CHO afforded RP = C(H)R′; this is an alternative route to phosphaalkenes.[620] The reaction of {(Me$_3$SiN)$_3$N}Li$_3$ with MoCl$_4$ or WCl$_4$ gave [{(Me$_3$SiN)$_3$N}MCl] (M = Mo or W).[621] The molybdenum(IV) halide complex afforded the derivatives [{(Me$_3$SiN)$_3$N}MoR] upon reaction with lithium alkyls and in some cases these eliminated H$_2$ under mild conditions to give alkyne complexes [{(Me$_3$SiN)$_3$N}Mo ≡ CR].[621,622] The tungsten analogues undergo similar elimination reactions.[623] Reaction of [{(Me$_3$SiN)$_3$N}MCl] (M = Mo or W) with LiEHPh (E = P or As) gave [{(Me$_3$SiN)$_3$N}M ≡ E] containing terminal M ≡ E triple bonds.[624,625] Although [{(Me$_3$SiN)$_3$N}W ≡ N], which is analogous to the molybdenum nitride/monodentate amide derivative described earlier, could be isolated by reaction with NaN$_3$,[625] the direct reaction of [{(Me$_3$SiN)$_3$N}M] (M = Mo or W) with dinitrogen is remarkable. Reduction of [{(Me$_3$SiN)$_3$N}MoCl] with magnesium powder afforded [{(Me$_3$SiN)$_3$N}MoN$_2$}$_2$Mg (thf)$_2$][628] which is an Mg^{2+} salt of the [(Me$_3$Si)$_3$N}]MoN$_2$]$^-$ ion which is best described as a diazenido species whose structural parameters resemble those in the [N$_3$N$_F$]Mo-N= N-Si(Pr^i)$_3$ (N$_3$N$_F$ = (C$_6$F$_5$NCH$_2$CH$_2$)$_3$N) and {[(Bu^{t}Me$_2$SiNCH$_2$CH$_2$)$_3$]Mo}$_2$(μ-N$_2$) which had previously been reported.[627,621] Reaction of the magnesium salt with FeCl$_2$ afforded the disproportionated product [Fe{N$_2$Mo[(Me$_3$SiN)$_3$N]}$_3$] (Figure 6.7).[626] However, the reaction is relatively complex and other products such as the end-on bound N$_2$ species {(Me$_3$SiN)$_3$N}MoN$_2$ were also observed. The latter has Mo-N$_2$, and MoN=N bond distances of 1.99 and 1.09 Å.[633]

The formation of relatively stable species of the type [(ArN)$_3$N]Mo−N = N−MoN(NAr)$_3$], in which further reduction of the N$_2$ moiety is not very favored, can be prevented by the use of larger [(ArN)$_3$N] triamido ligands. This led to the development of tris(amido)amine ligands

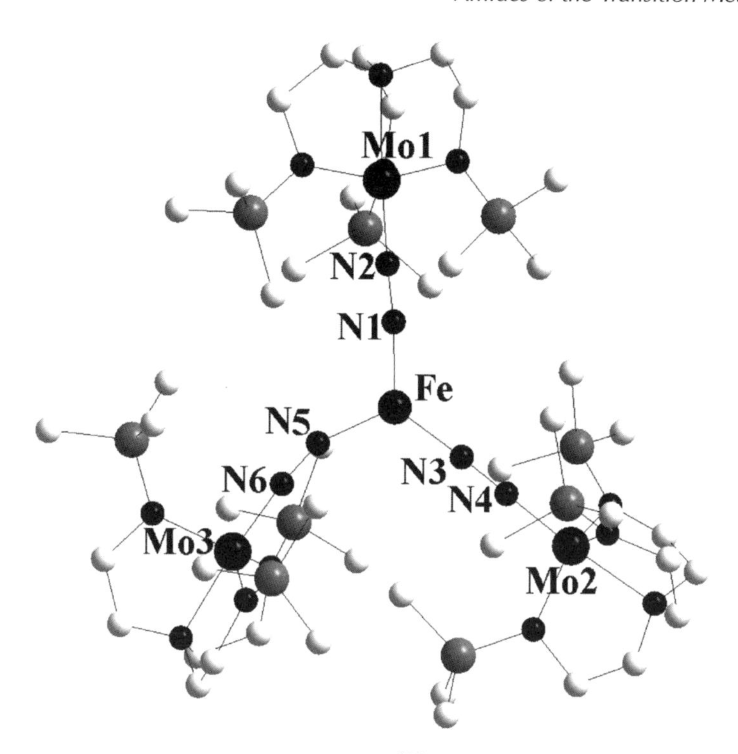

Figure 6.7 *View of $[Fe\{N_2Mo[(Me_3SiN)_3N]\}_3]$[626] (without H atoms) showing the three-coordinate iron atom and four-coordinate molybdenum atoms. Iron, molybdenum and nitrogen atoms are shown as black spheres, silicon atoms are grey and carbon atoms are given in white shadow. Selected bond lengths: Fe-N1 1.86, Fe-N3 1.84, Fe-N5 1.82, Mo1-N2 1.86, Mo2-N4 1.81, Mo3-N6 1.82 Å*

incorporating terphenyl substituents. In particular the use of the HIPT-substituted ligand in which the nitrogens carry terphenyls the central aryl of which is substituted at the 3 and 5 positions by bulky $-C_6H_2Pr^i_3$-2,4,6 groups is shown in **K**.[48] This has generated some remarkable results.

K [HIPT = $C_6H_3(C_6H_2Pr^i_3$-2',4',6')$_2$-3,5]

The hexaisopropylterphenyl or HIPT-substituted triamide when attached to molybdenum[629–632] or tungsten affords an approximately trigonal pocket in which N_2 or its reduced products are sterically protected. Dinitrogen was reduced catalytically in heptane by [Mo{(HITPN)$_3$N}] in the presence of [2,6-lutidinium][B(C$_6$H$_3$(CF$_3$)$_2$-3,5)$_4$] as a proton source and Cr(η^5-C$_5$Me$_5$)$_2$ as an electron source. Many of the intermediates in the reduction process including N_2, NNH, NNH_2, N–, NH and NH_3 complexes of the molybdenum species have been isolated and characterized. The catalytic properties of similarly substituted triamido-amine species have been studied.[633a] Several tungsten analogues of many of the molybdenum complexes described above have been synthesized,[633b] but the tungsten species do not behave as dinitrogen reduction catalysts under conditions similar to those used for molybdenum. A more recent report has disclosed the synthesis of tris(pyrrolyl-α-methyl)amines, which are related to the azatranes (**G**), and have been shown to form zirconium, molybdenum, and hafnium derivatives.[634]

The tripodal ligands of type **I** have been used with titanium(IV) and vanadium(V) to give four- or five-coordinate complexes with interesting properties.[598–601,635] For example, supplementary ligands, *e.g.*, a pyridine donor, can dissociate from the metal in [Ti(**I**)(py)(NBu^t)] to give (**I**)Ti(NBu^t) which can decoordinate its pyridinyl donor to give a highly reactive low (i.e. three) -coordinate Ti(IV) species which can induce C–N coupling and Me$_3$SiCl elimination reactions when reacted with unsaturated species.[599–636]

6.4 Other Chelating Amido Ligands

Numerous other polyamido ligands that feature three or more nitrogen donor sites are known. In these, the nitrogens can be incorporated as part of a ring as they are in 1,4,7 triazacyclononanes (**L**),[41] cyclams (**M**)[638] or the tris(aminobenzyl)-1,4,7-triazacyclononane species (**N**).[639]

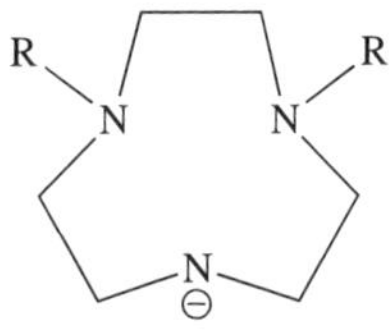

L

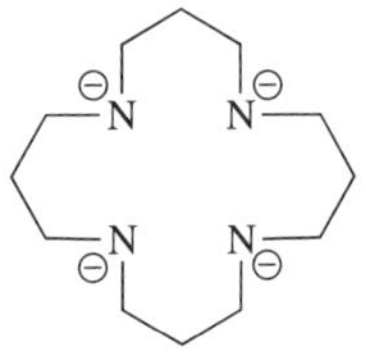

M

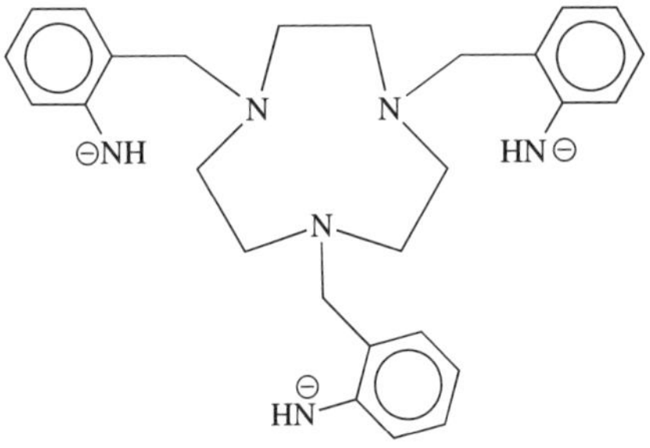

N

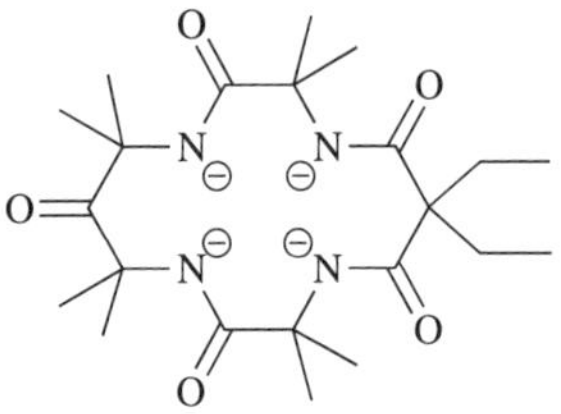

O

In addition, ligands of type **O** and related species have been used for the stabilization of middle and late high oxidation state transition metal complexes and the synthesis of oxidation catalysts.[637–639]

Acyclic, multidentate amido ligands permit greater flexibility in the coordination geometry of their complexes as exemplified by studies of diamidoamine and related ligands, a selection of which is illustrated by the structures **P–U**.

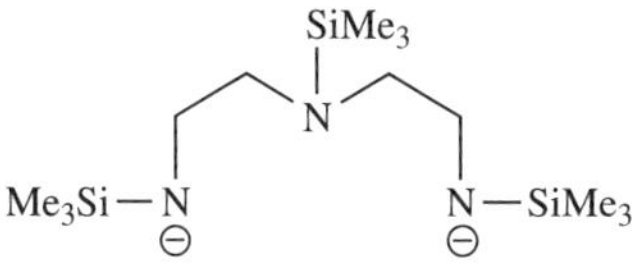

P 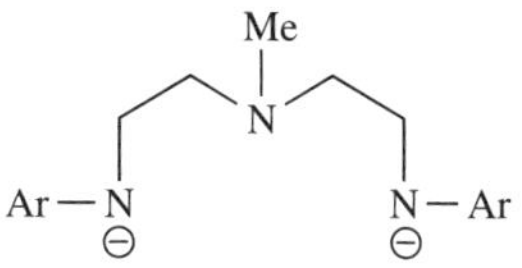 Q 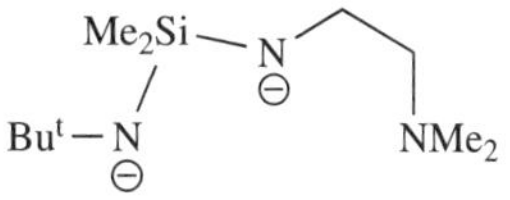 R

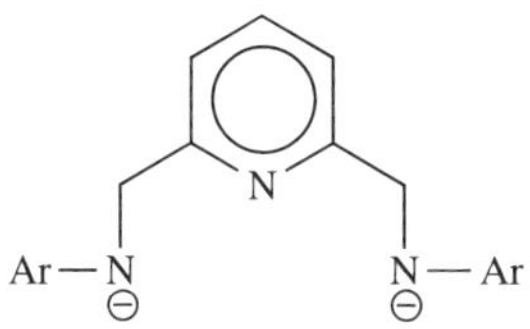

S 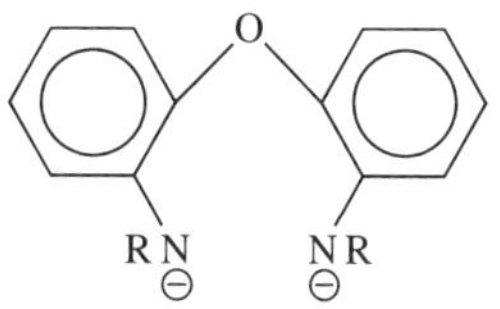 T 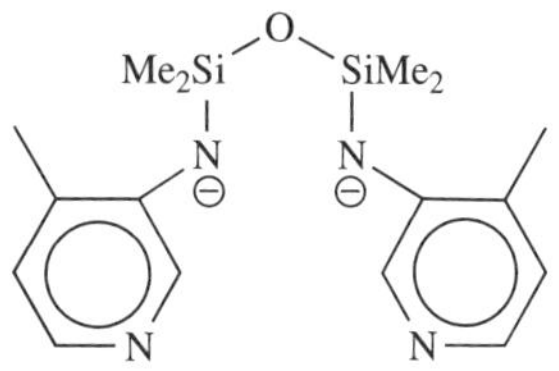 U

The chemistry of early transition metal complexes featuring these ligands has been reviewed in several recent articles.[30–33,35–37] As with the podand type ligands of type **G–J** the study of these ligands was justified by their possible applications in the synthesis of new transition metal complexes for olefin polymerization catalysis as part of the new generation of non-metallocene catalysts. The ligand **P** has been shown to bind so that the ligating nitrogens are mutually *cis* to each other in group 4 metal complexes,[639–645] whereas ligand **Q** binds in the two axial and one equatorial position of a distorted trigonal bipyramidal coordination sphere.[646] These and related compounds in which two amido nitrogens flank a neutral nitrogen or oxygen donor, as in **P**, **Q**, **S**, **S** and **U**, have been studied extensively.[647–653] Several of these complexes can polymerize 1-hexene in a living fashion.

The use of supporting amido ligands for transition metal based catalysts was intensified by reports of the use of bidentate amide donors of type **V**, with titanium and zirconium[654–657] (the related ligand **W** is also known).[658]

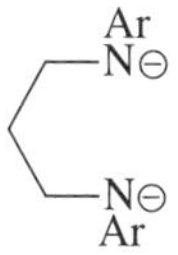

V 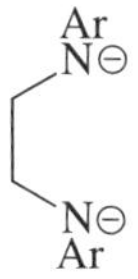 W 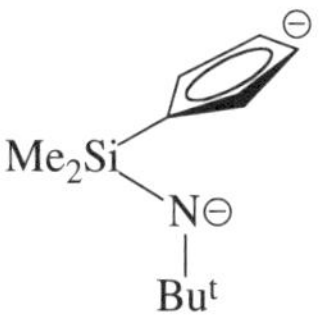 X 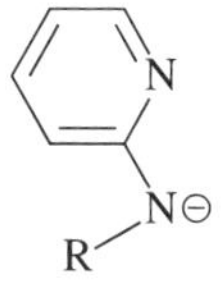 Y

In addition, chiral diamido ligands based on substituted 1,2-diaminocyclohexanes or diaminobinaphthalene have been prepared and tested for possible catalytic activity.[659,660] The use of amido ligands based on substituted 1,2-diaminobenzene and the reaction of their early transition metal complexes with unsaturated molecules has also been investigated.[661,662] The main impetus for the study of such ligands was to achieve greater control over the electronic and Lewis acidic properties of the olefin polymerization catalysts or precatalysts when bound to the group 4 metal. One way of doing this was to explore ligands other than the ubiquitous cyclopentadienyl and its various derivatives. Thus the linked cyclopentadienyl/amido hybrid ligand **X**[663–665] represents an intermediate stage between cyclopentadienyl and amido ligands. It was introduced by Bercaw and coworkers in the late 1980s. It was first used in organoscandium complexes for olefin polymerization and represented the initiation of the move to chelating amido groups as supporting ligands for catalysts. This development arose from the observation that the reactivity of the metal centre was greatly affected by the steric properties of the ligand. The introduction of the amido group as coligand renders the metal more open and more electron-deficient thereby promoting olefin insertion. The catalytic properties of these so-called 'constrained geometry' complexes have been investigated extensively because of their industrial importance and the area has been reviewed.[23] The amido pyridine ligand **Y** and related ligands have also been used to prepare a variety of transition metal derivatives.[666–668] Chiral ruthenium(II) complexes of the related chelating ligand, $\bar{N}(H)\{C(H)R\}_2\bar{N}Tos$ (R = H[669] or Ph[670]), have been used by Noyori and his group for asymmetric hydrogen transfer reactions between an alcohol or H_2 and a ketone.[669,670]

The ligands $[1,2\text{-}C_6H_4(NCH_2Bu^t_2)_2]^{2-}$ (≡ Lig^1),[673] $[1,3\text{- or }1,4\text{-}C_6H_4(NSiMe_3)_2]^{2-}$ (≡ Lig^2 or Lig^3),[671] $[1,2\text{-}C_6H_4(NSiMe_2R)_2]^{2-}$ [R = Me (Lig^4), R = CH=CH$_2$ (Lig^5)][672] and $[1,3\text{-}C_6H_4(CH_2NC_6H_3R_2\text{-}2,6)_2]^{2-}$ [R = Me (Lig^6), R = Pr^i (Lig^7)][258] have featured in a number of crystalline Zr(IV) complexes; Lig^2, Lig^3 and Lig^6 were the source of *m*- and *p*-bis[zirconyl(IV)amino]cyclophanes such as **Z**.[671,258]

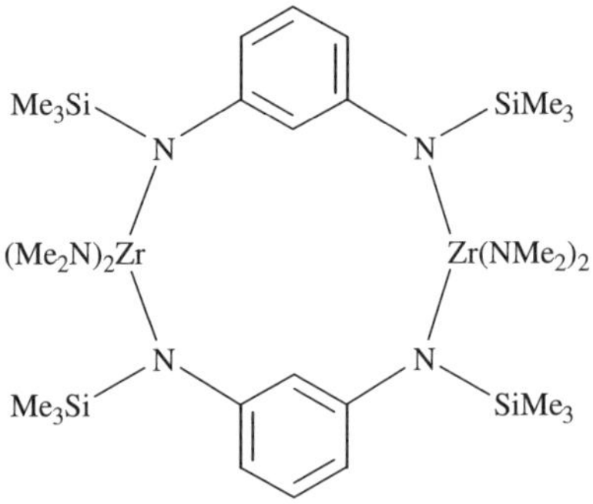

Z

References

1. M. H. Chisholm and I. P. Rothwell, in *Comprehensive Coordination Chemistry*, ed. Wilkinson, G.; Gillard, R. D.; McCleverty, J. A., Pergamon Oxford, 1987, vol. 2, p. 161. See also *Comprehensive Coordination Chemistry II*, ed. McCleverty, J. A.; Meyer, T. B.; Elsevier, Amsterdam, 2004, vols. 4–6; these chapters cover some aspects of transition metal amides.
2. M. H. Chisholm and D. H. Clark, *Comments Inorg. Chem.*, 1987, **6**, 23.

3. H. E. Bryndza and W. Tam, *Chem. Rev.*, 1988, **88**, 1163.
4. P. P. Power, *Comments Inorg. Chem.*, 1989, **8**, 177.
5. M. D. Fryzuk and C. D. Montgomery, *Coord. Chem. Rev.*, 1989, **95**, 1.
6. K. Dehnicke, *Chemiker-Zeitung*, 1990, **114**, 295.
7. M. D. Fryzuk, *Can. J. Chem.*, 1992, **70**, 2839.
8. D. M. Roundhill, *Chem. Rev.*, 1992, **92**, 1.
9. J. G. Verkade, *Acc. Chem. Res.*, 1993, **26**, 483.
10. P. P. Power, *Chemtracts: Inorganic Chemistry*, 1994, **6**, 181.
11. J. Barker and M. Kilner, *Coord. Chem. Rev.*, 1994, **133**, 219.
12. D. M. Hoffman, *Polyhedron*, 1994, **13**, 1169.
13. F. T. Edelmann, *Coord. Chem. Rev.*, 1994, **137**, 403.
14. C. Floriani, *Chem. Commun.*, 1996, 1257.
15. K. G. Caulton, *New. J. Chem.*, 1994, **18**, 25.
16. C. Floriani, *Pure Appl. Chem.*, 1996, **68**, 1.
17. R. R. Schrock, *Pure Appl. Chem.*, 1997, **69**, 2197.
18. R. R. Schrock, *Acc. Chem. Res.*, 1997, **30**, 9.
19. C. C. Cummins, *Chem. Commun.*, 1998, 1777.
20. C. C. Cummins, *Prog. Inorg. Chem.*, 1998, **47**, 685.
21. K. Dehnicke and F. Weller, *Coord. Chem. Rev.*, 1997, **158**, 103.
22. P. Mountford, *Chem. Soc. Rev.*, 1998, **27**, 105.
23. A. L. McKnight and R. M. Waymouth, *Chem. Rev.*, 1998, **98**, 2587.
24. J. F. Hartwig, *Angew. Chem., Int. Ed.*, 1998, **37**, 2046.
25. B. H. Yang and S. L. Buchwald, *J. Organomet. Chem.*, 1999, **576**, 125.
26. G. J. P. Britovsek, V. C. Gibson and D. F. Wass, *Angew. Chem., Int. Ed.*, 1999, **38**, 428.
27. P. L. Holland, R. A. Andersen and R. G. Bergman, *Comments Inorg. Chem.*, 1999, **21**, 115.
28. S. Alvarez, *Coord. Chem. Rev.*, 1999, **193–195**, 13.
29. O. Just and W. S. Rees, Jr., *Adv. Mater. Opt. Electron.*, 2000, **10**, 213.
30. L. H. Gade, *Chem. Commun.*, 2000, 173.
31. R. Kempe, *Angew. Chem., Int. Ed.*, 2000, **39**, 468.
32. C. F. Caro, M. F. Lappert and P. G. Merle, *Coord. Chem. Rev.*, 2001, **219–221**, 605.
33. L. H. Gade and P. Mountford, *Coord. Chem. Rev.*, 2001, **216–217**, 65.
34. L. Bourget-Merle, M. F. Lappert and J. R. Severn, *Chem. Rev.*, 2002, **102**, 3031.
35. L. H. Gade, *Acc. Chem. Res.*, 2002, **35**, 575.
36. R. Kempe, H. Noss and T. Irrgang, *J. Organomet. Chem.*, 2002, **647**, 12.
37. L. H. Gade, *J. Organomet. Chem.*, 2002, **661**, 85.
38. M. D. Fryzuk, *Modern Coordination Chemistry*, 2002, 187.
39. J. R. Fulton, A. W. Holland, D. J. Fox and R. G. Bergman, *Acc. Chem. Res.*, 2002, **35**, 44.
40. T. Schareina and R. Kempe, *Synthetic Methods of Organometallic and Inorganic Chemistry*, 2002, **10**, 1.
41. J. A. R. Schmidt, G. R. Giesbrecht, C. Cui and J. Arnold, *Chem. Commun.*, 2003, 1025.
42. V. C. Gibson and S. K. Spitzmesser, *Chem. Rev.*, 2003, **103**, 283.
43. S. Gambarotta and J. Scott, *Angew. Chem., Int. Ed.*, 2004, **43**, 5298.
44. R. H. Kempe, *Eur. J. Inorg. Chem.*, 2003, 791.
45. P. P. Power, *J. Organomet. Chem.*, 2004, **689**, 3904.
46. B. A. MacKay and M. D. Fryzuk, *Chem. Rev.*, 2004, **104**, 385.
47. P. L. Holland, *Can. J. Chem.*, 2005, **83**, 296.
48. R. R. Schrock, *Acc. Chem. Res.*, 2005, **38**, 955.
49. A. L. Odom, *Dalton Trans.*, 2005, 225.
50. S. L. Buchwald, C. Mauger, G. Mignani and U. Scholz, *Adv. Synth. Catal.*, 2006, **348**, 23.

51. (a) C. C. Cummins, *Angew. Chem., Int. Ed.*, 2006, **45**, 862; (b) J. S. Figueroa and C. C. Cummins, *Dalton Trans.*, 2006, 2161.
52. D. J. Mindiola, *Acc. Chem. Res.*, 2006, **39**, 813.
53. (a) E. A. MacLachlan and M. D. Fryzuk, *Organometallics*, 2006, **25**, 1530; (b) L.-C. Liang, *Coord. Chem. Rev.*, 2006, **250**, 1152.
54. T. B. Gunnoe, *Eur. J. Inorg. Chem.*, 2007, 1185.
55. M. F. Lappert, A. R. Sanger, R. C. Srivastava and P. P. Power, *Metal and Metalloid Amides: Synthesis, Structure, and Physical and Chemical Properties*, Horwood-Wiley, Chichester, 1980, p. 847.
56. T. Fjeldberg and R. A. Andersen, *J. Mol. Struct.*, 1985, **128**, 49.
57. R. Anwander, O. Runte, J. Eppinger, G. Gerstberger, E. Herdtweck and M. Spiegler, *J. Chem. Soc., Dalton Trans.*, 1998, 847.
58. M. Karl, K. Harms, G. Seybert, W. Massa, S. Fau, G. Frenking and K. Dehnicke, *Z. Anorg. Allg. Chem.*, 1999, **625**, 2055.
59. C. Meermann, P. Sirsch, K. W. Törnroos and R. Anwander, *Dalton Trans.*, 2006, 1041.
60. L. K. Knight, W. E. Piers and R. McDonald, *Organometallics*, 2006, **25**, 3289.
61. M. Karl, G. Seybert, W. Massa and K. Dehnicke, *Z. Anorg. Allg. Chem.*, 1999, **625**, 375.
62. L. K. Knight, W. E. Piers, P. Fleurat-Lessard, M. Parvez and R. McDonald, *Organometallics*, 2004, **23**, 2087.
63. (a) A. Haaland, K. Rypdal, H. V. Volden and R. A. Andersen, *J. Chem. Soc., Dalton Trans.*, 1992, 891; (b) M. E. Darre, T. Feinstein, S. Parsons, C. Pulham, D. W. H. Rankin and B. A. Smart, *Polyhedron*, 2006, **25**, 923.
64. M. A. Putzer, B. Neumüller and K. Dehincke, *Z. Anorg. Allg. Chem.*, 1998, **624**, 929.
65. O. Wagner, M. Jansen and H. P. Baldus, *Z. Anorg. Allg. Chem.*, 1994, **620**, 366.
66. A. J. Mountford, W. Clegg, S. J. Coles, R. N. Harrington, P. N. Horton, S. M. Humprhey, M. B. Hursthouse, J. A. Wright and S. J. Lancaster, *Chem. Eur. J.*, 2007, **13**, 4535.
67. D. L. Swartz II and A. L. Odom, *Organometallics*, 2006, **25**, 6125.
68. D. G. Dick, R. Rousseau and D. W. Stephan, *Can. J. Chem.*, 1991, **69**, 357.
69. G. Zi, H.-W. Li and Z. Xie, *Organometallics*, 2002, **21**, 3850.
70. H. Cai, W. H. Lam, X. Yu, X. Liu, Z.-Z. Wu, T. Chen, Z. Lin, X.-T. Chen, X.-Z. You and Z. Xue, *Inorg. Chem.*, 2003, **42**, 3008.
71. (a) H. Aneetha, F. Basuli, J. Bollinger, J. C. Huffman and D. J. Mindiola, *Organometallics*, 2006, **25**, 2402; (b) A. J. Mountford, S. J. Lancaster and S. J. Coles, *Acta. Crystallogr.*, 2007, **C63**, m401.
72. C. Boisson, J. C. Berthet, M. Ephritikhine, M. Lance and M. Nierlich, *J. Organomet. Chem.*, 1997, **531**, 115.
73. A. Martin, M. Mena, C. Yelamos, R. Serrano and P. R. Raithby, *J. Organomet. Chem.*, 1994, **467**, 79.
74. W. J. Sartain and J. P. Selegue, *Organometallics*, 1987, **6**, 1812.
75. W. J. Sartain and J. P. Selegue, *J. Am. Chem. Soc.*, 1985, **107**, 5818.
76. C. Airoldi, D. C. Bradley, H. Chudzynska, M. B. Hursthouse, K. M. A. Malik and P. R. Raithby, *J. Chem. Soc., Dalton Trans.*, 1980, 2010.
77. D. C. Bradley, H. Chudzynska, J. D. J. Backer-Dirks, M. B. Hursthouse, A. A. Ibrahim, M. Motevalli and A. C. Sullivan, *Polyhedron*, 1990, **9**, 1423.
78. Z. Duan, L. M. Thomas and J. G. Verkade, *Polyhedron*, 1996, **16**, 635.
79. A. R. Johnson, P. W. Wanandi, C. C. Cummins and W. M. Davis, *Organometallics*, 1994, **13**, 2907.
80. J. C. Peters, A. R. Johnson, A. L. Odom, P. W. Wanandi, W. M. Davis and C. C. Cummins, *J. Am. Chem. Soc.*, 1996, **118**, 10175.
81. E. Hollink, P. Wei and D. W. Stephan, *Can J. Chem.*, 2004, **82**, 1634.
82. T. Agapie, P. L. Diaconescu, D. J. Mindiola and C. C. Cummins, *Organometallics*, 2002, **21**, 1329.
83. A. Mendiratta, J. S. Figueroa and C. C. Cummins, *Chem. Commun.*, 2005, 3403.
84. A. Mendiratta and C. C. Cummins, *Inorg. Chem.*, 2005, **44**, 7319.

85. J. C. Peters, J.-P. F. Cherry, J. C. Thomas, L. Baraldo, D. J. Mindiola, W. M. Davis and C. C. Cummins, *J. Am. Chem. Soc.*, 1999, **121**, 10053.
86. B. R. Jagirdar, R. Murugavel and H.-G. Schmidt, *Inorg. Chim. Acta*, 1999, **292**, 105.
87. D. A. Straus, M. Kamigaito, A. P. Cole and R. M. Waymouth, *Inorg. Chim. Acta*, 2003, **349**, 65.
88. V. M. Visciglio, P. E. Fanwick and I. P. Rothwell, *Acta Crystallogr.*, 1994, **C50**, 896.
89. L. E. Turner, M. G. Thorn, P. E. Fanwick and I. P. Rothwell, *Organometallics*, 2004, **23**, 1576.
90. (a) D. L. Thorn, W. A. Nugent and R. L. Harlow, *J. Am. Chem. Soc.*, 1981, **103**, 357; (b) C. Lorber, R. Choukroun and L. Vendier, *Inorg. Chem.*, 2007, **46**, 3192; (c) S. Hamura, T. Oda, Y. Shimizu, K. Matsubara and H. Nagashima, *Dalton Trans.*, 2002, 1521.
91. C. J. Carmalt, A. C. Newport, S. A. O'Neill, I. Parkin, A. J. P. White and D. J. Williams, *Inorg. Chem.*, 2005, **44**, 615.
92. B. D. Ward, H. Risler, K. Weitershaus, S. Bellemin-Laponnaz, H. Wadepohl and L. H. Gade, *Inorg. Chem.*, 2006, **45**, 7777.
93. (a) H. Wang, H.-S. Chan and Z. Xie, *Organometallics*, 2006, **25**, 2569; (b) H. Shen, H.-S. Chan and X. Xie, *Organometallics*, 2007, **26**, 2694; (c) S. K. Mandal, P. M. Gurubasaravaj, H. W. Roesky, R. B. Oswald, J. Magull and A. Ringe, *Inorg. Chem.*, 2007, **46**, 7594.
94. G. Zi, H. W. Li and Z. Xie, *Organometallics*, 2002, **21**, 3850.
95. P. Shukla, J. A. Johnson, D. Vidovic, A. H. Cowley and C. D. Abernethy, *Chem. Commun.*, 2004, 360.
96. M. Galakhov, A. Martin, M. Mena, F. Palacris and P. R. Raithby, *Organometallics*, 1995, **14**, 131.
97. R. V. Bynum, W. E. Hunter, R. D. Rogers and J. L. Atwood, *Inorg. Chem.*, 1980, **19**, 2368.
98. Y. Takashima, Y. Nakayama, T. Hirao, H. Yasuda and A. Harada, *J. Organomet. Chem.*, 2004, **689**, 612.
99. R. K. Thomson, F. E. Zahariev, Z. Zhang, B. O. Patrick, Y. A. Wang and L. L. Schafer, *Inorg. Chem.*, 2005, **44**, 8680.
100. R. K. Minhas, L. Scoles, S. Wong and S. Gambarotta, *Organometallics*, 1996, **15**, 1113.
101. L. Scoles, R. Minhas, R. Duchateau, J. Jubb and S. Gambarotta, *Organometallics*, 1994, **13**, 4978.
102. B. Schwarze, W. Milius and W. Schnick, *Z. Naturforsch.*, 1997, **52B**, 819.
103. W. Schnick, R. Bettenhausen, B. Götze, H. A. Höppe, H. Huppertz, E. Irran, K. Köllisch, R. Lauterbach, M. Orth, S. Rannabauer, T. Schlieper, B. Schwarze and F. Wester, *Z. Anorg. Allg. Chem.*, 2003, **629**, 902.
104. D. M. Giolando, K. Kirschbaum, L. J. Graves and U. Bolle, *Inorg. Chem.*, 1992, **31**, 3887.
105. Y. Bai, M. Noltemeyer and H. W. Roesky, *Z. Naturforsch.*, 1991, **46B**, 1357.
106. A. V. Firth and D. W. Stephan, *Inorg. Chem.*, 1998, **37**, 4732.
107. J. L. Bennett and P. T. Wolczanski, *J. Am. Chem. Soc.*, 1997, **119**, 10696.
108. P. J. Sinnema, T. P. Spaniol and J. Okuda, *J. Organomet. Chem.*, 2000, **598**, 179.
109. M. Fan, E. N. Duesler and R. T. Paine, *Appl. Organomet. Chem.*, 2003, **17**, 549.
110. M. E. Gross and T. Siegrist, *Inorg. Chem.*, 1992, **31**, 4898.
111. A. M. Martins, J. R. Ascenso, C. G. de Azevedo, *et al. J. Chem. Soc., Dalton Trans.*, 2000, 4332.
112. K. Nomura and K. Fujii, *Macromolecules*, 2003, **36**, 2633.
113. M. G. Thorn, P. E. Fanwick and I. P. Rothwell, *Organometallics*, 1999, **18**, 4442.
114. B. Wrackmeyer, J. Weidinger and W. Milius, *Z. Anorg. Allg. Chem.*, 1998, **624**, 98.
115. K. Nomura and K. Fujii, *Organometallics*, 2002, **21**, 3042.
116. Y. E. Ovchinnikov, M. V. Ustinov, V. A. Igonin, Y. T. Struchkov, I. D. Kalikhman and M. G. Voronkov, *J. Organomet. Chem.*, 1993, **461**, 75.
117. Y. Bai, H. W. Roesky and M. Noltemeyer, *Z. Anorg. Allg. Chem.*, 1991, **595**, 21.
118. G. R. Giesbrecht, J. C. Gordon, D. L. Clark, C. A. Hijar, B. L. Scott and J. G. Watkin, *Polyhedron*, 2003, **22**, 153.
119. P. W. Wanandi, W. M. Davis, C. C. Cummins, M. A. Russell and D. E. Wilcox, *J. Am. Chem. Soc.*, 1995, **117**, 2110.
120. L. Scoles and S. Gambarotta, *Inorg. Chim. Acta*, 1995, **235**, 375.

121. N. Beydoun, R. Duchateau and S. Gambarotta, *J. Chem. Soc., Chem. Commun.*, 1992, 244.
122. A. R. Johnson, W. M. Davis and C. C. Cummins, *Organometallics*, 1996, **15**, 3825.
123. A. Mendiratta, C. C. Cummins, F. A. Cotton, S. A. Ibragimov, C. A. Murillo and D. Villagran, *Inorg. Chem.*, 2006, **45**, 4328.
124. E. Brady, J. R. Telford, G. Mitchell and W. Lukens, *Acta Crystallogr.*, 1995, **C51**, 558.
125. W. W. Lukens, Jr., M. R. Smith, III and R. A. Andersen, *J. Am. Chem. Soc.*, 1996, **118**, 1719.
126. J. Feldman and J. C. Calabrese, *J. Chem. Soc., Chem. Commun.*, 1991, 1042.
127. C. C. Cummins, C. P. Schaller, G. D. van Duyne, P. T. Wolczanski, A. W. E. Chan and R. Hoffmann, *J. Am. Chem. Soc.*, 1991, **113**, 2985.
128. R. Duchateau, S. Gambarotta, N. Beydoun and C. Bensimon, *J. Am. Chem. Soc.*, 1991, **113**, 8986.
129. R. Choukroun, P. Moumboko, S. Chevalier, M. Etienne and B. Donnadieu, *Angew. Chem., Int. Ed.*, 1998, **37**, 3169.
130. M. G. Fickes, W. M. Davis and C. C. Cummins, *J. Am. Chem. Soc.*, 1995, **117**, 6384.
131. C. P. Gerlach and J. Arnold, *Inorg. Chem.*, 1996, **35**, 5770.
132. C. P. Gerlach and J. Arnold, *Organometallics*, 1997, **16**, 5148.
133. F. Preuss, M. Steidel, M. Vogel, G. Overhoff, G. Hornung, W. Towae, W. Frank, G. Reiss and S. Muller-Becker, *Z. Anorg. Allg. Chem.*, 1997, **623**, 1220.
134. V. J. Murphy and H. Turner, *Organometallics*, 1997, **16**, 2495.
135. P. Berno and S. Gambarotta, *Angew. Chem., Int. Ed.*, 1995, **34**, 822.
136. S. R. Dubberley, B. R. Tyrrell and P. Mountford, *Acta Crystallogr.*, 2001, **C57**, 902.
137. (a) C. Lorber, R. Choukroun and L. Vendier, *Organometallics*, 2004, **23**, 1845; (b) C. Lorber, R. Choukroun and B. Donnadieu, *Inorg. Chem.*, 2002, **41**, 4217.
138. P. Berno and S. Gambarotta, *J. Chem. Soc., Chem. Commun.*, 1994, 2419.
139. N. Desmangles, S. Gambarotta, C. Bensimon, S. Davis and H. Zahalka, *J. Organomet. Chem.*, 1998, **562**, 53.
140. J. C. Peters, L. M. Baraldo, T. A. Baker, A. R. Johnson and C. C. Cummins, *J. Organomet. Chem.*, 1999, **591**, 24.
141. Z. Duan, M. Schmidt, V. G. Young, Jr., X. Xie, R. E. McCarley and J. G. Verkade, *J. Am. Chem. Soc.*, 1996, **118**, 5302.
142. M. Moore, K. Feghali and S. Gambarotta, *Inorg. Chem.*, 1997, **36**, 2191.
143. C. P. Gerlach and J. Arnold, *Organometallics*, 1996, **15**, 5260.
144. K. B. P. Ruppa, N. Desmangles, S. Gambarotta, G. Yap and A. L. Rheingold, *Inorg. Chem.*, 1997, **36**, 1194.
145. J.-I. Song, P. Berno and S. Gambarotta, *J. Am. Chem. Soc.*, 1994, **116**, 6927.
146. N. Desmangles, H. Jenkins, K. B. Ruppa and S. Gambarotta, *Inorg. Chim. Acta*, 1996, **250**, 1.
147. (a) P. Berno, R. Minhas, S. Hao and S. Gambarotta, *Organometallics*, 1994, **13**, 1052; (b) P. Berno and S. Gambarotta, *Organometallics*, 1994, **13**, 2569.
148. P. Berno, M. Moore, R. Minhas and S. Gambarotta, *Can. J. Chem.*, 1996, **74**, 1930.
149. M. Moore, S. Gambarotta, G. Yap, L. M. Liable-Sands and A. L. Rheingold, *Chem. Commun.*, 1997, 643.
150. M. Moore, S. Gambarotta and C. Bensimon, *Organometallics*, 1997, **16**, 1086.
151. A. L. Odom, C. C. Cummins and J. D. Protasiewicz, *J. Am. Chem. Soc.*, 1995, **117**, 6613.
152. A. L. Odom and C. C. Cummins, *Organometallics*, 1996, **15**, 898.
153. A. L. Odom and C. C. Cummins, *Polyhedron*, 1998, **17**, 675.
154. D. J. Mindiola and C. C. Cummins, *Angew. Chem., Int. Ed.*, 1998, **37**, 945.
155. A. L. Odom, D. J. Mindiola and C. C. Cummins, *Inorg. Chem.*, 1999, **38**, 3290.
156. K. B. P. Ruppa, K. Feghali, I. Kovacs, K. Aparna, S. Gambarotta, G. P. A. Yap and C. Bensimon, *J. Chem. Soc., Dalton Trans.*, 1998, 1595.
157. M. P. Coles, C. I. Dalby, V. C. Gibson, W. Clegg and M. R. J. Elsegood, *Polyhedron*, 1995, **14**, 2455.

158. A. A. Danopoulos, D. M. Hankin, G. Wilkinson, S. M. Cafferkey, T. K. N. Sweet and M. B. Hursthouse, *Polyhedron*, 1997, **16**, 3879.
159. K. C. Chew, W. Clegg, M. P. Coles, M. R. J. Elsegood, V. C. Gibson, A. J. P. White and D. J. Williams, *J. Chem. Soc., Dalton Trans.*, 1999, 2633.
160. E. W. Jandciu, P. Legzdins, W. S. McNeil, B. O. Patrick and K. M. Smith, *Chem. Commun.*, 2000, 1809.
161. (a) K. H. D. Ballem, V. Shetty, N. Etkin, B. O. Patrick and K. M. Smith, *Dalton Trans.*, 2004, 3431; (b) D. Rearden, I. Kovacs, K. B. P. Ruppa, K. Feghali, S. Gambarotta and J. Petersen, *Chem. Eur. J.*, 1997, **3**, 1482.
162. A. A. Danopoulos, G. Wilkinson, T. K. N. Sweet and M. B. Hursthouse, *J. Chem. Soc., Dalton Trans.*, 1996, 271.
163. R. S. Samuel, B. O. Patrick and K. M. Smith, *Can. J. Chem.*, 2004, **82**, 1788.
164. R. D. Koehn, G. Kociok-Koehn and M. Haufe, *Chem. Ber.*, 1996, **129**, 25.
165. R. Messere, M.-R. Spirlet, D. Jan, A. Demonceau and A. F. Noels, *Eur. J. Inorg. Chem.*, 2000, 1151.
166. J. J. H. Edema, S. Gambarotta and A. L. Spek, *Inorg. Chem.*, 1989, **28**, 811.
167. J. J. H. Edema, S. Gambarotta, A. Meetsma, A. L. Spek, W. J. J. Smeets and M. Y. Chiang, *J. Chem. Soc., Dalton Trans.*, 1993, 789.
168. H. Chen, R. A. Bartlett, M. M. Olmstead, P. P. Power and S. C. Shoner, *J. Am. Chem. Soc.*, 1990, **112**, 1048.
169. R. A. Bartlett, H. Chen and P. P. Power, *Angew. Chem., Int. Ed.*, 1989, **28**, 316.
170. G. A. Sim, D. I. Woodhouse and G. R. Knox, *J. Chem. Soc., Dalton Trans.*, 1979, 83.
171. E. W. Jandciu, J. Kuzelka, P. Legzdins, S. J. Rettig and K. M. Smith, *Organometallics*, 1999, **18**, 1994.
172. J. J. Ellison, P. P. Power and S. C. Shoner, *J. Am. Chem. Soc.*, 1989, **111**, 8044.
173. M. A. Putzer, A. Pilz, U. Müller, B. Neumüller and K. Dehnicke, *Z. Anorg. Allg. Chem.*, 1998, **624**, 1336.
174. M. A. Beswick, C. N. Harmer, P. R. Raithby, A. Steiner, K. L. Verhorevoort and D. S. Wright, *J. Chem. Soc., Dalton Trans.*, 1997, 2029.
175. M. P. Putzer, B. Neumüller, K. Dehnicke and J. Magull, *Chem. Ber.*, 1996, **129**, 715.
176. B. D. Murray and P. P. Power, *Inorg. Chem.*, 1984, **23**, 4584.
177. R. A. Andersen, K. Faegri, Jr., J. C. Green, A. Haaland, M. F. Lappert, W. P. Leung and K. Rypdal, *Inorg. Chem.*, 1988, **27**, 1782.
178. D. C. Bradley, M. B. Hursthouse, K. M. Abdul Malik and R. Moseler, *Trans. Met. Chem.*, 1978, **3**, 253.
179. A. Belforte, F. Calderazzo, U. Englert and J. Strähle, *J. Chem. Soc., Chem. Commun.*, 1989, 801.
180. (a) H. Chen, R. A. Bartlett, H. V. R. Dias, M. M. Olmstead and P. P. Power, *J. Am. Chem. Soc.*, 1989, **111**, 4338; (b) A. Panda, M. Stender, M. M. Olmstead, P. Klavins and P. P. Power, *Polyhedron*, 2003, **22**, 67.
181. D. K. Kennepohl, S. Brooker, G. M. Sheldrick and H. W. Roesky, *Z. Naturforsch.*, 1992, **47B**, 9.
182. D. C. Bradley, M. B. Hursthouse, A. A. Ibrahim, K. M. Abdul Malik, M. Motevalli, R. Möseler, H. Powell, J. D. Runnacles and A. C. Sullivan, *Polyhedron*, 1990, **9**, 2959.
183. M. Andruh, H. W. Roesky, M. Noltemeyer and H. G. Schmidt, *Z. Naturforsch.*, 1994, **49B**, 31.
184. A. Belforte, F. Calderazzo, U. Englert, J. Strähle and K. Wurst, *J. Chem. Soc., Dalton Trans.*, 1991, 2419.
185. (a) L. M. Carrella, W. Clegg, D. V. Graham, L. M. Hogg, A. R. Kennedy, J. Klett, R. E. Mulvey, E. Rentschler and L. Russo, *Angew. Chem., Int. Ed.*, 2007, **46**, 4662; (b) J. Garcia-Alvarez, A. R. Kennedy and R. E. Mulvey, *Angew. Chem., Int. Ed.*, 2007, **46**, 1105.
186. M. Niemeyer and P. P. Power, *Chem. Commun.*, 1996, 1573.
187. B. D. Murray and P. P. Power, *J. Am. Chem. Soc.*, 1984, **106**, 7011.
188. H. Chen, M. M. Olmstead, D. C. Pestana and P. P. Power, *Inorg. Chem.*, 1991, **30**, 1783.

189. M. Bochmann, A. K. Powell and X. Song, *Inorg. Chem.*, 1994, **33**, 400.
190. C. Weymann, A. A. Danoupoulos, G. Wilkinson, T. K. N. Sweet and M. B. Hursthouse, *Polyhedron*, 1996, **15**, 3605.
191. A. Panda, M. Stender, R. J. Wright, M. M. Olmstead, P. Klavins and P. P. Power, *Inorg. Chem.*, 2002, **41**, 3909.
192. C. C. Lu and J. C. Peters, *Inorg. Chem.*, 2006, **45**, 8597.
193. C. Soria Alvarez, A. Bashall, A. D. Bond, D. Cave, E. A. Harron, R. A. Layfield, M. E. G. Mosquera, M. McPartlin, J. M. Rawson, P. T. Wood and D. S. Wright, *Dalton Trans.*, 2003, 3002.
194. J. S. Duncan, T. M. Nazif, A. K. Verma and S. C. Lee, *Inorg. Chem.*, 2003, **42**, 1211.
195. S. L. Stokes, W. M. Davis, A. L. Odom and C. C. Cummins, *Organometallics*, 1996, **15**, 4521.
196. Y. Ohki, Y. Sunada and K. Tatsumi, *Chem. Lett.*, 2005, **34**, 172.
197. N. A. Eckert, J. M. Smith, R. J. Lachicotte, P. L. Holland and *Inorg Chem.*, 2004, **43**, 3306.
198. S. Takemoto, S. I. Ogura, H. Yo, Y. Hosokoshi, K. Kamikawa and H. Matsuzake, *Inorg. Chem.*, 2006, **45**, 4871.
199. C. W. Spicer, G. R. Potratz, W. M. Reiff and G. S. Girolami, Abstract of Papers 232nd ACS National Meeting San Francisco, CA, 2006,
200. M. M. Olmstead, P. P. Power and S. C. Shoner, *Inorg. Chem.*, 1991, **30**, 2547;
201. (a) R. A. Bartlett and P. P. Power, *J. Am. Chem. Soc.*, 1987, **109**, 7563. (b) H. Y. Au-Yeung, C. H. Lam, C.-K. Lam, W.-Y. Wong and H. K. Lee, *Inorg. Chem.*, 2007, **46**, 7695.
202. U. Siemeling, U. Vorfeld, B. Neumann and H.-G. Stammler, *Inorg. Chem.*, 2000, **39**, 5159.
203. D. J. Fox and R. G. Bergman, *Organometallics*, 2004, **23**, 1656.
204. Y. Ohki, Y. Takikawa, T. Hatanaka and K. Tatsumi, *Organometallics*, 2006, **25**, 3111.
205. U. Siemeling, U. Vorfeld, B. Neumann and H.-G. Stammler, *Organometallics*, 1998, **17**, 483.
206. J. J. Ellison, K. Ruhlandt-Senge, and P. P. Power, *Angew. Chem., Int. Ed.*, 1994, **33**, 1178.
207. R. A. Bartlett, J. J. Ellison, P. P. Power and S. C. Shoner, *Inorg. Chem.*, 1991, **30**, 2888.
208. H. Chen, M. M. Olmstead, D. C. Pestana and P. P. Power, *Inorg. Chem.*, 1991, **30**, 1783.
209. T. Komuro, H. Kawaguchi and K. Tatsumi, *Inorg. Chem.* 2002, **41**, 5083.
210. F. M. MacDonnell, K. Ruhlandt-Senge, J. J. Ellison, R. H. Holm and P. P. Power, *Inorg. Chem.*, 1995, **34**, 1815.
211. A. Murso and D. Stalke, *Dalton Trans.*, 2004, 2563.
212. A. K. Verma and S. C. Lee, *J. Am. Chem. Soc.*, 1999, **121**, 10838.
213. M. S. Hill and P. B. Hitchcock, *J. Organomet. Chem.*, 2004, **689**, 3163.
214. M. W. Bouwkamp, E. Lobkovsky and P. J. Chirik, *Inorg. Chem.*, 2006, **45**, 2.
215. H. Hope, M. M. Olmstead, B. D. Murray and P. P. Power, *J. Am. Chem. Soc.*, 1985, **107**, 712.
216. M. M. Olmstead, P. P. Power and G. Sigel, *Inorg. Chem.*, 1986, **25**, 1027.
217. (a) R. A. Bartlett, X. Feng, M. M. Olmstead, P. P. Power and K. J. Weese, *J. Am. Chem. Soc.*, 1987, **109**, 4851; (b) U. Koelle, B. Fuss, M. Belting and E. Raabe, *Organometallics*, 1986, **5**, 980.
218. H. Link and D. Fenske, *Z. Anorg. Allg. Chem.*, 1999, **625**, 1878.
219. H. Link, P. Reiss, S. Chitsaz, H. Pfistner and D. Fenske, *Z. Anorg. Allg. Chem.*, 2003, **629**, 755.
220. (a) J. Cámpora, P. Palma, D. del Rio, M. Mar Conejo, and E. Álvarez, *Organometallics*, 2004, **23**, 5653; (b) J. Cámpora, I. Matas, P. Palma, E. Álvarez, C. Griaff and A. Tiripiccio, *Organometallics*, 2007, **26**, 3840.
221. D. D. VanderLende, J. M. Boncella and K. A. Abboud, *Acta. Crystallogr.*, 1995, **C51**, 591.
222. P. L. Holland, R. A. Andersen, R. G. Bergman, J. Huang and S. P. Nolan, *J. Am. Chem. Soc.*, 1997, **119**, 2800.
223. P. L. Holland, R. A. Andersen and R. G. Bergman, *J. Am. Soc. Chem.*, 1996, **118**, 1092.
224. N. A. Eckert, E. M. Bones, R. J. Lachicotte and P. L. Holland, *Inorg. Chem.*, 2003, **42**, 1720.
225. E. Kogut, H. L. Wiencko, L. Zhang, D. E. Cordeau and T. H. Warren, *J. Am. Chem. Soc.*, 2005, **127**, 11248.
226. D. J. Mindiola and G. L. Hillhouse, *J. Am. Chem. Soc.*, 2001, **123**, 4623.

227. P. Reiss and D. Fenske, *Z. Anorg. Allg. Chem.*, 2000, **626**, 1317.
228. P. P. Power, K. Ruhlandt-Senge and S. C. Shoner, *Inorg. Chem.*, 1991, **30**, 5013.
229. H. Chen, M. M. Olmstead, S. C. Shoner and P. P. Power, *J. Chem. Soc., Dalton Trans.*, 1992, 451.
230. H. Hope and P. P. Power, *Inorg. Chem.*, 1984, **23**, 936.
231. S. Gambarotta, M. Bracci, C. Floriani, A. Chiesi-Villa and C. Guastini, *J. Chem. Soc., Dalton Trans.*, 1987, 1883.
232. (a) A. M. James, R. K. Laxman, F. R. Fronczek and A. W. Maverick, *Inorg. Chem.*, 1998, **37**, 3785; (b) P. Miele, J. D. Foulon, N. Hovnanian, J. Durand and L. Cot, *Eur. J. Inorg. Solid State Chem.*, 1992, **29**, 573.
233. R. P. Davies, S. Hornauer and P. B. Hitchcock, *Angew. Chem., Int. Ed.*, 2007, **46**, 5191.
234. L. A. Goj, E. D. Blue, S. A. Delp, T. B. Gunnoe, T. R. Cundari, A. W. Pierpont, J. L. Petersen and P. D. Boyle, *Inorg. Chem.*, 2006, **45**, 9032.
235. E. D. Blue, A. Davis, D. Conner, T. B. Gunnoe, P. D. Boyle and P. S. White, *J. Am. Chem. Soc.*, 2003, **125**, 9435.
236. M. Niemeyer, *Acta Crystallogr.*, 2001, **E57**, m491.
237. A. A. Danopoulos, G. Wilkinson, B. Hussain-Bates, and M. B. Hursthouse, *J. Chem. Soc., Dalton Trans.*, 1990, 2753.
238. P. B. Hitchcock, Q.-G. Huang, M. F. Lappert and X.-H. Wei, *J. Mater. Chem.*, 2004, **14**, 3266.
239. M. Westerhausen, M. Hartmann, A. Pfitzner and W. Schwarz, *Z. Anorg. Allg. Chem.*, 1995, **621**, 837.
240. W. A. Herrmann, F. C. Munck, G. R. J. Artus, O. Runte and R. Anwander, *Organometallics*, 1997, **16**, 682.
241. W. J. Evans, D. S. Lee and J. W. Ziller, *J. Am. Chem. Soc.*, 2004, **126**, 454.
242. (a) D. V. Gribkov, K. C. Hultzsch and F. Hampel, *Chem. Eur. J.*, 2003, **9**, 4796; (b) I. S. Edworthy, A. Blake, C. Wilson and P. L. Arnold, *Organometallics*, 2007, **26**, 3684.
243. M. H. Chisholm, C. E. Hammond and J. C. Huffman, *Polyhedron*, 1988, **7**, 2515.
244. R. Kempe, *Z. Krystallogr. New Cryst. Struct.*, 2000, **215**, 275.
245. K. Hagen, C. J. Holwill, D. A. Rice and J. D. Runnacles, *Inorg. Chem.*, 1988, **27**, 2032.
246. M. Polamo, I. Mutikainen and M. Leskela, *Z. Kristallogr. New Cryst. Struct.*, 1996, **211**, 641.
247. X. Yu and Z.-L. Xue, *Inorg. Chem.*, 2005, **44**, 1505.
248. Y. Bai, H. W. Roesky, M. Noltemeyer and M. Witt, *Chem. Ber.*, 1992, **125**, 825.
249. A. Antiñolo, G. S. Bristow, G. K. Campbell, A. W. Duff, P. B. Hitchcock, R. A. Kamarudin, M. F. Lappert, R. S. Norton, N. Sanjindeen, D. J. W. Winterborn, J. L. A. Atwood, W. E. Hunter and H. Zhang, *Polyhedron*, 1989, **8**, 601.
250. X. Liu, Z. Wu, Z. Peng, Y.-D. Wu and Z. Xue, *J. Am. Chem. Soc.*, 1999, **121**, 5350.
251. X. Liu, Z. Wu, H. Cai, *et al.*, *J. Am. Chem. Soc.*, 2001, **123**, 8011.
252. J.-S. M. Lehn and D. M. Hoffman, *Inorg. Chem.*, 2002, **41**, 4063.
253. H. Cai, W. H. Lam, X. Yu, X. Liu, Z.-Z. Wu, T. Chen, Z. Lin, X.-T. Chen, X.-Z. You and Z. Xue, *Inorg. Chem.*, 2003, **42**, 3008.
254. Z. Wu, J. B. Diminnie and Z. Xue, *Inorg. Chem.*, 1998, **37**, 6366.
255. R. Kempe, G. Hillebrand and A. Spannenberg, *Z. Krystallogr. New Cryst. Struct.*, 1997, **212**, 490.
256. Z. Wu, J. B. Diminnie and Z. Xue, *Inorg. Chem.*, 1998, **37**, 2570.
257. J. An, L. Urrieta, R. Williams, W. Tikkanen, R. Bau and M. Yousufuddin, *J. Organomet. Chem.*, 2005, **690**, 4376.
258. S. Daniéle, P. B. Hitchcock, M. F. Lappert, T. A. Nile and C. M. Zdanski, *J. Chem. Soc., Dalton Trans.*, 2002, 3980.
259. H. Qiu, H. Cai, J. B. Woods, Z. Wu, T. Chen, X. Yu and Z.-L. Xue, *Organometallics*, 2005, **24**, 4190.
260. J. R. Galsworthy, M. L. H. Green, N. Maxted and M. Müller, *J. Chem. Soc., Dalton Trans.*, 1998, 387.
261. X. Yu, S. Bi, I. A. Guzei, Z. Lin and Z.-L. Xue, *Inorg. Chem.*, 2004, **43**, 7111.

262. A. Kasani, S. Gambarotta and C. Bensimon, *Can. J. Chem.*, 1997, **75**, 1494.
263. Z. Ziniuk, I. Goldberg and M. Kol, *J. Organomet. Chem.*, 1997, **545–546**, 441.
264. A. Baunemann, Y. Kim, M. Winter and R. A. Fischer, *Dalton Trans.*, 2006, 121.
265. Z. Wu, J. B. Diminnie and Z. Xue, *J. Am. Chem. Soc.*, 1999, **121**, 4300.
266. R. Fandos, M. Martinez-Ripoll, A. Otero, M. J. Ruiz, A. Rodriguez and P. Terreros, *Organometallics*, 1998, **17**, 1465.
267. R. Wartchow and S. Doye, *J. Organomet. Chem.*, 1998, **566**, 287.
268. C. P. Schaller, C. C. Cummins and P. T. Wolczanski, *J. Am. Chem. Soc.*, 1996, **118**, 591.
269. W. A. Nugent and R. L. Harlow, *Inorg. Chem.*, 1979, **18**, 2030.
270. S. Brenner, R. Kempe and P. Arndt, *Z. Anorg. Allg. Chem.*, 1995, **621**, 2121.
271. M. H. Lee, J.-W. Hwang, Y. Kim, Y. Han and Y. Do, *Organometallics*, 2000, **19**, 5514.
272. J. Strauch, T. H. Warren, G. Erker, R. Frohlich and P. Saarenketo, *Inorg. Chim. Acta*, 2000, **300–302**, 810.
273. W. A. Herrmann, M. J. A. Morawietz and T. Priermeir, *J. Organomet. Chem.*, 1996, **506**, 351.
274. K. V. Axenov, M. Klinga, M. Leskelae, V. Kotov and T. Repo, *Eur. J. Inorg. Chem.*, 2004, 4702.
275. A. Vogel, T. Priermeier and W. A. Herrmann, *J. Organomet. Chem.*, 1997, **527**, 297.
276. R. R. Schrock, A. L. Casado, J. T. Goodman, L.-C. Liang, P. J. Bonitatebus and W. M. Davis, *Organometallics*, 2000, **19**, 5325.
277. H. Wang, H. S. Chan and Z. Xie, *Organometallics*, 2006, **25**, 2569.
278. J. C. Yoder, M. W. Day and J. E. Bercaw, *Organometallics*, 1998, **17**, 4946.
279. R. Thomas, A. Milanov, R. Bhakta, U. Patil, M. Winter, P. Ehrhart, R. Waser and A. Devi, *Chem. Vap. Deposition*, 2006, **12**, 295.
280. W. J. Sartain and J. P. Selegue, *Organometallics*, 1984, **3**, 1922.
281. M. R. Mason, B. N. Fneich and K. Kirschbaum, *Inorg. Chem.*, 2003, **42**, 6592.
282. G. M. Diamond, R. F. Jordan and J. L. Petersen, *Organometallics*, 1996, **15**, 4045.
283. W. A. Herrmann, N. W. Huber and J. Behm, *Chem. Ber.*, 1992, **125**, 1405.
284. S. A. A. Shah, H. Dorn, A. Voigt, H. W. Roesky, E. Parisini, H.-G. Schmidt and M. Noltemeyer, *Organometallics*, 1996, **15**, 3176.
285. S. A. A. Shah, H. Dorn, H. W. Roesky, E. Parisini, H.-G. Schmidt and M. Noltemeyer, *J. Chem. Soc., Dalton Trans.*, 1996, 4143.
286. G. Bai, D. Vidovic, H. W. Roesky and J. Magull, *Polyhedron*, 2004, **23**, 1125.
287. G. Bai, P. Mueller, H. W. Roesky and I. Usón, *Organometallics*, 2000, **19**, 4675.
288. M. M. Banaszak Holl and P. T. Wolczanski, *J. Am. Chem. Soc.*, 1992, **114**, 3854.
289. D. A. Kissounko, A. Epshteyn, J. C. Fettinger and L. R. Sita, *Organometallics*, 2006, **25**, 1076.
290. E. V. Avtomonov and K. A. Rufanov, *Z. Naturforsch.*, 1999, **54B**, 1563.
291. L. J. Procopio, P. J. Carroll and D. H. Berry, *J. Am. Chem. Soc.*, 1994, **116**, 177.
292. B. L. Rupert, J. Arnold and A. Krajete, *Acta Crystallogr.*, 2006, **E62**, m950.
293. M. Driess, J. Aust and K. Merz, *Eur. J. Inorg. Chem.*, 2002, 2961.
294. M. Polamo, I. Mutikainen and M. Leskelaa, *Acta Crystallogr.*, 1996, **C52**, 1348.
295. G. M. Diamond, R. F. Jordan and J. L. Petersen, *Organometallics*, 1996, **15**, 4030.
296. A. Milanov, R. Bhakta, A. Baunemann, H.-W. Becker, R. Thomas, P. Ehrhart, M. Winter, and A. Devi, *Inorg. Chem.*, 2006, **45**, 11008.
297. (a) A. Milanov, R. Bhakta, R. Thomas, P. Ehrhart, M. Winter, R. Waser and A. Devi, *J. Mater. Chem.*, 2006, **16**, 437; (b) M. Hellwig, A. Milanov, D. Barecca, J.-L. Deborde, R. Thomas, M. Winter, U. Kunze, R. A. Fischer and A. Devi, *Chem. Mater.*, 2007, **19**, 6077.
298. W. A. Herrmann, W. Baratta and E. Herdtweck, *Angew. Chem., Int. Ed.*, 1996, **35**, 1951.
299. (a) A. S. Batsanov, A. V. Churakov, J. A. K. Howard, A. K. Hughes, A. L. Johnson, A. J. Kingsley, I. S. Neretin and K. Wade, *J. Chem. Soc., Dalton Trans.*, 1999, 3867; (b) H. O. Davies, A. C. Jones, E. A. McKinnell, J. Raftery, C. A. Muryn, M. Afzaal, and P. O'Brien,

J. Mater. Chem., 2006, **16**, 2226; (c) J. Engering, E.-M. Peters and M. Jansen, *Z. Kristallogr. - New Cryst. Struct.*, 2003, **218**, 199.
300. D. M. Hoffman and S. P. Rangarajan, *Acta Crystallogr.*, 1996, **C52**, 1616.
301. P. Berno and S. Gambarotta, *Organometallics*, 1995, **14**, 2159.
302. D. J. Mindiola and C. C. Cummins, *Organometallics*, 2001, **20**, 3626.
303. D. J. Mindiola, K. Meyer, J.-P. F. Cherry, T. A. Baker and C. C. Cummins, *Organometallics*, 2000, **19**, 1622.
304. M. G. Fickes, A. L. Odom and C. C. Cummins, *Chem. Commun.*, 1997, 1993.
305. J. K. Brask, M. G. Fickes, P. Santrirutnugul, V. Dura-Vila, A. L. Odom and C. C. Cummins, *Chem. Commun.*, 2001, 1676.
306. J. S. Figueroa and C. C. Cummins, *Angew. Chem., Int. Ed.*, 2004, **43**, 984.
307. J. S. Figueroa and C. C. Cummins, *J. Am. Chem. Soc.*, 2004, **126**, 13916.
308. N. A. Piro, J. S. Figueroa, J. T. McKellar and C. C. Cummins, *Science*, 2006, **313**, 1276.
309. J. S. Figueroa, N. A. Piro, C. R. Clough and C. C. Cummins, *J. Am. Chem. Soc.*, 2006, **128**, 940.
310. J. S. Figueroa and C. C. Cummins, *J. Am. Chem. Soc.*, 2003, **125**, 4020.
311. J. S. Figueroa and C. C. Cummins, *Angew. Chem., Int. Ed.*, 2005, **44**, 4592.
312. A. Antiñolo, F. Carrillo-Hermosilla, J. Fernandez-Baeza, M. Lanfranchi, A. Lara-Sanchez, A. Otero, E. Palomares, M. A. Pellinghelli and A. M. Rodriguez, *Organometallics*, 1998, **17**, 3015.
313. M. J. Humphries, M. L. H. Green, R. E. Douthwaite and L. H. Rees, *J. Chem. Soc., Dalton Trans.*, 2000, 4555.
314. S. M. Pugh, D. J. M. Trösch, M. E. G. Skinner, L. H. Gade and P. Mountford, *Organometallics*, 2001, **20**, 3531.
315. A. Antiñolo, A. Otero, F. Urbanos, S. Garcia-Blanco, S. Martinez-Carrera and J. Sanz-Aparicio, *J. Organomet. Chem.*, 1988, **350**, 25.
316. M. Tayebani, K. Feghali, S. Gambarotta and C. Bensimon, *Organometallics*, 1997, **16**, 5084.
317. (a) D. M. Hoffman and S. P. Rangarajan, *Polyhedron*, 1993, **12**, 2899; (b) S. G. Bott, D. M. Hoffman and S. P. Rangarajan, *Inorg. Chem.*, 1995, **34**, 4335.
318. S. G. Bott, D. M. Hoffman and S. P. Rangarajan, *J. Chem. Soc., Dalton Trans.*, 1996, 1979.
319. K. Hagen, C. J. Holwill, D. A. Rice and J. D. Runnacles, *Inorg. Chem.*, 1992, **31**, 4733.
320. M. H. Chisholm, L. S. Tan and J. C. Huffman, *J. Am. Chem. Soc.*, 1982, **104**, 4879.
321. M. K. T. Tin, G. P. A. Yap and D. S. Richeson, *Inorg. Chem.*, 1999, **38**, 998.
322. M. G. Thorn, J. R. Parker, P. E. Fanwick and I. P. Rothwell, *Organometallics*, 2003, **22**, 4658.
323. H. Cai, X. Yu, T. Chen, X.-T. Chen, X.-Z. You and Z. Xue, *Can. J. Chem.*, 2003, **81**, 1398.
324. M. H. Chisholm, J. C. Huffman and L.-S. Tan, *Inorg. Chem.*, 1981, **20**, 1859.
325. I. A. Guzei, G. P. A. Yap and C. H. Winter, *Inorg. Chem.*, 1997, **36**, 1738.
326. M. K. T. Tin, G. P. A. Yap and D. S. Richeson, *Inorg. Chem.*, 1998, **37**, 6728.
327. M. K. T. Tin, N. Thirupathi, G. P. A. Yap and D. S. Richeson, *J. Chem. Soc., Dalton Trans.*, 1999, 2947.
328. N. Thirupathi, G. P. A. Yap and D. S. Richeson, *Organometallics*, 2000, **19**, 2573.
329. R. C. Mills, P. Doufou, K. A. Abboud and J. M. Boncella, *Polyhedron*, 2002, **21**, 1051.
330. M. A. Fox, J. A. K. Howard, A. K. Hughes, J. M. Malget and D. S. Yufit, *J. Chem. Soc., Dalton Trans.*, 2001, 2263.
331. A. S. Batsanov, P. A. Eva, M. A. Fox, J. A. K. Howard, A. K. Hughs, A. L. Johnson, A. M. Martin and K. Wade, *J. Chem. Soc., Dalton Trans.*, 2000, 3519.
332. A. Baunemann, Y. Kim, M. Winter and R. A. Fischer, *Dalton Trans.*, 2006, 121.
333. C. Bradley, M. B. Hursthouse, K. M. A. Malik and G. B. C. Vuru, *Inorg Chim Acta*, 1980, **44**, L5.
334. P. A. Fox, S. D. Gray, M. A. Bruck and D. E. Wigley, *Inorg. Chem.*, 1996, **35**, 6027.
335. D. C. Bradley, M. B. Hursthouse, K. M. A. Malik, A. J. Neilson and C. B. C. Vuru, *J. Chem. Soc., Dalton Trans.*, 1984, 1069.

336. (a) S. W. Schweiger, D. L. Tillison, M. G. Thorn, P. E. Fanwick and I. P. Rothwell, *J. Chem. Soc., Dalton Trans.*, 2001, 2401; (b) L. Scoles, K. B. P. Rupper and S. Gambarotta, *J. Am. Chem. Soc.*, 1996, **118**, 2529.
337. M. C. Burland, T. Y. Meyer and S. J. Geib, *Acta Crystallogr. Sect C*, 2003, **C59**, m46.
338. T. C. Jones, A. J. Nielson and C. E. F. Rickard, *J. Chem. Soc., Chem. Commun.*, 1984, 205.
339. A. Merkulov, S. Schmidt, K. Harrms and J. Sundermeyer, *Z. Anorg. Allg. Chem.*, 2005, **631**, 1810.
340. J. Gavenonis and T. D. Tilley, *Organometallics*, 2002, **21**, 5549.
341. S. P. Bloess and J. Nuss, *Z. Krystallogr. – New Cryst. Struct.*, 2006, **221**, 209.
342. A. Baunemann, D. Rische, A. Milanov, Y. Kim, M. Winter, C. Gemil and R. A. Fischer, *Dalton Trans.*, 2005, 3051.
343. R. E. Blake, Jr., D. M. Antonelli, L. M. Henling, W. P. Schaefer, K. I. Hardcastle and J. E. Bercaw, *Organometallics*, 1998, **17**, 718.
344. D. Profilet, P. E. Fanwick and I. P. Rothwell, *Polyhedron*, 1992, **11**, 1559.
345. D. C. Bradley, M. B. Hursthouse, A. J. Howes, A. N. d. M. Jelfs and M. Thorton-Pett, *J. Chem. Soc., Dalton Trans.*, 1991, 841.
346. A. Baunemann, M. Winter, K. Csapek, C. Gemel and R. A. Fischer, *Eur. J. Inorg. Chem.*, 2006, 4665.
347. (a) S. Suh and D. M. Hoffman, *Inorg. Chem.*, 1996, **35**, 5015; (b) F. A. Cotton, L. M. Daniels, C. A. Murillo and X. Wang, *J. Am. Chem. Soc.*, 1996, **118**, 12449.
348. M. H. Chisholm, C. E. Hammond and J. C. Huffman, *J. Chem. Soc., Chem. Commun.*, 1987, 1423.
349. Z. Gebeyehu, F. Weller, B. Neumueller and K. Dehnicke, *Z. Anorg. Allg. Chem.*, 1991, **593**, 99.
350. C. E. Laplaza, M. J. A. Johnson, J. Peters, A. L. Odom, E. Kim, C. C. Cummins, G. N. George and I. J. Pickering, *J. Am. Chem. Soc.*, 1986, **118**, 8623.
351. J. B. Greco, J. C. Peters, T. A. Baker, W. M. Davis, C. C. Cummins and G. Wu, *J. Am. Chem. Soc.*, 2001, **123**, 5003.
352. J. M. Blackwell, J. S. Figueroa, F. H. Stephens and C. C. Cummins, *Organometallics*, 2003, **22**, 3351.
353. J. C. Peters, A. L. Odom and C. C. Cummins, *Chem. Commun.*, 1997, 1995.
354. T. Agapie, P. L. Diaconescu and C. C. Cummins, *J. Am. Chem. Soc.*, 2002, **124**, 2412.
355. C. E. Laplaza, W. M. Davis and C. C. Cummins, *Angew. Chem., Int. Ed.*, 1995, **34**, 2402.
356. J.-P. F. Cherry, A. R. Johnson, L. M. Baraldo, Y.-C. Tsai, C. C. Cummins, S. V. Kryatov, E. V. Rybak-Akimova, K. B. Capps, C. D. Hoff, C. M. Haar and S. P. Nolan, *J. Am. Chem. Soc.*, 2001, **123**, 7271.
357. M. J. A. Johnson, A. L. Odom and C. C. Cummins, *Chem. Commun.*, 1997, 1523.
358. T. Agapie, A. L. Odom and C. C. Cummins, *Inorg. Chem.*, 2000, **39**, 174.
359. W. A. Nugent and R. L. Harlow, *Inorg. Chem.*, 1980, **19**, 777.
360. T. Chen, K. R. Sorasaenee, Z. Wu, J. B. Diminnie and Z. Xue, *Inorg. Chim. Acta*, 2003, **345**, 113.
361. L. N. Bochkarev, Y. E. Begantsova, V. I. Shcherbakov, N. E. Stolyarova, I. K. Grigorieva, I. P. Malysheva, G. V. Basova, A. L. Bochkarev, Y. P. Barinova, G. K. Fukin, E. V. Baranov, Y. A. Kurskii and G. A. Abakumov, *J. Organomet. Chem.*, 2005, **690**, 5720.
362. F. H. Stephens, J. S. Figueroa, P. L. Diaconescu and C. C. Cummins, *J. Am. Chem. Soc.*, 2003, **125**, 9264.
363. D. Fenske, A. Frankenau and K. Dehnicke, *Z. Anorg. Allg. Chem.*, 1989, **574**, 14.
364. S. K. Ignatov, N. H. Rees, S. R. Dubberley, A. G. Razuvaev, P. Mountford and G. I. Nikonov, *Chem. Commun.*, 2004, 952.
365. A. R. Johnson, W. M. Davis, C. C. Cummins, S. Serron, S. P. Nolan, D. G. Musaev and K. Morokuma, *J. Am. Chem. Soc.*, 1998, **120**, 2071.
366. Y.-C. Tsai, M. J. A. Johnson, D. J. Mindiola, C. C. Cummins, W. T. Klooster and T. F. Koetzle, *J. Am. Chem. Soc.*, 1999, **121**, 10426.
367. A. Sinha, R. R. Schrock, P. Müller and A. H. Hoveyda, *Organometallics*, 2006, **25**, 4621.
368. M. J. A. Johnson, P. M. Lee, A. L. Odom, W. M. Davis and C. C. Cummins, *Angew. Chem., Int. Ed.*, 1997, **36**, 87.

369. M. H. Chisholm, C. E. Hammond and J. C. Huffman, *Polyhedron*, 1988, **7**, 399.
370. J. C. Green, R. P. G. Parkin, X. Yan, A. Haaland, W. Scherer and M. A. Tafipolsky, *J. Chem. Soc., Dalton Trans.*, 1997, 3219.
371. C. E. Laplaza and C. C. Cummins, *Science*, 1995, **268**, 861.
372. T. Murahashi, C. R. Clough, J. S. Figueroa and C. C. Cummins, *Angew. Chem., Int. Ed.*, 2005, **44**, 2560.
373. A. L. Odom, P. L. Arnold and C. C. Cummins, *J. Am. Chem. Soc.*, 1998, **120**, 5836.
374. A. Mendiratta, C. C. Cummins, O. P. Kryatova, E. V. Rybak-Akimova, J. E. McDonough and C. D. Hoff, *Inorg. Chem.*, 2003, **42**, 8621.
375. Y.-C. Tsai, F. H. Stephens, K. Meyer, A. Mendiratta, M. D. Gheorghiu and C. C. Cummins, *Organometallics*, 2003, **22**, 2902.
376. D. J. Mindiola, Y.-C. Tsai, R. Hara, Q. Chen, K. Meyer and C. C. Cummins, *Chem. Commun.*, 2001, 125.
377. M. H. Chisholm, C. E. Hammond and J. C. Huffman, *Polyhedron*, 1989, **8**, 129.
378. M. H. Chisholm, C. E. Hammond and J. C. Huffman, *Polyhedron*, 1989, **8**, 1419.
379. K. Most, N. C. Moesch-Zanetti, D. Vidovic and J. Magull, *Organometallics*, 2003, **22**, 5485.
380. T. Adachi, D. L. Hughes, S. K. Ibrahim, S. Okamoto, C. J. Pickett, N. Yabanouchi and T. Yoshida, *J. Chem. Soc., Chem. Commun.*, 1995, 1081.
381. C. Laplaza, A. L. Odom, W. M. Davis, C. C. Cummins and J. D. Protasiewicz, *J. Am. Chem. Soc.*, 1995, **117**, 4999.
382. F. H. Stephens, J. S. Figueroa, C. C. Cummins, O. P. Kryatova, S. V. Kryatov, E. V. Rybak-Akimova, J. E. McDonough and C. D. Hoff, *Organometallics*, 2004, **23**, 3126.
383. M. H. Chisholm, D. A. Haitko, K. Folting and J. C. Huffman, *J. Am. Chem. Soc.*, 1981, **103**, 4046.
384. M. J. Chetcuti, M. H. Chisholm, K. Folting, J. C. Huffman and J. Janos, *J. Am. Chem. Soc.*, 1982, **104**, 4684.
385. M. J. Chetcuti, M. H. Chisholm, K. Folting, D. A. Haitko, J. C. Huffman and J. Janos, *J. Am. Chem. Soc.*, 1983, **105**, 1163.
386. T. W. Coffindaffer, I. P. Rothwell and J. C. Huffman, *Inorg. Chem.*, 1984, **23**, 1433.
387. T. W. Coffindaffer, I. P. Rothwell and J. C. Huffman, *Inorg. Chem.*, 1985, **24**, 1643.
388. R. H. Cayton, M. H. Chisholm, K. Folting, J. L. Wesemann and K. G. Moodley, *J. Chem. Soc., Dalton Trans.*, 1997, 3161.
389. M. H. Chisholm, K. Folting, J. C. Huffman and I. P. Rothwell, *Inorg. Chem.*, 1981, **20**, 1497.
390. M. H. Chisholm, D. A. Haitko, J. C. Huffman and K. Folting, *Inorg. Chem.*, 1981, **20**, 171.
391. M. H. Chisholm, K. Folting, J. C. Huffman and I. P. Rothwell, *Inorg. Chem.*,1981, **20**, 1854.
392. R. G. Abbott, F. A. Cotton and L. R. Falvello, *Inorg. Chem.*, 1990, **29**, 514.
393. M. H. Chisholm, I. P. Parkin, J. C. Huffman, E. M. Lobkovsky and K. Folting, *Polyhedron*, 1991, **10**, 2839.
394. M. H. Chisholm, H. T. Chiu, K. Folting and J. C. Huffman, *Inorg. Chem.*, 1984, **23**, 4097.
395. M. J. Chetcuti, M. H. Chisholm, H. T. Chiu and J. C. Huffman, *J. Am. Chem. Soc.*, 1983, **105**, 1060.
396. W. E. Buhro, M. H. Chisholm, K. Folting and J. C. Huffman, *J. Am. Chem. Soc.*, 1987, **109**, 905.
397. M. H. Chisholm, J. C. Huffman and J. W. Pasterczyk, *Inorg. Chem.*, 1987, **26**, 3781.
398. M. H. Chisholm, I. P. Pankin, W. I. E. Streib and K. S. Folting, *Polyhedron*, 1991, **10**, 2309.
399. M. H. Chisholm, J.-H. Huang, J. C. Huffman and I. P. Parkin, *Inorg. Chem.*, 1997, **36**, 1642.
400. M. H. Chisholm, K. Folting, W. E. Streib and D.-D. Wu, *Inorg. Chem.*, 1998, **37**, 50.
401. M. H. Chisholm, D. A. Haitko, J. C. Huffman and K. Folting, *Inorg. Chem.*, 1981, **20**, 2211.
402. P. Schollhammer, F. Y. Petillon, S. Poder-Guillou, J. Y. Saillard, J. Talarmin and K. W. Muir, *Chem. Commun.*, 1996, 2633.
403. A. Wlodarczyk, A. J. Edwards and J. A. McCleverty, *Polyhedron*, 1988, **7**, 103.
404. F. Liang, H. W. Schmalle and H. Berke, *Inorg. Chem.*, 2004, **43**, 993.
405. K. Hagen, C. J. Holwill, D. A. Rice and J. D. Runnacles, *Acta Chem. Scand.*, 1988, **A42**, 578.

406. D. M. Berg and P. R. Sharp, *Inorg. Chem.*, 1987, **26**, 2959.
407. M. H. Chisholm, T. A. Budzichowski, F. J. Feher and J. Ziller, *Polyhedron*, 1992, **11**, 1575.
408. S. Dietz, V. Allured and M. R. DuBois, *Inorg. Chem.*, 1993, **32**, 5418.
409. (a) C. R. Clough, J. B. Greco, J. S. Figueroa, P. L. Diaconescu, W. M. Davis and C. C. Cummins, *J. Am. Chem. Soc.*, 2004, **126**, 7742; (b) A. R. Fox, C. R. Clough, N. A. Piro and C. C. Cummins, *Angew. Chem., Int. Ed.*, 2007, **46**, 973.
410. D. Doufou, K. A. Abboud and J. M. Boncella, *Inorg. Chim. Acta*, 2003, **345**, 103.
411. H.-T. Chiu, S.-H. Chuang, G.-H. Lee and S.-M. Peng, *Polyhedron*, 1994, **13**, 2443.
412. A. J. Nielson, G. R. Clark and C. E. F. Rickard, *Aust. J. Chem.*, 1997, **50**, 259.
413. S. M. Holmes, D. F. SchaferII P. T. Wolczanski and E. B. Lobkovsky, *J. Am. Chem. Soc.*, 2001, **123**, 10571.
414. B. Schwarze, W. Milius and W. Schnick, *Chem. Ber.*, 1997, **130**, 701.
415. K. J. Ahmed, M. H. Chisholm, K. Folting and J. C. Huffman, *J. Am. Chem. Soc.*, 1986, **108**, 989.
416. K. J. Ahmed, M. H. Chisholm and J. C. Huffman, *Organometallics*, 1985, **4**, 1312.
417. F. A. Cotton, E. V. Dikarev and W.-Y. Wong, *Polyhedron*, 1997, **16**, 3893.
418. M. H. Chisholm, M. J. Hampden-Smith, J. C. Huffman and K. G. Moodley, *J. Am. Chem. Soc.*, 1988, **110**, 4070.
419. R. H. Cayton, M. H. Chisholm, M. J. Hampden-Smith, J. C. Huffman and K. G. Moodley, *Polyhedron*, 1992, **11**, 3197.
420. R. H. Cayton, M. H. Chisholm, K. Folting, J. L. Wesermann and K. G. Moodley, *J. Chem. Soc., Dalton Trans.*, 1997, 3161.
421. M. H. Chisholm, B. W. Eichorn, K. Folting and J. C. Huffman, *Inorg. Chim. Acta*, 1988, **144**, 193.
422. M. H. Chisholm, I. P. Parkin and J. C. Huffman, *Polyhedron*, 1997, **10**, 1215.
423. M. H. Chisholm, G. J. Gama and I. P. Parkin, *Polyhedron*, 1993, **12**, 961.
424. W. E. Buhro, M. H. Chisholm, K. Folting, J. C. Huffman, J. D. Martin and W. E. Streib, *J. Am. Chem. Soc.*, 1992, **114**, 557.
425. W. E. Buhro, M. H. Chisholm, K. Folting, J. C. Huffman, J. D. Martin and W. E. Streib, *J. Am. Chem. Soc.*, 1988, **110**, 6563.
426. T. A. Budzichowski, M. H. Chisholm, D. B. Tiedtke, N. E. Gruhn and D. L. Lichtenberger, *Polyhedron*, 1998, **17**, 705.
427. H. W. Roesky, N. Bertel, F. Edelmann, R. Herbst, E. Egert and G. M. Sheldrick, *Z. Naturforsch.*, 1986, **41B**, 1506.
428. M. H. Chisholm, K. Folting, S. T. Haubrich and J. D. Martin, *Inorg. Chim. Acta*, 1993, **213**, 17.
429. M. H. Chisholm, J. C. Gallucci and C. B. Hollandsworth, *J. Organomet. Chem.*, 2006, **691**, 93.
430. (a) M. H. Chisholm, S. T. Haubrich, J. C. Huffman and W. E. Streib, *J. Am. Chem. Soc.*, 1997, **119**, 1634; (b) M. H. Chisholm, K. S. Kramer, J. D. Martin, J. C. Huffman, E. B. Lobkovsky and W. E. Streib, *Inorg. Chem.*, 1992, **31**, 4469.
431. O. Coutelier, R. M. Gauvin, G. Nowogrocki, J. Trébosc, L. Delevoye and A. Mortreux, *Eur. J. Inorg. Chem.*, 2007, 5541.
432. M. H. Chisholm, R. M. Jansen and J. C. Huffman, *Organometallics*, 1992, **11**, 2305.
433. K. J. Ahmed, M. H. Chisholm, K. Folting and J. C. Huffman, *Inorg. Chem.*, 1985, **24**, 4039.
434. D. C. Bradley, M. B. Hursthouse and H. R. Powell, *J. Chem. Soc., Dalton Trans.*, 1989, 1537.
435. F. A. Cotton, E. V. Dikarev and W.-Y. Wong, *Inorg. Chem.*, 1997, **36**, 80.
436. T. A. Budzichowski, M. H. Chisholm, D. B. Tiedkte, J. C. Huffman and W. E. Streib, *Organometallics*, 1995, **14**, 2318.
437. M. H. Chisholm, E. F. Putilina, K. Folting and W. E. Streib, *J. Cluster Sci.*, 1994, **5**, 67.
438. W. E. Buhro, M. H. Chisholm, J. D. Martin, J. C. Huffman, K. Folting and W. E. Streib, *J. Am. Chem. Soc.*, 1989, **111**, 8149.
439. D. C. Bradley, H. M. Dawes, M. B. Hursthouse and H. R. Powell, *Polyhedron*, 1988, **7**, 2049.
440. P. J. Perez, P. S. White, M. Brookhart and J. L. Templeton, *Inorg. Chem.*, 1994, **33**, 6050.

441. P. J. Perez, L. Luan, P. White, M. Brookhart and J. L. Templeton, *J. Am. Chem. Soc.*, 1992, **114**, 7928.
442. P. G. Edwards, G. Wilkinson, M. B. Hursthouse and K. M. A. Malik, *J. Chem. Soc., Dalton Trans.*, 1980, 2467.
443. C. S. Masui and J. M. Mayer, *Inorg. Chim. Acta*, 1996, **251**, 325.
444. M. A. Dewey, A. M. Arif and J. M. Gladysz, *J. Chem. Soc., Chem. Commun.*, 1991, 712.
445. M. A. Dewey, D. A. Knight, A. Arif and J. M. Gladysz, *Chem. Ber.*, 1992, **125**, 815.
446. E. Hevia, J. Perez, V. Riera and D. Miguel, *Organometallics*, 2002, **21**, 1966.
447. E. Hevia, J. Perez, V. Riera and D. Miguel, *Organometallics*, 2003, **22**, 257.
448. M. Jahncke, A. Neels, H. Stoeckli-Evans and G. Süss-Fink, *J. Organomet. Chem.*, 1998, **565**, 97.
449. A. W. Kaplan, J. C. M. Ritter and R. G. Bergman, *J. Am. Chem. Soc.*, 1998, **120**, 6828.
450. G. C. Martin, G. J. Palenik and J. M. Boncella, *Inorg. Chem.*, 1990, **29**, 2027.
451. J. F. Hartwig, R. A. Andersen and R. G. Bergman, *Organometallics*, 1991, **10**, 1875.
452. R. E. Blake, Jr., R. H. Heyn and T. D. Tilley, *Polyhedron*, 1992, **11**, 709.
453. A. K. Burrell and A. J. Steedman, *Organometallics*, 1997, **16**, 1203.
454. J. M. Boncella, T. M. Eve, B. Rickman and K. A. Abboud, *Polyhedron*, 1998, **17**, 725.
455. K. N. Jayaprakash, T. B. Gunnoe and P. D. Boyle, *Inorg. Chem.*, 2001, **40**, 6481.
456. D. Conner, K. N. Jayaprakash, T. B. Gunnoe and P. D. Boyle, *Inorg. Chem.*, 2002, **41**, 3042.
457. D. Conner, K. N. Jayaprakash, T. B. Gunnoe and P. D. Boyle, *Organometallics*, 2002, **21**, 5265.
458. A. W. Holland and R. G. Bergman, *J. Am. Chem. Soc.*, 2002, **124**, 14684.
459. T. J. Crevier and J. M. Mayer, *J. Am. Chem. Soc.*, 1998, **120**, 5595.
460. A. G. Maestri, K. S. Cherry, J. J. Toboni and S. N. Brown, *J. Am. Chem. Soc.*, 2001, **123**, 7459.
461. T. Daniel, N. Mahr and H. Werner, *Chem. Ber.*, 1993, **126**, 1403.
462. W.-T. Wong and T.-S. Wong, *J. Organomet. Chem.*, 1997, **542**, 29.
463. V. A. Maksakov, I. V. Slovokhotova, A. V. Virovets, S. P. Babailov, V. P. Kirin and N. V. Podberezskaya, *Russ. J. Coord. Chem.*, 1998, **24**, 501.
464. H. Matsuzaka, T. Kamura, K. Ariga, Y. Watanabe, T. Okubo, T. Ishii, M. Yamashita, M. Kondo and S. Kitagawa, *Organometallics*, 2000, **19**, 216.
465. J.-J. Brunet, J.-C. Daran, D. Neibecker and L. Rosenberg, *J. Organomet. Chem.*, 1997, **538**, 251.
466. (a) C. Tejel, M. A. Ciriano, M. Bordonaba, J. A. Lopez, F. J. Lahoz and L. A. Oro, *Chem. Eur. J.*, 2002, **8**, 3128; (b) C. Tejel, M. A. Ciriano, J. A. Lopez, S. Jimmez, M. Bordonaba and L. A. Oro, *Chem. Eur. J.*, 2007, **13**, 2044.
467. P. Zhao, C. Krug and J. F. Hartwig, *J. Am. Chem. Soc.*, 2005, **127**, 12066.
468. Y.-W. Ge, F. Peng and P. R. Sharp, *J. Am. Chem. Soc.*, 1990, **112**, 2632.
469. (a) C. Tejel, M. A. Ciriano, M. Bordonaba, J. A. Lopez, F. J. Lahoz and L. A. Oro, *Inorg. Chem.*, 2002, **41**, 2348; (b) K. Ishwata, S. Kuwata and T. Ikariya, *Dalton Trans.*, 2007, 3606.
470. D. Rais and R. G. Bergman, *Chem. Eur. J.*, 2004, **10**, 3970.
471. J. Zhao, A. S. Goldman and J. F. Hartwig, *Science*, 2005, **307**, 1080.
472. M. Kanzelberger, X. Zhang, T. J. Emge, A. S. Goldman, J. Zhao, C. Incarvito and J. F. Hartwig, *J. Am. Chem. Soc.*, 2003, **125**, 13644.
473. A. L. Casalnuovo, J. C. Calabrese and D. Milstein, *Inorg. Chem.*, 1987, **26**, 971.
474. F. T. Ladipo and J. S. Merola, *Inorg. Chem.*, 1990, **29**, 4172.
475. A. W. Kaplan, J. C. M. Ritter and R. G. Bergman, *J. Am. Chem. Soc.*, 1998, **120**, 6828.
476. M. K. Kolel-Veetil, J. F. Curley, P. R. Yadav and K. J. Ahmed, *Polyhedron*, 1994, **13**, 919.
477. H. Matzuzaka, K. Ariga, H. Kase, T. Kamura, M. Kondo, S. Kitagawa and M. Yamasaki, *Organometallics*, 1997, **16**, 4514.
478. M. K. Kolel-Veetil, A. L. Rheingold and K. J. Ahmed, *Organometallics*, 1993, **12**, 3439.
479. M. D. Fryzuk and K. Joshi, *Organometallics*, 1989, **8**, 722.
480. M. K. Kolel-Veetil, M. Rahim, A. J. Edwards, A. L. Rheingold and K. J. Ahmed, *Inorg. Chem.*, 1992, **31**, 3877.
481. H. Von Armin, W. Massa, A. Zinn and K. Dehnicke, *Z. Naturforsch.*, 1991, **46B**, 992.

482. L. A. Villaneuva, K. A. Abboud and J. M. Boncella, *Organometallics*, 1994, **13**, 3921.
483. J. Ruiz, M. T. Martinez, C. Vincente, G. Garcia, G. Lopez, P. A. Chaloner and P. B. Hitchcock, *Organometallics*, 1993, **12**, 4321.
484. L. Barloy, R. M. Gauvin, J. A. Osborn, C. Sizun, R. Graff and N. Kyritsakas, *Eur. J. Inorg. Chem.*, 2001, 1699.
485. J. Ruiz, V. Rodriguez, G. Lopez, J. Casabo, E. Molins and C. Miravitlles, *Organometallics*, 1999, **18**, 1177.
486. G. Sanchez, J. L. Serrano, J. Garcia, G. Lopez, J. Perez and E. Molins, *Inorg. Chim. Acta*, 1999, **287**, 37.
487. S. Kannan, A. J. James and P. R. Sharp, *Inorg. Chim. Acta*, 2003, **345**, 8.
488. S. Kannan, A. J. James and P. R. Sharp, *Polyhedron*, 2000, **19**, 155.
489. G. Mann and J. F. Hartwig, *J. Am. Chem. Soc.*, 1996, **118**, 13109.
490. M. W. Hooper and J. F. Hartwig, *Organometallics*, 2003, **22**, 3394.
491. M. Yamashita and J. F. Hartwig, *J. Am. Chem. Soc.*, 2004, **126**, 5344.
492. Y.-J. Kim, J.-C. Choi and K. Osakada, *J. Organomet. Chem.*, 1995, **491**, 97.
493. M. Kretschmer and L. Heck, *Z. Anorg. Allg. Chem.*, 1982, **490**, 215.
494. L. Heck, M. Ardon, A. Bino and J. Zaap, *J. Am. Chem. Soc.*, 1988, **110**, 2691.
495. N. W. Alcock, P. Bergamini, T. J. Kemp, P. G. Pringle, S. Sostero and O. Traverso, *Inorg. Chem.*, 1991, **30**, 1594.
496. C. A. O'Mahoney, I. P. Parkin, D. J. Williams and J. D. Woollins, *Polyhedron*, 1989, **8**, 1979.
497. A. C. Albeniz, V. Calle, P. Espinet and S. Gomez, *Inorg. Chem.*, 2001, **40**, 4211.
498. S. Kannan, A. J. James and P. R. Sharp, *Polyhedron*, 2000, **19**, 155.
499. D. T. Eadie, A. Pidcock, S. R. Stobart, E. T. Brennan and T. S. Cameron, *Inorg. Chim. Acta*, 1982, **65**, L111.
500. P. B. Hitchcock, M. F. Lappert and L. J. M. Pierssens, *Chem. Commun.*, 1996, 1189.
501. P. Reiss and D. Fenske, *Z. Anorg. Allg. Chem.*, 2000, **626**, 2245.
502. M. A. Cinellu, G. Minghetti, M. V. Pinna, S. Stoccoro, A. Zucca and M. Manassero, *Eur. J. Inorg. Chem.*, 2003, 2304.
503. U. Grässle and J. Strähle, *Z. Anorg. Allg. Chem.*, 1985, **531**, 26.
504. H. N. Adams, U. Grässle, W. Hiller and J. Strähle, *Z. Anorg. Allg. Chem.*, 1983, **504**, 7.
505. S. D. Bunge, O. Just and W. S. Rees, Jr., *Angew. Chem., Int. Ed.*, 2000, **39**, 3082.
506. G. L. Wegner, A. Jockisch, A. Schier and H. Schmidbaur, *Z. Naturforsch.*, 2000, **55B**, 347.
507. O. C. Dermer and W. C. Fernelius, *Z. Anorg. Allg. Chem.*, 1934, **221**, 83.
508. A. L. Casalnuova, J. C. Calabrese and D. Milstein, *J. Am. Chem. Soc.*, 1988, **110**, 6738.
509. S. Park, A. L. Rheingold and D. M. Roundhill, *Organometallics*, 1991, **10**, 615.
510. M. G. Vale and R. R. Schrock, *Organometallics*, 1991, **10**, 1661.
511. K. N. Jayaprakash, D. Conner and T. B. Gunnoe, *Organometallics*, 2001, **20**, 5254.
512. D. Conner, K. M. Jayaprakash, T. R. Cundari and T. B. Gunnoe, *Organometallics*, 2004, **23**, 2724.
513. T. Büttner, J. Geier, G. Frisson, J. Harmer, C. Calle, A. Schweiger, H. Schönberg and H. Grützmacher, *Science*, 2005, **307**, 235.
514. D. C. Bradley, M. H. Chisholm, C. E. Heath and M. B. Hursthouse, *J. Chem. Soc., Chem. Commun.*, 1969, 1261.
515. V. Schomaker and D. P. Stevenson, *J. Am. Chem. Soc.*, 1941, **63**, 37.
516. R. Blom and A. Haaland, *J. Mol. Struct.*, 1985, **128**, 21.
517. S. J. Brois, *Trans. NY Acad Sci.*, 1969, **31**, 931.
518. K. Mislow, *Trans NY Acad Sci.*, 1973, **35**, 227.
519. J. C. Green, M. Payne, E. A. Seddon and R. A. Andersen, *J. Chem. Soc., Dalton Trans.*, 1982, 887.
520. M. H. Chisholm, A. H. Cowley and M. Lattman, *J. Am. Chem. Soc.*, 1980, **102**, 46.
521. B. E. Bursten, F. A. Cotton, J. C. Green, E. A. Seddon and G. G. Stanley, *J. Am. Chem. Soc.*, 1980, **102**, 4579.

522. (a) R. A. Andersen, D. B. Beach and W. L. Jolly, *Inorg. Chem.*, 1985, **24**, 4741; (b) D. B. Beach and W. L. Jolly, *Inorg. Chem.*, 1986, **25**, 875.
523. A. K. Baev and V. E. Mikhailov, *Russ. J. Phys. Chem.*, 1989, **63**, 949.
524. M. F. Lappert, D. S. Patil and J. B. Pedley, *J. Chem. Soc., Chem. Commun.*, 1975, 830.
525. F. A. Adedeji, K. J. Cavell, S. Cavell, J. A. Connor, G. Pilcher, H. A. Skinner and M. T. Zafarani-Moattar, *Faraday Trans. 1*, 1979, **75**, 603.
526. J. A. Connor, G. Pilcher, H. A. Skinner, M. H. Chisholm and F. A. Cotton, *J. Am. Chem. Soc.*, 1978, **100**, 7738.
527. D. C. Bradley and M. J. Hillyer, *Trans. Faraday Soc.*, 1966, **62**, 2374.
528. L. E. Schock and T. J. Marks, *J. Am. Chem. Soc.*, 1988, **110**, 7701.
529. A. A. Palacios, P. Alemany and S. Alvarez, *Inorg. Chem.*, 1999, **38**, 707.
530. S. L. Lee, F. Y. Li and T. W. Shia, *Bull. Inst. Chem. Acad. Sinica*, 1992, **39**, 69.
531. D. C. Bradley and M. H. Chisholm, *Acc. Chem. Res.*, 1976, **9**, 273.
532. S. A. Macgregor, *Organometallics*, 2001, **20**, 1860.
533. H. Jacobs and D. Peters, *J. Less Common Met.*, 1986, **119**, 99.
534. H. Jacobs, D. Peters and K. M. Hassiepen, *J. Less Common Met.*, 1986, **118**, 31.
535. H. Jacobs and T. J. Hennig, *Z. Anorg. Allg. Chem.*, 1998, **624**, 1823.
536. B. Frohling, G. Kreiner and H. Jacobs, *Z. Anorg. Allg. Chem.*, 1999, **625**, 211.
537. A. Tenten and H. Jacobs, *Z. Anorg. Allg. Chem.*, 1991, **604**, 113.
538. A. Tenten and H. Jacobs, *J. Alloys Compd.*, 1991, **177**, 193.
539. A. Tenten and H. Jacobs, *J. Less. Common Met.*, 1991, **170**, 145.
540. P. Andersen, H. Bengaard, J. Glerup, A. Gumm and S. Larsen, *Acta Chem. Scand.*, 1998, **52**, 1313.
541. H. Lueken, H. Schilder, H. Jacobs and U. Zachwieja, *Z. Anorg. Allg. Chem.*, 1995, **621**, 959.
542. M. T. Flood, R. F. Ziolo, J. E. Early and H. B. Gray, *Inorg. Chem.*, 1973, **12**, 2153.
543. G. Hillhouse and J. E. Bercaw, *J. Am. Chem. Soc.*, 1984, **106**, 5472.
544. V. V. Bekman and R. A. Goloseeva, *Zh. Neorg. Khim.*, 1981, **26**, 541.
545. J. R. Fulton, S. Sklenak, M. W. Bouwkamp and R. G. Bergman, *J. Am. Chem. Soc.*, 2002, **124**, 4722.
546. J. Haggin, *Chem. Eng. News*, 1993, **71**, 23.
547. D. C. Bradley, *Chem. Brit.*, 1975, **11**, 393.
548. P. G. Eller, D. C. Bradley, M. B. Hursthouse and D. W. Meek, *Coord. Chem. Rev.*, 1977, **24**, 1.
549. E. C. Alyea, J. S. Basi, D. C. Bradley and M. H. Chisholm, *Chem. Commun.*, 1968, 495.
550. H. Bürger and U. Wannagat, *Monatsh. Chem.*, 1964, **95**, 1099.
551. H. Bürger and U. Wannagat, *Monatsh. Chem.*, 1963, **94**, 1007.
552. H.-O. Fröhlich, *Z. Chem.*, 1980, **20**, 108.
553. H.-O. Fröhlich and W. Römhild, *Z. Chem.*, 1980, **20**, 154.
554. H.-O. Fröhlich and H. Francke, *Z. Chem.*, 1988, **28**, 413.
555. H.-O. Fröhlich and W. Roemhild, *Z. Chem.*, 1979, **19**, 414.
556. V. Brito, H.-O. Fröhlich and B. Müller, *Z. Chem.*, 1979, **19**, 28.
557. D. R. Manke and D. G. Nocera, *Inorg. Chem.*, 2003, **42**, 4431.
558. D. R. Manke and D. G. Nocera, *Inorg. Chim Acta*, 2003, **345**, 235.
559. T. H. Warren, R. R. Schrock and W. M. Davis, *Organometallics*, 1998, **17**, 308.
560. J. T. Patton, S. G. Feng and K. A. Abboud, *Organometallics*, 2001, **20**, 3399.
561. D. C. Bradley, M. B. Hursthouse, C. W. Newing and A. J. Welch, *J. Chem. Soc., Chem. Commun.*, 1972, 567.
562. K. J. Fisher, *Inorg. Nucl. Chem. Lett.*, 1973, **9**, 921.
563. G. S. Sigel, R. A. Bartlett, D. Decker, M. M. Olmstead and P. P. Power, *Inorg. Chem.*, 1987, **26**, 1773.
564. T. A. Chesnokova, E. V. Zhezlova, A. N. Kornev, Y. V. Fedotova, L. N. Zakharov, G. K. Fukin, Y. A. Kursky, T. G. Mushtina and G. A. Domrachev, *J. Organomet. Chem.*, 2002, **642**, 20.

565. M. M. Olmstead, G. Sigel, H. Hope, X. Xu and P. P. Power, *J. Am. Chem. Soc.*, 1985, **107**, 8087.
566. M. M. Olmstead, P. P. Power and G. A. Sigel, *Inorg. Chem.*, 1988, **27**, 580.
567. P. P. Power and S. C. Shoner, *Angew. Chem., Int. Ed.*, 1991, **30**, 330.
568. D. E. Gindelberger and J. Arnold, *Inorg. Chem.*, 1993, **32**, 5813.
569. C. E. Laplaza, A. R. Johnson and C. C. Cummins, *J. Am. Chem. Soc.*, 1996, **118**, 709.
570. J. K. Brask, V. Dura-Vila, P. L. Diaconescu and C. C. Cummins, *Chem. Commun.*, 2002, 902.
571. B. W. Howk, E. L. Little, S. L. Scott and G. M. Whitman, *J. Am. Chem. Soc.*, 1954, **76**, 1899.
572. G. P. Pez and J. E. Galle, *Pure Appl. Chem.*, 1985, **57**, 1917.
573. P. J. Walsh, A. M. Baranger and R. G. Bergman, *J. Am. Chem. Soc.*, 1992, **114**, 1708.
574. A. M. Baranger, P. J. Walsh and R. G. Bergman, *J. Am. Chem. Soc.*, 1993, **115**, 2753.
575. S. Y. Lee and R. G. Bergman, *Tetrahedron*, 1995, **51**, 4255.
576. R. L. Zuckerman, S. W. Krska and R. G. Bergman, *J. Organomet. Chem.*, 2000, **608**, 172.
577. R. L. Zuckerman and R. G. Bergman, *Organometallics*, 2000, **19**, 4795.
578. L. L. Anderson, J. Arnold and R. G. Bergman, *J. Am. Chem. Soc.*, 2005, **127**, 14542.
579. D. A. Watson, M. Chiu, R. G. Bergman and *Organometallics*, 2006, **25**, 4731.
580. T. E. Müller and M. Beller, *Chem. Rev.*, 1998, **98**, 675.
581. R. T. Ruck, R. L. Zuckerman, S. W. Krska and R. G. Bergman, *Angew. Chem., Int. Ed.*, 2004, **43**, 5372.
582. F. Basuli, H. Aneetha, J. C. Huffman and D. J. Mindiola, *J. Am. Chem. Soc.*, 2005, **127**, 17992.
583. H. Aneetha, F. Basuli, J. Bollinger, J. C. Huffman and D. J. Mindiola, *Organometallics*, 2006, **25**, 2402.
584. M. D. Fryzuk, J. B. Love, S. J. Rettig and V. G. Young, *Science*, 1997, **275**, 1445.
585. M. D. Fryzuk, J. B. Love and S. J. Rettig, *Chem. Commun.*, 1996, 2783.
586. M. D. Fryzuk, J. B. Love and S. J. Rettig, *Organometallics*, 1998, **17**, 846.
587. (a) M. D. Fryzuk, T. S. Haddad, M. Mylvaganam, D. H. McConville and S. J. Rettig, *J. Am. Chem. Soc.*, 1993, **115**, 2782; (b) E. A. MacLachlan, F. M. Hess, B. O. Patrick and M. D. Fryzuk, *J. Am. Chem. Soc.*, 2007, **129**, 10895.
588. (a) L.-C. Liang, J.-M. Lin and C.-H. Hung, *Organometallics*, 2003, **22**, 3007; (b) M. J. Ingelson, M. Pink, H. Fan and K. G. Caulton, *Inorg. Chem.*, 2007, **46**, 10321.
589. L. Fan, B. M. Foxman and O. V. Ozerov, *Organometallics*, 2004, **23**, 326.
590. O. V. Ozerov, C. Guo, V. A. Papkov and B. M. Foxman, *J. Am. Chem. Soc.*, 2004, **126**, 4792.
591. (a) B. C. Bailey, J. C. Huffman, D. J. Mindiola, W. Weng and O. V. Ozerov, *Organometallics*, 2005, **24**, 1390; (b) J. C. DeMott, F. Basuli, U. J. Kilgore, B. M. Foxman, J. C. Huffman, O. V. Ozerov and D. J. Mindiola, *Inorg. Chem.*, 2007, **46**, 6271.
592. L. Fan and O. V. Ozerov, *Chem. Commun.*, 2005, 4450.
593. W. Weng, C. Guo, R. Celenligil-Cetin, B. M. Foxman and O. V. Ozerov, *Chem. Commun.*, 2006, 197.
594. L. H. Gade and N. Mahr, *J. Chem. Soc., Dalton Trans.*, 1993, 489.
595. H. Memmler, L. H. Gade and J. W. Lauher, *Inorg. Chem.*, 1994, **33**, 3064.
596. M. Schubart, B. Findeis, L. H. Gade, W.-S. Li and M. McPartlin, *Chem. Ber.*, 1995, **128**, 329.
597. M. Veith, O. Schutt and V. Huch, *Z. Anorg. Allg. Chem.*, 1999, **625**, 1155.
598. S. Friedrich, M. Schubart, L. H. Gade, I. J. Scowen, A. J. Edwards and M. McPartlin, *Chem. Ber.*, 1997, **130**, 1751.
599. A. Bashall, P. E. Collier, L. H. Gade, M. McPartlin, P. Mountford, S. M. Pugh, S. Radojevic, M. Schubart, I. J. Scowen and D. J. M. Trösch, *Organometallics*, 2000, **19**, 4784.
600. S. M. Pugh, D. J. M. Trösch, D. J. Wilson, A. Bashall, F. G. N. Cloke, L. H. Gade, P. B. Hitchcock, M. McPartlin, J. F. Nixon and P. Mountford, *Organometallics*, 2000, **19**, 3205.
601. D. J. M. Trösch, P. E. Collier, A. Bashall, L. H. Gade, M. McPartlin, P. Mountford and S. Radojevic, *Organometallics*, 2001, **20**, 3308.
602. B. D. Ward, G. Orde, E. Clot, A. R. Cowley, L. H. Gade and P. Mountford, *Organometallics*, 2005, **24**, 2368.

603. S. Friedrich, H. Memmler, L. H. Gade, W.-S. Li and M. McPartlin, *Angew. Chem., Int. Ed.*, 1994, **33**, 676.
604. H. Memmler, U. Kauper, L. H. Gade, I. J. Scowen and M. McPartlin, *Chem. Commun.*, 1996, 1751.
605. B. Findeis, M. Schubart, C. Platzek, L. H. Gade, I. Scowen and M. McPartlin, *Chem. Commun.*, 1996, 219.
606. A. Schneider, L. H. Gade, M. Breuning, G. Bringmann, I. J. Scowen and M. McPartlin, *Organometallics*, 1998, **17**, 1643.
607. P. Mehrkhodavandi, P. J. Bonitatebus Jr. and R. R. Schrock, *J. Am. Chem. Soc.*, 2000, **122**, 7841.
608. P. Mehrkhodavandi and R. R. Schrock, *J. Am. Chem. Soc.*, 2001, **123**, 10746.
609. P. Mehrkhodavandi, R. R. Schrock and L. L. Pryor, *Organometallics*, 2003, **22**, 4569.
610. M. Tayebani, S. Conoci, K. Feghali, S. Gambarotta and G. P. A. Yap, *Organometallics*, 2000, **19**, 4568.
611. J. Scott, S. Gambarotta, G. Yap and D. G. Rancourt, *Organometallics*, 2003, **22**, 2325.
612. P. Crewdson, S. Gambarotta, G. P. A. Yap and L. K. Thompson, *Inorg. Chem.*, 2003, **42**, 8579.
613. C. C. Cummins, J. Lee, R. R. Schrock and W. D. Davis, *Angew. Chem., Int. Ed.*, 1992, **31**, 1501.
614. A. C. Filippou, S. Schneider and G. Schnakenburg, *Inorg. Chem.*, 2003, **42**, 6974.
615. C. C. Cummins, R. R. Schrock and W. M. Davis, *Inorg. Chem.*, 1994, **33**, 1448.
616. V. Christou and J. Arnold, *Angew. Chem., Int. Ed.*, 1993, **32**, 1450.
617. J. S. Freundlich, R. R. Schrock and W. M. Davis, *Organometallics*, 1996, **15**, 2777.
618. J. S. Freundlich, R. R. Schrock, C. C. Cummins and W. M. Davis, *J. Am. Chem. Soc.*, 1994, **116**, 6476.
619. C. C. Cummins, R. R. Schrock and W. M. Davis, *Angew. Chem., Int. Ed.*, 1993, **32**, 756.
620. J. S. Freundlich, R. R. Schrock and W. M. Davis, *J. Am. Chem. Soc.*, 1996, **118**, 3643.
621. K.-Y. Shih, R. R. Schrock and R. Kempe, *J. Am. Chem. Soc.*, 1994, **116**, 8804.
622. K.-Y. Shih, K. Totland, S. W. Seidel and R. R. Schrock, *J. Am. Chem. Soc.*, 1994, **116**, 12103.
623. R. R. Schrock, S. W. Seidel, N. C. Moesch-Zanetti, D. A. Dobbs, K.-Y. Shih and W. M. Davis, *Organometallics*, 1997, **16**, 5195.
624. N. C. Zanetti, R. R. Schrock and W. M. Davis, *Angew. Chem., Int. Ed.*, 1995, **34**, 2044.
625. N. C. Moesch-Zanetti, R. R. Schrock, W. M. Davis, K. Wanninger, S. W. Seidel and M. B. O'Donoghue, *J. Am. Chem. Soc.*, 1997, **119**, 11037.
626. M. B. O'Donoghue, N. C. Zanetti, W. M. Davis and R. R. Schrock, *J. Am. Chem. Soc.*, 1997, **119**, 2753.
627. M. Kol, R. R. Schrock, R. Kempe and W. M. Davis, *J. Am. Chem. Soc.*, 1994, **116**, 4382.
628. M. B. O'Donoghue, W. M. Davis and R. R. Schrock, *Inorg. Chem.*, 1998, **37**, 5149.
629. D. V. Yandulov and R. R. Schrock, *J. Am. Chem. Soc.*, 2002, **124**, 6252.
630. D. V. Yandulov, R. R. Schrock, A. L. Rheingold, C. Ceccarelli and W. M. Davis, *Inorg. Chem.*, 2003, **42**, 796.
631. D. V. Yandulov and R. R. Schrock, *Science*, 2003, **301**, 76.
632. D. V. Yandulov and R. R. Schrock, *Inorg. Chem.*, 2005, **44**, 1103.
633. (a) V. Ritleng, D. V. Yandulov, W. W. Weare, R. R. Schrock, A. S. Hock and W. M. Davis, *J. Am. Chem. Soc.*, 2004, **126**, 6150; (b) D. V. Yandulov and R. R. Schrock, *Can. J. Chem.*, 2005, **83**, 341.
634. K. M. Wampler and R. R. Schrock, *Inorg. Chem.*, 2007, **46**, 8463.
635. S. M. Pugh, A. J. Blake, L. H. Gade and P. Mountford, *Inorg. Chem.*, 2001, **40**, 3992.
636. A. J. Blake, P. E. Collier, L. H. Gade, M. McPartlin, P. Mountford, M. Schubart and I. J. Scowen, *Chem. Commun.*, 1997, 1555.
637. C. C. Cummins and R. R. Schrock, *Inorg. Chem.*, 1994, **33**, 395.
638. M. M. Olmstead, P. P. Power and M. Viggiano, *J. Am. Chem. Soc.*, 1983, **105**, 2927.
639. O. Schlager, K. Wieghardt and B. Nuber, *Inorg. Chem.*, 1995, **34**, 6456.
640. T. J. Collins, C. Slebodnick and E. S. Uffelman, *Inorg. Chem.*, 1990, **29**, 3433.

641. T. J. Collins, *Acc. Chem. Res.*, 1994, **27**, 279.
642. A. Ghosh, F. Tiago de Oliveira, T. Yano, T. Nishioka, E. S. Beach, I. Kinoshita, E. Muenck, A. D. Ryabov, C. P. Horwitz and T. J. Collins, *J. Am. Chem. Soc.*, 2005, **127**, 2505.
643. F. G. N. Cloke, P. B. Hitchcock and J. B. Love, *J. Chem. Soc., Dalton Trans.*, 1995, 25.
644. H. C. S. Clark, F. G. N. Cloke, P. B. Hitchcock, J. B. Love and A. P. Wainwright, *J. Organomet. Chem.*, 1995, **501**, 333.
645. J. B. Love, H. C. S. Clark, F. G. N. Cloke, J. C. Green and P. B. Hitchcock, *J. Am. Chem. Soc.*, 1999, **121**, 6843.
646. L.-C. Liang, R. R. Schrock, W. M. Davis and D. H. McConville, *J. Am. Chem. Soc.*, 1999, **121**, 5797.
647. F. J. Schattenmann, R. R. Schrock and W. M. Davis, *Organometallics*, 1998, **17**, 989.
648. R. R. Schrock, R. Baumann, S. M. Reid, J. T. Goodman, R. Stumpf and W. M. Davis, *Organometallics*, 1999, **18**, 3649.
649. R. Baumann and R. R. Schrock, *J. Organomet. Chem.*, 1998, **557**, 69.
650. J. T. Goodman and R. R. Schrock, *Organometallics*, 2001, **20**, 5205.
651. R. R. Schrock, A. L. Casado, J. T. Goodman, L.-C. Liang, P. J. Bonitatebus, Jr. and W. M. Davis, *Organometallics*, 2000, **19**, 5325.
652. R. R. Schrock, P. J. Bonitatebus, Jr., and Y. Schrodi, *Organometallics*, 2001, **20**, 1056.
653. R. R. Schrock, J. Adamchuk, K. Ruhland and L. P. H. Lopez, *Organometallics*, 2005, **24**, 857.
654. J. D. Scollard and D. H. McConville, *J. Am. Chem. Soc.*, 1996, **118**, 10008.
655. F. Guerin, D. H. McConville and N. C. Payne, *Organometallics*, 1996, **15**, 5085.
656. F. Guerin, D. H. McConville and J. J. Vittal, *Organometallics*, 1996, **15**, 5586.
657. J. D. Scollard, D. H. McConville and S. J. Rettig, *Organometallics*, 1997, **16**, 1810.
658. H. Chen, R. A. Bartlett, H. V. R. Dias, M. M. Olmstead and P. P. Power, *Inorg. Chem.*, 1991, **30**, 2487.
659. B. Tsuie, D. C. Swenson, R. F. Jordan and J. L. Petersen, *Organometallics*, 1997, **16**, 1392.
660. Z. J. Tonzetich and R. R. Schrock, *Polyhedron*, 2006, **25**, 469.
661. D. D. VanderLende, K. A. Abboud and J. M. Boncella, *Organometallics*, 1994, **13**, 3378.
662. E. A. Ison, C. O. Ortiz, K. Abboud and J. M. Boncella, *Organometallics*, 2005, **24**, 6310.
663. W. E. Piers, P. J. Shapiro, E. E. Bunel and J. E. Bercaw, *Synlett.*, 1990, 74.
664. P. J. Shapiro, E. Bunel, W. P. Schaefer and J. E. Bercaw, *Organometallics*, 1990, **9**, 867.
665. J. Okuda, *Chem. Ber.*, 1990, **123**, 1649.
666. A. Spannenberg, H. Fuhrmann, P. Arndt, W. Baumann and R. Kempe, *Angew. Chem., Int. Ed.*, 1999, **37**, 3363.
667. H. Kay Lee, Y. L. Wong, Z.-Y. Zhou, Z.-Y. Zhang, D. K. P. Ng and T. C. W. Mak, *J. Chem. Soc., Dalton Trans.*, 2000, 539.
668. H. K. Lee, C. H. Lam, S.-L. Li, Z.-Y. Zhang and T. C. W. Mak, *Inorg. Chem.*, 2001, **40**, 4691.
669. K. J. Haack, S. Hashiguchi, A. Fujii, T. Ikariya and R. Noyori, *Angew. Chem., Int. Ed.*, 1997, **36**, 285.
670. T. Ohkuma, N. Utsumi, K. Tsutsumi, K. Murata, C. Sandoval and R. Noyori, *J. Am. Chem. Soc.*, 2006, **128**, 8724.
671. S. Daniéle, P. B. Hitchcock and M. F. Lappert, *Chem. Commun.*, 1999, 1909.
672. S. Daniéle, P. B. Hitchcock, M. F. Lappert and P. G. Merle, *J. Chem. Soc., Dalton Trans.*, 2001, 13.
673. P. B. Hitchcock, Q. Huang, M. F. Lappert, X.-H. Wei and M. Zhou, *Dalton Trans.*, 2006, 2991.

7

Amides of Zinc, Cadmium and Mercury

7.1 Introduction

In the 1980 edition of this chapter, the total number of amido derivatives of all types for the group 12 metals was less than 200 characterized compounds. These were described in ca. 90 papers in the primary literature. Detailed structural knowledge of these compounds was especially scant however. For example, there existed only a single published X-ray crystal structure of a zinc amide; i.e. that of the dimer $[\{HZnNMe(CH_2)_2NMe_2\}_2]$,[1] which featured a chelating amido ligand and four-coordinate zinc centres. No detailed structure of a cadmium amide was known and only one crystal structure of a mercury species related to the amides – that of the acetamide $[Hg\{NHC(O)Me\}_2]$ – (which was stated to have the isomeric structure $[Hg\{N{=}C(OH)Me\}_2]$[2]) had been reported. ^1H NMR studies of this species in solution, however, indicated that it did in fact have the acetamido structure $[Hg\{NHC(O)Me\}_2]$, which displayed a tendency to polymerize to $[\{Hg(NC(O)Me)\}_n]$ with elimination of $MeC(O)NH_2$.[3] Of the earlier work the majority of the references concerned zinc derivatives and there was a much smaller number of well-characterized compounds known for mercury and only six citations for cadmium. This proportion of effort for each element is also perpetuated in the more recent and current work. There have been no major reviews that deal exclusively with amido derivatives of group 12 metals as a separate class; however, the amides have been reviewed to varying extents in more wide ranging treatises of the coordination complexes and organometallic derivatives of these elements.[4–13] As in the case of the transition metal amides (Chapter 6), this review of the group 12 amides is focused primarily on the synthesis, structure and characterization of derivatives of monodentate ligands although several derivatives of multidentate and/or π-delocalized ligands are also discussed.

Metal Amide Chemistry Michael Lappert, Andrey Protchenko, Philip Power and Alexandra Seeber

7.2 Neutral Homoleptic Zinc, Cadmium and Mercury Amides

Some indication of the degree of increase in our overall knowledge of group 12 metal amide compounds since 1980 can be gauged from the structural data presented in Table 7.1.[14–22] The data concern only monomeric, homoleptic zinc amides, which feature two-coordinate zinc and represent a relatively small fraction of the known number of zinc amide structures. These simple compounds were generally synthesized via the simple reaction of the lithium salt of the amide with zinc chloride. Data for the eleven compounds listed show that the N–Zn–N array is linear, or almost linear, in all cases (as illustrated in Figure 7.1) and that the Zn–N distances span the narrow range of ca. 1.81–1.86 Å. As is observed elsewhere in the Periodic Table for metal amides, these distances are shorter than that predicted from the sum of the covalent radii for Zn (1.25 Å) and nitrogen (0.73 Å). The shortening is probably a result of ionic effects[23,24] and the low metal coordination number rather than significant multiple character in the Zn–N bond. The longest Zn–N bond lengths are observed with the bulkier amido ligands $[Zn\{N(SiMePh_2)\}_2]$[16] and $[Zn\{N(SiMe_3)(SiBu^tPh_2)\}_2]$.[22]

In contrast, the number of known structures of homoleptic, monomeric two-coordinate amides of cadmium is currently limited to just one example. The structure of the compound $[Cd\{N(SiMe_3)_2\}_2]$ has been determined by electron diffraction and the data showed that it has a linear NCdN skeleton with a Cd–N bond length of 2.03(2) Å.[25] It is clear, however, that several other homoleptic cadmium amides (known and others as yet not synthesized) can be predicted to have monomeric structures in the solid state. For example ^{113}Cd NMR studies of amides such as $[Cd\{N(SiMe_3)_2\}_2]$ or $[Cd\{N(SiMe_2Ph)_2\}_2]$ indicate that they are

Table 7.1 Selected X-ray structural data for some two-coordinate monomeric zinc amides

Compound	Zn–N (Å)	N–Zn–N (°)	Reference
$[Zn\{N(SiMe_3)_2\}_2]$	1.824(14) (ged)	180	14
$[Zn\{N(SiMe_3)_2\}_2]$	1.833(11)	175.2(4)	15
$[Zn\{N(SiMePh_2)\}_2]$	1.850(3)	177.59(2)	16
Bu^t–N(Zn)–$SiMe_2$–O–$SiMe_2$–N(Zn)–Bu^t Bu^t–N(Zn)–$SiMe_2$–O–$SiMe_2$–N(Zn)–Bu^t	1.840(3), 1.847(3)	171.41	17
$[Zn\{N(Bu^t)SiMe_3\}_2]$	1.823(8), 1.812(8)	179.4(5)	18
$[Zn(NBu^t_2)_2]$	1.824(3), 1.831(3)	179.65(13)	19
$[Zn\{\overline{NCMe_2(CH_2)_3CMe_2}\}_2]$	1.822(2), 1.819(2)	179.41(9)	20
$[Zn\{N(SiMe_3)C_6H_3Me_2\text{-}2,6\}_2]$	1.817(2)	180	21
$[Zn\{N(SiMe_3)C_6H_3Pr^i_2\text{-}2,6\}_2]$	1.821(1)	180	21
$[Zn\{N(SiMe_3)C_6H_3Bu^t_2\text{-}2,5\}_2]$	1.839(4), 1.837(4)	179.6(2)	21
$[Zn\{N(SiMe_3)(SiBu^tPh_2)\}_2]$	1.853(2), 1.858(2)	167.9(1)	22
$[Zn\{N(SiMe_3)(1\text{-}Ad)\}_2]$	1.827(2)	177.46(6)	22

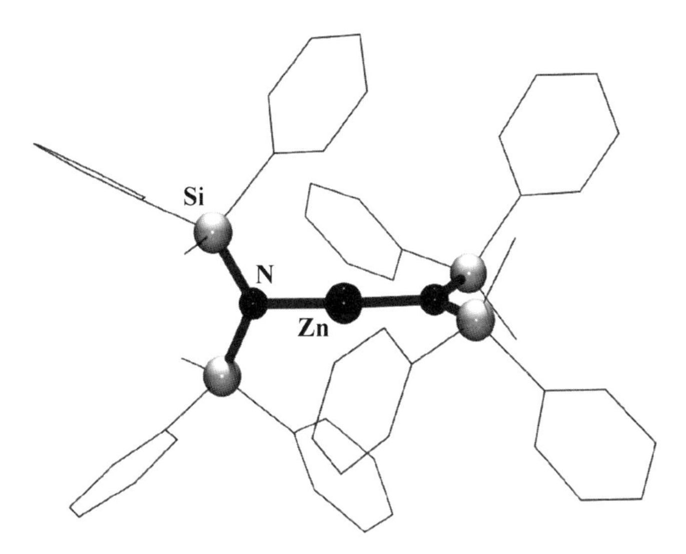

Figure 7.1 *Drawing of the structure of [Zn{N(SiMePh$_2$)$_2$}$_2$].[16] Some structural details are given in Table 7.1*

monomeric in hydrocarbon solution.[26,27] A small number of monomeric, homoleptic mercury(II) amides have been characterized by X-ray crystallography. Their structures include the unusual species [Hg{N(CF$_3$)TeF$_5$}$_2$] (Hg–N = 1.97(3) Å, N–Hg–N = 180°),[28] as well as [Hg{N(SiMe$_3$)R}$_2$] (R = Ph,[21] Hg–N = 2.024 Å avg., N–Hg–N = 174.1 (2)°; R = C$_6$H$_3$Me$_2$-2,6,[21] Hg–N = 2.017 Å avg., N–Hg–N = 176.1(1)°; R = C$_6$H$_3$Pr$^i_2$-2,6;[21] Hg–N = 2.008 Å avg., N–Hg–N = 174.7(4)°); R = 1-Ad,[22] Hg–N = 2.007 Å avg., N–Hg–N = 178.8(1)°); R = SiBu^{t}Ph$_2$,[22] Hg–N = 2.030 Å avg., N–Hg–N = 176.1(3)°) and [Hg(NSiMe$_2$(CH$_2$)$_2$SiMe$_2$)$_2$] (Hg–N = 2.000(2) Å, N–Hg–N = 180°).[29] Gas electron diffraction studies of [Hg{N(SiMe$_3$)$_2$}$_2$] show that it is monomeric with a linear N–Hg–N skeleton that has a Hg–N bond length of 2.01(2) Å.[30]

The key to the stabilization of two-coordinate monomeric structures is obviously ligand size; however, the metal also plays a role because in the case of zinc and cadmium the tendency to associate is more marked than it is for mercury. An example is provided by the monomeric structure observed for [Hg{N(SiMe$_3$)Ph}$_2$],[21] whereas a dimeric structure is seen for its zinc counterpart [(Zn{N(SiMe$_3$)Ph}$_2$)$_2$].[21] A further example involves the already mentioned structure of [Hg(NSiMe$_2$(CH$_2$)$_2$SiMe$_2$)$_2$],[29] which is monomeric, whereas the corresponding zinc and cadmium species [(M{NSiMe$_2$(CH$_2$)$_2$SiMe$_2$}$_2$)$_2$] (M = Zn or Cd) are amido-bridged dimers with metals that are bound to two bridging and a terminal amido ligand.[29] The dimeric structure of the zinc species may also be contrasted with that of [Zn{N(SiMe$_3$)$_2$}$_2$],[15] which is monomeric (cf. Table 7.1) but is very closely related sterically to [(Zn{NSiMe$_2$(CH$_2$)$_2$SiMe$_2$}$_2$)$_2$] whose slightly less crowded amido ligand is sufficient to permit dimerization.[29] Other examples of dimeric, homoleptic zinc amide derivatives of less bulky ligands include [{Zn(NPh$_2$)$_2$}$_2$],[31] [{Zn(NBu$^i_2$)$_2$}$_2$][19] and [Zn{N(CH$_2$Ph)$_2$}$_2$]$_2$.[32] They also dimerize through amido bridging to afford three-coordinate metals. The dibenzylamido derivative displays a concentration dependent monomer $\rightleftharpoons$ dimer equilibrium in solution.[32]

7.3 Ionic Metal Amides

7.3.1 Amidometallates

An ionic group 12 metal amide had not been well characterized at the time of the 1980 book. Homoleptic examples can be obtained by treatment of a zinc amide with one equivalent of an alkali metal or alkaline earth metal amide. For example, the reaction of $Zn\{N(SiMe_3)_2\}_2$ with $NaN(SiMe_3)_2$ gave the contact ion pair salt $[NaZn\{N(SiMe_3)_2\}_3]$ which, upon treatment with 12-crown-4, afforded the separated ion pair $[Na(12\text{-crown-}4)_2]$ $[Zn\{N(SiMe_3)_2\}_3]$.[33] The X-ray crystal structure of the anion consists of a zinc surrounded by three $-N(SiMe_3)_2$ groups. The geometry at zinc is planar with N–Zn–N angles that deviate only slightly from a mean of 120°. The average Zn–N bond length is 1.972 Å and the NSi_2 ligand planes subtend an average angle near 50° with respect to the ZnN_3 plane. The Zn–N distance is 0.14–0.15 Å longer than in $[Zn\{N(SiMe_3)_2\}_2]$ presumably as a result of the higher coordination number and anionic character of the zinc. Earlier work on reactions of $Ca\{N(SiMe_3)_2\}_2$ or $LiN(SiMe_3)_2$ with $Zn\{N(SiMe_3)_2\}_2$ in 1,2-dimethoxyethane (dme) yielded the salts $[Ca(Zn\{N(SiMe_3)_2\}_3)_2\text{•}3dme]$ or $[LiZn\{N(SiMe_3)_2\}_3\text{•}3dme]$. The former species yielded $[Ca(18\text{-crown-}6)][Zn\{N(SiMe_3)_2\}_3]_2$ upon treatment with 18-crown-6, which also exists as separate ion pairs in which the $[Zn\{N(SiMe_3)_2\}_3]^-$ ions presumably have a similar structure to that described above.[34]

7.3.2 Zincation Mediated by Amidozinc Complexes

Investigations by Mulvey and his group have uncovered very unusual structures associated with the mixed alkali metal-zinc reagents.[13a] For example, it has been found that a 1 : 1 mixture of $KN(SiMe_3)_2$ and $Zn\{N(SiMe_3)_2\}_2$ readily deprotonates toluene [Equation (7.1)] to afford

$$KN(SiMe_3)_2 + Zn\{N(SiMe_3)_2\}_2 \xrightarrow{PhMe} [KZn\{N(SiMe_3)_2\}_2(CH_2Ph)]_\infty + HN(SiMe_3)_2 \quad (7.1)$$

the benzyl product $[KZn\{N(SiMe_3)_2\}_2(CH_2Ph)]_\infty$ which was structurally characterized (Fig. 7.2),[35] whereas the analogous magnesium species affords the salt $[K(\eta^6\text{-PhMe})_2][Mg\{N(SiMe_3)_2\}_3]$ without toluene activation.[36] Changing the alkali metal counter cation in the zincate system also seems to suppress toluene deprotonation because it is possible to isolate the salt $[Rb(\eta^6\text{-PhMe})_3][Zn\{N(SiMe_3)_2\}_3]$ without toluene activation.[36]

The alkali-metal mediated activation of toluene prompted further work involving more basic amido ligands such as $\text{-}\overline{NCMe_2(CH_2)_3C}Me_2$.[37] In earlier work it had been reported by Kondo and coworkers that the use of the reagent $LiZnBu^t_2$ $(\overline{NCMe_2(CH_2)_3C}Me_2)$, which was synthesized in situ, resulted in high levels of chemo- and regioselectivity for the activation of aromatic ring species.[38–40] Mulvey and his group showed that it was possible to

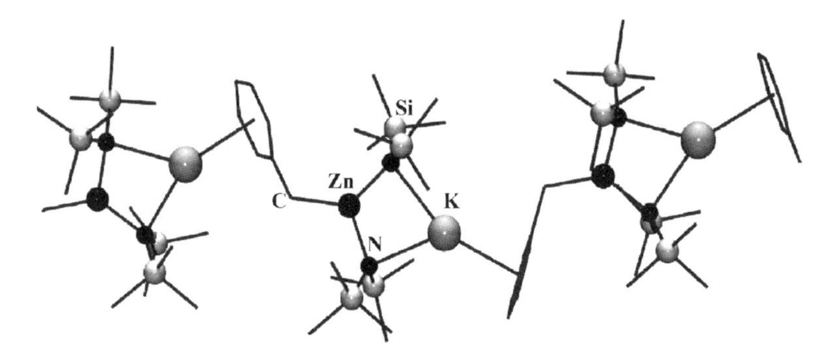

Figure 7.2 *Drawing of the structure of $[KZn\{N(SiMe_3)_2\}_2(CH_2Ph)]_\infty$, which arises from the facile deprotonation of solvent toluene by a 1 : 1 mixture of $KN(SiMe_3)_2$ and $Zn\{N(SiMe_3)_2\}_2$*[35]

crystallize the related sodium salt $[(tmeda)Na(\mu\text{-}Bu^t)(\mu\text{-}\overline{NCMe_2(CH_2)_3CMe_2})ZnBu^t]$ (tmeda = $Me_2NCH_2CH_2NMe_2$) whose crystal structure revealed that the zinc was bound to $\text{-}\overline{NCMe_2(CH_2)_3CMe_2}$ and two Bu^t ligands.[37] The zinc has a distorted trigonal planar coordination and is bridged to the Na^+ ion by a methyl moiety from one of the Bu^t groups as well as the amido ligand. This species reacted with benzene to form the phenyl derivative $[(tmeda)Na(\mu\text{-}Ph)(\mu\text{-}\overline{NCMe_2(CH_2)_3CMe_2})ZnBu^t]$ in which the phenyl group, formed by the deprotonation of benzene, displaced the bridging Bu^t moiety [Equation (7.2)].

$$[(tmeda)Na(\mu\text{-}Bu^t)(\mu\text{-}\overline{NCMe_2(CH_2)_3CMe_2})ZnBu^t] + C_6H_6 \longrightarrow$$
$$[(tmeda)Na(\mu\text{-}Ph)(\mu\text{-}NCMe_2(CH_2)_3CMe_2)ZnBu^t] + HCMe_3 \qquad (7.2)$$

This led to the description of such reaction as an 'alkali-metal-mediated zincation'. This was based on the fact it is a zincation (i.e. a zinc-hydrogen exchange) that proceeds only in the presence of an alkali metal compound (commonly an alkali metal amide) to afford a synergic metalation of a substrate. In more recent work it was shown that the originally used lithium salt $[(thf)Li(\mu\text{-}tmp)(\mu\text{-}Bu^t)ZnBu^t]$ and the related adduct $[Ph(Pr^i{}_2N)C(O)Li(\mu\text{-}tmp)\text{-}(\mu\text{-}Bu^t)ZnBu^t]$[41] crystallized with similar structures to their sodium counterparts. Other unusual reactions have also been uncovered. For example mixing $ZnBu^n{}_2$, and $Li\overline{NCMe_2(CH_2)_3CMe_2}$ in the presence of tmeda formed the salt $[(tmeda)Li(\mu\text{-}Bu^n)\text{-}(\mu\text{-}\overline{NCMe_2(CH_2)_3CMe_2})Zn(Bu^n)]$, which can activate ferrocene to form, inter alia, the salt $[Li(thf)_4][Zn\{Fe(\eta^5\text{-}C_5H_5)(\eta^5\text{-}C_5H_4)\}_3]$ in which three ferrocenes are σ-bonded to the zinc through a single carbon from one of the cyclopentadienide rings.[42] Investigation of the reaction of $[(tmeda)Na(\mu\text{-}Bu^t)(\mu\text{-}\overline{NCMe_2(CH_2)_3CMe_2})Zn(Bu^t)]$ with benzophenone showed that it efficiently butylates this molecule to yield the 4-Bu^t enolate salt $[(tmeda)Na(\mu\text{-}\overline{NCMe_2(CH_2)_3CMe_2})\{\mu\text{-}O\text{-}C(Ph){=}C_6H_5(H)Bu^t\}ZnBu^t]$.[43]

More recent work involving anisole as the aromatic substrate[44] has shown that this molecule is orthodeprotonated by $LiNCMe_2(CH_2)_3CMe_2$ and $ZnBu^t_2$ in the presence of thf to afford the products shown in Equation (7.3).

$$Li\bullet + ZnBu^t_2 \xrightarrow{2PhOMe,\ thf} (thf)Li(\mu\text{-}\bullet)(\mu\text{-}C_6H_4OMe)Zn(C_6H_4OMe)$$
$$Li\bullet + ZnBu^t_2 \xrightarrow{1PhOMe,\ thf} (thf)Li(\mu\text{-}\bullet)(\mu\text{-}C_6H_4OMe)ZnBu^t \quad (7.3)$$

$\bullet = NCMe_2(CH_2)_3CMe_2$

Without thf, anisole may solvate the lithium as in Equation (7.4).

$$Li\bullet + ZnBu^t_2 \xrightarrow{2PhOMe} (PhOMe)Li(\mu\text{-}\bullet)(\mu\text{-}C_6H_4OMe)Zn\text{–}Bu^t \quad (7.4)$$

$\bullet = NCMe_2(CH_2)_3CMe_2$

Treatment of $ZnMe_2$ with $LiNCMe_2(CH_2)_3CMe_2$ in the presence of tmeda afforded the conventional zincate product in accordance with Equation (7.5).

$$ZnMe_2 + Li\bullet \xrightarrow{tmeda} (tmeda)Li(\mu\text{-}\bullet)(\mu\text{-}CH_3)Zn\text{–}CH_3 \quad (7.5)$$

$\bullet = NCMe_2(CH_2)_3CMe_2$

In contrast, with use of $-N(SiMe_3)_2$ as the amide as in Equation (7.6), the 'inverse zincate' product was obtained.[45]

$$2ZnMe_2 + 2LiN(SiMe_3)_2 \xrightarrow{tmeda} ZnMe_2(tmeda) + 1/n\left[Li\{\mu\text{-}N(SiMe_3)_2\}_2Li\text{–}CH_3\text{–}Zn(tmeda)\text{–}CH_3\right]_n \quad (7.6)$$

More recent work on the zincate system has focused on the mechanism[41,46,47] of site directed zincation[38] as illustrated by Scheme 7.1.

The crystal structure of the lithium complex $[Pr^i_2N(Ph)COLi(\mu\text{-tmp})(\mu\text{-}Bu^t)ZnBu^t]$ shows that the amide solvates the lithium ion through oxygen with zincation of the aryl occurring via interaction of the ortho aryl hydrogen Bu^t or tmp group.[41] The crystal

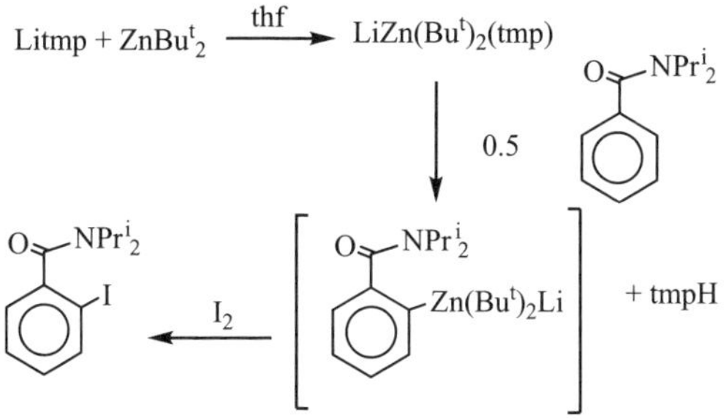

Litmp + $ZnBu^t_2$ —thf→ $LiZn(Bu^t)_2(tmp)$

0.5

I_2

$Zn(Bu^t)_2Li$

+ tmpH

__Scheme 7.1__ Proposed mechanism for synthesis of N,N′-diisopropyl-2-iodobenzamide via site directed zincation[38]

structure of the sodium salt $[(tmeda)Na(\mu\text{-}Pr^i{}_2NC(O)C_6H_4)(tmp)Zn(Bu^t)]$[46] (which corresponds to the proposed intermediate represented in brackets in Scheme 7.1) shows that Bu^tH is eliminated and the $Pr^i{}_2NC(O)C_6H_4$ and the tmp groups bridge the sodium and zinc centres with sodium being coordinated by oxygen from the $Pr^i{}_2NC(O)C_6H_4$ moiety. This suggests that activation occurs via interaction of the ortho-H with a zinc Bu^t substituent.[47] Other investigations have focused on the zincation of aromatic heterocycles[48] and multiple zincation of aromatics such as benzene[49a] or naphthalene.[49b] Deprotonative metallation reactions using amidozincate reagents have been recently reviewed.[13b] Finally, a recent report has shown that the addition of tmpMgCl·LiCl to $ZnCl_2$ affords $(tmp)_2Zn\cdot 2MgCl_2\cdot 2LiCl$ which zincates sensitive heterocycles such as 2-phenyl-1,3,4-oxadiazole that can be further derivatized by reaction CH_2CHCH_2Br.[50a] In addition, it has been shown that $Zn(tmp)_2$ by itself is a useful base for selective functionalization of C–H bonds.[50b]

7.3.3 Other Ionic Group 12 Metal Amido Salts

Another class of compounds that is closely related to the zincates are the so-called 'inverse crowns'.[12,43,51] These species are formed from mixtures of $[Zn\{N(SiMe_3)_2\}_2]$ (or

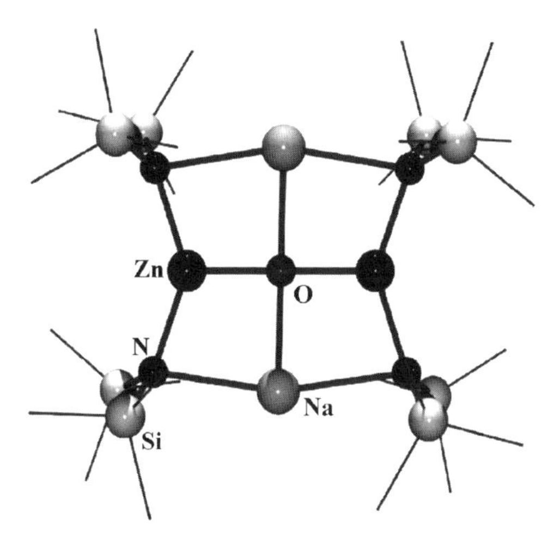

__Figure 7.3__ Drawing of the structure of $[Na_2Zn_2\{N(SiMe_3)_2\}_4(O)]$[52]

$[Mg\{N\text{-}(SiMe_3)_2\}_2]$) and $MN(SiMe_3)_2$ (M = Na or K) which can then interact with trace amounts of O_2/H_2O to afford products such as $[\{M_2Zn_2\{N(SiMe_3)_2\}_4(O_2)\}_\infty]$ or $[M_2Zn_2\{N\text{-}(SiMe_3)_2\}_4(O)]$ (Figure 7.3). The structure[52] is composed of an eight-membered ring of alternating four metal and four nitrogen atoms in which the metals are alternately alkali metal and zinc atoms. Central oxide or peroxide ions provide electroneutrality. Thus the conventional host-guest arrangement of ligand donor and guest acceptor is reversed since in this case the 'guest' is the anion O^{2-} or $O_2^{\ 2-}$ rather than a cation and the host metal centres behave as Lewis acids.

Other interesting reaction products have also been obtained from zinc amides. For example treatment of $Zn\{N(SiMe_3)_2\}_2$ with $[H(OEt_2)_2][B(C_6F_5)_4]$ yielded the unusual salt $[(Et_2O)_3Zn\{N(SiMe_3)_2\}][B(C_6F_5)_4]$. The cationic zinc centre is coordinated to three ethers as well as the amide and has a distorted tetrahedral coordination.[53] The Zn–N bond length is 1.907(3) Å and is significantly shorter than the 1.972 Å in $[Zn\{N(SiMe_3)_2\}_3]^-$ in spite of the higher coordination number of zinc. This species behaves as an efficient catalyst for the ring-opening polymerization of epoxides and ε-caprolactone. Similarly, reaction of $[RdadH][H_2N\{B(C_6F_5)_3\}_2]$ [Rdad = $(MeC{=}NC_6H_3R_2\text{-}2,6)_2$] with $M\{N(SiMe_3)_2\}_2$ (M = Zn or Cd) afforded the salts $[(Rdad)M\{N(SiMe_3)_2\}][H_2N\{B(C_6F_5)_3\}_2]$ and the structure of the zinc derivative (R = Ph) was determined.[54] A further unusual ionic species was isolated from the treatment of $\{Sb(NCy)_3\}_2Li_6$ with six equivalents of $Cd\{N(SiMe_3)_2\}_2$ which afforded the unique bimetallic cage species $[Sb(NCy)_3\{CdN(SiMe_3)_2\}_3]\bullet[Li\{N(SiMe_3)_2\}\text{-}(thf)]_2$ with a $SbCd_3N_3$ open cuboidal core structure.[55]

7.4 Lewis Base Complexes, Chelated Metal Amides and Heteroleptic Amido Complexes

There exist numerous other classes of group 12 metal amide species. Among the most prominent are those that incorporate either a chelating group or neutral donor co-ligands. Structurally characterized examples include the simple Lewis base complex $[Zn\text{-}(NPh_2)_2(thf)_2]$,[31] the chelates $[Zn\{N(CH_2CH_2NMe_2)_2\}_2]_2$,[56] $[Zn\{Bu^t_2P(Se)NPr^i\}_2]$,[57] $[Zn\{N(Bu^t)S(O)C_6H_4Me\text{-}4\}_2]$,[58] $[M\{N(8\text{-quinolyl})SiMe_3\}_2]$ (M = Zn or Cd),[59] $[Zn\{N\text{-}(SiMe_2Bu^t)(CH_2\text{-}2\text{-py})\}_2]$,[60] $[Cd\{N(Bu^t)Si(Me)_2O\}_2C_6H_4\text{-}1,2]_2$,[61] $[Zn\{N(H)(pm(OMe)_2\text{-}4,6)\}_2(tmeda)]$ (pm = pyrimidine),[62] LZn_2R_2 (L = dibenzofuran$(NCH_2CH_2NMe_2)_2$-4,6, R = Ph, $CH_2P(O)Me_2CH_2C(O)NMe_2$),[63] and $[(Zn\{N(PMes_2)\}_2C_6H_4\text{-}1,2)_2]$.[64] In addition there is a large range of heteroleptic metal amides. These often involve a combination of alkyl and amide ligands which arise from the use of zinc dialkyls in their synthesis via alkane elimination upon reaction with an amine. Examples include $[(MeZn\{N(H)Si\text{-}Pr^i_3\})_2]$,[65] $[(MeZn\{\mu\text{-}N(Cy)C_5H_4N\text{-}2\})_2]$,[66] $[\{MeZn\{NH(1\text{-}Ad)\}(thf)\}_2]$,[65] $[\{EtZn\{\mu\text{-}N\text{-}(H)Ph\}(thf)\}_3]$,[67] $[(EtZn\{\mu\text{-}N(H)C_6H_2Me_3\text{-}2,4,6\})_2]$,[67] $[(EtZn\{\mu\text{-}N(H)Bu^t\})_3]$,[67] $[(EtZn\text{-}\{2\text{-}N(SiMe_3)C_5H_3NMe\text{-}6\})_2]$,[68] $[(Me_3SiCH_2Zn\{\mu\text{-}N(H)C_6H_3Pr^i_2\text{-}2,6\})_2]$,[68] $[(EtZn\{N(H)\text{-}SiPr^i_3\})_2]$,[65] $[\{EtZn\{\mu\text{-}N(H)(1\text{-}naphthyl)\}(thf)\}_3]$,[69] $[(MeZn\{\mu\text{-}N(Me)(CH_2)_3NMe_2\})_2]$,[70] $[(MeZn\{\mu\text{-}N(Me)(CH_2)_2NMe_2\})_2]$,[70] $[(Zn(\mu\text{-}OCEt_3)\{N(SiMe_3)_2\})_2]$,[71] $[(MeZn\{\mu\text{-}N(CH_2\text{-}C_5H_4N\text{-}2)_2\})_2]$,[72] $[\{(\pm)\text{-}trans\text{-}1,2\text{-}C_6H_{10}(NSiMe_3)_2\}_2Zn_4Et_4]$ and $[\{(\pm)\text{-}trans\text{-}1,2\text{-}C_6H_{10}\text{-}(NSiMe_3)(NHSiMe_3)\}_2Zn_2Et_2]$,[73] $[EtZn\{N(C_6H_3Pr^i_2\text{-}2,6)(C_6H_4PPh_2\text{-}2)\}]$,[74] $[\{(Me_3Si)_3\text{-}CZn\}_2(\mu\text{-}Cl)(\mu\text{-}NC_4H_2Bu^t_2\text{-}2,5)]$.[75a] In addition, a series of zinc amides of formula

[{Zn(μ-tetraalkylguanidinate)N($SiMe_3$)$_2$}$_2$] (tetraalkylguanidinate = N=C(NR_2)-(NR'_2)(R/R'=Me/Et;R/NR'_2=Me/$\overline{N(CH_2)_3CH_2}$,Et/$\overline{N(CH_2)_3CH_2}$,Me/$\overline{N(CH_2)_4CH_2}$, Et/$\overline{N(CH_2)_4CH_2}$;NR_2=NR'_2= $\overline{N(CH_2)_3CH_2}$ have been synthesized and structurally characterized.[75b] The heteroleptic cadmium species [Cd(η^1-C_5Me_5){μ-N-($SiMe_3$)$_2$}$_2$][76] features terminal η^1-C_5Me_5 groups and bridging amido ligands.

Heteroleptic zinc amides can also arise from interactions with transition metal complexes. The reaction of [Zn{N($SiMe_3$)$_2$}$_2$] with MH_3L_3 (M = Rh, Ir; L_3 = (PMe_2Ph)$_3$ and MeC(CH_2PR_2)$_3$ (R = Me or Ph)) afforded the heterometallic hydrides of the type [L_3MH_2Zn{N($SiMe_3$)$_2$}].[77] The transition metal geometry is non-rigid and a crystal structure of [MeC(CH_2PPh_2)$_3$Rh(H)$_2$Zn{N($SiMe_3$)$_2$}] shows that the hydrides are terminally bound to rhodium and that the zinc has linear coordination being bonded to rhodium and a terminal –N($SiMe_3$)$_2$ group. The reaction of [MeC(CH_2PMe_2)$_3$RhH_2Zn{N-($SiMe_3$)$_2$}] with [RhH_3{MeC(CH_2PMe_2)$_3$}] led to H_2 elimination and the isolation of [{MeC(CH_2PMe_2)$_3$RhH}$_2$(μ-H)(μ-Zn{N($SiMe_3$)$_2$})] in which Zn{N($SiMe_3$)$_2$} bridges the rhodiums.[77]

Four crystalline zinc complexes, containing the 1,2-benzenebis(neopentylamido) ligand (L^{2-}), were prepared from [Li_2(L)] and $ZnCl_2$: [Li(thf)$_4$][Zn(L)($L^{\cdot -}$)] (**1**), [{Li(OEt_2)-(μ-L)Zn}$_2$(μ-L)], [Li(thf)$_4$][{Zn(μ-L)}$_3$(μ_3-Cl)] and [{Li(OEt_2)(μ-L)Zn(μ-L)Zn(LH)].[78] The most interesting is paramagnetic **1**, having an unpaired electron delocalized between two ligands (broad singlet with g_{iso} = 2.0034 in thf). This compound is related to the paramagnetic diazabutadiene complexes [Zn(Bu^t_2DAB)$_2$]$^+$[Zn(Bu^t_2DAB)$_2$]$^-$ (septet with g_{iso} = 2.0068 in thf attributed to the anion)[79] and [K(thf)$_3${Zn(Bu^t_2DAB)$_2$}]·thf (broad septet attributed to **2**).[80]

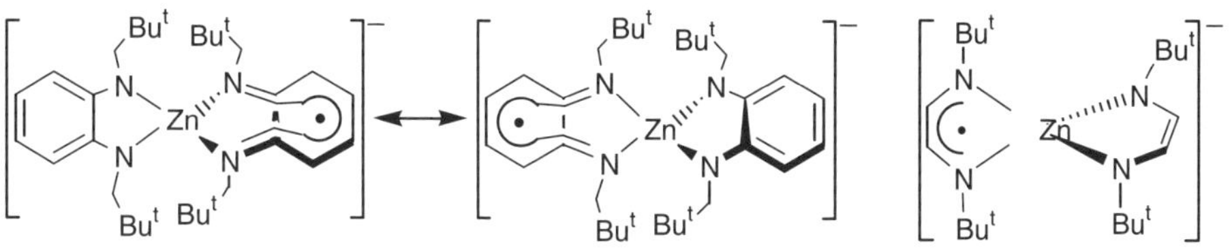

1 (anion) **2**

There are several other examples of zinc complexes that are closely related to amides but not within the scope of this survey. These include derivatives of diaminocyclophosphazanes,[81] arsenidoamines,[82] aminoiminophosphanes,[83] and other PN species,[84,85] carboxylic amides,[86] and chelating diimines,[87] triazenides,[88] amidinates,[88] or aminotroponiminates.[89,90] There is also a high level of interest in zinc diketiminates because of their application in the polymerization of lactides and the copolymerization of epoxides with carbon dioxide,[91–99] as well as their ability to stabilize three-coordination at zinc as in [(HC{C(Me)N$C_6H_3Pr^i_2$-2,6}$_2$)ZnNPr^i_2].[100] A bulky β-diketiminato ligand has also been shown to stabilize a Zn–Zn bond in [Zn_2({N($C_6H_3Pr^i_2$-2,6)CMe}$_2$CH)$_2$].[101]

Besides interest in their structural and chemical characteristics, there are several other motivations for current work on group 12 metal amides. Among the most prominent are their use as precursors for MOCVD,[100,102–105] their employment as synthons for other derivatives of these metals [58,106–109] and their use in the preparation of catalyst precursors.[52,53,91–99,110,111] Some of these studies have also provided important information on several fundamental compound classes; for example, synthetic and physical data for the simplest diorganoamides of zinc such as $[\{Zn(NR_2)_2\}_2]$ (R = Me, Pr^i or Bu^i),[70] or examples of well characterized thiolato (e.g. $[Zn_2\{\mu\text{-}N(SiMe_3)_2\}\{SC_6H_2(CF_3)_3\text{-}2,4,6\}_3]$ and $[Zn_3\{N(SiMe_3)_2\}_2(\mu\text{-}SC_6H_2Pr^i_3\text{-}2,4,6)_4]$[110]) or phosphido[112] complexes (see also Ref. 112 for the interesting amido/phosphido cluster $[\{ZnN(SiMe_3)_2\}_3\{(PMe)_2SiPr^i_2\}_2H]$) of zinc, cadmium and mercury,[108,112] or cationic zinc amides.[52,53]

Finally it is noteworthy that information on metal-nitrogen bond strengths in group 12 metal amides originally described in the 1980 edition has since been published in the primary literature.[113] In addition we draw attention to the fact that the structures of the parent amido derivatives $Zn(NH_2)_2$ and $Na_2Zn(NH_2)_4$ have been determined.[114] In $Zn(NH_2)_2$ the amido ions have a distorted cubic close packed arrangement in which a quarter of the tetrahedral voids are occupied by Zn^{2+} ions.

References

1. D. C. Bradley, *Inorg. Macromol. Rev.*, 1970, **1**, 141; N. A. Bell, P. T. Moseley, H. M. M. Shearer and C. B. Spencer, *Acta Crystallogr., Sect. B*, 1980, **36**, 2950.
2. B. Kamenar and D. Grdenić, *Inorg. Chim. Acta*, 1969, **3**, 25.
3. D. B. Brown and M. B. Robin, *Inorg. Chim. Acta*, 1969, **3**, 644.
4. R. H. Prince, Zinc and Cadmium; in *Comprehensive Coordination Chemistry* Eds. G. Wilkinson, R. D. Gillard and J. A. McCleverty Pergamon, Oxford, 1987, Volume 5, Chapter 56.1.
5. K. Brodersen and H.-U. Hummel, Mercury; in *Comprehensive Coordination Chemistry* Eds. G. Wilkinson, R. D. Gillard and J. A. McCleverty Pergamon, Oxford, 1987, Volume 5, Chapter 56.2.
6. S. J. Archibald, Zinc; in *Comprehensive Coordination Chemistry II* Eds. J. A. McCleverty and T. J. Meyer, Elsevier, Amsterdam, 2004, Chapter 6.8.
7. D. K. Breitinger, Cadmium and Mercury; in *Comprehensive Coordination Chemistry II*, Eds. J. A. McCleverty and T. J. Meyer Elsevier, Amsterdam, 2004, Chapter 6.9.
8. J. Boersma, Zinc and Cadmium; in *Comprehensive Organometallic Chemistry*, Eds. G. Wilkinson, F. G. A. Stone and E. W. Abel Pergamon, Oxford, 1982, Volume 2, Chapter 16.
9. J. L. Wardell, Mercury; in *Comprehensive Organometallic Chemistry*, Eds. G. Wilkinson, F. G. A. Stone and E. W. Abel Pergamon, Oxford, 1982, Volume 2 Chapter 17,
10. A. G. Davies, J. L. Wardell, Mercury; in *Comprehensive Organometallic Chemistry II* Eds. E. W. Abel, F. G. A. Stone and G. Wilkinson Pergamon/Elsevier, Amsterdam, 1995, Volume 3, Chapter 3.
11. P. O'Brien, Cadmium and Zinc; in *Comprehensive Organometallic Chemistry II* Eds. E. W. Abel, F. G. A. Stone and G. Wilkinson Pergamon/Elsevier, Amsterdam, 1995, Volume 3, Chapter 4.
12. R. E. Mulvey, *Chem. Commun.*, 2001, 1049.
13. (a) R. E. Mulvey, *Organometallics*, 2006, **25**, 1060. (b) R. E. Mulvey, F. Mongin, M. Uchiyama and Y. Kondo, *Angew. Chem., Int. Ed.*, 2007, **46**, 3802.
14. A. Haaland, K. Hedberg and P. P. Power, *Inorg. Chem.*, 1984, **23**, 1972.
15. G. Margraf, H.-W. Lerner, M. Bolte and M. Wagner, *Z. Anorg. Allg. Chem.*, 2004, **630**, 217.
16. P. P. Power, K. Ruhlandt-Senge and S. C. Shoner, *Inorg. Chem.*, 1991, **30**, 5013.
17. A. J. Elias, H.-G. Schmidt, M. Noltemeyer and H. W. Roesky, *Eur. J. Solid State Inorg. Chem.*, 1992, **29**, 23.
18. W. S. Rees, D. M. Green and W. Hesse, *Polyhedron*, 1992, **11**, 1697.

19. H. Schumann, J. Gottfriedsen and F. Girgsdies, *Z. Anorg. Allg. Chem.*, 1997, **623**, 1881.
20. W. S. Rees, O. Just, H. Schumann and R. Weimann, *Polyhedron*, 1998, **17**, 1001.
21. H. Schumann, J. Gottfriedsen, S. Dechert and F. Girgsdies, *Z. Anorg. Allg. Chem.*, 2000, **626**, 747.
22. Y. Tang, A. M. Felix, B. J. Boro, L. N. Zakharov, A. L. Rheingold and R. A. Kemp, *Polyhedron*, 2005, **24**, 1093.
23. V. Schomaker and D. P. Stevenson, *J. Am. Chem. Soc.*, 1941, **63**, 37.
24. R. Blom and R. Haaland, *J. Mol. Struct.*, 1985, **128**, 21.
25. E. C. Alyea, K. J. Fisher and T. Fjeldberg, *J. Mol. Struct.*, 1985, **127**, 325.
26. K. J. Fisher and E. C. Alyea, *Polyhedron*, 1984, **3**, 509.
27. E. C. Alyea and K. J. Fisher, *Polyhedron*, 1986, **5**, 695.
28. J. S. Thrasher, J. Nielsen, S. G. Bott, D. J. McClure, S. A. Morris and J. L. Atwood, *Inorg. Chem.*, 1988, **27**, 570.
29. O. Just, D. A. Gaul and W. S. Rees, *Polyhedron*, 2001, **20**, 815.
30. E. C. Alyea, K. J. Fisher and T. Fjeldberg, *J. Mol. Struct.*, 1985, **130**, 263.
31. M. A. Putzer, A. Dashti-Mommertz, B. Neumüller and K. Dehnicke, *Z. Anorg. Allg. Chem.*, 1998, **624**, 263.
32. D. R. Armstrong, G. C. Forbes, R. E. Mulvey, W. Clegg and D. M. Tooke, *J. Chem. Soc., Dalton Trans.*, 2002, 1656.
33. M. A. Putzer, B. Neumüller and K. Dehnicke, *Z. Anorg. Allg. Chem.*, 1997, **623**, 539.
34. M. Westerhausen, *Z. Anorg. Allg. Chem.*, 1992, **618**, 131.
35. W. Clegg, G. C. Forbes, A. R. Kennedy, R. E. Mulvey and S. T. Liddle, *Chem. Commun.* 2003, 406.
36. G. C. Forbes, A. R. Kennedy, R. E. Mulvey, B. A. Roberts and R. B. Rowlings, *Organometallics*, 2002, **21**, 5115.
37. P. C. Andrikopoulos, D. R. Armstrong, H. R. L. Barley, W. Clegg, S. H. Dale, E. Hevia, G. W. Honeyman, A. R. Kennedy and R. E. Mulvey, *J. Am. Chem. Soc.*, 2005, **127**, 6184.
38. Y. Kondo, M. Shilai, M. Uchiyama and T. Sakamoto, *J. Am. Chem. Soc.*, 1999, **121**, 3539.
39. T. Imahori, M. Uchiyama, T. Sakamoto and Y. Kondo, *Chem. Commun.*, 2001, 2450.
40. M. Uchiyama, T. Miyoshi, Y. Kajihara, T. Sakamoto, Y. Otani, T. Ohwada and Y. Kondo, *J. Am. Chem. Soc.*, 2002, **124**, 8514.
41. W. Clegg, S. H. Dale, E. Hevia, G. W. Honeyman and R. E. Mulvey, *Angew. Chem., Int. Ed.*, 2006, **45**, 2370.
42. H. R. L. Barley, W. Clegg, S. H. Dale, E. Hevia, G. W. Honeyman, A. R. Kennedy and R. E. Mulvey, *Angew. Chem., Int. Ed.*, 2005, **44**, 6018.
43. E. Hevia, G. W. Honeyman, A. R. Kennedy and R. E. Mulvey, *J. Am. Chem. Soc.*, 2005, **127**, 13106.
44. W. Clegg, S. H. Dale, A. M. Drummond, E. Hevia, G. W. Honeyman and R. E. Mulvey, *J. Am. Chem. Soc.*, 2006, **128**, 7434.
45. D. V. Graham, E. Hevia, A. R. Kennedy and R. E. Mulvey, *Organometallics*, 2006, **25**, 3297.
46. W. Clegg, S. H. Dale, R. W. Harrington, E. Hevia, G. W. Honeyman and R. E. Mulvey, *Angew. Chem., Int. Ed.*, 2006, **45**, 2374.
47. D. R. Armstrong, W. Clegg, S. H. Dale, E. Hevia, L. M. Hogg, G. W. Honeyman and R. E. Mulvey, *Angew. Chem., Int. Ed.*, 2006, **45**, 3775.
48. B. Conway, E. Hevia, A. R. Kennedy and R. E. Mulvey, *Chem. Commun.*, 2007, 2864.
49. (a) D. R. Armstrong, W. Clegg, S. H. Dale, D. V. Graham, E. Hevia, L. M. Hogg, A. R. Kennedy and R. E. Mulvey, *Chem. Commun.*, 2007, 598; (b) W. Clegg, S. H. Dale, E. Hevia, L. M. Hogg, G. W. Honeyman, R. E. Mulvey and C. T. O'Hara, *Angew. Chem., Int. Ed.*, 2006, **45**, 6548.
50. (a) S. H. Wunderlich and P. Knochel, *Angew. Chem. Int. Ed.*, 2007, **46**, 7685; (b) M. L. Hlavinka and J. Hagadorn, *Organometallics*, 2007, **26**, 4105.

51. A. R. Kennedy, R. E. Mulvey and R. B. Rowlings, *J. Am. Chem. Soc.*, 1998, **120**, 7816.
52. G. C. Forbes, A. R. Kennedy, R. E. Mulvey, R. B. Rowlings, W. Clegg, S. T. Liddle and C. C. Wilson, *Chem. Commun.*, 2000, 1759.
53. Y. Sarazin, M. Schormann and M. Bochmann, *Organometallics*, 2004, **23**, 3296.
54. M. D. Hannant, M. Schormann, D. L. Hughes and M. Bochmann, *Inorg. Chim. Acta*, 2005, **358**, 1683.
55. A. Bashall, M. A. Beswick, C. N. Harmer, M. McPartlin, M. A. Paver and D. S. Wright, *J. Chem. Soc., Dalton Trans.*, 1998, 517.
56. B. Luo, B. E. Kucera and W. L. Gladfelter, *Polyhedron*, 2006, **25**, 279.
57. M. Bochmann, G. C. Bwembya, M. B. Hursthouse and S. J. Coles, *J. Chem. Soc., Dalton Trans.*, 1995, 2813.
58. O. I. Guzyr, L. N. Markovskii, M. I. Povolotskii, H. W. Roesky, A. N. Chernega and E. B. Rusanov, *J. Mol. Struct.*, 2006, **788**, 89.
59. L. M. Engelhardt, P. C. Junk, W. C. Patalinghug, R. E. Sue, C. L. Raston, B. W. Skelton and A. H. White, *J. Chem. Soc., Chem. Commun.*, 1991, 930.
60. C. Koch, A. Malassa, C. Agthe, H. Görls, R. Biedermann, H. Krautscheid, and M. Westerhausen, *Z. Anorg. Allg. Chem.*, 2007, **633**, 375.
61. M. Veith and K. L. Woll, *Chem. Ber.*, 1993, **126**, 2383.
62. A. Bashall, J. M. Cole, F. Garcia, A. Primo, A. Rothenberger, M. McPartlin and D. S. Wright, *Inorg. Chim. Acta*, 2003, **354**, 41.
63. M. L. Hlavinka and J. R. Hagadorn, *Organometallics*, 2005, **24**, 4116.
64. F. Majuomo-Mbe, P. Lönnecke and E. Hey-Hawkins, *Organometallics*, 2005, **24**, 5287.
65. M. Westerhausen, T. Bollwein, A. Pfitzner, T. Nilges and H. J. Deiseroth, *Inorg. Chim. Acta*, 2001, **312**, 239.
66. S. J. Birch, S. R. Boss, S. C. Cole, M. P. Coles, R. Haigh, P. B. Hitchcock and A. E. Wheatley, *Dalton Trans.*, 2004, 3568.
67. M. M. Olmstead, W. J. Grigsby, D. R. Chacon, T. Hascall and P. P. Power, *Inorg. Chim. Acta*, 1996, **251**, 273.
68. L. M. Engelhardt, G. E. Jacobsen, W. C. Patalinghug, B. W. Skelton, C. L. Raston and A. H. White, *J. Chem. Soc., Dalton Trans.*, 1991, 2859.
69. M. G. Davidson, D. Elilio, S. L. Less, M. Avelino, P. R. Raithby, R. Snaith and D. S. Wright, *Organometallics*, 1993, **12**, 1.
70. M. A. Malik, P. O'Brien, M. Motevalli and A. C. Jones, *Inorg. Chem.*, 1997, **36**, 5076.
71. S. C. Goel, M. Y. Chiang and W. E. Buhro, *Inorg. Chem.*, 1990, **29**, 4646.
72. M. Westerhausen, A. N. Kniefel, I. Lindner and J. Grcic, *Z. Naturforsch.*, 2004, **59B**, 161.
73. D. Chakraborty and E. Y.-X. Chen, *Organometallics*, 2003, **22**, 769.
74. L.-C. Liang, W.-Y. Lee and C.-H. Hung, *Inorg. Chem.*, 2003, **42**, 5471.
75. (a) M. Westerhausen, M. Wieneke, H. Nöth, T. Seifert, A. Pfitzner, W. Schwarz, O. Schwarz and J. Weidlein, *Eur. J. Inorg. Chem.*, 1998, 1175; (b) T. L. Cleland and S. D. Bunge, *Polyhedron*, 2007, **26**, 5506.
76. C. C. Cummins, R. R. Schrock and W. M. Davis, *Organometallics*, 1991, **10**, 3781.
77. R. L. Geerts, J. C. Huffman, D. E. Westerberg, K. Folting and K. G. Caulton, *New. J. Chem.*, 1988, **12**, 455.
78. P. B. Hitchcock, M. F. Lappert and X.-H. Wei, *Dalton Trans.*, 2006, 1181.
79. M. G. Gardiner, G. R. Hanson, M. J. Henderson, F. C. Lee and C. L. Raston, *Inorg. Chem.*, 1994, **33**, 2456.
80. E. Rijnberg, B. Richter, K.-H. Thiele, J. Boersma, N. Veldman, A. L. Spek and G. van Koten, 1998, **37**, 56.
81. G. R. Lief, D. F. Moser, L. Stahl and R. J. Staples, *J. Organomet. Chem.*, 2004, **689**, 1110.

82. A. D. Bond, A. D. Hopkins, A. Rothenberger, R. Wolf, A. D. Woods and D. S. Wright, *Organometallics* **20**, 2001, 4454.
83. U. Wirringa, H. Voelker, H. W. Roesky, Y. Shermolovich, L. Markovski, I. Usón M. Noltemeyer and H.-G. Schmidt, *J. Chem. Soc., Dalton Trans.*, 1995, 1951.
84. A. Kasani, R. McDonald and R. G. Cavell, *Organometallics*, 1999, **18**, 3775.
85. M. S. Hill and P. B. Hitchcock, *J. Chem. Soc., Dalton Trans.*, 2002, 4694.
86. S. R. Boss, R. Haigh, D. J. Linton, P. Schooler, G. P. Shields and A. E. H. Wheatley, *Dalton Trans.*, 2003, 1001.
87. I. J. Blackmore, V. C. Gibson, P. B. Hitchcock, C. W. Rees, D. J. Williams and A. J. P. White, *J. Am. Chem. Soc.*, 2005, **127**, 6012.
88. N. Nimitsiriwat, V. C. Gibson, E. L. Marshall, P. Takolpuckdee, A. K. Tomov, A. J. P. White, D. J. Williams, M. R. J. Elsegood, and S. H. Dale, *Inorg. Chem.*, 2007, **46**, 9988.
89. M. T. Gamer and P. W. Roesky, *Eur. J. Inorg. Chem.*, 2003, 2145.
90. J.-S. Herrmann, G. A. Luinstra and P. W. Roesky, *J. Organomet. Chem.*, 2004, **689**, 2720.
91. M. Cheng, D. R. Moore, J. J. Reczek, B. M. Chamberlain, E. B. Lobkovsky and G. W. Coates, *J. Am. Chem. Soc.*, 2003, **123**, 8738.
92. J. Prust, A. Stasch, W. Zheng, H. W. Roesky, E. Alexopoulos, I. Usón, D. Böhler and T. Schuchardt, *Organometallics*, 2001, **20**, 3825.
93. D. R. Moore, M. Cheng, E. B. Lobkovsky and G. W. Coates, *J. Am. Chem. Soc.*, 2003, **125**, 11911.
94. J. Lewinsky, Z. Ochal, E. Bojarski, E. Tratkiewicz, I. Justyniak and J. Lipkowski, *Angew. Chem., Int. Ed.*, 2003, **42**, 4643.
95. S. D. Allen, D. R. Moore, E. B. Lobkovsky and G. W. Coates, *J. Organomet. Chem.*, 2003, **683**, 137.
96. A. P. Dove, V. C. Gibson, E. L. Marshall, A. J. P. White and D. J. Williams, *Dalton Trans.*, 2004, 570.
97. S. Y. Lee, S. J. Na, H. Y. Kwon, B. Y. Lee and S. O. Kang, *Organometallics*, 2004, **23**, 5382.
98. S. D. Allen, C. M. Byrne and G. W. Coates, ACS Symposium Ser., 2006, **921**, 116.
99. C. M. Byrne and G. W. Coates, *Polymer Preprints*, 2005, **46**, 162.
100. M. H. Chisholm, J. Gallucci and K. Phomphrai, *Inorg. Chem.*, 2002, **41**, 2785.
101. Y. Wang, B. Quillian, P. Wei, H. Wang, X.-J. Yang, Y. Xie, R. B. King, P. v. R. Schleyer, H. F. Schaefer III and G. H. Robinson, *J. Am. Chem. Soc.*, 2005, **127**, 11944.
102. E. Maile and R. A. Fischer, *Chem. Vap. Deposition*, 2005, **11**, 409.
103. M. A. Malik and P. O'Brien, *Polyhedron*, 1997, **16**, 3593.
104. W. S. Rees, D. M. Green, T. J. Anderson, E. Bretschneider, B. Pathangey, C. Park and J. Kim, *J. Electron. Mater.*, 1992, **21**, 361.
105. S. Suh, L. A. Minea, D. M. Hoffman, Z. Zhang and W.-K. Chu, *J. Mater. Sci. Lett.*, 2001, **20**, 115.
106. T. J. Boyle, S. D. Bunge, T. M. Alam, G. P. Holland, T. J. Headley and G. Avilucea, *Inorg. Chem.*, 2005, **44**, 1309.
107. M. A. Beswick, N. L. Cromhout, C. N. Harmer, J. S. Palmer, P. R. Raithby, A. Steiner, K. L. Verhorevoort and D. S. Wright, *Chem. Commun.*, 1997, 583.
108. S. C. Goel, M. Y. Chiang, D. J. Rauscher and W. E. Buhro, *J. Am. Chem. Soc.*, 1993, **115**, 160.
109. M. H. Chisholm and K. Phomphrai, *Inorg. Chim. Acta*, 2003, **350**, 121.
110. H. Grützmacher, M. Steiner, H. Pritzkow, L. Zsolnai, F. Huttner and A. Sebal, *Chem Ber.*, 1992, **125**, 2199.
111. M. H. Chisholm, J. C. Gallucci, H. Zhen and J. C. Huffman, *Inorg. Chem.*, 2001, **40**, 5051.
112. C. von Hanisch and O. Rubner, *Eur. J. Inorg. Chem.*, 2006, 1657.
113. I. E. Gümrükçüoglü, J. Jeffery, M. F. Lappert, J. B. Pedley and A. K. Rai, *J. Organomet Chem.*, 1988, **341**, 53.
114. B. Fröhling, G. Kreiner and H. Jacobs, *Z. Anorg. Allg. Chem.*, 1999, **625**, 211.

8

Amides of the Group 13 Metals

8.1 Introduction

In our 1980 book[1] there were ca. 230 references to the primary literature for amido derivatives of Al-Tl and the structure of about thirty compounds had been determined. Three-quarters of these references concerned aluminium amido species. The number of well-characterised complexes of these metals thus greatly exceeded those of the neighbouring main group 1 and group 2 elements. The detailed study of group 13 metal amido species had, in fact, blossomed during the 1960s and '70s and an understanding of their synthesis, properties and structures was developing rapidly. The area has continued to flourish and the more recent work has been driven by their application as precursors for III–V materials,[2–7] as well as a lively interest in multiple bonding between the group 13 metals and nitrogen,[8–11] which in itself is part of the theme of multiple bonding to heavier main group elements.[12] There is also a rapidly expanding development of complexes with the metals in lower ($< +3$) oxidation states, particularly those with metal–metal bonds and this has led to the synthesis of a series of unique 'metalloid' cluster compounds stabilised by amido and related ligands.[13–16] Currently, about thirty lower oxidation state group 13 metal derivatives of amido or closely related ligands have been structurally characterized, whereas none was known in 1980.

This survey is confined to an overview of group 13 amides containing M-NR_2 (R = H, alkyl, aryl, silyl, etc.) groups; derivatives where nitrogen is part of a delocalized or polydentate ligand, are not generally covered. However, as in the 1980 book, iminometallanes (metal imides) of formula $(RMNR')_n$ are treated because of their close relationship to amides and also because of the relevance of the lower derivatives (n = 1, 2 or 3) for possible M–N multiple bonding.

Metal Amide Chemistry Michael Lappert, Andrey Protchenko, Philip Power and Alexandra Seeber

8.1.1 Synthesis

There are three major synthetic routes to group 13 metal amides as exemplified by reactions of Equations (8.1–8.3) (M = Al-Tl, X = halogen, M′ = Li, Na or K).

$$MX_3 + 3M'NR_2 \longrightarrow M(NR_2)_3 + 3M'X \quad (8.1)$$

$$MR_3 + HNR'_2 \longrightarrow R_2MNR'_2 + RH \quad (8.2)$$

$$LiAlH_4 + [NR_3H]X \longrightarrow 1/n(H_2AlNR_2)_n + LiX + H_2 + RH \quad (8.3)$$

The simple, salt elimination reaction of Equation (8.1) has been employed for amides of all the group 13 metals. In addition, it is currently the only well-established route to M(I) metal amides where M = Ga or Tl. The alkane elimination route of eqn. (8.2) is generally employed only for M = Al or Ga. This synthetic approach is also used for the metal imides $(RMNR')_n$ where a primary amine H_2NR' is the reactant. The use of metal hydrides, of which Equation (8.3) is but one example, is limited mainly to aluminium and, to a lesser extent, gallium because of the decreased stability of the heavier metal hydrides.

Several other synthetic approaches, for example (i) transamination, (ii) the direct reaction between a primary or a secondary amine with a metal(III) halide, or (iii) the direct reaction of a metal (e.g. aluminium) and an amine in the presence or absence of hydrogen, as well as a number of other approaches (outlined in Chapter 4 of Ref. 1) have been known for many years and are occasionally employed. The procedure (iii), in particular, has been used for matrix isolation studies of the simplest group 13 metal amides (see below). However, it is the three general approaches of Equations (8.1–8.3) that are by far the most commonly used.

8.1.2 M–N Bonding (M = Al, Ga, In or Tl)

Boron-nitrogen π-bonding was well-established in 1980, but π-bonding involving the heavier elements Al-Tl was virtually unexplored. Since the group 13 metals are quite electropositive,[17] and nitrogen is amongst the most electronegative of the elements, the M–N bond has considerably more ionic character than B–N. The M–N π–bonding is thus weakened because of reduced orbital overlap. The polar nature of the M–N bond also confers high reactivity and group 13 metal amides are very susceptible to reaction with protic or unsaturated reagents. Thus the group 13 metal amides react readily with moisture or oxygen to form thermodynamically favored M–O bonded products.

Numerous group 13 metal amide structures are now known and the metal-nitrogen bonds are usually observed to be up to 0.2 Å shorter than calculated from the sum of the covalent radii (see below).[8–12] Moreover, planar geometries are almost invariably observed at nitrogen for terminal M-NR_2 moieties. The M–N shortening and the planar nitrogen coordination are suggestive of multiple bonding.[8,12] However, it is noteworthy that short M–N bonds and planar nitrogen geometries are seen even where there is little reason to expect significant multiple bonding, for example, in coordinatively saturated complexes. Nonetheless, in metal amide or related imide species with unsaturated (i.e. low) metal coordination, multiple bonding has to be considered a possibility.

8.1.3 Multiple Character in M–N (M = Al – Tl) Bonds

When a group 13 metal having a coordination number of three or lower is bound to a group 15 element substituted by one or two organic groups, overlap of the p-orbitals from the M and N atoms may occur, at least in principle, to generate multiple bonding. The simplest case is when there is a single amido substituent bound to a diorgano group 13 metal centre as in the amide in **A** or imide **B**.

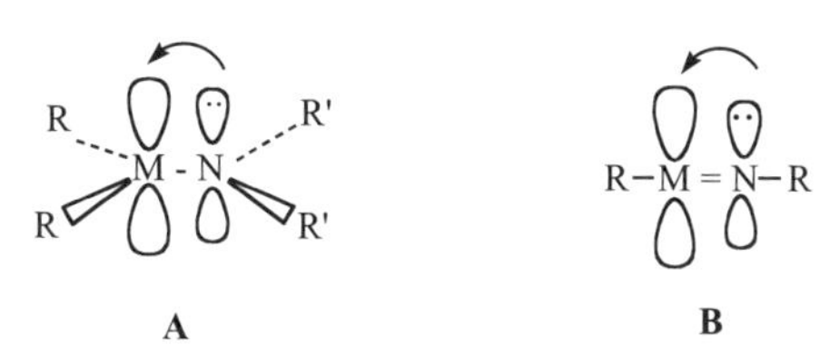

For **A**, the nitrogen coordination is invariably planar because of a low to zero inversion barrier which facilitates possible overlap of the M and N p-orbitals. The π-overlap in **B** is facilitated by a linear geometry. However, a further consideration in the case of **B** is the increasing stability of the lower oxidation state of the metal when descending the group.[17] This effect is most prominent in thallium and it implies an increasing stability of the monovalent unit R-M: which may lead to destabilization or distortion of structure **B**.

The synthesis of stable examples of **A** and **B** using large substituents and the subsequent study of the extent of their M–N π-bonding has been a major theme of group 13 element amides. These investigations have been paralleled by spectroscopic work on parent derivatives trapped in matrices at low temperature and by theoretical calculations.[18–27] There is a general consensus that in amidoalanes, -gallanes and -indanes, ionic resonance effects are dominant in shortening the M–N bond lengths and M–N π-bonding effects due to p-p orbital overlap are small and have a maximum energy value of ca. 10 $kcal{\cdot}mol^{-1}$. This view is consistent not only with the theoretical data but also with empirically predicted bond distances corrected for ionic effects.[28] The possibility of multiple M–N bonding via negative hyperconjugation has also been considered.[27] Other experimental observations support the lack of significant M–N multiple bonded character. For example, increasing the number of amido substituents on the group 13 element decreases the M–N bond length. This is the opposite to what is predicted on the basis of π-bonding because π-bonding to more than one amido group is expected to be weaker due to competition for overlap with the p-orbital on the metal. In addition, in related group 13 metal alkoxides and aryloxides, where the M–O bond has greater ionic character, the amount of shortening is not only greater[29] than in the amides but also exceeds the shortening attributable to the smaller size of oxygen relative to nitrogen. Thus, both factors favour ionic effects over the existence of significant π-bonding.

8.2 Aluminium Amides

8.2.1 Aluminium Parent Amides ($^{-}NH_2$ as Ligand)

The simplest aluminium amides have hydrogen substituents and the generation of species such as H_2AlNH_2, $HAlNH_2$ and $AlNH_2$ has been reported *via* the insertion of the metal atom into the N–H bond of NH_3.[18–21,23–25] These compounds have been trapped in frozen matrices at low temperatures and studied by IR and UV spectroscopy. For the simplest aluminium amide, H_2AlNH_2, calculations afforded an Al–N distance of 1.779 Å and an Al–N rotational barrier of 12.1 kcal mol^{-1}.[23] Similarly, in Me_2AlNH_2 the Al–N distance and rotation barrier were calculated to be 1.790 Å and 9.8 kcal mol^{-1}.[20] The latter species can be isolated as the trimer $[\{Me_2Al(\mu\text{-}NH_2)\}_3]$ as can its Bu^t substituted counterpart $[\{Bu^t{}_2Al(\mu\text{-}NH_2)\}_3]$.[30] Both feature hexagonal $(AlN)_3$ rings with Al–N bond lengths between 1.92 and 1.99 Å. The tetrahedral aluminium geometries in each complex are distorted to different extents and the Al_3N_3 ring in the methyl-substituted derivative has a skew-boat conformation while the *tert*-butyl derivative has a planar conformation. ^1H NMR studies showed that the trimer $[\{Me_2Al(\mu\text{-}NH_2)\}_2]$ is thermodynamically favoured over the dimer $[\{Me_2Al(\mu\text{-}NH_2)\}_2]$.[30b] The reaction of NH_3 with $[Al(SiMe_3)_3]$[31] or $[Al\{N(SiMe_3)_2\}_3]$[32] afforded the NH_2 bridged dimers $[\{(Me_3Si)_2Al(\mu\text{-}NH_2)\}_2]$ or $[(\{(Me_3Si)_2N\}_2Al(\mu\text{-}NH_2))_2]$ which have planar Al_2N_2 cores with bridging Al–N distances in the range 1.92–1.96 Å and an Al–N terminal distance of ca 1.85 Å in the latter complex. In addition, the reaction can produce the tetrametallic derivatives $[(\{(Me_3Si)_2N\}_2Al(NH_2)_2)_3Al]$[32] and $[\{(Me_3Si)_2Al(NH_2)_2\}_3Al]$[33] (Figure 8.1), in which six NH_2 groups surround the central aluminium in a distorted octahedral geometry (average Al–N bond lengths are near 2.02 Å).

Stable aluminium amides with terminal $-NH_2$ groups are rare. A simple example is $[Al(NH_2)_2(\{N(Dipp)C(Me)\}_2CH)]$,[34] which was obtained by ammonolysis of the corresponding dichloride; the $Al-NH_2$ bond lengths are 1.789(4) Å. The reaction of the $LiN(H)Pr^i$ with

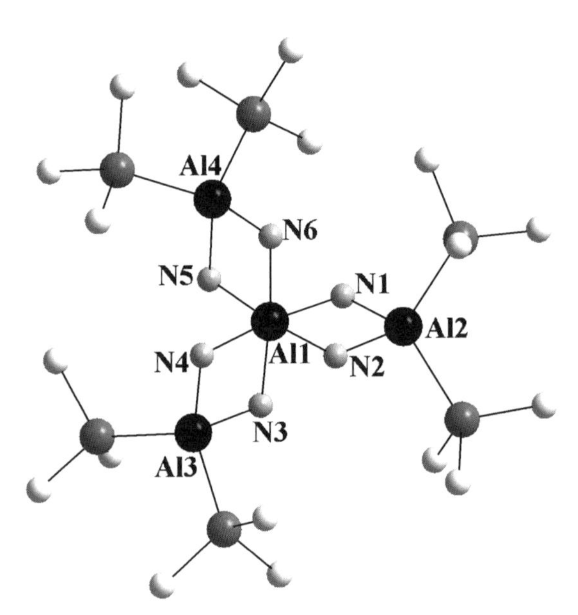

Figure 8.1 *Illustration of the structure of crystallized $[\{(Me_3Si)_2Al(NH_2)_2\}_3Al]$.[33] Aluminium atoms are shown as black spheres, silicon atoms are dark grey and nitrogen and carbon atoms are white. Selected bond lengths: Al1-N1 2.022(4), Al1-N2 2.017(5), Al1-N3 2.022(3), Al2-N1 1.935(4), Al3-N2 1.936(4), Al3-N3 1.923(5), Al1-Al2 2.911(3), Al1-Al3 2.905(2) Å*

$AlCl_3$ in the presence of H_2NPr^i afforded the unusual species $[AlCl_3(NH_2Pr^i)_2\{Al(NH_2)NH_3\}\{AlCl(NHPr^i)NPr^i\}]_2]$.[35] The Al–$NH_2$ distance is 1.880(6) Å.

8.2.2 Monomeric Aluminium Amides

Three-coordinate aluminium amides. Most organo substituted aluminium amides are associated through N–Al donor acceptor bonding to afford a coordination number of four or more at aluminium. The introduction of suitable bulky substituents at either the aluminium or nitrogen can prevent association and allow the isolation of monomeric compounds with three-coordination at the metal. For over two decades the only structurally characterized, three-coordinate aluminium amide was $[Al\{N(SiMe_3)_2\}_3]$ (Al–N = 1.78(2) Å).[36] Since the early 1990s, however, a number of other examples of three-coordinate aluminium amides have been synthesized and characterized. These include the tris(amides) $[Al\{NPr^i_2\}_3]$,[35,37] which has Al–N bond lengths in the range 1.793(4) to 1.801(4) Å,[37] and $[Al(tmp)_2\{N(H)Ph\}]$ (tmp = 2,2,6,6 –tetramethylpiperidido) in which Al–N is in the range 1.790(2)–1.813(2) Å.[38]

Several three-coordinate aluminium amido derivatives featuring either one or two amido ligands were obtained by reaction of the diorganoaluminium halide with a lithium amide. These include $[Bu^t_2AlNR_2]$ (R = Mes, $SiPh_3$), $[Bu_2^t\ Al\{N(R)SiPh_3\}]$ (R = C_6H_3-Pr^i_2-2,6 (Dipp), 1-adamantanyl, (1-Ad)) whose Al–N bond distances are in the range 1.8234(4) to 1.880(4) Å.[39] The variation in bond distances is mainly a consequence of the steric effects of the different substituents in these compounds although the bonds are all considerably shorter than the sum of the covalent radii of Al (1.3 Å) and sp^2 hybridized N (0.73 Å).[17] Barriers to Al–N rotation were observed for NR_2 = $NMes_2$, $N(SiPh_3)_2$ and $N(C_6H_3$–Pr^i_2-2,6)$SiPh_3$; only in the case of the last species, where the angle between the coordination planes at Al and N is small (i.e. 16.1°), could the barrier (9.9 kcal mol^{-1}) be attributed to Al–N π-bonding. A shorter Al–N bond length of 1.813(7) Å was observed in $[Mes_2AlN(SiMe_3)_2]$,[40] but in this derivative the angle between the Al and N coordination planes is 49°, thereby precluding significant Al–N p-pπ-bonding. Further shortening was observed in the three-coordinate primary monoamido derivative $[Trip_2Al\{N(H)Dipp)\}]$, [(Al–N = 1.784 (3) Å); Dipp = $C_6H_3Pr^i_2$-2,6, Trip = $C_6H_2Pr^i_3$-2,4,6].[41] The coordination planes at Al and N are almost coincident and an Al–N rotational barrier of 9.5 kcal mol^{-1} was estimated. A hydrogen elimination route was used to prepare $[Mes^*(H)AlN(SiMe_3)_2]$.[42]

Very bulky silylamido ligands have been used to stabilize the monoamides $[Me(Cl)AlN\{Si(Ph)Bu^t_2\}\{SiMeBu^t_2\}]$[43] (Al–N = 1.852(1) Å) and $[Me_2AlN\{SiPhBu^t_2\}\{SiMeBu^t_2\}]$ (Al–N = 1.869(2) Å).[43,44] The bis(amido) complexes $[MesAl\{N(SiMe_3)_2\}_2]$[37] (Fig. 8.2) and the $[MeAlN(Dipp)(CH_2)_3NDipp]$,[45] prepared by salt and alkane eliminations, have Al–N distances of 1.807 (avg.) and 1.763 (avg.) Å respectively. The latter may be the shortest known Al–N bond length for a stable amide. In addition, a series of bis(amido) alanes of formula $[(tmp)_2AlX]$ (X = Cl,[46] Br,[46] I,[46] ODipp,[38] SPh,[38] PPh_2,[38] $AsPh_2$,[38] Ph,[38] $Si(SiMe_3)_3$,[38] $Fe(CO)_2(\eta^5$-$C_5H_5)$,[38] $^1/_2Fe(\eta^5$-$C_5H_4)_2$[38]) having Al–N distances in the range 1.782(6) to 1.847(4) Å were prepared by salt elimination reactions. The bisamide $[ClAl\{N(Ph)(SiMePh_2)\}_2]$ (Al–N = 1.829(2) Å (avg.) was also prepared by this route.[47] Other aluminium amides such as $[Al\{N(H)Mes^*\}_3]$,[48] $[Mes^*Al\{N(H)Dipp\}_2]$,[49] and $[MeAl\{N(SiMe_3)_2\}_2]$[50] are also believed to have monomeric structures but they have not been characterized by X-ray crystallography.

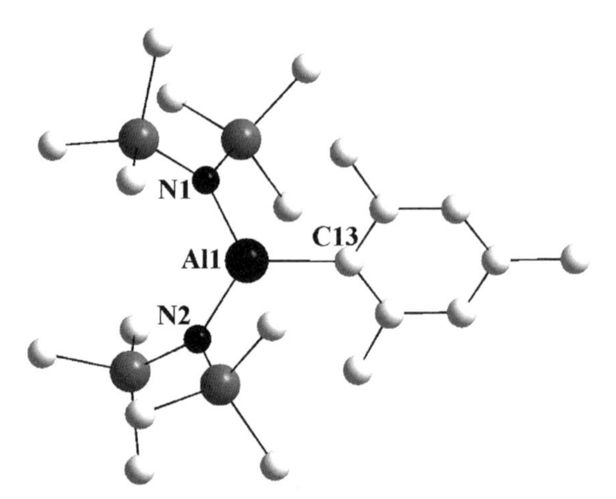

Figure 8.2 *Representation of the structure of [MesAl{N(SiMe$_3$)$_2$}$_2$].[37] Aluminium and nitrogen atoms are shown as black spheres and carbon atoms are white. Selected bond lengths: Al1-N1 1.804(2), Al1-N2 1.809(2), Al1-C13 1.970(3) Å*

Four-coordinate, monomeric aluminium amides. Monomeric, four-coordinate aluminium amido derivatives are now quite numerous, yet none had been structurally characterized before 1980. They may be divided into two broad categories: those which are Lewis base adducts and those in which four-coordination for aluminium is a result of complexation by a chelating ligand.

Several Lewis base-stabilized monomeric amidoalanes having a distorted tetrahedral metal coordination have been structurally characterized. A simple example is [Me$_3$N·Al{N-(H)Dipp}$_3$].[51] (Al—N$_{amido}$ 1.82–1.84 Å) and (Al—N$_{amino}$ 2.038(2) Å). Similar geometries are observed in the compounds [Me$_3$N·AlH{NH-2,6-Pr$^i_2$C$_6$H$_3$}$_2$],[51] [thf·Al{N(H)Si-(NMe$_2$)$_3$}$_3$][52] [Me$_3$N·AlH$_2$(tmp)][53] (where tmpH = 2,2,6,6-tetramethylpiperidine) and [Me$_3$N·AlH{N(Dipp)(CH$_2$)$_3$N(Dipp)}],[54] which also exhibit Al—N bond lengths to the amido group nitrogen atoms that are shorter than those to the amino nitrogen atoms by about 0.2 Å (av. 1.83 and 2.04 Å, respectively). With the exception of the thf adduct, these complexes were prepared by reaction of the amine with AlH$_3$·NMe$_3$. The adduct [Me$_3$N·Al-(Cl)H{N(SiMe$_3$)$_2$}] (Al—N = 1.823(4) Å) was prepared by salt elimination.[55] The phosphine oxide-stabilized amidoalane [Ph$_3$PO·Al{NMe$_2$}$_3$][56] has distorted tetrahedral geometry at both Al and P with average Al—N and Al—O distances near 1.81 and 1.84 Å, respectively. It was synthesized by direct reaction of [{Al(NMe$_2$)$_3$}$_2$] (see below) with Ph$_3$PO. A transamination reaction between piperidine and [{Al(NMe$_2$)$_3$}$_2$] afforded the adduct [Me$_2$NH·Al(piperdido)$_3$].[57] Several other Lewis bases have been used to form complexes to aluminium amides, as in [(thf)AlX(tmp)$_2$] (X = Cl or Br, tmp = 2,2,6,6-tetramethylpiperidido) and [(py)AlX(tmp)$_2$] (X = Cl or Br) [(py)AlI$_2$(tmp)] with Al—N distances in the range 1.809(5)–1.849(3) Å.[58]

Aluminium complexes of chelating amido ligands are perhaps the most common type of monomeric four-coordinate aluminium amide. A full listing of such complexes will not be provided and selected examples will suffice to illustrate synthetic approaches and structures.[59–72] A large variety of ligands (usually bidentate) has been used to obtain them – with the standard routes of alkane-, hydrogen- and salt- elimination reactions being most prominent.

A transamination route, as shown in Equation (8.4) ($SiR_3 = SiMe_3$, $Si(Bu^t)Me_2$),[59,60] has afforded complexes with unusual trigonal, monopyramidal coordination at aluminium.

$$[\{Al(NMe_2)_3\}_2] + 2N\{CH_2CH_2N(SiR_3)H\}_3 \longrightarrow 2Al\{N(SiR_3)CH_2CH_2\}_3N + 6HNMe_2 \quad (8.4)$$

The aluminium is coordinated by three silylamido nitrogens in an essentially trigonal planar fashion ($\sum{}^\circ Al = 359.7(3)$) with coordination by the amine nitrogen in an axial position. The Al–N(amido) bond lengths average 1.809(3) Å and the axial Al–N(amine) distance is 1.983(6) Å. The related podand derivatives $[\{HC(SiMe_2NR)_3\}Al{\cdot}thf]$ ($R = CH_2Ph$[61] or C_6H_4Me-4[62]), synthesized by salt elimination reactions, were reported recently.

Treatment of $Me_3N \cdot AlH_3$ with $H(Me_3Si)NCH_2CH_2N(SiMe_3)H$ afforded $[\{CH_2N(SiMe_3)_2]_2AlH\}_2]$ or $[\{CH_2N(SiMe_3)\}_2Al\{N(SiMe_3)\}\{CH_2CH_2N(H)SiMe_3\}]$ while the addition of this diamine to $[AlH_2(Cl) \cdot NMe_3]$ afforded $[\{Me_3SiN(H)CH_2CH_2N(SiMe_3)\}AlCl_2]$.[63] The complexes $[Me_2Al\{N(Ph)CH_2CH_2NH_2\}]$[69] and $[Me_2AlNCH_2C_5H_4N\text{-}2]$,[70] prepared by alkane elimination, have Al–N and Al–C bond lengths between 1.79–1.98 Å and 1.96–1.98 Å, respectively. The synthesis of the four-coordinate $(2\text{-}CyOC_6H_4NCy)AlMe_2$[71] was carried out by alkane elimination or by metathesis with the lithium salt of the ligand. Various synthetic methods were employed for the tridentate chelate complexes $(Me_3SiN\{NCH_2CH_2N(SiMe_3)\}_2)AlX$ [(X = Cl (salt elimination), H (chloride with $LiAlH_4$), Me (alkane elimination)). These were studied in connection with the ring-opening polymerization of propylene oxide and (*D,L*)-lactide. The Al–N (amide) and Al–N (amine) distances are in the range 1.811(1) to 1.836(2) Å and 1.998(1) to 2.019(2) Å.[72]

Higher coordinate monomeric aluminium amides. The distorted square pyramidal complexes $[MeAl\{OC(Me)C_6H_4NH\text{-}2\}_2]$,[73] $[ClAl\{NH(8\text{-}C_9H_6N)\}_2]$[74] and $[(Me_2N)Al\{N(Me)_2C(NPr^i)_2\}]$,[75] which were prepared by methane, salt, or amine elmination respectively, feature Me, Cl or NMe_2 groups in apical positions; the $Al{-}N_{amido}$ bond lengths are in the range 1.82 to 1.87 Å. The aluminium in $[Me_2Al(tedta)]$[76] (where tedtaH = tetraethyldiethylenetriamine) has distorted trigonal bipyramidal coordination geometry with the methyl groups in apical positions (mean Al–C distance is 1.99 Å); the tridentate ligand occupies the equatorial positions with shorter Al–N bond lengths to the amido nitrogen atom (1.84(1) Å) than to the amino nitrogen atom (2.350(8) Å). The hydrogen atoms in (volatile) $[H_2Al\{N(CH_2CH_2NMe_2)_2\}]$[77] occupy equatorial positions of the distorted trigonal bipyramidal aluminium atom (mean Al–H bond lengths are 1.54 Å); the tridentate amido ligand coordinates via the central amido nitrogen atom in equatorial and the two amino nitrogen atoms in axial positions (av. Al–N distances are 1.82 and 2.16 Å, respectively). With Me_3NMH_3 it formed $H_2[Al\{N(2\text{-}CH_2CH_2NMe_2)_2\}$ (M = Al, Ga).

The three bidentate ligands in the compound $[Al\{N(2\text{-}C_5H_4N)_2\}_3]$[74] are in a distorted octahedron around the aluminium atom. The aluminiums in $[AlCl_2\{N(C_5H_4N\text{-}2)(C_5H_4N\text{-}2)\}py_2]$, prepared as shown in Equation (8.5), and, also

$[Al\{N(C_5H_4N\text{-}2)_2\}\{N(C_5H_4N\text{-}2)(C_5H_4N\text{-}2)\}_2]$, have distorted octahedral geometry, with the dipyridyl ligand coordinating in a bidentate fashion in the former and in a tridentate mode in the latter.[78]

$$HN(C_5H_4N\text{-}2)_2 \xrightarrow{1.\ LiBu^n,\ 2.\ AlCl_3,\ 3.\ py} \qquad (8.5)$$

The use of macrocyclic ligands such as cyclam ([14]aneN_4) or the macrocyclic bis(amido) phosphine ligand $PhP(CH_2SiMe_2NSiMe_2CH_2)_2PPh$ (P_2N_2), has led to interesting products such as $[(Me_2AlCl)_2Al(N_4C_{10}H_{20})AlMe_2]$[79] or $[HAl(P_2N_2)]$,[80] but these complexes lie outside our scope.

8.2.3 Dimeric Aluminium Amides

Aluminum amide dimers constitute the largest class of aluminium amides. Only a selection of the large number of structurally characterized species can be discussed here. The majority feature either planar[31,49,51,63,64,81–96] or non-planar[31,51,65,82,96–104] four-membered $(AlN)_2$ or $(AlX)_2$ (X = an atom other than nitrogen) rings. The Al-$N_{bridging}$ and Al-$N_{terminal}$ bond lengths, generally range from 1.78–1.82 and 1.95–2.02 Å, respectively.

The simplest homoleptic, dimeric dialkylamide is $[\{Al(NMe_2)_3\}_2]$,[84,105] whose structure is illustrated in Figure 8.3. This compound was first reported in 1955[105] but it was not structurally characterized until 1993.[84] It was synthesized by reaction of three equivalents

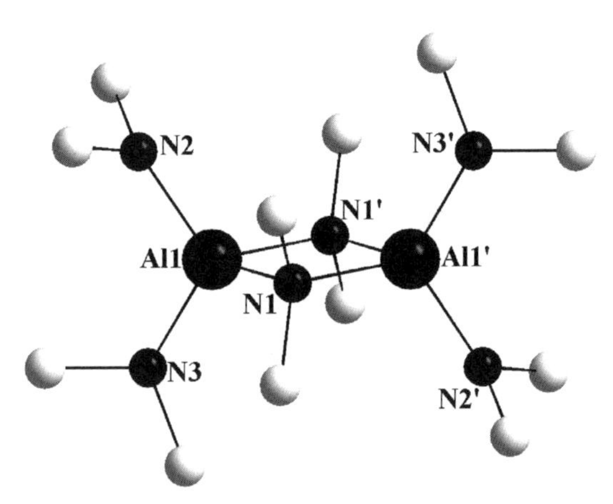

Figure 8.3 *Structural representation of $[\{Al(NMe_2)_3\}_2]$,[84] with the aluminium and nitrogen atoms shown as black spheres and carbon atoms in white. Selected bond lengths: Al1-N1 1.951 (2), Al1-N2 1.805(2), Al1-N3 1.799(2) Å*

of $LiNMe_2$ with $AlCl_3$.The aluminiums have distorted tetrahedral geometry and the bridging Al–N distances are considerably longer than the terminal Al–N bond lengths (av 1.96 and 1.80 Å, respectively). The related amides $Al(NEt_2)_3$ and $Al\{N(H)Pr^i\}_3$ may have similar dimeric structures.[35] Other simple dimeric amides are the parent amido derivatives $[R_2Al(\mu\text{-}NH_2)\}_2]$ ($R = SiMe_3$ or $N(SiMe_3)_2$) mentioned in Section 8.2.1.[31,32] The dimeric primary aluminium amide $[\{Me_2Al(\mu\text{-}NHPr^i)\}_2]$, synthesized by alkane elimination, was shown to exist as cis and trans isomers in a 2:1 ratio in the solid state and over a wide temperature range in solution.[106]

Non-planarity of the $(AlN)_2$ cores is usually a result of steric effects as in $[(AlMe_2\{\mu\text{-}N(R)C_6H_2Bu^t\text{-}2,4,6)_2\}$ ($R = SiMe_3$, CH_2Bu^t).[103] In $[\{(Me_3Si)_2N\}(Me_2N)Al(\mu\text{-}NMe_2)_2Al(H)\{N(SiMe_3)_2\}]$[99] the two aluminiums have different coordination environments (Figure 8.4). The bulky $^-N(SiMe_3)_2$ substituents are the most likely cause of the ring folding. The bidentate diamido ligands $C_6H_4\text{-}1,2\text{-}(NR)_2$ ($R = SiMe_3$ or CH_2Bu^t) produce fold angles of ca. 37–39° in the Al_2N_2 core of $[(Al\{(NR)_2C_6H_4\text{-}1,2\}Me(NR)_2)_2]$.[103a] Compounds $[AlMe_2\{N(R)C_6H_4{\cdot}N(H)R\text{-}1,2,3\}]$ and $[(AlMe_2)_2\{\mu\text{-}N(R)C_6H_4N(H)R\}]$ were also reported. Heating $[(AlMe_2\{\mu\text{-}N(H)C_6H_2Bu^t{}_3\text{-}2,4,6\})_2]$ afforded crystalline **1**.[103b]

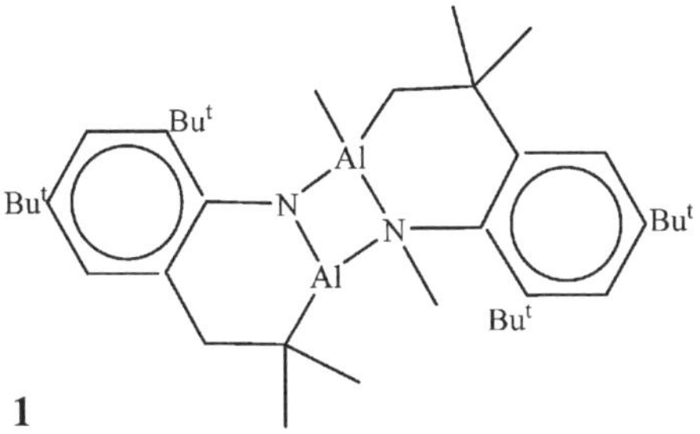

A dimeric amido complex **2** incorporating five-coordinate aluminium is produced by the reaction of $[\{Al(NMe_2)_3\}_2]$ with the azatrane $N\{CH_2CH_2N(Me)H\}_3$.[102]

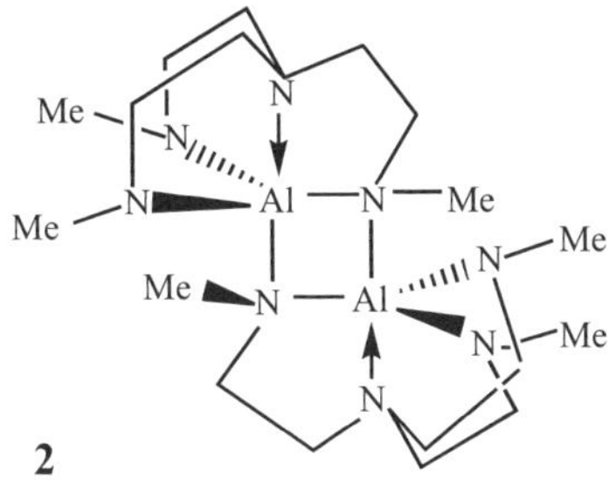

The terminal and bridging Al-N(amide) bond lengths are in the range 1.821(6)–1.844 (6) Å and 1.968(5)–2.049(5) Å, respectively, while the Al–N(amine) bonds are 2.124(6) and 2.160(6) Å.

Association may occur through bridging by atoms other than nitrogen in heteroleptic aluminium amides, particularly when the amido groups are sterically demanding. For example, in $[\{(\mu\text{-}H)Al(tmp)_2\}_2]$[98] and $[\{(\mu\text{-}H)Al\{N(SiMe_3)_2\}_2]$[99] the bridging Al–H bond lengths are in the range 1.67 to 1.78 Å. The halogen-bridged compounds $[\{Al(\mu\text{-}F)(tmp)_2\}_2]$[107] and $[\{Al(\mu\text{-}Cl)\{2,6\text{-}Pr^i{}_2C_6H_3N(SiMe_3)Dipp\}Cl\}_2]$,[108] feature folded,

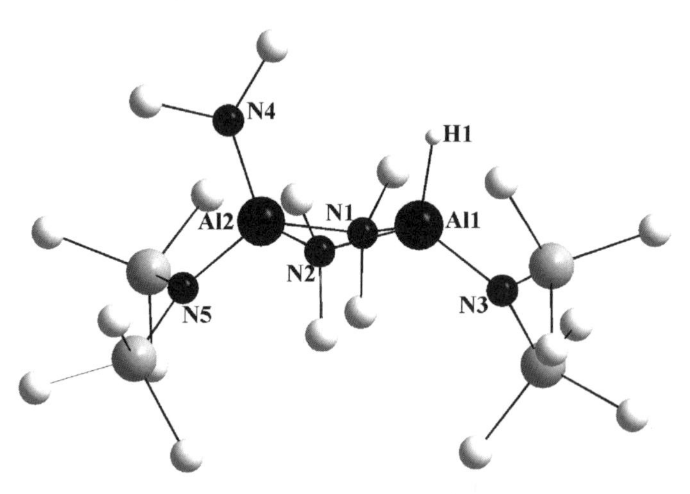

Figure 8.4 *Example of a folded $(AlN)_2$ ring in $[\{(Me_3Si)_2N\}(Me_2N)Al(\mu\text{-}NMe_2)_2Al(H)\{N(SiMe_3)_2\}]$.[99] Aluminium and nitrogen atoms are shown as black spheres, silicon atoms are grey and carbon atoms are white. Selected bond lengths: Al1-N1 1.948(7), Al1-N2 1.966(8), Al1-N3 1.836(4), Al1-H1 1.50, Al2-N1 1.985(8), Al2-N2 1.987(7), Al2-N4 1.782(8), Al2-N5 1.842(4) Å*

four-membered $(AlF)_2$ and $(AlCl)_2$ rings, respectively. These structures may be contrasted with those of the less crowded $[\{AlBr_2(\mu\text{-}NEt_2)\}_2]$[109] or $[\{Al(Cl)(NPr^i_2)(\mu\text{-}NEt_2)\}_2]$[35] where the bridging of the metals is via the amido group. Alkyl bridging is observed both in $[(MeAl(\mu\text{-}Me)\{N(Mes)(SiMe_2Mes)\})_2]$,[110] and in the sterically crowded dimer $[\{Me\text{-}\{Dipp(Me_3Si)N\}Al(\mu\text{-}Me)\}_2]$.[95] The asymmetric four-membered (AlNAlC) ring in $[Ph_2Al\{NBu^t)_2SiMe_2\}]_2$[68] is formed by the bridging coordination of one amido and one phenyl group (mean Al–N and Al–C distances in the ring are 1.99 and 2.18 Å, respectively). The two aluminiums have different coordination environments, with one having two terminal phenyl groups (Al–$C_{terminal}$ is 2.024(3) and 2.028(4) Å) and the second having one terminal phenyl and one terminal amido group (Al–C and Al–N are 2.004(5) and 1.824 (6) Å, respectively). Two different types of bridge are found in the X-ray crystal structure of $[(Me_2N)_2Al(\mu\text{-}NMe_2)(\mu\text{-}OBu^t)Al(NMe_2)_2]$.[96]

The reaction of Mes*AlH_2 with $Bu^tC{\equiv}N$ afforded $[(Mes^*Al\{N(Bu^t)CH_2\})_2]$ which contains a chair-shaped $(AlNC)_2$ ring in which the trigonal planar coordinated aluminiums are located above and below the $(NC)_2$ plane, with mean Al–N and Al–C bond lengths of 1.796(2) and 1.99(1) Å, respectively.[87] A six-membered Al_2N_4 ring structure is observed in $[LAlN(Me)NH\}_2]$ (L = HC{C(Me)N(Dipp)}$_2$), obtained by reaction of $Me(H)NNH_2$ with $LAlH_2$.[111]

8.2.4 Higher Aggregate Aluminium Amides

The most common higher structures are trimers as in $[\{Al(\mu\text{-}NH_2)R_2\}_3]$ (R = Me, Bu^t, see C.1).[30] The complex $[\{H_2AlNMe_2\}_3]$,[92] features a six-membered $(AlN)_3$ ring in a chair-like conformation (Al–N bond lengths are between 1.94 and 1.96 Å). The trimeric alane complex $[\{H_2Al(tmp)\}_3]$,[91] characterized by NMR and IR spectroscopy, was proposed to contain an $(AlH)_3$ ring with terminal amido and hydrogen ligands. Crystalline $[\{(Dipp(Me_3Si)N\}(F)Al(\mu\text{-}F)\}_3]$, synthesized by salt elimination, has an Al_3F_3 ring in which one of the F atoms deviates by ca. 0.6 Å from the almost planar Al_3F_2 moiety.[95] The $(AlN)_3$ chair-like ring in $[\{MeAl(\mu\text{-}PhNCH_2CH_2NH)\}_3]$,[69] synthesized by alkane elimination, has the ligands arranged in a propeller-like fashion. The aluminiums have distorted tetrahedral geometry with *endo*- and *exo*-cyclic Al–N distances of 1.83(2) and 1.92(2) Å. The chair-like structure in the trimetallic complex $[Me_8Al_3(tedta)]$[75] (tedtaH tetraethyldiethylenetriamine), obtained by alkane elimination, is formed by the tridentate coordination of the monoamido, bisamino tedta ligand in three different coordination modes. One aluminium bridges the central amido nitrogen and one amino nitrogen to form a planar $(AlN)_2$ ring ($Al{-}N_{amido}$ and $Al{-}N_{amino}$ bond lengths are 1.959(9) and 2.031(1) Å, respectively). The amido nitrogen in turn bridges to a second aluminium, which has distorted tetrahedral geometry completed by three methyl groups (Al–N and Al–C are 2.044(9) and ca. 1.99 Å, respectively). The remaining aluminium also has distorted tetrahedral geometry with a longer Al–N interaction with the remaining amino nitrogen atom (Al–N is 2.10(1) Å) and three methyl groups (Al–C bonds are between 1.98 and 2.00 Å). The reaction of AlR_3 (R = Me or Et) with ethylenediamine (en) in a 3 : 2 ratio affords $[\{RAl\{N(H)CH_2CH_2N(H)AlR_2\}_2]$. The structure of the methyl derivative features a central aluminium bound to an ethyl group and four amido nitrogens from two chelating en ligands. Two $AlMe_2$ moieties bridge the en nitrogens.[112]

The tetrametallic compounds $[\{Me_2Al\{NHCH_2\text{-}4\text{-}py\}AlMe_3\}_2]$, formed from $2AlMe_3$ and $[(Me_2Al\{N(H)CH_2\text{-}4\text{-}py\})_2]$.[83a] Treatment of $DippNH_2$ with $(Me_2AlF)_4$ affords the tetrametallic $(AlMe_2)_4(F)_2(NHDipp)_2$ which has an 8-membered $Al_4N_2F_2$ ring structure.[83b] The tetrametallic $[\{[CH_2N(SiMe_3)]_2AlH\}_2\{HAlN(SiMe_3)\text{-}CH_2CH_2NAlH_2\}]$[63] feature central, planar $(AlN)_2$ rings with mean Al–N bond lengths of 1.96 and 1.97 Å, respectively. An adamantanyl type compound $[Cl_2Al_4(NMe_2)_6(NMe)\text{-}O]$,[113] obtained adventitiously from $Al(NMe_2)_3$ and pmdeta in H_2O, has a similar structure to that in $[Al_4Cl_4\text{-}(NMe_2)_4(NMe)_2]$.[114] The Al_4N_5O scaffold consists of four bridgehead aluminums linked by NMe_2, an NMe and an oxo group. Two aluminiums carry chloro and two carry NMe_2 terminal groups. In $[Al_4Cl_4(NMe_2)_4(NMe)_2]$ the four bridgehead aluminiums are linked by four NMe_2 and two NMe groups and each chlorine is terminally bound to the metals (mean Al–O, $Al{-}N_{imido}$ and $Al{-}N_{amido}$ bond lengths are 1.70, 1.79 and 1.94 Å, respectively). The terminal amido and chloro ligands have average Al–N and Al–Cl distances of 1.79 and 2.16 Å, respectively (see Section B.6 for discussion of aluminium imides). The polyhedral $Al_4C_4N_4$ framework in the compound $[\{(\mu\text{-}AlH)(\mu_3\text{-}CH_2NBu^t)\}_4]$,[115] prepared from $AlH_3{\cdot}NMe_3$ and Bu^tNC, has a cage structure with C atoms in *exo*-positions that can be formally viewed as having been inserted into four Al–N bonds of an Al_4N_4 cube (mean Al–C and Al–N bond lengths are 1.99 and 1.97 Å, respectively). The core comprises six faces formed by two $(Al_2C_2N_2)$ rings in a boat conformation and four folded (Al_2CN_2) rings.

A ten-membered ring is observed in the compound $[L_2(AlMe_2)_4(thf)_2(toluene)_2]$[116] (**3**) (L = 1,3-bis(2,6-diisopropylanilino)squaraine). There are two bridging $AlMe_2$ groups (Al–N bond length is 1.916(2) Å) and two terminal $AlMe_2$ moeties that are bound to oxygen atoms located on external positions (Al-O distance is 1.823(2) Å) of the squaraine ligand. The treatment of $AlMe_3$ with various triaminosilanes afford several the cage structures one of which is illustrated in Equation (8.6) (R = $N(SiMe_3)Dipp$ or $N(SiMe_2\text{-}Pr^i)Dipp$).[117]

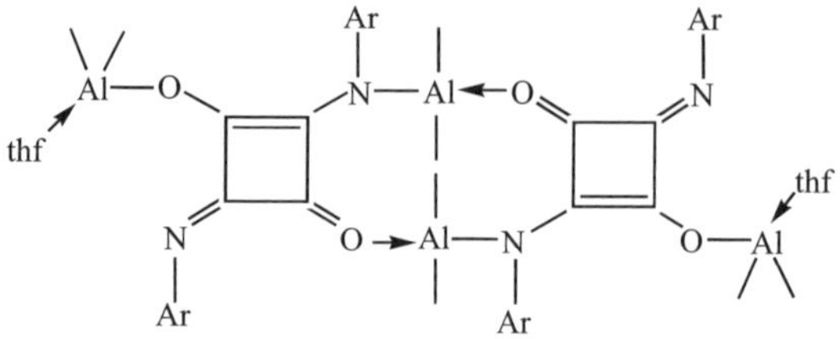

3

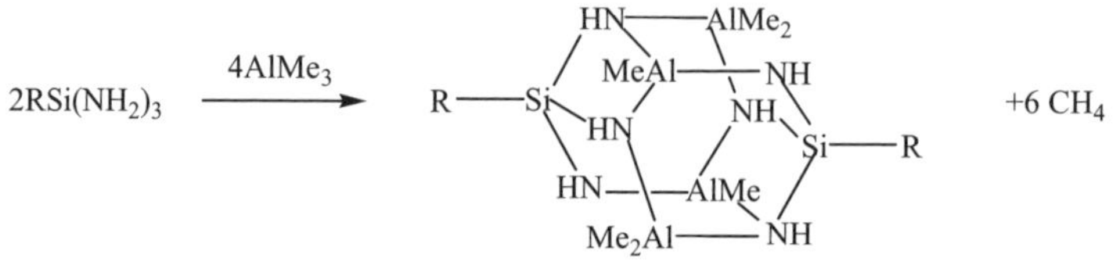

$$2RSi(NH_2)_3 \xrightarrow{4AlMe_3} [\{RSi(NH)_3\}_2(AlMe_2)_2(AlMe)_2] + 6\,CH_4 \quad (8.6)$$

The reaction of $AlCl_3$ with $LiNPr^i_2$ in the presence of H_2NPr^i yielded the heptametallic aggregate $[AlCl_3(H_2NPr^i)_2\{Al(NH_3)(NH_2)[AlCl(NHPr^i)(NPr^i)]_2\}_2]$,[34] as described in Section 8.2.1.

8.2.5 Heterometallic Aluminium Amides

A very large number of mixed metal aluminium amides has been reported. The majority are lithium-aluminium amide salts that exhibit a variety of different structures. Only a small number of these compounds are discussed here.[39,52,57,72,118–138] The simplest is $LiAl(NH_2)_4$ produced from the reaction of lithium and aluminium in liquid ammonia at 80 to 100 °C. The atomic arrangement of $LiAl(NH_2)_4$ has been studied by IR-spectroscopy and single crystal X-ray crystallography and was found to be a new variant of the $GaPS_4$-type structure.[118]

The commonly observed $(AlN)_2$ ring in aluminium amide complexes is also seen in dimers such as $[(Me_2Al\{NLi(thf)Bu^t\})_2]$ and $[(Bu^t_2Al\{NLi(thf)Ph\})_2]$.[119] In these the planar $(AlN)_2$ rings have distorted tetrahedral aluminium coordination and non-linear two-coordinate lithiums (mean Al–N and Li–N bond lengths are 1.88 and 1.84 Å, respectively). A similar structure is formed in the compound $[(Me_2Al\{NLi(thf)_2Ph\})_2]$,[120] in which the lithiums have distorted trigonal planar geometry (av. Li–N and Li–O distances are 2.02 and 1.93 Å, respectively) and the aluminiums have distorted tetrahedral geometry (av. Al–N and Al–C bond lengths are 1.89 and 1.98 Å, respectively).

The four-membered motif (AlN_2Li)[39,57,72,122–124] is found in several compounds, for example $[\{(2,6\text{-}Pr^i_2C_6H_3)NH\}_2Al\{\mu\text{-}NH(2,6\text{-}Pr^i_2C_6H_3)\}_2Li(thf)]$[124] and $[Mes_2Al\{NHBu^t\}_2Li(thf)]$,[119] which represent complexes with planar and non-planar (AlN_2Li) rings, respectively. The *endo*-cyclic Al–N and Li–N bond lengths in the two complexes are almost identical (av. 1.91 and 2.03 Å, respectively). The complex $[\{Li(Al(NHBu^t)_4\}_2]$[5] has a dimeric structure in which the lithium ions bridge the two $[Al(NHBu^t)_4]^-$ units to afford a distorted trigonal planar geometry for each lithium (mean Li–N distance is 2.08 Å).[125] The aluminium atoms have distorted tetrahedral geometry with longer bond lengths to the *endo*-cyclic nitrogen atoms (av. 1.90 Å) than to the *exo*-cyclic nitrogen atoms (1.795(4) Å).

A variety of more complex aggregate structures derived from a multidentate, mutlilithiated ligand as in [{cyclo-$(NHSiMe_2)_4AlLi(thf)_2\}_2$] or tridentate ligand as in [{$Me_2C\{CH_2N$-$(SiMe_3)\}_2AlH_2Li\}_2$][130a] or [{Dipp(Me_3Si)N}Si(NH_2)(NH)$_2AlMe_2$}{M(thf)}$_2$], (M = Li or Na)[130b] are also known.

The addition of trimethylaluminium to $LiN(SiMe_3)_2$ in toluene yielded the amido adduct [$LiN(SiMe_3)_2AlMe_3$],[131] which is formally an amido-aluminate salt; it comprises infinite chains of $LiN(SiMe_3)_2AlMe_3$ units linked by an interaction between the lithium ion and a methyl group from a neighbouring molecule (Li–C distance is 2.157(8) Å). The aluminium atoms have distorted tetrahedral coordination geometry with three methyl groups and one amido group (Al–C and Al–N bond lengths are 1.97–2.03 and 1.944(3) Å, respectively).

The addition of Li(tmp) to $AlBu^i_3$ in thf solution afforded 'Li$AlBu^i_3$tmp', which is an excellent reagent for the alumination of functionalized arenes.[132a] Mulvey and co-workers showed that the related sodium salt has the structure [(tmeda)Na(μ-tmp)(μ-Bu^i)$AlBu^i_2$][132b] and more recently demonstrated that the lithium salt could be crystallized as the solvate [{Pr^i_2N}(Ph)CO]Li(μ-tmp)(μ-Bu^i)$AlBu^i_2$].[133]

Separated ion pair structures for lithium aluminium amides are also commonly observed[118,121,123,134] as in [$Li(dme)_3$][Al(H){$N(CH_2Ph)_2\}_3$]][134] and [$Li(thf)_4$][Al{N-$(CH_2Ph)_2\}_4$]],[135] with mean Al–N bond lengths of 1.85 Å.

There are also several examples of heterometallic aluminium amide salts of heavier alkali metals. The reaction of [(Al(Cl)(μ-Cl){N(Dipp)$SiMe_3$})$_2$] with excess potassium or sodium phenylacetylide yielded the dimeric aluminum phenylethynyl complexes [(M-(thf){DippN($SiMe_3$)Al($CCPh)_3$})$_2$] (M = Na or K).[107] Each bridging potassium and sodium ion is π-coordinated by four phenylethynyl groups and one thf σ-donor. The sodium salt [Na-{(Dipp)($SiMe_3$)N}$AlEt_3$(thf)],[127] from [Na{Al(H)Et_3}] and HN(Dipp)$SiMe_3$, has an anionic framework of [(Dipp)($SiMe_3$)N}$AlEt_3$]$^-$ units connected *via* Na-C interactions (from η^4-Ph groups) to form a polymeric chain structure (Na···C distances are between 2.80 and 3.10 Å). The compound [{tmeda}Na(μ-tmp)(μ-Bu^i)$AlBu^i_2$],[132b] (mentioned above), synthesized from Na(tmp), Bu^i_3Al and tmeda, has a planar (Na···CAlN) ring; it is related to the corresponding alkali metal tmp derivatives of magnesium and zinc (see Chapters 3 and 7).

Mixed aluminium-gallium amide derivatives have also been reported, as in [$H_2Al\{N$-$(CH_2CH_2NMe_2)_2\}GaH_3$], prepared from $GaH_3{\cdot}NMe_3$ and [$H_2AlN(CH_2)_2NMe_2$]; it comprises a four-coordinate gallium atom with distorted tetrahedral geometry and a five-coordinate aluminium atom with distorted trigonal bipyramidal geometry.[77] The two metal centres are connected via a bridging amido nitrogen atom coordinating in apical position at the gallium atom and in an equatorial position at the aluminium atom (mean Ga–N and Al–N_{amido} bond lengths are 2.06 and 1.96 Å, respectively). The remaining tetrahedral positions at the gallium atom are occupied by hydrogens (Ga–H distances are between 1.43 and 1.44 Å). The trigonal bipyramidal geometry at the aluminium comprises two hydrogens in equatorial positions (Al–H distances are 1.44 to 1.45 Å) and axial amino nitrogen atoms from the tridentate amido ligand (Al–N_{amino} bond lengths are 2.17 to 2.23 Å).

Mixed aluminium-transition metals amido derivatives are also known. The reaction of $AlMe_3$ with [Mn{$N(SiMe_3)_2\}_2$(thf)] afforded [(Mn(μ-Me){$N(SiMe_3)_2AlMe_3$})$_2$],[136] which comprises manganese centered units dimerized via Mn(μ-Me)$_2$Mn interactions with

Mn–C distances of 2.201(11) and 2.285(11) Å. The aluminium atoms have distorted tetrahedral coordination geometry with Al-N and Al-C distances of 1.966(8) and 1.98–2.05 Å, respectively. The reaction of $[\{Al(NMe_2)_3\}_2]$ with $Mn_2(CO)_{10}$ or $Fe(CO)_5$ gave $[Mn_2(CO)_9\{C(NMe_2)OAl_2(NMe_2)_5\}]$ or $[Fe_2(CO)_8\{C(NMe_2)OAl(NMe_2)_2\}_2]$.[137] The manganese derivative features a planar $(AlN)_2$ ring (Al–N distances are 1.916(3) and 1.928(4) Å), in which one aluminium atom is in a terminal position with coordination of two amido groups (Al–N is 1.771(5) Å). The second aluminium is linked to the manganese centres by a bridging carbonyl unit, which coordinates end-on to the Mn–Mn bonded unit (Mn–Mn distance is 2.913(1) Å). The structure of the iron derivative is shown in Figure 8.5.

The complex $[(tmp)_2Al\{Fe(\eta^5\text{-}C_5H_5)(CO)_2\}]$ was synthesized by reaction of $(tmp)_2AlBr$ with $[NaFe[(\eta^5\text{-}C_5H_5)(CO)_2]$.[138] The aluminium atom has an unusual, slightly distorted trigonal planar geometry and has quite long Al–N bond lengths (1.847(4) and 1.862(4) Å); the Al–Fe distance is 2.45 Å.

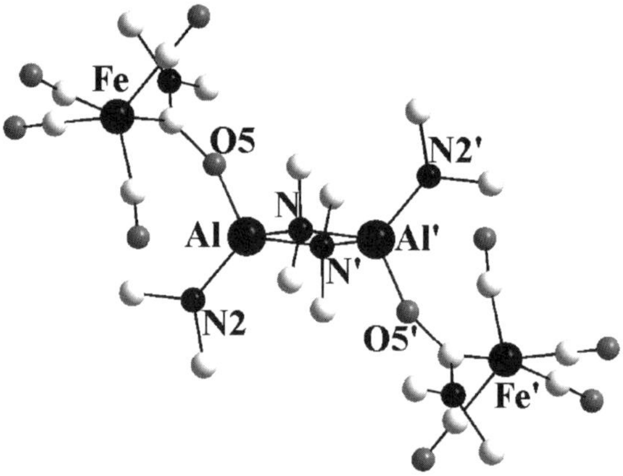

Figure 8.5 *Illustration of the structure of $[Fe_2(CO)_8\{C(NMe_2)OAl(NMe_2)_2\}_2]$[137] showing the planar $(AlN)_2$ core. Aluminium, iron and nitrogen atoms are shown as black spheres, oxygen atoms are grey and carbon atoms are white. Selected bond lengths: Al-N1 1.934(3), Al-N2 1.769 (2), Al-O5 1.73, Fe-$C_{carbonyl}$ 1.998(3), Fe-C(O) 1.77–1.80 Å*

8.2.6 Aluminium Imides (Iminoalanes)

These are compounds of the general formula $(RAlNR')_n$ (R = alkyl, aryl, hydrogen, halide; R′ = alkyl, aryl, silyl, hydrogen, n = 1, 2, 3, 4, 6, 7, 8) and related derivatives. They were first studied systematically during the 1960s and '70s and several examples had been structurally characterized by 1980. At that time, all the known compounds had three-dimensional aluminium-nitrogen cage structures composed of alternating AlR and NR′ units with aggregation numbers of 4, 6, 7 and 8. No lower aggregated imides, i.e. n = 1, 2 or 3, were known. In the ensuing period, particularly in the last two decades, there have been some notable advances, not only in the expansion and development of the known types of imides but also in the extension of the area to encompass the lower aggregates. Aspects of the synthesis, structure and reactivity of these compounds have been reviewed[139–142] and their simple derivatives have also been studied by computational methods.[143–147]

The major previously used routes to the higher cages (i.e. where $n \geq 4$) involved elimination reactions between aluminium alkyls and primary amines or the reaction of an aluminium hydride species, *e.g.*, $Me_3N{\cdot}AlH_3$ or $LiAlH_4$ with primary amines or primary ammonium halides. A number of other routes had also been used among which was the reaction between a nitrile and aluminium hydride, Equation (8.7).

$$Me_3N{\cdot}AlH_3 + RCN \longrightarrow 1/n(HAlNCH_2R)_n + NMe_3 \qquad (8.7)$$

This route has been recently employed to synthesize compounds such as $(HAlNCH_2R)_6$ ($R = Ph$, C_6H_4Me-4, $C_6H_4CF_3$-4),[148,149] $(XAlNCH_2Ad\text{-}1)_7$ ($X = H$ or F),[150] $(HAlNR)_6$ (R = 2-thiophenyl, $Fe(\eta^5\text{-}C_5H_5)(\eta^5\text{-}C_5H_4)$)[151] and related species such as $[(\{MeAl(\mu_2\text{-}F)\}_2N(Dipp))_4]$ whose structure consists of almost parallel octagonal Al_4F_3N rings in the $(MeAl)_4(\mu\text{-}F)_3(\mu_2\text{-}NDipp)$ moieties linked by two F and two NDipp groups that bridge the aluminiums in each ring.[152] In addition, more traditional routes involving elimination of alkanes from aluminium amides have yielded imides such as $[(MesNAlMe)_4]$,[88] $[\{Bu^tAlNSiPh_3\}_4]$,[153] $[MeAlNC_6H_4\text{-}4\text{-}F)_{4\text{ or }6}]$,[154] $[(MeAlNC_6F_5)_4]$,[155] or $[(MeAlNCH_2Ph)_6]$.[156]

The reaction of $AlMe_3$ and H_2NDipp ($Dipp = C_6H_3\text{-}Pr^i_2\text{-}2,6$) afforded the first (and currently the sole) example of trimeric iminoalane as shown in Equation (8.8).[157]

$$3AlMe_3 + 3H_2NDipp \longrightarrow (MeAlNDipp)_3 + 6CH_4 \qquad (8.8)$$

Its structure features a planar Al_3N_3 array with uniform Al–N bond lengths of 1.782(4) Å. It is an Al–N analogue of the well-known borazines and may be written in the 'alumazene' form

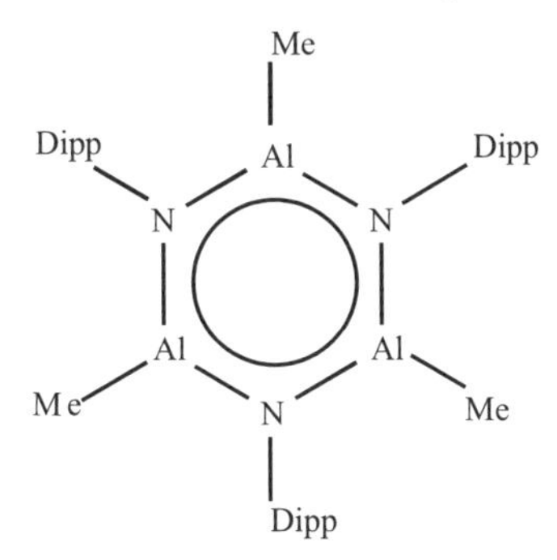

However, the π-character in this molecule is minimal because of the high polarity of the Al–N bond.[158–161] This is reflected in the fact that this molecule has an extensive chemistry that involves ready disruption of any potential aromatic character.[162–169] For example, it coordinates to metallocene trifluorides,[162,163] $OP(OMe)_3$,[164] $Me_3SiOP(O)O_2^{2-}$,[165,166] $OPPh_3$,[167] $MeS(O)_2O$[168] and more recently pyridine and 4-dimethylaminopyridine.[169a] One of these complexes is illustrated in Figure 8.6.[165] Treatment of $(MeAlNDipp)_3$ with Me_3SnF sequentially replaces the aluminium methyls with fluoride with formation of thf adducts as in $[\{(thf)(F)AlNDipp\}_3]$.[169b]

Dimeric and monomeric iminoalanes have also been stabilized with use of bulky substituents. For example, the reaction of $(AlCp^*)_4$ ($Cp^* = C_5Me_5$) with Me_3SiN_3 afforded the iminoalane illustrated in Equation (8.9)[170] which has a planar Al_2N_2 core whose average Al–N distance is ca. 1.81 Å. A simpler dimeric iminoalane $[\{\eta^1\text{-}C_5Me_5AlNSiBu^t_3\}_2]$[171] was obtained upon the reaction of $(AlCp^*)_4$ with $Bu^t_3SiN_3$. Other examples of dimers include $[(Mes^*AlNPh)_2]$ (Al–N = 1.824 Å (avg.)),[172] $[(Mes^*AlNSiPh_3)]$,[173]

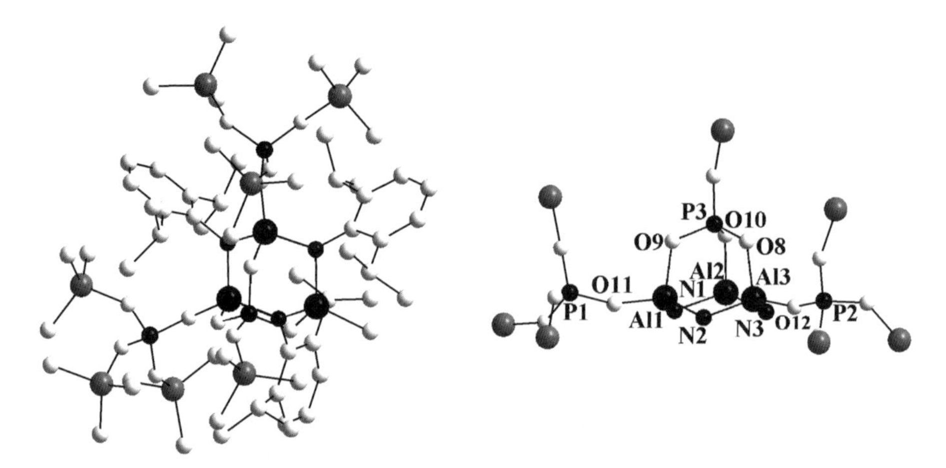

Figure 8.6 *Left: Structural representation of the trimeric complex [(MeAl)]{2,6-$Pr^i_2C_6H_3N$}$_3${Al-(OP($OSiMe_3$)$_3$)}$_2${$O_3POSiMe_3$}][165] with the ($Al_3N_3O_3P$) core highlighted in bold. Right: view of the adamantane-type core of [(MeAl)]{2,6-$Pr^i_2C_6H_3N$}$_3${Al(OP($OSiMe_3$)$_3$)}$_2${$O_3POSiMe_3$}] with carbon and hydrogen atoms removed. Aluminium, nitrogen and phosphorus atoms are shown as black spheres, silicon atoms are grey and oxygen and carbon atoms are white. Selected bond lengths: Al1-O9 1.846(5), Al2-O10 1.885(5), Al3-O8 1.817(5), Al1-N1 1.801(6), Al1-N3 1.812 (6), Al2-N1 1.853(6), Al2-N2 1.844(6), Al3-N2 1.809(6), Al3-N3 1.819(6) Å*

(Al–N = 1.842 Å (avg.)) and [{(η^5-C_5H_5)AlNDipp}$_2$] (Al–N = 1.804 Å avg.)[174] obtained by hydrocarbon or hydrogen elimination reactions. In addition, there are base-stabilized dimeric iminoalanes of formula [(Bu^t(thf)AlNSiMe$_3$)$_2$],[175] [{H(Me_3N)AlNDipp}$_2$][51] and [{(η^5-C_5Me_5)(dmap)AlNSiRR$'_2$}$_2$] (dmap = 4-dimethylaminopyridine; R, R$'$ = Et Pr^i; R = Bu^t, R$'$ = Me).[175] The thermolysis of [(Bu^t(thf)AlNSiMe$_3$)$_2$] at 150°C gave [(Bu^tAlN-SiMe$_3$)$_4$] in Equation (8.10).[176] However, this does not have the expected heterocubane structure but is comprised of three annulated four-membered rings.

$$(Cp^*Al)_4 + 4Me_3SiN_3 \xrightarrow{-4N_2} [(Me_3Si)_2N(\eta^6\text{-}^*Cp)Al\text{–}N(Al(Cp^*))(Al(N(SiMe_3)_2))N\text{–}Al(Cp^*)_2] \qquad (8.9)$$

$$2\{Bu^t(thf)AlNSiMe_3\}_2 \xrightarrow{150^\circ C} [(Bu^tAlNSiMe_3)_4] + 4\ thf \qquad (8.10)$$

The reaction of $[(Me_2NAlH_2)_3]$ with H_2NNMe_2 afforded $[Al_4(NHNMe_2)_4(NNMe_2)_4(NH_2NMe_2)_2]$ which has a similar ladder-like Al_4N_4 core framework.[177] The reactivity of $[\{H(Me_3N)AlNDipp\}_2]$ with iodine afforded interesting derivatives such as $[DippN\text{-}\{AlI_2(NMe_3)\}_2]$.[178] Other interesting iminoalane derivatives include the Al–N cyclopentadienide analogue Mes*Al(NPh)(AlMes*)(NPh)NPh[179] and the trilithio capped Al_4N_6 heteroadamantanyl iminoalane anion, $[(HAl)_4(NPh)_6\{Li(OEt_2)\}_3]$.[180]

No structure of a stable monomeric iminoalane is currently available. However, the β-diketiminate stabilized monomeric $[(L)]AlNC_6H_3(C_6H_2Pr^i_3\text{-}2,4,6)_2\text{-}2,6]$[181] (L = HC{C-(Me)N(Dipp)}$_2$) is believed to have a very similar structure to its gallium analogue (cf. Section 8.3.5) which has a very short Ga–N distance of 1.701 Å consistent with Ga–N multiple bonding. It readily underwent intramolecular additions to the Al–N bond to give crystalline products.[182] For an account of the chemistry of :Al(L), see Ref. 184.

8.2.7 Aluminium(I) Amides

The simplest, stable Al(I) amides are the tetrametallic species $[(Cp^*Al)_3AlN(SiMe_3)_2]$ (**5**)[185]

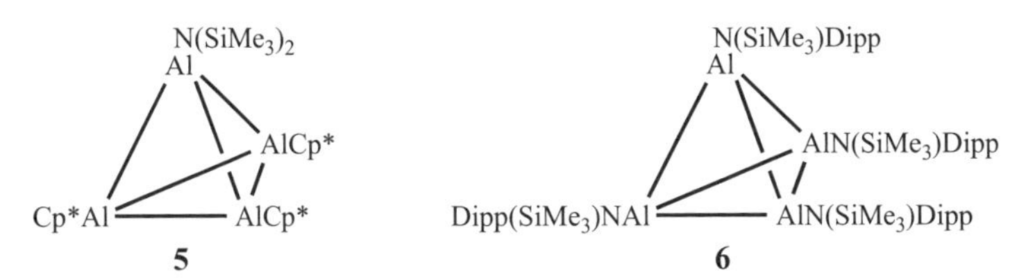

and $[Al_4\{N(SiMe_3)Dipp\}_4]$ (**6**).[186] Compound **5**, obtained by treatment of $(AlCp^*)_4$ with $Li\{N(SiMe_3)_2\}_2$ in the presence of tmeda, features a distorted tetrahedron of four aluminiums whose average Al(Cp*)-Al(amide) distances of 2.66 Å, are ca 0.1 Å shorter than the Al-(Cp*)-Al(Cp*) distances; the Al–N bond length is 1.847(2) Å. Compound **6** was synthesized by reduction of the iodide precursor (which has a dimeric iodide-bridged structure $[(\{Dipp(Me_3Si)N\}Al(\mu\text{-}I)(I))_2]$ with Na/K alloy.[186] The almost perfectly tetrahedral Al_4 core has an average Al–Al bond length near 2.62 Å and an average Al–N bond distance of 1.815 Å.

The most fascinating lower oxidation state aluminium amides are the series of anionic clusters[13–16,187] formula $[Al_7(NR_2)_6]^-$,[188] $[Al_{12}(NR_2)_6]^-$,[189] $[Al_{14}(NR_2)_6]^{2-}$,[190] $[Al_{69}(NR_2)_{18}]^{3-}$,[191] and $[Al_{77}(NR_2)_{20}]^{2-}$,[192] (R = SiMe$_3$) described by Schnöckel and co-workers. These were synthesized by the reaction of a metastable 'Al(I)X' (X = halogen) solution, generated at low temperatures, with an alkali metal salt $MN(SiMe_3)_2$. Under different conditions many stages in the aggregation and disproportion ($AlX \rightarrow Al(O) + AlX_3$) process can occur and the addition of $[N(SiMe_3)_2]^-$ or other anionic bulky ligands allow some of these aggregates, to be crystallized.[13–16] A distinguishing feature of these clusters is that they are 'metalloid' in that they have metal-metal bonding involving metal centres bonded only to other metal atoms. Their structures can be described in terms of sections from the structure of aluminium metal. Finally it is noteworthy that, of the anionic clusters, only $[Al_7\{N(SiMe_3)_2\}_6]^-$ corresponds to an aluminium oxidation state of +1. The others possess average metal oxidation states that are considerably lower. The Al_{77} cluster is paramagnetic having a delocalized electron.[192]

8.3 Gallium Amides

8.3.1 Introduction

In our 1980 book there were fewer than fifty references to the primary literature for gallium, indium and thallium amides. Since then the range of known compounds has undergone much expansion. As is the case for aluminium, interest in gallium amides owes much to their possible application in MOCVD for the synthesis of metal nitrides.[2–7] The synthesis, structures and reactivity of gallium amides, in many instances, parallel those of the corresponding aluminium species. However, their total number remains considerably lower than that of aluminium. Gallium amides, along with indium amides, were reviewed in 2001 and tables of structurally characterized compounds were provided.[11] Consequently, this section will focus only on major developments and more recent publications.

Contrary to the usual trend, the covalent radius of gallium is ca. 0.05 Å smaller that that of aluminium, which is reflected in Ga–Ga distances which are ca. 0.1 Å shorter than the Al–Al distances in metal-metal bonded compounds that carry the same substituents.[12–16,187] On this basis it might be expected that Ga–N bond lengths should be ca. 0.05 Å shorter than the corresponding Al–N bonds. However, experimentally determined Ga–N distances in amido derivatives are usually ca. 0.05 Å longer than corresponding Al–N distances. This may be illustrated by the data in Table 8.1 which features several pairs of aluminium and gallium amides that carry identical ligand sets (isoleptic).

It can be seen that the difference between the Al–N and Ga–N bond lengths range from ca. 0.05–0.10 Å and that the Al–N distances are always shorter. This finding supports the importance of ionic effects for the M–N bond lengths.[8,28] In essence, the lower electropositive character in gallium does not induce as great a degree of M–N shortening as that seen for aluminium.[28] Significantly, the differences between the less polar metal-carbon bond lengths is less pronounced and is in keeping with the lower ionic character of the M–C bond. For the M–Si bonds ionic effects are much lower and the smaller size of gallium

Table 8.1 Metal-nitrogen bond lengths in isoleptic aluminium and gallium amides

Compounds	M–N (Å)	M–C (Å)	Reference
$[Trip_2Al\{N(H)Dipp\}]^a$	1.784(3)	1.960(4) avg.	41
$[Trip_2Ga\{N(H)Dipp\}]$	1.847(12)	1.99(1) avg.	41
$[Bu^t_2Al\{N(1\text{-}Ad)SiPh_3\}]$	1.849(4)	2.02(5) avg.	39
$[Bu^t_2Ga(N(1\text{-}Ad)SiPh_3\}]$	1.924(2)	2.05(1) avg.	41
$[(Me_3Si)_3SiAl(tmp)_2]^b$	1.846(2)	$2.513(6)^c$	38
$[(Me_3Si)_3SiGa(tmp)_2]$	1.908(3) 1.913(2)	$2.468(1)^c$	193
$[Al\{N(SiMe_3)_2\}_3]$	1.78(2)		36
$[Ga\{N(SiMe_3)_2\}_3]$	1.868(1), 1.870(6)		37, 82, 194
$[Al_2(\mu\text{-}NMe_2)_2(NMe_2)_4]$ (Fig. 8.3)	1.803(3) (terminal) 1.965(14) (bridging)		84
$[Ga_2(\mu\text{-}NMe_2)_2(NMe_2)_4]$	1.855(6) (terminal) 2.013(7) (bridging)		84

^{a}Trip = $C_6H_2Pr^i_3$-2,4,6, Dipp = $C_6H_3Pr^i_2$-2,6
^{b}tmp = 2,2,6,6-tetramethylpiperidido;
^{c}M-Si distances

reasserts itself which is reflected in the fact that the Ga–Si bond[193] is ca. 0.05 Å shorter than that of Al–Si.[38] In effect, the more electronegative character of gallium in comparison to aluminium results in less shortening due to the smaller ionic correction in the less polar bonds formed by gallium. It is interesting to note that theory predicts that Ga–N π-bonding should be slightly stronger than Al–N π-bonding.[23,26] However, there are no experimental data to substantiate this prediction.

The simplest gallium amides have been generated via direct reaction of the metal with ammonia and trapped in matrices at low temperature. The structure of the primary amide H_2GaNH_2 is planar with a calculated Ga–N distance of 1.8221 Å (cf. 1.7790 Å for H_2AlNH_2).[23] Other calculations afford similar values.[26,27] Under ambient conditions H_2GaNH_2 is a trimer and displays a six-membered $(GaN)_3$ ring in a chair conformation with a mean Ga-N distance near 1.99 Å.[195,196] A very similar structure is observed for the related parent amido derivative $[\{Me_2Ga(\mu\text{-}NH_2)\}_3]$ which has Ga–N distances in the range of 1.93–2.05 Å.[197] However, the Bu^t substituted trimer $[\{Bu^t{}_2Ga(\mu\text{-}NH_2)\}_3]$ (Ga–N = 2.02 Å (avg.))[198] has a planar $(Ga–N)_3$ ring. An unusual parent amide of formula $[Ga(NH)_2F_2{\cdot}NH_3]$ is formed by the reaction of NH_4F with gallium metal in the presence of In. The structure consists of layers of corner bridged $Ga(NH_3)_2F_4$ and $Ga(NH_2)F_4$ octahedra (one F^- being common) with Ga–N distances of 1.943(3) and 1.901(6) Å, respectively.[199] The primary gallium amide derivative $[HC\{C(Me)N(Dipp)\}_2Ga(NH_2)_2]$[200] containing terminal NH_2 groups is stabilized by the bulky β-diketiminate ligand.

8.3.2 Monomeric Gallium Amides

Numerous examples of monomeric gallium amides have been synthesized and characterized. They were obtained for the most part by salt elimination reactions. The simplest compounds involve three-coordinate gallium that carries one, two, or three amido substituents. For the three-coordinate monoamides, several, such as $[Mes^*Ga(Cl)\{N(H)Ph\}]$, (Ga–N = 1.832 (10) Å avg.),[37] $[Mes^*Ga(Cl)\{N(SiMe_3)_2\}]$ (Ga–N = 1.867(10) Å),[201] $Trip_2Ga\{N(H)Dipp\}$ (Ga–N = 1.847(12) Å avg.),[27] $Trip_2GaNPh_2$ (Ga–N = 1.878(7) Å)[41] and $[Mes^*{}_2Ga\{N(H)\text{-}Ph\}]$ (Ga–N = 1.874(4) Å)[37] have almost coincident gallium and nitrogen coordination planes. Restricted rotation around the Ga–N bond was observed by VT NMR spectroscopy in some of these, but it was impossible to say if this was due to π-bonding or steric hindrance. It is noteworthy, however, that some monoamides, $[Bu^t{}_2Ga\{N(R)SiPh_3\}]$ (R = Bu^t, Ga–N 1.906(5) Å; R = 1-adamantanyl, Ga–N = 1.924(2) Å)[27] or $[Et_2Ga\{N(Bu^t)BMes_2\}]$ (Ga–N 1.937(3) Å)[202] have high (70–90°) torsion angles between the Ga and N coordination planes. All these complexes have significantly longer Ga–N distances than in those where the planes are coincident. While this is consistent with the absence of Ga–N π-bonding, it may be argued that greater steric hindrance causes both the high torsion angle and longer bonds. The complex $[Et_2Ga\{N(Bu^t)BMes_2\}]$ has the longest Ga–N distance probably because of competitive bonding by the $BMes_2$ group. Several monomeric bisamido complexes are also known including the tmp derivatives $[tmp_2GaX]$ (X = Ph,[203] OPh,[203] PBu$^t{}_2$,[204] and Si$(SiMe_3)_3$[203]), the P–P bonded $[(\{tmp_2Ga\{Bu^tP\}\})_2]$[204] as well as $[Mes^*Ga\{N(H)Ph\}_2]$,[37] $[ClGa\{N(SiMe_3)_2\}_2]$[37] and $[ClGa\{N(Dipp)SiMe_3\}\{N(Dipp)SiMe_2N(Dipp)SiMe_3\}$[205] that have Ga–N bond lengths in the range 1.83–1.91 Å. The tmp ligand also supports three-coordinate geometry and a Ga–Ga bond in $[(tmp)_2GaGa(tmp)_2]$ (Ga–Ga = 2.52(1) Å, Ga–N = 1.90(4) Å)[206] as does the bidentate ligand $[Me_2C\{CH_2N(SiMe_3)\}_2]^-$ (L^-) in [LGaGaL] (Ga–Ga = 2.385(1) Å, Ga–N = 1.832(3) Å).[134] The three-coordinate, monomeric tris(amido)gallium derivatives, the silylamide $[Ga\{N(SiMe_3)_2\}_3]$[37,82,194] (cf. Table 8.1) and the dialkylamide $[Ga(NCy_2)_3]$, (Ga–N = 1.836(6) Å avg.)[134] have been structurally characterized.

With slightly less bulky substituents, monomeric gallium amide complexes with one or more Lewis base donors are readily formed. Among the simplest of these are $[H_2Ga\{N(SiMe_3)_2\}(quin)]$ (quin = quinuclidine),[207] $\{HGa\{N(SiMe_3)_2\}_2(quin)]$,[207] and $[HGa\{N(H)\,Dipp\}_2(quin)]$,[207] synthesized by addition of the lithium amide to $H_2GaCl(quin)$ or $[HGaCl_2(quin)]$. Each has distorted tetrahedral metal coordination, as have $[Cl_2Ga\{N(H)SiMe_3\}(thf)]$ (Ga-N = 2.026(3) Å)[208] and $[Cl_2Ga\{N(SiMe_3)_2\}(quin)]$.[209] The Ga–N bond lengths in the four-coordinate complexes are generally in the range of 1.86–2.02 Å. The complex $[Ga\{N(H)Dipp\}_3(py)]$ has a Ga–N(amido) bond length of 1.877(6) Å (avg.) and a Ga–N(py) of 2.051(7) Å.[47] The silylated azatrane complex $[Ga\{N(SiMe_3)CH_2CH_2\}_3N]$ is believed to have trigonal monopyramidal coordination like its aluminium congener.[102]

Higher-coordinate monomeric gallium amides are exemplified by $[H_2Ga\{N(CH_2CH_2NMe_2)_2\}]$,[77,210] which has a distorted trigonal bipyramidal geometry at gallium with the tripodal ligand coordinating *via* the amido group in equatorial position (Ga–N bond length is 1.874(2) Å) and the two amino groups coordinating in axial positions (Ga–N distance is 2.2826(16) Å). The two hydride donors coordinate in the remaining equatorial positions with Ga–H bond lengths of 1.47(2) Å. Similar structures are observed in the compounds $[Cl_2Ga\{N(CH_2CH_2NEt_2)_2\}]$,[66] $[Cl_2Ga\{N(CH_2CH_2NMe_2)_2\}]$,[77] and $[(N_3)_2Ga\{N(CH_2CH_2NEt_2)_2\}]$,[66] where the tridentate ligand occupies one equatorial (mean $Ga–N_{amide}$ are 1.84, 1.85 and 1.84 Å, respectively) and two axial positions (average $Ga–N_{amine}$ are 2.29, 2.20 and 2.26 Å, respectively) and the remaining equatorial positions of the distorted trigonal bipyramidal geometry are occupied by Cl (average Ga-Cl distance in both complexes is 2.21 Å) and azido (mean Ga–N distance is 1.93 Å) groups, respectively. These compounds were studied primarily because of their use as MOCVD precursors from GaN.

8.3.3 Associated Gallium Amides

Most gallium amides are dimers which feature a planar $(GaN)_2$ core with Ga–N bond lengths in the range 1.96–2.10 Å.[86,91,211–244] As is the case for the monomers many of these compounds have been listed and described in a review.[11] As a result this discussion will focus on fundamental trends and the more recent developments.

The synthetic methods used usually involve salt, alkane, or hydrogen elimination. The simplest monodialkylamido species is $[\{H_2Ga(\mu\text{-}NMe_2)\}_2]$[242] which was formed by elimination of hydrogen from the adduct $Me_2(H)N{\cdot}GaH_3$ at room temperature. Its structure, which was determined by electron diffraction, afforded a Ga–N distance of 2.027(4) Å. The structure may be contrasted with that of its aluminium analogue which is trimeric.[92] The higher (Al > Ga) aggregation maybe attributed to the larger size of aluminium. The difference in the M–N bond lengths (cf. Table 8.1) is typical.

Large differences between the structures of dimeric aluminium and gallium amides are unusual as exemplified by the structure of the simplest homoleptic gallium dialkylamide $[\{Ga(NMe_2)_3\}_2]$ (cf. Table 8.1) whose structure closely resembles that of its aluminium analogue in that it has a planar Ga_2N_2 core, tetrahedral coordination at each gallium and terminal and bridging M–N distances that are ca. 0.05 Å longer than the corresponding

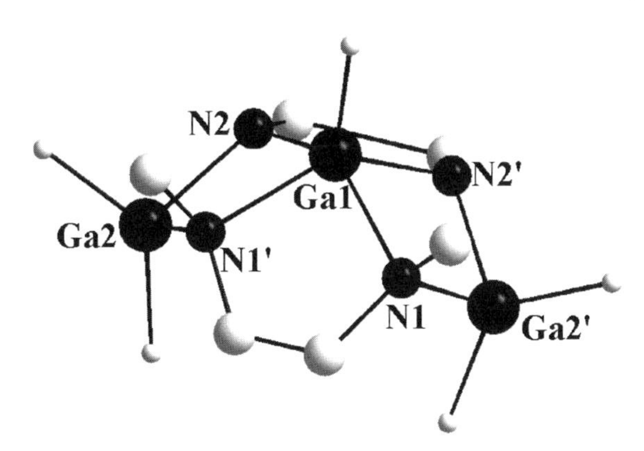

Figure 8.7 *Illustration of the trimeric [$H_5Ga_3\{(NMeCH_2)_2\}_2$].[245] Gallium and nitrogen atoms are shown as black spheres, carbon and hydrogen atoms are white. Selected bond lengths: Ga1-N1 2.082(8), Ga1-N2 2.08(1), Ga2-N1′ 1.962(8), Ga2-N2′ 1.979(8) Å*

Al–N bonds.[84] The dimeric hydride [$\{HGa(NMe_2)_2\}_2$] was prepared from of $HGaCl_2$(quin) and $LiNMe_2$. It features bridging and terminal amide groups and may be converted to the imide (H or NH) or $(HGaNMe)_n$ under suitable conditions.[218]

The primary trisamide derivative [$\{Ga\{\mu\text{-}N(H)Bu^t\}\{N(H)Bu^t\}_2$], synthesized by salt elimination, also has a dimeric structure,[194] although in this case the Ga_2N_2 core has a folded conformation for steric reason with distorted tetrahedral gallium coordination and bridging and terminal Ga–N bond lengths of 2.008(8) and 1.813(10)/1.904(9) Å respectively. The tris(amido) azatrane [$(Ga\{N(CH_2CH_2NMe)_3\})_2$] is also dimeric but the galliums are five-coordinate. However, its structure is very similar to that of its aluminium analogue **2** (Section B.3).[95] The dimeric amides [$(Cl_2Ga\{\mu\text{-}N(H)R\})_2$] (R = Bu^t, $SiMe_3$, $SiMe_2Ph$) were prepared by heating the adducts $Cl_3Ga{\cdot}NH_2R$.[239]

There are a number of amides that contain two or more different types of gallium coordination environments. An example is [$(H_2Ga\{N(CH_2CH_2NMe_2)_2\})(GaH_3)$][210] which is formed by complexation of GaH_3 to the amido nitrogen of the tridentate amido ligand in [$H_2Ga\{N(CH_2CH_2NMe_2)_2\}$] (see Section 8.3.2).[210] The Ga-N(amide) and GaH_3-N(amide) distances are 1.922(2) and 2.058(2) Å, respectively. In [$H_2Ga\{N(CH_2CH_2NMe_2)_2\}$] the Ga–N(amide) bond is 1.874(2) Å. A similar conplexation of a gallane $GaEt_3$ occurs in [$\{Et_2Ga(N\{CH_2(2\text{-py})\}_2)\}(GaEt_3)$], although in this instance the structure of the original amide [$\{Et_2Ga(N\{CH_2(2\text{-py})\}_2)\}_2$] is dimeric.[243]

The trimer [$\{H_2Ga(\mu\text{-}NH_2)\}_3$],[197] which has a chair configuration was discussed in Section 8.3.1. The six-membered ring in [$H_2Ga\{N(H)Me\})_3$] [213] has a 'skew-boat' conformation with Ga–N bond lengths near 1.97 Å. An unusual trimetallic complex [$H_5Ga_3\{N(Me)CH_2CH_2N(Me)\}_2$][245] was formed from $GaH_3{\cdot}NMe_3$ and HN(Me)-CH_2CH_2N(Me)H (Figure 8.7).

8.3.4 Heterometallic Gallium Amides

The synthesis and structure of these complexes follow trends established for aluminium, although the number of compounds is lower. The majority are lithium salts of anionic gallium amides and are listed in a recent review.[11] A straightforward example is the solvent-separated ion pair $[Li(thf)_4][Ga\{N(CH_2Ph)\}_4]$,[134] obtained by salt elimination, with Ga–N bond lengths in the range 1.89–1.93 Å. More often, the lithium ion is associated with the amidogallate as exemplified by $[Cy_2Ga\{NHBu^t\}_2Li(thf)_2]$ which contains a planar $LiGaN_2$ core in which the Ga–N bond length is 1.99 Å. The structure of $[Bu^n{}_2Ga\{N(H)(C_6H_3Me_2\text{-}2,6)\}_2Li(OEt_2)]$ (Ga–N = 2.009 Å avg.) is similar.[229] The amidogallate $[\{LiN(SiMe_3)_2GaMe_3\}_n]$[132] has a structure that is very similar to its aluminium congener; there are infinite chains of $LiN(SiMe_3)_2GaMe_3$ units linked by Li–CH_3 interactions. The Ga–N bond length is 2.038(4) Å (cf. 1.944(3) Å for the aluminium analogue).

8.3.5 Iminogallanes (Gallium Imides)

In 1980 well-characterized iminogallanes were hardly known. The only exception was $[(MeGaNMe)_6(Me_2Ga\{N(H)Me\})_2]$,[246] which consists of an hexagonal $(MeGaNMe)_6$ prism in which two opposite Ga–N edges connecting the hexagonal $(MeGaNMe)_3$ faces are each bound to a $Me_2GaN(H)Me$ moiety. Since that time the synthesis and structures of several gallium imido species with cage structures have been published. One of the major considerations guiding their study is their potential use as precursors for gallium nitride,[247] but, there is also significant interest in obtaining lower aggregates of the general formula $(RGaNR')_n$ (n = 1, 2 or 3) where Ga–N multiple bonding is possible at least in principle.

The simplest gallium imides are $[(HGaNH)_n]$,[247,248] and $[(Ga(NH)_{1.5})_n]$,[249,250] which have polymeric structures, were studied primarily as precursors for gallium nitride. They were prepared either by the action of various Lewis bases on $[(H_2GaNH_2)_3]$, or from ammonia and $[\{Ga(NMe_2)_3\}_2]$. Several gallium imido cages have been characterized by X-ray crystallography. The highest aggregate is $[(PhGaNMe)_7]$[251] and is formed by thermolysis of $[\{Ph_2GaN(H)Me\}_2]$ in dodecane at 220 °C. Its structure is derived from an hexagonal prismatic structure $(PhGaNMe)_6$ in which opposite hexagonal faces are capped by GaPh and NMe moieties. It features Ga–N distances in the range of 1.93 to 2.00 Å. Three hexamers, $[(MeGaNC_6H_4\text{-}4\text{-}F)_6]$,[148] $[(MeGaNBu^i)_6]$[252] and $[(EtGaNEt)_6]$,[253] obtained by hydrocarbon elimination from the amide precursor, have been structurally characterized. All have an hexagonal prismatic structure and Ga–N bonds in the range of 1.937 to 1.971 Å. Similarly, the tetramers $[(MeGaNR)_4]$ (R = C_6F_5,[254] Bu^t,[255] $SiMe_3$[82]) and $[(PhGaNPh)_4]$,[251] were synthesized by elimination from amide precursors. They possess Ga_4N_4 cubic cores with Ga-N distances that range from 1.952 to 2.039 Å. There are several mixed ligand cage species, including $[(PhGa)_4(NHBu^i)_4(NBu^i)_2]$,[255] $[(PhGa)_7(NHMe)_4(NMe)_5]$,[256] and $[\{F_5C_6(H)N\}Ga(MesGa)_3(NC_6F_5)_4]$ as well as $[Li(Ga\{NBu^t\}\{NHBu^t\}_2)]_2[Li\{NHBu^t\}]$[257] as shown in Figure 8.8.

No trimeric gallium imide similar to $[(MeAlNDipp)_3]$ has been isolated. Attempts to synthesize its gallium analogue led to activation of a CH_3 group from the iso-propyl substituent of the Dipp ligand to afford $[(MeGa\{\mu\text{-}N(H)C_6H_3Pr^i\text{-}6\text{-}CHMeCH_2\text{-}2\})_2]$.[88] The dimer $[\{\eta^1\text{-}C_5Me_5)GaNC_6H_3Me_2\text{-}2,6\}_2]$,[258] featuring three-coordinate galliums, was obtained from $Ga(\eta^5\text{-}C_5Me_5)$ and the azide. The Ga_2N_2 core is planar with Ga–N bond lengths of 1.850–1.870 Å. Monomeric gallium imides were stabilized with use of extremely bulky substituents and were obtained by the reactions given by Equation (8.11).[181,259–260]

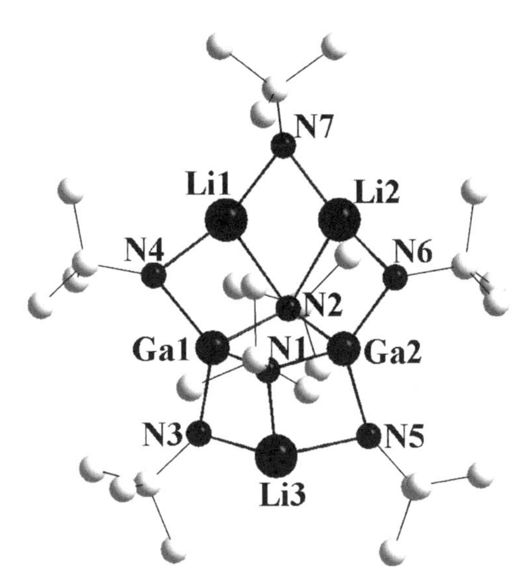

Figure 8.8 *View of [{LiGa(NBu^t)(NHBu^t)$_2$}$_2${LiNHBu^t}][257] highlighting the central metal-nitrogen framework. Gallium, lithium and nitrogen atoms are shown as black spheres and carbon atoms are white. Selected bond lengths: Ga1-N1 1.933(3), Ga1-N2 1.970(3), Ga2-N1 1.938(3), Ga2-N2 1.972(3), Ga1-N3 1.938(3), Ga1-N4 1.930(3), Ga2-N5 1.943(3), Ga2-N6 1.931(3), Li1-N2 2.239(8), Li1-N4 2.031(9), Li2-N2 2.245(8), Li2-N6 2.054(8), Li3-N1 2.033(8), Li3-N3 2.086(8), Li3-N5 2.069(8) Å*

$$\mathrm{RGa + N_3R' \longrightarrow RGaNR' + N_2} \quad (8.11)$$

$$\mathrm{R = C_6H_3Dipp_2\text{-}2,6;\ R' = C_6H_3(C_6H_2Me_2\text{-}2,6\text{-}Bu^t\text{-}4)_2\text{-}2,6}$$

$$\mathrm{R = H\{C(Me)N(Dipp)\}_2;\ R' = C_6H_3Trip_2\text{-}2,6}$$

The first example was obtained by reacting the gallium(I) β-diketiminate[259,260a] with the bulky azide $N_3C_6H_3(C_6H_2Me_2\text{-}2,6\text{-}Bu^t\text{-}4)_2\text{-}2,6$. The imido product has a short Ga–N distance of 1.742(3) Å with a bending angle at nitrogen of 134.6(3)°. However, the coordination geometry at the three-coordinate gallium is pyramidalized ($\sum^\circ = 351.69$) showing that multiple Ga–N bonding is weakened. Subsequently, the use of a large terphenyl substituent at gallium permitted the formation of the product [2,6-$Dipp_2H_3C_6$-$GaNC_6H_3(C_6H_2Me_2\text{-}2,6\text{-}Bu^t\text{-}4)_2$-2,6] in which both gallium and nitrogen are two-coordinate.[260b] The gallium-N(imide) distance was extremely short (1.701(3) Å) with a planar C(ipso)GaNC(ipso) core arrangement as well as bent geometries at gallium, 148.2° and nitrogen 141.7°. The short distance is consistent with considerable multiple bond character. However, the *trans*-bending observed in the core is indicative of reduced overlap efficiency and as a result a weakening of the Ga–N bond. Similarly, the amido-imido complex [{$(Me_3Si)(2,6\text{-}Mes_2H_3C_6)N$}$GaNC_6H_3Mes_2$-2,6], obtained by treatment of the Ga(I) amide [Ga{$N(SiMe_3)C_6H_3Mes_2$-2,6}] with $N_3C_6H_3Mes_2$-2,6, also possesses a planar NGaNC(ipso) core with Ga–N(imide) and Ga–N(amide) distances of 1.743(5) and 1.862(5) Å, respectively.[253]

8.3.6 Gallium Amides in Low (<+3) Oxidation States

The Ga–Ga bonded, formally gallium(II), amides $[(tmp)_2GaGa(tmp)_2]$ (Ga–Ga = 2.52(1) and Ga–N = 1.90(4) Å) and [LGaGaL] ($L = Me_2C\{CH_2N(SiMe_3)\}_2$; Ga–Ga = 2.385(1) and Ga–N = 1.832(3) Å, avg.) were mentioned in Section 8.3.2. These dimers were synthesized by alkali metal reduction of the corresponding bis(amido)metal halide.[134,206]

The compound $[:Ga\{N(Dipp)C(Me)\}_2CH]$ was the first neutral gallium(I) species closely related to a gallium(I) amide to be characterized.[259–261] Its bonding and reactivity, which lie outside our coverage, have been discussed in a review.[261] Similarly, the related guanidinate $[:Ga\{N(Dipp)\}_2CH]$[262] and the anionic gallium heterocycles of formula $[:\overline{GaN(R)CHCHN}(R)]^-$, R = Bu^t[263] or Dipp,[264] as well as their Ga–Ga bonded derivatives fall outside our scope, cf. Ref. 265.

The simplest gallium(I) amide is the monomer $[:Ga\{N(SiMe_3)C_6H_3Mes_2\text{-}2,6\}]$ which was synthesized by the reaction of 'GaI' with $LiN(SiMe_3)C_6H_3Mes_2\text{-}2,6$.[260b] It features an essentially one coordinate gallium with a long Ga–N bond of 1.980(2) Å. An interesting sidelight on gallium(I) amides such as $:GaNR_2$ is that they are isomers of the imide RGaNR. Surprisingly, calculations on model species HGaNH and MeGaNMe show that their $GaNH_2$ and $GaNMe_2$ isomers have considerably greater stability; for example, ca 36.7 kcal mol^{-1} in the case of methyl substituted $GaNMe_2$,[260] which was attributed to weak Ga–N multiple bonding in $MeGaNMe_2$ and the low energy of the gallium lone pair in $:GaNR_2$.

In a similar manner to aluminium, gallium can form numerous 'metalloid' clusters stabilized by various bulky ligands. These include $[Ga_{22}\{N(SiMe_3)_2\}_{10}]^{2-}$,[266] $[Ga_{22}\{N(SiMe_3)_2\}_{10}Br_{10}]^{2-}$,[267] $[Ga_{23}\{N(SiMe_3)_2\}_{11}]$,[268] and $[Ga_{84}\{N(SiMe_3)_2\}_{20}]^{4-}$.[269]

8.4 Indium Amides

8.4.1 Introduction

In 1980 only a handful of indium amides had been characterized and only one X-ray crystal structure had been published.[1] During the ensuing period the investigation of indium amides has been spurred by their possible applications in technology.[270,271] In addition, there has been an interest in low coordinate and oxidation state compounds. Both indium amides[11] and low oxidation state indium derivatives[272] have been recently reviewed and the extensive range of compounds listed and discussed. The simple amido derivatives $HInNH_2$ and $InNH_2$ have been trapped in frozen matrices[23–27] and charaterized both spectroscopically and computationally.[18–27]

8.4.2 Monomeric Indium(III) Amides

The neutral parent amido species $[In(NH_2)_3]$ was synthesized by reaction of KNH_2 with InI_3 in liquid ammonia.[273] Its manner of polymerization in the solid is unknown because diffraction studies were inconclusive as a result of crystal twinning. It was shown to decompose cleanly and quantitatively at moderate temperature to yield InN with liberation of ammonia.[273]

The isolation of neutral monomeric and stable indium amides in the solid state is only feasible with use of bulky substituents, and the synthesis is usually accomplished by a salt elimination route. For example, $[Bu^t_2In\{N(Dipp)SiPh_3]$[274] has three-coordinate indium and nitrogen centres with a relatively small torsion angle of 15.5° between their coordination planes. The In–N bond length of 2.104(3) Å is ca. 0.25 Å longer than that in similar gallium compounds such as $[Bu^t_2Ga\{N(1\text{-}Ad)SiPh_3\}]$ (Ga–N = 1.847(12) Å)[41] which reflects the much larger size of indium in comparison to gallium.[17] Very similar In–N bond lengths of 2.103(6) and 2.094(4) Å were observed in $[Mes^*In\{N(SiMe_3)_2\}_2]$.[201] Interestingly, more structurally characterized monomeric trisamides of indium are known than for any other group 13 metal. Examples include $[In\{N(SiMe_3)_2\}_3]$ (In = 2.049 (1) Å),[274] $[In(NCy_2)_3]$ (In–N = 2.035(11) Å avg.),[275] $[In(tmp)_3]$ (In–N = 2.074(8) Å avg.),[276] $[In\{N(H)Mes^*\}_3]$, (In–N = 2.068(7) Å avg.)[48] and the hydrazide $[In\{N(SiMe_3)NMe_2\}_3]$ (In–N = 2.078(3) Å).[277]

Higher metal coordination numbers in monomeric indium amides are achieved by use of less bulky ligands in combination with neutral Lewis bases as in $[In\{N(Ph)SiMe_3\}_3(OEt_2)]$,[278] $[In(NPh_2)_3(py)]$[278] or $[In\{N(Bu^t)SiMe_3\}_3(pyNMe_2\text{-}4)]$,[278] which have In-N distances in the range 2.09–2.12 Å. In addition, there is the chelated species $[In\{N(H)Dipp\}_2(NPr^i)_2CBu^t\}]$[279] which has In–N(amido) distances that average 2.12 Å and In–N (amidinate) bonds of 2.16 Å. Treatment of the latter with $Ti(NMe_2)_4$ resulted in elimination of two equivalents of $HNMe_2$ and the formation of $[\{Bu^tC(NPr^i)_2\}In(\mu\text{-}NDipp)_2Ti(NMe_2)_2]$.[279] The four-coordinate In(III) amides $[InCl_{2\text{-}n}\{N(H)Dipp\}\{(NDipp)CH\}_2CPh]_n]$ have ben prepared.[104]

Higher metal coordination numbers are also observed in the distorted square pyramidal indium(III) amides $[MeIn\{N(Me)(2\text{-}py)\}_2]$[280] and $[MeIn\{2\text{-}(N(CH_2Ph)_{amide}(2\text{-}py)\}_2]$[281] in which the bidentate ligands occupy equatorial positions (mean In–N bond lengths are 2.14 and 2.16 Å, respectively; mean In–N(py) bond lengths are 2.34 Å) and the methyls have apical positions with In–C distances of 2.146(2) and 2.169(2) Å, respectively. A distorted trigonal bipyramidal geometry is observed in $[In\{N(H)Dipp\}_3(py)_2]$,[48] in which the py groups have axial positions (In–N bond lengths are 2.421(4) and 2.460(4) Å) and the three amido ligands are in equatorial positions with shorter In–N distances between 2.11 and 2.12 Å.

8.4.3 Associated Indium Amides

The majority of indium amides that have been structurally characterized are dimers that are bridged via amido ligands to afford an $(InN)_2$ core,[156,282–297] as exemplified in $[(Me_2In\{\mu\text{-}N(Me)Ph)\}_2]$,[220] in which the $(InN)_2$ ring is planar with In–N = 2.282 Å. Folded (InN) rings are observed in $[\{Bu^t_2In(\mu\text{-}NEt_2)\}_2]$[286] and $[\{(Bu^t_2In(\mu\text{-}NBu^n)_2]$[286] which have average In–N bond distances near 2.269 Å. These derivatives were obtained by an unusual photo-induced route involving the reaction of $InBu^t_3$ with Me_3SnNEt_2 or $Me_3SnNBu^n_2$. The azatrane $[In\{N(CH_2CH_2NMe)_3\}_2]$[293] was synthesized via a transamination reaction from $[\{In(NEt_2)_3\}_2]$.[294] Unlike its aluminium or gallium congeners (Figure 8.9), the dimer is non-symmetric; the ring In–N bond lengths vary widely from 2.06 to 2.46 Å, the bond lengths to the four terminal amido nitrogens have a wide range (1.96 to 2.35 Å) and the In–N bonds to amino nitrogen atoms (2.099(2) and 2.327(3) Å) complete the distorted square pyramidal geometries of each metal.

The dimer $[\{In(NEt_2)_3\}_2]$ was synthesized by reaction of $LiNEt_2$ with $InCl_3$.[294] It was characterized by NMR and mass spectrometry. Oddly, the detailed synthesis and structure

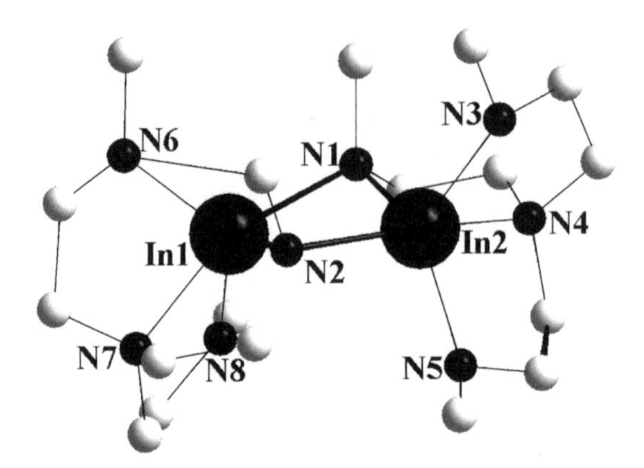

Figure 8.9 Representation of the dimeric indium azatrane derivative [In{N(CH$_2$CH$_2$NMe)$_3$)]$_2$[293] highlighting the central (InN)$_2$ ring in bold. Indium and nitrogen atoms are shown as black spheres and carbon atoms are white. Selected bond lengths: In1-N1 2.063(3), In1-N2 2.393(3), In1-N6 2.083(3), In1-N7 2.099(2), In1-N8 2.350(3), In2-N1 2.456(3), In2-N2 2.224(3), In2-N3 2.208(3), In2-N4 2.327(3), In2-N5 1.960(3) Å

of the simplest diorganoamido indium(III) species [{In(NMe$_2$)$_3$}$_2$] has not been elaborated in the literature although it has been mentioned in a patent.[295] The dimeric compound [{ClIn(μ-N(Me)(SiMe$_2$)$_2$NMe)}$_2$] [296] features an (InN)$_2$ ring as part of a tricyclic ladder structure. The indium has distorted tetrahedral geometry with one terminal chloro ligand (In–Cl bond length is 2.350(3) Å) and three bridging nitrogen atoms. The central (InN)$_2$ core has relatively long In–N bond lengths (2.204(8) Å) compared to those to the two outer six-membered (InNCNSiN) rings (2.036(7) Å). Dimerization can also occur through halide bridging as in [(Mes(μ-Cl)In{N(SiMe$_3$)$_2$})$_2$][289] with In–Cl bond lengths between 2.57 and 2.61 Å. The aryl groups and the amido ligands each coordinate to one of the distorted tetrahedral indium centres with mean In–C and In–N bond lengths of 2.09 and 2.02 Å. The species [{Me$_2$In(NC$_4$Me$_4$)}$_2$] also exists as dimers with dimerization occurring through In–C contacts (2.575 Å) with a ring carbon of the pyrrole.[297]

Several higher aggregate indium amide structures have been characterized, including [MeIn(MeIn{N(Me)(CH$_2$)$_2$NMe})$_2$];[280] the three indium atoms have either distorted square pyramidal (In–N distances between 2.27 and 2.28 Å, mean In–C 2.21 Å) or tetrahedral (mean In–N and In–C bond lengths are 2.21 and 2.19 Å, respectively) coordination environments. The trinuclear macrocyclic structure observed in the compound [{(Me$_2$In)$_2$(NC$_5$H$_4$CMeNNC(S)NC$_6$H$_5$)$_2$}(InMe)·thf][298] is derived from a thiosemicarbazone. The tricyclic In$_3$N$_4$ core in the heteroleptic complex [In$_3$Br$_4$(NBu^t)(NHBu^t)$_3$][299] contains two different indium atom coordination environments. The structure is comprised of two InBr units (In–Br bond lengths are 2.510(1) and 2.729(1) Å) and one InBr$_2$ fragment (In–Br bond length is 2.462(1) Å) that construct a triangle of indium atoms (In–In distances are 3.12 and 3.24 Å) bridged by amido groups (In–N bond lengths are between 2.14 and 2.34 Å).

A polymeric chain structure is observed in [Et$_2$In(pyr)]$_n$[300] (Hpyr = pyrrole; n = 2 in the vapour); the distorted trigonal bipyramidal geometry at each indium features equatorial coordination of the ethyl (In–C bond lengths are 2.141(5) and 2.143(5) Å) and amido groups (In–N bond length is 2.166(4) Å) and relatively short intermolecular interactions with neighbouring indium atoms (In–In distances are between 2.95 and 3.06 Å) in axial positions.

8.4.4 Heterometallic Indium Amides

Several mixed metal complexes of indium amides incorporating lithium ions have been reported. The compound $[\{LiIn(HNMeSiMe_2NMe_2)_2(MeNSiMe_2NMe)\}_2]$[296] has an adamantane-type $Li_2In_2Si_2N_4$ core, in which the $[MeNSiMe_2NMe]^2$ ligands act bridges within the cage and the monodeprotonated $[HN(Me)SiMe_2NMe]^-$ ligands are involved in four six-membered rings that flank the macrocycle (*exo*-cyclic In–N bond lengths 2.10–2.11 Å, *endo*-cyclic In–N distances 2.16–2.17 Å). The complex $[(Li(pyNMe_2\text{-}4)(In\{NMe(SiMe_3)\}_4)]$ was synthesized by the reaction of $LiNMe(SiMe_3)$ with $InCl_3$ to form the intermediate $Li[In\{NMe(SiMe_3)\}_4]$ as the monoetherate, prior to the addition of *p*-(dimethylamino)pyridine ($pyNMe_2$-4). The structure consists of a central, planar (LiNInN) cycle with mean In–N and Li–N bond lengths of 2.17 and 2.00 Å, respectively. The indium has distorted tetrahedral geometry completed by the coordination of two terminal amido ligands (In–N distances are 2.05–2.08 Å). The solvent separated ion pair salt $[Li(dme)_3][InBr_2\{N(SiMe_3)_2\}_2]$[301] was obtained from $InBr_3$ and $LiN(SiMe_3)_2$. In the compound $[(thf)_3Li(\mu\text{-}Cl)(Cl)_2In\{N(SiMe_3)(Dipp)\}]$[302] the In–Cl and Li–Cl distances are 2.36–2.42 and 2.392(4) Å, respectively, and the In–N distance is 2.054(2) Å. The anionic units in both sets of separated ion pairs $[Li(thf)_4][InCl\{NPh_2\}_3]$[278] and $[Li(thf)_4][In\{N(CH_2Ph)_2\}_4]$[275] contain distorted tetrahedral indium centres with In–N bond lengths between 2.02–2.15 and 2.08–2.12 Å, respectively.

The structure of $[Cs(\eta^6\text{-}PhMe)_3(\mu\text{-}F)In\{N(SiMe_3)_2\}_3]$[303] features a distorted tetrahedral indium connected to a caesium ion *via* a bridging μ_2-F ligand with In–F and Cs–F bond lengths of 2.065(2) and 2.777(2) Å, respectively (Figure 8.10). The In–N bond lengths are 2.12 and 2.13 Å. The caesium has additional Cs-arene interactions with three toluene solvent molecules (average Cs-C_{aryl} distances are ca 3.81 Å).

The reaction of $[\{(\eta^5\text{-}C_5H_5)(CO)_3Mo\}_2InCl]$ with $LiN(SiMe_3)_2$ afforded $[\{(\eta^5\text{-}C_5H_5)(CO)_3Mo\}_2In\{N(SiMe_3)_2\}]$,[304] which features a trigonal planar indium geometry and an In–N bond length of 2.123(3) Å and two In–Mo distances of 2.818(1) and 2.833(1) Å. Similarly, the reaction of $[\{(\eta^5\text{-}C_5H_5)(CO)_3Mo\}InCl_2]$ with $LiN(SiMe_3)_2$ gave $[\{(\eta^5\text{-}C_5H_5)(CO)_3Mo\}In\{N(SiMe_3)_2\}_2]$,[304] with comparable In–N and In–Mo bond lengths (2.116(3) and 2.794(1) Å, respectively).

8.4.5 Iminoindanes (Indium Imides)

No well-characterized indium imido derivative was known in 1980. In the past decade several examples have been synthesized and their structures determined. At present, however, their structural diversity does not match those of aluminium or gallium. The highest degree of aggregation currently known is four.

Among the first to be described was the iminoindane complex $[\{MeIn(NC_6H_4F\text{-}4)(thf)\}_4]$[154] prepared as shown in Scheme 8.1. The framework consists of a distorted cube with In and N atoms at alternating corners (In–N bond lengths are between 2.19 and 2.41 Å). The distorted trigonal pyramidal coordination geometry at each indium atom is

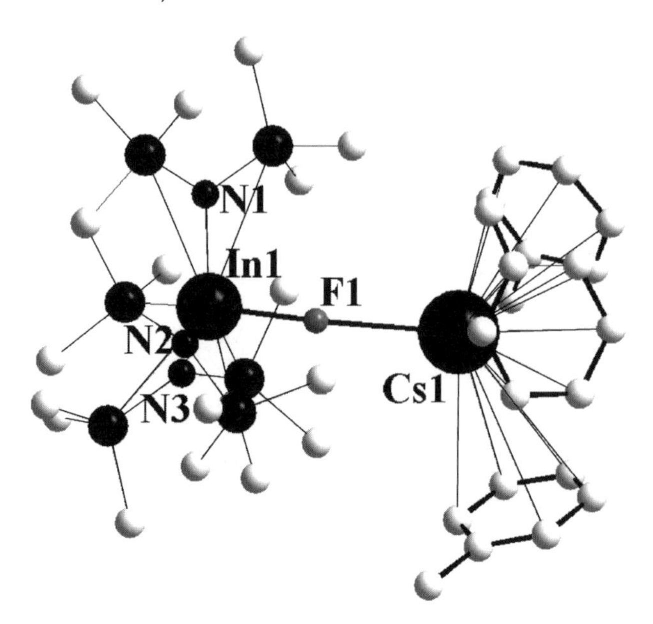

Figure 8.10 *Representation of the crystal structure of* $[Cs((\eta^6\text{-}PhMe)_3(\mu\text{-}FIn\{N(SiMe_3)_2\}_3)]$.[303] *Indium, caesium, nitrogen and silicon atoms are shown as black spheres, fluorine atoms are grey and carbon atoms are white. Selected bond lengths: In1-N1 2.121(3), In1-N2 2.128(3), In1-N3 2.127(3), In1-F1 2.065(2), Cs1-F1 2.777(2) Å*

completed by one methyl group and one thf molecule (In–C and In–O distances are 2.117 (7) and 2.732(3) Å, respectively).

The imide $[(MeInNC_6F_5)_4]$[254] was synthesized by a similar method. The tetrameric cubanes $[In_4X_4(\mu\text{-}NBu^t)_4]$ (where X = Cl, Br or I),[299] formed from the corresponding indium trihalide, InX_3 and $LiHNBu^t$, have In_4N_4 heterocubane structures that are similar to their aluminium and gallium counterparts. The chloride $[In_4Cl_4(NBu^t)_4]$ is shown in Figure 8.11.

There are no lower aggregated indium imides, but there is an example of a monomer – $[(2,6\text{-}Dipp_2H_3C_6)In\{NC_6H_3(C_6H_2Me_2\text{-}2,6\text{-}Bu^t\text{-}4)_2\text{-}2,6\}]$[181] which has a planar *trans*-bent C(ipso)InNC(ipso) core with a short In–N bond of 1.928(3) Å and bending angles of 142.2° at indium and 134.9° at nitrogen.

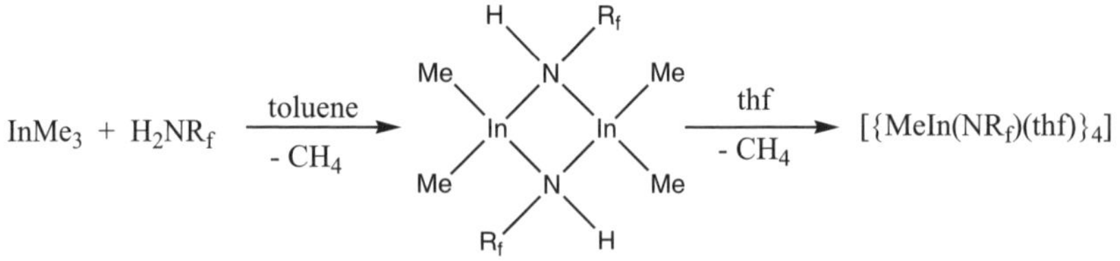

$InMe_3 + H_2NR_f$ —(toluene, $-CH_4$)→ $[Me_2In(\mu\text{-}N(H)R_f)]_2$ —(thf, $-CH_4$)→ $[\{MeIn(NR_f)(thf)\}_4]$

H, R_f, Me, N, Me, In, In, Me, N, Me, R_f, H

Scheme 8.1 *Synthesis of iminoindane complexes* ($R_f = C_6H_4F\text{-}4$)[254]

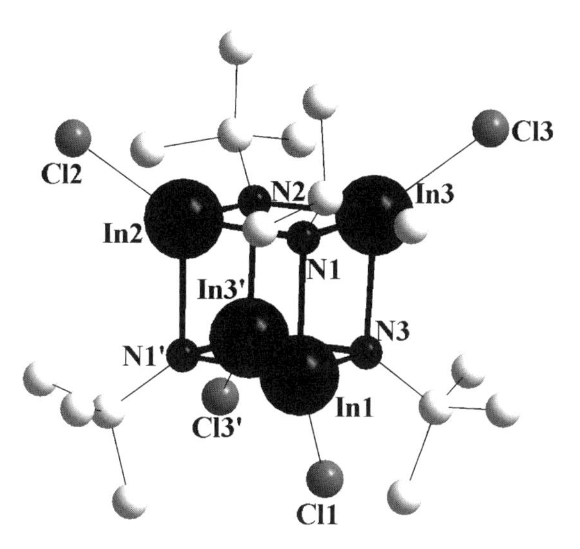

Figure 8.11 *Illustration of the In_4N_4 heterocubane structure in $[In_4Cl_4(NBu^t)_4]$[299] highlighting the cubane in bold. Indium and nitrogen atoms are shown as black spheres, chlorine atoms are grey and carbon atoms are black. Selected bond lengths: In1-N1 2.189(3), In1-N3 2.177(5), In1-Cl1 2.325(2), In2-N1 2.176(3), In2-N2 2.187(4), In2-Cl2 2.320(2), In3-N2 2.187(3), In3-N2 2.178(3), In3-N3 2.176(3), In3-Cl3 2.319(1) Å*

8.4.6 Indium Amides in Oxidation States <+3

The chemistry of indium complexes of all types in metal oxidation states lower than +3 has been comprehensively reviewed.[272] Few lower oxidation state mononuclear amido complexes of indium are well characterized, however, and no structure has been reported for an In(I) amide. The compound $\{In\{N(SiMe_3)_2\}_n$, which is unstable,[305] has been characterized NMR spectroscopy but its structure is unknown. The structures of several In(I) complexes, related to amides but outside our current scope, have been described. Like its aluminium and gallium counterparts, the β-diketiminate derivative $[:In\{N(Dipp)C(Me)\}_2CH]$[306] has been characterized, as has the closely related species $[:In\{N(Dipp)C(CF_3)\}_2CH]$.[307] These feature V-shaped, two-coordination at the metal. The less bulky $[(In\{N(Mes)C(Me)\}_2\text{-}CH)_2]$[308a] and $[(In\{C_6H_3\text{-}2,6\text{-}Me_2)C(Me)\}_2CH]$[308b] are dimeric with long In–In bonds of 3.1967(4) and 3.3400(5) Å. The amidinate $[In\{N(Dipp)\}_2CBu^t\}]$[309] has a structure in which the indium is coordinated by one of the ligand nitrogens and η^6 by one of the Dipp substituents. The guanadinate $[In\{N(Dipp)_2CNCy_2\}]$ has also been reported.[262]

There are a small number of indium(II) amide structures that contain In–In bonds. These have been obtained with use of multidentate ligands. They were generally synthesized by reduction of the ligand indium halide precursor. Among such compounds are: (i) the halogeno derivatives $[\{In(Cl)(MesNCHCHNMes)\}_2]$[310a] (In–In = 2.7280(9) Å) and $[\{In\text{-}(Cl)(\{N(Dipp)C(Me)\}_2CH)\}]$[310b] (In–In) = 2.8243(7) Å; (ii) [LIn-InL](L^{2-} = $\bar{N}(Bu^t)$-$Si(Me)(\mu\text{-}NBu^t)_2Si(Me)\bar{N}Bu^t)$[311] in which the indiums are each coordinated by two amido and one amino nitrogens from the ligand (In–N(amido) = 2.12(1) and 2.18(1) Å; In–N (amino) = 2.298(7) Å) with In–In = 2.768(1) Å; (iii) the imido-phosphazane $[(Bu^tNP)_2\text{-}(Bu^tN)_2InIn(NBu^t)_2(PNBu^t)_2]$ (In–In = 2.7720(4) Å)[312a] and $[(In\{PP(CH_2SiMe_2NSi\text{-}$

$Me_2NCH_2)_2PPh\})_2]$[312b] (In–In = 2.7168(12) Å) and (iv) the chelate complex (thf)LInInL-(thf) (L = 1,8-$(NSiMe_3)_2C_{10}H_6$) (In–In = 2.7237(6) Å)[313a] and LInInL (L = η^3-$\{Me_3SiN$-$(CH_2\}_3(\mu$-In)CMe} (In–In = 2.807(1) Å.[313b]

8.5 Thallium Amides

8.5.1 Introduction

The structures of only two thallium amides, $[(Me_2TlNMe_2)_2]$[314] and $[Tl\{N(SiMe_3)_2\}_3]$,[315] both of which were published in the late 1970s, had been reported in the previous edition of this book. Interest in the field has since increased considerably and the range of compounds has been extended to amido derivatives of Tl(I). A review dedicated to thallium(I) amides was published in 2003.[316] The clustering of Tl(I) centers by weakly attractive metal–metal interactions[317] as well as metal–arene interactions are characteristic of their structural chemistry and give rise to finite or infinite aggregates in the solid state. The major focus of thallium amide work since 1980, and particularly over the last decade, has shifted to derivatives of Tl(I). This is not only because of the application such derivatives as amide transfer agents but also because of interest in weak metal-metal interactions as a result of dispersive forces.

8.5.2 Thallium(I) Amides

Thallium is unique among the group 13 metals in that it forms the widest variety of M(I) derivatives, as a result of the increased stability of the lower oxidation state.[316] The structural characterization of thallium(I) amides in the solid state began only within the past decade and the number of structures has been limited in part due to their extreme air sensitivity and thermal lability. Thallium(I) amides often have thallium-thallium,[317] and in some cases, thallium-arene interactions that result in finite or infinite aggregates. Thallium (I) amides were precedented in the literature by the structures of related species such as the benztriazolate,[318] 1,3-diphenyltriazenide,[319] 1,5-di-tolylpentaazadienide[319] and pyrazolylborate.[320] The first structures were of the multidentate diaminodisiletidinede or (tris-t-butylamido) methylsilane ligands (see below).

The first well characterized thallium(I) derivative of a monodentate amide ligand was the silylamide $[\{TlN(SiMe_3)_2\}_2]$ obtained from $LiN(SiMe_3)_2$ and TlCl in toluene. It is a centrosymmetric dimer with an almost perfectly square planar Tl_2N_2 core as shown in Figure 8.12.[321]

There are weak exocyclic Tl—Tl interactions (Tl-Tl = 3.935(1) Å) between the dimers (cf. Tl—Tl = 3.650(1) Å) within the dimer).

In the gas phase there is dissociation to $TlN(SiMe_3)_2$ monomers in which the Tl–N distance is 2.15(1) Å.[322] The use of the more bulky amide ligand -N(Me)($C_6H_3Mes_2$-2,6) led to the monomer [TlN(Me)($C_6H_3Mes_2$-2,6)]; Tl–N bond lengths = 2.379(3) and 2.348(3) Å in the two different molecules of the asymmetric unit.[323] The Tl is essentially one-coordinate with only weak interactions between Tl and one of the flanking rings of the $C_6H_3Mes_2$-2,6-substituent (Tl—centroid 3.026(2) Å avg.). It is noteworthy that the Tl—N distance is quite close to that observed in the dimer $[\{TlN(SiMe_3)_2\}_2]$. Employment of the bulky amide N($SiMe_3$)Dipp afforded the tetramer $[\{TlN(SiMe_3)Dipp\}_4]$[324] as shown in Figure 8.13.

Monomeric thallium(I) amides and related species have also be obtained by using chelating amido or related ligands, as in the β-diketiminato complexes [Tl({NDippC-(R)}$_2$CH)] (R = CH_3 or CF_3)[307] and TlL (L = $\{N(SiMe_3)C(Ph)\}_2CH$ or {N(Dipp)C-

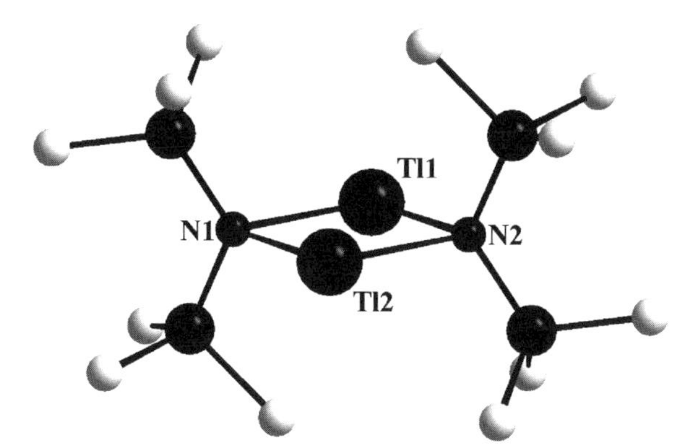

Figure 8.12 An illustration of the planar $(TlN)_2$ ring in the compound $[(Tl\{N(SiMe_3)_2\})_2]$.[321] Thallium, nitrogen and silicon atoms are shown as black spheres and carbon atoms are white. Selected bond lengths: Tl1-N1 2.581(7), Tl1-N2 2.576(7) Å

$(H)_2CPh\}$.[325] In $[Tl\{N(quinolinyl\text{-}8)_2\}]$ the tridentate amido ligand prevents aggregation via close thallium-thallium contacts.[326]

The reaction of $[(Bu^tNP)_2\{Bu^tNLi(thf)\}_2]$ with TlCl in a PhMe/thf mixture yielded the dimeric thallium(I) amide $[(Bu^tNP)_2(Bu^tNTl)_2]$,[312] which has a heterocubane type structure (Tl. . .Tl distance is 3.546(3) Å) with Tl–N bond lengths between 2.55 and 2.62 Å. The silylamido ligand in the dimeric complex $[\{(Me_2Si)_2(NBu^t)_4\}Tl_2]$[327] results in a

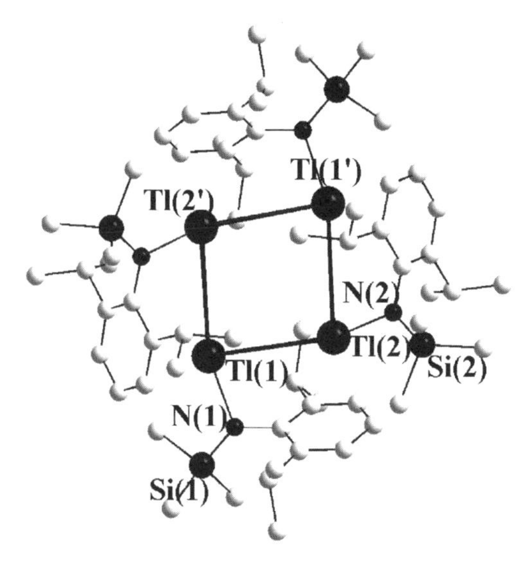

Figure 8.13 Illustration of the tetrameric $[Tl\{N(SiMe_3)(C_6H_3Pr^i{}_2\text{-}2,6)\}]_4$[324] highlighting the Tl–Tl interactions as bold lines. Thallium, nitrogen and silicon atoms are shown as black spheres and carbon atoms are white. Selected bond lengths: Tl(l)-N(l) 2.303(6), Tl(2)-N(2) 2.307(6) Å, Tl-Tl interactions 4.06 Å avg., Tl-centroid = ca. 3.11 Å

bowl-shaped structure with the two-coordinate thallium(I) ions located on the rim (Tl—N bond lengths are between 2.52 and 2.73 Å). The hexanuclear complex $[\{H_3CC\{CH_2N(Tl)SiMe_3\}_3\}_2]$[328] comprises two trimetallated fragments connected *via* close Tl—Tl interactions lying between 3.41 and 3.80 Å. The individual units are formed by the coordination of each nitrogen of the tridentate tris(amido) ligand in a bridging fashion between two thallium (I) ions in a distorted adamantane-type framework (Tl—N bond lengths range from 2.34 to 2.55 Å). Conversely, the puckered $(TlN)_2$ rings in $[((2\text{-}C_5H_4N)C(CH_3)\{CH_2N(Tl)\text{-}SiMe_3\}_2)_2]$[329] interact in a head-to-tail fashion with considerably shorter Tl—Tl distances (3.500(2) Å). Due to the nature of the tridentate ligand, the two thallium ions are in different coordination environments. One has two-coordinate thallium involved solely in the formation of the $(TlN)_2$ ring (*intra*-cyclic Tl—N bond lengths are between 2.41 and 2.47 Å), while the second thallium ion is three-coordinate with an additional interaction with the pyridyl nitrogen atom (Tl—N distance is 2.468(15) Å). The metal-metal interaction occurs between the three-coordinate thallium ion of one molecule and the two-coordinate thallium ion of a neighbouring dimer.

The octanuclear complex $[(Tl\{N(SiMe_3)_2{=}C(H)N(CH)N(CH){=}N(SiMe_3)\})_8]$[330] is formed from four dimeric units, each consisting of two thallium(I) ions and two amido ligands. The thalliums have four different coordination environments, with coordination numbers ranging from three to five and Tl—N bond lengths between 2.49 and 2.75 Å and are related *via* an inversion centre. Short metal-metal interactions lead to cluster formation, with Tl—Tl distances between 3.65 and 3.89 Å. The polymeric structure of the compound $[MeSi\{SiMe_2N(Tl)Bu^t\}_3]$[331] is formed by the linking of monomeric units via interaction between two of the thallium ions (Tl—Tl distance is 3.673(2) Å). The monomeric units comprise two thallium ions and two nitrogen atoms atoms in a $(TlN)_2$ ring with the third thallium ion being coordinated in a 'pendant' position along the ligand framework (Tl—N bond lengths are between 2.23 and 2.54 Å). The effect of the 'availability' of the thallium ion on the crystal packing has also been investigated using the two compounds $[CH_2\{CH_2N(Tl)SiMe_3\}_2]$ and $[CH_2\{CH_2N(Tl)Si(Bu^t)Me_2\}_2]$.[332] Both compounds form puckered $(TlN)_2$ rings (Tl—N bond lengths are between 2.41–2.48 and 2.40–2.48 Å, respectively), but steric shielding of the larger butyl group prevented the formation of metal-metal interactions (shortest Tl—Tl distance is > 4.07 Å). The crystal structure of the former compound is, however, governed by two-fold metal-metal interactions that facilitate a three-dimensional network with intra- and inter-strand Tl—Tl distances of 3.697(3) and 3.775(3) Å, respectively. The conformation of the tripodal tris(amido) ligand in the complex $[(C_6H_5)C\{CH_2N(Tl)SiMe_3\}_3]$[333] results in each amido nitrogen atom coordinating to a different thallium(I) ion to form a trimetalated fragment (Tl—N bond lengths lie between 2.41 and 2.75 Å) in which Tl-arene interactions are facilitated by the proximity of the thallium ions to the phenyl substituents. The short distance between a thallium and the ipso carbon of a phenyl group (2.87(3) Å) is noteworthy and the remaining Tl-arene distances lie in the range 3.34 to 3.74 Å. The fragments are connected 'tail-to-tail' into a dimeric structure by the formation of a planar $(TlN)_2$ ring, which are polymerized into infinite chains via intermolecular thallium-thallium interactions (Tl—Tl distances are 3.7471(15) Å). Thermolysis of the Tl(I) amide derived from the tripodal amine $MeSi\{SiMe_2NHBu^t\}_3$ afforded the dimetallated diamido amine $[MeSi\{SiMe_2N(Tl)Bu^t\}_2\}SiMe_2NHBu^t]$ in which a Tl_2N_2 core structure is formed with one of the thalliums weakly coordinated to the remaining nitrogen.[334]

8.5.3 Thallium(III) Derivatives

A number of thallium(III) derivatives related to amides containing the $TlMe_2$ unit have been isolated. For example, compounds featuring the stable Me_2Tl^+ moiety bound to various two coordinate nitrogens in delocalized rings as in $[Me_2Tl(DABRhd)]$[335] (HDABRhd = *p*-dimethylaminobenzylidenerhodanine) and $[Me_2Tl(DABTd)(dmso)]$[336] (HDABTd = 5-(4′-dimethylaminobenzylidene)-2-thiohydantoin), whose thallium has distorted octahedral coordination geometry with various two-coordinate nitrogens in delocalized rings having relatively long Tl−N distances of 2.660(6) and 2.687(9) Å, respectively, were characterized. The long bond lengths reflect the higher coordination number at the metal centre. Similarly, the thallium(III) atoms in $[Me_2Tl(2\text{-thiouracil})]$[337] and $[Me_2Tl(PyRd)]_2$[338] (HPyRd = 5-(2-pyridylmethylenerhodanine)) are linked to neighbouring monomeric units *via* longer contacts to either an oxygen or a sulfur atom (Tl−O and Tl−S distances are 2.72 (3) and 3.530(2) Å, respectively). A helical polymer is formed in the compound $[Me_2Tl\{N\text{-}(NC_2H_4)_2\}]$,[339] where the tridentate ligand bridges the five-coordinate thallium(III) ions (Tl−N bond lengths lie between 2.48 and 2.69 Å). Infinite helical chains are observed in $[Me_2Tl(HDAPTSC)]$[340] ($H_2DAPTSC$ = 2,6-diacetylpyridinebis(thiosemicarbazone)), where the monodeprotonated thiosemicarbazone acts as a tetradentate ligand (mean Tl−C, Tl−N and Tl−S distances are 2.14, 2.62 and 2.69 Å, respectively) and the distorted square pyramidal coordination at the thallium ion is completed by a sulfur atom from a neighbouring ligand. Dideprotonation of the ligand facilitates a rare example of a monophenylthallium(III) compound, $[PhTl(DAPTSC)]\cdot C_3H_6O$,[340] where the distorted pentagonal pyramidal thallium geometry with short $Tl-N_{amide}$ bond lengths between 2.51 and 2.56 Å.

The dimeric thallium(III) amide $[(\{bis(\text{pentafluorophenyl})\}Tl(dpa))]_2$[341] (where Hdpa = 2,2′-dipyridylamine) comprises distorted trigonal bipyramidal thallium(III) ions that are coordinated by two pentafluorophenyl groups (Tl−C distances are 2.12(4) and 2.16 (4) Å), two nitrogen atoms from one ligand ($Tl-N_{amide}$ and $Tl-N_{pyr}$ distances are 2.27(3) and 2.57(4) Å, respectively) and one pyridyl nitrogen atom from a bridging ligand (Tl−N distance is 2.46(3) Å).

8.5.4 Thallium Amides in Mixed Oxidation States

The mixed valence compound $[(CH_2\{CH_2NSiMe_3\}_2)_2Tl^{III}Tl^{I}]$ was synthesized from the reaction of $[H_2C\{CH_2N(Li)SiMe_3\}_2]_2$ with $TlCl_3$.[342] The structure comprises a planar $(TlN)_2$ core with the two thalliums having different coordination geometries. The thallium (I) is two-coordinate (mean Tl−N bond length is 2.54 Å), thallium(III) has distorted tetrahedral geometry with two intra-cyclic bonds (average Tl−N distance is 2.23 Å) as well as two shorter interactions with an additional amido nitrogen atom from each ligand system (Tl−N distances are 2.11(3) and 2.12(3) Å). The in situ formed lithium amide, from $HC\{SiMe_2NH(p\text{-Tol})\}_3$ and Bu^nLi with TlCl, yielded the mixed valence compound $[HC\{SiMe_2NH(p\text{-Tol})\}_3(Tl^{III}Bu^n)(Tl^{I})]$.[343] The tripodal ligand coordinates to the two thallium ions in a heterotopic fashion, with bidentate coordination to the two-coordinate thallium(I) ion and tridentate complexation of the four-coordinate thallium(III) ion. The remaining site on the thallium(III) ion is occupied by the butyl substituent, with a Tl−C bond length of 2.188(9) Å. Coordination of two of the donor nitrogen atoms to the thallium(I) ion effects a lengthening of the two corresponding Tl(III)-N bonds (2.321(7) and 2.326(7) Å) compared

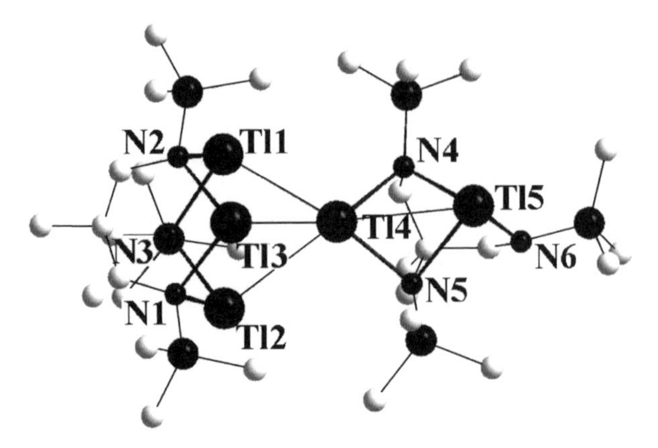

Figure 8.14 Representation of $[Tl_5(H)\{H_3CC(CH_2NSiMe_3)_3\}_2]$[313b] highlighting the central Tl_5 core in bold. Thallium, nitrogen and silicon atoms are shown as black spheres and carbon atoms are white. Selected bond lengths: Tl1-Tl2 3.707(3), Tl1-Tl3 3.837(3), T12-Tl3 3.756(3), T14-Tl1 3.409(3), T14-Tl2 3.541(3), T14-Tl3 3.479(3), Tl4-Tl5 3.403(3), Tl1/4-N1/5 2.39(1)-2.49(1), T15-hyphen;N6 2.62(2) Å

to the third Tl(III)-N bond (2.155(7) Å). The thallium(I)-nitrogen bond lengths are much longer (2.694(7) and 2.721(8) Å), which may be due to the steric congestion at the coordination sites whereby the two thallium ions are positioned in extremely close proximity to one another (Tl–Tl distance is 3.3620(9) Å). The molecular units are joined into weakly associated dimers by Tl-arene interactions with one tolyl group of a neighbouring molecule (Tl-centroid distance is 3.093 Å) as well as metal-metal contacts (intermolecular Tl–Tl distances are 3.7615(13) Å). An unique central Tl_5 core is observed in the mixed-valence compound $[Tl_5(H)\{H_3CC(CH_2NSiMe_3)_3\}_2]$,[313b] formed from aggregation of a dimetallated and a trimetallated fragment and whose structure was described as a 'spiked tetrahedron' (see Figure 8.14). $[H_3CC\{CH_3N(Tl)SiMe_3\}_3][H_3CC\{CH_2NSiMe_3\}_3(H)(Tl)(Li(thf)] \cdot (toluene)$[328] has comparable parameters.

8.5.5 Heterometallic Derivatives

The mixed metal thallium(I) amide $[C_{10}H_6(N\{Li(thf)_2\}SiMe_3)(N\{Tl\}SiMe_3)]$[344] is dimeric, with the bis(amido) ligands coordinating to both lithium and thallium metal ions via both nitrogen atoms, forming a (LiNTlN) ring (Li–N and Tl–N bond lengths are between 1.97–2.02 and 2.47–2.50 Å, respectively). Dimerization of the two heterometallic units occurs through Tl-arene interactions between the thallium ions and the naphthalene rings of a neighbouring molecule (Tl-centroid distance is 3.5106 Å). No Tl–Tl interaction is observed (Tl-Tl distance is 3.982(2) Å). The dimeric structure observed in the compound $[\{(2\text{-}C_5H_4N)C(CH_2N(Li)SiMe_3)\}\{CH_2N(Tl)SiMe_3\}]_2$[330] comprises a central $Li_2Tl_2N_4$ macrocycle with average Li–N and Tl–N bond lengths of 2.02 and 2.47 Å, respectively. The thallium(I) ions are two-coordinate, whereas the lithium ions are three-coordinate due to coordination by a terminal pyridyl nitrogen atom. The reaction of the thallium(I) amide $[CH_2\{CH_2N(Tl)SiMe_3\}_2]$ with InCl yielded the mixed metal compound $[\{CH_2(CH_2NSiMe_3)_2\}_2In^{III}Tl^{I}]$.[342] The ligands bridge the two metal ions with one amido nitrogen atom to form a planar (InNTlN) ring with In–N and Tl–N distances of 2.207(14) and 2.601(14) Å. The thallium(I) is two-coordinate and indium(III) is four-coordinate with the remaining amido nitrogen from each ligand completing the distorted tetrahedral geometry (In–N distances are 2.027(14) and 2.039(14) Å).

References

1. M. F. Lappert, P. P. Power, A. R. Sanger and R. C. Srivastava, *Metal and Metalloid Amides: Synthesis, Structure and Physical and Chemical Properties*, Horwood-Wiley, Chichester, 1980, p. 847.
2. A. H. Cowley and R. A. Jones, *Angew. Chem., Int. Ed.*, 1989, **28**, 1208.
3. I. R. Grant, in *Chemistry of Aluminium, Gallium, Indium and Thallium*, ed. A. J. Downs, Blackie, Glasgow, 1993, p. 292.
4. (a) Strite and H. Morko, *J. Vac. Sci. Technol. B*, 1992, **10**, 1237; (b) F. C. Sauls and L. V. Interrante, *Coord. Chem. Rev.*, 1993, **128**, 193; (c) D. M. Hoffman, *Polyhedron*, 1994, **13**, 1169; (d) D. A. Neumayer and J. G. Ekerdt, *Chem. Mater.*, 1996, **8**, 9.
5. A. R. Barron, *CVD of Nonmetals*, ed. W. S. Rees. Jr. VCH, Weinheim, 1996, p. 300.
6. A. C. Jones and P. O'Brien, *CVD of Compound Semiconductors*, VCH, Weinheim, 1997, p. 296.
7. J. A. Jegier and W. L. Gladfelter, *Coord. Chem. Rev.*, 2000, **206–207**, 631.
8. P. J. Brothers and P. P. Power, *Adv. Organomet. Chem.*, 1996, **39**, 1.
9. A. J. Downs, *Coord. Chem. Rev.*, 1999, **189**, 59.
10. C. -C. Chang and M. A. Ameerunisha, *Coord. Chem. Rev.*, 1999, **189**, 199.
11. C. J. Carmalt, *Coord. Chem. Rev.*, 2001, **223**, 217.
12. P. P. Power, *Chem. Rev.*, 1999, **99**, 3463.
13. H. Schnöckel and A. Schnepf, *Adv. Organomet. Chem.*, 2001, **47**, 235.
14. H. Schnöckel and H. Kohlein, *Polyhedron*, 2002, **21**, 489.
15. A. Schnepf and H. Schnöckel, *Angew. Chem., Int. Ed.*, 2002, **41**, 3532.
16. H. Schnöckel, *Dalton Trans.*, 2005, 3131.
17. A. J. Downs, in *Chemistry of Aluminium, Gallium, Indium and Thallium*, (ed. A. J. Downs), Blackie, Glasgow, 1993, p. 1.
18. A. J. Downs, H. J. Himmel and L. Manceron, *Polyhedron*, 2002, **21**, 473.
19. D. D. Randall and K. L. Jaffrey, *J. Phys. Chem.*, 1994, **98**, 8930.
20. J. Müller, *J. Am. Chem. Soc.*, 1996, **118**, 6370.
21. D. V. Lanzisera and L. Andrews, *J. Phys. Chem. A*, 1997, **101**, 5082.
22. T. P. Hamilton and A. W. Shaikh, *Inorg. Chem.*, 1997, **36**, 754.
23. H. J. Himmel, A. J. Downs and T. M. Greene, *Chem. Commun.*, 2000, 871.
24. H. J. Himmel, A. J. Downs and T. M. Greene, *J. Am. Chem. Soc.*, 2000, **122**, 9793.
25. B. Gaertner and H. J. Himmel, *Inorg. Chem.*, 2002, **41**, 2496.
26. W. H. Fink, P. P. Power and T. L. Allen, *Inorg. Chem.*, 1997, **36**, 1431.
27. B. L. Kormos and C. J. Cramer, *Inorg. Chem.*, 2003, **42**, 6691.
28. The simplest such correction is derived from electronegativity differences using formulae given in: (a) V. Schomaker and D. P. Stevenson, *J. Am. Chem. Soc.*, 1941, **63**, 37; (b) and R. Blom and A. Haaland, *J. Mol. Struct.*, 1985, **129**, 1. Although they are empirical, the corrections are grounded in the fact that the ionic correction for the shortening corresponds roughly to the ionic component of the wavefunction which increases with the increasing electronegativity difference between the bonding atoms. (c) See also: A. Haaland, in *Coordination Chemistry of Aluminium*, ed. G. H. Robinson, VCH, New York, 1993, Chapter 1, for a discussion of Al-ligand distances in molecular compounds.

29. M. A. Petrie, M. M. Olmstead and P. P. Power, *J. Am. Chem. Soc.*, 1991, **113**, 8704.
30. (a) L. V. Interrante, G. A. Sigel, M. Garbauskas, C. Hejna and G. A. Slack, *Inorg. Chem.*, 1989, **28**, 252; (b) F. C. Sauls, L. V. Interrante and Z. Jiang, *Inorg. Chem.*, 1990, **29**, 2989.
31. J. F. Janik, E. N. Duesler and R. T. Paine, *Inorg. Chem.*, 1987, **26**, 4341.
32. K. J. Paciorek, L. J. H. Nakahara, L. A. Moferkamp, C. George, J. L. Flippen-Anderson, R. Gilardi and W. R. Schmidt, *Chem. Mater.*, 1991, **3**, 82.
33. J. F. Janik, E. N. Duesler and R. T. Paine, *Inorg. Chem.*, 1988, **27**, 4335.
34. V. Jancik, L. W. Pineda, J. Pinkas, H. W. Roesky, D. Neculai, A. M. Neculai and R. Herbst-Irmer, *Angew. Chem., Int. Ed.*, 2004, **43**, 2142.
35. C. -C. Chang, M.-D. Li, M. Y. Chiang, S. M. Peng, Y. Wang and G.-H. Lee, *Inorg. Chem.*, 1997, **36**, 1955.
36. G. M. Sheldrick and W. S. Sheldrick, *J. Chem. Soc. (A)*, 1969, 2297.
37. P. J. Brothers, R. J. Wehmschulte, M. M. Olmstead, K. Ruhlandt-Senge, S. R. Parkin and P. P. Power, *Organometallics*, 1994, **13**, 2792.
38. K. Knabel, I. Krossing, H. Nöth, H. Schwenk-Kirchner, M. Schmidt-Amelunxen and T. Seifert, *Eur. J Inorg. Chem.*, 1998, 1095.
39. M. A. Petrie, K. Ruhlandt-Senge and P. P. Power, *Inorg. Chem.*, 1993, **32**, 1135.
40. D. A. Atwood and D. R. Rutherford, *Main Group Chem.*, 1996, **1**, 431.
41. K. M. Waggoner, K. Ruhlandt-Senge, R. J. Wehmschulte, X. He, M. M. Olmstead and P. P. Power, *Inorg. Chem.*, 1993, **32**, 2557.
42. R. J. Wehmschulte and P. P. Power, *Inorg. Chem.*, 1998, **37**, 2106.
43. J. Niesman, U. Klingebiel, C. Röpken, M. Noltemeyer and R. Herbst-Irmer, *Main Group Chem.*, 1998, **2**, 297.
44. J. Niesman, U. Klingebiel, M. Noltemeyer and R. Boese, *Chem. Commun.*, 1997, 365 .
45. D. Chakraborty and E. Y. X. Chen, *Organometallics*, 2002, **21**, 1438.
46. I. Krossing, H. Nöth, C. Tacke and M. Schmidt, *Chem. Ber. Recl.*, 1997, **130**, 1047.
47. S. D. Waezsada, C. Rennekamp, H. W. Roesky, C. Röpken and E. Parisini, *Z. Anorg. Allg. Chem.*, 1998, **624**, 987.
48. J. S. Silverman, C. J. Carmalt, A. H. Cowley, R. D. Culp, R. A. Jones and B. G. McBurnett, *Inorg. Chem.*, 1999, **38**, 296.
49. R. J. Wehmschulte and P. P. Power, *J. Am. Chem. Soc.*, 1996, **118**, 791.
50. H. Krause and J. Weidlein, *Z. Anorg. Allg. Chem.*, 1988, **563**, 116.
51. T. Bauer, S. Schulz, H. Hupfer and M. Nieger, *Organometallics* 2002, **21**, 2931.
52. J. S. Bradley, F. Cheng, S. J. Archibald, R. Suppilit, R. Rovai, C. W. Lehmann, C. Krüger and F. Lefebvre, *Dalton Trans.*, 2003, 1846.
53. J. L. Atwood, G. A. Koutsantonis, F. C. Lee and C. L. Raston, *J. Chem. Soc., Chem. Commun.*, 1994, 91.
54. H. P. Zhu, J. F. Chai, H. W. Roesky, *et al.*, *Eur. J. Inorg. Chem.*, 2003, 3113.
55. M. G. Gardiner, G. A. Koutsantonis, S. M. Lawrence, F. C. Lee and C. L. Raston, *Chem. Ber.*, 1996, **129**, 545.
56. J. Pinkas, H. Wessel, Y. Yang, M. L. Montero, M. Noltemeyer, M. Froba and H. W. Roesky, *Inorg. Chem.*, 1998, **37**, 2450.
57. D. R. Armstrong, F. J. Craig, A. R. Kennedy and R. E. Mulvey, *J. Organomet. Chem.*, 1998, **550**, 355.
58. I. Krossing, H. Nöth and H. Schwenk-Kircher, *Eur. J. Inorg. Chem.*, 1998, 927.
59. J. Pinkas, B. Gaul and J. Verkade, *J. Am. Chem. Soc.*, 1993, **115**, 3925.
60. J. Pinkas, T. Wang, R. A. Jacobson and J. G. Verkade, *Inorg. Chem.*, 1994, **33**, 4202.
61. H. Zhu and E. Y.-X. Chen, *Organometallics*, 2007, **26**, 5395.
62. H. Zhu and E. Y.-X. Chen, *Inorg. Chem.*, 2007, **46**, 1481.

63. M. G. Gardiner, G. A. Koutsantonis, S. M. Lawrence and C. L. Raston, *Inorg. Chem.*, 1996, **35**, 5696.
64. J. L. Atwood, S. M. Lawrence and C. L. Raston, *J. Chem. Soc., Chem. Commun.*, 1994, 73.
65. M. G. Gardiner, S. M. Lawrence and C. L. Raston, *Inorg. Chem.*, 1996, **35**, 1349.
66. M. J. Zaworotko and J. L. Atwood, *Inorg. Chem.*, 1980, **19**, 268.
67. H. Sussek, O. Stark, A. Devi, H. Pritzkow and R. A. Fischer, *J. Organomet. Chem.*, 2000, **602**, 29.
68. M. Veith, H. Lange, O. Recktenwald and W. Frank, *J. Organomet. Chem.*, 1985, **294**, 273.
69. S. J. Trepanier and S. N. Wang, *J. Chem. Soc., Dalton Trans.*, 1995, 2425.
70. V. C. Gibson, C. Redshaw, A. J. P. White, and D. J. Williams, *J. Organomet. Chem.*, 1998, **550**, 453.
71. P. Haquette, S. Dagorne, R. Welter and G. Jaouen, *J. Organomet.Chem.*, 2003, **682**, 240.
72. N. Emig, H. Nguyen, H. Krautscheid, R. Reau, J. B. Cazaux and G. Bertrand, *Organometallics*, 1998, **17**, 3599.
73. J. Lewínski, J. Zachara, T. Kopéc and Z. Ochal, *Polyhedron*, 1997, **16**, 1337.
74. L. M. Engelhardt, M. G. Gardiner, C. Jones, P. C. Junk, C. L. Raston and A. H. White, *J. Chem. Soc., Dalton Trans.*, 1996, 3053.
75. A. P. Kenney, G. P. A. Yap, D. S. Richeson and S. T. Barry, *Inorg. Chem.*, 2005, **44**, 2926.
76. S. J. Trepanier and S. N. Wang, *Organometallics*, 1994, **13**, 2213.
77. B. Luo, B. E. Kucera and W. L. Gladfelter, *Dalton Trans.*, 2006, 4491.
78. M. Pfeiffer, A. Murso, L. Mahalakshmi, D. Moigno, W. Kiefer and D. Stalke, *Eur. J Inorg. Chem.*, 2002, 3222.
79. G. H. Robinson, M. F. Self, S. A. Sangokoya and W. T. Pennington, *J. Am. Chem. Soc.*, 1989, **111**, 1520.
80. M. D. Fryzuk, G. R. Giesbrecht, S. J. Rettig and G. P. A. Yap, *J. Organomet. Chem.*, 1999, **591**, 63.
81. S. J. Schauer, W. T. Pennington and G. H. Robinson, *Organometallics*, 1992, **11**, 3287.
82. S. Kühner, R. Kuhnle, H.-D. Hausen and J. Weidlein, *Z. Anorg. Allg. Chem.*, 1997, **623**, 25.
83. (a) S. J. Trepanier and S. N. Wang, *Organometallics*, 1996, **15**, 760; (b) C. Rennekamp, A. Stasch, P. Müller, H. W. Roesky, M. Noltemeyer, H.-G. Schmidt and I. Usón, *J. Fluor. Chem.*, 2000, **102**, 17.
84. K. M. Waggoner, M. M. Olmstead and P. P. Power, *Polyhedron*, 1990, **9**, 257.
85. C. Gordon, P. Shukla, A. H. Cowley, J. N. Jones, D. W. Keogh and B. L. Scott, *Chem. Commun.*, 2002, 2710.
86. E. K. Styron, C. H. Lake, C. L. Watkins, L. K. Krannich, C. D. Incarvito and A. L. Rheingold, *Organometallics*, 2000, **19**, 32536.
87. R. J. Wehmschulte and P. P. Power, *Inorg. Chem.*, 1998, **37**, 6906.
88. K. M. Waggoner and P. P. Power, *J. Am. Chem. Soc.*, 1991, **113**, 3385.
89. M. G. Gardiner, S. M. Lawrence and C. L. Raston, *J. Chem. Soc., Dalton Trans.*, 1996, 4163.
90. D. C. Bradley, I. S. Harding, I. A. Maia and M. Motevalli, *J. Chem. Soc., Dalton Trans.*, 1997, 2969.
91. M. M. Andrianarison, M. C. Ellerby, I. B. Gorrell, P. B. Hitchcock, J. D. Smith and D. R. Stanley, *J. Chem. Soc., Dalton Trans.*, 1996, 211.
92. K. Ouzounis, H. Riffel, H. Hess, U. Kohler and J. Weidlein, *Z. Anorg. Allg. Chem.*, 1983, **504**, 67.
93. K. Schmid, H. D. Hausen, K. W. Klinkhammer and J. Weidlein, *Z. Anorg. Allg. Chem.*, 1999, **625**, 945.
94. J. D. Fisher, P. J. Shapiro, G. P. A. Yap and A. L. Rheingold, *Inorg. Chem.*, 1996, **35**, 271.
95. S. D. Waezsada, F. -Q. Liu, E. F. Murphy, H. W. Roesky, M. Teichert, I. Usón, H.-G. Schmidt, T. Albens, E. Parisini and M. Noltemeyer, *Organometallics*, 1997, **16**, 1260.
96. M. H. Chisholm, V. F. DiStasi and W. E. Streib, *Polyhedron*, 1990, **9**, 253.
97. M. Niemeyer and M. Hoffmann, *Acta Crystallogr., Sect. E: Struct. Rep.*, 2005, **61**, M1334.
98. C. Klein, H. Nöth, M. Tacke and M. Thomann, *Angew.Chem., Int. Ed.*, 1993, **32**, 886.

99. J. F. Janik, R. L. Wells, A. L. Rheingold and I. A. Guzei, *Polyhedron* 1998, **17**, 4101.
100. D. Chakraborty and E. Y. X. Chen, *Organometallics*, 2003, **22**, 769.
101. G. H. Robinson, S. A. Sangokoya, F. Moise and W. T. Pennington, *Organometallics*, 1988, **7**, 1887.
102. J. Pinkas, T. Wang, R. A. Jacobson and J. G. Verkade, *Inorg. Chem.*, 1994, **33**, 5244.
103. (a) J. -P. Bezombes, B. Gehrhus, P. B. Hitchcock, M. F. Lappert and P. G. Merle, *Dalton Trans.*, 2003, 1821; (b) P. B. Hitchcock, H. A. Jasim, M. F. Lappert and H. D. Williams, *Polyhedron*, 1990, **9**, 245.
104. Y. Cheng, D. J. Doyle, P. B. Hitchcock and M. F. Lappert, *Dalton Trans.*, 2006, 4499.
105. E. Wiberg and A. May, *Z. Naturforsch.*, 1955, **10B** 681.
106. S. Amirkhalili, P. B. Hitchcock, A. D. Jenkins, J. Z. Nyathi and J. D. Smith, *J. Chem. Soc., Dalton Trans.*, 1981, 377.
107. M. Schiefer, H. Hatop, H. W. Roesky, H.-G. Schmidt and M. Noltemeyer, *Organometallics*, 2002, **21**, 1300.
108. D. R. Armstrong, R. R. Davies, D. J. Linton, R. Snaith, P. Schooler and A. E. H. Wheatley, *J. Chem. Soc., Dalton Trans.*, 2001, 2838.
109. T. Y. Her, C. C. Cheng, J. U. Ysai, Y. Y. Lai, L. K. Liu, H. C. Chang and J. H. Chen, *Polyhedron*, 1993, **12**, 731.
110. J. R. Gardinier and F. P. Gabbaï, *New J. Chem.*, 2001, **25**, 1567.
111. S. Shravan Humar, S. Singh, F. Hongjun, H. W. Roesky, D. Vidore and J. Magull, *Organometallics*, 2004, **23**, 6327.
112. Z. Jiang, L. V. Interrante, D. Kwon, F. S. Tham and R. Kullnig, *Inorg. Chem.*, 1991, **30**, 995.
113. G. Leggett, M. Motevalli and A. C. Sullivan, *J. Organomet. Chem.*, 2000, **598**, 36.
114. U. Thewalt and I. Kawada, *Chem. Ber.*, 1970, **103**, 2754.
115. W. Zheng, A. Stasch, J. Prust, H. W. Roesky, F. Cimpoesu, M. Noltemeyer and H.-G. Schmidt, *Angew. Chem., Int. Ed.*, 2001, **40**, 3461.
116. J. Rong, Y. Peng, H. W. Roesky, J. Li, D. Vidovic and J. Magull, *Eur. J. Inorg. Chem.*, 2003, 3110.
117. C. Rennekamp, P. Müller, J. Prust, H. Wessel, H. W. Roesky and I. Usón, *Eur. J. Inorg. Chem.*, 2000, 1861.
118. H. Jacobs, K. Jaënichen, C. Hadenfeldt and R. Juza, *Z. Anorg. Allg. Chem.*, 1985, **531**, 125.
119. D. Rutherford and D. A. Atwood, *J. Am. Chem. Soc.*, 1996, **118**, 11535.
120. D. A. Atwood and D. Rutherford, *Organometallics*, 1996, **15**, 436.
121. M. J. Harvey, M. Proffitt, P. Wei and D. A. Atwood, *Chem. Commun.*, 2001, 2094.
122. S. S. Al-Juaid, C. Eaborn, I. B. Gorrell, S. A. Hawkes, P. B. Hitchcock and J. D. Smith, *J. Chem. Soc., Dalton Trans.*, 1998, 2411.
123. M. G. Gardiner, C. L. Raston, B. W. Skelton and A. H. White, *Inorg. Chem.*, 1997, **36**, 2795.
124. M. A. Beswick, N. Choi, C. N. Harmer, M. McPartlin, M. E. G. Mosquera, P. R. Raithby, M. Tombul and D. S. Wright, *Chem. Commun.*, 1998, 1383.
125. J. S. Silverman, C. J. Carmalt, D. A. Neumayer, A. H. Cowley, B. G. McBurnett and A. Decken, *Polyhedron*, 1998, **17**, 977.
126. M. G. Gardiner, S. M. Lawrence and C. L. Raston, *Inorg. Chem.*, 1995, **34**, 4652.
127. M. L. Montero, H. Wessel, H. W. Roesky, M. Teichert and I. Usón, *Angew. Chem., Int. Ed.*, 1997, **36**, 629.
128. J. K. Brask, T. Chivers and G. P. A. Yap, *Chem. Commun.*, 1998, 2543.
129. N. Kocher, C. Selinka, D. Leusser, D. Kost, I. Kalikham and D. Strähle, *Z. Anorg. Allg. Chem.*, 2004, **630**, 1777.
130. (a) G. Linti, W. Köstler and A. Rodig, *Z. Anorg. Allg. Chem.*, 2002, **628**, 1319; (b) P. Böttcher, H. W. Roesky, M. G. Walawalkar and H.-G. Schmidt, *Organometallics* 2001, **20**, 790.
131. M. Niemeyer and P. P. Power, *Organometallics*, 1995, **14**, 5488.

132. (a) M. Uchiyama, Y. Naka, Y. Matsumoto and T. Ohwada, *J. Am. Chem. Soc.*, 2004, **126**, 10526; (b) J. Garcia-Álvarez, D. V. Graham, A. R. Kennedy, R. E. Mulvey and S. Waetherstone, *Chem. Commun.*, 2006, 3208.
133. J. Garcia-Álvarez, E. Hevia, A. R. Kennedy, J. Klett and R. E. Mulvey, *Chem. Commun.*, 2007, 2402.
134. J. Pauls and B. Neumüller, *Inorg. Chem.*, 2001, **40**, 121.
135. J. Pauls and B. Neumüller, *Z. Anorg. Allg. Chem.*, 2001, **627**, 583.
136. M. Niemeyer and P. P. Power, *Chem. Commun.*, 1996, 1573.
137. J. F. Janik, E. N. Duesler and R. T. Paine, *J. Organomet. Chem.*, 1987, **323**, 149.
138. B. N. Anand, I. Krossing and H. Nöth, *Inorg. Chem.*, 1997, **36**, 1979.
139. M. Cesari and S. Cucinella, in *The Chemistry of Inorganic Mono- and Heterocycles, Volume 1* (eds. I. Haiduc and D. B. Sowerby), Academic, London, 1987, Chapter 6, p. 167.
140. W. Uhl and H. W. Roesky, in *Molecular Clusters of the Main Group Elements*, (eds. M. Driess and H. Nöth), Wiley-VCH, Weinheim, 2004, p. 369.
141. M. Veith, *Chem. Rev.*, 1990, **90**, 3.
142. A. Y. Timoshkin, *Coord. Chem. Rev.*, 2005, **249**, 2094.
143. H. J. Himmel, A. J. Downs, J. C. Green and T. M. Green, *J. Chem. Soc., Dalton Trans.*, 2001, 535.
144. A. Y. Timoshkin, *Inorg. Chem. Commun.*, 2003, **6**, 274.
145. R. D. Davy and H. F. Schaefer, *Inorg. Chem.*, 1998, **37**, 2291.
146. A. Timoshkin and G. Frenking, *Inorg. Chem.*, 2003, **42**, 60.
147. A. Y. Timoshkin and H. F. Schaefer, *Inorg. Chem.*, 2004, **43**, 3080.
148. N. Dastagiri Reddy, H. W. Roesky, M. Noltemeyer and H.-G. Schmidt, *Inorg. Chem.*, 2002, **41**, 2374.
149. N. Dastagiri Reddy, S. Kumar, H. W. Roesky, D. Vidovic, J. Magull, M. Noltemeyer and H.-G. Schmidt, *Eur. J. Inorg. Chem.*, 2003, 442.
150. Y. Peng, J. F. Rong, D. Vidovic, H. W. Roesky, T. Labahn, J. Magull, M. Noltemeyer and H.-G. Schmidt, *J. Fluor. Chem.*, 2004, **125**, 951.
151. S. S. Kumar, N. Dastagiri Reddy, H. W. Roesky, D. Vidovic, J. Magull and R. F. Winter, *Organometallics*, 2003, **22**, 3348.
152. H. Wessel, H. S. Park, P. Müller, H. W. Roesky and I. Usón, *Angew. Chem., Int. Ed.*, 1999, **38**, 813.
153. D. M. Choquette, M. J. Timm, J. L. Hobbs, T. M. Nicholson, M. M. Olmstead and R. P. Planalp, *Inorg. Chem.*, 1993, **32**, 2600.
154. C. Schnitter, S. D. Waezsada, H. W. Roesky, M. Teichert, I. Usón and E. Parisini, *Organometallics*, 1997, **16**, 1197.
155. T. Belgardt, S. D. Waezsada, H. W. Roesky, H. Gornitzka, L. Häming and D. Stalke, *Inorg. Chem.*, 1994, **33**, 6247.
156. E. K. Styron, C. H. Lake, D. H. Powell, L. K. Krannich and C. L. Watkins, *J. Organomet. Chem.*, 2002, **649**, 78.
157. K. M. Waggoner, H. Hope and P. P. Power, *Angew. Chem., Int. Ed.*, 1988, **27**, 1699.
158. W. H. Fink and J. C. Richards, *J. Am. Chem. Soc.*, 1991, **113**, 3393.
159. N. Matsunaga, T. R. Cundari, M. W. Schmidt and M. S. Gordon, *Theor. Chim. Acta*, 1992, **83**, 52.
160. P. Schleyer, H. Jiao, N. J. R. van Eikena Hommes, V. G. Malkin and O. L. Malkin, *J. Am. Chem. Soc.*, 1997, **119**, 12669.
161. J. J. Engelberts, R. W. A. Havenith, J. H. van Lenthe, L. W. Jenneskens and P. Fowler, *Inorg. Chem.*, 2005, **44**, 5266.
162. H. Wessel, C. Rennekamp, H. W. Roesky, M. L. Montero, P. Muller and I. Usón, *Organometallics*, 1998, **17**, 1919.
163. H. Wessel, M. L. Montero, C. Rennekamp, H. W. Roesky, P. Yu and I. Usón, *Angew. Chem., Int. Ed.*, 1998, **37**, 843.

164. J. Pinkas, H. Wessel, Y. Yang, M. L. Montero, M. Noltemeyer, M. Frola and H. W. Roesky, *Inorg. Chem.*, 1998, **37**, 2450.
165. J. Pinkas, J. Löbl, D. Dastych, M. Necas and H. W. Roesky, *Inorg. Chem.*, 2002, **41**, 6914.
166. J. Pinkas, J. Löbl and H. W. Roesky, *Phosphorus, Sulfur and Silicon*, 2004, **179**, 759.
167. J. Löbl, M. Necas and J. Pinkas, *Main Group Chem.*, 2006, **5**, 79.
168. J. Löbl, J. Pinkas, H. W. Roesky, W. Plass and H. Görls, *Inorg. Chem*, 2006, **45**, 6571.
169. (a) J. Löbl, A. Y. Timoshkin, T. Cong, M. Necas, H. W. Roesky and J. Pinkas, *Inorg. Chem.*, 2007, **46**, 5678; (b) H. Hohmeister, H. Wessel, P. Lobinger, H. W. Roesky, P. Müller, I. Usón, H.-G. Schmidt, M. Noltemeyer and J. Magull, *J. Fluor. Chem.*, 2003, **120**, 59.
170. S. Schulz, L. Häming, R. Herbst-Irmer, H. W. Roesky and G. M. Sheldrick, *Angew. Chem., Int. Ed.*, 1994, **33**, 969.
171. S. Schulz, A. Voigt, H. W. Roesky, L. Hamig and R. Herbst-Irmer, *Organometallics*, 1996, **15**, 5252.
172. R. J. Wehmschulte and P. P. Power, *Inorg. Chem.*, 1996, **118**, 791.
173. R. J. Wehmschulte and P. P. Power, *Inorg. Chem.*, 1998, **37**, 6906.
174. J. D. Fischer, P. J. Shapiro, G. P. A. Yap and A. L. Rheingold, *Inorg. Chem.*, 1996, **35**, 271.
175. S. Schulz, F. Thomas, W. M. Priesmann and M. Nieger, *Organometallics*, 2006, **25**, 1392.
176. W. Uhl, J. Molter and R. Koch, *Eur. J. Inorg. Chem.*, 1999, 2021.
177. J. S. Silverman, C. D. Abernethy, R. A. Jones and A. H. Cowley, *Chem. Commun.*, 1999, 1645.
178. T. Bauer, S. Schulz and M. Nieger, *Z. Anorg. Allg. Chem.*, 2004, **630**, 1807.
179. R. J. Wehmschulte and P. P. Power, *Inorg. Chem.*, 1996, **35**, 2717.
180. K. W. Henderson, A. R. Kennedy, A. E. McKeown and R. E. Mulvey, *Angew. Chem., Int. Ed.*, 2000, **39**, 3879.
181. N. J. Hardman, C. Cui, H. W. Roesky, W. H. Fink and P. P. Power, *Angew. Chem., Int. Ed.*, 2001, **40**, 2172.
182. H. Zhu, J. Chai, V. Chandrasekhar, H. W. Roesky, J. Magull, D. Vidovic, H-G. Schmidt, M. Noltemeyer, P. P. Power and W. A. Merrill, *J. Am. Chem. Soc.*, 2004, **126**, 9472.
183. C. Cui, H. W. Roesky, H.-G. Schmidt, M. Noltemeyer, H. Hao and F. Cimpoesu, *Angew. Chem., Int. Ed.*, 2000, **39**, 4274.
184. H. W. Roesky, *Inorg. Chem.*, 2004, **43**, 7284.
185. H. Sitzmann, M. F. Lappert, C. Dohmeier, C. Üffing and H. Schnöckel, *J. Organomet. Chem.*, 1998, **561**, 203.
186. M. Scheifer, N. Dastagiri Reddy, H. W. Roesky and D. Vidovic, *Organometallics*, 2003, **22**, 3637.
187. G. Linti, H. Schnöckel, W. Uhl and N. Wiberg, in *Molecular Clusters of the Main Group Elements*, (eds., M. Driess and H. Nöth), Wiley-VCH, Weinheim, 2004, p. 145.
188. A. Purath, R. Köppe and H. Schnöckel, *Angew. Chem., Int. Ed.*, 1999, **38**, 2926.
189. A. Purath, R. Köppe and H. Schnöckel, *Chem. Commun.*, 1999, 1933.
190. H. Köhlein, G. Stößer, E. Baum, E. Möllhausen, U. Huniar and H. Schnöckel, *Angew. Chem., Int. Ed.*, 2000, **39**, 799.
191. H. Köhnlein, A. Purath, C. Klemp, E. Baum, I. Krossing, G. Stößer and H Schnöckel, *Inorg. Chem.*, 2001, **40**, 4830.
192. A. Ecker, H. Weckert and H. Schnöckel, *Nature*, 1997, **387**, 389.
193. R. Frey, G. Linti and K. Polborn, *Chem. Ber.*, 1994, **127**, 101.
194. D. A. Atwood, V. O. Atwood, A. H. Cowley, R. A. Jones, J. L. Atwood and S. G. Bott, *Inorg. Chem.*, 1994, **33**, 3251.
195. J. P. Campbell, J. W. Hwang, V. G. Young, R. B. Von Dreele, C. J. Cramer and W. L. Gladfelter, *J. Am. Chem. Soc.*, 1998, **120**, 521.
196. J. W. Hwang, S. A. Hanson, D. Britton, J. F. Evans, K. F. Jensen and W. L. Gladfelter, *Chem. Mater.*, 1990, **2**, 342.

197. M. J. Almond, M. G. B. Drew, C. E. Jenkins and D. A. Rice, *J. Chem. Soc., Dalton Trans.*, 1992, 5.
198. (a) D. A. Atwood, A. H. Cowley, P. R. Harris, R. A. Jones, S. U. Koschmieder, C. M. Nunn, J. L. Atwood and S. G. Bott, *Organometallics*, 1993, **12**, 24; (b) D. A. Atwood, A. H. Cowley and R. A. Jones, *Organometallics* 1993, **12**, 236.
199. M. Roos and G. Meyer, *Z. Anorg. Allg. Chem.*, 1999, **625**, 1839.
200. V. Jancik, L. W. Pineda, A. C. Stückl, H. W. Roesky and R. Herbst-Irmer, *Organometallics*, 2005, **24**, 1511.
201. W. P. Leung, C. M. Y. Chan, B. M. Wu and T. C. W. Mak, *Organometallics*, 1996, **15**, 5179.
202. G. Linti, *J. Organomet. Chem.*, 1994, **465**, 79.
203. G. Linti, R. Frey and K. Polborn, *Chem. Ber.*, 1994, **127**, 1387.
204. G. Linti, R. Frey, H. Köstler and H. Schwenk, *Chem. Ber.*, 1997, **130**, 663.
205. G. Linti and R. Frey, *Z. Anorg. Allg. Chem.*, 1997, **623**, 531.
206. G. Linti, R. Frey and M. Schmidt, *Z. Naturforsch.*, 1994, **49B** 958.
207. B. Luo, V. G. Young and W. L. Gladfelter, *Inorg. Chem.*, 2000, **39**, 1705.
208. N. L. Pickett, O. Just, D. G. Van Derveer and W. S. Rees, Jr., *Acta Crystallogr., Sect. C: Cryst. Struct. Commun.*, 2000, **56**, 560.
209. B. Luo, V. G. Young and W. L. Gladfelter, *J. Organomet. Chem.*, 2002, **649**, 268.
210. B. Luo, B. E. Kucera and W. L. Gladfelter, *Chem. Commun.*, 2005, 3463.
211. B. Luo and W. L. Gladfelter, *Chem. Commun.*, 2000, 825.
212. E. S. Schmidt, A. Jockisch and H. Schmidbaur, *J. Chem. Soc., Dalton Trans.*, 2000, 1039.
213. S. Marchant, C. Y. Tang, A. J. Downs, T. M. Greene, H. J. Himmel and S. Parsons, *Dalton Trans.*, 2005, 3281.
214. B. Luo, M. Pink and W. L. Gladfelter, *Inorg. Chem.*, 2001, **40**, 307.
215. B. Luo and W. L. Gladfelter, *Inorg. Chem.*, 2002, **41**, 590.
216. L. Grocholl, S. A. Cullison, J. J. Wang, D. C. Swenson and E. G. Gillan, *Inorg. Chem.*, 2002, **41**, 2920.
217. D. A. Atwood, R. A. Jones, A. H. Cowley, S. G. Bott and J. L. Atwood, *J. Organomet. Chem.*, 1992, **434**, 143.
218. B. Luo and W. L. Gladfelter, *J. Organomet. Chem.*, 2004, **689**, 666.
219. D. A. Atwood, V. O. Atwood, D. F. Carriker, A. H. Cowley, F. P. Gabbai, R. A. Jones, M. R. Bond and C. J. Carrano, *J. Organomet. Chem.*, 1993, **463**, 29.
220. O. T. Beachley, M. J. Noble, M. R. Churchill and C. H. Lake, *Organometallics* 1992, **11**, 1051.
221. O. T. Beachley, M. J. Noble, M. R. Churchill and C. H. Lake, *Organometallics*, 1998, **17**, 3311.
222. O. T. Beachley, M. J. Noble, M. R. Churchill and C. H. Lake, *Organometallics*, 1998, **17**, 5153.
223. L. H. van Poppel, S. G. Bott and A. R. Barron, *Polyhedron*, 2003, **22**, 9.
224. D. A. Atwood, R. A. Jones, A. H. Cowley, S. G. Bott and J. L. Atwood, *Polyhedron*, 1991, **10**, 1897.
225. B. J. Bae, J. E. Park, Y. Kim, J. T. Park and I. H. Suh, *Organometallics*, 1999, **18**, 2513.
226. P. Jutzi, B. Neumann, G. Reumann and H. G. Stammler, *Organometallics*, 1999, **18**, 2037.
227. S. J. Schauer, C. H. Lake, C. L. Watkins, L. K. Krannich and D. H. Powell, *J. Organomet. Chem.*, 1997, **549**, 31.
228. S. J. Schauer, C. H. Lake, C. L. Watkins and L. K. Krannich, *Organometallics*, 1996, **15**, 5641.
229. S. T. Barry and D. S. Richeson, *J. Organomet. Chem.*, 1996, **510**, 103.
230. W. R. Nutt, K. J. Murray, J. M. Gulick, J. D. Odom, Y. Ding and L. Lebioda, *Organometallics*, 1996, **15**, 1728.
231. D. A. Neumayer, A. H. Cowley, A. Decken, R. A. Jones, V. Lakhotia and J. G. Ekerdt, *J. Am. Chem. Soc.*, 1995, **117**, 5893.
232. J. T. Park, Y. Kim, J. Kim, K. Kim and Y. Kim, *Organometallics*, 1992, **11**, 3330.
233. B. S. Lee, W. T. Pennington and G. H. Robinson, *Inorg. Chim. Acta*, 1991, **190**, 173.

234. W. R. Nutt, J. S. Blanton, F. O. Kroh and J. D. Odom, *Inorg. Chem.*, 1989, **28**, 2224.
235. P. L. Baxter, A. J. Downs, D. W. H. Rankin and H. E. Robertson, *J. Chem. Soc., Dalton. Trans.*, 1985, 807.
236. W. R. Nutt, J. A. Anderson, J. D. Odom, M. M. Williamson and B. H. Rubin, *Inorg. Chem.*, 1985, **24**, 159.
237. W. R. Nutt, R. E. Stimson, M. F. Leopold and B. H. Rubin, *Inorg. Chem.*, 1982, **21**, 1909.
238. P. I. Arvidsson, G. Hilmersson and O. Davidsson, *Chem. Eur. J.*, 1999, **5**, 23485.
239. C. J. Carmalt, J. D. Mileham, A. J. P. White and D. J. Williams, *Dalton Trans.*, 2003, 4255.
240. M. R. Kopp, T. Krauter, A. Dashti-Mommertz and B. Neumüller, *Z. Naturforsch.*, 1999, **54B** 627.
241. O. T. Beachley, D. B. Rosenblum, M. R. Churchill, D. G. Churchill and L. M. Krajkowski, *Organometallics*, 1999, **18**, 2543.
242. P. L. Baxter, A. J. Downs, D. W. H. Rankin and H. Robertson, *J. Chem. Soc., Dalton Trans.*, 1985, 807.
243. M. Westerhausen, A. N. Kneifel, P. Mayer and H. Nöth, *Z. Anorg. Allg. Chem.*, 2004, **630**, 2013.
244. W. R. Nutt, J. A. Anderson, J. D. Odom, M. W. Williamson and B. H. Rubin, *Inorg. Chem.*, 1985, **24**, 159.
245. J. L. Atwood, S. G. Bott, C. Jones and C. L. Raston, *Inorg. Chem.*, 1991, **30**, 4868.
246. S. Amirkhalili, P. B. Hitchcock, J. D. Smith and J. G. Stamper, *J. Chem. Soc., Dalton Trans.*, 1979, 1206.
247. J. A. Jegier, S. McKennan and W. L. Gladfelter, *Chem. Mater.*, 1999, **10**, 2041.
248. J. A. Jegier, S. McKennan and W. L. Gladfelter, *Inorg. Chem.*, 1999, **38**, 2726.
249. J. F. Janik and R. L. Wells, *Chem. Mater.*, 1996, **8**, 2706.
250. R. J. Jouet, A. P. Purdy, R. L. Wells and J. F. Janik, *J. Cluster Sci.*, 2002, **13**, 469.
251. B. Luo and N. L. Gladfelter, *Inorg. Chem.*, 2002, **41**, 590.
252. K. Schmid, M. Niemeyer and J. Weidlein, *Z. Anorg. Allg. Chem.*, 1999, **625**, 186.
253. B. Luo and W. L. Gladfelter, *J. Cluster Sci.*, 2002, **13**, 461.
254. T. Belgardt, H. W. Roesky, M. Noltemeyer and H. G. Schmidt, *Angew. Chem., Int. Ed.*, 1993, **32**, 1101.
255. F. Cordeddu, H.-D. Hausen and J. Weidlein, *Z. Anorg. Allg. Chem.*, 1996, **622**, 573.
256. B. Luo and W. L. Gladfelter, *Inorg. Chem.*, 2002, **41**, 6249.
257. J. K. Brask, T. Chivers, G. Schatte and G. P. A. Yap, *Organometallics*, 2000, **19**, 5683.
258. P. Jutzi, B. Neumann, G. Reumann and H.-G. Stammler, *Organometallics*, 1999, **18**, 2037.
259. N. J. Hardman, B. E. Eichler and P. P. Power, *Chem. Commun.*, 2000, 1491.
260. (a) R. J. Wright, A. D. Phillips, T. L. Allen, W. H. Fink and P. P. Power, *J. Am. Chem. Soc.*, 2003, **125**, 1694; (b) R. J. Wright, M. Brynda, J. C. Fettinger, A. R. Betzer and P. P. Power, *J. Am. Chem. Soc.*, 2006, **128**, 12498.
261. N. J. Hardman, A. D. Phillips and P. P. Power, Group 13 Chemistry from Fundamentals to Applications, ACS Symposium Series 822, (eds; P. J. Shapiro and D. A. Atwood), American Chemical Society, Washington D.C., Chapter 2.
262. C. Jones, P. C. Junk, J. A. Platts and A. Stasch, *J. Am. Chem. Soc.*, 2006, **128**, 2206.
263. E. S. Schimdt, A. Jockisch and H. Schmidbaur, *J. Am. Chem. Soc.*, 1999, **121**, 9758.
264. R. J. Baker, R. D. Farley, C. Jones, M. Kloth and D. M. Murphy, *Dalton Trans.*, 2002, 3844.
265. R. J. Baker and C. Jones, *Coord. Chem. Rev.*, 2005, **249**, 1857.
266. A. Schnepf, G. Stößer and H. Schnöckel, *Angew. Chem., Int. Ed.*, 2002, **41**, 1882.
267. A. Schnepf, R. Köppe, F. Wechert and H. Schnöckel, *Chem. Eur. J.*, 2004, **10**, 1977.
268. J. Hartig, A. Stößer, P. Hauser and H. Schnöckel, *Angew. Chem., Int. Ed.*, 2007, **46**, 1658.
269. A. Schnepf and H. Schnöckel, *Angew. Chem., Int. Ed.*, 2001, **40**, 711.
270. D. A. Neumayer and J. G. Ekerdt, *Chem. Mater.*, 1996, **8**, 9.
271. D. M. Hoffman, *Polyhedron*, 1994, **13**, 1169.
272. J. A. Pardoe and A. J. Downs, *Chem. Rev.*, 2007, **107**, 2.

273. A. P. Purdy, *Inorg. Chem.*, 1994, **33**, 282.
274. M. A. Petrie, K. Ruhlandt-Senge, H. Hope and P. P. Power, *Bull. Soc. Chim. Fr.*, 1993, **130**, 851.
275. J. Pauls, S. Chitsaz and B. Neumüller, *Z. Anorg. Allg. Chem.*, 2001, **627**, 1723.
276. R. Frey, V. D. Gupta and G. Linti, *Z. Anorg. Allg. Chem.*, 1996, **622**, 1060.
277. B. Luo, C. J. Cramer and W. L. Gladfelter, *Inorg. Chem.*, 2003, **42**, 3431.
278. J. Kim, S. G. Bott and D. M. Hoffman, *Inorg. Chem.*, 1998, **37**, 3835.
279. J. T. Patton, M. M. Bokota and K. A. Abboud, *Organometallics*, 2002, **21**, 2145.
280. M. Arif, D. C. Bradley, D. M. Frigo, M. B. Hursthouse and B. Hussain, *J. Chem. Soc., Chem. Commun.*, 1985, 783.
281. Y. L. Zhou and D. S. Richeson, *Organometallics*, 1995, **14**, 3558.
282. O. T. Beachley, C. Bueno, M. R. Churchill, R. B. Hallock and R. G. Simmons, *Inorg. Chem.*, 1981, **20**, 2423.
283. I. Grafova, R. Vivani, A. Grafov and F. Benetollo, *J. Organomet. Chem.*, 2004, **689**, 3000.
284. K. A. Aitchison, J. D. J. Backerdirks, D. C. Bradley, M. M. Faktor, D. M. Frigo, M. B. Hursthouse, B. Hussain and R. L. Short, *J. Organomet. Chem.*, 1989, **366**, 11.
285. O. T. Beachley, D. J. Macrae, M. R. Churchill, A. Y. Kovalevsky and E. S. Robards, *Organometallics*, 2003, **22**, 3991.
286. D. L. Freeman, J. D. Odom, W. R. Nutt and L. Lebioda, *Inorg. Chem.*, 1997, **36**, 2718.
287. R. Hillwig, K. Harms and K. Dehnicke, *J. Organomet. Chem.*, 1995, **501**, 327.
288. B. Neumüller, *Z. Naturforsch.*, 1990, **45B**, 1559.
289. B. Neumüller, *Z. Naturforsch.*, 1991, **46B**, 753.
290. B. Neumüller, *Chem. Ber.*, 1989, **122**, 2283.
291. E. K. Styron, S. J. Schauer, C. H. Lake, C. L. Watkins and L. K. Krannich, *J. Organomet. Chem.*, 1999, **585**, 266.
292. A. M. Arif, D. C. Bradley, H. Dawes, D. M. Frigo, M. B. Hursthouse and B. Hussain, *J. Chem. Soc., Dalton Trans.*, 1987, 2159.
293. P. L. Shutov, S. S. Karlov, K. Harms, O. K. Poleshchuk, J. Lorberth and G. S. Zaitseva, *Eur. J. Inorg. Chem.*, 2003, 1507.
294. G. Rossetto, N. Brianese, A. Camporese, M. Porchia and R. Bertoncello, *Main Group Met. Chem.*, 1991, **14**, 113.
295. T. Ryorin, H. Saito, Y. M. Tanetani, T. Yuasa and M. Harada, *Jpn. Kokai Tokkyo Koho*, 2007, JP2007077245.
296. J. Kim, S. G. Bott and D. M. Hoffman, *J. Chem. Soc., Dalton Trans.*, 1999, 141.
297. J. Tödtmann, W. Schwarz, J. Weidlein and A. Haaland, *Z. Naturforsch.*, 1993, **48B**, 1437.
298. C. Paek, S. O. Kang, J. Ko and P. J. Carroll, *Organometallics*, 1997, **16**, 4755.
299. T. Grabowy and K. Merzweiler, *Z. Anorg. Allg. Chem.*, 2000, **626**, 736.
300. M. Porchia, F. Benetollo, N. Brianese, G. Rossetto, P. Zanella and G. Bombieri, *J. Organomet. Chem.*, 1992, **424**, 1.
301. C. Jones, P. C. Junk and N. A. Smithies, *Main. Group Met. Chem.*, 2003, **26**, 35.
302. J. Prust, P. Müller, C. Rennekamp, H. W. Roesky and I. Usón, *J. Chem. Soc., Dalton Trans.*, 1999, 2265.
303. M. R. Kopp and B. Neumüller, *Z. Anorg. Allg. Chem.*, 1998, **624**, 361.
304. T. Grabowy and K. Merzweiler, *Z. Anorg. Allg. Chem.*, 1999, **625**, 2045.
305. N. Wiberg, R. Blank, K. Amelunxen, N. Nöth, H. Schnöckel, E. Baum and A. Purath, *Eur. J. Inorg. Chem.*, 2002, 341.
306. M. S. Hill and P. B. Hitchcock, *Chem. Commun.*, 2004, 1818.
307. M. S. Hill, P. B. Hitchcock and R. Pongtavornpinyo, *Dalton Trans.*, 2005, 273.
308. (a) M. S. Hill, P. B. Hitchcock and R. Pongtavornpinyo, *Dalton Trans.*, 2007, 731; (b) M. S. Hill, P. B. Hitchcock, and R. Pongtovornipinyo, *Angew. Chem. Int. Ed.*, 2005, **44**, 4231.
309. C. Jones, P. C. Junk, J. A. Platts, D. Rothmann and A. Stasch, *Dalton Trans.*, 2005, 2479.

310. (a) R. J. Baker, R. D. Farley, C. Jones, M. Kloth and D. M. Murphy, *Chem. Commun.*, 2002, 1196; (b) M. S. Stender and P. P. Power, *Polyhedron*, 2002, **21**, 525.
311. M. Veith, F. Goffing, S. Becker and V. Huch, *J. Organomet. Chem.*, 1991, **406**, 105.
312. (a) L. Grocholl, I. Schranz, L. Stahl and R. J. Staples, *Inorg. Chem.*, 1998, **37**, 2496; (b) M. D. Fryzuk, G. R. Giesbrecht, S. J. Rettig, and G. P. A. Yap, *J. Organomet. Chem.*, 1999, **591**, 63.
313. (a) K. W. Hellman, C. H. Galka, I. Rüdenauer, L. H. Gade, I. J. Scowen and M. McPartlin, *Angew. Chem., Int. Ed.*, 1998, **37**, 1948; (b) K. W. Hellman, L. H. Gade, A. Steiner, D. Stalke and F. Möller, *Angew. Chem. Int. Ed.*, 1997, **36**, 160.
314. K. Mertz, W. Schwartz, B. Elerivein, J. Weidlein, H. Hess and H. J. Hanson, *Z. Anorg. Allg. Chem.*, 1977, **429**, 99.
315. P. Krommes and J. Lorberth, *J. Organomet. Chem.*, 1977, **131**, 415.
316. L. H. Gade, *Dalton Trans.*, 2003, 267.
317. P. Pyykkö, *Chem. Rev.*, 1997, **97**, 597.
318. J. Reedjik, G. Roelofsen, A. R. Siedle and A. L. Spek, *Inorg. Chem.*, 1979, **18**, 1947.
319. J. Beck and J. Strähle, *Z. Naturforsch.*, 1986, **41B**, 1381.
320. A. H. Cowley, R. L. Geerts, C. M. Nunn and S. Trofimenko, *J. Organomet. Chem.*, 1989, **365**, 19.
321. K. W. Klinkhammer and S. Henkel, *J. Organomet. Chem.*, 1994, **480**, 167.
322. A. Haaland, D. J. Shorokhov, H. V. Volden and K. W. Klinkhammer, *Inorg. Chem.*, 1999, **38**, 1118.
323. R. J. Wright, M. Brynda and P. P. Power, *Inorg. Chem.*, 2005, **44**, 3368.
324. S. D. Waezsada, T. Belgardt, M. Noltemeyer and H. W. Roesky, *Angew. Chem., Int. Ed.*, 1994, **33**, 1351.
325. Y. Cheng, P. B. Hitchcock, M. F. Lappert and M. Zhou, *Chem. Commun.*, 2005, 752.
326. J. C. Peters, S. B. Harkins, S. D. Brown and M. W. Day, *Inorg. Chem.*, 2001, **40**, 5083.
327. M. Veith, F. Goffing and V. Huch, *Chem. Ber.*, 1988, **121**, 943.
328. K. W. Hellmann, L. H. Gade, R. Fleischer and T. Kottke, *Chem. Eur. J.*, 1997, **3**, 1801.
329. C. H. Galka, D. J. M. Trosch, M. Schubart, L. H. Gade, S. Radojevic, I. J. Scowen and M. McPartlin, *Eur. J. Inorg. Chem.*, 2000, 2577.
330. W. M. Boesveld, P. B. Hitchcock, M. F. Lappert and H. Nöth, *Angew. Chem., Int. Ed.*, 2000, **39**, 222.
331. K. W. Hellmann, L. H. Gade, I. J. Scowen and M. McPartlin, *Chem. Commun.*, 1996, 2515.
332. K. W. Hellmann, L. H. Gade, R. Fleische and D. Stalke, *Chem. Commun.*, 1997, 527.
333. C. H. Galka and L. H. Gade, *Inorg. Chem.*, 1999, **38**, 1038.
334. C. H. Galka, P. Renner and L. H. Gade, *Inorg. Chem. Commun.*, 2001, **4**, 332.
335. J. S. Casas, A. Macias, N. Playa, A. Sanchez, J. Sordo and J. M. Varela, *Polyhedron* 1992, **11**, 2231.
336. J. S. Casas, A. Castinèiras, N. Playá, A. Sánchez, J. Sordo, J. M. Varela and E. M. Vazquez-Lopez, *Polyhedron* 1999, **18**, 3653.
337. M. S. Garciatasende, M. I. Suarez, A. Sánchez, J. S. Casas, J. Sordo, E. E. Castellano and Y. P. Mascarenhas, *Inorg. Chem.*, 1987, **26**, 3818.
338. J. S. Casas, E. E. Castellano, A. Macias, N. Playa, A. Sánchez, J. Sordo, J. M. Varela and J. Zukerman-Schpector, *Inorg. Chim. Acta*, 1995, **238**, 129.
339. H. Gornitzka and D. Stalke, *Eur. J. Inorg. Chem.*, 1998, 311.
340. J. S. Casas, E. E. Castellano, J. Ellena, M. S. Garcia-Tasende, A. Sanchez, J. Sordo, E. M. Vazquez-Lopez and M. J. Vidarte, *Z. Anorg. Allg. Chem.*, 2003, **629**, 261.
341. G. B. Deacon, S. J. Faulks, B. M. Gatehouse and A. J. Jozsa, *Inorg. Chim. Acta*, 1977, **21**, L1.
342. K. W. Hellmann, A. Bergner, L. H. Gade, I. J. Scowen and M. McPartlin, *J. Organomet. Chem.*, 1999, **573**, 156.
343. C. H. Galka and L. H. Gade, *Chem. Commun.*, 2001, 899.
344. K. W. Hellman, C. H. Galka, L. H. Gade, A. Steiner, D. S. Wright, T. Köttke and D. Stalke, *Chem. Commun.*, 1998, 549.

9

Subvalent Amides of Silicon and the Group 14 Metals

9.1 Introduction

In our 1980 book, the synthesis and characterisation of amides of these elements were discussed in 37 pages of text and 83 pages of tabulated data for 1161, 186, 274 and 66 individual amides of Si, Ge, Sn and Pb, respectively; there were 396 (Si), 66 (Ge), 142 (Sn) and 19 (Pb) bibliographic citations.[1] All the silicon entries referred to Si^{IV} amides and the majority of those for the heavier members of the group also were of compounds in the +4 oxidation state. However, 8 (Ge), 9 (Sn) and 3 (Pb) citations were for monomeric bivalent amides; the X-ray structures of $Ge(TMP)_2$[2] and $\overline{SnN(Bu^t)Si(Me)_2N}Bu^t$[3] had been published. The trivalent state was represented by solution EPR spectral data on three Ge^{III} and four Sn^{III} amides. Part 2 of our book had major sections on reactions of the tetravalent Group 14 metal amides, covering insertion or metathetical exchange reactions, and experiments with metal hydrides, Lewis acids or bases; the main thrust was on such amides of silicon(IV) and tin(IV), somewhat less on those of Ge^{IV} and very little of Pb^{IV} amides.[1]

This survey is restricted to thermally robust (rather than transient) *subvalent* amides of the Group 14 elements; thus advances in the chemistry of compounds in the +4 oxidation state are omitted, but some reviews which mention the topic (reviews on compounds in the +2 oxidation state also included) are available;[4] more specialised surveys are cited later.

The seminal discovery relating to thermally robust bivalent Group 14 metal amides related to reports of the crystalline monomeric bis(trimethylsilyl)amides of Ge, Sn and Pb;[5] the tin compound $Sn[N(SiMe_3)_2]_2$ had been prepared independently but was erroneously assigned as the Sn–Sn-bonded dimer.[6] These M^{II} amides were thermochromic, becoming colourless at $-196\,^{\circ}C$ and reverting to coloured compounds as the temperature was raised, distilling as red-brown vapours.[5] The tendency to redness was most marked for Pb and least for Ge. These diamagnetic compounds were formulated as bent singlets, as confirmed in the

Metal Amide Chemistry Michael Lappert, Andrey Protchenko, Philip Power and Alexandra Seeber

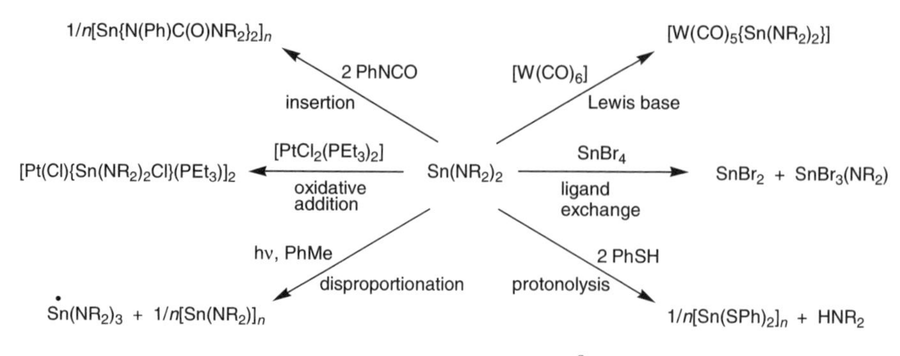

Scheme 9.1 *(R = SiMe₃)*[8]

pre-1980 period by the electron diffraction study of $[Sn\{N(SiMe_3)_2\}_2]$,[7] and the X-ray structures of crystalline $Ge(TMP)_2$[2] and $\overline{SnN(Bu^t)Si(Me)_2N}Bu^t$ (which also exists as a dimer).[3] The compounds were shown to undergo insertion, oxidative addition, disproportionation; and Lewis base, ligand exchange and protonolysis reactions,[5,8] as illustrated for the tin compound in Scheme 9.1.[8]

The photochemical synthesis and characterisation (by EPR solution spectra at 20 °C in C_6H_{14}) of the first stable trivalent germanium and tin amides $M^{III}[N(SiMe_3)_2]_3$ was first reported in 1974; in solution they were shown to be persistent for more than three months at ambient temperature.[9] An alternative synthesis from a tetravalent precursor for the Ge^{III} amide is shown in Equation (9.1) (X = Cl or Br).[10]

$$4[Ge\{N(SiMe_3)_2\}_3X] + (\mathrm{EtNCH_2CH_2NEt})C{=}C(\mathrm{EtNCH_2CH_2NEt}) \xrightarrow[C_6H_{14}]{h\nu} 4\dot{G}e[N(SiMe_3)_2]_3 + 2\left[(\mathrm{EtNCH_2CH_2NEt})C{-}X\right]X\downarrow \qquad (9.1)$$

A wider range of monomeric bivalent Group 14 metal amides $M[N(Bu^t)SiMe_3]_2$, $M'[N(GeMe_3)_2]_2$, $M'[N(GeEt_3)_2]_2$ and $M'[N(GePh_3)_2]_2$ (M = Ge, Sn, Pb; M′ = Ge, Sn) was prepared.[11] Photolysis in C_6H_{14} at ambient temperature afforded $M'[N(GeMe_3)_2]_3$ and $M'[N(GeEt_3)_2]_3$, while the other Ge^{II} and Sn^{II} amides were unreactive and the Pb^{II} amides decomposed, depositing a lead mirror.[11] The stability of solutions of the M^{III} amides decreased in the sequence $M'[N(SiMe_3)_2]_3 \gg M'[N(Bu^t)SiMe_3]_3 < M'[N(GeMe_3)_2]_3 \approx M'[N(GeEt_3)_2]_3$.[11] A detailed EPR spectroscopic study of these compounds was published.[12]

9.2 Subvalent Amidosilicon Compounds

9.2.1 Introduction

The first bis(amino)silylene **1** was described in 1992, but was stable only up to 77 K.[13a] The transient $Si(NPr^i_2)_2$ was identified by trapping reactions and by electronic spectra at low temperatures.[13b] The major breakthrough in the chemistry of stable silylenes was the report in 1994 by Denk, West, Haaland and their collaborators of the synthesis and electron diffraction study of **2**.[14] This was followed in 1995 by the preparation and X-ray structure of **3**.[15] Related compounds **4** (1996),[16] **5** (1998),[17] and **6** (2001)[18] were subsequently described. Each of **2–6** was obtained by the reduction of a dichloro- or dibromosilane precursor with K, Na/K alloy or KC_8.

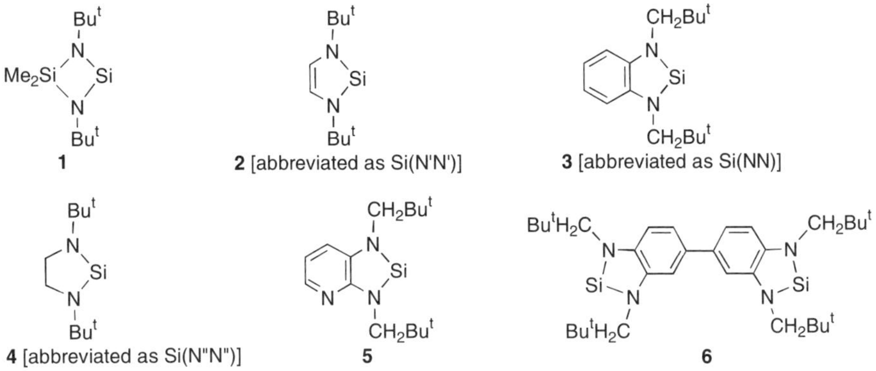

1 **2** [abbreviated as Si(N'N')] **3** [abbreviated as Si(NN)]

4 [abbreviated as Si(N"N")] **5** **6**

The chemistry of bis(amino)silylenes, particularly of **2** and **3** and to a lesser extent **4**, has been studied in some detail and has relatively recently been comprehensively reviewed;[19,20,21] hence only an outline of these researches will be featured in Sections 9.2.2 and 9.2.3.

9.2.2 Bis(amino)silylenes: pre-2001[19,20]

The brief account in this section is based mainly on Ref. 20 but also Ref. 19, having 78 and 47 literature citations, respectively, the majority being from the groups of West and Gehrhus/Lappert. Molecular structures were established by electron diffraction for **2** and X-ray diffraction for **3** and **4**. In solution, **4** was in dynamic equilibrium, Equation (9.2); the tetramer **7** was X-ray-characterised.

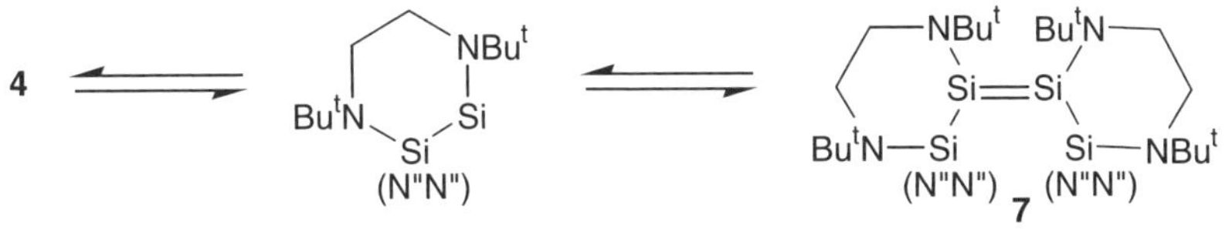

(9.2)

The electronic structures of **2–4** were probed by their He(I) and He(II) photoelectron spectra. Calculations reported on isodesmic hydrogenation reactions of model compounds were consistent with significant π-delocalisation in **2** and **3**; support for that view came inter alia from chemical shift tensors for **2–4**, determined from $^{29}Si\{^1H\}$ CPMAS-NMR spectra.

In their reaction chemistry, **2–4** may be described as nucleophilic silylenes, consistent with their $^{29}Si\{^1H\}$ NMR spectroscopic chemical shifts and their $n(Si) \rightarrow 3p_\pi(Si)$ visible band being at much higher and at lower frequency, respectively, than in the dialkylsilylene $\overline{SiC(SiMe_3)_2(CH_2)_2C}(SiMe_3)_2$.

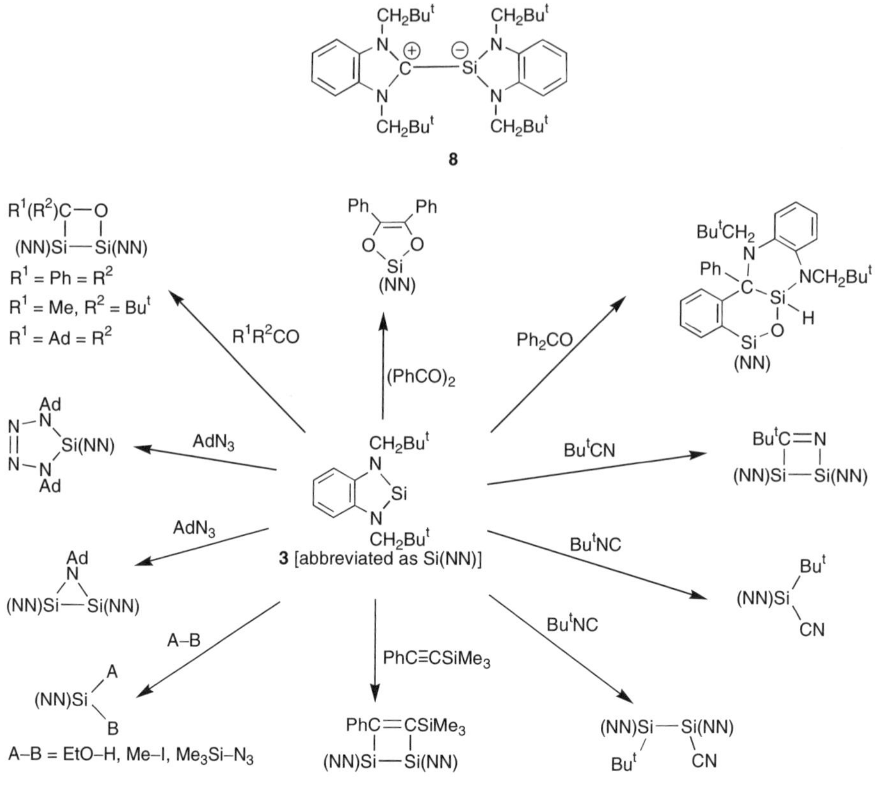

Scheme 9.2[20]

The reactions of **2–4** were classified as follows: (i) nucleophilic addition reactions of a bis(amino)silylene to an unsaturated organic compound; (ii) insertion of a bis(amino)-silylene into a compound containing an O–H, C–Cl, C–Br, C–I or B–C bond; (iii) insertion of a bis(amino)silylene into a compound having an M–N (M = Li, Na, K), Li–C, Li–Si or M′(II)–X (M′ = Ge, Sn, Pb; X = C, N, O, Cl) bond; (iv) insertion of a silylene into a transition metal–X bond; (v) further (*cf.*, i and ii) oxidative addition of a chalcogen or Bu^tNC to a bis(amino)silylene; (vi) bis(amino)silylenes as reducing reagents; (vii) bis-(amino)silylenes as ligands in d- or f-block chemistry; and (viii) an adduct **8** with the bis-(amino)carbene analogue of **3**. The reactions of **2–4** described in references 19 and 20 are summarised succinctly in Schemes 9.2–9.8.

9.2.3 Bis(amino)silylenes: 2001–2004

This account is based on the 2004 review by Hill and West which has 77 literature citations, 21 of which precede the year 2000.[21] Also the matter there covered is dealt with in this section only in outline; we deemed it appropriate to cite references given there to the post-2000 period; the relevant original bibliography has also been consulted.

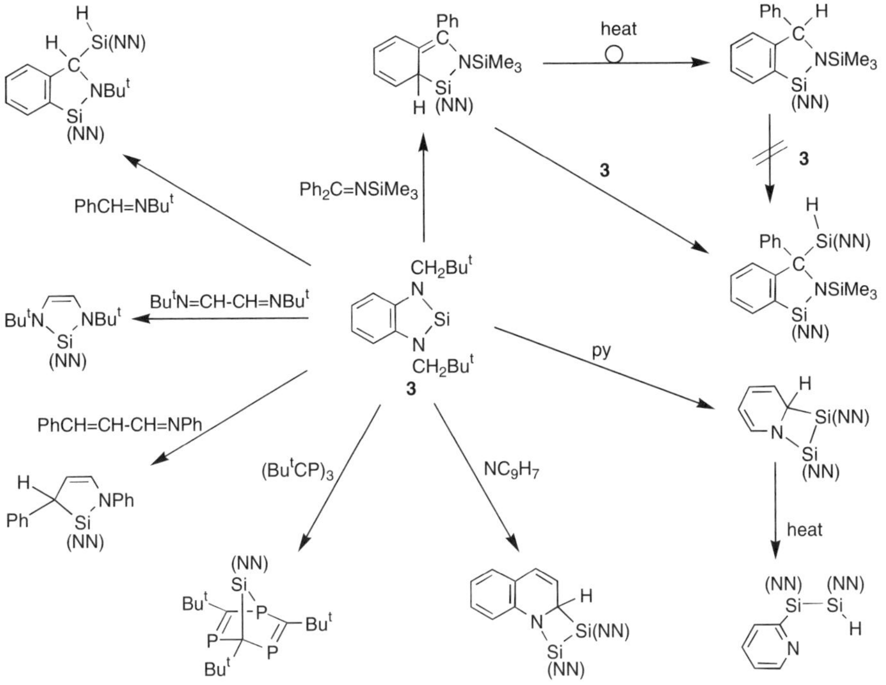

Scheme 9.3 [20]

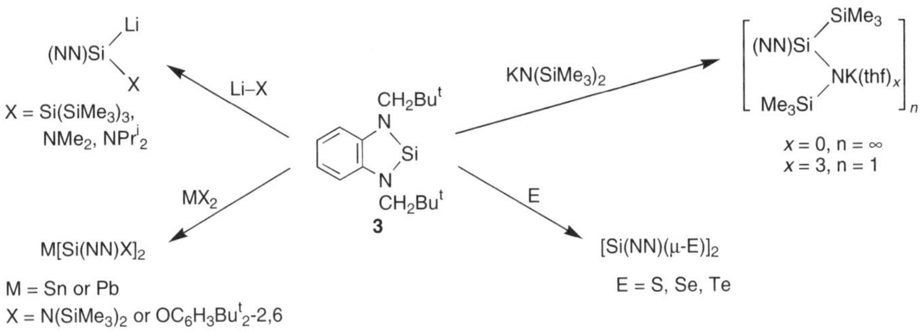

Scheme 9.4 [20]

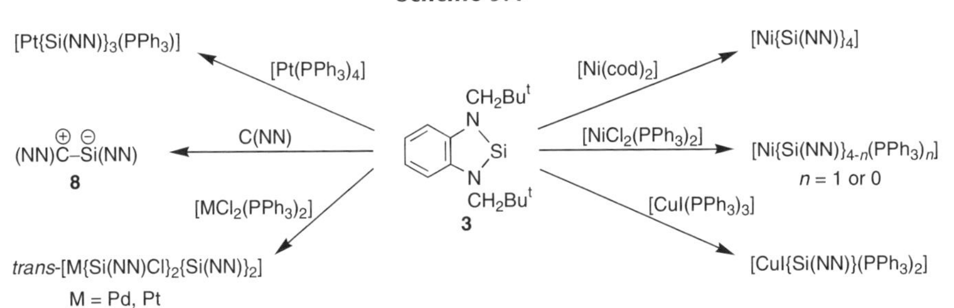

Scheme 9.5 [20]

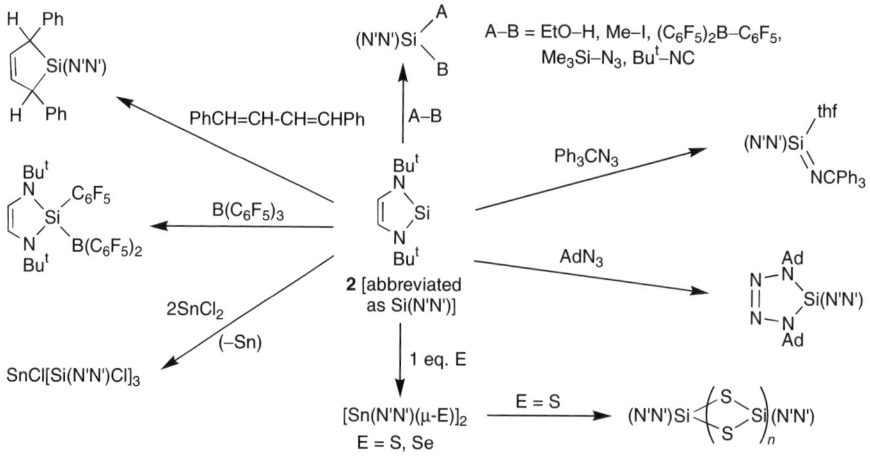

Scheme 9.6 [19,20]

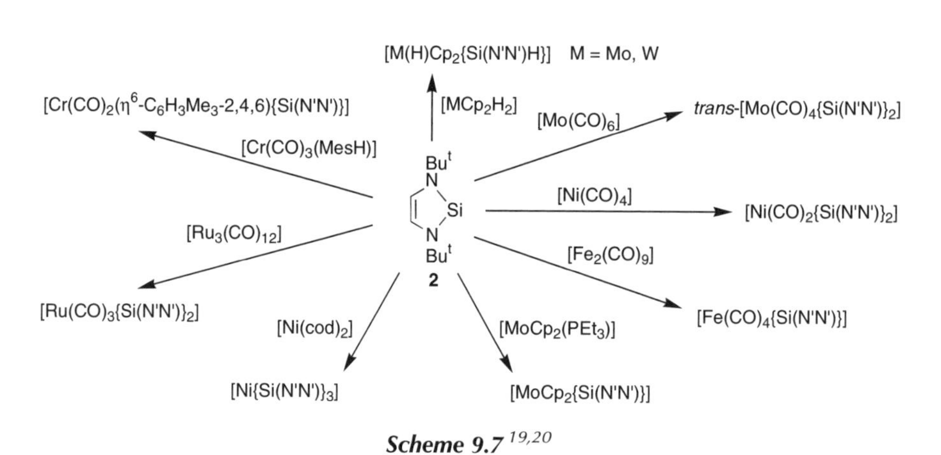

Scheme 9.7 [19,20]

The compound $Si[N(SiMe_3)_2]_2$, obtained by KC_8 reduction of $SiBr_2[N(SiMe_3)_2]_2$ in thf at $-78\,°C$, was stable at $< -20\,°C$, $\delta[^{29}Si\{^1H\}] = +223.9$ ppm for the silylene Si nucleus; with ROH it afforded $SiH(OR)[N(SiMe_3)_2]_2$ (R = Me, Ph); DFT calculations were used to compute $\delta[^{29}Si]$, the molecular structure of the silylene, and to calculate its stability relative to its isoleptic Group 14 metal(II) amides (Sn > Ge ≫ Si).[22]

The low temperature structure of the silylene Si(N′N′) (**2**) confirmed its monomeric nature and showed that the ring SiNCCN is planar.[23] Reduction of 9,9-dichloro-10-methyl-9-sila-10-azaanthracene with $LiC_{10}H_8$ yielded a transient silylene, as established by a trapping experiment.[24] Reduction of **9** using an excess of K yielded successively the potassium salts of the tetrameric radical anion **10** and the dianion **11**, Equation (9.3).[25]

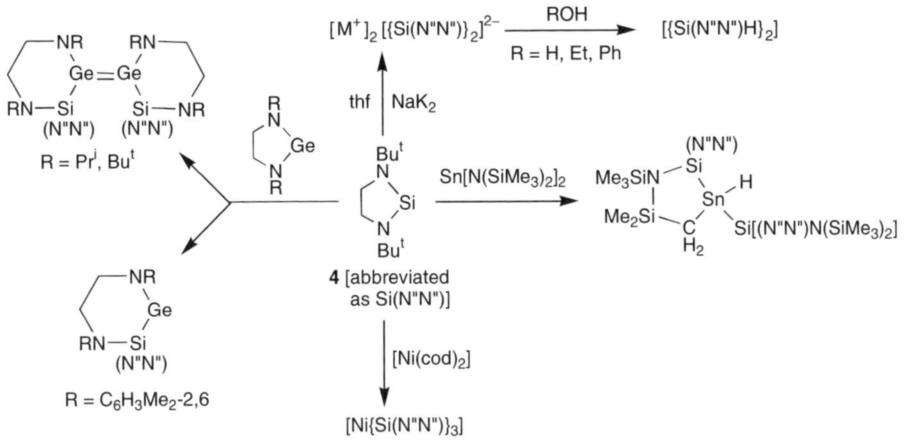

Scheme 9.8 [19,20,21]

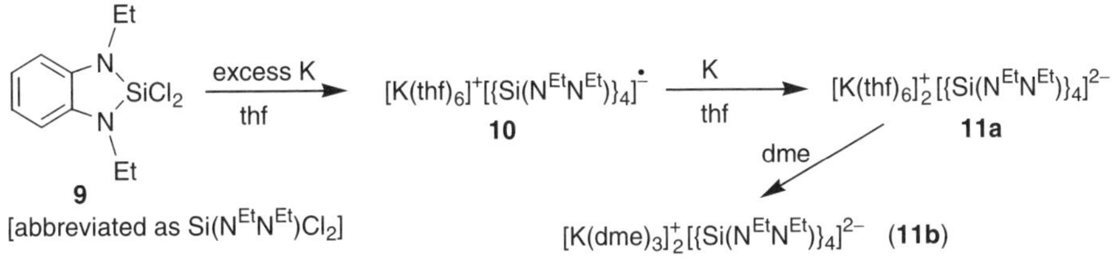

(9.3)

The X-ray structures of the paramagnetic crystalline solvent-separated ion pair and the crystalline salt **11b** (K^+ is weakly η^2-C_6H_4-bound) showed the Si_4 ring in each to be a square with Si–Si bond lengths of 2.347(2) and 2.284(2) Å, respectively. These data, together with similar findings on the isobutyl analogues of **9–11**, demonstrate the importance of sterically demanding *N*-substituents R for the stabilisation of such silylenes o-$C_6H_4\{N(R)\}_2Si$.[25]

Evidence for aromatic π-delocalisation in Si(N′N′) (**2**) included (i) the greatly enhanced (Raman) C=C stretching mode in **2** over non-aromatic reference compounds;[26] and comparisons of (ii) the enthalpies (**2** > **4**) of the isodesmic reactions for **2**(or **4**) + $SiH_4 \rightarrow$ Si(N′N′ or N″N″)H_2 + SiH_2,[27a] or (iii) singlet-triplet splitting energies (**4** > **2**).[27a,27b] The ^{29}Si-NMR spectral chemical shift tensors of 21 transient silylenes, including $Si(NH_2)X$ (X = H, Me) have been computed, suggesting that there is a correlation between λ_{max} and the isotropic chemical shift.[28] A DFT study of **2**, **3** and **5** (and related compounds) and their radical anions led inter alia to calculations of reduction potentials of the silylenes ($E^0 = -2.97$, -2.88 and -2.69 V vs SCE, respectively).[29] Experimental data gave -2.69 V for **2** but two reduction waves for the saturated analogue **4** (-1.75 and -2.35 V).[30]

Radical species including **10**[25] and **12**[31] [R = $MCp(CO)_3$ (M = Mo, W), $Re(CO)_5$, TEMPO, $P(O)(OPr^i)_2$, CH_2Ph] were generated, for example, for R = $Re(CO)_5$ by irradiation of **2** with $[Re_2(CO)_{10}]$. Treatment of Si(N′N′) (**2**) with Bu^{t}Cl gave Si(N′N′)(Cl)Bu^t, and with

CCl_4, $CHCl_3$, CH_2Cl_2, or $PhCH_2Cl$ gave the appropriate Cl(N′N′)Si–Si(N′N′)$CR^1R^2R^3$ via a proposed chloride-bridged intermediate (N′N′)Si$\cdots$Cl$CR^1R^2R^3$; from bromobenzene, Si(N′N′)(Br)Ph and Br(N′N′)Si–Si(N′N′)Ph were obtained.[32]

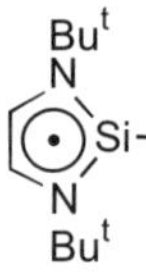

12

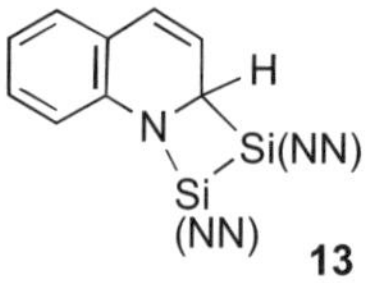

13

Treatment of Si(NN) (**3**) with pyridine (see also Scheme 9.3) has been studied in detail; with quinoline, the product was **13**.[33] The X-ray-characterised stable disilacyclopropane was prepared as shown in Equation (9.4).[34] Further (see Scheme 9.4) insertion reactions of **3** into Li–X bonds have been reported, generating the adducts **14** [X, L_n = Me, $(OEt_2)_2$;

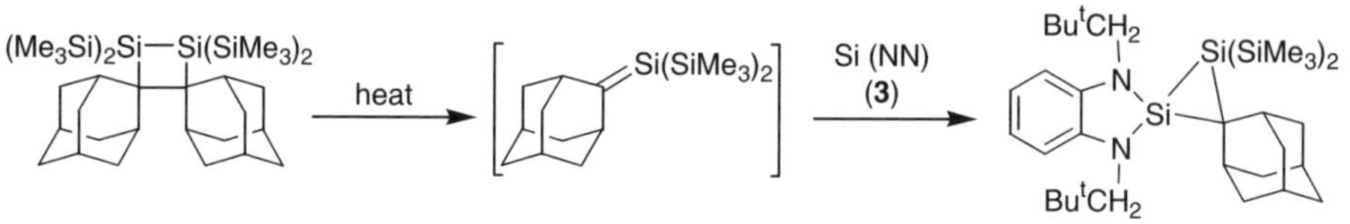

(9.4)

Bu^t, $(thf)_3$; $CH(SiMe_3)_2$, $(thf)_2$].[35] The azatrisilacyclobutane **15** was obtained from **3** and the crystalline 1-azaallyllithium compound **16**.[36] While **3** with $(Bu^tCP)_3$ gave the (1 + 4) cycloadduct (Scheme 9.3),[37] silylene **4** reacted as the dimer to give the related adduct **17**.[38] Treatment of **3** with the phosphirene **18a**, **18b** or **18c** gave **19a**, **19b** or **20a** (with **20b**), respectively; in the formation of **19b**, the intermediate crystalline zwitterionic compound **21** was isolated.[39] For the reactions of **3** with Bu^{t}CN, AdCN, Bu^{t}NC, AdN_3 and Me_3SiN_3,[40] see Scheme 9.2.

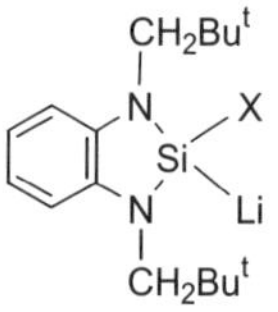

14

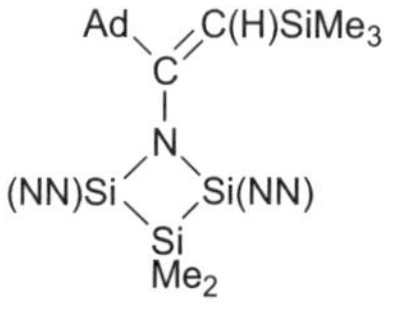

15

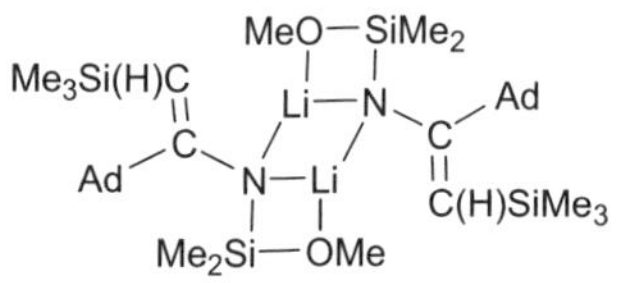

16

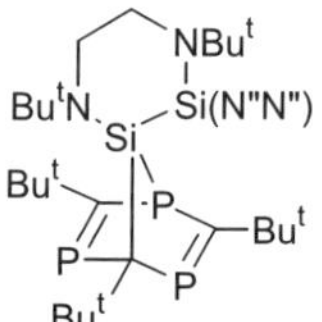

17

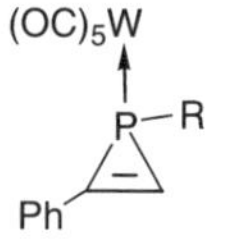

18a R = Ph
18b R = Me

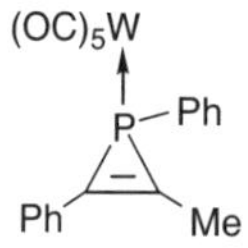

18c

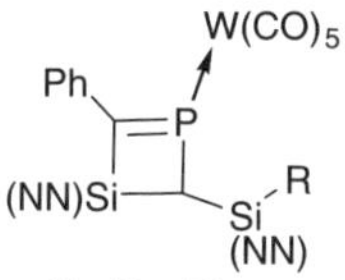

19 a R = Ph
19 b R = Me

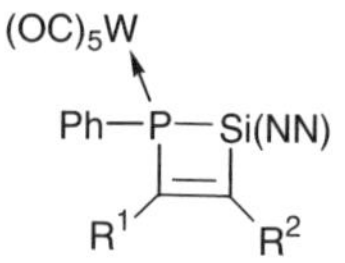

20a R^1 = Ph, R^2 = Me
20b R^1 = Me, R^2 = Ph

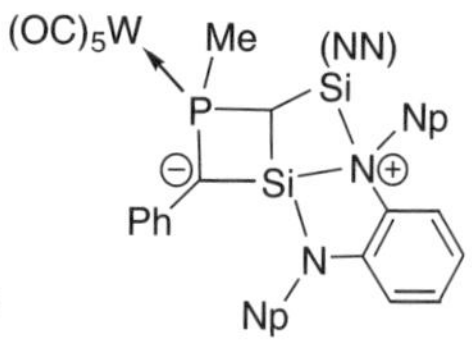

21

Although calculations indicated that electron transfer to Si(N′N′) (**2**) was more favourable than for the saturated analogue Si(N″N″) (**4**), reduction with K gave no anionic product with **2**, whereas **4** with 1K/thf gave the transient salt of the disilyl dianion $[\{Si(N''N'')\}_2]^{2-}$ and with 2K/thf gave the more robust salt of the dianion $[Si(N''N'')]^{2-}$. The former could be trapped with H_2O as the dihydride, while the latter slowly deprotonated thf at 25 °C yielding the monoanionic $[Si(N''N'')H]^-$ which was trapped with an electrophile.[30] These results may be compared with those for **3**, Equation (9.3).[25]

From **2** and $[P(NCy_2)_2][AlCl_4]$ the Si^{IV} compound $Si(N'N')Cl\{P(NCy_2)_2\}$ was isolated; the formation of the salt $[Si(N'N')\{P(NCy_2)_2\}][AlCl_4]$ as an intermediate was supported by *ab initio* calculations.[41] The complex $[\{Pd(PPh_3)(\mu\text{-}Si(N'N'))\}_2]$, obtained from **2** and $[Pd(PPh_3)_4]$, was a catalyst for Suzuki[42] and Stille[21] cross-coupling. The highly labile Pd-$(\mathbf{2})_3$ and Pd$(\mathbf{4})_4$, derived from $Pd(PBu^t_3)_2$ and **2** or **4**, were trapped as the crystalline complexes $[\{Pd(\mathbf{2})(\mu\text{-}\mathbf{2})\}_2]$ and $[\{Pd(\mathbf{4})(\mu\text{-}\mathbf{4})\}_2]$.[43] Details have appeared[44] of reactions of the silylene **3** with various Ni^0, Ni^{II}, Pt^0, Pd^{II}, Pt^{II} or Cu^I substrates, as summarised in Scheme 9.5; these were examples of (i) insertion of Si(NN) (**3**) into M–Cl (M = Pd, Pt) bonds, (ii) ligation of **3** to Ni^0, Ni^{II}, Pt^0, Pd^{II}, Pt^{II} or Cu^I complexes, (iii) a combination of processes i and iii, and (iv) reductive dechlorination of a Ni^{II} dichloride. The new data included (a) X-ray structures of $[Ni\{Si(NN)\}_4]$, *trans*-$[Pd\{Si(NN)Cl\}_2\{Si(NN)\}_2]$, $[Pt(PPh_3)\{Si(NN)\}_3]$, $[CuI(PPh_3)_2\{Si(NN)\}]$, and (b) a study of the fluxional behaviour of *trans*-$[Pd\{Si(NN)\text{-}Cl\}_2\{Si(NN)\}_2]$ in $CDCl_3$ solution.[44] Reaction of $[\{RuCp^*Cl\}_4]$ with **2** gave $[RuCp^*(Cl)\text{-}\{Si(N'N')\}]$ which with $HexSiH_3$ yielded the crystalline complex **22**, while $[RuCp^*(NCMe)_3]$ [OTf] and **2** yielded a salt tentatively formulated as **23**.[45] Well characterised crystalline

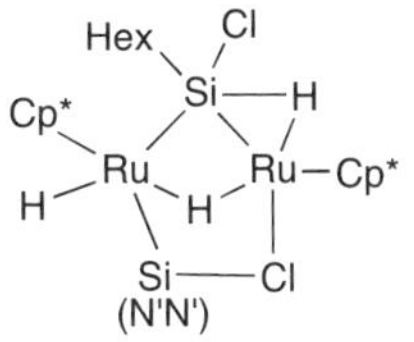

22 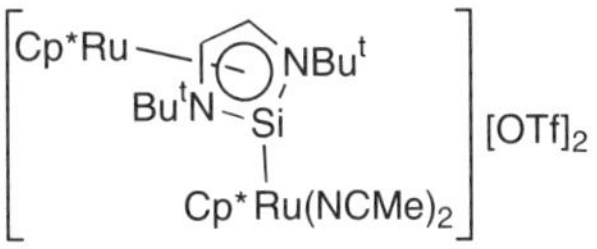 **23** 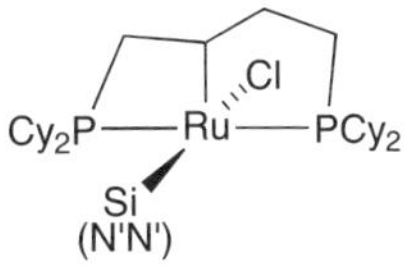 **24**

complexes $[LnCp_3\{Si(NN)\}]$ were obtained from $[LnCp_3]$ (Ln = Y, Yb) and **3** in PhMe, whereas $LaCp_3$ was unreactive; in PhMe-d_8 each of the 1:1 adducts dissociated (Y > Yb);[46a] the complex $[SmCp^*_2\{Si(N'N')\}]$ was obtained similarly from $[SmCp^*_2]$ and was not accessible from $[SmCp^*_2(thf)]$ and **2**.[46b] The crystalline complex **24** was obtained from $[L(Cl)Ru(\mu\text{-}Cl)_3Ru(L)(N_2)]$ and **2** $[L = Cy_2P(CH_2)_4PCy_2]$; treatment of **24** with H_2 containing a trace of H_2O gave $O\{Si(N'N')H\}_2$, and CO displaced Si(N'N') from **24**.[47] The compound $[MoCp_2\{Si(N'N')\}]$ was prepared from $[MoCp_2(PEt_3)]$ and Si(N'N').[48] A computational study of methyl migration from a Pd^{II} centre to its coordinated carbene, silylene or germylene E(N'N') indicated a low barrier for E = Si or Ge; the possible implications for catalysis were discussed.[48] The chemistry of transition metal complexes of acyclic base-stabilised silylenes has been reviewed.[49]

9.2.4 Bis(amino)silylenes: Post-2004

The colourless, monomeric *rac*-bis(amino)silylene $SiN(Bu^t)\{C(H)Me\}_2NBu^t$ (**25**) was prepared by KC_8 reduction in thf/NEt_3 of the *rac*-dibromosilane precursor. At ambient temperature, unlike its methyl-free analogue **4** (for the low temperature monomeric X-ray structure of **4**, see Ref. 50), it showed no tendency to oligomerise.[51] Attempts to make a thermally stable Et or Bu^i analogue of **3** failed, although presumably they were transient intermediates along the pathway to the salts **10** and **11**, Equation (9.3) (a preliminary

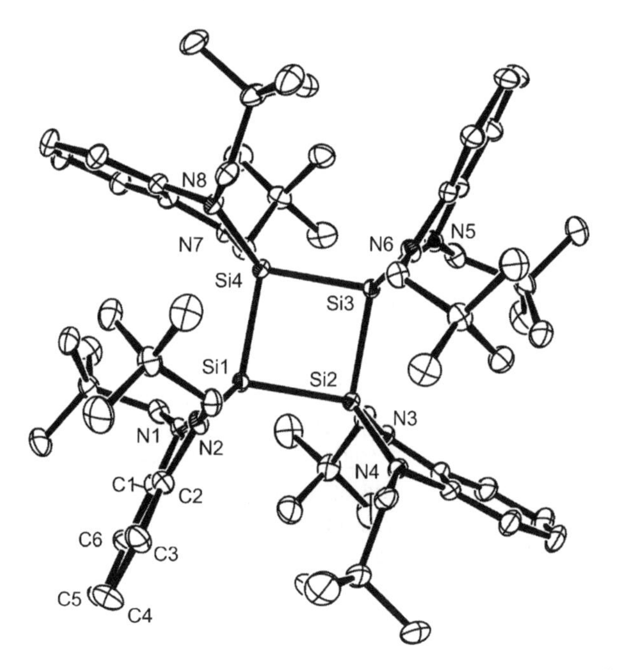

Figure 9.1 *Structure of the anion of* $[Na(thf)_6]_2[c\text{-}\{Si(NN)\}_4]$[52]

communication is Ref. 25) [52]. The X-ray structures of $[K(thf)_6][c\text{-}\{Si(N^{Et}N^{Et})\}_4]$ (**10**),[25] $[K(dme)_3]_2[c\text{-}\{Si(N^{Et}N^{Et})\}_4]$ (**11b**),[25] $[Na(thf)_6]_2[c\text{-}\{Si(NN)\}_4]$ (anion, Figure 9.1), $[Na(thf)_6][Na(thf)_5][c\text{-}\{Si(NN)\}_3]$ (anion, Figure 9.2) and $[\{Si(NN)X\}_2]$ ($X = Cl, SiMe_3$) were reported.[52]

Treatment of $[Cl_2Si\{N(CH_2Bu^t)\}_2C_6H_3\text{–}3,4]_2$ with KC_8 in thf furnished the crystalline bis(silylene) **6**; the molecule has a two-fold rotation axis, the two $C_6H_3N_2Si$ planes are twisted about the C(1)–C(1′) axis by 30.3(1)° and in C_6D_6 $\delta[^{29}Si\{^1H\}] = 96.6$ ppm.[18]

The relative reactivity [**2** > C(N′N′)] of the silylene **2** and its carbene analogue C(N′N′) towards Pd^0 complexes has been studied by NMR spectroscopy and DFT calculations.[53]

The silylenes Si(N′N′) (**2**) and Si(N″N″) (**4**) were studied by cyclic voltammetry (CV) and computationally (as well as electron affinities and ionisation potentials – poor correlations).[54] In thf or 1,2-$C_6H_4Cl_2$, CV measurements showed irreversible waves both for oxidation (E_p: **2** > **4**) and reduction (E_p: **4** > **2**); furthermore the silylenes were more readily oxidised or reduced than the isoleptic germylenes.[54] Treatment of the *rac*-silylene **25** with oxygen gas, MeOH or H_2O in hexane at −78 °C afforded the cyclodisiloxane $[\overline{Si(N(Bu^t)\{C(H)Me\}_2N}Bu^t)(\mu\text{-}O)]_2$, methoxyhydrosilane $[\overline{Si(N(Bu^t)\{C(H)Me\}_2N}Bu^t)\text{-}H(OMe)]$ or hydrodisiloxane $[Si(\overline{N(Bu^t)\{C(H)Me\}_2N}Bu^t)H]_2O$, respectively.[51] Some free radical oxidation reactions of **2** are summarised in Scheme 9.9 [TEMPO = $\overline{ONCMe_2(CH_2)_3C}Me_2$]; in the TEMPO reaction, suggested intermediates were $Si(N'N')(OTMP)_2$ and Si(N′N′)O via the EPR-characterised (N′N′)SiO(TMP), while $(N'N')SiSi(N'N')OBu^t$ may have been on the pathway to the $(Bu^tO)_2$-trisilane.[55]

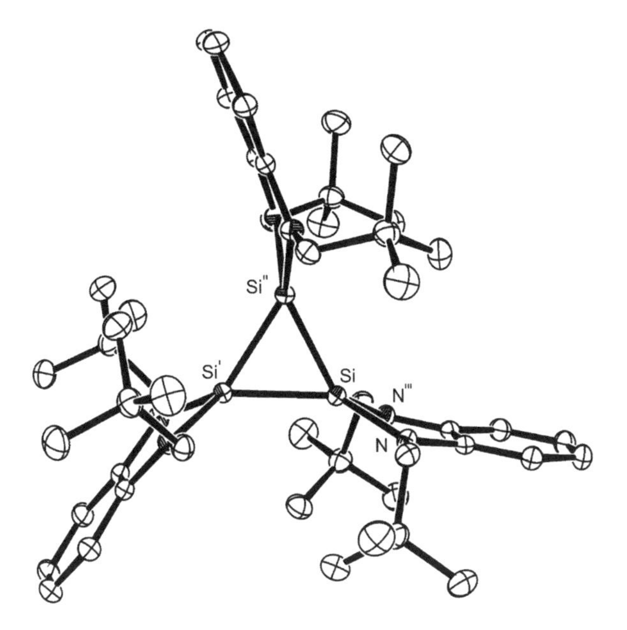

Figure 9.2 *Structure of the anion of [Na(thf)$_6$][Na(thf)$_5$][c-{Si(NN)}$_3$]*[52]

As shown in Scheme 9.10, the silylene **3** underwent oxidative additions with alkali metal (M) bis(trimethylsilyl)amides to give new alkali metal amides, rather than metal tris(amino)silyls M[Si(NN){N(SiMe$_3$)$_2$}](thf)$_x$ which, however, were suggested to have been intermediates.[56] Whereas M[N{Si(NN)(SiMe$_3$)}(SiMe$_2$R)](thf)$_x$ with R = Me underwent a further reaction with **3** to yield M[N{Si(NN)(SiMe$_3$)}$_2$], for R = Ph the double addition compound was the sole product in the **3**/Li[N(SiMe$_3$)(SiMe$_2$R)] system.[56]

Related reactions of **3** with LiNR$_2$ (R = Me, Pr^i) afforded the first tris(amino)silylmetal complexes, the X-ray-characterised [Li{Si(NN)NR$_2$}(thf)$_3$]; **3** with Li[N(SiMe$_3$)R′] gave the crystalline amides [Li{N(Si(NN)SiMe$_3$)R′}(thf)$_2$] (R′ = Bu^t, C$_6$H$_3$Me$_2$-2,6) and with R′ = C$_6$H$_3$Me$_2$-2,6 the crystalline [Li{Si(NN)N(SiMe$_3$)(C$_6$H$_3$Me$_2$-2,6)}(thf)$_2$] was isolated at low temperature.[57]

Crystalline Na-Si(NN) derivatives were prepared as shown in Scheme 9.11: the silylenoid [Si(NN)OMe]$^-$ (**26**), the dianion [(NN)Si-Si(NN)]$^{2-}$ (**27**) and the radical anion [c-{Si-(NN)}$_3$]$^-$ (**28**); X-ray structures are illustrated in Figures 9.3, 9.4 and 9.5,

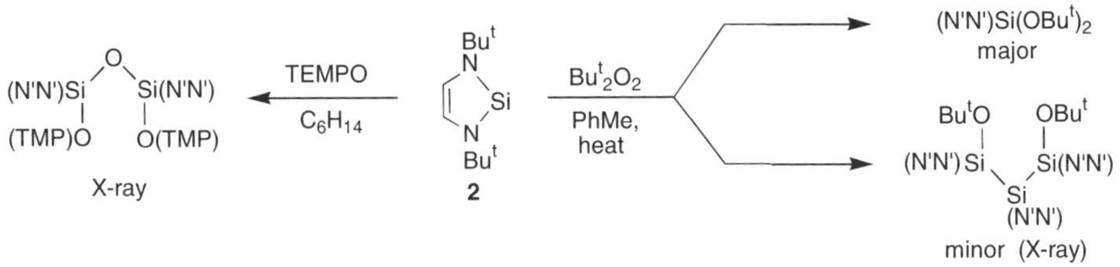

(N'N')Si(O)Si(N'N') with (TMP)O, O(TMP) — X-ray ← TEMPO, C$_6$H$_{14}$ — **2** (Bu^{t}N–Si–NBu^t) — Bu$^t_2$O$_2$, PhMe, heat → (N'N')Si(OBu^t)$_2$ major; Bu^{t}O(N'N')Si–Si(N'N')–Si(N'N')OBu^t minor (X-ray)

Scheme 9.9[55]

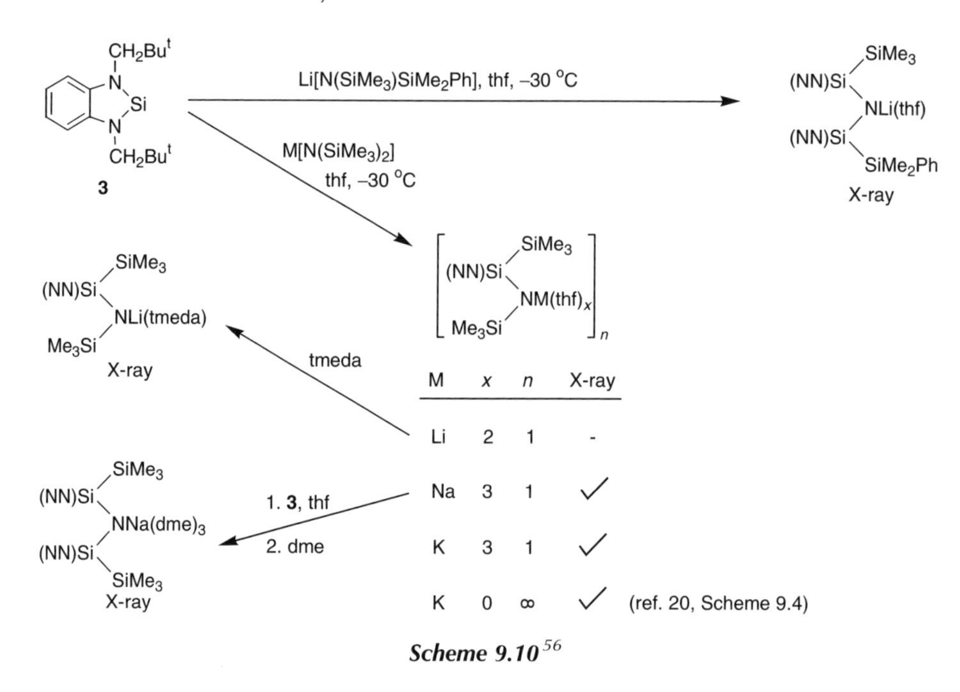

Scheme 9.10[56]

respectively.[58] The EPR spectrum of $[Na(thf)_4][c\text{-}\{Si(NN)\}_3]$ in dme was consistent with the single electron being delocalised over the Si_3 ring.

Further reactions (cf. Schemes 9.2 and 9.6) of bis(amino)silylenes with azides are summarised in Equation (9.5) and Scheme 9.12.[50] The isolation of **29** was taken as evidence for a transient iminosilane (N′N′)Si=NR or (N″N″)Si=NR in the pathway to the silatetrazolines of Equation (9.5) and Scheme 9.12 (R = Ph, Ph_3C, C_6H_4Me-4, $SiPh_3$).

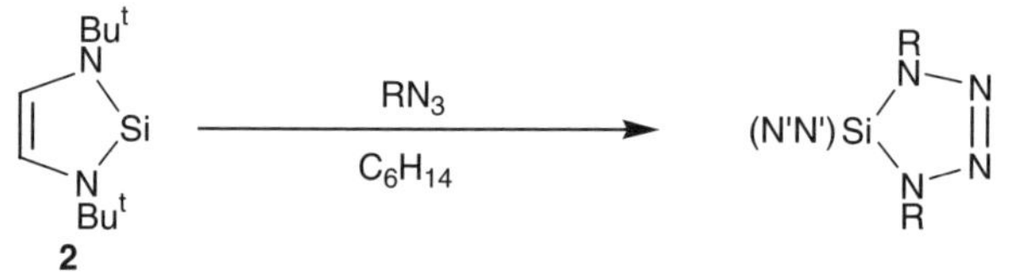

$$\mathbf{2} \xrightarrow[C_6H_{14}]{RN_3} (N'N')Si[N(R)N{=}NN(R)] \tag{9.5}$$

CH2But / N / Si / N / CH2But **3**; 1. NaOMe, thf 2. Et2O → [thf(Et2O)Na(OMe)Si(NN)]2 **26**; Na, thf → [Na(thf)4][c-{Si(NN)}3] **28**; excess Na, thf → (NN)Si(Na(thf)2)–Si(NN)(Na(thf)2) **27**; **28** excess Na, thf → **27**

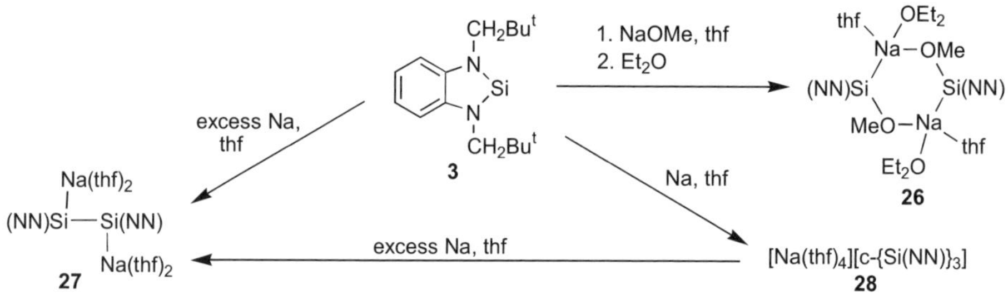

Scheme 9.11[58]

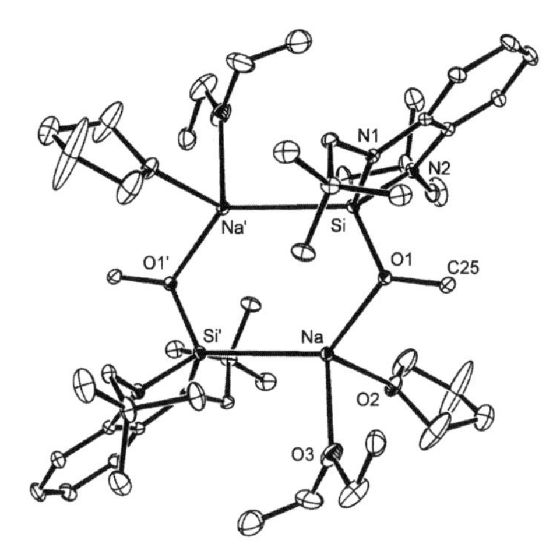

Figure 9.3 *Molecular structure of the centro-symmetric **26** [puckered $(SiNaO)_2$ ring]*[58]

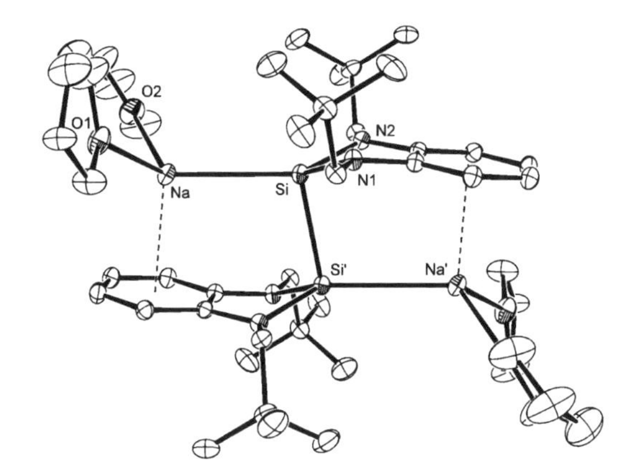

Figure 9.4 *Molecular structure of the centrosymmetric **27***[58]

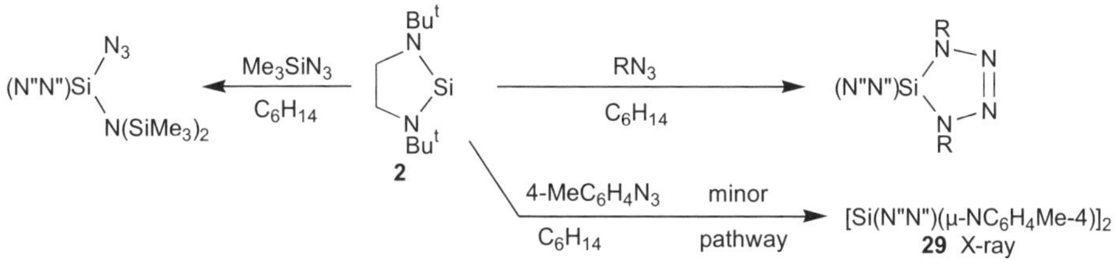

$(N''N'')Si(N_3)\{N(SiMe_3)_2\}$ $\xleftarrow[C_6H_{14}]{Me_3SiN_3}$ **2** $\xrightarrow[C_6H_{14}]{RN_3}$ $(N''N'')Si(N(R)N{=}NN(R))$

2 $\xrightarrow[C_6H_{14}]{4\text{-}MeC_6H_4N_3}$ (minor pathway) $[Si(N''N'')(\mu\text{-}NC_6H_4Me\text{-}4)]_2$ **29** X-ray

Scheme 9.12[50]

Treatment of the stannylene $Sn[\{N(CH_2Bu^t)\}_2C_{10}H_6\text{-}1,8]$ with Si(NN) (**3**) gave the crystalline bis(silyl)stannylene **30** which underwent fluxional processes in thf-d_8 or PhMe-d_8 and in the latter or C_6D_6 it finally fragmented yielding **31**, **3** and Sn metal.[59] The crystalline complexes **32** and **33** were obtained as shown in Scheme 9.13.[60]

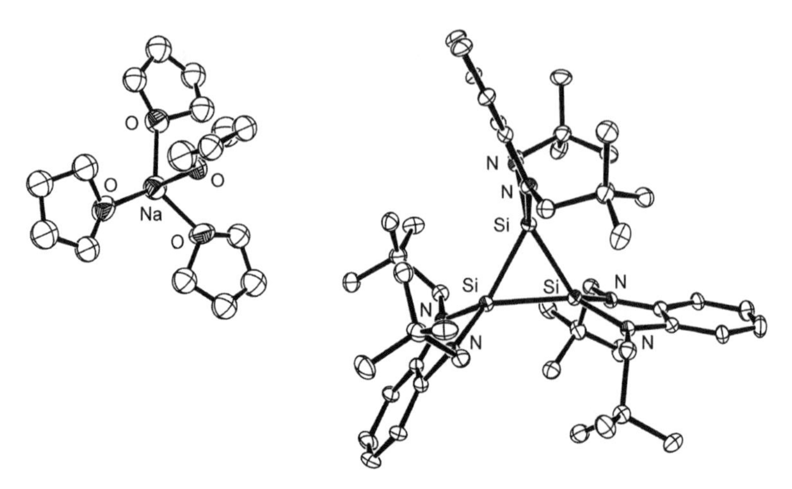

Figure 9.5 *Molecular structure of the salt* ***28****; the anion lies on a site of* $\bar{6}$ *symmetry;*[58] *(cf.*[53] *Figure 9.2)*

Si(NN) (**3**) underwent oxidative additions with various compounds A–B (Br–Br, Cl–$SiCl_3$, Br–$SiBr_3$, Cl–$SiCl_2Ph$, Cl–$SiCl_2Me$) affording the 1 : 1 Si(NN)(A)B and in some cases also ($SiCl_4$, $SiBr_4$, $SiCl_3Ph$) the 2 : 1 (NN)(A)Si–Si(NN)B adducts;[61] heating the disilane $[(NN)(Cl)Si]_2$ gave the trisilane $[(NN)(Cl)Si]_2SiCl_2$.[61] By contrast, **3** behaved as a reducing agent towards MCl_4 (M = Ge, Sn), furnishing $Si(NN)Cl_2$ and the appropriate metal(II) chloride.[61]

The oxidative addition reactions of a silylene with a halocarbon have already been discussed: **3** with MeI, Scheme 9.2; **2** with Bu^tCl, $PhCH_2Cl$, CH_2Cl_2, $CHCl_3$, CCl_4 or PhBr;[32] the formation of 1 : 1 [e.g. Si(N′N′)(Cl)R] or of 2 : 1 [e.g. (N′N′)(Cl)Si–Si(N′N′)R] adducts in such reactions with the generalised chlorocarbon ClR was originally postulated to proceed via an intermediate $(N'N')Si^{+\cdot\cdot\cdot}ClR^{-}$, and for the 2 : 1 adduct by a second

Scheme 9.13 *[R = $CH(SiMe_3)_2$]*[60]

$$\begin{array}{c} {}^{+}\mathrm{Si(N'N')} \\ / \\ \mathrm{(N'N')Si{-}{-}{-}{-}ClR^{-}} \\ \mathbf{34} \end{array}$$

intermediate **34**.[32] A computational study of reactions of **2** with XCH_3 or $ClCR_3$ concluded that such halophilic intermediates were energetically unlikely;[62a] the formation of the 2 : 1 adducts was postulated to follow the pathway $Si(N'N') + ClCR_3 \longrightarrow T.S.1 \longrightarrow (N'N')Si(Cl)\text{-}CR_3 \xrightarrow{Si(N'N')} T.S.2 \longrightarrow (N'N')(Cl)Si\text{–}Si(N'N')CR_3$, and a radical pathway was excluded.[62b] The notion that the 2 : 1 adduct was formed by insertion of Si(N′N′) (**2**) into the Si–Cl bond of the 1 : 1 adduct was inconsistent with the experimental observation[32] that **2** even with a large excess of $CHCl_3$ only yielded the 2 : 1 adduct.[63a] Moreover, while the earlier study employed restricted DFT calculations,[62] the radical mechanism was consistent with later computations.[63a,63b]

A detailed study by Gehrhus and coworkers of the reactions of Si(NN) (**3**) with RCl ($R = Pr^n$, Bu^t, Ph, c-C_3H_5, c-$C_3H_5CH_2$), RBr (R = Ph, c-$C_3H_5CH_2$), R_2CCl_2 (R = H, Me), R_2CBr_2 (R = H, Me) or $CHCl_3$ were reported.[64] In general the first formed product at ambient temperature was the disilane (NN)(X)Si–Si(NN)R′. Exceptions were (i) PhCl, which gave Si(NN)(Cl)Ph but required heating, (ii) CH_2Br_2 or CMe_2Br_2 (Scheme 9.14) and (iii) c-$C_3H_5CH_2X$ (Scheme 9.15). When heated, the disilane (NN)(X)Si–Si(NN)R′ gave the silane Si(NN)(X)R′ (R′X = Pr^nCl, Bu^tCl, Bu^tBr, PhBr, c-C_3H_5Br), $[(NN)(X)Si]_2CH_2$ (R′X = CH_2Cl_2), or successively **35** and $[(NN)(X)Si]_2CMe_2$.[64]

None of the 1 : 1 adducts with **3** gave the disilane, whereas the reverse reaction proved to be viable. The reactions of Scheme 9.15 provide good evidence for a radical mechanism.[64]

West and coworkers have extended their earlier[32] studies on reactions of the silylene **2** with a halocarbon RX.[65] Thus Si(N′N′) (**2**) or Si(N″N″) (**4**) with RX gave either 1 : 1 or 2 : 1 adducts or a mixture of the two. The 2 : 1 were favoured from **2** for RX = CCl_4, $CHCl_3$,

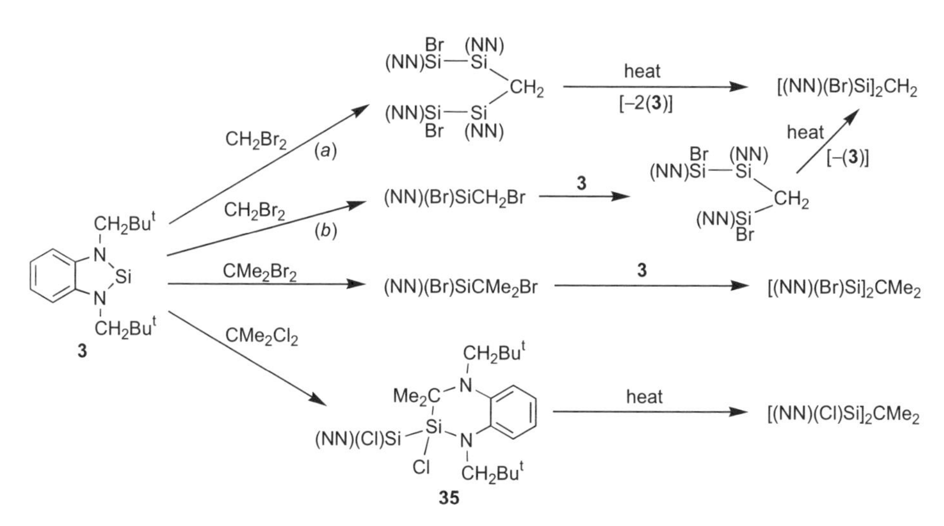

Scheme 9.14 *(a: CH_2Br_2 added to **3**; b: **3** added to excess CH_2Br_2)*[64]

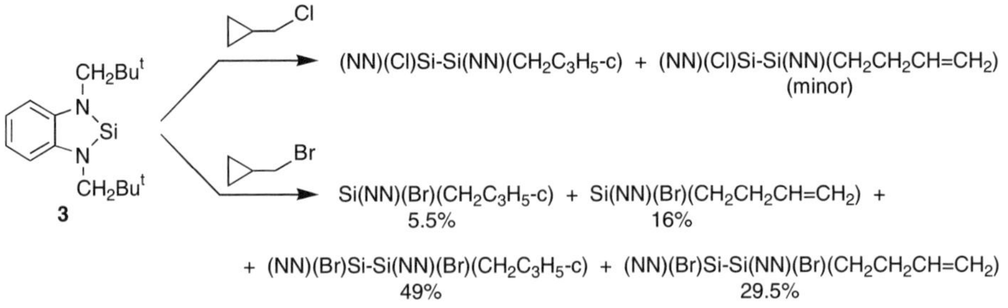

$(NN)(Cl)Si\text{-}Si(NN)(CH_2C_3H_5\text{-}c)$ + $(NN)(Cl)Si\text{-}Si(NN)(CH_2CH_2CH{=}CH_2)$ (minor)

$Si(NN)(Br)(CH_2C_3H_5\text{-}c)$ 5.5% + $Si(NN)(Br)(CH_2CH_2CH{=}CH_2)$ 16% +

+ $(NN)(Br)Si\text{-}Si(NN)(Br)(CH_2C_3H_5\text{-}c)$ 49% + $(NN)(Br)Si\text{-}Si(NN)(Br)(CH_2CH_2CH{=}CH_2)$ 29.5%

Scheme 9.15 [64]

CH_2Cl_2, $PhCH_2Cl$ or Bu^nCl; the 1 : 1 for RX = Bu^tCl or a bromoarene; while a mixture of the two was obtained for RX = Pr^iCl, Pr^iBr, n-$C_6H_{13}Br$ or PhBr. From **4**, the 2 : 1 adduct was the major product for RX = CCl_4, $CHCl_3$, CH_2Cl_2 or $PhCH_2Cl$; but a mixture of the two was found for RX = Pr^iCl, Bu^nCl, Bu^tCl, Pr^iBr, n-$C_6H_{13}Br$ or PhBr. Reaction of **2** or **4** with C_2Cl_6 gave $Si(N^xN^x)Cl_2$ and $(N^xN^x)(Cl)Si{-}Si(N^xN^x)Cl$. In the light of the calculations of Ref. 63, a radical mechanism was favoured, halogen atom abstraction from RX giving $\dot{S}i(N^XN^X)X + \dot{R}$ as initiators.[65] The formation of the dichlorides from C_2Cl_6 was attributed to the fragmentation of $CCl_3\dot{C}Cl_2$ giving $CCl_4 + \dot{C}l$.[65]

Heteroleptic amidosilylenes are extremely rare. Thus, reaction of the salt $[Si(\eta^5\text{-}C_5Me_5)]$-$[B(C_6F_5)_4]$ with $LiN(SiMe_3)_2$ led to a reactive intermediate $(C_5Me_5)SiN(SiMe_3)_2$, which yielded the crystalline disilene $[Si(\eta^1\text{-}C_5Me_5)\{N(SiMe_3)_2\}]_2$ with an Si=Si bond length of

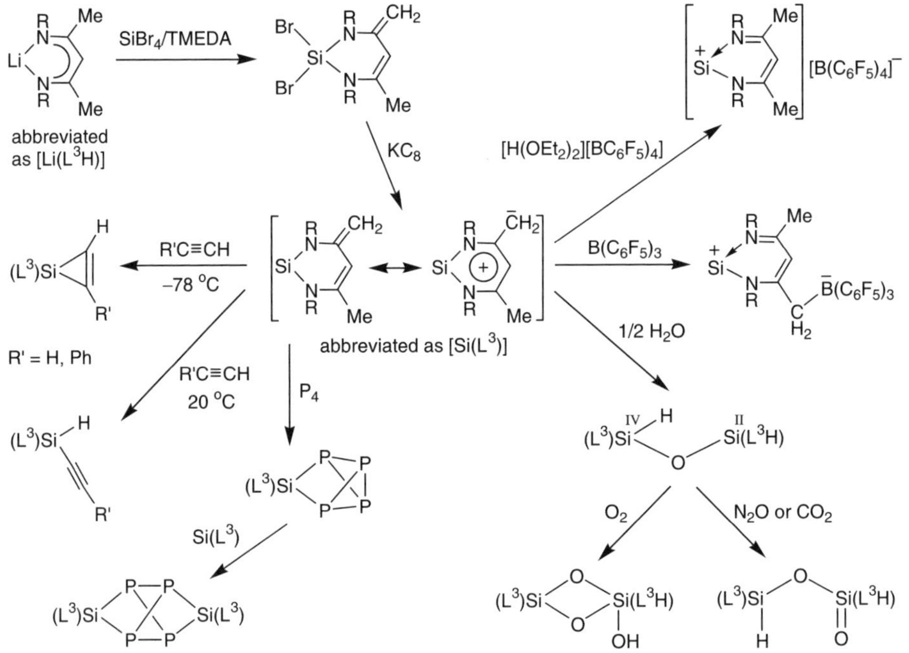

Scheme 9.16 *(R = $C_6H_3Pr^i{}_2$-2,6)*

2.1683(5) Å.[66] The diiron complex [{Fe(Cp)(CO)}$_2$(μ-CO){μ-SiN(SiMe$_3$)SiMe$_2$CH$_2$}] (Cp = η^5-C$_5$H$_5$), obtained by a photochemical reaction of [Fe(Cp)(CO)$_2$(SiMe$_3$)] with SiH$_3${N(SiMe$_3$)$_2$}, has a bridging alkylamidosilylene ligand.[67] Several heteroleptic three-coordinate silylenes [Si(L^1)X] (L^1 = [PhC(NBu^t)$_2$]$^-$; X = Cl, NMe$_2$, OBu^t, OPr^i, PPr$^i_2$) have recently been prepared by the reduction of the appropriate Si^{IV} dichloro compound with K; using the less bulky benzamidinato ligand, [PhC(NSiMe$_3$)$_2$]$^-$ (≡ L^2), led to the disproportionation products [Si(L^2)$_2$Cl$_2$] and Si.[68]

A new type of six-membered *N*-heterocyclic stable silylene has been reported recently.[69] The yellow crystalline [Si(L^3)] (L^3 = N(R)C(Me)CHC(=CH$_2$)NR, R = C$_6$H$_3$Pr$^i_2$-2,6) was obtained by the reduction of [Si(L^3)Br$_2$] with potassium-graphite (for the synthesis of the [Ge(L^3)] analogue, see Equation (9.7) of Section 9.3.2). In contrast to the Ge, Sn and Pb β-diketiminatometal(II) halides (Sections 9.3.2 and 9.3.4), the Si^{II} compound Si{(N(R)C(Me))$_2$CH}Br was not available; synthesis and selected reactions of [Si(L^3)] are shown in Scheme 9.16.[69,70]

9.3 Amidometal(II) Chemistry [Ge(II), Sn(II), Pb(II)]

9.3.1 Introduction

It is convenient to consider amides of Ge(II), Sn(II) and Pb(II) together. Published data related to the tin compounds exceed those for germanium, while lead(II) amides have been the least studied. A comprehensive review of this field is not available, but surveys on specific aspects exist; in chronological order they have dealt with (i) 'Heavy atom main group 4 analogues of carbenes, radicals and alkenes,' [including the synthesis, structures and physical properties of M{N(SiMe$_3$)$_2$}$_2$, some of their oxidative addition reactions and their role in transition metal chemistry];[71] (ii) 'Unsaturated molecules containing main group metals,' [chemistry based on SnN(Bu^t)Si(Me)$_2$NBu^t and its Ge and Pb analogues];[72] (iii) 'The role of group 14 element carbene analogues in transition metal chemistry,' [including M{N(SiMe$_3$)$_2$}$_2$];[73] (iv) 'Recent advances in the chemistry of bivalent compounds of Ge, Sn and Pb,' [including aspects of amidogermanium(II) and -tin(II) chemistry with ligands $\bar{N}$(SiMe$_3$)$_2$, TMP$^-$, 1,2-C$_6$H$_4$($\bar{N}$R)$_2$ (R = SiMe$_3$, CH$_2$Bu^t)];[74] (v) 'Metal centred molecules and their application in chemistry and physics,' [including materials based on monocyclic and polycyclic silylamides of Ge(II), Sn(II) and Pb(II)];[75] and (vi) '*N*-Heterocyclic germylenes and related compounds,' [including chemistry based on GeN(Bu^t)Si(Me)$_2$NBu^t and Ge[{N(SiMe$_3$)}$_2$C$_6$H$_4$-1,2].[76]

9.3.2 Homoleptic Metal(II) Amides: Synthesis, Structures and Physical Properties

In our 1980 book, the homoleptic Group 14 metal(II) amides M[N(R^1)R^2]$_2$ there reported were those with M = Ge and R^1 = R^2 = SiMe$_3$, SiEt$_3$, GeMe$_3$, GeEt$_3$, GePh$_3$; Ge[N(Bu^t)SiMe$_3$]$_2$ and Ge[NC(Me)$_2$(CH$_2$)$_3$CMe$_2$]$_2$ [≡ Ge(TMP)$_2$]; M = Sn and R^1 = R^2 = Me, Pr^i, Bu^t, Ph, SiMe$_3$, SiEt$_3$, GeMe$_3$, GeEt$_3$, GePh$_3$; Sn[N(Bu^t)SiMe$_3$]$_2$,

$Sn(TMP)_2$, $\overline{SnN(Bu^t)Si(Me)_2N}Bu^t$ (**1′**, the Sn analogue of the silylene **1**) and $\overline{SnN(SiMe_3)(CH_2)_nN}SiMe_3$ ($n = 2$, 3, 4); and M = Pb and $R^1 = R^2 = SiMe_3$; Pb[N-$(Bu^t)SiMe_3]_2$ and $Pb(TMP)_2$.[1] Since that time the field has grown: new compounds are listed in Tables 9.1–9.3.[77–101] The majority of these were prepared by a LiCl- or more rarely $MgCl_2$-elimination procedure from $GeCl_2$·dioxane-1,4, $SnCl_2$ or $PbCl_2$ and the appropriate lithium amide (shown as 'A' in Tables 9.1–9.3) or magnesium diamide ('A′' in Table 9.1). The amine-elimination method ('B' and 'B′' in Tables 9.1 and 9.2) is outlined in Scheme 9.17. Syntheses starting from $GeCl_4$ ['C' in Table 9.1, Equation (9.6)] or $PbCl_4$ ('B' in Table 9.3) have also been explored, the alkali metal amide behaving both as a ligand transfer and a reducing agent. A dehydrochlorinative procedure is shown in Equation (9.7),[95] and an insertion reaction in Equation (9.8).[101]

1. 4Na, Et_2O; 2. $GeCl_4$; R = C_6H_4Ph-2 → 4NaCl + Ge(NR)₂ (9.6)

R = $C_6H_3Pr^i_2$-2,6 ; + $Li[N(SiMe_3)_2]$ → LiCl + $HN(SiMe_3)_2$ + (9.7)

R = CH_2Bu^t ; + 2 Si (9.8)

The molecular structures of homoleptic Group 14 metal(II) amides have received much attention. Structures also became available for a number of the compounds known before 1979: $M[N(SiMe_3)_2]_2$ (M = Ge, Sn, Pb; Table 9.4),[110,111] $Sn(TMP)_2$,[111] [Sn-$(NMe_2)(\mu\text{-}NMe_2)]_2$ (crystalline[106a] and gaseous in equilibrium with the monomer at 385 K[106b]), and **37**.[99] The data of Table 9.4 illustrate several features for the series M[N-$(SiMe_3)_2]_2$ which, however, may be generalised for the wider class of compounds M[N-$(R)R']_2$.[111] (i) The N–M–N angles decrease in the sequence Ge > Sn > Pb, consistent with Bent's rule. (ii) The M–N bond lengths decrease in the reverse sequence.

Table 9.1 *Homoleptic germanium(II) amides*

Compound	Method[b]	Reference
$Ge[N(H)C_6H_2Bu^t_3\text{-}2,4,6]_2$[a]	A	77b
$Ge[N(SiMe_2OBu^t)_2]_2$[a]	A	78
$Ge[N(SiMe_2Pr^i)_2]_2$[a]	A (from $GeBr_2$)	79
$Ge[N(R)SiMe_3]_2$ $R = C_6H_4Me\text{-}2$, $C_6H_4F\text{-}2$, $C_6H_3Me_2\text{-}2,6$, $C_6H_3Pr^i_2\text{-}2,6$, $C_6H_2Me_3\text{-}2,4,6$	A	80
Ge(NBu^t)₂SiMe₂ ring: $Ge[N(Bu^t)]_2SiMe_2$	A	81
Ge–N(R)CH₂CH₂N(R) ring; R = Me, Pr[i], Bu[t],[a] Ph, $C_6H_3Me_2\text{-}2,6$	A (Bu^t) A′	81 (Me, Ph) 82 (Bu^t) 83
Ge–N(R)CH=CHN(R) ring; R = Bu[t],[a] CH_2Bu^t,[a] $C_6H_2Me_3\text{-}2,4,6$	A	82 (Bu^t) 84 (CH_2Bu^t) 85
Ge–N($C_6H_2Me_3$-2,4,6)C(Me)=C(Me)N($C_6H_2Me_3$-2,4,6) ring[a]	B	86
Ge–N(Bu[t])–Si(Me)(NBu[t])₂Si(Me)–N(Bu[t]) ring	A	87
Ge–N(R)C₆H₄N(R) ring; R = $SiMe_3$, CH_2Bu^t[a]	A	84 (CH_2Bu^t) 88 ($SiMe_3$)
Ge–N(CH_2Bu^t)–pyridine–N(CH_2Bu^t) ring[a]	A	84
Ge–N(Pr[i])–naphthalene–N(Pr[i]) ring[a]	A, B′	89

(continued)

Table 9.1 *(Continued)*

Compound	Method[b]	Reference
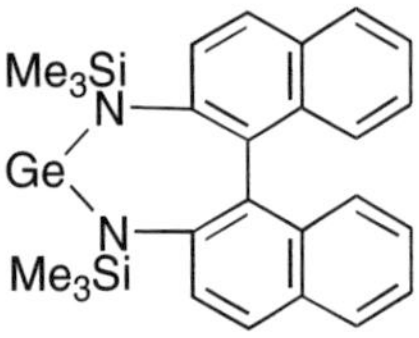	A	90
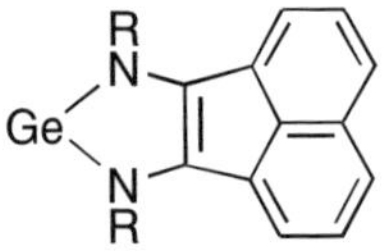 $R = C_6H_4Ph\text{-}2,^a\ C_6H_3Pr^i_2\text{-}2,6,^a\ C_6H_3Bu^t_2\text{-}2,5$	A′ C (C_6H_4Ph-2)	91
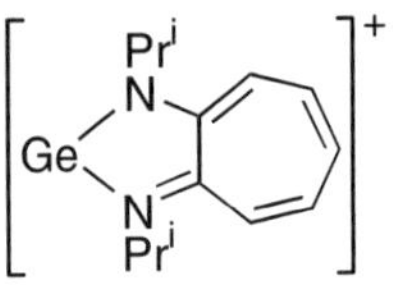 [ ]+ [a]	A	92
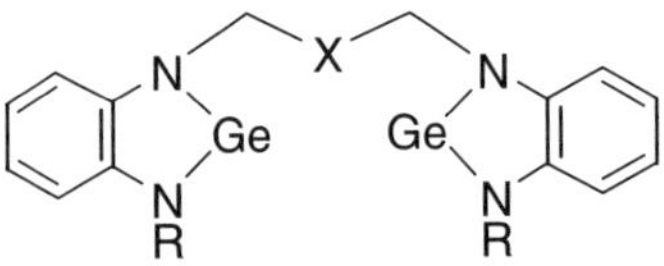 $R = CH_2Bu^t$ and $X = CMe_2, (CH_2)_2,^a (CH_2)_3,^a C_6H_4\text{-}1,2,^a C_6H_4\text{-}1,3,^a C_5H_3N\text{-}2,6$ (**36**)[a] $R = CH_2CH_3$ and $X = (CH_2)_2,^a (CH_2)_3$	A, B′	93a, 93b
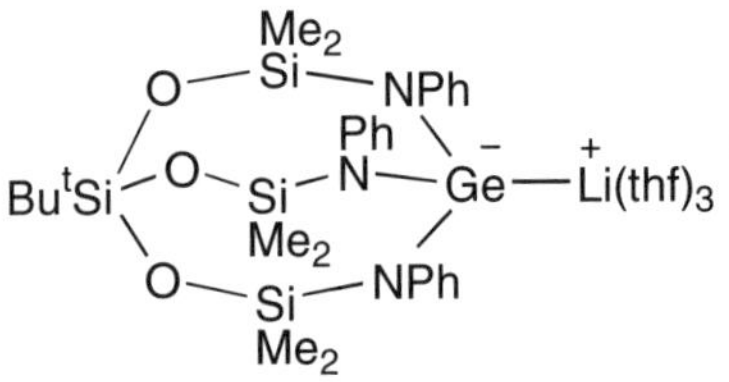 [a]	A	94
 [a]	D	95

[a]Molecular structure by X-ray diffraction

[b]Methods:

A: salt elimination from $GeCl_2$·dioxane-1,4 + 2Li[N(R)R′] or + Li[N(R)-X-N(R)]Li

A′: salt elimination from $GeCl_2$·dioxane-1,4 + $Mg[N(R)R']_2$

B and B′: amine elimination (Scheme 9.17)

C: salt elimination and reduction [Equation (9.6)]

D: dehydrochlorination [Equation (9.7)]

Table 9.2 *Homoleptic tin(II) amides*

Compound	Method[b]	Reference
$Sn[N(H)C_6H_2Bu^t_3\text{-}2,4,6]_2$[a]	A	77a, 77b
$Sn[N(SiMe_2Ph)_2]_2$[a]	A	96
$Sn[N(R)SiMe_3]_2$		
$R = C_6H_3Pr^i_2\text{-}2,6,$[a] $C_6H_2Me_3\text{-}2,4,6$[a] 1-adamantyl[a]	A	96 ($C_6H_3Pr^i_2$-2,6) 97a, 97b
$Sn[N(SiR^1_2R^2)SiMe_3]_2$		
$R^1 = Me$, $R^2 = Bu^t$;[a] $R^1 = Ph$, $R^2 = Bu^t$;[a]	A	97a
$R^1 = SiMe_3 = R^2$[a]		97c
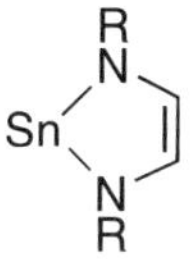 $R = Bu^t, Pr^i, C_6H_2Me_3\text{-}2,4,6$[a]	B B′	86 (Pr^i) 98
$Sn[N(SiMe_3)(CH_2)_nNSiMe_3]$ $n = 3$ or 4[a] (dimer, **37**)	A	99
$\{SnN(Bu^t)[SiR_2]NBu^t\}_2$ $[SiR_2] = SiCH_2(CH_2)_nCH_2$ ($n = 1$,[a] 2 or 3), $SiCH_2CH{=}CHCH_2$[a]	A	100
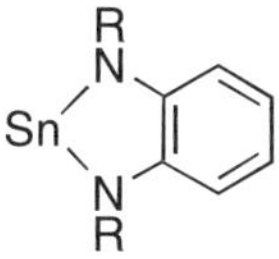 $R = SiMe_3, SiMe_2Bu^t, CH_2Bu^t$[a] Me (dimer),[a] $(CH_2)_2NMe_2$,[a] $(CH_2)_2OMe$[a]	A, B′	101 102a, 102b
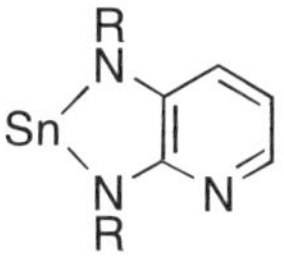 $R = CH_2Bu^t$	A	17
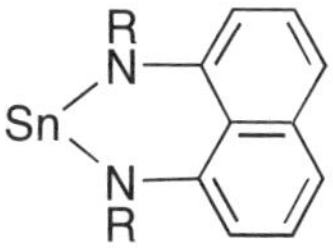 $R = Pr^i$,[a] CH_2Bu^t, $SiMe_3$[a]	B′	103 (CH_2Bu^t) 59 ($SiMe_3$) 104 (Pr^i)

(continued)

Table 9.2 *(Continued)*

Compound	Method[b]	Reference
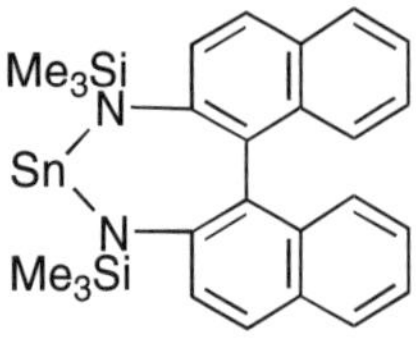	A	90
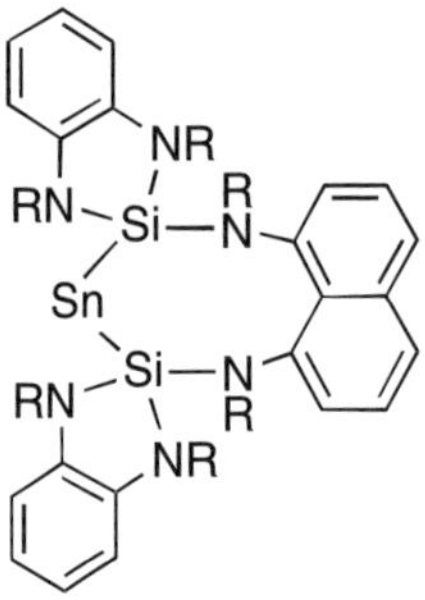 R = CH$_2$Bu^t [a]	C	59
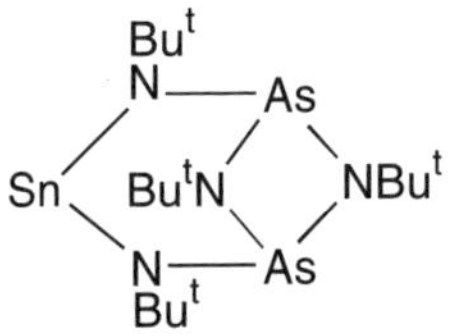	A	105
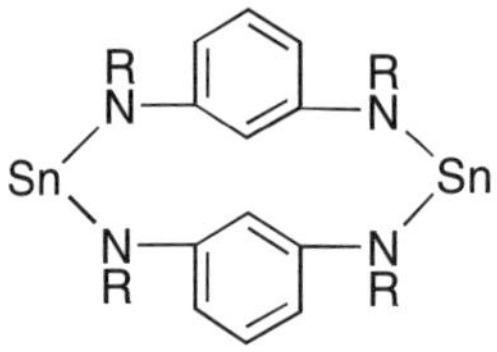 R = SiMe$_3$ [a]	A (+ligand redistribution)	136
[Sn(NMe$_2$)(μ-NMe$_2$)]$_2$ [a]	A	1, 106
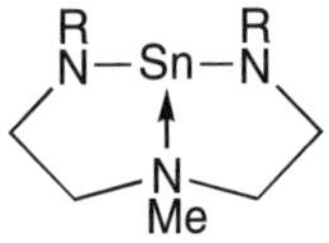 R = Pr^i, SiMe$_3$ [a]	A	107
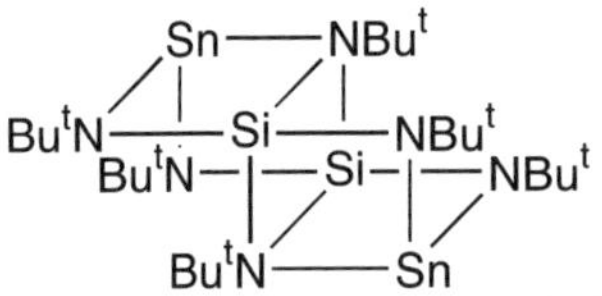 [a]	A	108

[a]Molecular structure by X-ray diffraction
[b]Methods:
A: salt elimination from $SnCl_2 + 2Li[N(R)R']$ or $+ Li[N(R)\text{-}X\text{-}N(R)]Li$
B and B′: amine elimination (Scheme 9.17)
C: insertion [Equation (9.8)]

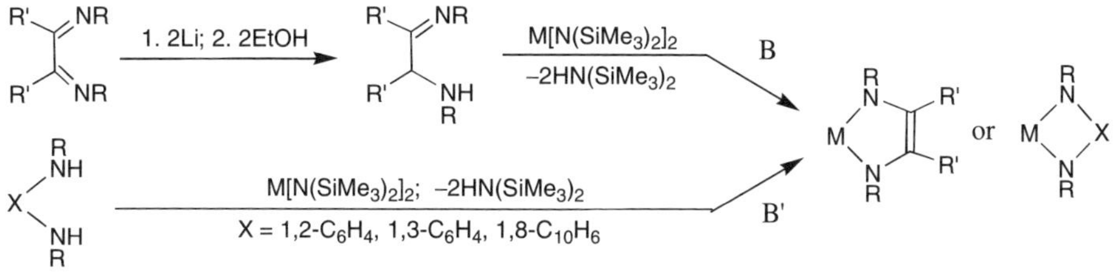

Scheme 9.17

(iii) Although the M–N bond lengths for a given amide are similar in both the solid and the vapour, the N–M–N angles are markedly smaller in the vapour; this is due to the difference in the relative conformations of the two ligands in the two phases; in the crystal not only intra- but also intermolecular contacts need to be minimised.[111]

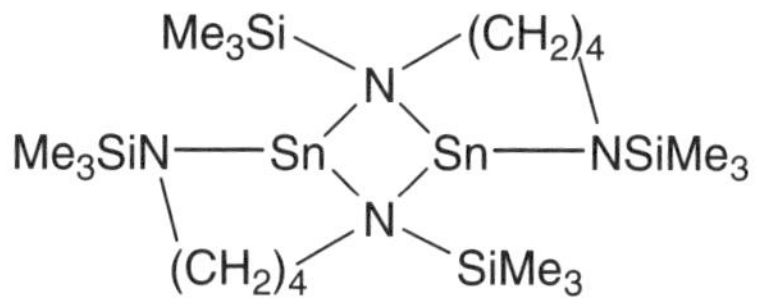

37

Table 9.3 *Homoleptic lead(II) amides*

Compound	Method[b]	Reference
$[Pb(NMe_2)(\mu\text{-}NMe_2)]_2$	A	106
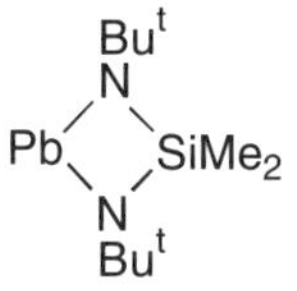 $Pb[N(Bu^t)]_2SiMe_2$ (cyclic)	A	81
$\{PbN(Bu^t)[SiR_2]NBu^t\}_2$; $[SiR_2] = SiCH_2(CH_2)_nCH_2$ ($n = 1$,[a] 2 or 3), $SiCH_2CH{=}CHCH_2$	A	100
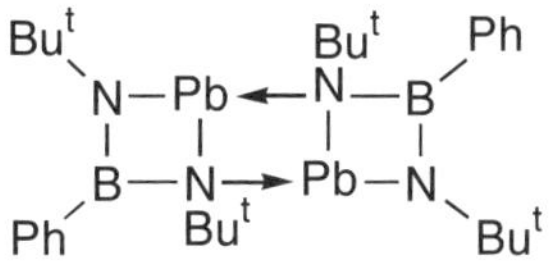 $[Pb\{N(Bu^t)B(Ph)N(Bu^t)\}]_2$ [a]	A B	109

[a]Molecular structure by X-ray diffraction
[b]Methods:
A: salt elimination from $PbCl_2 + 2Li[N(R)R']$ or $+ Li[N(R)\text{-}X\text{-}N(R)]Li$
B: salt elimination and reduction from $PbCl_4 + 2Li[N(Bu^t)B(Ph)N(Bu^t)]Li$

Table 9.4 *Some structural parameters for crystalline and gaseous Group 14 metal(II) amides*

	M–N Bond length (Å)		N–M–N Bond angle (°)	
M in $M[N(SiMe_3)_2]_2$	Crystal[a]	Gas[b]	Crystal[a]	Gas[b]
Ge	1.876(5)[c]	1.89(1)	107.1(2)[c]	101(1.5)
Sn	2.09(1)	2.09[d]	104.7(2)	96[d]
Pb	2.24(2)	2.20(2)	103.6(7)	91(2)

[a]At 140 K.[110] [b]At 380 K.[110] [c]Data from Ref. 111 [at 173(2) K]. [d]Data from Ref. 7.

The majority of the crystalline amides listed in Tables 9.1–9.3 are V-shaped Group 14 metal-centred monomers. Exceptions are $Bu^tSi[OSi(Me)_2N(Ph)]_3\overline{Ge}–Li^+(thf)_3$ which has a long Ge–Li contact of 2.904(12) Å,[94] the three-coordinate tin(II) amides $[Sn(NMe_2)(\mu\text{-}NMe_2)]_2$,[106] $Sn[\{N(R)(CH_2)_2\}_2NMe]$,[107] **37**,[99] $Sn_2Si_2(NBu^t)_6$,[108] $[Sn\{N(Me)\}_2C_6H_4\text{-}1,2]_2$,[102b] $[\{\overline{SnN(Bu^t)[SiR_2]N}Bu^t\}_2]$ ($[SiR_2] = \overline{SiCH_2(CH_2)C}H_2$ or $\overline{SiCH_2CH{=}CHC}H_2$)[100] and the three-coordinate lead(II) amides[81,100,106,109] listed in Table 9.3. The trigonal metal(II) amides have a stereochemically active lone pair of electrons at M; e.g. the sum of the three N-Sn-N angles of $[Sn(NMe_2)(\mu\text{-}NMe_2)]_2$ is 280.4°.[106] The bis-(germylene) **36** (Table 9.1) with the pyridinediyl-2,6 linker has a short intramolecular Ge···Ge contact of 3.041(5) Å,[93b] which was not observed in the phenylene-bridged compounds.[93a,93b]

Computational studies on isoleptic silylenes and germylenes $\overline{EN(R)CH_2CH_2N}R$ (e.g. **4**, E = Si) and $\overline{EN(R)CH{=}CHN}R$ (e.g. **2**, E = Si) (R = H, Bu^t) have been reported;[54,112a,112b] it was concluded that there is significant p_π-p_π-delocalisation for the latter compounds. The relationship between stability, acid-base and spin properties, nucleophilicity and electrophilicity in a series of silylenes was studied by conceptual density functional theory.[112c] Electronic structures of $\overline{EN(R)CH{=}CHN}R$ (E = Si, Ge, Sn, Pb) and their group 13, 15 and 16 analogues were analysed using various quantum chemical methods.[112d]

As discussed in Section 9.2.4, the stable silylenes **2** and **4** have also been studied by cyclic voltammetry (CV); similar procedures were carried out on the isoleptic germanium(II) amides.[54] The latter were less readily oxidised or reduced than **2** and **4**; the oxidation proceeded more readily for the saturated compounds. Of the four compounds, using 1,2-$C_6H_4Cl_2$ and $[NBu_4][ClO_4]$ as the electrolyte, only $\overline{GeN(Bu^t)CH{=}CHN}Bu^t$ showed a single reversible oxidation couple, whereas the saturated analogue exhibited a reversible couple only at its second oxidation step.[54]

Core excitation spectroscopy of such cyclic diaminocarbenes, -silylenes and -germylenes $\overline{EN(Bu^t)(CH_n)_2N}Bu^t$ ($n = 1, 2$), using inner shell electron energy loss spectroscopy and *ab initio* calculations, revealed significant π-delocalisation and in the unsaturated compounds ($n = 1$) aromatic stabilisation;[113] this view was reinforced by the Raman spectra of the latter (E = Si, Ge), the π-delocalisation being the greater for E = Ge.[114] The variable temperature ($77 \leq T \leq 155$) ^{119}Sn Mössbauer spectra of $Sn[N(SiMe_3)_2]_2$, $Sn[\overline{NCH_2C}H_2]_2$ and $[Sn(NMe_2)(\mu\text{-}NMe_2)]_2$ have been recorded;[115a] those of $Sn[(NPr^i)_2C_{10}H_6\text{-}1,8]$ ($98 \leq T \leq 255$) were used to study the Sn-ligand bond flexibility in this compound, which has an additional Sn-naphthyl π-interaction in the solid state.[115b]

The kinetics and mechanism of the thermal decomposition of $\overline{GeN(Bu^t)CH{=}CHNBu^t}$ between 140 and 440 °C leading to pure germanium, has been studied.[116a] Its chemical vapour deposition on a silicon wafer produced pure germanium at p-type silicon sites.[116b]

9.3.3 Protonolyses of Homoleptic Metal(II) Amides

Pre-1980 reports on protonolyses of homoleptic metal(II) amides were on $Sn(NR_2)_2$ with 2HA yielding SnA_2 [$R = SiMe_3$ and $A = OEt$, $OC_6H_2Bu^t_2$-2,6-Me-4, OAc, Cl, Cp, PBu^t_2, PPh_2, $AsPh_2$, SPh; R = Me and A = OEt or $A_2 = (OCH_2)_2NMe$] and on $Pb[N(SiMe_3)_2]_2$ which with 2ArOH gave $Pb(OAr)_2$ ($Ar = C_6H_2Bu^t_2$-2,6-Me-4).[1]

Treatment of $M[N(SiMe_3)_2]_2$ with excess of Bu^tOH afforded the homoleptic tert-butoxides of tin(II)[117] and lead(II),[119] the former characterised in the vapour phase as *trans*-$[Sn(OBu^t)(\mu\text{-}OBu^t)]_2$.[117] From Bu^t_3COH these amides gave $Sn(OCBu^t_3)_2$[117] and $Pb(OCBu^t_3)_2(HOCBu^t_3)$;[119] the X-ray-characterised isoleptic Ge(II) alkoxide was likewise obtained from $Ge[N(SiMe_3)_2]_2$.[117] The cluster compound $Pb_6O_4(OPr^i)_4$ was isolated from $Pb[N(SiMe_3)_2]_2$ and Pr^iOH.[119] The crystalline $Sn[N(SiMe_3)_2](OAr)$ was prepared from $Sn[N(SiMe_3)_2]_2$ and ArOH, but was also available from $Sn[N(SiMe_3)_2]_2$ and $Sn(OAr)_2$ or $[Sn(\mu\text{-}Cl)\{N(SiMe_3)_2\}]_2$ and LiOAr ($Ar = C_6H_2Bu^t_2$-2,6-Me-4).[120] The reaction of $Ge[N(SiMe_3)_2]_2$ and 2ArOH produced the crystalline aryloxides $[Ge(\mu\text{-}OAr)(OAr)]_2$ ($Ar = C_6H_2Me_3$-2,4,6, $C_6H_3Pr^i_2$-2,6) or $[Ge(OAr)_2]$ ($Ar = C_6H_3Ph_2$-2,6, C_6HPh_4-2,3,5,6), while a similar reaction with a thiol gave the germanium(IV) derivative $[Ge(H)(SC_6H_2Pr^i_3\text{-}2,4,6)_3]$.[121] Depending on stoichiometry, the compounds $M[N(SiMe_3)_2](OAr')$ (M = Ge, Sn) and $M(OAr')_2$ (M = Ge, Sn, Pb) were produced from the appropriate $M[N(SiMe_3)_2]_2$ and $Ar'OH[Ar' = C_6H_2(CH_2NMe_2)_3\text{-}2,4,6]$.[118a,118b] Reacting $\overline{MN(Bu^t)Si(Me)_2NBu^t}$ with Bu^tOH gave **38** ($X = OBu^t$; M = Ge, Sn).[122]

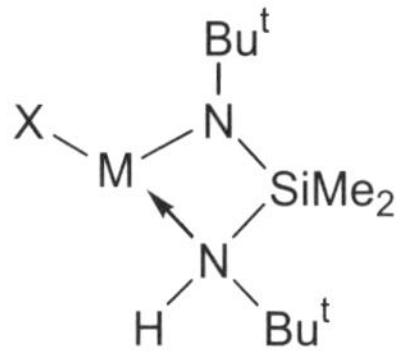

38

The crystalline metal(II) thiolates $M(SAr)_2$ (M = Ge, Sn, Pb) [also formed from MCl_2 + $2Li(SAr)(OEt_2)$ (M = Ge, Sn; $Ar = C_6H_2Bu^t_3$-2,4,6)], $M(SAr')(\mu\text{-}SAr')_2M(\mu\text{-}SAr')_2M(SAr')$ (M = Ge, Sn; $Ar' = C_6H_3Pr^i_2$-2,6) were prepared from $M[N(SiMe_3)_2]_2$/2ArSH or 2Ar′SH.[123] The compound $Sn(SAr)_2$ was alternatively isolated from equimolar portions of the reagents, showing that the heteroleptic compound $Sn[N(SiMe_3)_2](SAr)$ is unstable with respect to its decomposition products $Sn[N(SiMe_3)_2]_2$ and $Sn(SAr)_2$.[124] Nevertheless, the crystalline $[Pb\{N(SiMe_3)_2\}\{\mu\text{-}SC(SiMe_3)_3\}]_2$ was X-ray-characterised; in C_6D_6 at ambient temperature it also disproportionated.[124] From equimolar portions of $(Me_3Si)_3CSH$ and $Sn[N(SiMe_3)_2]_2$ in C_6H_{14} at ambient temperature $Sn[SC(SiMe_3)_3]_2$ was obtained; while $Ge[N(SiMe_3)_2]_2$ in PhMe under reflux gave the crystalline $[Ge(CH_2Ph)\{N(SiMe_3)_2\}(\mu\text{-}S)]_2$; in the absence of the thiol, there was no reaction.[124] Treatment of $Sn[N(SiMe_3)_2]_2$ with $PhP(C_6H_3R\text{-}3\text{-}SH\text{-}2)_2$ ($R = H$, $SiMe_3$) or $2HOS(O)_2CF_3$ gave the crystalline **39**[125] or **40**,[126] respectively; the latter was also available from $[Sn(\mu\text{-}Cl)\{N(SiMe_3)_2\}]_2$ and 2AgOTf.[126]

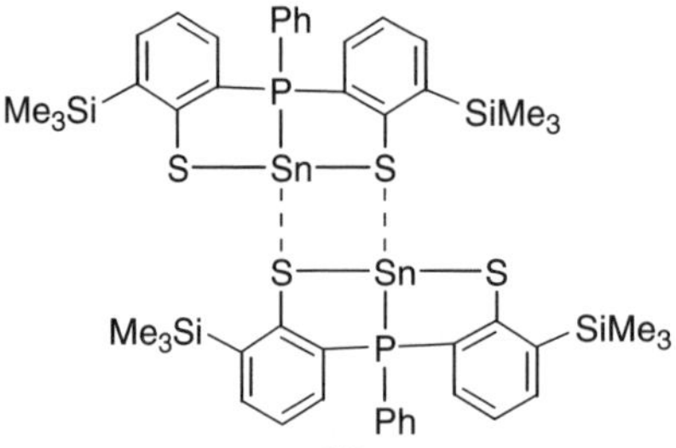

39

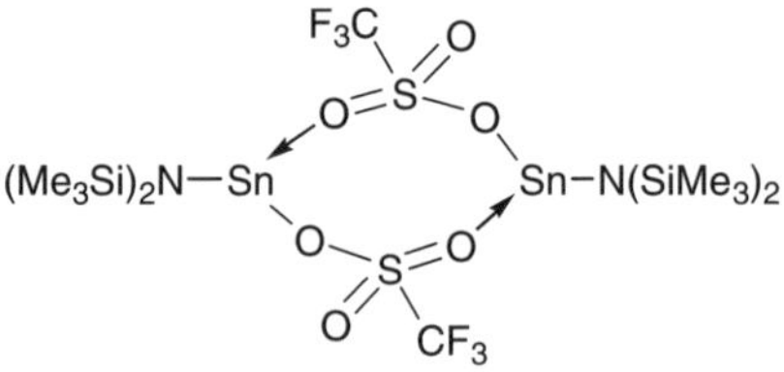

40

The reactions of $\overline{SnN(Bu^t)Si(Me)_2NBu^t}$ (**1′**), or its isoleptic Ge and Pb compounds, with Bu^tNH_2 have been studied by Veith and his coworkers (for an early review, see Ref. 72). The crystalline 1 : 1 adduct $[(\mathbf{1'})\cdot NH_2Bu^t]$[72,127] is clearly the first-formed product.[127] Next are the Me_2Si-containing compounds **38** [X = N(H)Bu^t] and **41** (M = Sn)[128] followed by **42** (M = Sn)[128] (X-ray[129]) and the cubane **43** (M = Sn).[128] The crystalline compounds **42** (M = Ge)[81] and **43** (M = Ge, Pb)[81] were likewise prepared and characterised, as were the mixed cubanes **43** ($M_4 = Sn_3Ge$, Sn_3Pb, Sn_2Pb_2, $SnPb_3$).[81] The cubanes $[Sn(NPr^i)]_4$ and $[Sn(NNMe_2)]_4$ were obtained from **1′** and Pr^iNH_2 and Me_2NNH_2 respectively,[130] and the silylimido-cubanes $[Sn(NSiR^1{}_2R^2)]_4$ from **1′** and $R^1{}_2R^2SiNH_2$ ($R^1{}_2R^2 = Me_3$, Et_3, Ph_3, Me_2Bu^t, $Bu^t{}_2H$).[131] Treatment of $\overline{PbN(Bu^t)Si(Me)_2NBu^t}$ with $Me_2Si(NHMe)_2$ yielded $[\overline{PbN(Me)Si(Me)_2N(Me)Si(Me)_2NMe}]_n$.[81] Hydrolysis of **42** (M = Sn), but not **43** (M = Sn), furnished **44**.[132a,132b] Reaction of **1′** with HX gave **38** (X = OBu^t, SBu^t;[122] Cl, Br, I;[133] η^5-C_5H_5, η^5-C_9H_7[134]); the lead analogue of **1** (unlike the isoleptic Ge compound) behaved similarly towards cyclopentadiene or C_9H_8.[134]

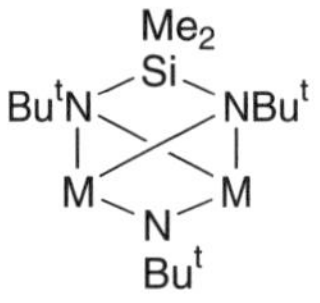

41 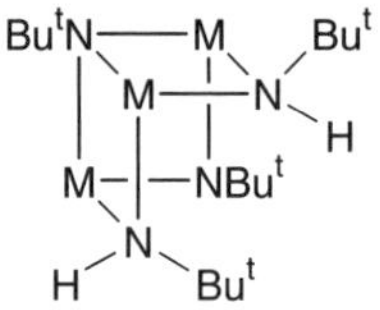 **42** 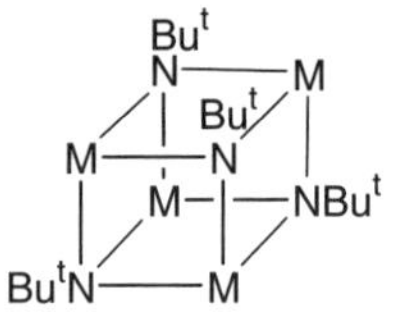 **43** 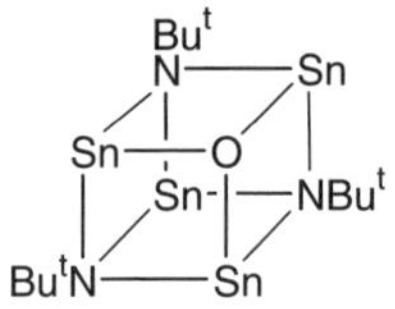 **44**

The crystalline germanium imides, the planar $[Ge(\mu\text{-}NC_6H_2Bu^t{}_3\text{-}2,4,6)]_2$ and $[Ge(\mu\text{-}NC_6H_3Pr^i{}_2\text{-}2,6)]_3$, were obtained by thermolysis of $Ge[N(H)C_6H_2Bu^t{}_3\text{-}2,4,6]_2$[77b] and from $Ge[N(SiMe_3)_2]_2/H_2NC_6H_3Pr^i{}_2\text{-}2,6$,[135a] respectively. Treatment of $Sn[N(SiMe_3)_2]_2$ with $C_6H_4[N(H)SiMe_3]_2\text{-}1,3$, $C_{10}H_6[N(H)R]_2\text{-}1,8$, or RN=CHCH$_2$N(H)R gave **45**,[136] **46** (R = $SiMe_3$, CH_2Bu^t, Pr^i; Table 9.2),[59,103,104] or successively **47** and $\overline{SnN(R)CH{=}CHNR}$ (R = Bu^t, $C_6H_2Me_3$-2,4,6);[98] the Ge or Pb analogues of **46** (R = Pr^i; Table 9.1),[89] or **47** and $\overline{PbN(R)CH{=}CHNR}$ (R = Bu^t),[98] were likewise obtained from the appropriate bis-(trimethylsilyl)amidometal substrate.

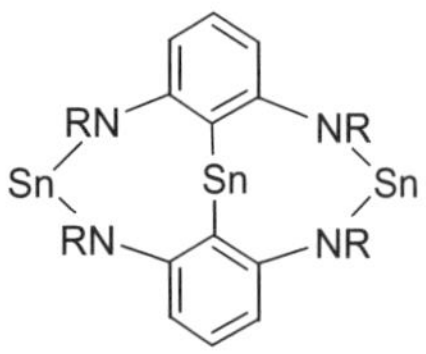

45 (R = $SiMe_3$) 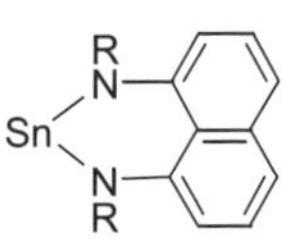 **46** (R = $SiMe_3$, CH_2Bu^t, Pr^i) 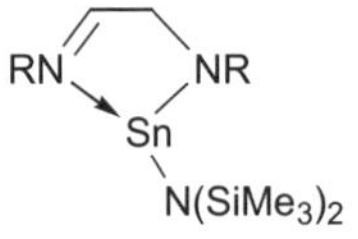 **47** (R = Bu^t, $C_6H_2Me_3$-2,4,6)

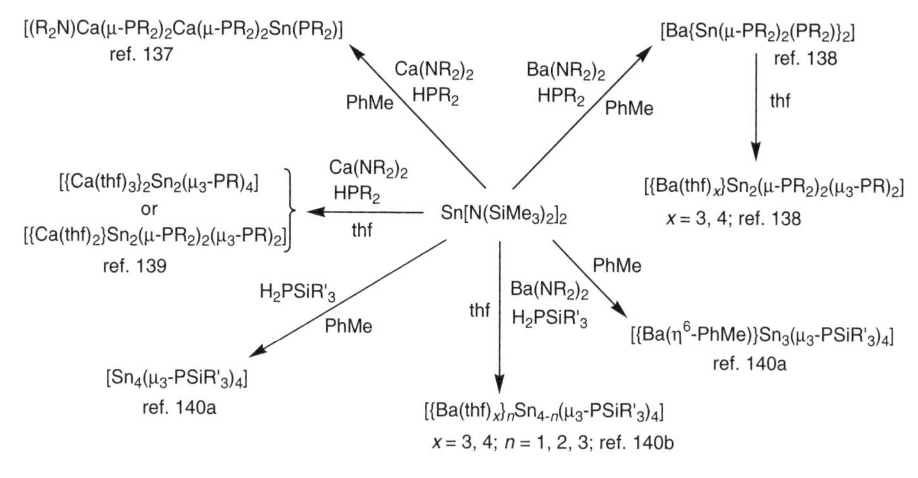

Scheme 9.18 *(R = SiMe$_3$, R′ = Bu^t)*

Treatment of $M[N(SiMe_3)_2]_2$ with $HP(SiMe_3)_2$ yielded $[M\{P(SiMe_3)_2\}\{\mu\text{-}P(SiMe_3)_2\}]_2$ (M = Sn, Pb).[135b]

Westerhausen, *et al.*, have converted $Sn[N(SiMe_3)_2]_2$ into a number of crystalline tin(II)-containing cluster compounds, as summarised in Scheme 9.18.[137–140a,140b]

Wright and coworkers have used $Sn(NMe_2)_2$ as a source of Sn-N- and Sn-P-containing compounds. Thus $Sn(NMe_2)_2$ with various primary amines RNH_2 was converted into (i) successively $[Sn(\mu\text{-}NMe_2)\{N(H)R\}]_2$, **48** and then $[Sn(\mu\text{-}NR)]_4$ (cf. **43**) (R = $C_6H_3Pr^i_2$-2,6 or $C_6H_2Me_3$-2,4,6);[141] (ii) **49** or $[Sn(\mu\text{-}NR)]_4$ [R = C_6H_4Me-4, $C_6H_3(OMe)$-2-Me-4, $C_6H_3(OMe)$-4-Me-2, $C_6H_3(OMe)_2$-2,4 or 3,4, or $N(CH_2)_2OCH_2CH_2$;[142] (iii) $[Sn(\mu\text{-}NR)]_4$ (R = c-C_6H_{11} or 2-$CH_2C_5H_4N$);[143] and (iv) $Sn_7(NR)_8$ (**50**) (R = 2-C_5H_3(Me-5)N[144a] or methyl- or methoxypyrimidinyl).[144b] The reaction of $4Sn(NMe_2)_2$ and successively

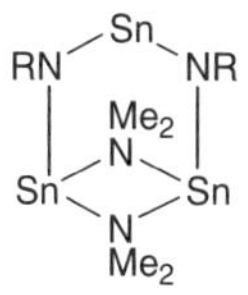

48 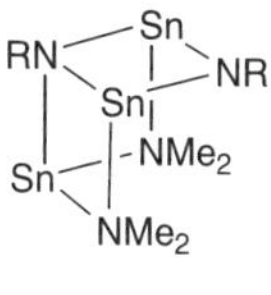 **49** [R = $C_6H_3(OMe)$-2-Me-6] 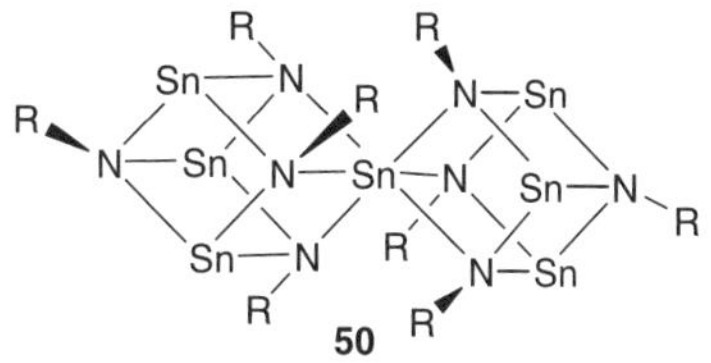 **50**

$4Bu^tNH_2$ and $3Li[N(H)C_{10}H_7\text{-}1]$ or $6Li[P(H)C_6H_{11}\text{-}c]$ gave the crystalline salt $[Li(thf)_4]$-[**51**] (R = 1-$C_{10}H_7$) or $[\{Sn_2(PC_6H_{11}\text{-}c)_3\}_2\{Li(thf)\}_4]$ ($[Li(thf)]_4$[**52**], containing a $Sn_4P_6Li_4$ core).[145] The crystalline salt $[Li(thf)]_4$[**52**] or the mixed metal compound **53** were obtained from the tin(II) amide and Li[P(H)R] (R = c-C_6H_{11} or Bu^t) or Li[P(H)R′] (R′ = $C_6H_2Me_3$-2,4,6).[146] The 1 : 2 or 1 : 3 stoichiometric reactions of $Sn(NMe_2)_2$ with M-$[P(H)Bu^t]$ (M = Li, Na, K) in thf gave the crystalline $[Li(thf)]_4$[**52**] (R = Bu^t), $[Na(thf)_6]$[**54**], and $[\{Sn_4(PBu^t)_5\}K_2(thf)_5]$ [containing a $\{P\cdots K(thf)_2\cdots thf\cdots K(thf)_2\cdots P\}$ unit].[147] A review paper described some of these compounds and also referred to compounds containing the dianion $[Sn_2(\mu\text{-}PR)_2(\mu\text{-}P_2R_2)]^{2-}$ (R = c-C_6H_{11}, $C_6H_2Me_3$-2,4,6) (*cf.*, **53**); the formation of the P(R)–P(R) bond was favoured when the counter cation implicated a heavier alkali metal.[148]

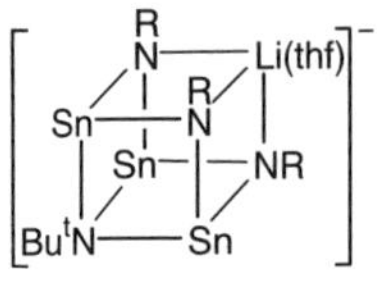

From $Sn[N(SiMe_3)_2]_2$ and H_2PR in the presence or absence of $SnCl_2$, the crystalline cluster compounds **55** [$R = Si(C_6H_2Pr^i_3\text{-}2,4,6)Pr^i_2$] and **56** ($R = SiPr^i_3$) were obtained.[149] Treatment of $Pb[N(SiMe_3)_2]_2$ with $Ph_2PCH_2C_5H_4N\text{-}2$ gave **57**.[150]

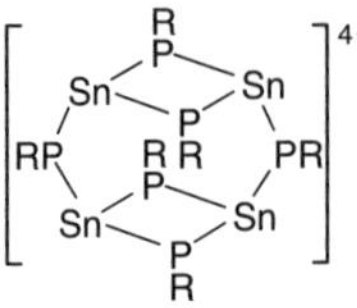

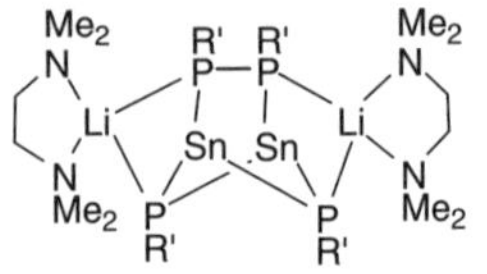

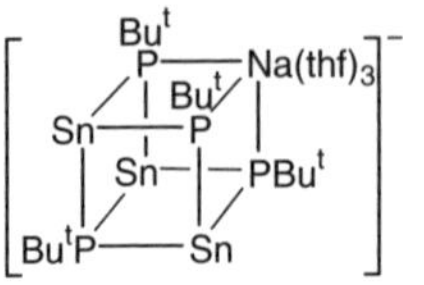

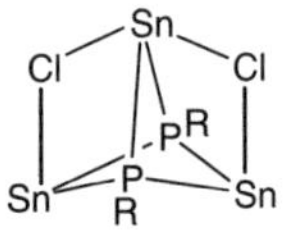

9.3.4 Heteroleptic Metal(II) Amides

In Section 9.3.3 the following heteroleptic group 14 metal(II) amides have already been described: **38**, **40–49**, **51**, **57**, $[Pb\{N(SiMe_3)_2\}\{\mu\text{-}SC(SiMe_3)_3\}]_2$ and $Sn[N(SiMe_3)_2](OC_6H_2Bu^t_2\text{-}2,6\text{-}Me\text{-}4)$, which has also been prepared from $Sn[N(SiMe_3)_2]_2 + Sn(OC_6H_2Bu^t_2\text{-}2,6\text{-}Me\text{-}4)_2$ or $[Sn(\mu\text{-}Cl)\{N(SiMe_3)_2\}]_2 + 2Li(OC_6H_2Bu^t_2\text{-}2,6\text{-}Me\text{-}4)$.[120] The compound $Sn[N(SiMe_3)_2](TMP)$, the first prochiral heteroleptic tin(II) amide, was obtained from $[Sn(\mu\text{-}Cl)\{N(SiMe_3)_2\}]_2 + 2Li(TMP)$.[120] Table 9.5 provides a summary of other heteroleptic metal(II) amides and their syntheses [as items 'A' to 'G′', etc., in footnote 'b' ('Methods') of Table 9.5; further procedures are outlined in Schemes 9.19–9.24 and Equations (9.9)[151c] and (9.10)[170]].

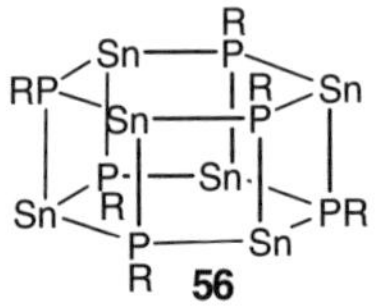

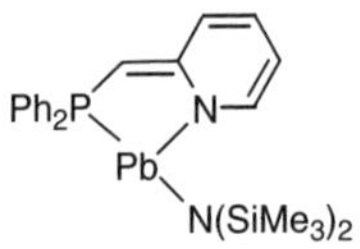

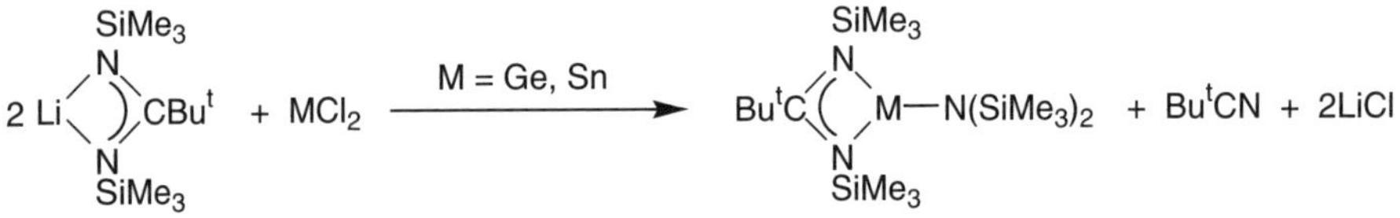

$$2\,Li[(NSiMe_3)_2CBu^t] + MCl_2 \xrightarrow{M = Ge,\ Sn} Bu^tC(NSiMe_3)_2M{-}N(SiMe_3)_2 + Bu^tCN + 2LiCl \quad (9.9)$$

$$4\,Me_2Si(NBu^t)(N(H)Bu^t)SnBr\ (\mathbf{38}\ (M = Sn, X = Br)) + 2LiBr + 8Bu^tNH_2 \longrightarrow [Sn_4Br_6\{N(H)Bu^t\}_4Li_2\{N(H)_2Bu^t\}_4] + 4Me_2Si[N(H)Bu^t]_2 \quad (9.10)$$

While crystalline $[Sn\{C_6H_3(NMe_2)_2\text{-}2,6\}\{N(SiMe_3)_2\}]$ has the Sn^{II} atom in a three-coordinate environment, in solution there is only a single ^1H NMR spectral signal for the two NMe_2 groups.[157] Whereas the crystalline lead compound $[M\{N(SiMe_3)_2\}$-

Table 9.5 *Heteroleptic Ge^{II}, Sn^{II} and Pb^{II} amides*

Compound	Method[b]	Reference
$M[N(SiMe_3)_2][\{N(R)\}_2CR']$		
$M=Sn$; $R=C_6H_{11}$-c, $R'=Me$[a]	B[151a,151b]	151a
$M=Sn, Ge$; $R=C_6H_{11}$-c, $R'=Bu^t$[a]		151b
$M=Sn, Ge$; $R=SiMe_3$, $R'=Bu^t$[a]	Equation (9.9)[151c]	151c
$M[N(SiMe_3)_2](OC_6H_2(CH_2NMe_2)_3$-2,4,6)	E	118a
$M=Sn, Ge, Pb$	B	118b
	G	
$[Sn\{N(SiMe_3)_2\}(\mu\text{-}OBu^t)]_2$[a]	G	152
	G′	
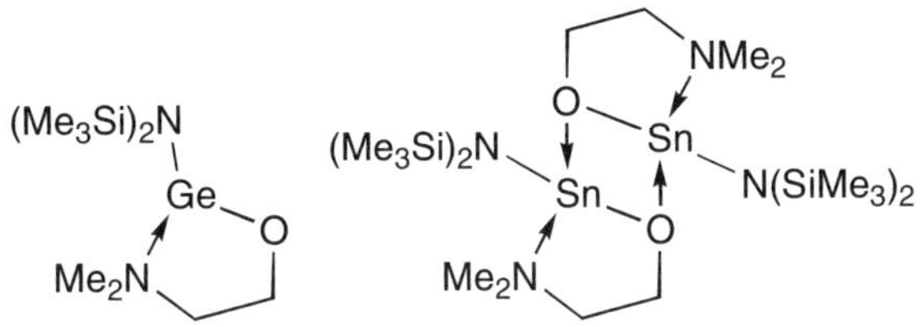 $(Me_3Si)_2N$–Ge(←Me_2N)–O chelate; $[(Me_3Si)_2N)Sn(\mu\text{-}O)(←NMe_2)]_2$ dimer	G	153a
$[Sn\{N(SiMe_3)_2\}(\mu\text{-}OBu^t)_2Sn(OSiMe_3)]$	Scheme 9.19	154
$[Pb\{N(SiMe_3)_2\}(\mu\text{-}OSiMe_3)]_2$[a]	Scheme 9.20	155
$Sn\{N(SiMe_3)_2\}R$		
$R=C_6H_3Trip_2$-2,6[a] ($Trip=C_6H_2Pr^i{}_3$-2,4,6),	D[156]	156 ($C_6H_3Trip_2$-2,6)
$C_6H_3(NMe_2)_2$-2,6	B[157]	157 [$C_6H_3(NMe_2)_2$-2,6]
$[Pb\{N(SiMe_3)_2\}\{\mu\text{-}SC(SiMe_3)_3\}]_2$[a]	E	124
$[Sn\{N(SiMe_3)_2\}NCS)]$[a]	G	158
$Sn\{N(SiMe_3)_2\}X$	B	120
$X=TMP$, $OC_6H_2Bu^t{}_2$-2,6-Me-4[a]	G	
$Ge\{N(R)SiMe_3\}X$	A	80
$R=C_6H_3Me_2$-2,6, $C_6H_3Pr^i{}_2$-2,6		
$X=Cl, Br$		
$Ge[N(SiMe_3)_2](C_6H_3(CH_2NEt_2)_2$-2,6)[a]	B	159
$[Sn\{N(SiMe_3)_2\}(\mu\text{-}Cl)]_2$[a]	(pre-1979[1])	160a, 160b
$[Sn\{N(C_6H_3Pr^i{}_2\text{-}2,6)SiMe_3\}(\mu\text{-}Cl)]_2$[a]	A	162
	G	
$[Sn\{(N(C_6H_3Pr^i{}_2\text{-}2,6)C(Me))_2CH\}(NMe_2)]$[a]	B, E	163
$[Pb\{(N(C_6H_3Pr^i{}_2\text{-}2,6)C(Me))_2CH\}\{N(SiMe_3)_2\}]$[a]	B, E	164a, 164b
$[Sn(TMP)(\mu\text{-}X)]_2$	G	160a, 160b
$X=F$,[a] Cl[a]		161
$[M(Cl)(\mu\text{-}NR_2)]_2$		153b
$M=Ge$, $R=Et$;[a]	A[c]	
$M=Sn$, $R=Me$[a]	G	
$[M\{N(SiMe_3)_2\}]_2(X)$	B	165
$X=[N(SiMe_3)]_2C_6H_4$-1,4, $M=Ge, Sn$;[a]		
$X=[N(SiMe_3)]_2C_6H_{10}$-1,4, $M=Sn$[a]		
$[M(\mu\text{-}NH_2)R]_2$	B (M = Pb)	166
$M=Ge$, $R=C_6H_3(C_6H_3Pr^i{}_2\text{-}2',6')_2$-2,6;[a]	F (M = Ge, Sn)	

(*continued*)

Table 9.5

Compound	Method[b]	Reference
M = Sn, Pb, R = $C_6H_3(C_6H_2Pr^i_3\text{-}2',4',6')_2\text{-}2,6$[a]		
$[M\{N(H)R\}\{\mu\text{-}N(H)R\}_2Li(thf)]$	C	167
M = Sn, R = $C_6H_4OMe\text{-}4$[a]	(Scheme 9.20)	
$[Sn\{N(C_6H_2Me_3\text{-}2,4,6)SiMe_3\}(\mu\text{-}Cl)]_3$[a]	A	97b
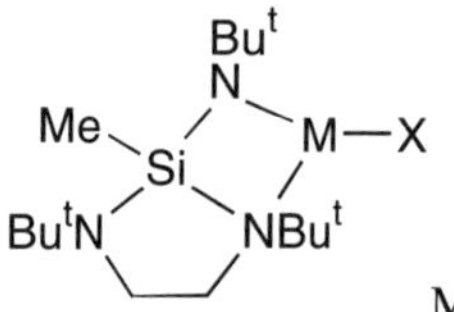 M = Sn; X = Cl M = Ge; X = Cl, $N(SiMe_3)_2$	A B	105 174
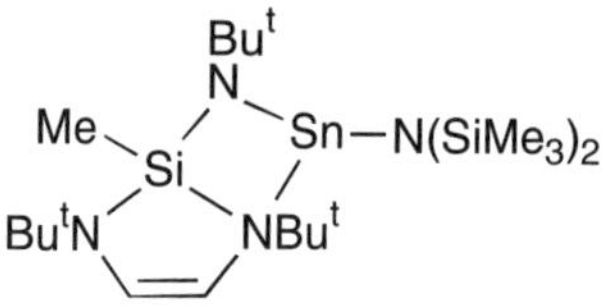 [a]	A	175
$[Ge\{N(H)R\}]_2(\mu\text{-}NR)$, R = $C_6H_2Bu^t_3\text{-}2,4,6$	E′	77b
$[\{Li(pmdeta)\}_2(\mu\text{-}Cl)]^+[Sn\{N(R)SiMe_2\}_3CH]^-$,[a]	A	176a, 176b
$[Li(thf)_3]^+[Sn\{N(R)SiMe_2\}_3CH]^-$,[a] R = $C_6H_4Me\text{-}4$		
$[Li(thf)]^+[Sn\{N(R)SiMe_2\}_3CMe]^-$, R = Bu^t	A	176b
$[Li(thf)]^+[Sn\{N(R)CH_2\}_3C(Me)]^-$, R = $SiMe_3$		
$[Li(thf)_n]^+[Sn\{N(CHMePh)SiMe_2\}_3CH]^-$ $n = 0, 1$[a]	A	177
$[Li(thf)_4]^+[Pb\{N(R)SiMe_2\}_3CH]^-$, R = $C_6H_4Me\text{-}4$	C	176b
$[Li(thf)]^+[Pb\{N(R)SiMe_2\}_3CMe]^-$, R = Bu^t[a]		
$[Pb\{N(R)CH_2\}_2C(Me)(CH_2O)Li]_2$, R = $SiMe_3$[a]		
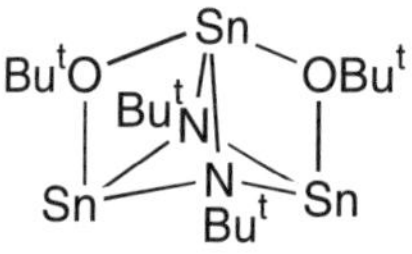 [a]	Scheme 9.22	178
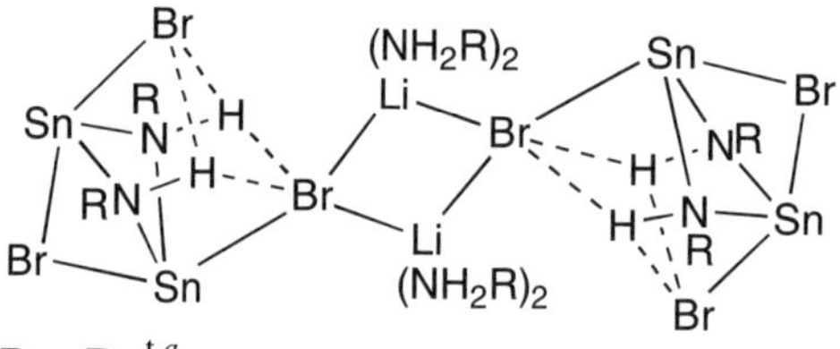 R = Bu^t[a]	Equation (9.10)	179
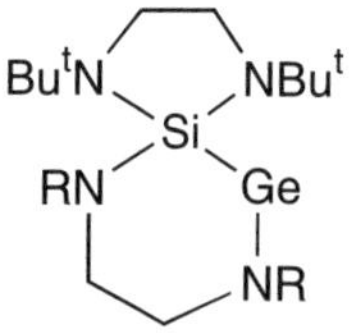 R = $C_6H_3Me_2\text{-}2,6$[a]	Scheme 9.8	83

Table 9.5

Compound	Method[b]	Reference
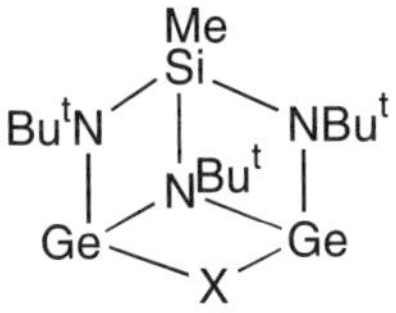 $X = N(SiMe_3)_2$,[a] Cl,[a] PPh_2 [a]	A (X = Cl) then NaX [X = $N(SiMe_3)_2$, PPh_2]	168
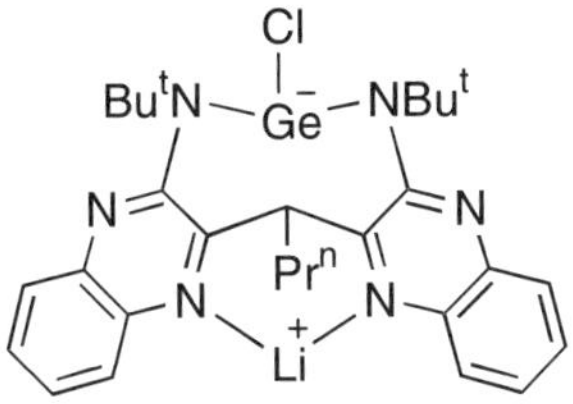 [a]	A[d]	169
$Pb[C_6H_2Bu^t\text{-}4\text{-}\{P(O)(OEt)_2\}_2\text{-}2,6][NPr^i_2]$	B	170
$Sn(Cp)\{N(SiMe_3)_2\}(\mu\text{-}Cp)Li(pmdeta)$	C	171
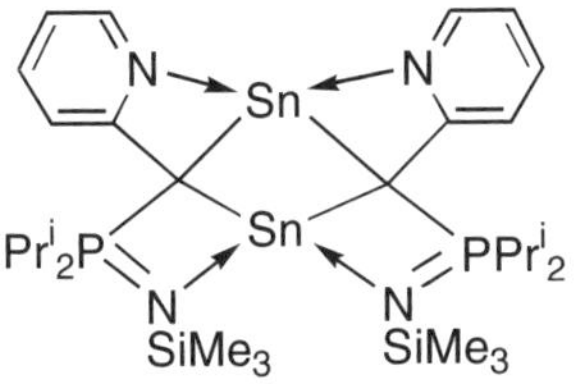 (*R*- and *S*-) [a]	Scheme 9.23	172
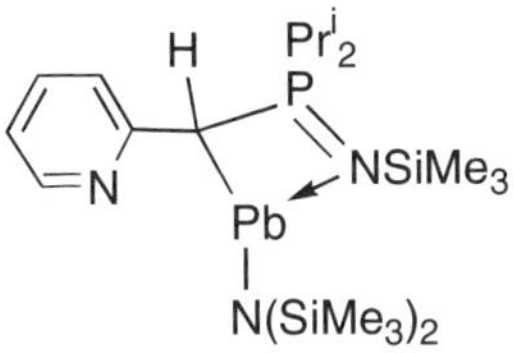 [a]		
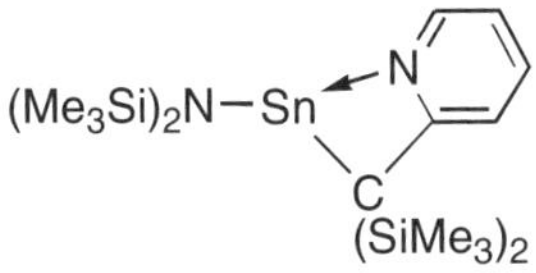 [a]	D	173a 173b
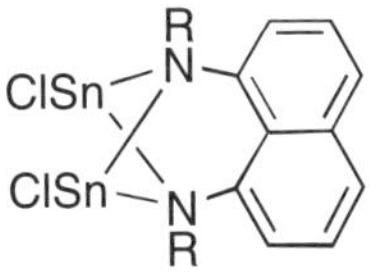 $R = SiMe_3$ [a]	G	59
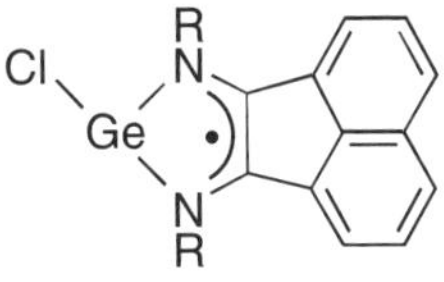 $R = C_6H_4Ph\text{-}2$,[a] $C_6H_3Pr^i_2\text{-}2,6$,[a] $C_6H_3Bu^t_2\text{-}2,5$ [a]	Scheme 9.24	180a

(*continued*)

Table 9.5

Compound	Method[b]	Reference
		180b
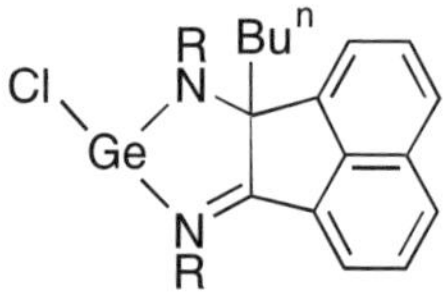 Ge chelate (Cl–Ge bonded to two N(R) groups of a Bu^n-substituted acenaphthene-derived amido/imino ligand); $R = C_6H_3Pr^i_2$-2,6[a]	Scheme 9.24	181

[a]Molecular structure by X-ray diffraction
[b]Methods:
A: salt elimination from $MCl_2 + M'[N(R)R']$ ($M' = Li$ or Na)
B: salt elimination from $M(L)Hal + M'[N(R)R']$ ($M' = Li$ or Na); or $M[N(R)R']Hal + LiR''$
C: LiCp elimination from $MCp_2 + Li[N(R)R']$
D: $Li[N(R)R']$ elimination from $M[N(R)R']_2 + LiR''$
E: $HN(R)R'$ elimination from $M[N(R)R']_2 + HA$; E′: from $2Ge[N(H)R]_2$
F: $[NH_4]Cl$ eliminatiom from $M(Cl)R + 2NH_3$
G: metathetical ligand exchange from $M[N(R)R']_2 + MX_2$; G′: CuX instead of MX_2
[c]Et_3GeNEt_2 was a source of an amide (elimination of Et_3GeCl)
[d]Amidolithium precursor obtained in situ from 2-N(H)Bu^t-3-Cl-quinoxaline + 2LiBu^n

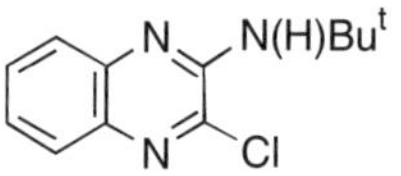

$\{\mu\text{-}SC(SiMe_3)_3\}]_2$ ($M = Pb$) was accessible, attempts to make the tin analogue led to the redistributed products $Sn[N(SiMe_3)_2]_2$ and $[Sn\{SC(SiMe_3)_3\}_2]_\infty$; and $Ge[N(SiMe_3)_2]_2$ with $HSC(SiMe_3)_3$ in toluene gave the crystalline *cis*-$[Ge(CH_2Ph)\{N(SiMe_3)_2\}$-$(\mu\text{-}S)]_2$.[124] The compound $Sn[N(SiMe_3)_2](TMP)$ is the only mononuclear heteroleptic tin(II) bis(amide).[120] Crystalline *cis*-$[Sn(\mu\text{-}Cl)(TMP)]_2$ in $[^2H_8]$toluene underwent rapid *cis* ⇌ *trans* isomerisation at ambient temperature, but distinct *cis* and *trans* isomers were shown to coexist below 192 K;[160a] the mechanism of this intramolecular isomerisation was suggested to implicate a T-shaped transition state.[161] The crystalline compounds *trans*-$[Ge(Cl)(\mu\text{-}NEt_2)]_2$ and *trans*-$[Sn(Cl)(\mu\text{-}NMe_2)]_2$[153b] are noteworthy for having a bridging amide (rather than Cl) in contrast to dimeric complexes with bulkier amido ligands such as TMP, $N(SiMe_3)_2$ or $N(C_6H_3Pr^i_2\text{-}2,6)SiMe_3$. Whereas the central MNM′N′ ring is planar for M = Sn, in the Ge analogue the ring is puckered, and the former rather than the latter has close intermolecular Sn···Cl contacts of 3.424(1) Å.[153b]

Although treatment of $SnCl_2$ with $Sn[N(SiMe_3)_2]_2$ gave the metathetical ligand exchange product $[Sn(\mu\text{-}Cl)\{N(SiMe_3)_2\}]_2$,[1] $SnCl_2$ with $[\{SnN(SiMe_3)_2\}_2\{\mu\text{-}C_6H_4(NSiMe_3)_2\text{-}1,4\}]$ led to the redox products Sn and $[SnCl_2\{N(SiMe_3)_2\}]_2\{\mu\text{-}C_6H_4(NSiMe_3)_2\text{-}1,4\}$.[165] The compound $SnCp\{N(SiMe_3)_2\}(\mu\text{-}Cp)Li(pmdeta)$ in C_6D_6 was in equilibrium

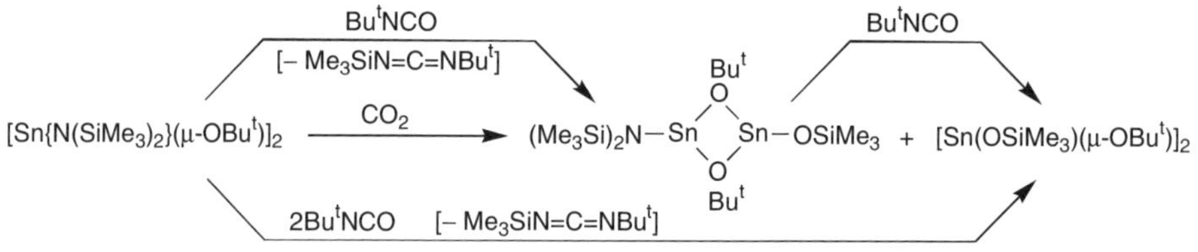

$[Sn\{N(SiMe_3)_2\}(\mu\text{-}OBu^t)]_2$ —(CO_2)→ $(Me_3Si)_2N{-}Sn(\mu\text{-}OBu^t)_2Sn{-}OSiMe_3$ + $[Sn(OSiMe_3)(\mu\text{-}OBu^t)]_2$
(upper path: Bu^tNCO [$- Me_3SiN{=}C{=}NBu^t$], then Bu^tNCO; lower path: $2Bu^tNCO$ [$- Me_3SiN{=}C{=}NBu^t$])

Scheme 9.19[154]

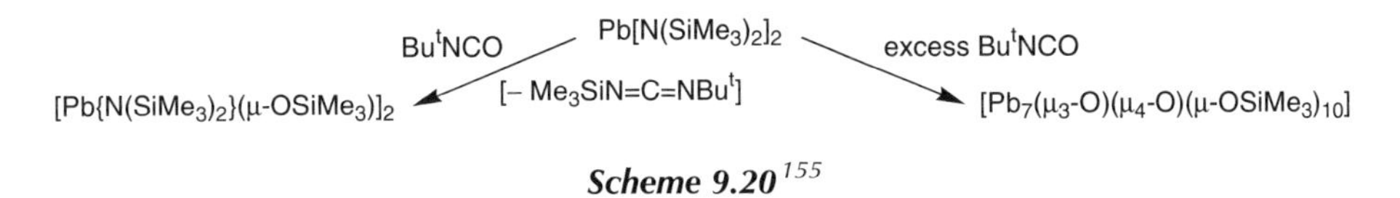

Scheme 9.20[155]

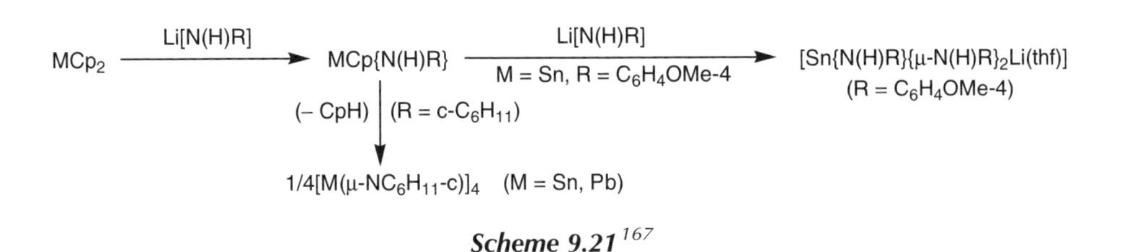

Scheme 9.21[167]

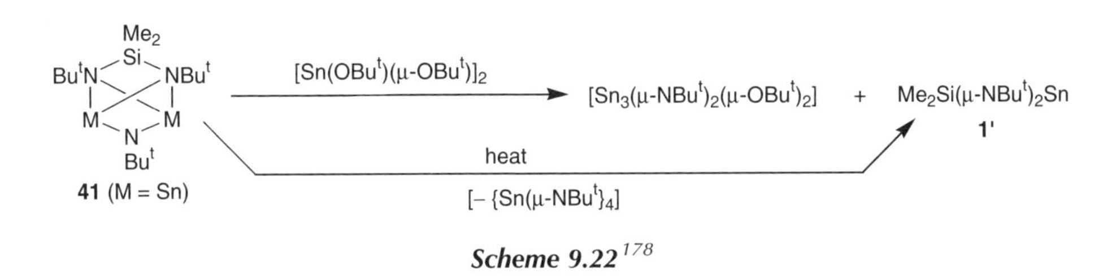

Scheme 9.22[178]

with $SnCp\{N(SiMe_3)_2\}$ and LiCp(pmdeta).[171] The ^1H NMR spectrum of $Sn\{C(SiMe_3)_2(C_5H_4N\text{-}2)\}[N(SiMe_3)_2]$ in $[^2H_8]$ toluene at 193 K showed four distinct $SiMe_3$ signals; two of these merged at 214 ± 2 K and the other two at 368 ± 2 K, attributed to restricted rotation about the Sn–C and Sn–NSi_2 bonds, respectively.[173b]

9.3.5 Metathetical Exchange Reactions

A metathetical exchange reaction of a tin(II) amide, leading to $Sn[N(SiMe_3)_2](OAr)$,[120] was described in Section 9.3.3. Such transformations also affording other heteroleptic Group 14 metal(II) amides have featured in Section 9.3.4, being listed in Table 9.5 (reactions there designated as of type D[156,173a,173b] and G or G′[59,118a,118b,120,152,153,158,162]) and in Schemes 9.22,[178] 9.23,[172] and 9.24.[180a,180b,181]

The X-ray-characterised zwitterionic Ge^{IV} amide **58** was obtained as shown in Scheme 9.25.[182]

The first example of an $[NR_2]^-/[R']^-$ exchange reaction at a Group 14 metal(II) site was in the 1976 synthesis of the first germanium(II) alkyl $Ge[CH(SiMe_3)_2]_2$ from $Ge[N(SiMe_3)_2]_2$ and $2Li[CH(SiMe_3)_2]$; the same procedure was effective for the isoleptic tin(II) dialkyl.[183] Further such $[NR_2]^-/[R']^-$ displacements from $M[N(SiMe_3)_2]_2$ and LiR′ have led to $[Sn\{C_6H_3(C_6H_2Pr^i_3\text{-}2',4',6')_2\text{-}2,6\}\{N(SiMe_3)_2\}]$,[156] $[Ge(C_6H_2Pr^i_3\text{-}2,4,6)_2]_2$,[184]

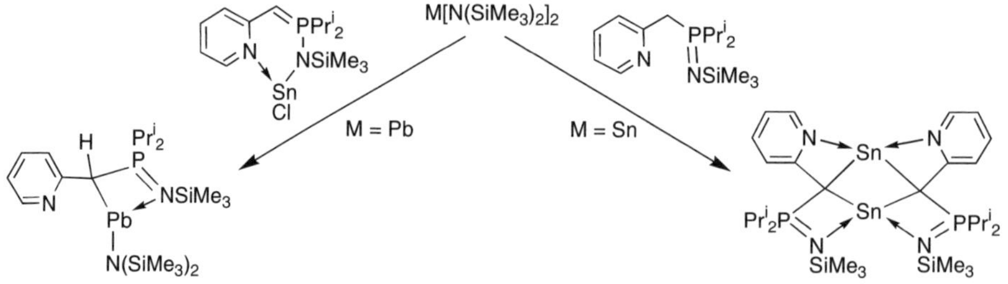

M[N(SiMe3)2]2
PPri2
NSiMe3
N
Sn
Cl
M = Pb
M = Sn
PPri2
NSiMe3
H Pri2
P
NSiMe3
Pb
N(SiMe3)2
N Sn N
Sn
Pri2P PPri2
N N
SiMe3 SiMe3

Scheme 9.23[172]

GeCl2·dioxane
Bun
R
N
Li(OEt2)
N
R
NR
(or Na+L−)
NR
Bun
R
N
Ge
Cl
N
R
R = C6H3Pri2-2,6 (ref. 181)
R
N
Ge
N
R
Cl
R = C6H3Pri2-2,6 (ref. 180a)
C6H4Ph-2 (ref. 180b)
C6H3But2-2,5 (ref. 180b)

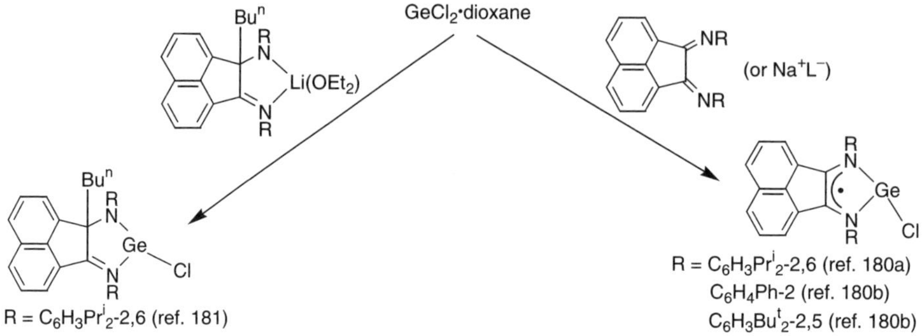

Scheme 9.24[180a,180b,181]

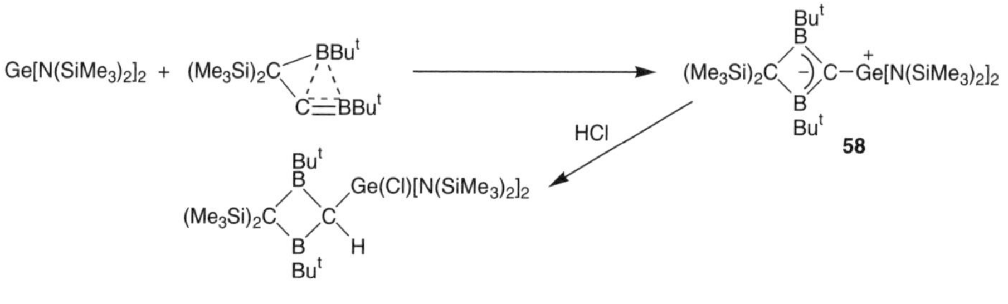

Ge[N(SiMe3)2]2 + (Me3Si)2C BBut C BBut
But B (Me3Si)2C C−Ge[N(SiMe3)2]2 B But
58
HCl
But B (Me3Si)2C C Ge(Cl)[N(SiMe3)2]2 B H But

Scheme 9.25[182]

$Sn(C_6H_2Bu^t{}_3\text{-}2,4,6)_2$,[185] PbR_2[186] and $Pb(R')R''$[186] [$R = C_6H_2Pr^i{}_3$-2,4,6; $R' = C_6H_2$-$\{CH(SiMe_3)_2\}_3$-2,4,6; $R'' = R$, R' or $CH(SiMe_3)_2$] with elimination in each case of $Li[N(SiMe_3)_2]$. The reaction of $M[N(SiMe_3)_2]_2$ (M = Sn, Pb), $[Sn(\mu\text{-}F)\{N(SiMe_3)_2\}]_2$ or $Sn(TMP)_2$ with AgOCN gave the appropriate silver(I) amide $[\{Ag(\mu\text{-}NR_2)\}_4]$.[187] The crystalline tetranuclear heteroleptic lead amide **59** was prepared as shown in Equation (9.11).[188]

(9.11)

Treatment of $Sn[N(SiMe_3)_2]R$ with $Si[\{N(CH_2Bu^t)\}_2C_6H_4\text{-}1,2]$ [$\equiv$ Si(NN), **3**] gave the tin(II) silyl compound $Sn(R)[Si(NN)\{N(SiMe_3)_2\}]$, demonstrating that $N(SiMe_3)_2 > R$ in migratory aptitude [$R = C_6H_3(NMe_2)_2\text{-}2,6$].[189]

Metathetical exchange of $M[N(SiMe_3)_2]_2$ (M = Sn or Pb) with $2K[Si(SiMe_3)_3]$ gave the metal(II) hypersilyl compounds $M[Si(SiMe_3)_3]_2$, which were monomeric in solution while in the solid state the lead compound was a monomer and the tin analogue was a Sn–Sn-bonded dimer.[190a] With the less bulky silyl reagent $K[Si(SiMe_3)_2Et]$, the crystalline products were $[K(OEt_2)_2Pb\{Si(SiMe_3)_2Et\}_3]$ and the first stable plumbyl radical $[Pb\{Si(SiMe_3)_2Et\}_3]$.[190b]

9.3.6 Reactions with Heterocumulenes

The synthesis of carbodiimides, and particularly of asymmetrically substituted compounds, from a Group 14 metal(II) amide and a heterocumulene, has been developed by Sita *et al.* The first (1996)[191] example was the conversion of $Sn[N(SiMe_3)_2]_2$ by CO_2 into $Me_3SiN{=}C{=}O$, which with further tin(II) amide gave $C[N(SiMe_3)]_2$,[191] Scheme 9.26 [reactions at ambient temperature in C_5H_{12}; $R = Bu^t$, C_6H_{11}, Pr^i, Bu^n, Ph, $C_6H_3R'_2\text{-}2,6$ ($R' = Me$, Pr^i)[192]]. While $Sn[N(SiMe_2Ph)_2]_2$ reacted with CO_2 in C_5H_{12} at 30 psi and 50 °C yielding $RN{=}C{=}O$ (4.4 parts), $C(NR)_2$ (1 part) and $[Sn(OR)(\mu\text{-}OR)]_2$ ($R = SiMe_2Ph$),[96] $Sn[N(C_6H_3Pr^i_2\text{-}2,6)SiMe_3]_2$ proved to be unreactive to both CO_2 and $R'N{=}C{=}O$; $Sn[N(SiMe_2Ph)_2]_2$ with various isocyanates $4\text{-}XC_6H_4NCO$ gave the asymmetrical carbodiimides at a rate related (Hammett plots) to the nucleophilicity of the isocyanate (X = OMe, Me, F, Cl, CF_3).[96] Various compounds $Sn[N(SiMe_3)Y]_2$ reacted with CO_2 at rates favouring the more congested tin amide (reactivity: $Y = SiPh_2Bu^t > SiMe_2Bu^t >$ 1-adamantyl).[97a]

Further examples of metal(II) amide/(CO_2 or Bu^tNCO) reactions are illustrated in Schemes 9.19[154] and 9.20.[155]

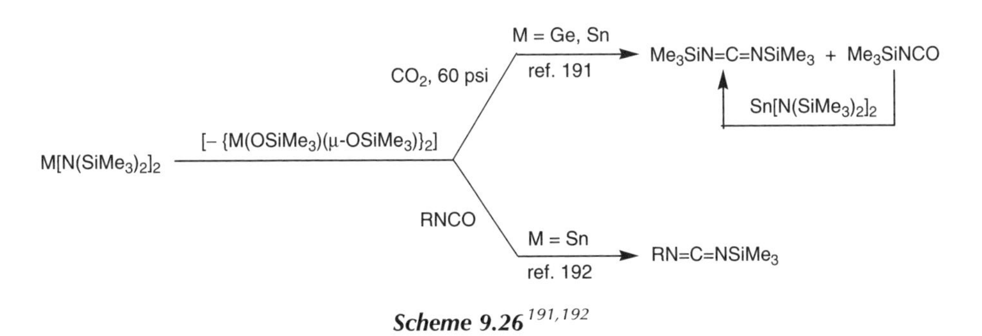

Scheme 9.26[191,192]

9.3.7 Oligomeric Metal(II) Imides

The title compounds are of formula $[M(\mu\text{-}NR)]_n$ (often M = Sn). Most are cubanes or contain a cubanoid cluster, such as $[Sn(\mu\text{-}NR)]_4$, $[Sn_3M(\mu\text{-}NR)_4]$[81] or $[Sn_4(\mu\text{-}NR)_3(\mu\text{-}O)]$,[132] and have nitrogen atoms in a four-coordinate environment, and being outside the scope of this review are dealt with only briefly. However, there are exceptions, containing three-coordinate nitrogen atoms. Such examples are the crystalline complexes **60** {prepared from $[Li(\mu\text{-}N(H)C_6H_2Bu^t_3\text{-}2,4,6)(thf)]_2$ and $GeCl_2$·diox, or from $Ge[N(H)C_6H_2Bu^t_3\text{-}2,4,6]_2$}[77] and **61** {obtained from $Ge[N(SiMe_3)_2]_2$ and $2,6\text{-}Pr^i_2C_6H_3NH_2$};[193a,193b] the $(Ge{-}N)_{2\text{ or }3}$ rings in **60** or **61** are planar. By contrast to the above reaction leading to **61**, $M[N(SiMe_3)_2]_2$ (M = Sn, Pb) with RNH_2 gave the appropriate cubane $[M(\mu\text{-}NR)]_4$ [M = Sn or Pb and R = $C_6H_3Pr^i_2$-2,6; or M = Sn and R = $B(C_6H_2Me_3\text{-}2,4,6)_2$].[193a]

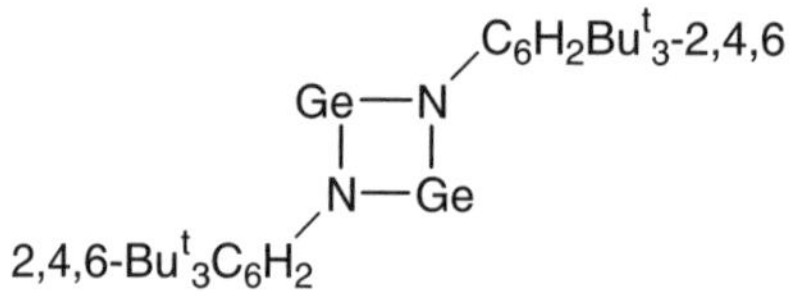

60

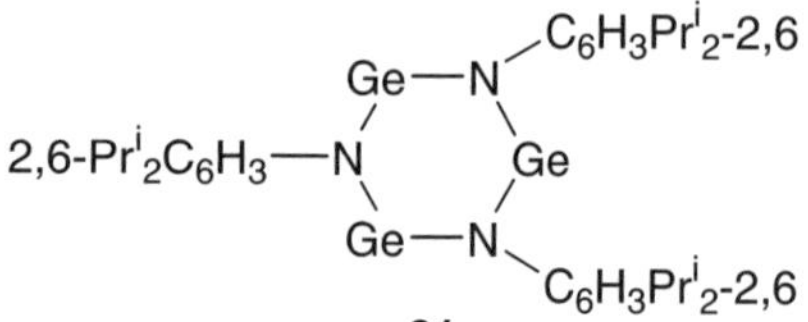

61

Cubanoid metal imides (most mentioned in Section 9.3.3) are the compounds **43**, **44**,[132a,132b,81] **50**,[144a,144b] **51**,[145] and others of formula $[Sn(\mu\text{-}NR)]_4$,[130,131,132a,132b,141,142,143,193a,194] $[M(\mu\text{-}NC_6H_{11}\text{-}c)]_4$ (M = Sn, Pb),[167] $[M(\mu\text{-}NPh)]_4$ (M = Ge, Sn, Pb),[197] $[\{M(\mu\text{-}NBu^t)\}_4(AlCl_3)_2]$ (M = Ge, Sn),[196] and $[\{Pb_4(\mu\text{-}NBu^t)_3(\mu\text{-}O)\}Ti(OPr^i)_4]$.[195] The crystalline compounds $[M(\mu\text{-}NR)]_4$ (M = Ge, Sn, Pb; R = $SiMe_3$ or $GeMe_3$) and $[Sn(\mu\text{-}NSnMe_3)]_4$ were prepared from equivalent portions of MCl_2 and the appropriate $Li[N(R)(SnMe_3)]$, with elimination of LiCl and Me_3SnCl.[198a,198b] Heterometallic anionic $[Li(thf)_4][\mathbf{51}]$ (R = Bu^t)[199a] and neutral $[\{Mg(thf)\}Sn_3(\mu\text{-}NBu^t)_4]$[199a] cubanoids were readily prepared by the metallation of **42** (M = Sn) with 2 LiBu or $MgBu_2$, respectively; their reactions (and those of $[Sn(\mu\text{-}NC_6H_3Pr^i_2\text{-}2,6)]_4$) with heavier chalcogens were studied.[199a,199b]

9.3.8 Metal(II) Amides based on 1,4-Diazabutadienes or a Related Compound

Compounds based on the $\bar{N}(R)CH{=}CH\bar{N}(R)$ skeleton have featured in Tables 9.1 and 9.2 (Refs. 17, 82–86, 88, 91, 92, 98, 101, 102) and in Schemes 9.17 and 9.24 (Refs. 180a, 180b, 181). Further examples are the crystalline compounds **62** [made from $GeCl_2\{N(Bu^t)CH{=}CHNBu^t\}$ + $GeCl_2$·diox][200] and **63** (M = Ge, Sn, Pb), obtained from the carbene $C[\{N(CH_2Bu^t)\}_2C_6H_4\text{-}1,2]$ and the isoleptic heavier Group 14 metal analogue;[201] the latter has an almost planar central C and a pyramidal M atom environment.[201]

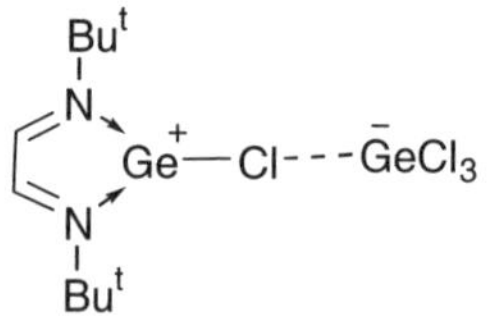

62

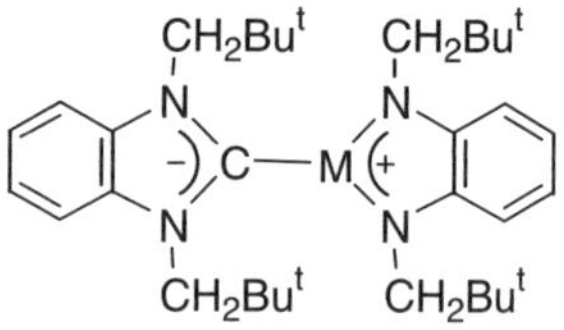

63

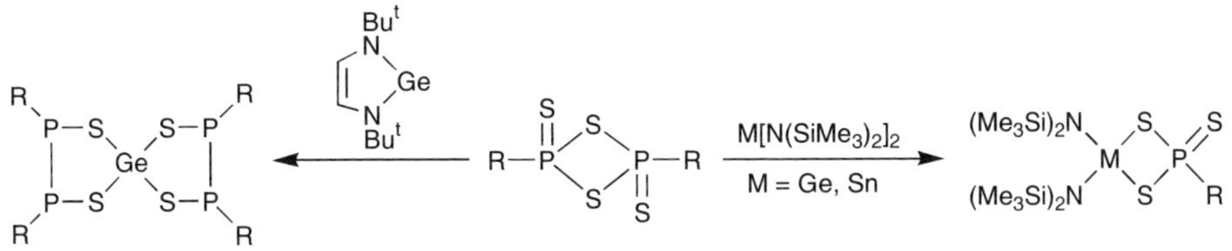

Scheme 9.27 *(R = C_6H_4OMe-4)*[202]

An unusual 1,4-diazabutadiene displacement, using Lawesson's reagent, is shown inter alia in Scheme 9.27.[202]

9.3.9 Oxidative Addition and Redox Reactions

The first (1976) oxidative addition reactions of a Group 14 metal(II) amide related to the $Sn[N(SiMe_3)_2]_2$/RX system (RX = MeBr, MeI, PhI), Equation (9.12); they were discovered by the Sussex group.[203a] Evidence for a free radical mechanism (kinetics, stereochemistry) was presented, involving $\cdot Sn[N(SiMe_3)_2]_2X$ and R· in the initiation step.[203a] When bromobenzene was the reagent, a trace of EtBr proved to be a catalyst, and using thf rather than benzene as the reaction medium gave $SnBr_2[N(SiMe_3)_2]_2$ as a significant by-product.[203b]

$$Sn[N(SiMe_3)_2]_2 + RX \longrightarrow Sn[N(SiMe_3)_2]_2(R)X \qquad (9.12)$$

Reactions of Equation (9.12) were extended to yield similar products (70–80%) from $Sn[N(SiMe_3)_2]_2$ and RCl (R = Bu^n, CH_2Cl, $CHCl_2$, CCl_3), RBr [R = Pr^n, Bu^t, Ph, $N(SiMe_3)_2$], or RI (R = Me, Et, Pr^i, Bu^n, Ph); likewise, $M(TMP)_2$ and RC(O)−Cl (R = Bu^t, Ph) or $CF_3C(O)-OC(O)CF_3$ gave the appropriate 1 : 1 adduct, as also was the case for the reaction between this anhydride with $M[N(SiMe_3)_2]_2$, $M(NBu^t{}_2)_2$ (M = Ge, Sn),[204] or $Sn[N(SiMe_3)_2](TMP)$.[120] This heteroleptic tin(II) diamide reacted similarly with MeI or (+)-$EtCH(Me)CH_2-Br$, in the case of the latter without significant diastereoselectivity.[120] Treatment of $Ge[N(SiMe_3)_2]_2$ with Br−CN gave the crystalline $GeBr(CN)[N(SiMe_3)_2]_2$.[205] The dimeric germanium(II) imide **60** with MeI furnished $\{GeI(Me)(\mu\text{-}NC_6H_2Bu^t{}_3\text{-}2,4,6)\}_2$.[77b]

Banaszak Holl and coworkers have discovered an interesting regioselective C−H activation reaction of certain ethers and alkanes (abbreviated as RH), using the $M[N(SiMe_3)_2]_2$/PhI system. These culminate in the formation of the appropriate $M(I)[N(SiMe_3)_2]_2R$ with elimination of C_6H_6; reactions with M = Ge are summarised in Scheme 9.28.[206a] High dilution conditions were crucial; optimal results were achieved by slow ambient temperature addition of $Ge[N(SiMe_3)_2]_2$ (0.02 *M* in RH) to an equimolar amount of PhI in RH; oxidative addition yielding $Ge(I)[N(SiMe_3)_2]_2Ph$ was a concentration-dependent side reaction. The proposed mechanism involved R·, Ph·, $\cdot Ge[N(SiMe_3)_2]_2R$ and $\cdot Ge[N(SiMe_3)_2]_2Ph$ intermediates in the chain propagation sequence.

Such C−H activation reactions were extended to the related Sn^{II} systems.[206b] The rates of addition for this chemistry were 4 to 6 times slower for the latter compared with the Ge reactions, in order to avoid the competing formation of $Sn(I)[N(SiMe_3)_2]_2Ph$. Thus, similar

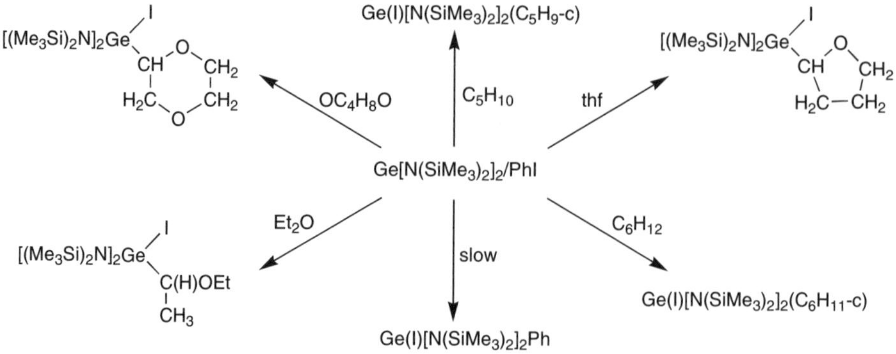

$Ge(I)[N(SiMe_3)_2]_2(C_5H_9\text{-}c)$

$[(Me_3Si)_2N]_2Ge(I)$–CH(OC_4H_8O ring) ← OC_4H_8O — C_5H_{10} — thf → $[(Me_3Si)_2N]_2Ge(I)$–CH(OC_4H_7 ring)

$Ge[N(SiMe_3)_2]_2/PhI$

$[(Me_3Si)_2N]_2Ge(I)C(H)(OEt)CH_3$ ← Et_2O — slow — C_6H_{12} → $Ge(I)[N(SiMe_3)_2]_2(C_6H_{11}\text{-}c)$

$Ge(I)[N(SiMe_3)_2]_2Ph$

Scheme 9.28 [206a]

$Sn(I)[N(SiMe_3)_2]_2R$ compounds were obtained from the tin(II) amide/PhI/RH with RH = thf, C_6H_{12}, Et_2O or 1,4-dioxane, as well as others with R = CH_2OBu^t (from $MeOBu^t$) and a mixture of C_5H_{11} isomers (ca. 6.2:1, secondary to primary).

Making use of an earlier finding that Sn^{IV} compounds containing Sn–CH_2R' bonds were suitable for Stille cross-coupling reactions,[207] one-pot C–H activation and cross-coupling products were isolated in modest yields (21% from $MeOBu^t$ and 5% from n-C_5H_{12} based on Sn^{II} amide) as shown in Equation (9.13) (R = CH_2OBu^t or n-C_5H_{11}).[206b]

$$Sn[N(SiMe_3)_2]_2 + PhI + RH \xrightarrow[\text{2. } [NBu^n_4]F,\ 3\ \text{mol}\ \%[Pd(PPh_3)_4]\ \ PhI,\ \text{dioxane, reflux}]{\text{1. RH}} Ph\text{–}R \qquad (9.13)$$

The cyclic diaminogermylenes $\overline{GeN(Bu^t)XN}Bu^t$ (X = CH_2CH_2 or CH=CH) with TEMPO yielded 1 : 2 $O\overline{NC(Me)_2(CH_2)_3C}Me_2$ adducts.[55] The unsaturated germylene and its isoleptic C^{II}, Si^{II} and Sn^{II} compounds gave the EPR-characterised radicals $[\overline{M\{N(Bu^t)CH{=}CHN}Bu^t\}R]^{\cdot}$ [R = TEMPO, $OC_6H_2Bu^t_2$-2,6-Me-4, OC_6Me_4-2,3,5,6-OH-4, Bu^t, $SiMe_3$, $WCp(CO)_3$, $Re(CO)_5$]; together with EPR spectral data, DFT calculations on model compounds showed that spin delocalisation to the five-membered ring decreased in the sequence M = Ge > Si > C, implicating the importance of the zwitterionic structure **64**.[208]

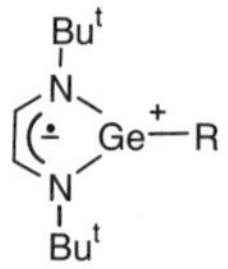

64

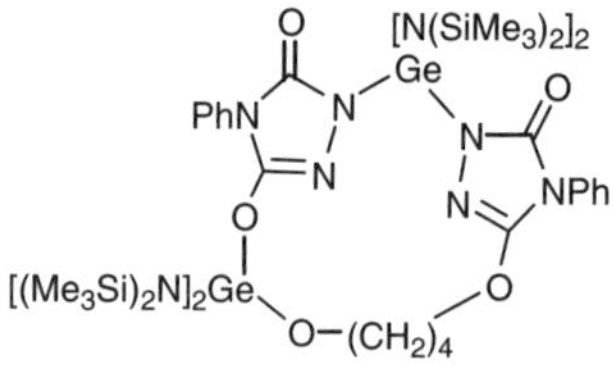

65

$Ge[N(SiMe_3)_2]_2$ in C_6H_6 at 25 °C with H_2 was converted into $Ge(H)_2[N(SiMe_3)_2]_2$ using $Ni[Ge\{N(SiMe_3)_2\}_2](PEt_3)_2$ (2 mol %) as catalyst.[209a] Treatment of this Ge^{II} amide with 4-phenyl-1,2,4-triazoline-3,5-dione in thf yielded the amide **65**.[209b] The compound $Bu^tSi[OSi(Me)_2N(Ph)]_3\bar{G}e{-}\overset{+}{L}i(thf)_3$ (cf. Table 9.1) and H_2O or O_2 gave $Bu^tSi[OSi(Me)_2N(Ph)]_3GeH$ or $[Bu^tSi\{OSi(Me)_2N(Ph)\}_3Ge(\mu_3\text{-}O)Li]_2$, respectively.[94]

Veith *et al.*, were the pioneers (1979 *et seq.*) in the oxidative addition of a chalcogen to a Group 14 metal(II) amide. Thus, crystalline compounds **66**,[210] Ge(L)(=S) (**67**) or $[Ge(L)(\mu\text{-}O)]_2$,[87] and Ge(L′)(=X) (**68**)[105] were obtained from $\overline{SnN(Bu^t)Si(Me)_2N}Bu^t$ (**1′**) and O_2,[210] Ge(L) and S_8 or O_2,[87] and Ge(L′) and O_2, S_8 or Me_3SiN_3 (with loss of N_2).[105] The compound $\overline{MN(Bu^t)Si(Me)_2N}Bu^t$ (M = Ge, Sn) and chalcogen (E) gave the crystalline complexes $\overline{M\{N(Bu^t)Si(Me)_2N}Bu^t\}(\mu\text{-}E)]_2$; [M = Ge with E = S, Se, Te; M = Sn with E = O (cf.[210] **66**), S, Se, Te];[211] $[\overline{Ge\{N(Bu^t)Si(Me)_2N}Bu^t\}(\mu\text{-}O)]_3$ was obtained from the Ge^{II} amide and Me_3NO.[211]

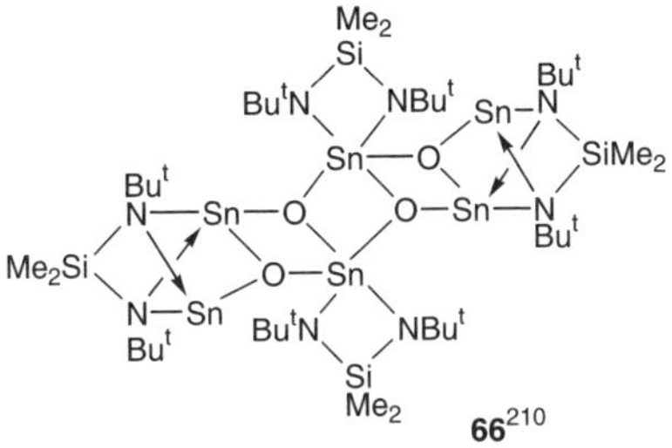

66[210] 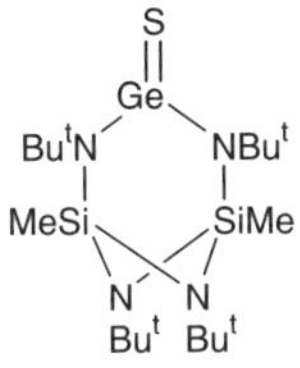 **67**[87] 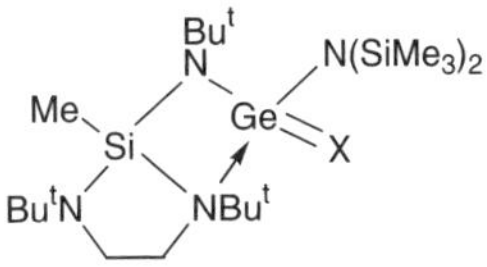 **68** (X = O, S, $NSiMe_3$)[105]

Experiments at Sussex on chalcogen [or $C_2(COOMe)_2$] reactions with $M[N(SiMe_3)_2]_2$, yielding crystalline heteroleptic M^{IV} diamides (M = Ge, Sn) are shown in Scheme 9.29; the compound $[Sn(L)(\mu\text{-}O)]_3$[90] was formed upon a slow aerial oxidation of Sn(L) (≡ **69**, cf.[90] Table 9.2)]. The heteroleptic β-diketiminatotin(II) amide $Sn(L')[N(SiMe_3)_2]$ with chalcogen or dibenzoylmethane gave $Sn(L')[N(SiMe_3)_2](=E)$[216] [L′ = $\{N(SiMe_3)C(Ph)\}_2CH$, E = S, Se] or $Sn(L')[\{OC(Ph)\}_2CH]$, respectively.[216] Compounds $[Ge(L'')(\mu\text{-}E)]_2$ were prepared from E (S, Se) and $Ge\{N(Pr^i)\}_2C_{10}H_8\text{-}1,8$ (≡ GeL″) (cf. Table 9.1).[89]

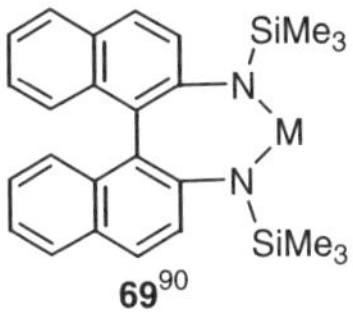

69[90]

The crystalline compounds $M[\{N(C_6H_{11}\text{-}c)\}_2CMe][N(SiMe_3)_2](EPh)_2$ (M = Ge, Sn; E = S or Se) were obtained from PhEEPh and the appropriate compound $M[\{N(C_6H_{11}\text{-}c)\}_2CMe][N(SiMe_3)_2]$.[217]

Bis(amido)germanium(IV)-containing block copolymers, obtained from a Ge^{II} amide have been reported by Kobayashi *et al.* These are the poly[bis(amido)germanium(IV) quinolates] $[-Ge\{N(SiMe_3)_2\}_2OC_6H_2R_2\text{-}2,5\text{-}O\text{-}4-]_n$ (R = H, Bu^t, Cl or Ph),[218a,218b] or their $Ge[N(Bu^t)SiMe_3]_2$ analogues,[218b] the corresponding enolates $[-Ge\{N(SiMe_3)_2\}_2\{\text{-C=C-}(CH_2)_m\text{-O-}\}]_n$[219a,219b] or the $Ge[\{N(SiMe_3)\}_2C_6H_4\text{-}1,2]$

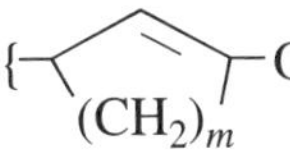

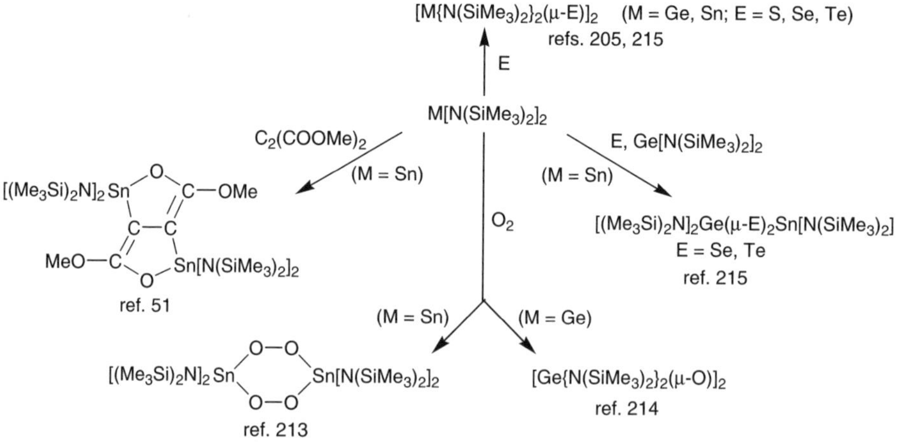

Scheme 9.29

analogues,[219b] and the related alkenes $[-Ge\{N(SiMe_3)_2\}_2\{-C(R)=C(H)-\}_m]_n$ (R = Bu^n, C_5H_{11}, C_6H_{13}, Ph),[220a] obtained from the appropriate Ge^{II} amide and 1,4-benzoquinone,[218a,218b] cyclic α,β-unsaturated ketone,[219a,219b] or RC≡CH.[220a] The corresponding $[\{MN(SiMe_3)_2\}_2\{(NSiMe_3)_2C_6H_4\text{-}1,4\}]$-containing polymer was prepared from M(Cl)-$[N(SiMe_3)_2]$ and successively $Li_2[(NSiMe_3)_2C_6H_4\text{-}1,4]$ and EtBr (M = Sn) or EtI (M = Ge).[220b] Treatment of $Ge[N(SiMe_3)_2]_2$ with 2-vinylpyridine or 1,4-naphthoquinone afforded compound **70**[221a] or **71**,[221b] respectively.

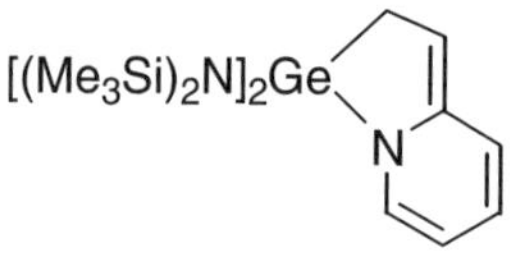

70[221a] 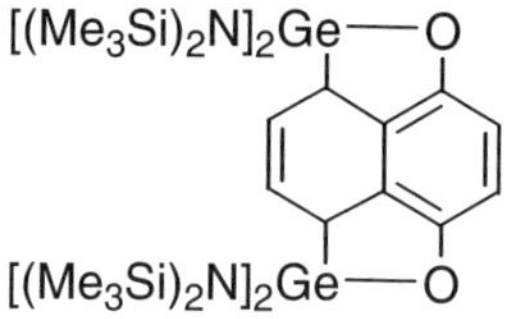 **71**[221b]

Veith's group has studied the oxidative addition or redox reaction of a phosphorus(III) chloride or $M'X_4$ (M' = Ge, Sn; X = Cl, Br, I) with $MN(Bu^t)Si(Me)_2NBu^t$ [M = Ge, Sn (**1'**), Pb]. Thus, **1'** with $M'X_4$ yielded $M'[N(Bu^t)Si(Me)_2NBu^t]X_2$ (X = Cl, Br, I and M' = Si[222a] or Ge[222b]). From $MN(Bu^t)Si(Me)_2NBu^t$, the compounds $P[Ge(Cl)\{N(Bu^t)Si(Me)_2NBu^t\}]_3$, $Sn(Cl)_2[N(Bu^t)Si(Me)_2NBu^t]$ (**72**), or $P(Cl)[N(Bu^t)Si(Me)_2NBu^t]$ were obtained using $1/3PCl_3$ (M = Ge) or PCl_3 (M = Sn, Pb).[223b,223c] Compound **1'** and $RPCl_2$ yielded **72** and $1/n(PR)_n$ (R = Ph and $n = 5$; or R = $C_6H_2Me_3$-2,4,6 and $n = 2$).[223a] Compounds $P(Cl)[Ge(Cl)\{N(Bu^t)Si(Me)_2NBu^t\}](R)$ were isolated from the Ge analogue of **1** and $RPCl_2$.[223b] Further redox reactions are shown in Equations (9.14),[224] (9.15),[225] (9.16),[165] (9.17),[226] and (9.18).[227]

$$3Pb[N(SiMe_3)_2]_2 + 2MoCl_5 \xrightarrow{Et_2O} 2Pb(Cl)[N(SiMe_3)_2]_3 + 2MoCl_3 + PbCl_2 \quad (9.14)$$

$$Sn[N(SiMe_3)_2]_2 + Yb \xrightarrow[2.\ C_6H_{14}/dme]{1.\ thf} [Yb\{N(SiMe_3)_2\}_2(dme)] + Sn \qquad (9.15)$$

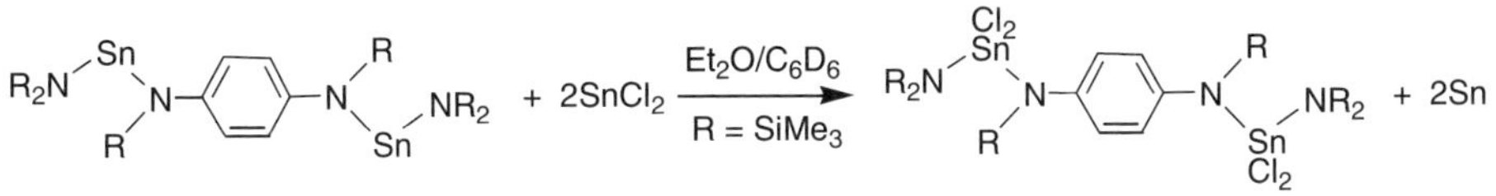

$$R_2N\text{-}Sn\text{-}N(R)\text{-}C_6H_4\text{-}N(R)\text{-}Sn\text{-}NR_2 + 2SnCl_2 \xrightarrow[R = SiMe_3]{Et_2O/C_6D_6} R_2N\text{-}SnCl_2\text{-}N(R)\text{-}C_6H_4\text{-}N(R)\text{-}SnCl_2\text{-}NR_2 + 2Sn \qquad (9.16)$$

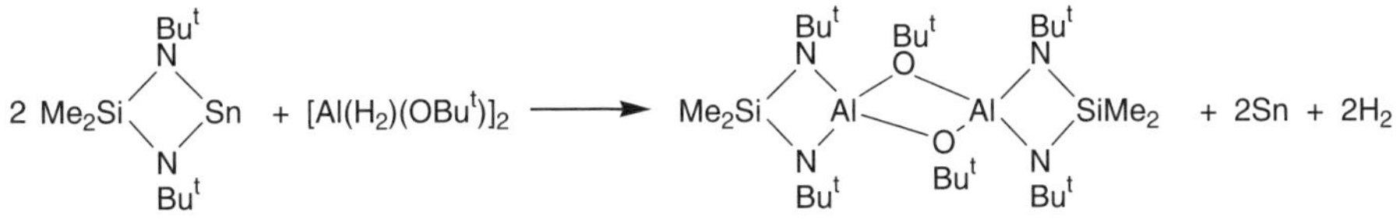

$$2\ Me_2Si(\mu\text{-}NBu^t)_2Sn + [Al(H_2)(OBu^t)]_2 \longrightarrow Me_2Si(\mu\text{-}NBu^t)_2Al(\mu\text{-}OBu^t)_2Al(\mu\text{-}NBu^t)_2SiMe_2 + 2Sn + 2H_2 \qquad (9.17)$$

$$6Sn[N(SiMe_3)_2]_2 + 12NaR \xrightarrow[R = SiBu^t_3]{MeOBu^t} (SnR)_6 + 3R_2 + 12NaN(SiMe_3)_2 \qquad (9.18)$$

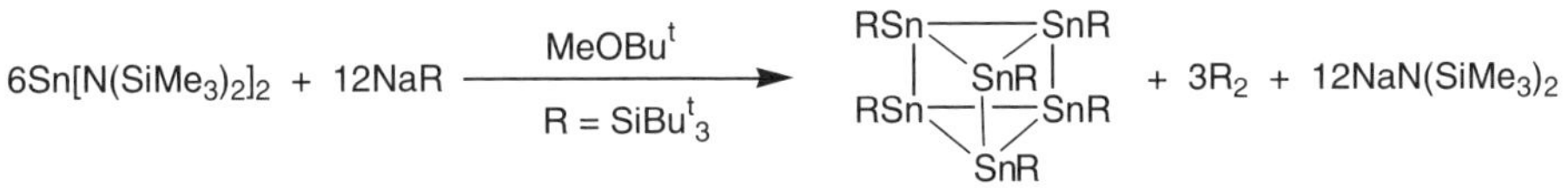

The tin amide $Sn[N(SiMe_3)_2]_2$ with the appropriate Group 2 metal (M) in 1,2-dimethoxyethane (dme) yielded $M[N(SiMe_3)_2]_2(dme)_n$ ($M = Ca$, Sr and $n = 1$, or $M = Ba$ and $n = 2$), whilst a similar reaction in toluene furnished $M[N(SiMe_3)_2]_2$ ($M = Mg$, Ca, Sr, Ba).[228] Nanoscaled tin and lead particles aligned in alumina tubes were obtained from the molecular precursors $M(\mu\text{-}NBu^t)_2SiMe_2$ ($M = Sn$, Pb) or $Pb[N(SiMe_3)_2]_2$ and $[Al(H)_2(OBu^t)]_2$.[229a,229b]

The cyclopropenylidenemetal compounds $\overline{C(NPr^i_2)=C(NPr^i_2)C}=M[N(SiMe_3)_2]_2$ were obtained from $[\{C(NPr^i_2)\}_2C(Cl)](OTf)$ and successively $LiBu^n$ and $M[N(SiMe_3)_2]_2$ ($M = Ge$, Sn, Pb).[230] Reactions between the bis(amino)silylene **4** and the isoleptic Ge^{II}

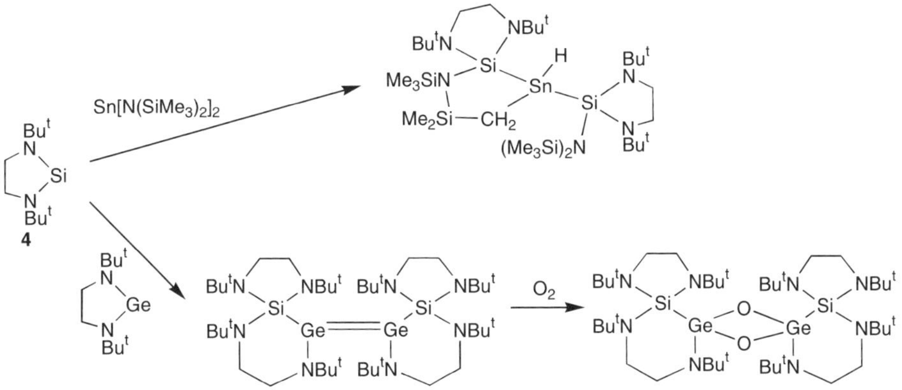

Scheme 9.30[231]

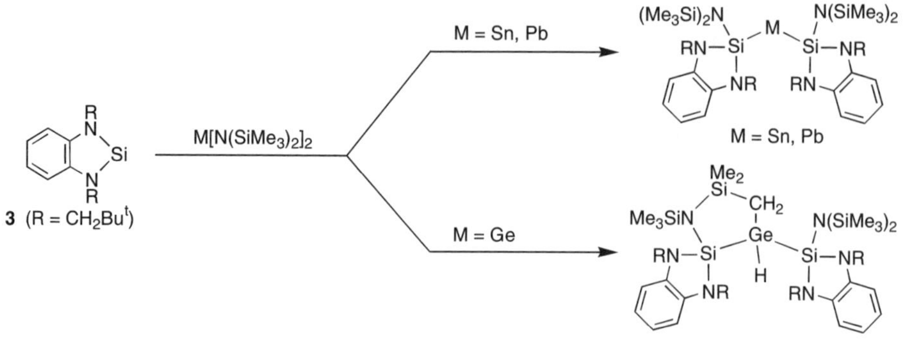

3 (R = CH_2Bu^t)
$M[N(SiMe_3)_2]_2$
M = Sn, Pb
$(Me_3Si)_2N$ M $N(SiMe_3)_2$
RN—Si Si—NR
NR RN
M = Sn, Pb
M = Ge
Me_2 Si CH_2
Me_3SiN Ge $N(SiMe_3)_2$
RN—Si H Si—NR
NR RN

Scheme 9.31 [232]

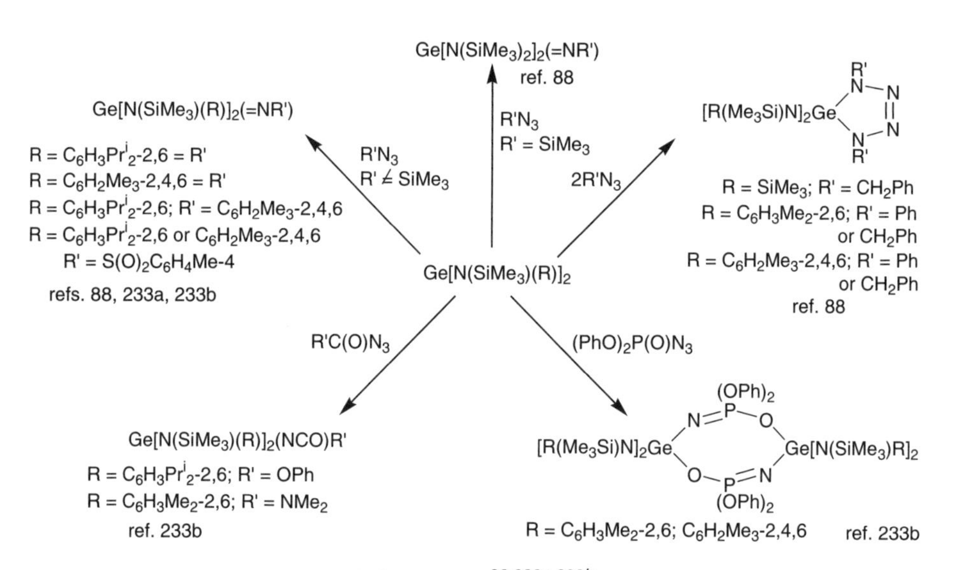

Scheme 9.32 [88,233a,233b]

compound or the Sn^{II} amide $Sn[N(SiMe_3)_2]_2$ are illustrated (cf. Scheme 9.8) in Scheme 9.30,[231] and those between **3** and $M[N(SiMe_3)_2]_2$ (M = Ge, Sn, Pb) in Scheme 9.31.[232]

The first (1989–1992) oxidative additions of an azide to a Group 14 metal(II) amide were reported by Meller, *et al.*, Equation (9.19) (R = Me, OBu^t)[88] and Scheme 9.32.[88,233a,233b] The formation of **68** was cited in Section 9.3.9.[105] Other examples are illustrated in Schemes 9.33[234] and 9.34;[235] compounds **73**[175] and **74**[236] were prepared from $Sn\{N(SiMe_3)_2\}[N(Bu^t)\{Si(Me)N(Bu^t)CH{=}CHNBu^t\}] + 2PhN_3$,[175] and the appropriate Ge^{II} amide + Me_3SiN_3,[236] respectively.

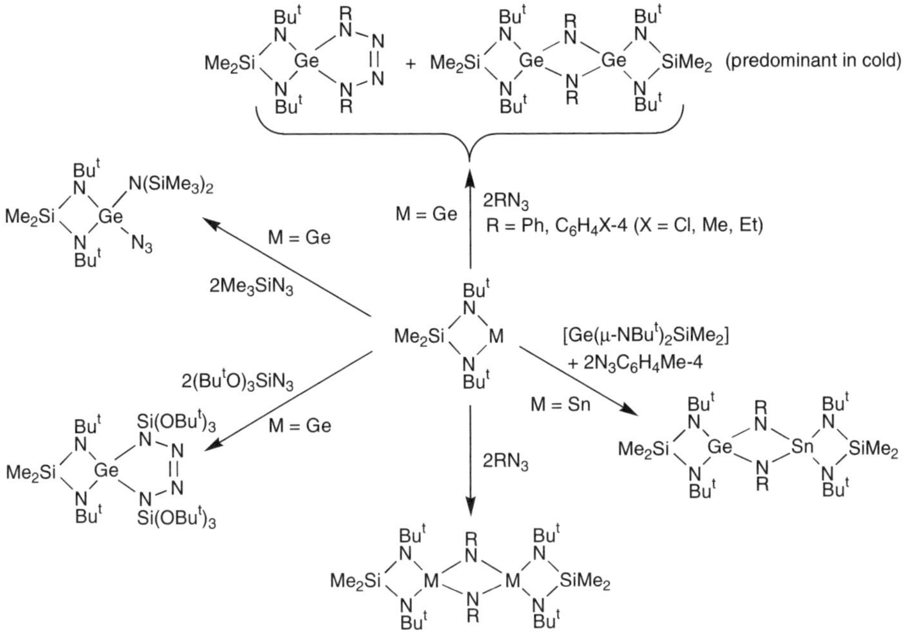

Bu^t R Bu^t R Bu^t
N N–N N N N
Me_2Si Ge ‖ + Me_2Si Ge Ge $SiMe_2$ (predominant in cold)
N N–N N N N
Bu^t R Bu^t R Bu^t

Bu^t
N $N(SiMe_3)_2$
Me_2Si Ge
N N_3
Bu^t

M = Ge

$2RN_3$

R = Ph, C_6H_4X-4 (X = Cl, Me, Et)

M = Ge

$2Me_3SiN_3$

Bu^t
N
Me_2Si M
N
Bu^t

$[Ge(\mu\text{-}NBu^t)_2SiMe_2]$

+ $2N_3C_6H_4Me$-4

$2(Bu^tO)_3SiN_3$

M = Sn

Bu^t $Si(OBu^t)_3$
N N–N
Me_2Si Ge ‖
N N–N
Bu^t $Si(OBu^t)_3$

M = Ge

Bu^t R Bu^t
N N N
Me_2Si Ge Sn $SiMe_2$
N N N
Bu^t R Bu^t

$2RN_3$

Bu^t R Bu^t
N N N
Me_2Si M M $SiMe_2$
N N N
Bu^t R Bu^t

M = Sn; R = Ph, C_6H_4X-4, $C(O)C_6H_4Me$-4, $S(O)_2C_6H_4Me$-4

M = Ge; R = $C(O)C_6H_4X$-4 (X = Me, OMe), $S(O)_2C_6H_4Me$-4

Scheme 9.33 [234]

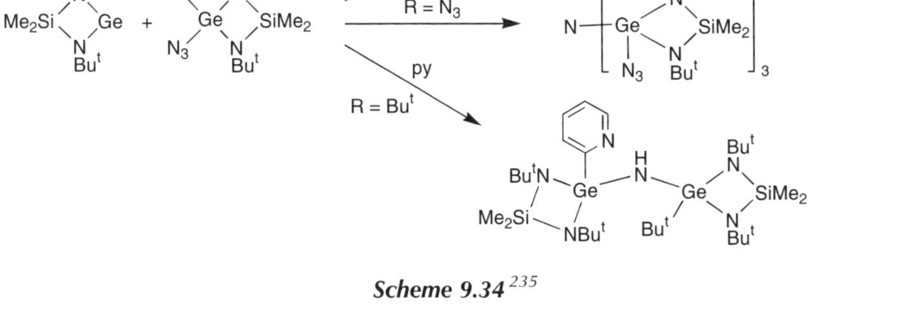

Me_2
Si
Bu^tN NBu^t
Bu^t Me Bu^t
N Ge N
Me_2Si Ge—N N—Ge $SiMe_2$
N N=N Me N
Bu^t Bu^t

R = Me

Bu^t Bu^t
N R N
Me_2Si Ge + Ge $SiMe_2$
N N_3 N
Bu^t Bu^t

R = N_3

Bu^t
N
N—[Ge $SiMe_2$]$_3$
N
N_3 Bu^t

py

R = Bu^t

N
H Bu^t
Bu^tN—Ge—N—Ge N
Me_2Si $SiMe_2$
NBu^t Bu^t N
Bu^t

Scheme 9.34 [235]

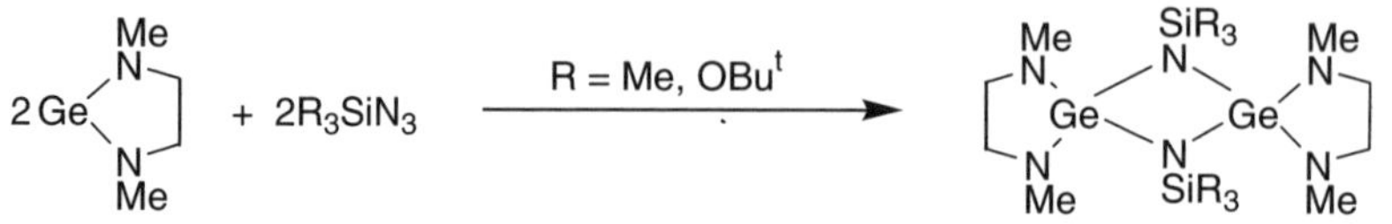

(9.19)

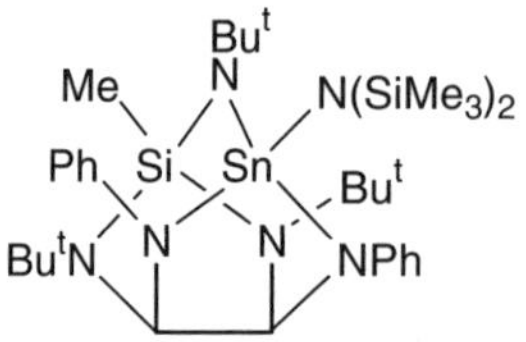

73[175]

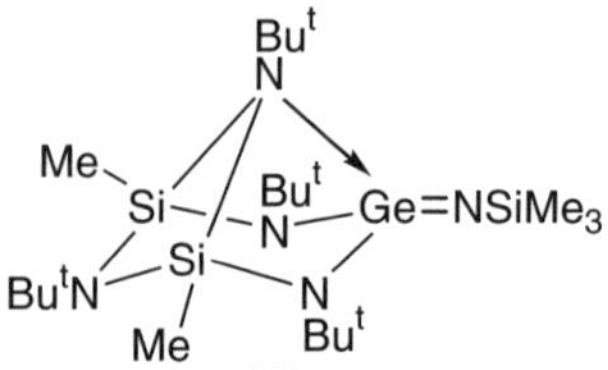

74[236]

9.3.10 Reactions with Transition Metal Complexes

The first example (1974; from Sussex) of the reaction of a Group 14 metal(II) amide with a transition metal complex was that between $Sn(NR_2)_2$ ($R = SiMe_3$) and $[FeCp(CO)_2Me]$ furnishing $[FeCp(CO)_2\{Sn(NR_2)_2Me\}]$;[5a] for an early (1986) review, see Ref. 71. The types of reaction of a transition metal complex and such an amide may be classified as shown in Table 9.6 (see a 1990 review)[73] and itemised in Table 9.7 and Schemes 9.35,[73,249] 9.36 and 9.37.[73,250]

In a majority of its reactions with a transition metal complex $L_nM'{-}X$ the amidometal(II) (M^{II}) amide behaves as an *M*-centred nucleophile. Thus, it may simply form a donor (M)-acceptor (M′) adduct, or it may displace a leaving group X from $L_nM'{-}X$. In the latter case, X may be anionic (e.g. Cl^-) or neutral (e.g. CO or PR'_3); thus in these instances the M^{II} amide behaves towards $L_nM'{-}X$ like a tertiary phosphine or carbon monoxide. Such reactions correspond to Type 1 or 2 behaviour (see Table 9.6). However, the M′–M-containing adduct may only be of transient existence (or be a transition state) and may undergo a 1,2-shift of Cl^- from M′ to M to generate a complex having an M′–M(X) core; such reactions are designated as being of Type 3 and are similar to a carbene-like (or CO-like) insertion into an M′–X bond. The metal(II) amide may function (cf. Type 7) not only as an *M*-centred ligand but also as a reducing agent, exemplified in Equation (9.20),[247a] or in

Table 9.6 *The role of M^{II} amides in transition metal (M′) chemistry*[73]

	Reaction product [from $M(NRR')_2$]	Role
Type 1	$L_nM'{-}M(NRR')_2$	$M(NRR')_2$ as ligand
Type 2	$L_nM'\{\mu{-}M(NRR')_2\}M'L_n$	Bridging $M(NRR')_2$ ligand
Type 3	$L_nM'{-}\{M(NRR')_2X\}$	$[M(NRR')_2X]^-$ as ligand
Type 4	$[M(M'L_n)_2]$ (see Scheme 9.34)	$M(NRR')_2$ as *N*-centred nucleophile
Type 5	$L_nM'{-}N(RR')$	$M(NRR')_2$ as $[NRR']^-$ transfer reagent
Type 6	(see Scheme 9.36)	Oxidative addition of C–H in $M(NRR')_2$ and cyclometallation
Type 7	$[M'L_m]$	$M(NRR')_2$ as dechlorinating (reducing) agent

Table 9.7 *Transition metal (M′) complexes containing M′–M^{II}–N< bonds (M = Ge, Sn or Pb) derived from an M^{II} amide (R = SiMe$_3$)*

Complex	M′ precursor	Type	Ref.
[ScCp$_2$(Me){Sn(NR$_2$)$_2$}]	[ScCp$_2$ (μ-Me)$_2$AlMe$_2$]	1	237
[M′(CO)$_5${Sn(NR$_2$)$_2$}] (M′ = Mo or W; M = Ge or Sn)	[M′(CO)$_6$]	1	237
trans-[M′(CO)$_4${Sn(NR$_2$)$_2$}$_2$] (M′ = Mo or W; M = Ge or Sn)	[M′(CO)$_4$(nbd)]	1	237
[M′(CO)$_5${Sn(μ-NBu^t)$_2$SiMe$_2$}] (M′ = Cr, Mo)	[M′(CO)$_6$]	1	238
cis-[M′(CO)$_4${Sn(μ-NBu^t)$_2$SiMe$_2$}$_2$] (M′ = Mo)	[M′(CO)$_6$]	1	238
cis-[Mo(CO)$_4${Ge(NCH$_2$Bu^t)CH=CHNCH$_2$Bu^t}$_2$]	[Mo(CO)$_4$-(NCEt)$_2$]	1	239
fac-[Mo(CO)$_3${Ge(NCH$_2$Bu^t)CH=CHNCH$_2$Bu^t}$_3$][a]	[Mo(CO)$_4$-(NCEt)$_2$]	1	239
[Mn(CO)$_5${SnBr(NR$_2$)$_2$}]	[MnBr(CO)$_5$]	3	237
[FeCp(CO)$_2${Sn(NR$_2$)$_2$X}] (X = F, I or Me)	[FeCp(X)(CO)$_2$]	3	5a, 237
[FeCp(CO)$_2$(M{(μ-NBu^t)$_2$SiMe$_2$}Me)] (M = Ge,[a] Sn)	[FeCp(Me)-(CO)$_2$]	3	240a
[{FeCp(CO)$_2$}$_2${μ-M(μ-NBu^t)$_2$SiMe$_2$}] (M = Ge,[a] Sn[a])	[{FeCp(CO)$_2$}$_2$]	2	240a
[Fe(CO)$_4${μ-Sn(μ-NBu^t)$_2$SiMe$_2$}]$_2$	[Fe$_2$(CO)$_9$]	2	240b
[Fe(CO)$_4${Ge(NBu^t)$_4$(SiMe)$_2$}][a] (*cf.* **74**)	[Fe$_2$(CO)$_9$]	1	236
[Co$_2$(CO)$_6${Ge(N(SiMe$_2$Pr^i)$_2$)$_2$}$_2$][a]	[Co$_2$(CO)$_8$]	1	79
X; Ph$_2$P; PPh$_2$; (OC)$_3$Fe; Pt(PR′$_3$); Sn[(μ-NBu^t)$_2$SiMe$_2$] — X = CH$_2$, NH and R′ = Ph; X = CH$_2$ and R′ = C$_6$H$_4$Me-4	Scheme 9.37	2	241
Ph$_2$P; PPh$_2$; (OC)$_3$Fe; Pt—Ge[(μ-NBu^t)$_2$SiMe$_2$]; Ge[(μ-NBu^t)$_2$SiMe$_2$]	Scheme 9.37	1, 2	241
Ph$_2$P; PPh$_2$; (OC)$_3$Fe; M′Me; (MeO)$_2$Si; O; Me; M[(μ-NBu^t)$_2$SiMe$_2$] — M′ = Pd, Pt, M = Sn; M′ = Pd, M = Ge[a]	Scheme 9.37	1	241
[Me$_2$Si(μ-NBu^t)$_2$Sn → M′(μ-Cl)$_2$M′ ← Sn(μ-NBu^t)$_2$SiMe$_2$] (Bu^t, N, Sn, M′, Cl, Cl, Cl, Cl, M′, Sn, N, Bu^t, SiMe$_2$) — M′ = Cr, Fe, Co[a]	M′Cl$_2$	1′	243

(continued)

Table 9.7 *(Continued)*

Complex	M′ precursor	Type	Ref.
$[Rh(\eta\text{-}PhMe)(\eta\text{-}C_2H_4)\{SnCl(NR_2)_2\}]$[a]	$[\{Rh(C_2H_4)\text{-}(\mu\text{-}Cl)\}_2]$	3	73
$[Rh(\eta\text{-}ArH)(\eta\text{-}alkene)\{SnCl(NR_2)_2\}]$ (alkene = trans-hex-2-ene, ArH = PhMe; alkene = C_8H_{14},[a] ArH = C_6H_6, PhMe[a] or $C_6H_3Me_3$-2,4,6)	$[\{Rh(alkene)\text{-}(\mu\text{-}Cl)\}_2]$	3	248
$[(Rh(\mu\text{-}Cl)\{Sn(NR_2)_2\}_2)_2]$	$[\{Rh(alkene)\text{-}(\mu\text{-}Cl)\}_2]$	1	248
$[Rh(cod)L\{SnMe(NR_2)_2\}]$ L = $\overline{CN(Me)(CH_2)_2N}Me$	[RhMe(cod)L]	3	73
$[Rh(cod)L\{SnCl(NR_2)_2\}]$[a] L = $\overline{CN(Me)(CH_2)_2N}Me$	[RhCl(cod)L]	3	73
$[Rh(cod)(PEt_3)\{SnCl(NR_2)_2\}]$[a]	$[RhCl(cod)\text{-}(PEt_3)]$	3	73
$[Rh(cod)(\{Sn(NR_2)_2\}_2(\mu\text{-}Cl))]$	$[\{Rh(\mu\text{-}Cl)\text{-}(cod)\}_2]$	1	73
$[(Rh(cod)\{\mu\text{-}Sn(NR_2)_2\}\{SnCl(NR_2)_2\})_2]$	$[\{Rh(\mu\text{-}Cl)\text{-}(cod)\}_2]$	2, 3	73
cis-$[RhCl\{Ge(NR_2)_2\}(PPh_3)_2]$[a]	$[RhCl(PPh_3)_3]$	1	73
$[Rh(CO)(dppe)\{SnCl(NR_2)_2\}]$	$[RhCl(CO)\text{-}(dppe)]$	3	73
cis-$[RhCl(PPh_3)_2\{Sn(NR_2)_2\}]$[a]	$[RhCl(PPh_3)_3]$	1	73
trans-$[Rh(CO)(PPh_3)_2\{SnCl(NR_2)_2\}]$	$[RhCl(CO)\text{-}(PPh_3)_2]$	3	73
cis-$[Rh(PPh_3)_2(\{Sn(\mu\text{-}NBu^t)_2SiMe_2\}_2(\mu\text{-}Cl))]$[a]	$[RhCl(PPh_3)_3]$	1	244
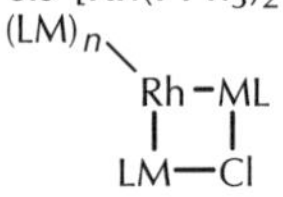 $(LM)_n$–Rh–ML, LM–Cl (four-membered ring Rh–ML–Cl–LM) ML = $M\{(\mu\text{-}NBu^t)_2SiMe_2\}$; M = Ge (n = 2),[a] 243 Sn (n = 3)[a] 244	$[RhCl(PPh_3)_3]$	1, 1″	243, 244
$[Ir\{\overline{SnCl(NR_2)N(R)SiMe_2CH_2}\}(cod)H\{Sn(NR_2)_2\}]$	Scheme 9.35	1, 3, 6	73
$[\{\overline{CH_2Me_2SiN(R)(NR_2)Ge}\}HIr(\mu\text{-}Cl)_2\text{-}\{\overline{Ge(NR_2)N(R)SiMe_2CH_2}\}IrH\{Ge(NR_2)_2\}]$[a]	Scheme 9.35	1, 2, 6	249
$[Ir\{\overline{GeCl(NR_2)N(R)SiMe_2CH_2}\}(CO_2)H\{Ge(NR_2)_2\}]$[a]	Scheme 9.35	1, 3, 6	249
$[Ni(CO)_{4-n}\{\overline{GeN(Bu^t)(CH_2)_2N}Bu^t\}_n]$ (n = 1, 2,[a] 3)	$[Ni(CO)_4]$	1	82
$[Ni\{Ge(NR_2)_2\}(PR_3)_2]$ R = Ph,[a] Et	$[Ni(cod)_2]/PR_3$	1	209a, 245c
$[Ni\{Ge(NPr^i)_2C_{10}H_6\text{-}1,8\}_4]$[a]	$[Ni(cod)_2]$	1	89
$[Ni\{Sn(\mu\text{-}NBu^t)_2SiMe_2\}_4]$[a]	$[Ni(cod)_2]$	1	242a
$[NiCp\{(Sn(\mu\text{-}NBu^t)_2SiMe_2)_2(\mu\text{-}Cp)\}]$[a]	$[NiCp_2]$	1	242b
$[NiCp\{Ge(\mu\text{-}NBu^t)_2SiMe_2\}\{(Ge(\mu\text{-}NBu^t)_2SiMe_2)Cp\}]$[a]	$[NiCp_2]$	1, 3	242b
$[M'\{(Sn(\mu\text{-}NBu^t)_2SiMe_2)_2(\mu\text{-}X)\}_2]$ (M′ = Ni, X = Br[a]; M′ = Pd, Pt; X = Cl[a])	$M'X_2$	1	243
cis-$[Pd(\eta\text{-}C_3H_5)Cl\{M(NR_2)_2\}]$ (M = Sn, Pb)	$[\{Pd(C_3H_5)\text{-}(\mu\text{-}Cl)\}_2]$	1	237
trans-$[Pd(CNBu^t)_2\{SnCl(NR_2)_2\}_2]$[a]	$[PdCl_2(CN\text{-}Bu^t)_2]$	3	73

Table 9.7 *(Continued)*

Complex	M′ precursor	Type	Ref.
$[\{Pd(PPh_3)\}_2\{\mu\text{-}Sn(\mu\text{-}NBu^t)_2SiMe_2\}_3]$[a]	$[Pd(PPh_3)_4]$	1	243
$[Pd\{Ge(NR_2)_2\}(PR_3)_2]$ R = Ph,[a] Et[a]	$[Pd(PPh_3)_4]$, $[Pd(ox)(PEt_3)_2]$	1	245d
$[(Pd\{Ge(NR_2)_2\}(\mu\text{-}dppe))_2]$[a]	$[Pd(ox)(dppe)]$	1	245d
$[Pt\{Ge(NR_2)_2\}(PEt_3)_2]$[a]	$[Pt(ox)(PEt_3)_2]$	1	245a
$[Pt\{M(NR_2)_2\}(PPh_3)_2]$ (M = Ge, Sn)	$[Pt(PPh_3)_3]$	1	73
cis-$[Pt(cod)\{SnCl(NR_2)_2\}_2]$	$[PtCl_2(cod)]$	3	237
$[(Pt(\mu\text{-}Cl)\{MCl(NR_2)_2\}(PEt_3))_2]$ (M = Ge, Sn[a] or Pb; *cis* or *trans*[a])	$[\{PtCl(\mu\text{-}Cl)\text{-}(PEt_3)\}_2]$	3	237, 246
$[M'\{M(NR_2)_2\}_3]$ (M′ = Pd, M = Ge, Sn; M′ = Pt, M = Ge, Sn[a])	$[Pt(cod)_2]$, $[PdCl_2(cod)]$	1, 7	246, 247a
$[(M'(CO)\{\mu\text{-}M(NR_2)_2\})_3]$ (M′ = Pd, M = Ge, Sn[a]; M′ = Pt, M = Ge, Sn[a])	$[Pt(cod)_2]$/CO, $[PdCl_2(cod)]$/CO	2, 7	247a, 247b
$[Cu(\{N(R')C(Me)\}_2CH)\{Ge(NR_2)_2\}]$[a] R′ = $C_6H_3Pr^i_2$-2,6	$[Cu(\{N(R')C\text{-}(Me)\}_2CH)]$	1	85
$[Cu(\{N(R')C(Me)\}_2CH)\{\overline{Ge(NR'')CH{=}CHNR''}\}]$[a] R′ = $C_6H_3Pr^i_2$-2,6; R″ = $C_6H_2Me_3$-2,4,6	$[Cu(\{N(R')C\text{-}(Me)\}_2CH)]$	1	85

[a]Molecular structure by X-ray diffraction

the formation of $PbCl\{N(SiMe_3)_2\}_3$ from the Pb^{II} amide and $MoCl_5$ [Equation (9.14), Section 9.3.9].[224] Rarer pathways, still based on *M*-centred nucleophilicity of the metal(II) amide, involve its additional role (Type 5) as an amide transfer agent, as in Equation (9.21),[73] or (Type 6) as a substrate for oxidative addition and/or cyclometallation (Scheme 9.35).[73,249] Finally, the metal(II) amide may use its *N*-centred nucleophilicity (Type 4), as illustrated in Scheme 9.36[73,250] and Equation (9.22).[240b]

$$[PdCl_2(cod)] + 4Sn[NR_2]_2 \xrightarrow{R = SiMe_3} [Pd\{Sn(NR_2)_2\}_3] + SnCl_2(NR_2)_2 \qquad (9.20)$$

$$2[Ni(acac)_2] + Sn[NR_2]_2 \xrightarrow{R = SiMe_3} [\{Ni(acac)(\mu\text{-}NR_2)\}_2] + Sn(acac)_2 \qquad (9.21)$$

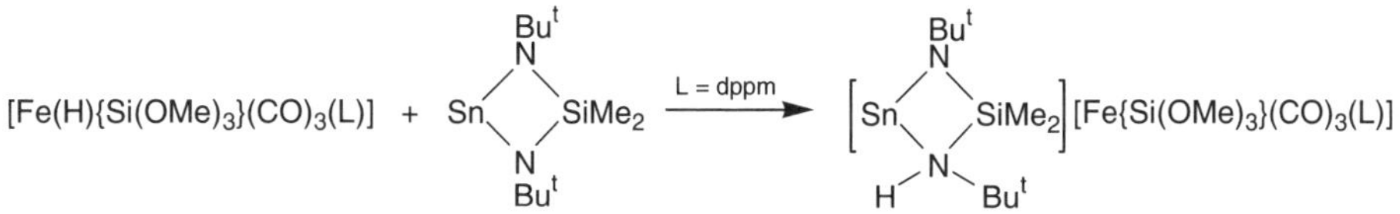

$$[Fe(H)\{Si(OMe)_3\}(CO)_3(L)] + \overline{Sn(\mu\text{-}NBu^t)_2SiMe_2} \xrightarrow{L = dppm} [\overline{Sn(\mu\text{-}NBu^t)(\mu\text{-}N(H)Bu^t)SiMe_2}][Fe\{Si(OMe)_3\}(CO)_3(L)] \qquad (9.22)$$

Some of the complexes listed in Table 9.7 (R = $SiMe_3$) which have a bond or bonds between a transition metal (M′) and a Group 14 metal(II) (M^{II}) amide have been shown to undergo further reactions. The first such reactions (1985, from Sussex) are shown in

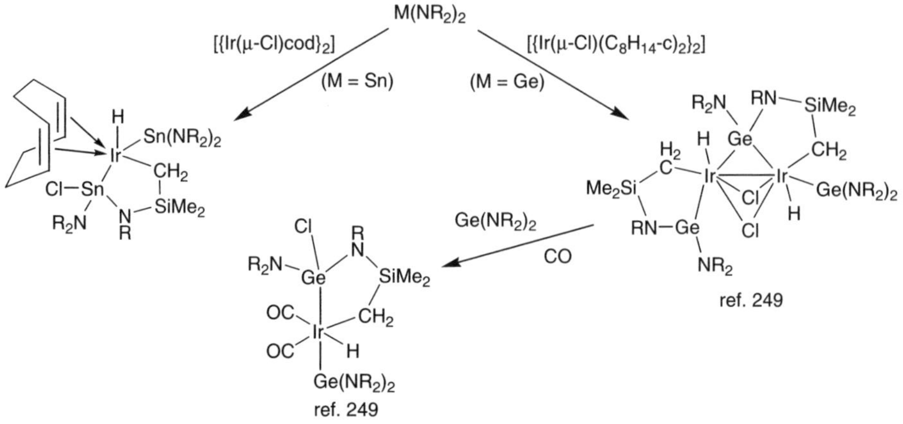

$M(NR_2)_2$

$[\{Ir(\mu\text{-}Cl)cod\}_2]$ (M = Sn) → $[Ir(H)\{Sn(NR_2)_2\}(cod)\{CH_2SiMe_2N(R)Sn(Cl)NR_2\}]$

$[\{Ir(\mu\text{-}Cl)(C_8H_{14}\text{-}c)_2\}_2]$ (M = Ge) → $[\{Ir(H)(\mu\text{-}Cl)\}_2\{\mu\text{-}Ge(NR_2)N(R)SiMe_2CH_2\}_2\{Ge(NR_2)_2\}]$ (ref. 249)

→ $Ge(NR_2)_2$, CO → $[Ir(H)(CO)_2\{Ge(NR_2)_2\}\{Ge(Cl)(NR_2)N(R)SiMe_2CH_2\}]$ (ref. 249)

Scheme 9.35 *(R = SiMe₃, cod = cycloocta-1,5-diene, C₈H₁₄-c = cyclooctene)*[73,249]

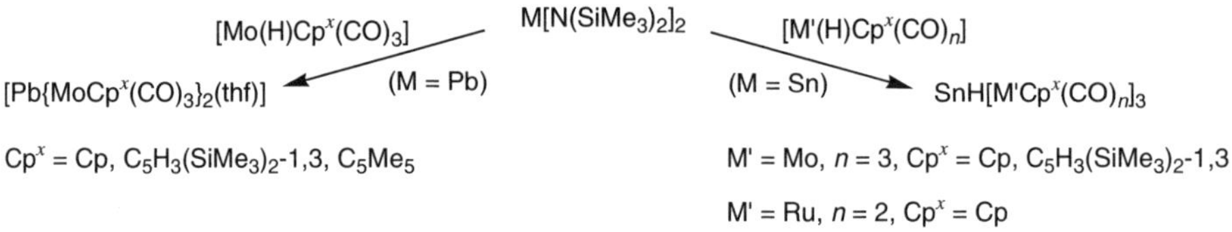

$M[N(SiMe_3)_2]_2$

$[Mo(H)Cp^x(CO)_3]$ (M = Pb) → $[Pb\{MoCp^x(CO)_3\}_2(thf)]$

Cp^x = Cp, $C_5H_3(SiMe_3)_2$-1,3, C_5Me_5

$[M'(H)Cp^x(CO)_n]$ (M = Sn) → $SnH[M'Cp^x(CO)_n]_3$

M' = Mo, n = 3, Cp^x = Cp, $C_5H_3(SiMe_3)_2$-1,3

M' = Ru, n = 2, Cp^x = Cp

Scheme 9.36 [73,250]

Equation (9.23) (cf. Table 9.7).[247a,247b] The crystalline products have a planar $[MM'(CO)]_3$ core; and in thf they underwent reversible electrochemical 1-electron reduction which for the Pt/Sn product was shown by EPR spectroscopy (subsequent to controlled potential electrolysis) to retain its $(PtSn)_3$ framework.[247b]

$$3[M'\{M(NR_2)_2\}_3] + 3CO \longrightarrow [\{M'(CO)\}_3\{\mu\text{-}M(NR_2)_2\}_3] \qquad (9.23)$$

From crystalline *cis*-$[Pt(H)\{Ge(NR_2)_2H\}(PEt_3)_2]$ (Table 9.8) (which in solution was in equilibrium with the *trans* isomer)[245b] and $Ge(CHR_2)_2$, the products were $[Pt\{Ge(CHR_2)_2\}(PEt_3)_2]$ and $Ge(H)_2(NR_2)_2$.[209a] The compound $[Pd\{C(H)_2OGe(NR_2)_2S\}]$ (Table 9.8) with CO yielded the redox products $Ge(NR_2)_2SC(O)C(H)_2O$ and Pd^0 adducts.[249] Some metathetical exchange and addition reactions are summarised in Scheme 9.38;[92] and Scheme 9.39 relates to some copper chemistry.[85]

The most detailed studies of reactions of M'–M^{II}(amide) compounds were carried out by Banaszak Holl and coworkers (1995–2004), as summarised in Table 9.8. These dealt with

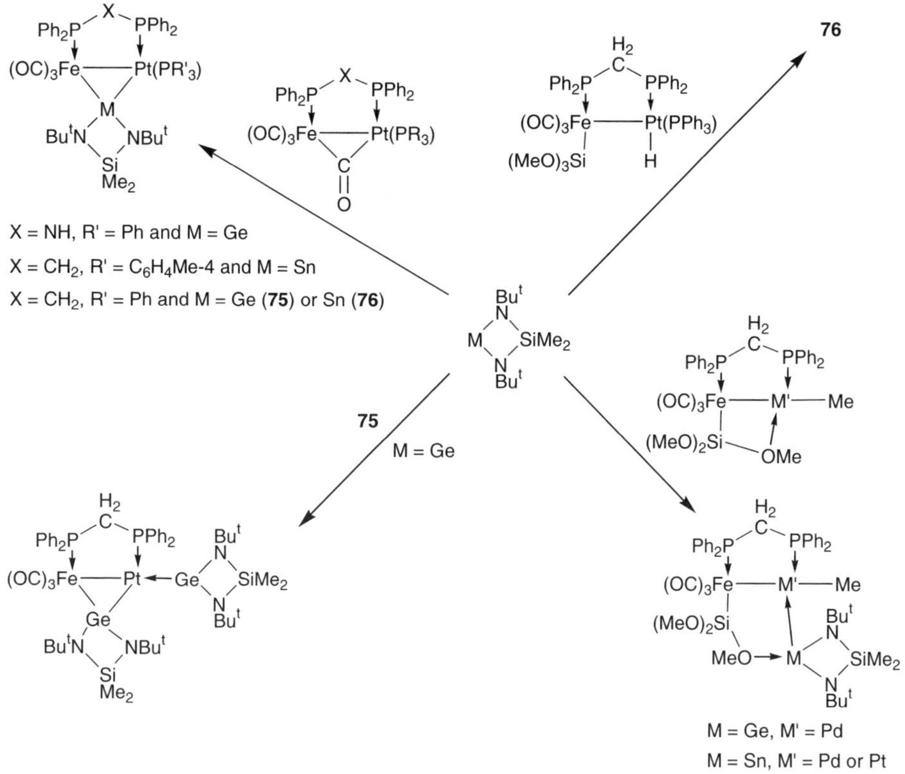

X
Ph_2P PPh_2
$(OC)_3Fe$ $Pt(PR'_3)$
M
Bu^tN NBu^t
Si
Me_2

X
Ph_2P PPh_2
$(OC)_3Fe$ $Pt(PR_3)$
C
O

H_2
C
Ph_2P PPh_2
$(OC)_3Fe$ $Pt(PPh_3)$
$(MeO)_3Si$ H

76

X = NH, R' = Ph and M = Ge

X = CH_2, R' = C_6H_4Me-4 and M = Sn

X = CH_2, R' = Ph and M = Ge (**75**) or Sn (**76**)

Bu^t
N
M $SiMe_2$
N
Bu^t

H_2
C
Ph_2P PPh_2
$(OC)_3Fe$ M' Me
$(MeO)_2Si$ OMe

75

M = Ge

H_2
C
Ph_2P PPh_2 Bu^t
$(OC)_3Fe$ Pt Ge N $SiMe_2$
Ge N Bu^t
Bu^tN NBu^t
Si
Me_2

H_2
C
Ph_2P PPh_2
$(OC)_3Fe$ M' Me
$(MeO)_2Si$ Bu^t N
MeO M $SiMe_2$
N Bu^t

M = Ge, M' = Pd

M = Sn, M' = Pd or Pt

Scheme 9.37 [241]

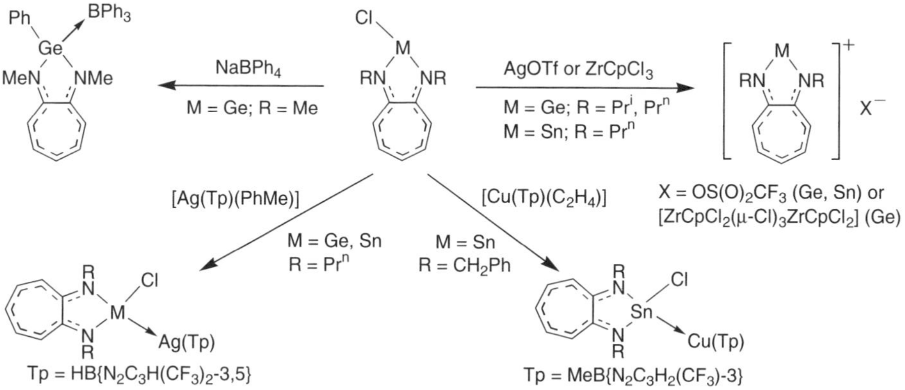

Ph BPh_3
Ge
MeN NMe

$NaBPh_4$

M = Ge; R = Me

Cl
M
RN NR

AgOTf or $ZrCpCl_3$

M = Ge; R = Pr^i, Pr^n

M = Sn; R = Pr^n

M
RN NR
X^-

X = $OS(O)_2CF_3$ (Ge, Sn) or $[ZrCpCl_2(\mu\text{-}Cl)_3ZrCpCl_2]$ (Ge)

[Ag(Tp)(PhMe)]

M = Ge, Sn

R = Pr^n

$[Cu(Tp)(C_2H_4)]$

M = Sn

R = CH_2Ph

R
N Cl
M
N Ag(Tp)
R

Tp = $HB\{N_2C_3H(CF_3)_2\text{-}3,5\}$

R
N Cl
Sn
N Cu(Tp)
R

Tp = $MeB\{N_2C_3H_2(CF_3)\text{-}3\}$

Scheme 9.38 [92]

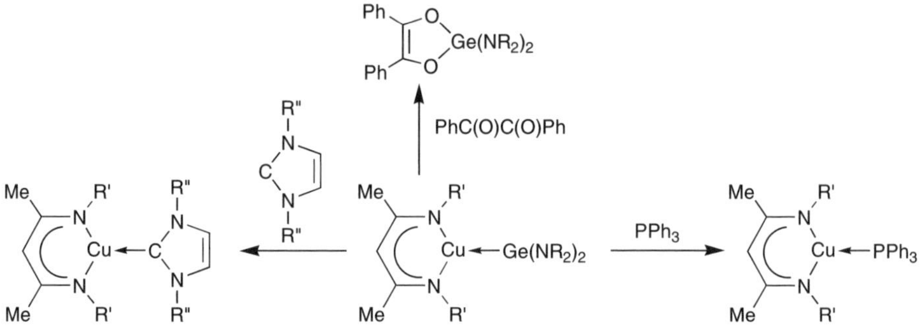

Scheme 9.39 *(L = R′ = $C_6H_3Pr^i_2$-2,6; R″ = $C_6H_2Me_3$-2,4,6)* [85]

cycloaddition reactions of the heavier Group 10 metal(0) complexes $[M'\{Ge(NR_2)_2\}L_2]$ [M′ = Pd or Pt, L = PEt_3 or PPh_3 or $L_2 = (Ph_2PCH_2)_2$ (≡ dppe)] leading to $M'^{II}/Ge^{IV}(NR_2)_2$-containing products [the majority of which had M′–Ge bond(s)].

9.4 Dimeric Metal(III) Imides: Biradicaloid Compounds

There is much current interest in four-membered ring compounds containing heteroatoms that are isoelectronic with a cyclobuta-1,3-dienediide. The first thermally robust parent homocyclic dianion $[\{C(SiMe_3)\}_4]^{2-}$ was only reported in 2000;[256a] the first neutral heterocyclic analogue $[P(C_6H_2Bu^t_3\text{-}2,4,6)\{\mu\text{-}C(Cl)\}]_2$, containing a Group 14 element (C) dates from 1995.[256b] Extensions to the 4th and 5th quantum group elements Ge[257] and Sn,[258] respectively, were published in 2004.

The crystalline, dark-violet compound **77** was obtained by Power and coworkers as shown in Equation (9.24),[257] derived from the remarkable germanium alkyne analogue RGeGeR [R = $C_6H_3(C_6H_3Pr^i_2\text{-}2',6')_2$-2,6]. The $(GeNSi)_2$ core of **77** is planar and the aryl ligands are arranged in a transoid fashion. Its diamagnetism is evident both from its

$$RGeGeR + 2Me_3SiN_3 \xrightarrow{n\text{-}C_6H_{14}} [RGe(\mu\text{-}NSiMe_3)]_2\ (\mathbf{77}) + 2N_2 \quad (9.24)$$

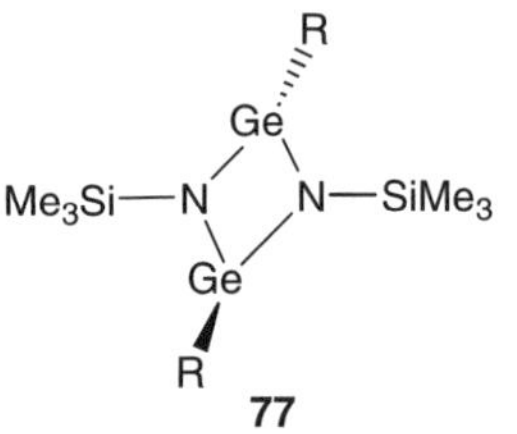

normal ^1H and ^{13}C NMR spectroscopic signals in toluene-d_8 and its failure to show an EPR spectral signal in the range 77–300 K. DFT calculations on the model compound $[GeMe(\mu\text{-}NSiH_3)]_2$ showed similar geometric features to those observed for crystalline **77**. The UV-visible spectrum in n-C_6H_{14} revealed an absorption maximum corresponding to the large HOMO-LUMO gap of 54.9 kcal mol^{-1}, close to the value calculated for the model analogue. The bonding was described as 'non-Kekulé, singlet biradicaloid'.

Table 9.8 *Pd^{II} and Pt^{II} complexes derived from [M{Ge(NR$_2$)$_2$}L$_2$] (M = Pd, Pt; R = SiMe$_3$; L = PEt$_3$, PPh$_3$ or L$_2$ = dppe)*

M^{II} complex	Precursor(s)	Ref.
cis-[L$_2$M(O–O)Ge(NR$_2$)$_2$]	M = Pd and L = PEt$_3$, PPh$_3$ or $^1/_2$dppe M = Pt and L = PEt$_3$[a]	245c 245b
cis-[L$_2$M(μ-O)$_2$Ge(NR$_2$)$_2$]	[L$_2$M(O–O)Ge(NR$_2$)$_2$], hν M = Pd and L = PEt$_3$, PPh$_3$ or $^1/_2$dppe M = Pt and L = PEt$_3$[a]	245c 245b
cis-[(Et$_3$P)$_2$Pt(OCH$_2$OO)Ge(NR$_2$)$_2$]	[(Et$_3$P)$_2$Pt(O–O)Ge(NR$_2$)$_2$] + H$_2$CO	254
cis-[(Et$_3$P)$_2$Pt(OS(O$_2$)O)Ge(NR$_2$)$_2$][a]	[(Et$_3$P)$_2$Pt(O–O)Ge(NR$_2$)$_2$] + SO$_2$	245b
cis-[(Et$_3$P)$_2$Pd(μ-S)Ge(NR$_2$)$_2$]	L = PEt$_3$[a] L$_2$ = dppe[a]	254
cis-[(Ph$_3$P)$_2$Pd(μ-S)$_2$Ge(NR$_2$)$_2$][a]	2COS	254
[(R$_2$N)$_2$Ge(μ-S)$_2$Pd(Et$_3$P)...(PEt$_3$)Pd(μ-S)$_2$Ge(NR$_2$)$_2$][a] (dimer, Pd←S)	COS	254
cis-[(Et$_3$P)$_2$Pd(S)(OS–O)Ge(NR$_2$)$_2$][a]	[(Et$_3$P)$_2$Pd(μ-S)Ge(NR$_2$)$_2$] + SO$_2$	255
[(Et$_3$P)Pd{S–Ge(NR$_2$)$_2$–O}$_2$S][a]	[(Et$_3$P)$_2$Pd(S)(OS–O)Ge(NR$_2$)$_2$], C$_6$H$_6$, reflux or 2PPh$_3$, 70 °C [− Pd(PPh$_3$)$_n$]	255

(continued)

Table 9.8 *(Continued)*

M^{II} complex	Precursor(s)	Ref.
cis-$[(Et_3P)_2Pd\{SGe(NR_2)_2OCH_2\}]$ 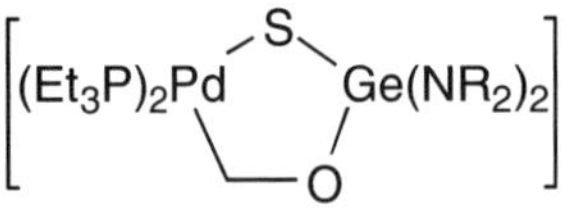	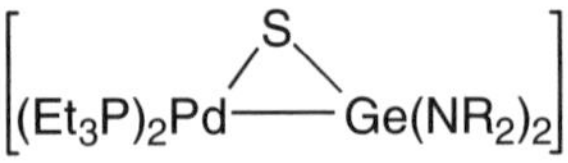 $[(Et_3P)_2Pd\{SGe(NR_2)_2\}] + (CH_2O)_3$	253
cis-$[(dppe)Pd\{SGe(NR_2)_2OCH_2\}]$[a] 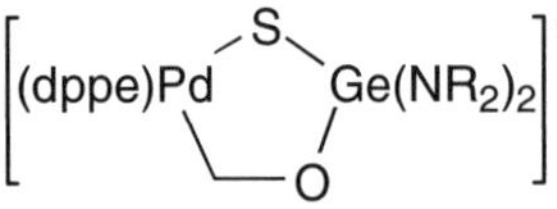	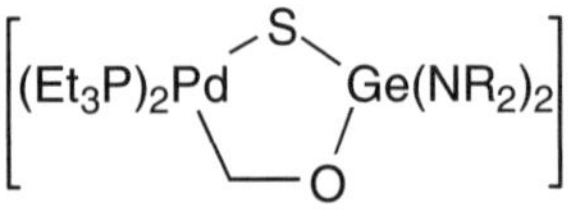 $[(Et_3P)_2Pd\{SGe(NR_2)_2OCH_2\}]$ + dppe	253
$[Pt(H)_2\{Ge(NR_2)_2\}(PEt_3)_2]$	H_2	245a
cis-$[Pt(H)\{Ge(NR_2)_2H\}(PEt_3)_2]$[a]	$[Pt(H)_2\{Ge(NR_2)_2\}(PEt_3)_2]$	245a
cis-$[(Et_3P)_2Pt\{C(O)OGe(NR_2)_2\}]$[a] 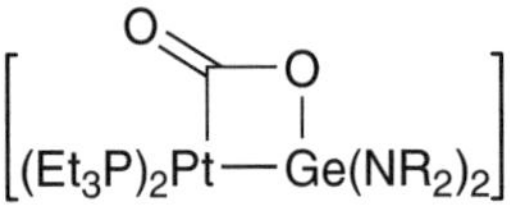	CO_2	245a
cis-$[(Et_3P)_2Pt\{CH_2OGe(NR_2)_2\}]$[a] 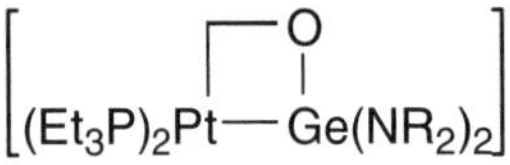	H_2CO	245a, 252
cis-$[(Et_3P)_2Pt\{N(2\text{-}XC_6H_4)OGe(NR_2)_2\}]$ X = H,[a] Me 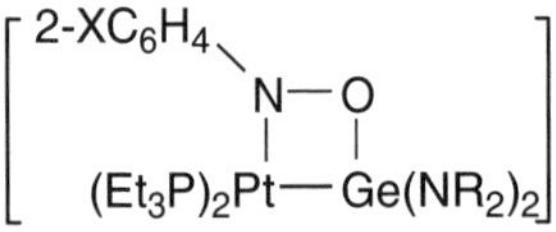	$2\text{-}XC_6H_4NO$	251
cis-$[(Et_3P)_2Pt\{N(Ph)C(O)N(Ph)OGe(NR_2)_2\}]$[a] 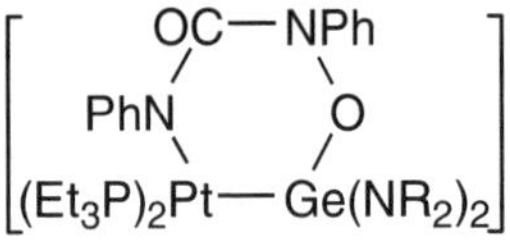	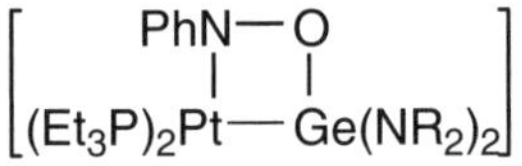 $[(Et_3P)_2Pt\{N(Ph)OGe(NR_2)_2\}]$ + PhNCO	251
cis-$[(Et_3P)_2Pt\{S(O_2)N(C_6H_4X\text{-}2)OGe(NR_2)_2\}]$ X = H,[a] Me 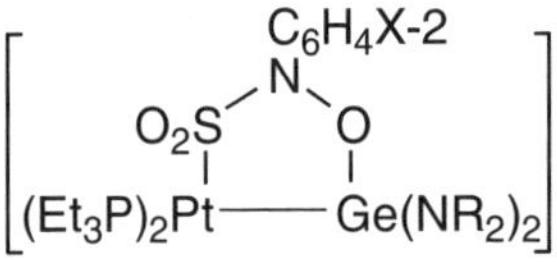	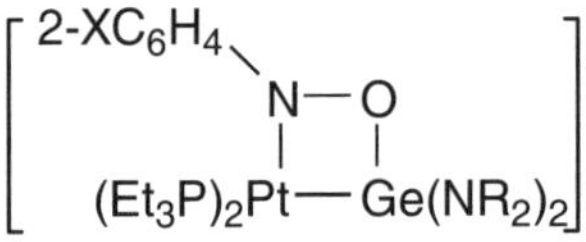 $[(Et_3P)_2Pt\{N(2\text{-}XC_6H_4)OGe(NR_2)_2\}] + SO_2$	251
cis-$[(Et_3P)_2Pt\{OCH_2N(C_6H_4X\text{-}2)OGe(NR_2)_2\}]$ X = H,[a] Me 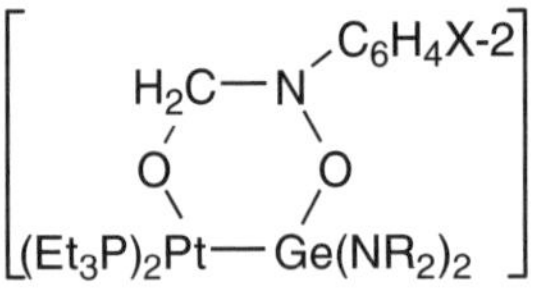	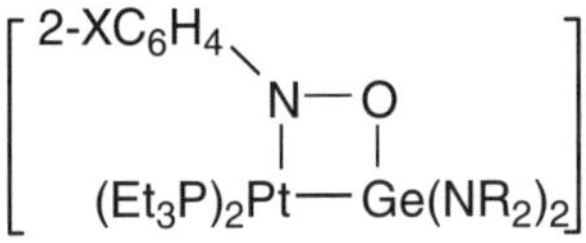 $[(Et_3P)_2Pt\{N(2\text{-}XC_6H_4)OGe(NR_2)_2\}] + H_2CO$	251

[a]Molecular structure by X-ray diffraction

The colourless, crystalline 1,3-diaza-2,4-distannacyclobuta-1,3-dienediide **78** was prepared by the Sussex group, as illustrated in Equation (9.25).[258] The synthesis is remarkable in the light of prior observations on Sn(X)X′/AgY systems, which led to (i) metathetical ligand exchange as in $[Sn\{N(SiMe_3)_2\}(\mu\text{-Cl})]_2 + AgOTf \rightarrow$ **40** (Section 9.3.3),[126] or $Sn[N(SiMe_3)_2]_2 + AgOCN$ (Section 9.3.5);[187] (ii) $Sn^{II} \rightarrow Sn^{IV}$ oxidative addition, as in $Sn[CH(SiMe_3)_2]_2 + 2AgOCN \rightarrow Sn[CH(SiMe_3)_2]_2(NCO)_2$;[259] or (iii) 1 : 1-adduct formation, as with $Sn[CH(SiMe_3)_2]_2 + AgNCS$.[259] It was suggested that each of these reactions proceeds by a common intermediate (or end-product, iii) the 1 : 1-adduct. For Sn(X)X′ being $[Sn\{N(SiMe_3)_2\}(\mu\text{-Cl})]_2$ and Y = OCN, this is followed by a redox reaction yielding $Sn(Cl)[N(SiMe_3)]$ (which dimerises to yield **78**) and $Ag + Me_3SiOCN$ (which is transformed into the carbodiimide + CO_2).

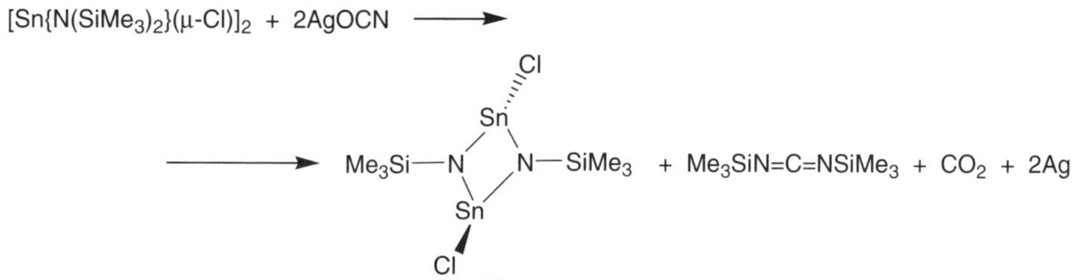

$[Sn\{N(SiMe_3)_2\}(\mu\text{-Cl})]_2$ + 2AgOCN ⟶

⟶ **78** + $Me_3SiN{=}C{=}NSiMe_3$ + CO_2 + 2Ag

(9.25)

Monomeric, diamagnetic **78** has a planar centrosymmetric $(SnN)_2$ ring with the *trans*-Si atoms only 6° out of this plane; the *trans*-Cl ligands are disposed orthogonally. The Sn···Sn′ separation of 3.398 Å shows there is no transannular bonding [likewise, the Ge···Ge′ distance of 2.755 Å precludes such an interaction in **77**[257]].[258] The X-ray structure of **78** revealed long range Sn···Cl″ contacts of 3.29 Å [*cf.*, Sn–Cl, 2.470(2) Å] between neighbouring molecules. The CPMAS NMR ^{13}C, ^{29}Si and ^{119}Sn chemical shifts were found at lower frequency than those recorded in $PhMe/C_6D_6$; and the ^{119}Sn signal was particularly sensitive to solvent attributed to a Sn←donor solvent interaction which at the limit was believed to disrupt the virtual dimer in the crystal. Computational studies (which did not take into account the intermolecular contacts) showed that the singlet state is substantially favoured over the triplet. While the calculated and experimental Sn′–Sn–Cl angle showed considerable disparity, there was good agreement for the remaining geometrical parameters. The singlet state was computed to be favoured over the triplet by $> 57\ \mathrm{kJ\ mol^{-1}}$.

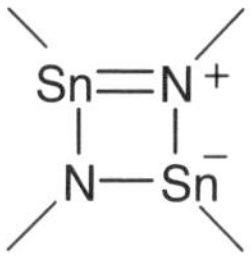

78′

The valence bond structure of **78** shown in **78′** (this is one of four resonance structures) is consistent with the calculated MOs, in which 3 of the 4 π MOs (2 almost non-bonding, and a bonding) are filled and the antibonding MO is empty. This MOs arise from the 4 ring atoms: ++ and +− on both the N (combinations of p-orbitals) and Sn (combinations of sp^3 hybrids) atoms. The three filled MOs are the ++ combinations of the Sn atoms and both the *N*-centred orbitals.[258] This analysis, corresponding to a delocalised π-bonded ring, is similar to that proposed for $[P(C_6H_2Bu^t_3\text{-}2,4,6)\{\mu\text{-}C(Cl)\}]_2$.[256b]

9.5 Higher-Nuclearity Group 14 Metalloid Clusters having Amido Ligands

The term 'metalloid cluster' was coined by Schnöckel and coworkers in the context of Al chemistry to designate those clusters in which the number of metal-metal contacts exceeds the number of metal-ligands contacts and by the presence of metal atoms which participate exclusively in metal-metal interactions. Reviews of such Group 13 and 14 metal clusters have appeared inter alia in Refs. 260a and 260b, respectively. With three exceptions, $[Ge_8\{N(SiMe_3)_2\}_6]$ (**79**)[261] and $[Sn_9\{Sn(NRR')\}_6]$ [$R = C_6H_3Pr^i_2\text{-}2,6$ and $R' = SiMe_3$ (**80**) or $SiMe_2Ph$ (**81**)],[162] previous Group 14 metalloid clusters $[M_x(\text{Ligand})_y]$ ($x > y$) have involved bulky aryl or silyl ligands, as in $[Sn_8(SiBu^t_3)_6]$ and $[Sn_{10}\{C_6H_3(C_6H_2Me_3\text{-}2',4',6')_2\text{-}2,6\}_3]^-$; the latter, apart from **80**, was of the highest nuclearity.

The dark red, crystalline cluster **79** was obtained by Schnepf and Köppe from the reaction of the transiently stable Ge^IBr (from liquid Ge and HBr under high vacuum at 1550 °C) with $Na[N(SiMe_3)_2]$ in toluene at −78 °C.[261] The cluster core is a distorted cube of Ge atoms, Figure 9.6 (the $SiMe_3$ groups attached to the nitrogen atoms joined to Ge2 to Ge6 and Ge8 are omitted), having Ge1−Ge7, Ge1−Ge2 (Ge4 or Ge6) and Ge2−Ge3 (Ge4 or Ge8) distances of 5.175(1), 2.499–2.503, and 2.661–2.672 Å, respectively. The intracuboid bond angles Ge2−Ge1−Ge6 (or Ge2−Ge1−Ge4) and Ge3−Ge4−Ge5 (or Ge2−Ge3−Ge4) are ca. 81.5 and 75.6°, respectively. The $Ge{-}N(SiMe_3)_2$ bond lengths of 1.869 ± 0.003 Å are similar to the 1.876(5) Å in $Ge[N(SiMe_3)_2]_2$[110] (Table 9.4). Compound **79** is EPR silent, but

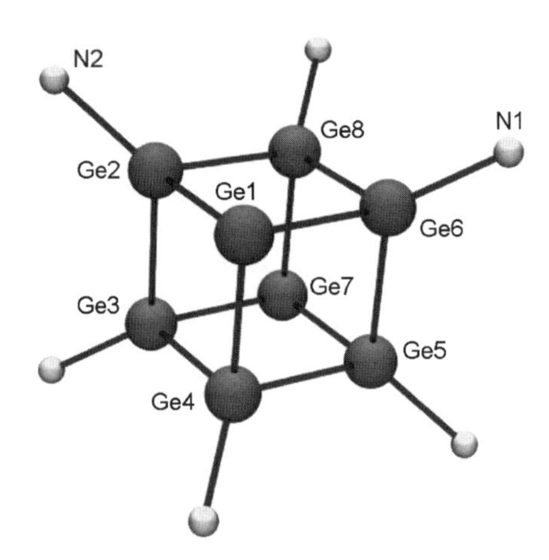

Figure 9.6[261]

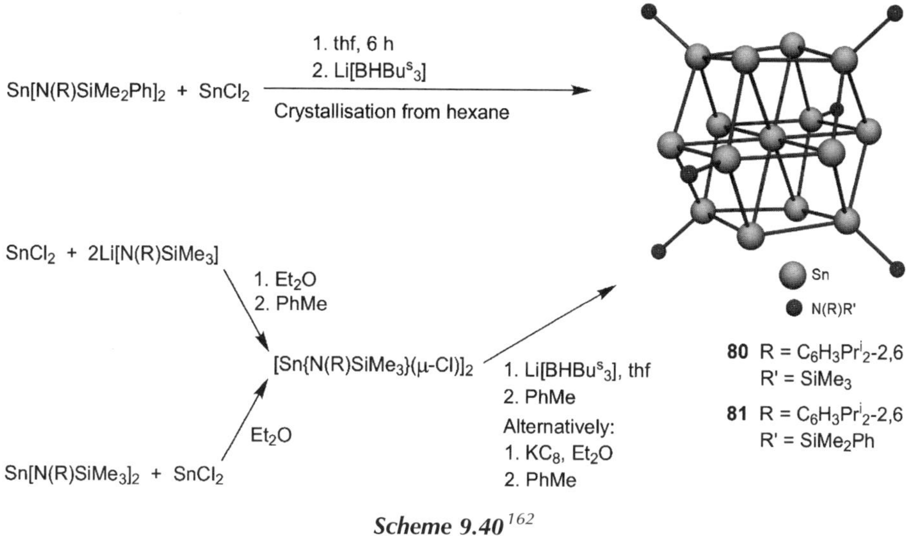

Scheme 9.40[162]

can be described as a singlet diradical (or diradicaloid) with a calculated energy difference between its singlet and triplet states of 96.5 kJ mol^{-1}.[261]

The black (or dark red in thin layers) crystalline clusters **80** and **81** were obtained in low yield, as shown in Scheme 9.40, by groups at Sussex and Davis.[162] These are the first body-centred clusters of a Group 14 element. Single crystals of **80** were isolated in four different space groups, but the molecular structure deduced from each is essentially identical and closely similar to that of **81**.

The metalloid character of **80** and **81** is revealed in the Sn–Sn interactions between the central tin atom (Sn1) and the eight 'naked' tin atoms (Sn_A) of the body-centred (Sn1)$(Sn_A)_8$ cube. These Sn1–Sn_A distances are closer than those (Sn1–Sn_B) between Sn1 and the amido-bearing tin atoms (Sn_B); average Sn–Sn bond lengths are listed in Table 9.9. The structural motif (bcc arrangement) of the [Sn1$(Sn_A)_8$] core in **80** and **81** is similar to that of metallic structures (such as that of α-iron) but for elemental tin only at pressures > 45 GPa.[262] Temperature-dependent ^{119}Sn Mössbauer spectral measurements (at Jerusalem) revealed two (rather than three) different tin sites in a ratio of 6 (corresponding to Sn_B) to 9 (Sn1 + 8Sn_A), and the isomer shift of the ^{119}Sn atoms in the Sn_9 core was almost identical to that for β-tin.[162] DFT calculations on the metal cluster [Sn_9{$Sn(NH_2)$}$_6$] reasonably reproduced the experimental Sn–Sn distances of Table 9.9.[162]

Table 9.9 *Average Sn–Sn bond lengths (Å) in clusters **80** and **81***[162]

Compound	Sn1–Sn_A	Sn1–Sn_B	Sn_A–Sn_B	Sn_A–Sn_A
80[a]	3.18 ± 0.02	3.33 ± 0.04	2.99 ± 0.02	3.67 ± 0.06
81	3.15 ± 0.03	3.41 ± 0.02	3.02 ± 0.04	3.64 ± 0.04

[a] As for **81**, these data are for the $P2_1/n$ crystal

References

1. M. F. Lappert, P. P. Power, A. R. Sanger and R. C. Srivastava, *Metal and Metalloid Amides*, Horwood-Wiley, Chichester, 1980.
2. M. F. Lappert, M. J. Slade, J. L. Atwood and M. J. Zavorotko, *J. Chem. Soc., Chem. Commun.*, 1980, 621.
3. M. Veith, *Z. Naturforsch., Teil B*, 1978, **33**, 1, 7.
4. (a) P. Neugebauer, B. Jaschke and U. Klingebiel, Recent developments in the chemistry of compounds with silicon-nitrogen bonds; in *The Chemistry of Organic Silicon Compounds* (eds. Z. Rappoport and Y. Apeloig), John Wiley & Sons, Inc., New York, 2001, vol. 3, ch. 6, pp. 429–467; (b) K. W. Klinkhammer, *The Chemistry of Organic Germanium, Tin and Lead Compounds* (ed. Z. Rappoport), John Wiley & Sons, Inc., New York, 2001, vol. 2, part 1, ch. 4; (c) M. Veith, *Chem. Rev.*, 1990, **90**, 3; (d) M. Veith and O. Recktenwald, *Topics Current Chem.*, 1982, **104**, 1; (e) J. Barrau, J. Escudié and J. Satgé, *Chem. Rev.*, 1990, **90**, 283; (f) M. Driess and H. Grützmacher, *Angew. Chem., Int. Ed.*, 1996, **35**, 828; (g) W. Petz, *Chem. Rev.*, 1986, **86**, 1019. (h) N. N. Zemlyanskii, I. V. Borisova, M. S. Nechaev, V. N. Khrustalev, V. V. Lunin, M. Yu. Antipin and Yu. A. Ustynyuk, *Russ. Chem. Bull.*, 2004, **53**, 980; (i) O. Kühl, *Coord. Chem. Rev.*, 2004, **248**, 411; (j) M. Weidenbruch, *Coord. Chem. Rev.*, 1994, **130**, 275; (k) N. Tokitoh and R. Okazaki, *Coord. Chem. Rev.*, 2000, **210**, 251; (l) J. Parr, Germanium, tin and lead; in *Comprehensive Coordination Chemistry II* (eds. J. A. McCleverty and T. J. Meyer), Elsevier, Oxford, 2004, vol. 3 (ed. G. F. R. Parkin), ch. 3.7, pp. 545–608; (m) J. Barrau and G. Rima, *Coord. Chem. Rev.*, 1998, **178**, 593; (n) M. Weidenbruch, *Eur. J. Inorg. Chem.*, 1999, 373; (o) C. E. Holloway and M. Melnik, *Main Group Met. Chem.*, 1997, **20**, 399 and 1998, **21**, 371; (p) C. E. Holloway and M. Melnik, *Main Group Met. Chem.*, 2003, **26**, 155.
5. (a) D. H. Harris and M. F. Lappert, *J. Chem. Soc., Chem. Commun.*, 1974, 895; (b) P. J. Davidson, D. H. Harris and M. F. Lappert, *J. Chem. Soc., Dalton Trans.*, 1976, 2268.
6. C. D. Schaeffer and J. J. Zuckerman, *J. Am. Chem. Soc.*, 1974, **96**, 7160.
7. M. F. Lappert, P. P. Power, M. J. Slade, L. Hedberg, K. Hedberg and V. Schomaker, *J. Chem. Soc., Chem. Commun.*, 1979, 369.
8. M. F. Lappert and P. P. Power, *Adv. Chem. Ser. (Am. Chem. Soc.)*, 1976, **157**, 70.
9. J. D. Cotton, C. S. Cundy, D. H. Harris, A. Hudson, M. F. Lappert and P. W. Lednor, *J. Chem. Soc., Chem. Commun.*, 1974, 651.
10. I. Gümrükçü, A. Hudson, M. F. Lappert, M. J. Slade and P. P. Power, *J. Chem. Soc., Chem. Commun.*, 1980, 776.
11. M. J. S. Gynane, D. H. Harris, M. F. Lappert, P. P. Power, P. Rivière and M. Rivière-Baudet, *J. Chem. Soc., Dalton Trans.*, 1977, 2004.
12. (a) A. Hudson, M. F. Lappert and P. W. Lednor, *J. Chem. Soc., Dalton Trans.*, 1976, 2369; (b) M. F. Lappert and P. W. Lednor, *Adv. Organomet. Chem.*, 1976, **14**, 345.
13. (a) M. Veith, E. Werle, R. Lisowsky, R. Köppe and H. Schnöckel, *Chem. Ber.*, 1992, **125**, 1375; (b) S. Tsutsui, K. Sakamoto and M. Kira, *J. Am. Chem. Soc.*, 1998, **120**, 9955.
14. M. Denk, R. Lennon, R. Hayashi, R. West, A. V. Belyakov, H. P. Verne, A. Haaland, M. Wagner and N. Metzler, *J. Am. Chem. Soc.*, 1994, **116**, 2691.
15. B. Gehrhus, M. F. Lappert, J. Heinicke, R. Boese and D. Bläser, *J. Chem. Soc., Chem. Commun.*, 1995, 1931.
16. (a) R. West and M. Denk, *Pure Appl. Chem.*, 1996, **68**, 785; (b) M. Haaf, T. A. Schmedake, B. J. Paradise and R. West, *Can. J. Chem.*, 2000, **78**, 1526.
17. J. Heinicke, A. Oprea, M. K. Kindermann, T. Kárpáti, L. Nyulászi and T. Veszprémi, *Chem. Eur. J.*, 1998, **4**, 541.
18. B. Gehrhus, P. B. Hitchcock and M. F. Lappert, *Z. Anorg. Allg. Chem.*, 2005, **631**, 1383.

19. M. Haaf, T. A. Schmedake and R. West, *Acc. Chem. Res.*, 2000, **33**, 704.
20. M. F. Lappert and B. Gehrhus, *J. Organomet. Chem.*, 2001, **617–618**, 209.
21. N. J. Hill and R. West, *J. Organomet. Chem.*, 2004, **689**, 4165.
22. G.-H. Lee, R. West and T. Müller, *J. Am. Chem. Soc.*, 2003, **125**, 8114.
23. D. F. Moser, I. A. Guzei and R. West, *Main Group Met. Chem.*, 2001, **24**, 811.
24. M. E. Lee, H. M. Cho, C. H. Kim and W. Ando, *Organometallics*, 2001, **20**, 1472.
25. B. Gehrhus, P. B. Hitchcock and L. Zhang, *Angew. Chem., Int. Ed.*, 2004, **43**, 1124.
26. L. A. Leites, S. S. Bukalov, M. Denk, R. West and M. Haaf, *J. Mol. Struct.*, 2000, **550**, 329.
27. (a) J. Oláh and T. Veszprémi, *J. Organomet. Chem.*, 2003, **686**, 112; (b) J. Oláh, F. De Proft, T. Veszprémi and P. Geerlings, *J. Phys. Chem. A*, 2004, **108**, 490.
28. T. Müller, *J. Organomet. Chem.*, 2003, **686**, 251.
29. L. Pause, M. Robert, J. Heinicke and O. Kühl, *J. Chem. Soc., Perkin Trans. 2*, 2001, 1383.
30. R. West, T. A. Schmedake, M. Haaf, J. Becker and T. Müller, *Chem. Lett.*, 2001, **30**, 68.
31. B. Tumanskii, P. Pine, Y. Apeloig, N. J. Hill and R. West, *J. Am. Chem. Soc.*, 2004, **126**, 7786.
32. D. F. Moser, T. Bosse, J. Olson, J. L. Moser, I. A. Guzei and R. West, *J. Am. Chem. Soc.*, 2002, **124**, 4186.
33. B. Gehrhus and P. B. Hitchcock, *J. Organomet. Chem.*, 2004, **689**, 1350.
34. B. Gehrhus and P. B. Hitchcock, *Organometallics*, 2004, **23**, 2848.
35. X. Cai, B. Gehrhus, P. B. Hitchcock, M. F. Lappert and J. C. Slootweg, *J. Organomet. Chem.*, 2002, **643–644**, 272.
36. F. Antolini, B. Gehrhus, P. B. Hitchcock and M. F. Lappert, *Angew. Chem., Int. Ed.*, 2002, **41**, 2568.
37. S. B. Clendenning, B. Gehrhus, P. B. Hitchcock and J. F. Nixon, *Chem. Commun.*, 1999, 2451.
38. S. B. Clendenning, B. Gehrhus, P. B. Hitchcock, D. F. Moser, J. F. Nixon and R. West, *J. Chem. Soc., Dalton Trans.*, 2002, 484.
39. J. C. Slootweg, F. J. J. de Kanter, M. Schakel, A. W. Ehlers, B. Gehrhus, M. Lutz, A. M. Mills, A. L. Spek and K. Lammertsma, *Angew. Chem., Int. Ed.*, 2004, **43**, 3474.
40. B. Gehrhus, P. B. Hitchcock and M. F. Lappert, *Z. Anorg. Allg. Chem.*, 2001, **627**, 1048.
41. R. Pietschnig, *Chem. Commun.*, 2004, 546.
42. A. Fürstner, H. Krause and C. W. Lehmann, *Chem. Commun.*, 2001, 2372.
43. W. A. Herrmann, P. Härter, C. W. K. Gstöttmayr, F. Bielert, N. Seeboth and P. Sirsch, *J. Organomet. Chem.*, 2002, **649**, 141.
44. A. G. Avent, B. Gehrhus, P. B. Hitchcock, M. F. Lappert and H. Maciejewski, *J. Organomet. Chem.*, 2003, **686**, 321.
45. J. M. Dysard and T. D. Tilley, *Organometallics*, 2000, **19**, 4726.
46. (a) X. Cai, B. Gehrhus, P. B. Hitchcock and M. F. Lappert, *Can. J. Chem.*, 2000, **78**, 1484; (b) W. J. Evans, J. M. Perotti, J. W. Ziller, D. F. Moser and R. West, *Organometallics*, 2003, **22**, 1160.
47. D. Amoroso, M. Haaf, G. P. A. Yap, R. West and D. E. Fogg, *Organometallics*, 2002, **21**, 534.
48. D. S. McGuinness, B. F. Yates and K. J. Cavell, *Organometallics*, 2002, **21**, 5408.
49. M. Okazaki, H. Tobita and H. Ogino, *Dalton Trans.*, 2003, 493.
50. N. J. Hill, D. F. Moser, I. A. Guzei and R. West, *Organometallics*, 2005, **24**, 3346.
51. W. Li, N. J. Hill, A. C. Tomasik, G. Bikzhanova and R. West, *Organometallics*, 2006, **25**, 3802.
52. B. Gehrhus, P. B. Hitchcock, R. Pongtavornpinyo and L. Zhang, *Dalton Trans.*, **2006**, 1847.
53. A. Zeller, F. Bielert, P. Härter, W. A. Herrmann and T. Strassner, *J. Organomet. Chem.*, 2005, **690**, 3292.
54. A. Dhiman, T. Müller, R. West and J. Y. Becker, *Organometallics*, 2004, **23**, 5689.
55. A. Naka, N. J. Hill and R. West, *Organometallics*, 2004, **23**, 6330.
56. F. Antolini, B. Gehrhus, P. B. Hitchcock, M. F. Lappert and J. C. Slootweg, *Dalton Trans.*, 2004, 3288.
57. B. Gehrhus, P. B. Hitchcock and M. Parruci, *Dalton Trans.* 2005, 2720.

58. F. Antolini, B. Gehrhus, P. B. Hitchcock and M. F. Lappert, *Chem. Commun.*, 2005, 5112.
59. A. G. Avent, C. Drost, B. Gehrhus, P. B. Hitchcock and M. F. Lappert, *Z. Anorg. Allg. Chem.*, 2004, **630**, 2090.
60. E. Ionescu, B. Gehrhus, P. B. Hitchcock, M. Nieger and R. Streubel, *Chem. Commun.*, 2005, 4842.
61. B. Gehrhus, P. B. Hitchcock and H. Jansen, *J. Organomet. Chem.*, 2006, **691**, 811.
62. (a) M.-D. Su, *J. Am. Chem. Soc.*, 2003, **125**, 1714; (b) M.-D. Su, *Chem. Phys. Lett.*, 2003, **374**, 385.
63. (a) H. Joo and M. L. McKee, *J. Phys. Chem. A*, 2005, **109**, 3728; (b) R.-E. Li, J.-H. Sheu, and M.-D. Su, *Inorg. Chem.*, 2007, **46**, 9245.
64. M. Delawar, B. Gehrhus and P. B. Hitchcock, *Dalton Trans.*, 2005, 2945.
65. D. F. Moser, A. Naka, I. A. Guzei, T. Müller and R. West, *J. Am. Chem. Soc.*, 2005, **127**, 14730.
66. P. Jutzi, A. Mix, B. Rummel, W. W. Schoeller, B. Neumann and H.-G. Stammler, *Science*, 2004, **305**, 849.
67. M. Hirotsu, T. Nishida, H. Sasaki, T. Muraoka, T. Yoshimura and K. Ueno, *Organometallics*, 2007, **26**, 2495.
68. C.-W. So, H. W. Roesky, P. M. Gurubasavaraj, R. B. Oswald, M. T. Gamer, P. G. Jones and S. Blaurock, *J. Am. Chem. Soc.*, 2007, **129**, 12049.
69. M. Driess, S. Yao, M. Brym, C. van Wüllen and D. Lentz, *J. Am. Chem. Soc.*, 2006, **128**, 9628.
70. (a) M. Driess, S. Yao, M. Brym and C. van Wüllen, *Angew. Chem., Int. Ed.*, 2006, **45**, 6730; (b) S. Yao, M. Brym, C. van Wüllen and M. Driess, *Angew. Chem., Int. Ed.*, 2007, **46**, 4159; (c) S. Yao, Y. Xiong, M. Brym and M. Driess, *J. Am. Chem. Soc.*, 2007, **129**, 7268; (d) Y. Xiong, S. Yao, M. Brym and M. Driess, *Angew. Chem., Int. Ed.*, 2007, **46**, 4511; (e) S. Yao, C. van Wüllen, X.-Y. Sun and M. Driess, *Angew. Chem., Int. Ed.*, 2008, **47**, 3250.
71. M. F. Lappert, *Silicon, Germanium, Tin and Lead Compounds*, 1986, **9**, 129.
72. M. Veith, *Angew. Chem., Int. Ed.*, 1987, **26**, 1.
73. M. F. Lappert and R. S. Rowe, *Coord. Chem. Rev.*, 1990, **100**, 267.
74. M. F. Lappert, *Main Group Metal Chemistry*, 1994, **17**, 183.
75. M. Veith, A. Rammo, S. Faber and B. Schillo, *Pure Appl. Chem.*, 1999, **71**, 401.
76. O. Kühl, *Coord. Chem. Rev.*, 2004, **248**, 411.
77. (a) B. Çetinkaya, P. B. Hitchcock, M. F. Lappert, M. C. Misra and A. J. Thorne, *J. Chem. Soc., Chem. Commun.*, 1984, 148; (b) P. B. Hitchcock, M. F. Lappert and A. J. Thorne, *J. Chem. Soc., Chem. Commun.*, 1990, 1587.
78. M. Veith and A. Rammo, *Z. Anorg. Allg. Chem.*, 2001, **627**, 662.
79. A. Schnepf, *Z. Anorg. Allg. Chem.*, 2006, **632**, 935.
80. A. Meller and C.-P. Gräbe, *Chem. Ber.*, 1985, **118**, 2020.
81. M. Veith and M. Grosser, *Z. Naturforsch., Teil B*, 1982, **37**, 1375.
82. W. A. Herrmann, M. Denk, J. Behm, W. Scherer, F.-R. Klingan, H. Bock, B. Solouki and M. Wagner, *Angew. Chem., Int. Ed.*, 1992, **31**, 1485.
83. A. Schäfer, W. Saak and M. Weidenbruch, *Z. Anorg. Allg. Chem.*, 1998, **624**, 1405.
84. O. Kühl, P. Lönnecke and J. Heinicke, *Polyhedron*, 2001, **20**, 2215.
85. J. T. York, V. G. Young, Jr. and W. B. Tolman, *Inorg. Chem.*, 2006, **45**, 4191.
86. T. Gans-Eichler, D. Gudat, K. Nättinen and M. Nieger, *Chem. Eur. J.*, 2006, **12**, 1162.
87. M. Veith, S. Becker and V. Huch, *Angew. Chem., Int. Ed.*, 1989, **28**, 1237.
88. J. Pfeiffer, W. Maringgele, M. Noltemeyer and A. Meller, *Chem. Ber.*, 1989, **122**, 245.
89. P. Bazinet, G. P. A. Yap, and D. S. Richeson, *J. Am. Chem. Soc.*, 2001, **123**, 11162.
90. C. Drost, P. B. Hitchcock and M. F. Lappert, *J. Chem. Soc., Dalton Trans.*, 1996, 3595.
91. I. L. Fedushkin, A. A. Skatova, V. A. Chudakova, N. M. Khvoinova, A. Yu. Baurin, S. Dechert, M. Hummert and H. Schumann, *Organometallics*, 2004, **23**, 3714.
92. (a) H. V. R. Dias and Z. Wang, *J. Am. Chem. Soc.*, 1997, **119**, 4650; (b) A. E. Ayers and H. V. R. Dias, *Inorg. Chem.*, 2002, **41**, 3259; (c) H. V. R. Dias, Z. Wang and V. K. Diyabalange, *Inorg. Chem.*, 2005, **44**, 7322.

93. (a) A. V. Zabula, F. E. Hahn, T. Pape and A. Hepp, *Organometallics*, 2007, **26**, 1972; (b) F. E. Hahn, A. V. Zabula, T. Pape and A. Hepp, *Eur. J. Inorg. Chem.*, 2007, 2405.
94. M. Veith, O. Schütt and V. Huch, *Angew. Chem., Int. Ed.*, 2000, **39**, 601.
95. M. Driess, S. Yao, M. Brym and C. van Wüllen, *Angew. Chem., Int. Ed.*, 2006, **45**, 4349.
96. J. R. Babcock, L. Liable-Sands, A. L. Rheingold and L. R. Sita, *Organometallics*, 1999, **18**, 4437.
97. (a) Y. J. Tang, A. M. Felix, V. W. Manner, L. N. Zakharov, A. L. Rheingold, B. Moasser and R. A. Kemp, *ACS Symposium Series*, 2006, **917**, 410; (b) Y. Tang, A. M. Felix, L. N. Zakharov, A. L. Rheingold and R. A. Kemp, *Inorg. Chem.*, 2004, **43**,7239; (c) M. Westerhausen, J. Greul, H.-D. Hausen, and W. Schwarz, *Z. Anorg. Allg. Chem.*, 1996, **622**, 1295.
98. T. Gans-Eichler, D. Gudat and M. Nieger, *Angew. Chem., Int. Ed.*, 2002, **41**, 1888.
99. J. R. Babcock, C. Incarvito, A. L. Rheingold, J. C. Fettinger and L. R. Sita, *Organometallics*, 1999, **18**, 5729.
100. S.-J. Kim, Y.-J. Lee, S. H. Kim, J. Ko, S. Cho and S. O. Kang, *Organometallics*, 2002, **21**, 5358.
101. H. Braunschweig, B. Gehrhus, P. B. Hitchcock and M. F. Lappert, *Z. Anorg. Allg. Chem.*, 1995, **621**, 1922.
102. (a) F. E. Hahn, L. Wittenbecher, M. Kühn, T. Lügger and R. Fröhlich, *J. Organomet. Chem.*, 2001, **617–618**, 629; (b) F. E. Hahn, L. Wittenbecher, D. Le Van and A. V. Zabula, *Inorg. Chem.*, 2007, **46**, 7662.
103. C. Drost, P. B. Hitchcock and M. F. Lappert, *Angew. Chem., Int. Ed.*, 1999, **38**, 1113.
104. P. Bazinet, G. P. A. Yap, G. A. DiLabio and D. S. Richeson, *Inorg. Chem.*, 2005, **44**, 4616.
105. M. Veith, A. Rammo and M. Hans, *Phosphorus, Sulfur Silicon Relat. Elem.*, 1994, **93–94**, 197.
106. (a) M. M. Olmstead and P. P. Power, *Inorg. Chem.*, 1984, **23**, 413; (b) B. Beagley, N. G. Scott and D. Schmidling, *J. Mol. Struct.*, 1990, **221**, 15.
107. J.-L. Fauré, H. Gornitzka, R. Réau, D. Stalke and G. Bertrand, *Eur. J. Inorg. Chem.*, 1999, 2295.
108. M. Veith and R. Lisowsky, *Angew. Chem., Int. Ed.*, 1988, **27**, 1087.
109. A. Heine, D. Fest, D. Stalke, C. D. Habben, A. Meller and G. M. Sheldrick, *J. Chem. Soc., Chem. Commun.*, 1990, 742.
110. T. Fjeldberg, H. Hope, M. F. Lappert, P. P. Power and A. J. Thorne, *J. Chem. Soc., Chem. Commun.*, 1983, 639.
111. R. W. Chorley, P. B. Hitchcock, M. F. Lappert, W.-P. Leung, P. P. Power and M. M. Olmstead, *Inorg. Chim. Acta*, 1992, **198–200**, 203.
112. (a) C. Heinemann, W. A. Herrmann and W. Thiel, *J. Organomet. Chem.*, 1994, **475**, 73; (b) C. Boehme and G. Frenking, *J. Am. Chem. Soc.*, 1996, **118**, 2039; (c) J. Oláh, T. Veszprémi, F. De Proft and P. Geerlings, *J. Phys. Chem. A*, 2007, **111**, 10815; (d) H. M. Tuononen, R. Roesler, J. L. Dutton and P. J. Ragogna, *Inorg. Chem.*, 2007, **46**, 10693.
113. J. F. Lehmann, S. G. Urquhart, L. E. Ennis, A. P. Hitchcock, K. Hatano, S. Gupta and M. K. Denk, *Organometallics*, 1999, **18**, 1862.
114. L. A. Leites, S. S. Bukalov, A. V. Zabula, I. A. Garbuzova, D. F. Moser and R. West, *J. Am. Chem. Soc.*, 2004, **126**, 4114.
115. (a) K. C. Molloy, M. P. Bigwood, R. H. Herber and J. J. Zuckerman, *Inorg. Chem.*, 1982, **21**, 3709; (b) I. Nowik, H. A. Spinney, D. S. Richeson and R. H. Herber, *J. Organomet. Chem.*, 2007, **692**, 5680.
116. (a) S. Vepřek, J. Prokop, F. Glatz, R. Merica, F. R. Klingan and W. A. Herrmann, *Chem. Mater.*, 1996, **8**, 825; (b) *Idem, Proc. Electrochem. Soc.*, 1996, **96**, 247.
117. T. Fjeldberg, P. B. Hitchcock, M. F. Lappert, S. J. Smith and A. J. Thorne, *J. Chem. Soc., Chem. Commun.*, 1985, 939.
118. (a) J. Barrau, G. Rima and T. El Amraoui, *Inorg. Chim. Acta*, 1996, **241**, 9; (b) J. Barrau, G. Rima and T. El Amraoui, *J. Organomet. Chem.*, 1998, **561**, 167.

119. R. Papiernik, L. G. Hubert-Pfalzgraf and M. C. Massiani, *Polyhedron*, 1991, **10**, 1657.
120. H. Braunschweig, R. W. Chorley, P. B. Hitchcock and M. F. Lappert, *J. Chem. Soc., Chem. Commun.*, 1992, 1311.
121. C. S. Weinert, A. E. Fenwick, P. E. Fanwick and I. P. Rothwell, *Dalton Trans*. 2003, 532.
122. M. Veith, P. Hobein and R. Rösler, *Z. Naturforsch., Teil B*, 1989, **44**, 1067.
123. P. B. Hitchcock, M. F. Lappert, B. J. Samways and E. L. Weinberg, *J. Chem. Soc., Chem. Commun.*, 1983, 1492.
124. P. B. Hitchcock, H. A. Jasim, R. E. Kelly and M. F. Lappert, *J. Chem. Soc., Chem. Commun.*, 1985, 1776.
125. N. Froelich, P. B. Hitchcock, J. Hu, M. F. Lappert and J. R. Dilworth, *J. Chem. Soc., Dalton Trans.*, 1996, 1941.
126. P. B. Hitchcock, M. F. Lappert, G. A. Lawless, G. M. de Lima and L. J.-M. Pierssens, *J. Organomet. Chem.*, 2000, **601**, 142.
127. M. Veith, G. Schlemmer and M.-L. Sommer, *Z. Anorg. Allg. Chem.*, 1983, **497**, 157.
128. M. Veith, M.-L. Sommer, and D. Jäger, *Chem. Ber.*, 1979, **112**, 2581.
129. M. Veith, *Z. Naturforsch., Teil B*, 1980, **35**, 20.
130. M. Veith and G. Schlemmer, *Chem. Ber.*, 1982, **115**, 2141.
131. M. Veith, M. Opsölder, M. Zimmer and V. Huch, *Eur. J. Inorg. Chem.*, 2000, 1143.
132. (a) M. Veith and H. Lange, *Angew. Chem., Int. Ed.*, 1980, **19**, 401; (b) M. Veith and O. Recktenwald, *Z. Naturforsch., Teil B*, 1983, **38**, 1054.
133. M. Veith, M. Jarczyk and V. Huch, *Chem. Ber.*, 1988, **121**, 347.
134. M. Veith, M. Olbrich, W. Shihua and V. Huch, *J. Chem. Soc., Dalton Trans.*, 1996, 161.
135. (a) R. A. Bartlett and P. P. Power, *J. Am. Chem. Soc.*, 1990, **112**, 3660; (b) S. C. Goel, M. Y. Chiang, D. J. Rauscher and W. E. Buhro, *J. Am. Chem. Soc.*, 1993, **115**, 160.
136. H. Braunschweig, C. Drost, P. B. Hitchcock, M. F. Lappert and L. J.-M. Pierssens, *Angew. Chem., Int. Ed.* 1997, **36**, 261.
137. M. Westerhausen, M. M. Enzelberger and W. Schwarz, *J. Organomet. Chem.*, 1995, **491**, 83.
138. M. Westerhausen, H.-D. Hausen, and W. Schwarz, *Z. Anorg. Allg. Chem.*, 1995, **621**, 877.
139. M. Westerhausen and W. Schwarz, *Z. Anorg. Allg. Chem.*, 1996, **622**, 903.
140. (a) M. Westerhausen, M. Krofta, N. Wiberg, J. Knizek, H. Nöth and A. Pfitzner, *Z. Naturforsch. Teil B*, 1998, **53**, 1489; (b) M. Westerhausen, M. Krofta, S. Schneiderbauer and H. Piotrowski, *Z. Anorg. Allg. Chem.*, 2005, **631**, 1391.
141. R. E. Allan, M. A. Beswick, G. R. Coggan, P. R. Raithby, A. E. H. Wheatley and D. S. Wright, *Inorg. Chem.*, 1997, **36**, 5202.
142. A. Bashall, N. Feeder, E. A. Harron, M. McPartlin, M. E. G. Mosquera, D. Sáez and D. S. Wright, *J. Chem. Soc., Dalton Trans.*, 2000, 4104.
143. R. E. Allan, M. A. Beswick, A. J. Edwards, M. A. Paver, M. -A. Rennie, P. R. Raithby and D. S. Wright, *J. Chem. Soc., Dalton Trans.*, 1995, 1991.
144. (a) F. Benevelli, E. L. Doyle, E. A. Harron, N. Feeder, E. A. Quadrelli, D. Sáez and D. S. Wright, *Angew. Chem., Int. Ed.*, 2000, **39**, 1501; (b) D. R. Armstrong, F. Benevelli, A. D. Bond, N. Feeder, E. A. Harron, A. D. Hopkins, M. McPartlin, D. Moncrieff, D. Sáez, E. A. Quadrelli, A. D. Woods and D. S. Wright, *Inorg. Chem.*, 2002, **41**, 1492.
145. R. E. Allan, M. A. Beswick, N. L. Cromhout, M. A. Paver, P. R. Raithby, A. Steiner, M. Trevithick and D. S. Wright, *Chem. Commun.*, 1996, 1501.
146. A. D. Bond, A. Rothenberger, A. D. Woods and D. S. Wright, *Chem. Commun.*, 2001, 525.
147. F. García, J. P. Hehn, R. A. Kowenicki, M. McPartlin, C. M. Pask, A. Rothenberger, M. L. Stead and D. S. Wright, *Organometallics*, 2006, **25**, 3275.
148. F. García, A. D. Hopkins, C. M. Pask, A. D. Woods and D. S. Wright, *J. Mater. Chem.*, 2004, **14**, 3093.
149. M. Driess, S. Martin, K. Merz, V. Pintchouk, H. Pritzkow, H. Grützmacher and M. Kaupp, *Angew. Chem., Int. Ed.*, 1997, **36**, 1894.

150. A. Murso, M. Straka, M. Kaupp, R. Bertermann and D. Stalke, *Organometallics*, 2005, **24**, 3576.
151. (a) S. R. Foley, C. Bensimon and D. S. Richeson, *J. Am. Chem. Soc.*, 1997, **119**, 10359; (b) S. R. Foley, G. P. A. Yap and D. S. Richeson, *Organometallics*, 1999, **18**, 4700; (c) S. R. Foley, Y. Zhou, G. P. A. Yap and D. S. Richeson, *Inorg. Chem.*, 2000, **39**, 924.
152. M. J. McGeary, K. Folting and K. G. Caulton, *Inorg. Chem.*, 1989, **28**, 4051.
153. (a) V. N. Khrustalev, I. A. Portnyagin, N. N. Zemlyansky, I. V. Borisova, M. S. Nechaev, Yu. A. Ustynyuk, M. Yu. Antipin and V. Lunin, *J. Organomet. Chem.*, 2005, **690**, 1172; (b) V. N. Khrustalev, I. V. Glukhov, I. V. Borisova and N. N. Zemlyansky, *Appl. Organomet. Chem.*, 2007, **21**, 551.
154. X. Rimo and L. R. Sita, *Inorg. Chim. Acta*, 1998, **270**, 118.
155. C. S. Weinert, I. A. Guzei, A. L. Rheingold and L. R. Sita, *Organometallics*, 1998, **17**, 498.
156. L. Pu, M. M. Olmstead, P. P. Power and B. Schiemenz, *Organometallics*, 1998, **17**, 5602.
157. C. Drost, P. B. Hitchcock, M. F. Lappert and L. J.-M. Pierssens, *Chem. Commun.*, 1997, 1141.
158. R. W. Chorley, P. B. Hitchcock and M. F. Lappert, *J. Chem. Soc., Dalton Trans.*, 1992, 1451.
159. C. Bibal, S. Mazières, H. Gornitzka and C. Couret, *Organometallics*, 2002, **21**, 2940.
160. (a) R. W. Chorley, P. B. Hitchock, B. S. Jolly, M. F. Lappert and G. A. Lawless, *J. Chem. Soc., Chem. Commun.*, 1991, 1302; (b) R. W. Chorley, D. Ellis, P. B. Hitchock and M. F. Lappert, *Bull. Soc. Chim. Fr.*, 1992, **129**, 599.
161. D. A. Dixon, A. J. Arduengo and M. F. Lappert, *Heteroatom Chem.*, 1991, **2**, 541.
162. M. Brynda, R. Herber, P. B. Hitchcock, M. F. Lappert, I. Nowik, P. P. Power, A. V. Protchenko, A. Růžička and J. Steiner, *Angew. Chem., Int. Ed.*, 2006, **45**, 4333.
163. A. P. Dove, V. C. Gibson, E. L. Marshall, H. S. Rzepa, A. J. P. White and D. J. Williams, *J. Am. Chem. Soc.*, 2006, **128**, 9834.
164. (a) M. Chen, J. R. Fulton, P. B. Hitchcock, N. C. Johnstone, M. F. Lappert and A. V. Protchenko, *Dalton Trans.* 2007, 2770. (b) S. Yao, S. Block, M. Brym and M. Driess, *Chem. Commun.*, 2007, 3844.
165. H. Braunschweig, P. B. Hitchcock, M. F. Lappert and L. J.-M. Pierssens, *Angew. Chem., Int. Ed.*, 1994, **33**, 1156.
166. C. Stanciu, S. S. Hino, M. Stender, A. F. Richards, M. M. Olmstead and P. P. Power, *Inorg. Chem.*, 2005, **44**, 2774.
167. R. E. Allan, M. A. Beswick, M. K. Davies, P. R. Raithby, A. Steiner and D. S. Wright, *J. Organomet. Chem.*, 1998, **550**, 71.
168. M. Veith and M. Zimmer, *Z. Anorg. Allg. Chem.*, 1996, **622**, 1471.
169. O. Kühl, P. Lönnecke and J. Heinicke, *New J. Chem.*, 2002, **26**, 1304.
170. K. Jurkschat, K. Peveling and M. Schürmann, *Eur. J. Inorg. Chem.*, 2003, 3563.
171. M. A. Paver, C. A. Russell, D. Stalke and D. S. Wright, *J. Chem. Soc., Chem. Commun.*, 1993, 1349.
172. W.-P. Leung, K.-W. Wong, Z.-X. Wang and T. C. W. Mak, *Organometallics*, 2006, **25**, 2037.
173. (a) L. M. Engelhardt, B. S. Jolly, M. F. Lappert, C. L. Raston and A. H. White, *J. Chem. Soc., Chem. Commun.*, 1988, 336; (b) B. S. Jolly, M. F. Lappert, L. M. Engelhardt, A. H. White and C. L. Raston, *J. Chem. Soc., Dalton Trans.*, 1993, 2653.
174. M. Veith and A. Rammo, *Z. Anorg. Allg. Chem.*, 1997, **623**, 861.
175. M. Veith, B. Schillo and V. Huch, *Angew. Chem., Int. Ed.*, 1999, **38**, 182.
176. (a) K. W. Hellmann, P. Steinert and L. H. Gade, *Inorg. Chem.*, 1994, **33**, 3859; (b) K. W. Hellmann, L. H. Gade, O. Gevert and P. Steinert, *Inorg. Chem.*, 1995, **34**, 4069.
177. H. Memmler, U. Kauper, L. H. Gade, D. Stalke and J. W. Lauher, *Organometallics*, 1996, **15**, 3637.
178. M. Veith and W. Frank, *Angew. Chem., Int. Ed.*, 1984, **23**, 158.
179. M. Veith, P. Hobein and V. Huch, *J. Chem. Soc., Chem. Commun.*, 1995, 213.

180. (a) I. L. Fedushkin, N. M. Khvoinova, A. Yu. Baurin, G. K. Fukin, V. K. Cherkasov and M. P. Bubnov, *Inorg. Chem.*, 2004, **43**, 7807; (b) I. L. Fedushkin, N. M. Khvoinova, A. Yu. Baurin, V. A. Chudakova, A. A. Skatova, V. K. Cherkasov, G. K. Fukin and E. A. Baranov, *Russ. Chem. Bull.*, 2006, **55**, 74.
181. I. L. Fedushkin, M. Hummert and H. Schumann, *Eur. J. Inorg. Chem.*, 2006, 3266.
182. H. Meyer, G. Baum, W. Massa and A. Berndt, *Angew. Chem., Int. Ed.*, 1987, **26**, 798.
183. D. E. Goldberg, D. H. Harris, M. F. Lappert and K. M. Thomas, *J. Chem. Soc., Chem. Commun.*, 1976, 261.
184. H. Schäfer, W. Saak and M. Weidenbruch, *Organometallics*, 1999, **18**, 3159.
185. M. Weidenbruch, J. Schlaefke, A. Schäfer, K. Peters, H. G. von Schnering and H. Marsmann, *Angew. Chem., Int. Ed.*, 1994, **33**, 1846.
186. N. Kano, K. Shibata, N. Tokitoh and R. Okazaki, *Organometallics*, 1999, **18**, 2999.
187. P. B. Hitchcock, M. F. Lappert and L. J.-M. Pierssens, *Chem. Commun.*, 1996, 1189.
188. W.-P. Leung, Q.W.-Y. Ip, S.-Y. Wong and T. C. W. Mak, *Organometallics*, 2003, **22**, 4604.
189. C. Drost, B. Gehrhus, P. B. Hitchcock and M. F. Lappert, *Chem. Commun.*, 1997, 1845.
190. (a) K. W. Klinkhammer and W. Schwarz, *Angew. Chem., Int. Ed.*, 1995, **34**, 1334; (b) C. Förster, K. W. Klinkhammer, B. Tumanskii, H.-J. Krüger and H. Kelm, *Angew. Chem., Int. Ed.*, 2007, **46**, 1156.
191. L. R. Sita, J. R. Babcock and R. Xi, *J. Am. Chem. Soc.*, 1996, **118**, 10912.
192. J. R. Babcock and L. R. Sita, *J. Am. Chem. Soc.*, 1998, **120**, 5585.
193. (a) H. Chen, R. A. Bartlett, H. V. R. Dias, M. M. Olmstead and P. P. Power, *Inorg. Chem.*, 1991, **30**, 3390; (b) R. A. Bartlett and P. P. Power, *J. Am. Chem. Soc.*, 1990, **112**, 3660.
194. Y. J. Tang, L. N. Zakharov, A. L. Rheingold and R. A. Kemp, *Inorg. Chim. Acta*, 2006, **359**, 775.
195. R. Papiernik, L. G. Hubert-Pfalzgraf, M. Veith and V. Huch, *Chem. Ber. Recl.*, 1997, **130**, 1361.
196. M. Veith and W. Frank, *Angew. Chem., Int. Ed.*, 1984, **23**, 158.
197. W. J. Grigsby, T. Hascall, J. J. Ellison, M. M. Olmstead and P. P. Power, *Inorg. Chem.*, 1996, **35**, 3254.
198. (a) J. F. Eichler, O. Just and W. S. Rees, Jr. *ACS Symposium Series*, 2006, **917**, 122; (b) J. F. Eichler, O. Just and W. S. Rees, Jr. *Inorg. Chem.*, 2006, **45**, 6706.
199. (a) T. Chivers and D. J. Eisler, *Angew. Chem., Int. Ed.*, 2004, **43**, 6686; (b) D. J. Eisler and T. Chivers, *Chem. Eur. J.*, 2006, **12**, 233.
200. R. West, D. F. Moser, I. A. Guzei, G.-H. Lee, A. Naka, W. Li, A. Zabula, S. Bukalov and L. Leites, *Organometallics*, 2003, **22**, 4604.
201. B. Gehrhus, P. B. Hitchcock and M. F. Lappert, *J. Chem. Soc., Dalton Trans.*, 2000, 3094.
202. C. J. Carmalt, J. A. C. Clyburne, A. H. Cowley, V. Lomeli and B. G. McBurnett, *Chem. Commun.*, 1998, 243.
203. (a) M. J. S. Gynane, M. F. Lappert, S. J. Miles and P. P. Power, *J. Chem. Soc., Chem. Commun.*, 1976, 256; (b) M. J. S. Gynane, M. F. Lappert, S. J. Miles and P. P. Power, *J. Chem. Soc., Chem. Commun.*, 1978, 192.
204. M. F. Lappert, M. C. Misra, M. Onyszchuk, R. S. Rowe, P. P. Power and M. J. Slade, *J. Organomet. Chem.*, 1987, **330**, 31.
205. S. N. Nikolaeva, E. V. Avtonomov, J. Lorberth and V. S. Petrosyan, *Z. Naturforsch., Teil B*, 1998, **53**, 9.
206. (a) K. A. Miller, J. M. Bartolin, R. M. O'Neill, R. D. Sweeder, T. M. Owens, J. W. Kampf, M. M. Banaszak Holl and N. J. Wells, *J. Am. Chem. Soc.*, 2003, **125**, 8986; (b) J. M. Bartolin, A. Kavara, J. W. Kampf and M. M. Banaszak Holl, *Organometallics*, 2006, **25**, 4738.
207. A. Herve, A. L. Rodriguez and E. Fouquet, *J. Org. Chem.*, 2005, **70**, 1953.
208. B. Tumanskii, P. Pine, Y. Apeloig, N. J. Hill and R. West, *J. Am. Chem. Soc.*, 2005, **127**, 8248.
209. (a) K. E. Litz, J. E. Bender, J. W. Kampf and M. M. Banaszak Holl, *Angew. Chem., Int. Ed.*, 1997, **36**, 496; (b) A. C. Gottfried, J. Wang, E. E. Wilson, L. W. Beck, M. M. Banaszak Holl and J. W. Kampf, *Inorg. Chem.*, 2004, **43**, 7665.

210. M. Veith and O. Recktenwald, *Z. Anorg. Allg. Chem.*, 1979, **459**, 208,
211. M. Veith, M. Nötzel, L. Stahl and V. Huch, *Z. Anorg. Allg. Chem.*, 1994, **620**, 1264.
212. P. B. Hitchcock, H. A. Jasim, M. F. Lappert, W.-P. Leung, A. K. Rai and R. E. Taylor, *Polyhedron*, 1991, **10**, 1203.
213. R. W. Chorley, P. B. Hitchcock and M. F. Lappert, *J. Chem. Soc., Chem. Commun.*, 1992, 525.
214. D. Ellis, P. B. Hitchcock and M. F. Lappert, *J. Chem. Soc., Dalton Trans.*, 1992, 3397.
215. P. B. Hitchcock, E. Jang and M. F. Lappert, *J. Chem. Soc., Dalton Trans.*, 1995, 3179.
216. P. B. Hitchcock, J. Hu, M. F. Lappert and J. R. Severn, *Dalton Trans.*, 2004, 4193.
217. S. R. Foley, G. P. A. Yap, and D. S. Richeson, *J. Chem. Soc., Dalton Trans.*, 2000, 1663.
218. (a) S. Kobayashi, S. Iwata, M. Abe and S. Shoda, *J. Am. Chem. Soc.*, 1990, **112**, 1625; (b) S. Kobayashi, S. Iwata, M. Abe and S. Shoda, *J. Am. Chem. Soc.*, 1995, **117**, 2187.
219. (a) S. Kobayashi, S. Iwata, K. Yajima, K. Yagi and S. Shoda, *J. Am. Chem. Soc.*, 1992, **114**, 4929; (b) S. Shoda, S. Iwata, K. Yajima, K. Yagi, Y. Ohnishi and S. Kobayashi, *Tetrahedron*, 1997, **53**, 15281.
220. (a) S. Kobayashi and S. Cao, *Chem. Lett.*, 1993, 25; (b) S. Kobayashi and S. Cao, *Chem. Lett.*, 1994, 941.
221. (a) S. Iwata, S. Shoda and S. Kobayashi, *Organometallics*, 1995, **14**, 5533; (b) R. D. Sweeder, R. L. Gdula, B. J. Ludwig, M. M. Banaszak Holl and J. W. Kampf, *Organometallics*, 2003, **22**, 3222.
222. (a) M. Veith and R. Lisowsky, *Z. Anorg. Allg. Chem.*, 1988, **560**, 59; (b) M. Veith and E. Werle, *Z. Anorg. Allg. Chem.*, 1992, **609**, 19.
223. (a) M. Veith, V. Huch, J.-P. Majoral, G. Bertrand and G. Manuel, *Tetrahedron Lett.*, 1983, 4219; (b) M. Veith, M. Gouygou and A. Detemple, *Phosphorus, Sulfur Silicon Relat. Elem.*, 1993, **75**, 183; (c) M. Veith, M. Grosser and V. Huch, *Z. Anorg. Allg. Chem.*, 1984, **513**, 89.
224. D. M. Smith, B. Neumüller and K. Dehnicke, *Z. Naturforsch., Teil B*, 1998, **53**, 1074.
225. B. Çetinkaya, P. B. Hitchcock, M. F. Lappert and R. G. Smith, *J. Chem. Soc., Chem. Commun.*, 1992, 932.
226. M. Veith, O. Schütt, J. Blin, S. Becker, J. Frères and V. Huch, *Z. Anorg. Allg. Chem.*, 2002, **628**, 138.
227. N. Wiberg, H.-W. Lerner, H. Nöth and W. Ponikwar, *Angew. Chem., Int. Ed.*, 1999, **38**, 1103.
228. M. Westerhausen, *Inorg. Chem.* 1991, **30**, 96.
229. (a) M. Veith, S. Mathur, P. König, C. Cavelius, J. Biegler, A. Rammo, V. Huch, H. Shen and G. Schmid, *C. R. Chim.*, 2004, **7**, 509; (b) M. Veith, J. Frères, P. König, O. Schütt, V. Huch and J. Blin, *Eur. J. Inorg. Chem.*, 2005, 3699.
230. H. Schumann, M. Glanz, F. Girgsdies, F. E. Hahn, M. Tamm and A. Grzegorzewski, *Angew. Chem., Int. Ed.*, 1997, **36**, 2232.
231. A. Schäfer, W. Saak, M. Weidenbruch, H. Marsmann and G. Henkel, *Chem. Ber. Recl.*, 1997, **130**, 1733.
232. B. Gehrhus, P. B. Hitchcock and M. F. Lappert, *Angew. Chem., Int. Ed.*, 1997, **36**, 2514.
233. (a) A. Meller, G. Ossig, W. Maringgele, D. Stalke, R. Herbst-Irmer, S. Freitag and G. M. Sheldrick, *J. Chem. Soc., Chem. Commun.*, 1991, 1123; (b) A. Meller, G. Ossig, W. Maringgele, M. Noltemeyer, D. Stalke, R. Herbst-Irmer, S. Freitag and G. M. Sheldrick, *Z. Naturforsch., Teil B*, 1992, **47**, 162.
234. B. Klein and W. P. Neumann, *J. Organomet. Chem.*, 1994, **465**, 119.
235. M. Veith, E. Werle and V. Huch, *Z. Anorg. Allg. Chem.*, 1993, **619**, 641.
236. M. Veith, S. Becker and V. Huch, *Angew. Chem., Int. Ed.*, 1990, **29**, 216.
237. M. F. Lappert and P. P. Power, *J. Chem. Soc., Dalton Trans.*, 1985, 51.
238. M. Veith, H. Lange, K. Bräuer and R. Bachmann, *J. Organomet. Chem.*, 1981, **216**, 377.
239. O. Kühl, P. Lönnecke and J. Heinicke, *Inorg. Chem.*, 2003, **42**, 2836.

240. (a) M. Veith, L. Stahl and V. Huch, *Organometallics*, 1993, **12**, 1914; (b) P. Braunstein, M. Veith, J. Blin and V. Huch, *Organometallics*, 2001, **20**, 627; (c) P. Braunstein, V. Huch, C. Stern and M. Veith, *Chem. Commun.*, 1996, 2041.
241. M. Knorr, E. Hallauer, V. Huch, M. Veith and P. Braunstein, *Organometallics*, 1996, **15**, 3868.
242. (a) M. Veith, L. Stahl and V. Huch, *J. Chem. Soc., Chem. Commun.*, 1990, 359; (b) M. Veith and L. Stahl, *Angew. Chem., Int. Ed.*, 1993, **32**, 106.
243. M. Veith, A. Müller, L. Stahl, M. Nötzel, M. Jarczyk and V. Huch, *Inorg. Chem.*, 1996, **35**, 3848.
244. M. Veith, L. Stahl and V. Huch, *Inorg. Chem.*, 1989, **28**, 3278.
245. (a) K. E. Litz, K. Henderson, R. W. Gourley and M. M. Banaszak Holl, *Organometallics*, 1995, **14**, 5008; (b) K. E. Litz, M. M. Banaszak Holl, J. W. Kampf and G. B. Carpenter, *Inorg. Chem.*, 1998, **37**, 6461; (c) J. E. Bender, A. J. Shusterman, M. M. Banaszak Holl and J. W. Kampf, *Organometallics*, 1999, **18**, 1547; (d) Z. T. Cygan, J. E. Bender, K. E. Litz, J. W. Kampf and M. M. Banaszak Holl, *Organometallics*, 2002, **21**, 5373.
246. T. A. K. Al-Allaf, C. Eaborn, P. B. Hitchcock, M. F. Lappert and A. Pidcock, *J. Chem. Soc., Chem. Commun.*, 1985, 548.
247. (a) P. B. Hitchcock, M. F. Lappert and M. C. Misra, *J. Chem. Soc., Chem. Commun.*, 1985, 863; (b) G. K. Campbell, P. B. Hitchcock, M. F. Lappert and M. C. Misra, *J. Organomet. Chem.*, 1985, **289**, C1.
248. S. M. Hawkins, P. B. Hitchcock and M. F. Lappert, *J. Chem. Soc., Chem. Commun.*, 1985, 1592.
249. S. M. Hawkins, P. B. Hitchcock, M. F. Lappert and A. K. Rai, *J. Chem. Soc., Chem. Commun.*, 1986, 1689.
250. P. B. Hitchcock, M. F. Lappert and M. J. Michalczyk, *J. Chem. Soc., Dalton Trans.*, 1987, 2635.
251. K. E. Litz, J. W. Kampf and M. M. Banaszak Holl, *J. Am. Chem. Soc.*, 1998, **120**, 7484.
252. K. E. Litz, J. E. Bender, R. D. Sweeder, M. M. Banaszak Holl and J. W. Kampf, *Organometallics*, 2000, **19**, 1186.
253. Z. T. Cygan, J. W. Kampf and M. M. Banaszak Holl, *Organometallics*, 2004, **23**, 2370.
254. Z. T. Cygan, J. W. Kampf and M. M. Banaszak Holl, *Inorg. Chem.*, 2003, **42**, 7219.
255. Z. T. Cygan, J. W. Kampf and M. M. Banaszak Holl, *Inorg. Chem.*, 2004, **43**, 2057.
256. (a) A. Sekiguchi, T. Matsuo and H. Watanabe, *J. Am. Chem. Soc.*, 2000, **122**, 5652; (b) E. Niecke, A. Fuchs, F. Baumeister, M. Nieger and W. W. Schoeller, *Angew. Chem., Int. Ed.*, 1995, **34**, 555.
257. C. Cui, M. Brynda, M. M. Olmstead and P. P. Power, *J. Am. Chem. Soc.*, 2004, **126**, 6510.
258. H. Cox, P. B. Hitchcock, M. F. Lappert and L. J.-M. Pierssens, *Angew. Chem., Int. Ed.*, 2004, **43**, 4500.
259. P. B. Hitchcock, M. F. Lappert and L. J.-M. Pierssens, *Organometallics*, 1998, **17**, 2686.
260. (a) H. Schnöckel, *Dalton Trans.*, 2005, 3131;(b) N. Wiberg and P. P. Power; in *Molecular Clusters of the Main Group Elements* (eds. M. Driess and H. Nöth), Wiley-VCH, Weinheim, 2004, ch. 2.3.
261. A. Schnepf and R. Köppe, *Angew. Chem., Int. Ed.*, 2003, **42**, 911.
262. S. Desgreniers, Y. K. Vohra and A. L. Ruoff, *Phys. Rev. B*, 1989, **39**, 10359.

10

Amides of the Group 15 Metals (As, Sb, Bi)

10.1 Introduction

In our 1980 book, the synthesis and characterisation of the amides (excluding those in which the ligand is NH_2^-, NH^{2-} or N^{3-}) of arsenic (8 pp.) and antimony or bismuth (1 page) was described; tabulated data were presented in $13^1/_2$ pages (mostly As), covering 305 (As), 35 (Sb) and 8 (Bi) individual compounds.[1] There were 96 (As), 14 (Sb) and 6 (Bi) bibliographic citations. The vast majority of arsenic compounds were of three-coordinate As^{III}. Other As^{III} amides included four- and six-membered ring compounds (oligomers of As^{III} imides) such as $[As(Me)(\mu\text{-}NPh)]_2$ and $[As(Cl)(\mu\text{-}NMe)]_3$. Arsenic(V) compounds were the four- or five-coordinate complexes $[As\{N(H)Ph\}_3(O)]$ and $[As(F)\{N(H)Pr\}(OMe)_3]$, respectively. X-Ray diffraction data were only then available on two compounds: $As_4(NMe)_6$ and $[As(Cl)(\mu\text{-}NMe)]_3$.

In Part 2 of the book,[1] reference was made to RNCO insertions into $M^{III}\text{–}NR_2$ (M = As, mainly; but also M = Sb or Bi). Protolytic cleavage of $M^{III}\text{–}NR_2$ bonds was reported for a number of As^{III} but very few Sb^{III} or Bi^{III} amides. Metathetical exchange reactions (principally NMe_2/Cl exchanges) were established for a number of As^{III} amides but just two Sb^{III} or one Bi^{III} amide. $As(NMe_2)_3$ ($\equiv$ L) was shown to be a ligand in $[Ni(CO)_3(L)]$; $[As(Cl)(\mu\text{-}NBu^t)]_2$ was obtained by thermal reductive elimination from $[As(Cl)_3({=}NBu^t)]$.

The majority of amides (or imides) of arsenic, antimony and bismuth (collectively designated as of 'M'), have M in the +3 oxidation state, although several examples of As^V and Sb^V amides or imides have been reported; but only three papers have dealt with Bi^V compounds. The imides considered to be within the scope of this chapter are those having trivalent nitrogen as, for instance, in $[As(\mu\text{-}NR)X]_2$. There are very few examples of $As^I\text{–}N$ compounds, such as the bis(imido)arsenic(I) cation **1**,[2a] and only one of a thermally stable

Metal Amide Chemistry Michael Lappert, Andrey Protchenko, Philip Power and Alexandra Seeber

As^{II} amide, $As[N(SiMe_3)_2]_2$.[3] General reviews (1982–2004) of aspects of the organic chemistry of arsenic, antimony and bismuth have included a few examples of amidometal(III) (mainly As) compounds.[4–6] Bi- and tricyclic bis(amido)cyclophosph(III)azane compounds of As, Sb, Bi (M in **2**) have featured in a review (which had greater emphasis on the more prolific M = P analogues).[7] The use of aminoarsines, such as $As(NMe_2)_3$ and $As(NEt_2)R_2$, as $\bar{N}Me_2$ or $\bar{N}Et_2$ synthons has been examined.[8] Early reviews have dealt with pentacoordinate arsenic(V) compounds, including five amides,[9a,9b] and of some amidoarsenic(III) precursors.[9b]

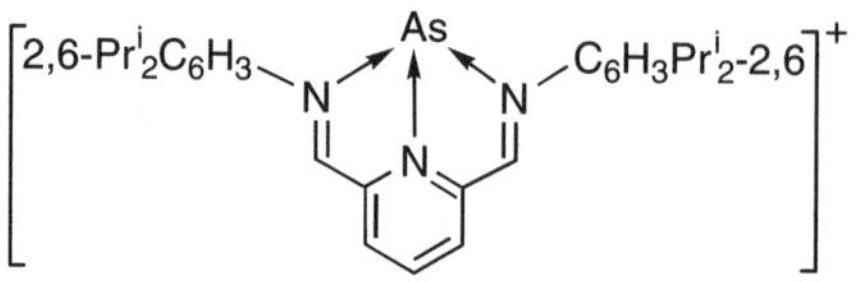

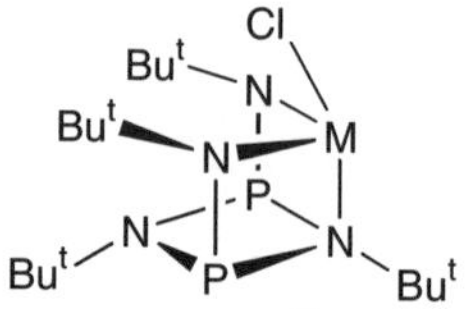

Wright and collaborators have reviewed the synthesis and structures of salts containing anions such as **3–5** or their derivatives (also As or Bi analogues) in articles entitled 'Structure and bonding in organometallic anions of heavy Group 14 and 15 elements;'[10a] 'Synthetic applications of p-block metal dimethylamido reagents:'[10b] and 'Toroidal main group macrocycles: new opportunities for cation and anion coordination' (e.g. in **6**).[10c] Chivers and coworkers published a review entitled 'Imido analogues of common oxo anions: a new episode in the chemistry of cluster compounds;'[11a] and a comprehensive survey (96 references) of 'The chemistry of pnictogen(III)-nitrogen ring systems' (M = P, As, Sb, Bi), including four- (e.g. **7**) and six- (e.g. **8**) membered ring compounds and their derivatives as well as **3–5** and macrocycles.[11b]

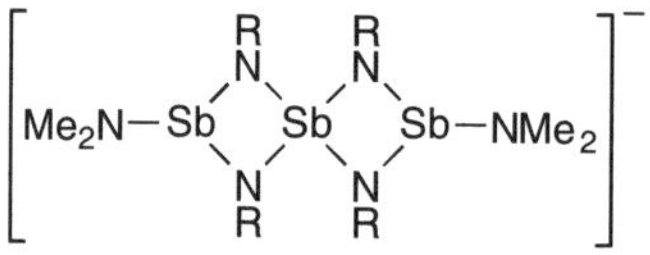

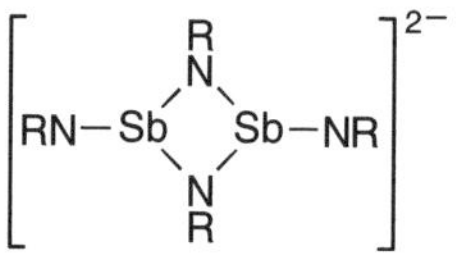

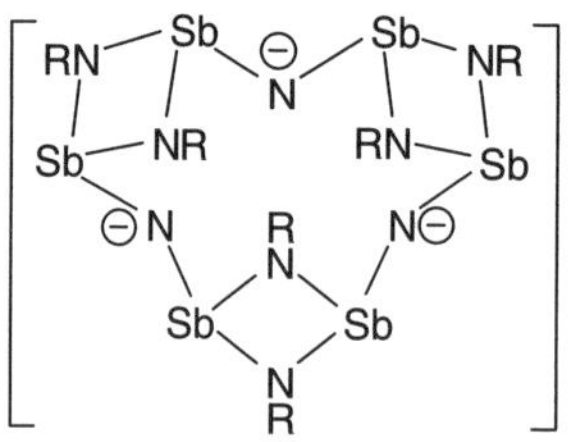

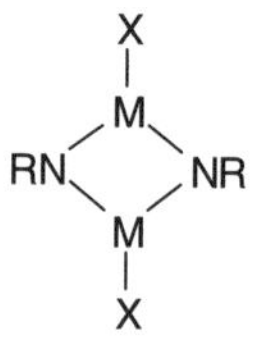

An article on 'Coordination complexes of bismuth(III) involving organic ligands with pnictogene or chalcogene donors' has a brief (9 refs.) section on amidobismuth(III) compounds.[12]

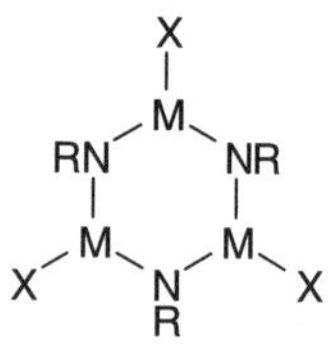

10.2 Mononuclear Group 15 Metal(III) Amides

10.2.1 Introduction

In earlier work,[1] no molecular structural data were available on mononuclear M^{III} amides, but this has changed markedly in the last three (and particularly last two) decades. Another significant development has been the use of the title compounds, and especially the homoleptic dimethylamides $M(NMe_2)_3$, as reagents,[8] or synthons for the preparation of bridging binuclear imides **7** (Section 10.3) or cluster imides (e.g. **3–7**) (Section 10.4), or $Bi[N(SiMe_3)_2]_3$ as a precursor for atomic layer deposition.[19c] Heteroleptic amidometal chlorides have featured as precursors to cationic metal(III) amides (Section 10.2.4).

10.2.2 Synthesis, Structures and Protolyses of Metal(III) Amides

The first (1980) homoleptic Bi^{III} amide, $Bi[N(SiMe_3)_2]_3$, was prepared from $BiCl_3$ and $3Li[N(SiMe_3)_2]$;[3] later workers used the sodium[14a,19a] (also[19a] for the Sb^{III} analogue from $SbCl_3$) or potassium[19b] bis(trimethylsilyl)amide. Similar salt elimination reactions have been used for the synthesis of other $M^{III}L_3$ compounds [L = NMe_2 (M = As, Sb, Bi), NR_2 (M = Bi and R = Et, Pr^i, $SiMe_2R'$ {R' = Et, Bu^n, $CH{=}CH_2$}), $N(Bu^t)CH_2CH_2NMe_2$ (M = Bi), NPh_2 (M = Bi), $N(H)C_6H_3Me_2$-2,6 (M = Sb), $N(H)C_6H_2Bu^t{}_3$-2,4,6 (M = Sb, Bi), NC_4H_4 (M = As), and $N(SiMe_3)C_5H_4N$-2 (M = Sb, Bi)],[13b,14b,15,17a,19c,20,21] as well as the cage compounds **9a** and **9b** (M = Sb, Bi).[17b,17c] The compound $As[N(H)Bu^t]_3$ was obtained from $As(NMe_2)_3/3Bu^tNH_2$ by $HNMe_2$ elimination,[16a] and **10** from $M(NR'_2)_3$ (M = Sb, R' = Et; M = Bi, R' = Me) and $N\{CH_2CH_2N(R)H\}_3$.[16b]

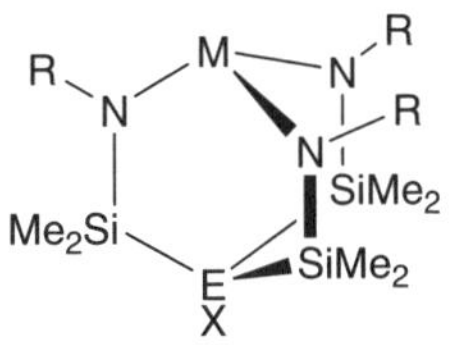

9a R = Bu^t, EX = CH, M = Sb or Bi

9b R = C_6H_4Me-4, EX = SiMe, M = Sb

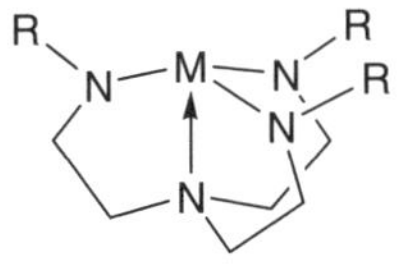

10 R = Me, $SiMe_3$; M = Sb, Bi

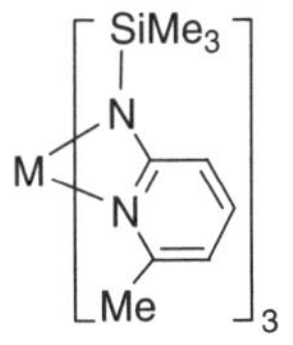

11 M = Sb, Bi

Selected available structural data are assembled in Table 10.1. The first such information for each of the heavier Group 15 metal(III) homoleptic amides was for tris(morpholinato)-arsenic (1980),[20] $Sb[N(H)C_6H_2Bu^t{}_3$-2,4,6$]_3$ (1996),[17a] and $Bi(NPh_2)_3$ (1989).[15] Each of the compounds of Table 10.1 is trigonal pyramidal at M [the sum of the angles subtended at M ranges from 287.9° for $Bi[N(H)C_6H_2Bu^t{}_3$-2,4,6$]_3$[17a] to 312° for gaseous $As(NMe_2)_3$[13a,13b]] and generally trigonal planar at the nitrogen atoms. For the C_s-symmetrical $M(NMe_2)_3$ compounds, however, only two of the nitrogen atoms are nearly planar while the third is distinctly pyramidal.[13a,13b,14b] The crystalline distorted octahedral compounds **11** (M = Sb, Bi) (with *fac*-amido centres) were obtained from the appropriate MCl_3 and $[Li\{N(SiMe_3)$-$C_5H_3N(Me$-6)-2$\}(OEt_2)]_2$.[22]

Tris(pyrrolido)arsenic, unlike the isoleptic robust P analogue, decomposed at ambient temperature, while the Sb and Bi compounds proved to be inaccessible from the $MCl_3/3Li$-(NC_4H_4) system.[21] Whereas $MCl_3/3Li[N(H)C_6H_2Bu^t{}_3$-2,4,6$]$ furnished the thermally

Table 10.1 Some structural parameters for homoleptic Group 15 metal(III) amides

Compound	M–N (Å)	N–M–N′ (°)	Ref.
$As(NMe_2)_3$[a]	1.849(5), 1.870(5)[a]	100.3(9), 111.2(9)[a]	13a 13b
$Sb(NMe_2)_3$[a]	2.022(12), 2.043(12)[a]	94.7(8), 107.3(8)[a]	13a, 13b
$Bi(NMe_2)_3$	2.180(21), 2.189(18)	96.2(9), 98.3(5)	14b
$Sb[N(H)C_6H_2$-Bu^t_3-2,4,6$]_3$	2.041(6), 2.048(6)	85.6(2), 104.4(3)	17a
$Bi[N(H)C_6H_2$-Bu^t_3-2,4,6$]_3$	2.14(2), 2.179(14), 2.214(13)	82.7(6), 106.7(6)	17a
$Bi(NPh_2)_3$	2.16(3), 2.17(2), 2.26(3) 2.12(2), 2.21(4), 2.28(2)	96(1), 100(1), 100(1) 97(1), 99(1), 101(1)	15
$Bi[N(SiMe_3)_2]_3$	2.199(8) to 2.272(8)	100.3(3) to 104.3(3)	19c
As(N)$_3$	1.84(2), 1.86(2), 1.87(3)	94.7(7), 96.3(7), 98.9(7)	21
As(N O)$_3$	1.853(3), 1.866(4)	94.1, 94.1, 109.9	20
9a (M = Sb)	2.051(3), 2.052(3), 2.062(3)	98.8(1), 99.1(1), 99.3(1)	17b
9a (M = Bi)	2.149(7), 2.161(7), 2.182(7)	97.1(3), 97.2(3), 97.5(3)	17b
9b	2.050(3), 2.053(3), 2.062(2)	96.7(1), 97.9(1), 99.6(1)	17c
10 (M = Sb)	2.049(2) [3.200(2) – transannular]	100.91(7)	16b
10 (M = Bi)	2.167(5) [3.021(4) – transannular]	103.5(2)	16b

[a]By gas electron diffraction

robust (CD_2Cl_2, reflux) homoleptic Sb and Bi amides,[17] with $AsCl_3$ the product was $As[N(H)C_6H_2Bu^t_3$-2,4,6$](=NC_6H_2Bu^t_3$-2,4,6);[23a] and MCl_3 with $3Li[N(H)C_6H_3R_2$-2,6] afforded $[M\{N(H)C_6H_3R_2$-2,6$\}(\mu$-$NC_6H_3R_2$-2,6$)]_2$ (M = Sb, R = Me;[18] M = Bi, R = Pr^i[24]). The crystalline 24-membered macrocycle $Sb_{12}(\mu$-$NPh)_{18}$ [i.e. hexameric $\{Sb(\mu$-$NPh)\}_2(\mu$-NPh)] was obtained from the transient $Sb[N(H)Ph]_3$, prepared from $SbCl_3$ and 3Li[N(H)-Ph].[18a] $As(NMe_2)_3$ with $(AlHMe_2)_3$[25b] or $(AlMe_3)_2$[25a] yielded $AlMe_2(NMe_2)$.

Compounds $M[N(R^1)R^2]_3$ (like heteroleptic amido analogues; *cf.*[8] for an early review) are useful starting materials, as summarised for their reactions with protic reagents in Table 10.2; noteworthy are the reactions of $Sb(NEt_2)_3$[30b] or $Bi(NMe_2)_3$[16c] with (1 or 2) HN-$(C_6F_5)(C_5H_4N$-2) yielding Sb^{III} or Bi^{III} products containing the $[N(C_6F_4NEt_2$-2$)(C_5H_4N$-2$)]^-$ or $[N\{C_6F_3(NMe_2)_2$-2,6$\}(C_5H_4N$-2$)]^-$ ligand. For reactions with $M[N(H)R]_n$ (e.g. M = Li and $n = 1$) see Sections 10.3 and 10.4.

10.2.3 Synthesis, Structures and Reactions of Heteroleptic Mononuclear Bis(amido)-metal(III) Compounds

The following mononuclear bis(amido)metal(III) compounds were reported in the post-1978 period: $M(Cl)[N(SiMe_3)_2]_2$ (M = As, Sb),[3] $As(CF_3)[N(SiMe_3)_2]_2$,[35] **12** (M = As, Sb, Bi),[36a] $As(C_6H_4X)[N(H)Bu^t]_2$ (X = H, *p*-Me, *p*-Br, *o*-OMe, *o*-Cl),[37] **13** (R = Et,[38] Me,[39] Bu^t[40]), **14** (M = As, Sb),[41] **15** (M = Sb, Bi),[42] **16** (M = As, Sb; R = Bu^t, $C_6H_2Me_3$-2,4,6; X = Cl, Br);[44a,44b] $SbF(NEt_2)\{N(C_6F_4NEt_2-2)(C_5H_4N-2)\}$;[30b] $MF\{N(C_6F_5)(C_5H_4N$-2$)\}$-$\{N(C_6F_4NR_2$-2$)(C_5H_4N$-2$)\}$ (M = Sb, R = Et;[30b] M = Bi, R = Me[16c]); the Br analogue of **17**[45] and $Sb[N(R)SiMe_3]_2X$ (R = Bu^t, $SiMe_3$; X = Cl, Br).[46]

Table 10.2 *Protolytic cleavage reactions of homoleptic Group 15 metal(III) amides*

$M[N(R)R']_3$	HA	Product	Ref.
$M(NMe_2)_3$ (M = As, Sb)	3 $HESi(SiMe_3)_3$ (E = Se, Te)	$M[ESi(SiMe_3)_3]_3$ (M = As, E = Se; M = Sb, E = Se, Te)	26
$Sb(NMe_2)_3$	H_2NR	$[Sb(NMe_2)(\mu\text{-}NR)]_2$ R = $C_6H_2(OMe)_3$-3,4,5,[27a] C_5H_3N-2-Me-5,[27a] $C_6H_3Pr^i{}_2$-2,6,[27b] c-C_6H_{11}[27c]	27a,27b, 27c
$Sb(NMe_2)_3$	$[P(NH_2)(\mu\text{-}NBu^t)_2P\{N(H)Bu^t\}]$	$[Sb(NMe_2)(\mu\text{-}NR)]_2$ R = $P(\mu\text{-}NBu^t)_2P\{N(H)Bu^t\}$	28
$Sb(NEt_2)_3$	$HN(C_6F_5)(C_5H_4N\text{-}2)$	$[SbF(NEt_2)\{N(C_6F_4NEt_2\text{-}2)(C_5H_4N\text{-}2)\}]$	30b
$Sb(NEt_2)_3$	2 $HN(C_6F_5)(C_5H_4N\text{-}2)$	$[SbF\{N(C_6F_4NEt_2\text{-}2)(C_5H_4N\text{-}2)\}\text{-}\{N(C_6F_5)(C_5H_4N\text{-}2)\}]$	30b
$Sb(NEt_2)_3$	3 $HN(C_6F_5)(C_5H_4N\text{-}2)$	$Sb[N(C_6F_5)(C_5H_4N\text{-}2)]_3$	30b
$Sb(NMe_2)_3$	1,8-$C_{10}H_6[N(H)Pr^i]_2$	$Sb(NMe_2)[\{N(Pr^i)\}_2C_{10}H_6\text{-}1,8]$	41a
$Sb(NMe_2)_3$	$[N(H)Pr^i]_2C{=}NPr^i$	$Sb[\{N(Pr^i)\}_2CNPr^i][\{N(Pr^i)\}_2C\{N(H)Pr^i\}]$	27d
$Sb(NMe_2)_3$	$(HO)_3SiR$	$[SbO_3SiR)]_4$ R = $N(SiMe_3)C_6H_3Pr^i{}_2$-2,6	29
$Bi(NMe_2)_3$	$(HO)_3SiR$	$[Bi_{12}(O_3SiR)_8(\mu_3\text{-}O)_4Cl_4(thf)_8]$ R = $N(SiMe_3)C_6H_3Pr^i{}_2$-2,6	29
$Bi(NMe_2)_3$	$HN(C_6F_5)(C_5H_4N\text{-}2)$	$BiF_2[N\{C_6F_3(NMe_2)_2\text{-}2,6\}(C_5H_4N\text{-}2)]$	16c
$Bi(NMe_2)_3$	2 $HN(C_6F_5)(C_5H_4N\text{-}2)$	$BiF\{N(C_6F_4NMe_2\text{-}2)(C_5H_4N\text{-}2)\}\text{-}\{N(C_6F_5)(C_5H_4N\text{-}2)\}$	16c
$Bi(NMe_2)_3$	3 $HN(C_6F_5)(C_5H_4N\text{-}2)$	$Bi[N(C_6F_5)(C_5H_4N\text{-}2)]_3$	16c
$Bi[N(R^1)R^2]_3$ $R^1 = Me = R^2$; $R^1 = SiMe_3 = R^2$; $R^1 = Me$, $R^2 = SiMe_3$	3 HCp	$BiCp_3$	30a
8Bi$[N(SiMe_3)_2]_3$	8 (Bu^t / OH)$_8$, 4 H_2O	$[Bi_8O_4(Bu^tC8)_2]$ (Bu^tC8 = *p*- Bu^{t}calix[8]arene)	31
$Bi[N(SiMe_3)_2]_3$	3 HOR	$Bi(OR)_3$ R = Et, Pr^i, Bu^t, Am^t, $(CH_2)_2NMe_2$, $CH(Me)CH_2NMe_2$, $C(Me)_2CH_2OMe$	32
$Bi[N(SiMe_3)_2]_3$	3 $HOCH_2Bu^t$	$[Bi(\mu\text{-}OR)(OR)_2(HOR)]_2$ R = CH_2Bu^t	33
$Bi[N(SiMe_3)_2]_3$	3 $HOSiPh_3$	$Bi(OSiPh_3)_3$	34
$Bi[N(SiMe_3)_2]_3$	3 $HESi(SiMe_3)_3$ (E = Se, Te)	$Bi[ESi(SiMe_3)_3]_3$	26
$M[N(SiMe_3)_2]_3$ (M = Sb, Bi)	3 $HEC_6H_2R_3$-2,4,6 (E = S, Se)	$M[EC_6H_2R_3\text{-}2,4,6]_3$ R = Me, Pr^i, Bu^t	19a

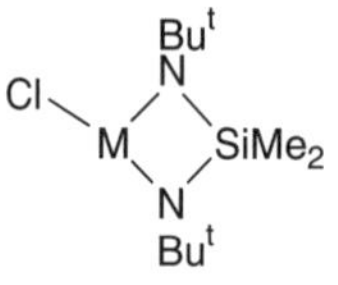

Whereas the $MCl_2Ph/2Li[N(H)Bu^t]$ reaction for M = As yielded $As(C_6H_5)[N(H)Bu^t]_2$,[37] for the heavier pnictogens the product was the dinuclear imide $[M(C_6H_5)(\mu\text{-}NBu^t)]_2$ (M = Sb, Bi);[43] likewise (see Section 10.2.2), $BiCl_3$ with $2Li[N(H)C_6H_3Pr^i{}_2\text{-}2,6]$ gave $[Bi\{N(H)C_6H_3Pr^i{}_2\text{-}2,6\}(\mu\text{-}NC_6H_3Pr^i{}_2\text{-}2,6)]_2$.[24] Apart from the four- (**17**) and five- (**15**) coordinate complexes, the bis(amido)metal(III) chlorides are trigonal pyramidal at M. X-Ray structural data are available for **12** (M = As, Sb, Bi),[36] **13** ($R = Bu^t$),[40] **14** (M = As, Sb),[41b] $Sb(NMe_2)[\{N(Pr^i)\}_2C_{10}H_6\text{-}1,8]$,[41a] **15** (M = Sb, Bi),[42] **16** ($R = Bu^t$ and MX = AsCl, SbBr; or R = Mes, MX = SbCl),[44a,44b] and **17**.[45] The antimony atom in **15** is in a distorted trigonal bipyramidal environment, with the axial positions occupied by Cl and a nitrogen atom of the pyridyl moiety.[42] (In contrast to **15**, the corresponding bismuth compound is an asymmetric $(\mu\text{-}Cl)_2$ dimer [42].)

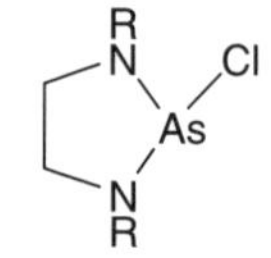

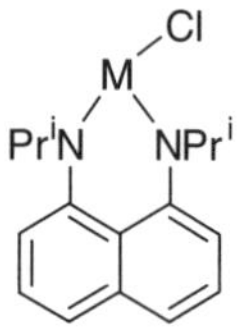

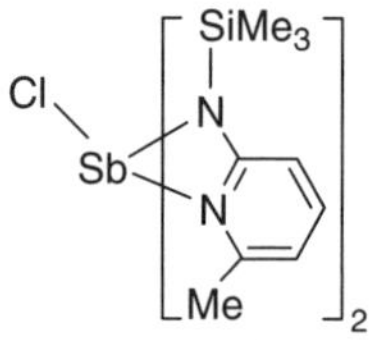

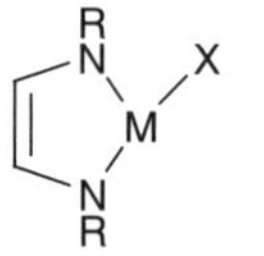

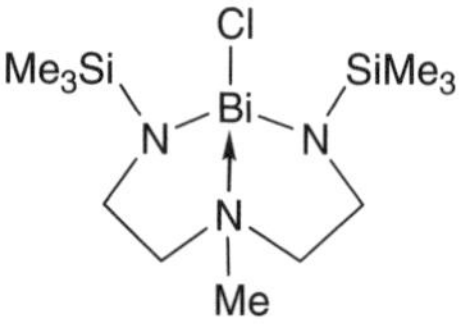

The above bis(amido)metal chlorides were generally [but see Equation (10.1)[47]] prepared from the appropriate MCl_3 by elimination of (i) LiCl,[3,35,36a,38,40,42,43a,43b,45,46] (ii) $[HNEt_3]Cl$,[37,41,44a,44b] or (iii) $SnCl_2$ [for **12**,[36] using $Sn(\mu\text{-}NBu^t)_2SiMe_2$ as reagent].

$$+ \ Sb(Cl)(NMe_2)_2 \longrightarrow \ + \ 2HNMe_2 \qquad (10.1)$$

The bis(amido)metal(III) chlorides have been used extensively as sources of bis(amido)-metal(III) salts (see Section 10.2.5).[39,41,44a,44b,45,48,49] The chloride **16** (M = Sb, $R = Bu^t$) was converted into the iodide by treatment with Me_3SiI.[44a] The crystalline constrained geometry complexes $MCl\{C_5Me_4Si(Me)_2NBu^t\}$ (M = As, Sb) were prepared from MCl_3 and the corresponding chloro-Mg precursor.[53b] The As–N bonds of $AsMe(NMe_2)_2$ and an Sb–N bond of $SbCl(NMe_2)_2$ were readily cleaved upon reaction with $(AlHMe_2)_3$[25b] or $(AlMe_3)_2$[25a] or Bu^tNH_2,[50] respectively. The crystalline compound **18** was obtained from $As(C_6H_5)[N(H)Bu^t]_2$ and 2LiBu; with Se, the corresponding $As^V{=}Se$ compound was obtained.[51] An M′/Bi exchange reaction converted **17** with $M'Cl_3$ (M′ = Al, Ga) into its isoleptic Al or Ga compound and $BiCl_3$.[45] Reduction of a toluene solution of

$AsCl[N(SiMe_3)_2]_2$ with the electron-rich olefin $\overline{EtN(CH_2)_2N(Et)C}=\overline{CN(Et)(CH_2)_2NEt}$ gave the persistent radical $^{\bullet}As[N(SiMe_3)_2]_2$.[3] The chloroantimony(III) compound **16** ($M = Sb$, $R = Bu^t$, $X = Cl$) with $SbCl_5$ gave $Bu^tN{=}CHCH{=}NBu^t + 2SbCl_3$, possibly via $SbCl_3$ + a transient Sb^V compound **16** ($M = Sb$, $R = Bu^t$, $X = Cl_3$).[47] The intramolecular (at the metal) $[NR_2]^-/F^-$ displacement reactions in Sb^{III} ($R = Et$)[30b] and Bi^{III} ($R = Me$)[16c] compounds involving the ligands $[NR_2]^-/[N(C_6F_5)(C_5H_4N\text{-}2)]^-$ are noteworthy [also in bis(amido) (Section 10.2.3) complexes]; hydrolysis afforded $HN(C_6F_4NR_2\text{-}2)(C_5H_4N\text{-}2)$ ($R = Et$,[30b] Me[16c]).

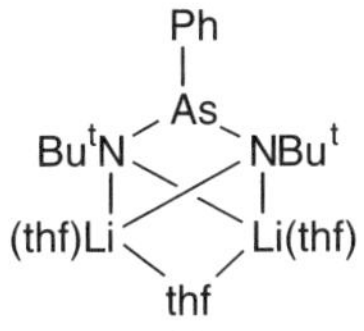

Ph, As, Bu^tN, NBu^t, (thf)Li, Li(thf), thf

18

Reactions of MCl_3 with $n Li_2[(NBu^t)_2BPh]$ yielded crystalline compounds: (i) with $n = 1$, $[MCl\{(NBu^t)_2BPh\}]$ (M = As, Sb, Bi); (ii) with $n = 3/2$, $[M_2\{(NBu^t)_2BPh\}_3]$ (M = Sb, Bi); and (iii) with $n = 2$, $Li[M\{(NBu^t)_2BPh\}_2]$ (M = As, Sb, Bi);[43b] their structures illustrate the versatile coordinating ability of the diamido ligand $[(NBu^t)_2BPh]^{2-}$.

10.2.4 Synthesis, Structures and Reactions of Heteroleptic Mononuclear Amidometal(III) Compounds

The following mononuclear mono(amido)metal(III) complexes were reported in the post-1978 period: the thermally labile $Sb[N(Bu^t)SiMe_3]X_2$ (X = Cl, Br)[46] and $As(\eta^5\text{-}C_5Me_5)Cl(NMe_2)$,[53a] and the robust $MCl_2[N(SiMe_3)_2]$ (M = As, Sb),[3] $As(CF_3)_2[N(SiMe_3)_2]$,[35] $As(\eta^5\text{-}C_5Me_5)F[N(SiMe_3)_2]$,[54] $As(\eta^5\text{-}C_5Me_5)Cl[N(R^1)R^2]$ ($R^1 = SiMe_3 = R^2$;[54] $R^1 = H$, $R^2 = Bu^t$[54]), $AsCl_2[N(SiMe_3)C_5H_3N(Me\text{-}6)\text{-}2]$,[22] $AsCl_2[N(AsCl_2)C_5H_3N(Me\text{-}6)\text{-}2]$,[22] $AsCl_2[N(R)C_6H_2Me_3\text{-}2,4,6]$ (R = H, $SiMe_3$),[52] $M[N(R)\{B(Cl)(TMP)\}]X_2$ (**19**) (M = As, Sb; R = Bu^t, $C_6H_3Pr^i_2\text{-}2,6$; X = F, Cl, Br),[55] $BiF_2[N\{C_6F_3(NMe_2)_2\text{-}2,6\}(C_5H_4N\text{-}2)]$,[16c]

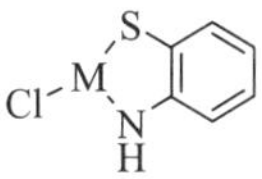

S, M, Cl, N, H (**20**) (M = Bi, X = Me;[58a] M = As, X = Cl [58b,58c]). Their syntheses from as precursor MX_3, $AsCl_2R$ ($R = \eta^5\text{-}C_5Me_5$,[53a,54] CF_3[35]) {or $As(\eta^5\text{-}C_5Me_5)F_2 + Na[N(SiMe_3)_2]$ for $As(\eta^5\text{-}C_5Me_5)F[N(SiMe_3)_2]$ [54]} involved elimination of LiCl,[3,32,52,35,46] LiBr,[46,58a] HCl,[54,58b,58c] NaCl,[54] or Me_3SiCl.[53] Alternative methods [for **19**,[55] **21**,[52] and **23** (via **22**) [22]] are summarised in Equations (10.2),[55](10.3),[52] and (10.4)[54] and Scheme 10.1;[22] compound **21** was originally obtained from $AsCl_3 + 2Li[N(H)C_6H_2Bu^t_3\text{-}2,4,6]$ in hot toluene.[23b]

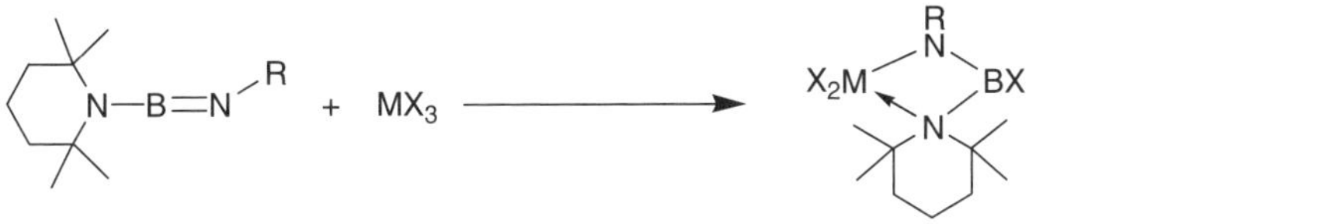

$$\text{TMP-N-B{=}N-R} + MX_3 \longrightarrow X_2M(\mu\text{-NR})(\leftarrow \text{TMP-N})BX \qquad (10.2)$$

19 (M = As, Sb; R = Bu^t, $C_6H_3Pr^i_2$-2,6; X = F, Cl, Br)

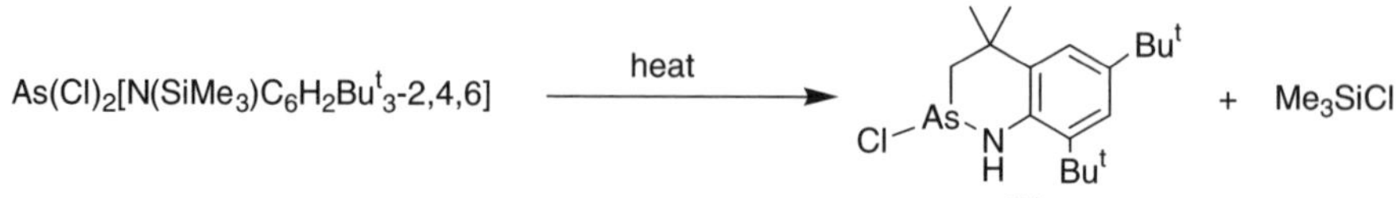

$$As(Cl)_2[N(SiMe_3)C_6H_2Bu^t_3\text{-}2,4,6] \xrightarrow{\text{heat}} \mathbf{21} + Me_3SiCl \qquad (10.3)$$

$$As(\eta^5\text{-}C_5Me_5)(Cl)[N(SiMe_3)_2] \xrightarrow{[Co(\eta^5\text{-}C_5H_5)_2F]} As(\eta^5\text{-}C_5Me_5)(F)[N(SiMe_3)_2] \qquad (10.4)$$

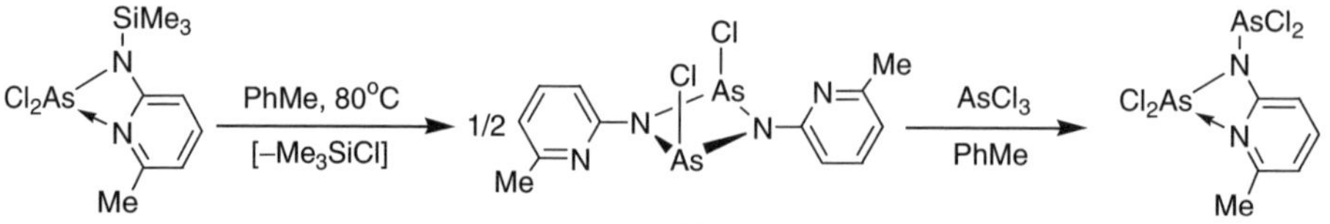

PhMe, 80°C; [–Me_3SiCl]; 1/2 **22**; $AsCl_3$, PhMe; **23**

Scheme 10.1[22]

X-Ray structural data are available for the three-coordinate M^{III} amides $As(\eta^5\text{-}C_5Me_5)Cl[N(R^1)R^2]$ ($R^1 = SiMe_3 = R^2$; $R^1 = H$, $R^2 = Bu^t$),[54] $AsCl_2[N(R)C_6H_2Me_3\text{-}2,4,6]$ (R = H, $SiMe_3$),[52] **20** (M = As, X = Cl)[58c] and **21**;[52] the four-coordinate crystalline pnictogene(III) complexes **19** (M = As or Sb, R = Bu^t, X = Cl)[55] and **23**[22] have the distorted trigonal bipyramidal M atom with the atoms (for **23**) N_{amido} and Cl and the lone pair in equatorial sites.

The cyclodimeric metal(III) imides **22**,[22] $[Sb(X)(\mu\text{-}NBu^t)]_2$ (X = Cl, Br)[46] and $[As(\eta^5\text{-}C_5Me_5)(\mu\text{-}NBu^t)]_2$[54] were obtained from $AsCl_2[N(SiMe_3)C_5H_3N(Me\text{-}6)\text{-}2]$ (Scheme 10.1),[22] the labile $Sb[N(Bu^t)SiMe_3]X_2$,[46] or $As(\eta^5\text{-}C_5Me_5)Cl[N(H)Bu^t]$ [54] by elimination of Me_3SiCl[20,46] or Me_3SiBr,[46] or by treatment of $As(\eta^5\text{-}C_5Me_5)Cl[N(H)Bu^t]$ with $Na[N(SiMe_3)_2]$ or Me_3SnNMe_2,[54] respectively. Another Me_3SiCl elimination reaction yielded the cyclic amidoarsenic(III) chloride **21**, Equation (10.3).[52] Reacting the labile compound $As(\eta^5\text{-}C_5Me_5)Cl(NMe_2)$ with Al_2Cl_6 in CH_2Cl_2 yielded the salt $[As(\eta^5\text{-}C_5Me_5)(NMe_2)][AlCl_4]$.[53a] Oxidation of the compound $As(CF_3)_2[N(SiMe_3)_2]$ with Cl_2 furnished the cyclodimeric arsenic(V) imide $[As(CF_3)_2Cl(\mu\text{-}NSiMe_3)]_2$.[35,56] The Cl^- ligand was displaced by F^- by addition of $[Co(\eta^5\text{-}C_5H_5)_2F]$ to $As(\eta^5\text{-}C_5Me_5)Cl[N(SiMe_3)_2]$, Equation (10.4).[54] The azides $M(Me)_2N_3$ (M = Sb, Bi) were obtained from $Sb(Me)_2NMe_2$ or $BiMe_2[N(SiMe_3)_2]$ and hydrazoic acid, HN_3.[57] Interaction of $As(Me)_2NMe_2$ and $As(H)Me_2$ or $(AlHMe_2)_3$ gave the diarsine $[AsMe_2]_2$ or the mixture of $[Al(Me)_2NMe_2]_2$, $[AsMe_2]_2$ and $As(H)Me_2$,[25a,25b] respectively; while addition of BH_3(thf) showed that the nitrogen rather than the arsenic atom of $As(Me)_2NMe_2$ was its Lewis base site.[59b] Treatment of $Sb(Cl)_2NMe_2$ with EtOH furnished $[Sb(Cl)_2(NHMe_2)(\mu\text{-}OEt)]_2$.[50]

10.2.5 Bis(amido)Metal(III) Salts

Table 10.3 lists the cationic bis(amido)metal(III) compounds which have been reported. The first of these, in 1981, described by Cowley and coworkers, was $[As(\eta^5\text{-}C_5Me_5)(NMe_2)][AlCl_4]$.[53a] An isoleptic series $[M(\mu\text{-}NBu^t)_2SiMe_2][AlCl_4]$ (M = As, Sb, Bi) was provided by Veith and coworkers in 1988;[48] this included the thus far sole example of a cationic amidobismuth compound, which has close cation-anion Bi ··· Cl contacts. It is evident from

Table 10.3 *Cationic Group 15 metal(III) amides*

Compound	Method[b]	Reference
[Bu^t N M $SiMe_2$ N Bu^t] $[GaCl_4]$ M = As, Sb,[a] Bi[a]	A	48
[Et N As N Et] [OTf]	A′	38
$[As(NEt_2)_2]$ (X = OTf, $AlCl_4$)	A′, A	38
[Bu^t N As N Bu^t]$_2$ $[GeCl_5][Cl]$	B	40
[Bu^t N Sb N Bu^t] [X] (X = OTf,[a] $SbCl_4$[a])	A or A′	47
[R N As N R] $[SnCl_5(thf)]$[a] R = $C_6H_2Me_3$-2,4,6	C	2a
[R N As N R] $[SnCl_5(thf)]$[a] R = $C_6H_3Pr^i_2$-2,6	D	62
[H N As X] $[AlCl_4]$[a] X = S, NH	A	58b, 58c
$[As(\eta^5-C_5Me_5)(NMe_2)][AlCl_4]$	A	53
[R N M N R] [X] M = As, R = Pr^i, X = $GaCl_4$[a]	A	41a, 41b
M = As, R = Ph, X = $GaCl_4$[a]	A	41a
M = Sb{N(H)Me_2}, X = OTf[a]	E	41a, 41b
[Me N As N Me] $[GaCl_4]$[a]	A	49

(continued)

Table 10.3 (*Continued*)

Compound	Method[b]	Reference
[As(N(R)As(Cl))₂NR]-type cyclic cation: R Cl / N—As / As NR / N—As / R Cl [$GaCl_4$] $R = C_6H_3Pr^i_2$-2,6,[a] $C_6H_3Me_2$-2,6	F	60
Bu^{t}N—SiMe$_2$ / As / Me / Me / Me / Me [$AlCl_4$][a]	A	53b
Bu^t / N / Sb / SiMe$_2$ / Me / Me / Me / Me [$AlCl_4$][a]	A	53b
Bu^t Me / N—Si / Sb Bu^{t}N NBu^t / N—Si / Bu^t Me [$InCl_4$][a]	G	61
NMe / Me N—As / As—NMe / MeN [$M'Cl_4$]$_2$ $M' = Al$,[a] Ga[a]	A	39

[a]Molecular structure by X-ray diffraction
[b]Methods:
A: addition of the appropriate Lewis acid $M'Cl_3$ ($M' = Al$, Ga, Sb) to the bis(amido)metal chloride;
A′: addition of Me_3SiOTf (with elimination of Me_3SiCl) (or AgOTf with elimination of AgCl) to the bis(amido)metal chloride; B - G: Equations (10.5)–(10.10), respectively

Table 10.3, that widely used precursors to these two-coordinate amidometal(III) salts have been the corresponding three-coordinate metal chlorides, methods A and A′ of Table 10.3, and Equation (10.9);[60] while those of B, C, D, F and G [cf. Equations (10.5),[40] (10.6),[2a] (10.7),[62] (10.9)[60] and (10.10)[61]] also probably implicate such an intermediate. Method E [cf. Equation (10.8)], a dimethylamine-elimination procedure,[41a,41b] may have some wider generality.

X-Ray structural data are now available on 15 of the cationic compounds, Table 10.3. For these, the following general features emerge: (i) the four- and five-membered pnictogene-containing rings are planar; (ii) where good quality data are available on cationic compounds and their chloride precursors, the M–N bond lengths are somewhat longer for the former, as illustrated in Table 10.4. The six-membered ring of crystalline $[\overline{As\{N(Me)(CH_2)_3N}Me\}][GaCl_4]$ is boat-shaped.[49] The cationic unit of crystalline $[\overline{As\{N(Me)(CH_2)_2N}Me\}]_2[M'Cl_4]_2$ ($M' = Al$, Ga) has a step-like fused tricyclic skeleton,

Table 10.4 *Comparative M–N bond lengths (Å) for two pairs of compounds*

Compound	M–N (Å)	Reference
$ClAs[N(Bu^t)CH]_2$ [a]	1.8045(12), 1.8057(12)	44a
$[As[N(R)CH]_2][OTf]$ R = $C_6H_2Me_3$-2,4,6	1.8377(14), 1.8271(13)	44a
$ClSb[N(Bu^t)CH]_2$ [b]	1.998(4), 2.000(4)	44b
$[Sb[N(Bu^t)CH]_2][OTf]$	2.018(2), 2.020(2)	44a

[a]There are no close intermolecular As···Cl contacts
[b]There are secondary Sb···Cl contacts, 3.862 Å

the centre of crystallographic symmetry being the centre of the internal four-membered As---NAs---N ring, the As–N, As–N_{bridge} and As···N_{bridge} bond lengths being 1.763(8), 1.955(7) and 2.103(8) Å, respectively (M′ = Al); the dication is fluxional in solution.[39] The tricyclic cation of $[Sb\{N(Bu^t)Si(Me)(\mu\text{-}NBu^t)_2Si(Me)NBu^t\}][InCl_4]$ has the positive charge centred on the antimony atom, with the Sb–NBu^t, Sb–NBu^t and Sb←NBu^t bond lengths being 2.256(7), 2.275(7) and 2.366(6) Å, respectively.[61]

Treatment of $[As\{N(Me)(CH_2)_nNMe\}][GaCl_4]$ (n = 2 or 3) with 2,3-dimethylbuta-1,3-diene yielded the crystalline salts **24** and **25**, in which the positive charge on the cation was *N*-, rather than *As*-, centred.[49] The Lewis acid character of the arsenic atom of $[As\{N(R)(CH_2)_2NR\}][OTf]$ (R = Me,[63a] Et [63b,38]) and $[As(NEt_2)_2][OTf]$,[63b,38] with respect to coordination by the pyridyl nitrogen atom of $4\text{-}Me_2NC_5H_4N$ or PBu_3, was probed by measuring equilibrium constants.

$[GaCl_4]^-$

24 n = 2, **25** n = 3

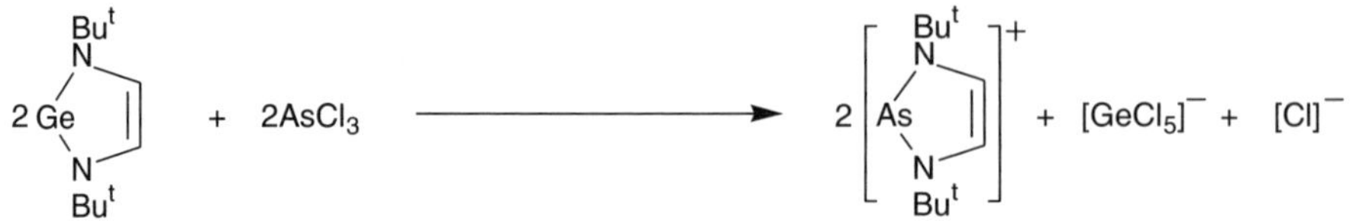

(10.5)[40]

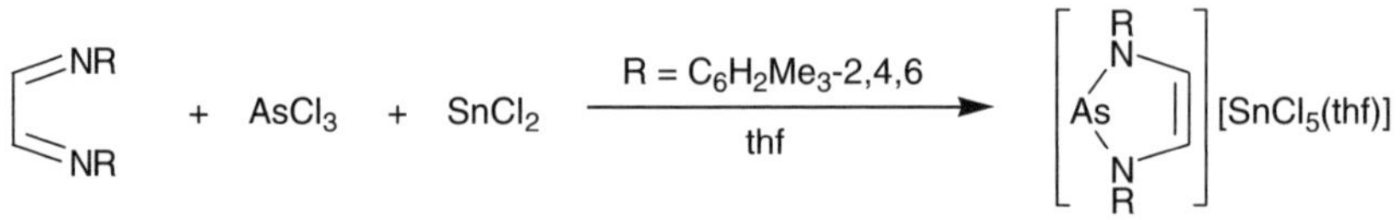

(10.6)[2a]

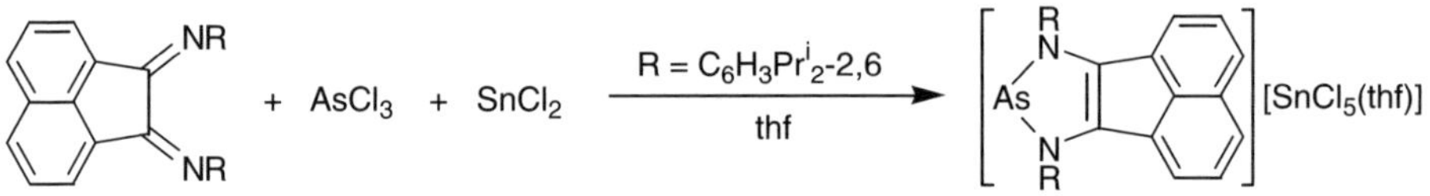

(10.7)[62]

(10.8)[41a,4

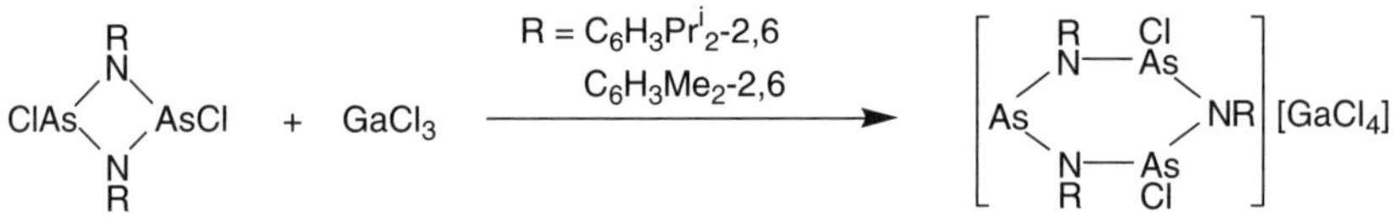

(10.9)[60]

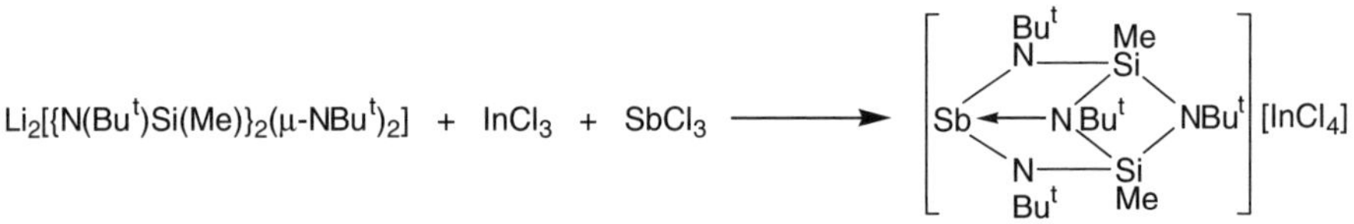

(10.10)[61]

10.3 Oligomeric Group 15 Metal Imides

The compounds considered in this section have bridging imido groups $(NR)^{2-}$ between two Group 15 metal centres, the most prolific being the binuclear compounds $[M(X)(\mu\text{-}NR)]_2$, Section 10.3.1. Alternatively, the ligands may be regarded as di(metallato)amides and hence are included in this survey.

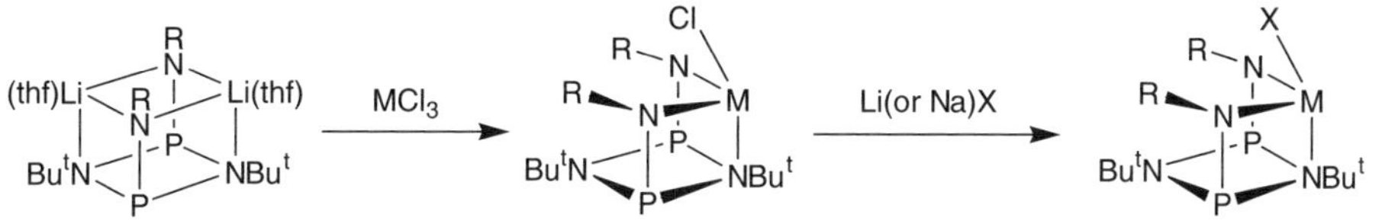

Scheme 10.2

The cage imides of Scheme 10.2 have just a single heavier group 15 metal component [M/R = Sb/Bu^t (**2**), Sb/Ph*, Bi/Bu^t (**2**); M/R/X = Sb/Bu^t/{N_3^*, OPh*, $N(SiMe_3)_2^*$}]; those marked with an asterisk were X-ray-characterised.[63c]

10.3.1 Binuclear and Oligomeric Group 15 Metal(III) Imides

The vast majority of these compounds have the metal in the oxidation state +3, i.e. they are the cyclopnict(III)azanes, $[M(X)(\mu\text{-NR})]_2$ (M = As, Sb, Bi). Chivers and coworkers have provided an excellent review.[11b] Table 10.5 lists the 40 such compounds reported to the end of 2006, together with a summary as to their method of preparation and, for 25, structures. Where appropriate, the latter refer inter alia to their geometry (*cis*- and/or *trans*-X_2) and conformation (*exo*/*exo* or *endo*/*endo*); these features are exemplified in **26**[70] and **27**.[70] The nitrogen environment is distorted trigonal planar; hence, the designations '*cis*/*trans*' refer to the relative disposition of the substituents X at the two atoms M. Selected geometric data on the thus far the only complete set of an isoleptic series of compounds $[M\{N(H)R\}(\mu\text{-NR})]_2$ (R = $C_6H_3Pr^i{}_2$-2,6 and M = As,[66] Sb,[66] Bi [24]) is presented in Table 10.6. Only a single case of an X-ray-characterised pair of *cis*/*trans* isomers has been reported and selected data for these are shown in Table 10.7. The four-membered $(MN)_2$ ring is generally planar, but a few examples (Tables 10.5–10.7) of puckered rings have been observed. The crystalline *trans*-complexes are prevalent, but in solution both forms are sometimes observed.[18,27b,65,68]

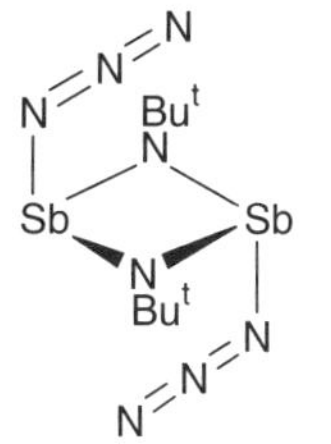

26 *trans/endo/endo*[70] 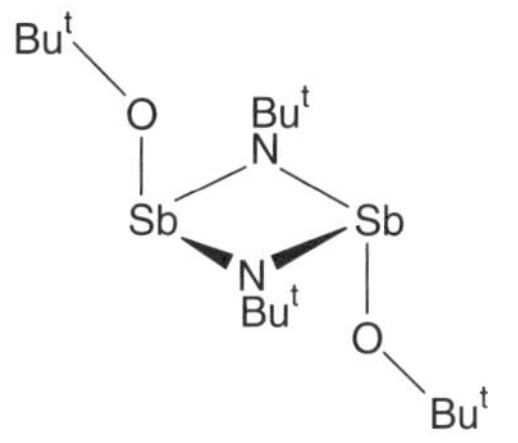 **27** *trans/exo/exo*[70]

Among the few reaction types of cyclodipnict(III)azanes are the nucleophilic (Cl^-/X^-) displacement reactions of $[M(Cl)(\mu\text{-NR})]_2$ (A in Table 10.5; Refs. 19b, 24, 43, 52, 65 and 68–72). Treatment of $[As\{N(H)R\}(\mu\text{-NR})]_2$ (R = $C_6H_3Pr^i{}_2$-2,6 or Cy) with the appropriate Group 12 metal bis(trimethylsilyl)amide or $ZnMe_2$ gave $M'[N(R)AsNR]_2$ (R = $C_6H_3Pr^i{}_2$-2,6; M′ = Zn, Cd)[67a] or **28**,[67b] respectively. The base adduct $[SbCl\{N(H)Me_2\}(\mu\text{-NBu}^t)]_2$ was obtained from $SbCl(NMe_2)_2$ and Bu^tNH_2.[50]

Reaction of the *in-situ* prepared $[Sb(NMe_2)(\mu\text{-NCy})]_2$ with $LiNH_2$ in thf, or Li[N(H)Cy] yielded the lithium salt of the trianion **29**[73a] (*cf.*, **6**) or $[\{Sb_2(NCy)_4\}_2Li_4]$

Table 10.5 *Cyclopnict(III)azanes $[M(X)(\mu\text{-}NR)]_2$ (M = As, Sb, Bi)*

M	X	R	Method/precursors	Reference (X-ray)[a]
As	Cl (*cis*)	Bu^t	A, C/$AsCl_3$, 2Li[$N(SiMe_3)_2$]	69[a] (almost planar)
As	C_6H_4Br-4 (*trans, exo/exo*)	Ph	B/As(C_6H_4Br-4)Cl_2, $3PhNH_2$, NEt_3	72[a]
As	C_6H_4R'	Bu^t	B/As(C_6H_4R')Cl_2, $3Bu^tNH_2$	37
	R′ = H, Br-4, OMe-2			
As	C_6H_4R'	C_6H_4R''	B/As(C_6H_4R')Cl_2, ($R''C_6H_4$)NH_2, NEt_3	72
	R′ = H	R″ = Br-4	B	72[a]
	H	OMe-2	B	72
	H	Cl-4	B	72
	Br-4	H	B	72
	OMe-2	H	B	72
	Me-4	H	B	72
As	Cl (*trans*)	$C_6H_2(CF_3)_3$-2,4,6	A,B/2$AsCl_3$, 4K[N(H)$C_6H_2(CF_3)_3$-2,4,6]	71[a]
As	η^3-C_5Me_5 (*cis, exo/exo*)	Me	B/As(C_5Me_5)Cl_2, $4MeNH_2$	54[a] (puckered)
As	η^3-$C_5Pr^i_4H$ (*trans, exo/exo*)	Me	B/As($C_5Pr^i_4H$)Cl_2, $4MeNH_2$	54[a] (planar)
As	Cl (*trans*)	$C_6H_2Bu^t_3$-2,4,6	B/$AsCl_2$[N(H)$C_6H_2Bu^t_3$-2,4,6], NEt_3	52[a] (planar)
As	Cl (*cis*)	C_5H_3N-2-Me-6	C/$AsCl_2$[N($SiMe_3$)C_5H_3N-2-Me-6]	22[a] (planar)
As	N(H)$C_6H_3Pr^i_2$-2,6 (*cis, exo/exo*)	$C_6H_3Pr^i_2$-2,6	A,B/$AsCl_3$, 3Li[N(H)$C_6H_3Pr^i_2$-2,6]	66[a] (puckered)
Sb	N(H)$C_6H_3Pr^i_2$-2,6 (*trans, exo/exo*)	$C_6H_3Pr^i_2$-2,6	A,B/$SbCl_3$, 2Li[N(H)$C_6H_3Pr^i_2$-2,6]	66[a] (planar)
Bi	N(H)$C_6H_3Pr^i_2$-2,6 (*trans, exo/exo*)	$C_6H_3Pr^i_2$-2,6	A,B/$BiCl_3$, 3K[N(H)$C_6H_3Pr^i_2$-2,6]	24[a] (planar)
Sb	Br	Bu^t	C/$SbBr_2$[N($SiMe_3$)Bu^t]	46
Sb	Cl (*cis*)	Bu^t	C/$SbCl_2$[N($SiMe_3$)Bu^t]	65,[a] 46 (2 puckered, 1 planar)
Sb	N(H)Bu^t	Bu^t	A/[SbCl(μ-NBu^t)]$_2$, 2Li[N(H)Bu^t]	65
Sb	N(H)$C_6H_3Me_2$-2,6 (*cis, exo/exo*)	Bu^t	A/[SbCl(μ-NBu^t)]$_2$, 2Li[N(H)$C_6H_3Me_2$-2,6]	65[a] (puckered)
Sb	N(H)$C_6H_3Pr^i_2$-2,6 (*cis, exo/exo*) (*trans, exo/exo*)	Bu^t	A/[SbCl(μ-NBu^t)]$_2$, 2Li[N(H)$C_6H_3Pr^i_2$-2,6]	65[a] (puckered) (planar)
Sb	Bu^t (*trans, exo/exo*)	Bu^t	A/[SbCl(μ-NBu^t)]$_2$, 2LiBu^t	68[a]
	OMe		A/[SbCl(μ-NBu^t)]$_2$, 2NaOMe	68
	OPh		A/[SbCl(μ-NBu^t)]$_2$, 2NaOPh	68
	OBu^t		A/[SbCl(μ-NBu^t)]$_2$, 2NaOBu^t	68
	$N(SiMe_3)_2$		A/[SbCl(μ-NBu^t)]$_2$, 2Li[$N(SiMe_3)_2$]	68
	$P(SiMe_3)_2$		A/[SbCl(μ-NBu^t)]$_2$, 2Li[$P(SiMe_3)_2$]	68
	Me		A/[SbCl(μ-NBu^t)]$_2$, 2LiMe	68

(*continued*)

Table 10.5 *(Continued)*

M	X	R	Method/precursors	Reference (X-ray)[a]
Sb (**26**)	N_3 (*trans, endo/endo*)	Bu^t	A/[SbCl(μ-NBu^t)]$_2$, 2NaN_3	70[a]
Sb (**27**)	OBu^t (*trans, exo/exo*)	Bu^t	A/[SbCl(μ-NBu^t)]$_2$, 2$KOBu^t$	70[a]
Sb	Ph (*trans*)	Bu^t	A/$SbCl_2$(Ph), 2Li[N(H)Bu^t]	43[a]
Bi	Ph (*trans*)	Bu^t	A/$BiCl_2$(Ph), 2Li[N(H)Bu^t]	43[a]
Sb	OTf (*trans, exo/exo*)	$C_6H_2Bu^t_3$-2,4,6	A/[SbCl(μ-$NC_6H_2Bu^t_3$-2,4,6)]$_2$, 2AgOTf	52[a] (planar)
Sb	N(H)$C_6H_3Me_2$-2,6 (*trans, exo/exo*)	$C_6H_3Me_2$-2,6	D/Sb[N(H)$C_6H_3Me_2$-2,6]$_3$	18[a]
Sb	NMe_2 (*trans*)	$C_6H_2(OMe)_3$-3,4,5	D/2Sb$(NMe_2)_3$, 2[3,4,5-$(MeO)_3C_6H_2NH_2$]	27a[a] (planar)
Sb	NMe_2 (*trans*)	C_5H_3N-2-Me-5	D/2Sb$(NMe_2)_3$, 2[5-Me-2-$NC_5H_3NH_2$]	27a[a] (planar)
Bi	N$(SiMe_3)_2$ (*trans*)	$SiMe_3$	A, B/$BiCl_3$, 3K[N$(SiMe_3)_2$] (minor product)	19b[a] (planar)

[a]Molecular structure by X-ray diffraction
[b]Methods:
A: M′Cl elimination (M′ = Li, Na, K, Ag);
B: HCl elimination;
C: Me_3SiCl or Me_3SiBr elimination;
D: NH(R′)R″ elimination; there are steric limitations; e.g. Sb[N(H)$C_6H_2Bu^t_3$-2,4,6]$_3$ in CD_2Cl_2 was unchanged after several hours at 60 °C.[17]

Table 10.6 *Comparative selected geometric data for crystalline* $[M\{N(H)R\}(\mu\text{-}NR)]_2$ *(M = As (cis), Sb (trans), Bi (trans); R =* $C_6H_3Pr^i_2$*-2,6)*

M	M–N (Å)[a] (endocyclic)	M–N (Å)[a] (exocyclic)	N-M-N (°)[a] (endocyclic)	M-N-M (°)[a] (endocyclic)	ΣM (°)	Ring conformation	Ref.
As	1.881	1.838	79.0	97.8	280.5	Puckered[b] (*exo/exo*)	66
Sb	2.064	2.032	77.7	102.3	274	Planar (*exo/exo*)	66
Bi	2.158	2.164	78.5	101.5	268.9	Planar (*exo/exo*)	24

[a] Average.
[b] Fold angle about As···As vector: 150.5°

(**30**),[73b] respectively. Compound **31** was obtained by interaction of $[As(Cl)(\mu\text{-}NBu^t)]_2$ and $[As\{N(Bu^t)Li\}(\mu\text{-}NBu^t)]_2$.[74] The product from $[Sb\{N(H)Bu^t\}(\mu\text{-}NBu^t)]_2$, *trans*-$[Sb\{N(H)\text{-}C_6H_3Pr^i_2\text{-}2,6\}(\mu\text{-}NBu^t)]_2$ or its *cis*-isomer with respectively $2NaBu^n$, $2LiBu^n$ or $2LiBu^n$ was the cubane $[Sb_2Na_2(\mu\text{-}NBu^t)_4]$, $[Sb_2Li_2(\mu\text{-}NBu^t)_2(\mu\text{-}NC_6H_3Pr^i_2\text{-}2,6)_2]$ or the tricyclic compound **32** [with $(SbBu^n)_n$ as possible coproduct], respectively.[65] The macrocycles $[\{Sb(\mu\text{-}NR)\}_2(\mu\text{-}NR)]_6$ [R = Ph, from $SbCl_3/3Li[N(H)Ph]$;[18a] R = C_6H_4OMe-2, from $Sb(NMe_2)_3/Sn\{N(H)R\}\{\mu\text{-}N(H)R\}Li(thf)_2$[18b]] contain six dimeric imidoantimony rings. From $[M(NMe_2)(\mu\text{-}NR)]_2$ and Li[N(H)R], the cubane (M = Bi, R = Bu^t) $[Bi_2\{Li(thf)\}_2(\mu\text{-}NBu^t)_4]$ and (M = Sb, R = Cy) a product with interlocked 'broken cubes' $[\{Sb_2(NCy)_4\}_2Li_4]$ (**30**) was obtained.[82e] Further examples are discussed in Section 10.5.

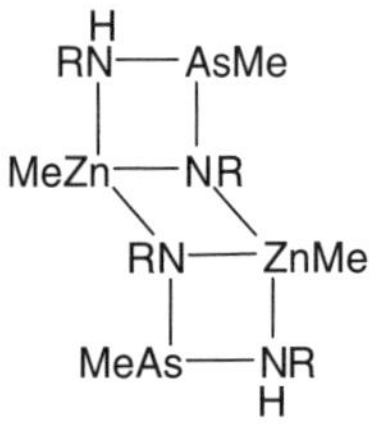

28 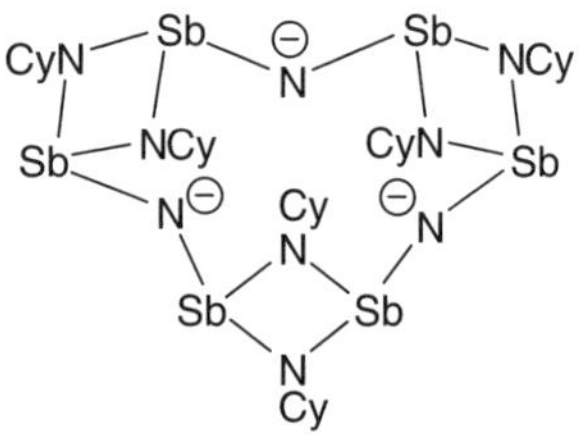 **29** 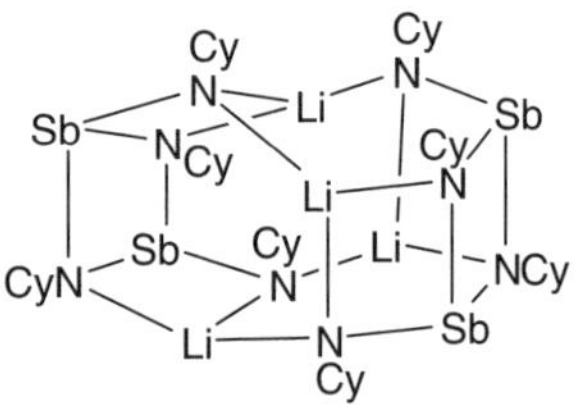 **30**

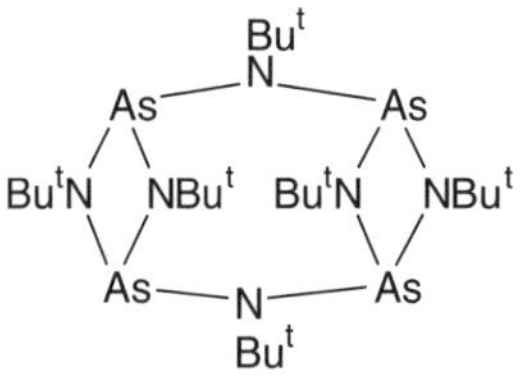

31 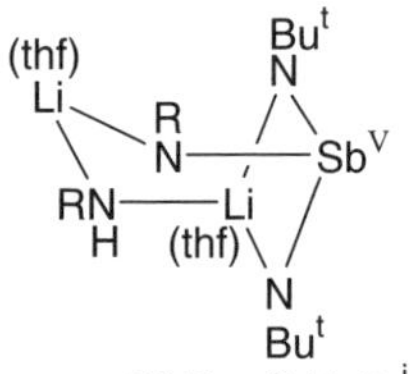 **32** R = $C_6H_3Pr^i_2$-2,6

10.3.2 Binuclear Group 15 Metal(V) Imides

The first cyclopnict(V)azane, reported by Roesky and coworkers in 1983, was the arsenic compound **33**, obtained by oxidative addition of Cl_2 and concomitant trimethyl(chloro)-silane elimination from a bis(trimethylsilyl)amidoAs(III) precursor, Equation (10.11). Each As atom in crystalline **33** is in a trigonal bipyramidal environment (Cl and N_{ax}, axial); there is a rhomboidal $(AsN)_2$ core, with the As–N_{ax} longer than the As–N_{eq} bonds [a feature which is replicated in the $(Sb^VN)_2$ compounds **34**,[75] **35**,[64] and **36**,[64] but differs from the $(M^{III}N)_2$ compounds of Section 10.3.1].[56] Thermolysis of **33** yielded the crystalline compounds $[As(CF_3)_2(\mu\text{-}N)]_x$ ($x = 3$ or 4).[35]

Table 10.7 *Comparative selected geometric data for crystalline cis-* and *trans*-[Sb{N(H)R}-(μ-NBu^t)]$_2$ *(R = $C_6H_3Pr^i_2$-2,6)*[65]

	M–N (Å)[a] (endocyclic)	M–N (Å)[a] (exocyclic)	N-M-N (°)[a] (endocyclic)	M-N-M (°)[a] (endocyclic)	ΣM (°)	Ring conformation
cis	2.038	2.055	78.1	100.5	271.4	Puckered[b] (R's *exo*/*exo*)
trans	2.031	2.058	84.3	95.7	273.2	Planar (R's *exo*/*exo*)

[a]Average. [b]Fold angle about Sb···Sb vector: 161°

$$2As(CF_3)_2[N(SiMe_3)_2] + 2Cl_2 \longrightarrow \mathbf{33} + 2Me_3SiCl \quad (10.11)$$

33

Five crystalline cyclodistib(V)azanes **34**,[75] **35**,[64] and **36**[64] have been reported: compound **34** by dimethylamine elimination from $Sb(NMe_2)_2Ph_3$ and $Ph(CH_2)_2NH_2$,[75] and the compounds **35** and **36** by LiCl elimination from the appropriate compounds $Sb(Cl)_2Ph_2(X)$ and 2Li[N(H)R].[64] Each of the crystalline compounds **34–36** has the Sb atoms in a distorted trigonal bipyramidal environment [axial: N and Ph (**35**) or N(H)R (**36**)]. In **35**, the two benzyls are mutually transoid,[64] whereas in **34** the two groups $Ph(CH_2)_2$ are cisoid;[75] the imido pairs and N(H)R pairs in **36** are mutually transoid.[64]

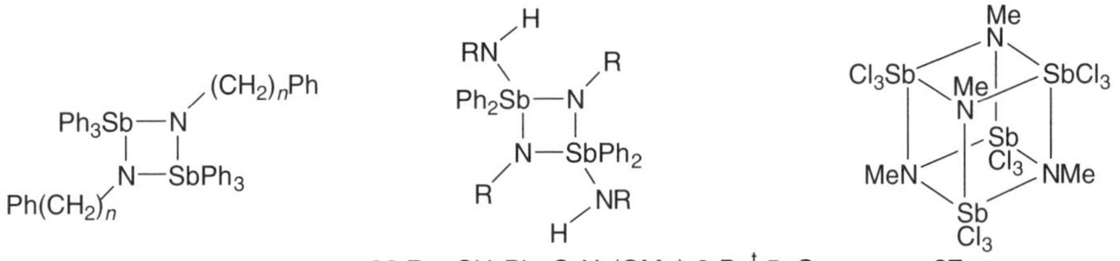

34 *n* = 2, **35** *n* = 1   **36** R = CH_2Ph, C_6H_4(OMe)-2-Bu^t-5, Cy   **37**

Interaction of $SbCl_5$ and an equimolar portion of $N(Me)(SiMe_3)_2$ gave the crystalline distorted cubane **37**, having Sb–N bond lengths 2.151 to 2.189 Å, Sb–N–Sb angles 99.6 to 100.8° and N–Sb–N angles 78.2 to 79.1°.[76] Compound **35** is included in this section, because although it is not a binuclear Group 15 metal(V) imide, it may be regarded as a dimer thereof.

10.4 Mononuclear Group 15 Metal(V) Amides

The title compounds are rare (see also **32**[65]). A requirement appears to be that the metal should have an adequate number of electron-withdrawing substituents. The simplest member to have been reported appears to be $Sb(NMe_2)_2Ph_3$, obtained as a solid and characterised by microanalysis, IR and ^{1}HNMR spectroscopy; it was prepared from $SbCl_2Ph_3 + 2LiNMe_2$, and was used as a precursor to **34** (Section 10.3.2).[75] The compounds $SbX(Me)_3\{N(C_6F_5)(C_5H_4N\text{-}2)\}$ (X = Cl or Br) and $Sb(Me)_3\{N(C_6F_5)(C_5H_4N\text{-}2)\}_2$

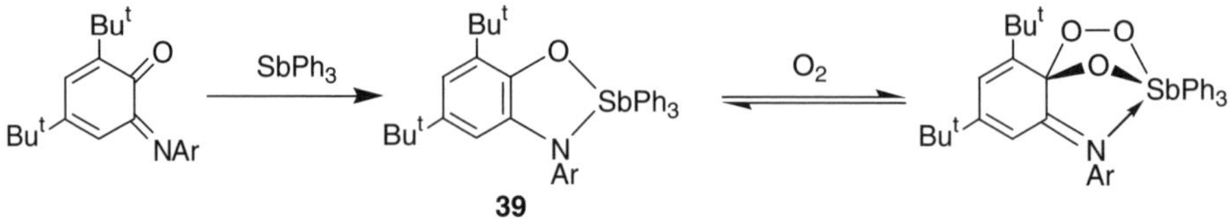

Scheme 10.3 $(Ar = C_6H_3Pr^i_2\text{-}2,6)$

were prepared from SbX_2Me_3 and (1 or 2) $Li[N(C_6F_5)(C_5H_4N\text{-}2)]$.[16c] The crystalline distorted trigonal bipyramidal complex $[SbCl_2(Me)\{N(SiMe_3)_2\}_2]$ [av. Sb–N 1.991 Å, Cl–Sb–Cl 175°] was obtained from $Sb(Me)\{N(SiMe_3)_2\}_2$ and SO_2Cl_2.[47a] The crystalline distorted trigonal bipyramidal As^V compound **38** (O atoms axial; pseudo-2-fold axis through As-C_{ipso})[77b] was prepared from $As(O)(OH)_2Ph$ and $2\text{-}HOC_6H_4NH_2$ by

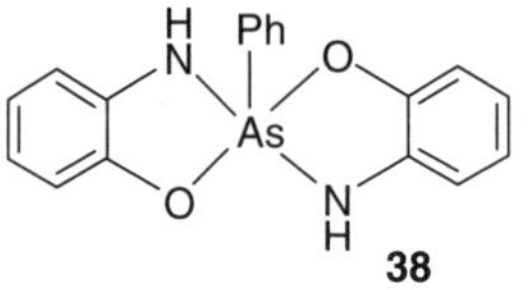

38

azeotropic distillation in benzene.[77a] The *N*-succinimidometal(V) compounds $M\{NC(O)(CH_2)_2CO\}(Ph)_3X$ [M = As, Bi and X = Br; M = Sb, X = Cl; M = Sb, Bi and X = $\{NC(O)(CH_2)_2CO\}$]] and the phthalimidometal(V) complexes $M(Br)\{NC(O)\text{-}C_6H_4CO\text{-}2\}R_3$ (M = As, Sb, Bi and R = Me, Ph) were synthesised either by (i) an oxidative addition of the appropriate *N*-halosuccinimide or *N*-bromophthalimide to MPh_3 or MMe_3, or (ii) the metathetical exchange reaction between MBr_2Ph_3 and $Ag\{NC(O)(CH_2)_2CO\}$.[79] A hydrogen chloride elimination procedure yielded the bis(imidazolo)metal(V) compounds $M(NC_3H_3N)_2(R)_2R'$ (M = As, Sb; R = Ph, C_6F_5; R′ = C_6F_5, C_6H_4Me-4) from the appropriate reagents $MCl_2(R)_2R'$, 2HNC_3H_3N and $2NEt_3$.[80] A Me_3SiCl elimination reaction yielded $AsCl\{N(Me)C(O)NC_6H_4CF_3\text{-}3\}_2$ from $AsCl_5$ and the substituted urea O=C-$\{N(Me)SiMe_3\}\{N(C_6H_4CF_3\text{-}3)SiMe_3\}$.[78a] The strong Lewis acid $SbCl_4\{N(SO_2F)_2\}$, obtained from $SbCl_5$ and $N(H)(SO_2F)_2$, formed *O*-centred adducts with $P(O)Ph_3$ and $S(O)Me_2$.[78b]

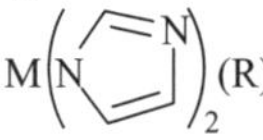

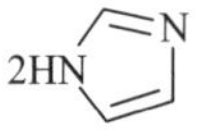

The distorted tetragonal pyramidal **39** [one C_{ipso} apical; Sb–N, 2.041(3) Å], prepared by Cherkasov *et al.*, as shown in Scheme 10.3, is remarkable in that it is unique among main group metal complexes in reversibly binding O_2 [Sb–N, 2.425(3) Å in the peroxo adduct].[81]

10.5 Group 15 Metal(III) Macrocyclic Imides

Reviews by Wright and coworkers have dealt with aspects of this theme, especially with poly(imidoantimonates) containing anions of the types **3–5** (Section 10.1) (e.g. R = Cy), generated by stepwise metallation of $Sb(NMe_2)_3$ with Li[N(H)R] and, to a limited extent, with bismuth analogues derived from $Bi(NMe_2)_3$.[10a,10b,10c] Examples include $M[Sb_3(NCy)_4(NMe_2)_2]$ (M = Li,[82a,82f] Na[82f]) and $K[Sb_3(NCy)_4\{N(H)Cy\}_2]$[82b] (cf. **3**), $M_4[Sb_2(NCy)_4]$ (M = Cu,[82b,82d] Ag[82b]) (cf. **4**) and $Pb_3[Sb(NCy)_3]_2$[82b] (cf. **5**). A very recent survey by Chivers and coworkers covers much of this ground, including mixed metal macrocycles such as certain cubanes, **28**,[67b] **30**,[73b] or **32**[65] (see Section 10.3.1), **40** [face-to-face dimers (inter-locked 'broken' cubanes) of the mixed metal cubanes $\{M_2M'_2(\mu_3\text{-}NCy)_4\}$ (M = As, Sb; M′ = Li, Na)].[27c,73b,82c,83] Compound $[K(\eta^6\text{-}PhMe)]$-$[Sb_3(\mu_3\text{-}NCy)_3(\mu\text{-}NCy)(OBu^t)_2]$ (schematically shown in **41**) was obtained from Li-$[Sb_3(NCy)_4(NMe_2)_2]$ and successively $KOBu^t$ and $2Bu^tOH$.[82f,83b] The same Sb_3 precursor afforded $Li[Sb_3(\mu_3\text{-}NCy)_3(\mu\text{-}NCy)\{S\text{-}2\text{-}Me(imidazole)\}_2]$ or $[Sb(\mu_3\text{-}NCy)_3(OC_5H_4N\text{-}2)]_4$ upon treatment with the appropriate thiol or 2-hydroxypyridine.[82f] The reaction of $As(NMe_2)_3$ with $3Li[P(H)R]$ gave $Li[As(PR)_3]$ ($R = Bu^t$, 1-Ad), which upon refluxing in toluene yielded the Zintl compound As_7Li_3.[84a] The analogous Sb Zintl compound was obtained by the low temperature (30–40 °C) thermolysis of the single-source precursor $[Li_6\{Sb(PCy)_3\}_2]\cdot 6N(H)Me_2$.[84b] The bis(imido)organoarsenate **42**, incorporating $LiBu^n$, was obtained from $AsCl_3$, RNH_2, NEt_3 and $LiBu^n$.[85] The imido(amido)bismuth compound **43** was prepared from $BiCl_3$ and $3Li[N(H)R]$.[86]

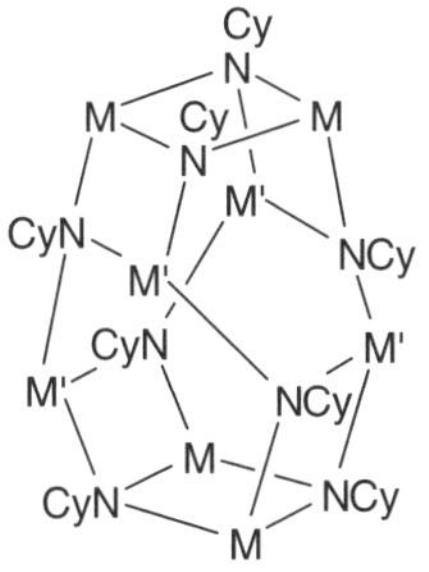

40 (M = As, Sb; M′ = Li, Na)

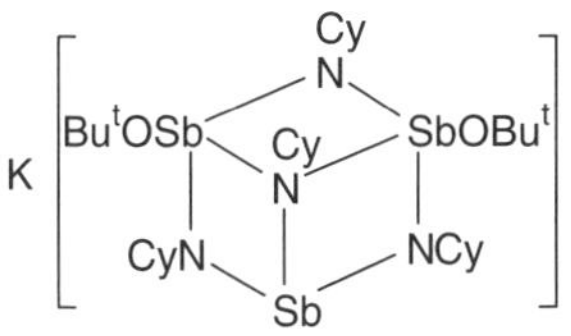

41

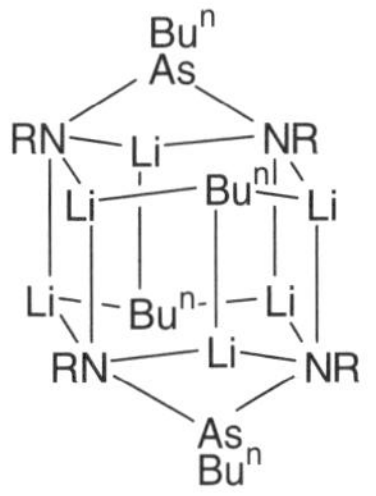

42 $R = C_6H_4(OPh)\text{-}2$

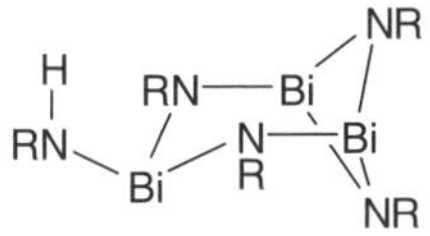

43 $R = C_6H_3Me_2\text{-}2,6$

10.6 Miscellaneous Group 15 Metal-Nitrogen Compounds

Photolytic reduction of a toluene solution of a chlorobis(amido)arsane, as shown in Equation (10.12), yielded the corresponding persistent As^{II} amide, characterised by its solution EPR spectrum at 300 K as a 1 : 1 : 1 : 1 quartet, $g_{av} = 2.008$, $a(^{75}As) = 31.2$ G, $t_{1/2}$ ca. 15 days.[3]

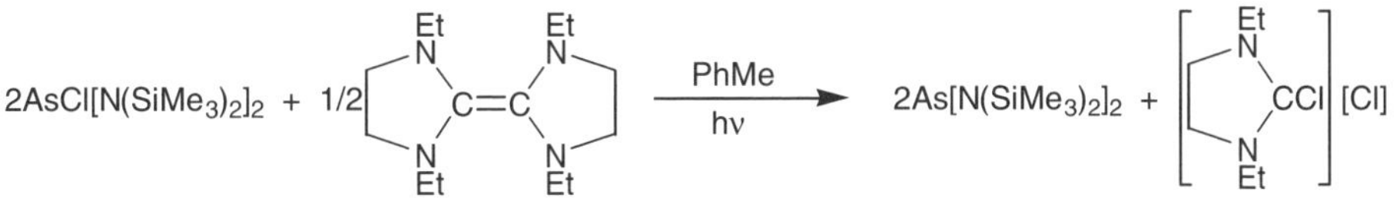

$$2AsCl[N(SiMe_3)_2]_2 + 1/2\,(\text{tetraethyl bis(imidazolidinylidene)}) \xrightarrow[h\nu]{PhMe} 2As[N(SiMe_3)_2]_2 + [\text{EtN(CH}_2)_2\text{NEt-CCl}][Cl] \quad (10.12)$$

Another product (also from the Sussex group) is the two-coordinate crystalline arsenic compound, the red As^{III} amido(imide) **44**, obtained from $AsCl_3$ and $3Li[N(H)C_6H_2Bu^t_3$-

$2,4,6]$.[23a] The As–N and As=N bond lengths are 1.745(7) and 1.714(7) Å, respectively, and the N–As–N angle is 98.8(3)°. A ^{1}H NMR spectral spin saturation-transfer experiment showed that in C_6D_6 solution **44** underwent slow prototropy, involving H^+ 1,3- N→N shifts.[23a] Compound **44** with $(AlMe_3)_2$ afforded $AlMe_2[N(H)C_6H_2Bu^t{}_3\text{-}2,4,6]$.[23b] The yellow dimeric imido complex $[As(Cl)(\mu\text{-}NC_6H_2Bu^t{}_3\text{-}2,4,6)]_2$ in Et_2O slowly changed colour to red,[52] which may have been due to its dissociation into the solvated monomer.

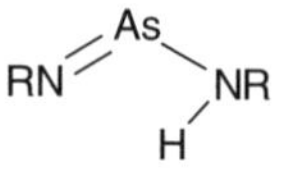

44 R = $C_6H_2Bu^t{}_3$-2,4,6

The fused bicyclic M^I (M = As, Sb) compounds **45** (R = Bu^t, 1-adamantyl) are neither amido- nor imido-metal complexes (however, alternative valence bond schemes with an amido N–M bond, **45′** and **45″**, may be implicated), but they are included here since they were derived from an amine $N(H)\{CH_2C(O)R\}_2$ (with MCl_3 and NEt_3 in thf);[87a] the X-ray structures of four of these (R = Bu^t, 1-Ad) were determined: M–N bond lengths 1.839(3) Å (**45a**) and 2.064(3) Å (**45b**) for the Bu^t compounds.[87b] These *tert*-butyl compounds **45a** and **45b** were shown to be *M*-centred ligands in *trans*-$[Pt(Me)(\mathbf{45b})(PPh_3)_2][SbF_6]$[88a] and $[\{Mn(\eta^5\text{-}C_5H_5)(CO)_2(\mathbf{45a}\text{ or }\mathbf{45b})]$,[88b] prepared by treatment with *trans*-$[Pt(Me)(OCMe_2)(PPh_3)_2][SbF_6]$[88a] and $[\{Mn(\eta^5\text{-}C_5H_5)(CO)_2(thf)]$,[88b] respectively.

The crystalline compound $[As\{N(SiMe_3)_2\}_2]_2CH_2$ (av. As–N 1.895 Å, av. N–As–N′ 107°) was obtained from $(AsCl_2)_2CH_2$ and $4LiN(SiMe_3)_2\}_2$.[89b] Its analogue, $[As(NMe_2)_2]_2CH_2$, with $1,2\text{-}C_6H_4(NH_2)_2$ and LiBu yielded **46**, the first compound containing the diazaarsolyl anion (av. As–N 1.788 Å).[89a] Similarly, the antimony analogue of **46** was prepared using $Sb(NMe_2)_3$ as the precursor, which with pmdeta oligomerises as the tetramer, av. Sb–N 2.02 Å in $[C_6H_4N_2SbLi(pmdeta)]_4$.[89c]

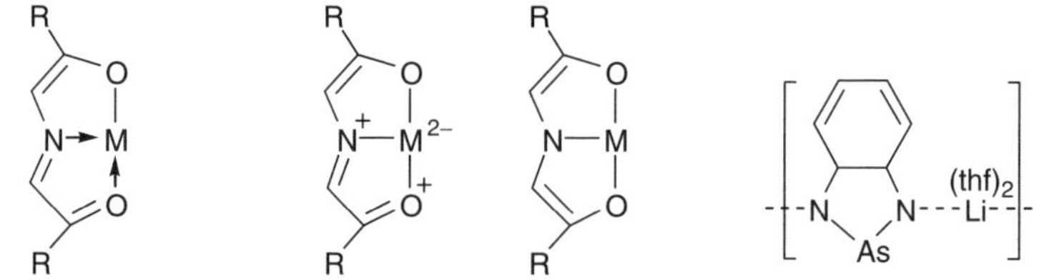

45a (M = As)
45b (M = Sb)
45'
45"
46

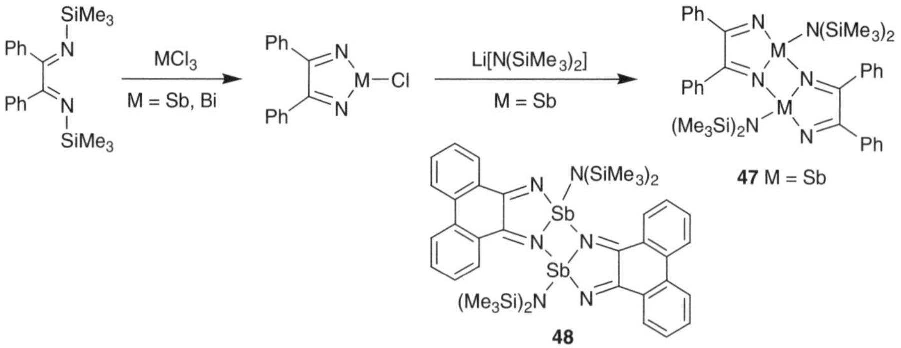

MCl_3, M = Sb, Bi; $Li[N(SiMe_3)_2]$, M = Sb

47 M = Sb

48

Scheme 10.4

The crystalline As^I salts [**1**][$SnCl_5$(thf)] and [**1**][As_2I_8] (the structure of the cation **1** is shown in Section 10.1) were synthesised from $C_5H_3N\{C(Me){=}NAr\}_2$-2,6 (Ar = $C_6H_3Pr^i_2$-2,6) and $AsCl_3/SnCl_2$ and AsI_3, respectively.[2a] The As–N(Ar) and As–N(py) bond lengths in the former are 2.095(4) and 1.862(5) Å, respectively.[2a]

The first 1,3,2-diazastiboles and -bismoles were prepared from *N,N′*-bis(trimethylsilyl)-1,4-diazabutadiene derivatives (derived from phenanthraquinone or benzil) and the appropriate MCl_3 (M = Sb, Bi), as shown in Scheme 10.4 for the benzil derivative; the crystalline compounds **47** and **48** are dimers.[90]

References

1. M. F. Lappert, P. P. Power, A. R. Sanger and R. C. Srivastava, *Metal and Metalloid Amides*, Horwood-Wiley, Chichester, 1980.
2. (a) G. Reeske and A. H. Cowley, *Chem. Commun.*, 2006, 1784; (b) B. D. Ellis and C. L. B. Macdonald, *Coord. Chem. Rev.*, 2007, **251**, 936.
3. M. J. S. Gynane, A. Hudson, M. F. Lappert, P. P. Power and H. Goldwhite, *J. Chem. Soc., Dalton Trans.*, 1980, 2428.
4. (a)W. Levason and G. Reid, *Arsenic, antimony and bismuth; in Comprehensive Coordination Chemistry* II (eds. J. A. McCleverty and T. J. Meyer,). Vol. 3 (ed. G. F. R. Parkin), Ch. 3, Elsevier, Oxford, 2004; (b) C. A. McAuliffe, *Arsenic, antimony and bismuth; in Comprehensive Coordination Chemistry I* (eds. G. Wilkinson, R. D. Gillard and J. A. McCleverty,), Vol. 3, Ch. 28, Pergamon, Oxford, 1987.
5. (a) J. L. Wardell, *Arsenic antimony and bismuth; in Comprehensive Organometallic Chemistry II* (eds. E. W. Abel, F. G. A. Stone and G. Wilkinson), Vol. 2 (ed. A. G. Davies), Ch. 8, Pergamon, Oxford, 1995; (b) J. L. Wardell, *Arsenic, antimony and bismuth; in Comprehensive Organometallic Chemistry I* (eds. G. Wilkinson, F. G. A. Stone and E. W. Abel,), Vol. 2, Ch. 13, Pergamon, Oxford, 1982.
6. (a) D. B. Sowerby, *Structural Chemistry of Organic Compounds Containing Arsenic, Antimony and Bismuth,* in *Organic Arsenic, Antimony and Bismuth Compounds* (ed. S. Patai), Ch. 2, John Wiley & Sons, Inc., New York, 1994; (b) S. M. Godfrey, C. A. McAuliffe, A. G. Mackie and R. G. Pritchard, *Chemistry of Arsenic, Antimony and Bismuth* (ed. N.C. Norman), Blackie, London, 1998.
7. L. Stahl, *Coord. Chem. Rev.*, 2000, **210**, 203.
8. F. Kober, *Synthesis*, 1982, 173.
9. (a) R. Bohra and H. W. Roesky, *Adv. Inorg. Chem. Radiochem.*, 1984, **28**, 201; (b) H. W. Roesky, *J. Organomet. Chem.*, 1985, **281**, 69.
10. (a) M. A. Paver, C. A. Russell and D. S. Wright, *Angew. Chem., Int. Ed.*, 1995, **34**, 1545; (b) M. A. Beswick and D. S. Wright, *Coord. Chem. Rev.*, 1998, **176**, 373; (c) E. L. Doyle, L. Riera and D. S. Wright, *Eur. J. Inorg. Chem.*, 2003, 3279.
11. (a) J. K. Brask and T. Chivers, *Angew. Chem., Int. Ed.*, 2001, **40**, 3960; (b) M. S. Balakrishna, D. J. Eisler and T. Chivers, *Chem. Soc. Rev.*, 2007, **36**, 650.
12. G. G. Briand and N. Burford, *Adv. Inorg. Chem.*, 2000, **50**, 285.
13. (a) A. V. Belyakov, P. E. Baskakova, A. Haaland, L. K. Krannich and O. Swang, *J. Gen. Chem. USSR*, 1997, **67**, 245. [*Zh. Obsh. Khim.*, 1997, **67**, 263]; (b) P. E. Baskakova, A. V. Belyakov, T. Colacot, L. K. Krannich, A. Haaland, H. V. Volden and O. Swang, *J. Mol. Struct.*, 1998, **445**, 311.
14. (a) C. J. Carmalt, N. A. Compton, R. J. Errington, G. A. Fisher, I. Moenandar and N. C. Norman, *Inorg. Synth.*, 1996, **31**, 98; (b) W. Clegg, N. A. Compton, R. J. Errington, G. A. Fisher, M. E. Green, D. C. R. Hockless and N. C. Norman, *Inorg. Chem.*, 1991, **30**, 4680.
15. W. Clegg, N. A. Compton, R. J. Errington, N. C. Norman and N. Wishart, *Polyhedron*, 1989, **8**, 1579.

16. (a) A. Bashall, A. D. Bond, A. D. Hopkins, S. J. Kidd, M. McPartlin, A. Steiner, R. Wolf, A. D. Woods and D. S. Wright, *J. Chem. Soc., Dalton Trans.*, 2002, 343; (b) P. L. Shutov, S. S. Karlov, K. Harms, D. A. Tyurin, A. V. Churakov, J. Lorberth and G. S. Zaitseva, *Inorg. Chem.*, 2002, **41**, 6147; (c) P. L. Shutov, S. S. Karlov, K. Harms, M. V. Zabalov, J. Sundermeyer, J. Lorberth and G. S. Zaitseva, *Eur. J. Inorg. Chem.*, 2007, 5684.
17. (a) N. Burford, C. L. B. Macdonald, K. N. Robertson and T. S . Cameron, *Inorg. Chem.*, 1996, **35**, 4013; (b) M. R. Mason, S. S. Phulpagar, M. S. Mashuta and J. F. Richardson, *Inorg. Chem.*, 2000, **39**, 3931; (c) L. H. Gade, B. Findeis, O. Gevert and H. Werner, *Z. Anorg. Allg. Chem.*, 2000, **626**, 1030.
18. (a) R. Bryant, S. C. James, J. C. Jeffery, N. C. Norman, A. G. Orpen and U. Weckenmann, *J. Chem. Soc., Dalton Trans.*, 2000, 4007; (b) M. A. Beswick, M. K. Davies, M. A. Paver, P. R. Raithby, A. Steiner and D. S. Wright, *Angew. Chem., Int. Ed.*, 1996, **35**, 1508.
19. (a) M. Bochmann, X. Song, M. B. Hursthouse and A. Karaulov, *J. Chem. Soc., Dalton Trans.*, 1995, 1649; (b) W. J. Evans, D. B. Rego and J. W. Ziller, *Inorg. Chim. Acta*, 2007, **360**, 1349; (c) M. Vehkamäki, T. Hatanpää, M. Ritala and M. Leskelä, *J. Mater. Chem.*, 2004, **14**, 3191.
20. C. Rømming and J. Songstad, *Acta Chem. Scand.* 1980, **34A**, 365.
21. J. L. Atwood, A. H. Cowley, W. E. Hunter and S. K. Mehrotra, *Inorg. Chem.*, 1982, **21**, 1354.
22. C. L. Raston, B. W. Skelton, V.-A. Tolhurst and A. H. White, *J. Chem. Soc., Dalton Trans.*, 2000, 1279.
23. (a) P. B. Hitchcock, M. F. Lappert, A. K. Rai and H. D. Williams, *J. Chem. Soc., Chem. Commun.*, 1986, 1633; (b) P. B. Hitchcock, H. A. Jasim, M. F. Lappert and H. D. Williams, *J. Chem. Soc., Chem. Commun.*, 1986, 1634.
24. U. Wirringa, H. W. Roesky, M. Noltemeyer and H.-G. Schmidt, *Inorg. Chem.*, 1994, **33**, 4607.
25. (a) L. K. Krannich, C. L. Watkins and D. K. Srivastava, *Polyhedron*, 1990, **9**, 289; (b) C. L. Watkins, L. K. Krannich, C. J. Thomas and D. K. Srivastava, *Polyhedron*, 1994, **13**, 3299.
26. S. R. Wuller, A. L. Seligson, G. P. Mitchell and J. Arnold, *Inorg. Chem.*, 1995, **34**, 4854.
27. (a) A. J. Edwards, M. A. Paver, M.-A. Rennie, P. R. Raithby, C. A. Russell and D. S. Wright, *J. Chem. Soc., Dalton Trans.*, 1994, 2963; (b) M. A. Beswick, C. N. Harmer, A. D. Hopkins, M. A. Paver, P. R. Raithby and D. S. Wright, *Polyhedron*, 1998, **17**, 745; (c) A. Bashall, M. A. Beswick, C. N. Harmer, A. D. Hopkins, M. McPartlin, M. A. Paver, P. R. Raithby and D. S. Wright, *J. Chem. Soc., Dalton Trans.*, 1998, 1389; (d) P. J. Bailey, R. O. Gould, C. N. Harmer, S. Pace, A. Steiner and D. S. Wright, *Chem. Commun.*, 1997, 1161.
28. M. A. Beswick, B. R. Elvidge, N. Feeder, S. J. Kidd and D. S. Wright, *Chem. Commun.*, 2001, 379.
29. U. N. Nehete, H. W. Roesky, V. Jancik, A. Pal and J. Magull, *Inorg. Chim. Acta*, 2007, **360**, 1248.
30. (a) J. Lorberth, W. Massa, S. Wocadlo, I. Sarraje, S.-H. Shin and X.-W. Li, *J. Organomet. Chem.*, 1995, **485**, 149; (b) P. L. Shutov, S. S. Karlov, K. Harms, D. A. Tyurin, J. Sundermeyer, J. Lorberth and G. S. Zaitseva, *Eur. J. Inorg. Chem.*, 2004, 2498.
31. L. Liu, L. N. Zakharov, A. L. Rheingold and T. A. Hanna, *Chem. Commun.*, 2004, 1472.
32. W. A. Herrmann, N. W. Huber, R. Anwander and T. Priermeier, *Chem. Ber.*, 1993, **126**, 1127.
33. T. J. Boyle, D. M. Pedrotty, B. Scott and J. W. Ziller, *Polyhedron*, 1998, **17**, 1959.
34. M.-C. Massiani, R. Papiernik, L. G. Hubert-Pfalzgraf and J.-C. Daran, *Polyhedron*, 1991, **10**, 437.
35. R. Bohra, H. W. Roesky, J. Lucas, M. Noltemeyer and G. M. Sheldrick, *J. Chem. Soc., Dalton Trans.*, 1983, 1011.
36. (a) M. Veith and B. Bertsch, *Z. Anorg. Allg. Chem.*, 1988, **557**, 7; (b) A.-M. Caminade, M. Veith, V. Huch and W. Malisch, *Organometallics*, 1990, **9**, 1798.
37. G. I. Kokorev, R. Z. Musin, Sh.-Kh. Badrutdinov, F. G. Khalitov, A. B. Platonov and F. D. Yambushev, *J. Gen. Chem. USSR*, 1989, **59**, 1376.
38. C. Payrastre, Y. Madaule, J. G. Wolf, T. C. Kim, M. R. Mazières, R. Wolf and M. Sanchez, *Heteroatom Chem.*, 1992, **3**, 157.
39. N. Burford, T. M. Parks, B. W. Royan, B. Borecka, T. S. Cameron, J. F. Richardson, E. J. Gabe and R. Haynes, *J. Am. Chem. Soc.*, 1992, **114**, 8147.

40. C. J. Carmalt, V. Lomeli, B. G. McBurnett and A. H. Cowley, *Chem. Commun.*, 1997, 2095.
41. (a) H. A. Spinney, I. Korobkov, G. A. DiLabio, G. P. A. Yap and D. S. Richeson, *Organometallics*, 2007, **26**, 4972; (b) H. A. Spinney, I. Korobkov and D. S. Richeson, *Chem. Commun.*, 2007, 1647.
42. C. L. Raston, B. W. Skelton, V.-A. Tolhurst and A. H. White, *Polyhedron*, 1998, **17**, 935.
43. (a) G. G. Briand, T. Chivers and M. Parvez, *Can. J. Chem.*, 2003, **81**, 169; (b) J. Konu, M. S. Balakrishna, T. Chivers and T.W. Swaddle, *Inorg. Chem.*, 2007, **46**, 2627.
44. (a) T. Gans-Eichler, D. Gudat and M. Nieger, *Heteroatom Chem.*, 2005, **16**, 327; (b) D. Gudat, T. Gans-Eichler and M. Nieger, *Chem. Commun.*, 2004, 2434.
45. J.-L. Fauré, H. Gornitzka, R. Réau, D. Stalke and G. Bertrand, *Eur. J. Inorg. Chem.*, 1999, 2295.
46. N. Kuhn and O.J. Scherer, *Z. Naturforsch., Teil B*, 1979, **34**, 888.
47. (a) W. Kolondra, W. Schwartz and J. Weidlein, *Z. Naturforsch., Teil B*, 1985, **40**, 872; (b) H. Krause and J. Weidlein, *Z. Anorg. Allg. Chem.*, 1988, **563**, 116.
48. M. Veith, B. Bertsch and V. Huch, *Z. Anorg. Allg. Chem.*, 1988, **559**, 73.
49. N. Burford, C. L. B. Macdonald, T. M. Parks, G. Wu, B. Borecka, W. Kwiatkowski and T. S. Cameron, *Can. J. Chem.*, 1996, **74**, 2209.
50. A. J. Edwards, N. E. Leadbeater, M. A. Paver, P. R. Raithby, C. A. Russell and D. S. Wright, *J. Chem. Soc., Dalton Trans.*, 1994, 1479.
51. G. G. Briand, T. Chivers and M. Parvez, *J. Chem. Soc., Dalton Trans.*, 2002, 3785.
52. N. Burford, T. S. Cameron, C. L. B. Macdonald, K. N. Robertson, R. Schurko, D. Walsh, R. McDonald and R. E. Wasylishen, *Inorg. Chem.*, 2005, **44**, 8058.
53. (a) S. G. Baxter, A. H. Cowley and S. K. Mehrotra, *J. Am. Chem. Soc.*, 1981, **103**, 5573; (b) R. J. Wiacek, C. L. B. Macdonald, J. N. Jones, J. M. Pietryga and A. H. Cowley, *Chem. Commun.*, 2003, 430.
54. E. V. Avtomonov, K. Megges, X. Li, J. Lorberth, S. Wocadlo, W. Massa, K. Harms, A. V. Churakov and J. A. K. Howard, *J. Organomet. Chem.*, 1997, **544**, 79.
55. A. Brandl and H. Nöth, *Chem. Ber.*, 1988, **121**, 1321.
56. H. W. Roesky, R. Bohra and W. S. Sheldrick, *J. Fluorine Chem.*, 1983, **22**, 199.
57. J. Müller, U. Müller, A. Loss, J. Lorberth, H. Donath and W. Massa, *Z. Naturforsch., Teil B*, 1985, **40**, 1320.
58. (a) T. Klapötke, *Polyhedron*, 1987, **6**, 1593; (b) N. Burford, B. W. Royan and P. S. White, *J. Am. Chem. Soc.*, 1989, **111**, 3746; (c) N. Burford, T. M. Parks, B. W. Royan, J. F. Richardson and P. S. White, *Can. J. Chem.*, 1992, **70**, 703.
59. (a) L. K. Krannich, R. K. Kanjolia and C. L. Watkins, *J. Ind. Chem. Soc.*, 1985, **62**, 795; (b) R. K. Kanjolia, L. K. Krannich and C. L. Watkins, *J. Chem. Soc., Dalton Trans.*, 1986, 2345; (c) V. K. Gupta, L. K. Krannich and C. L. Watkins, *Inorg. Chem.*, 1986, **25**, 2553.
60. N. Burford, J. C. Landry, M. J. Ferguson and R. McDonald, *Inorg. Chem.*, 2005, **44**, 5897.
61. M. Veith, F. Goffing and V. Huch, *Z. Naturforsch., Teil B*, 1988, **43**, 846.
62. G. Reeske, C. R. Hoberg, N. J. Hill and A. H. Cowley, *J. Am. Chem. Soc.*, 2006, **128**, 2800.
63. (a) C. Payrastre, J. G. Wolf, T. C. Kim and M. Sanchez, *Bull. Soc. Chim. Fr.*, 1992, **129**, 157;(b) C. Payrastre, Y. Madaule and J. G. Wolf, *Tetrahedron Lett.*, 1990, **31**, 1145; (c) D. F. Moser, I. Schranz, M. C. Gerrety, L. Stahl and R. J. Staples, *J. Chem. Soc., Dalton Trans.*, 1999, 751.
64. M. C. Copsey, S. B. Gallon, S. K. Grocott, J. C. Jeffery, C. A. Russell and J. M. Slattery, *Inorg. Chem.*, 2005, **44**, 5495.
65. D. J. Eisler and T. Chivers, *Inorg. Chem.*, 2006, **45**, 10734.
66. N. Burford, T. S. Cameron, K.-C. Lam, D. J. LeBlanc, C. L. B. Macdonald, A. D. Phillips, A. L. Rheingold, L. Stark and D. Walsh, *Can. J. Chem.*, 2001, **79**, 342.
67. (a) U. Wirringa, H. W. Roesky, M. Noltemeyer and H.-G. Schmidt, *Angew. Chem., Int. Ed.*, 1993, **32**, 1628; (b) A. D. Bond, A. D. Hopkins, A. Rothenberger, R. Wolf, A. D. Woods and D. S. Wright, *Organometallics*, 2001, **20**, 4454.
68. B. Ross, J. Belz and M. Nieger, *Chem. Ber.*, 1990, **123**, 975.
69. R. Bohra, H. W. Roesky, M. Noltemeyer and G. M. Sheldrick, *Acta Crystallogr., Sect. C: Cryst. Struct. Commun.*, 1984, **C40**, 1150.

70. D. C. Haagenson, L. Stahl and R. J. Staples, *Inorg. Chem.*, 2001, **40**, 4491.
71. J.-T. Ahlemann, H. W. Roesky, R. Murugavel, E. Parisini, M. Noltemeyer, H.-G. Schmidt, O. Müller, R. Herbst-Irmer, L. N. Markovskii and Yu. G. Shermolovich, *Chem. Ber.*, 1997, **130**, 1113.
72. G. I. Kokorev, I. A. Litvinov, V. A. Naumov, Sh. Kh. Batrutdinov and F. D. Yambushev, *J. Gen. Chem. USSR*, 1989, **59**, 1381.
73. (a) F. García, D. J. Linton, M. McPartlin, A. Rothenberger, A. E. H. Wheatley and D. S. Wright, *J. Chem. Soc., Dalton Trans.*, 2002, 481; (b) R. A. Alton, D. Barr, A. J. Edwards, M. A. Paver, P. R. Raithby, M.-A. Rennie, C. A. Russell and D. S. Wright, *J. Chem. Soc., Chem. Commun.*, 1994, 1481.
74. M. Veith, A. Rammo and M. Hans, *Phosphorus, Sulfur Silicon Relat. Elem.* 1994, **93–94**, 197.
75. A. J. Edwards, M. A. Paver, P. Pearson, P. R. Raithby, M.-A. Rennie, C. A. Russell and D. S. Wright, *J. Organomet. Chem.*, 1995, **503**, C29.
76. W. Neubert, H. Pritzkow and H. P. Latscha, *Angew. Chem., Int. Ed.*, 1988, **27**, 287.
77. (a) R. O. Day and R. R. Holmes, *Inorg. Chem.*, 1980, **19**, 3609; (b) R. O. Day, J. M. Holmes, A. C. Sau, J. R. Devillers, R. R. Holmes and J. A. Deiters, *J. Am. Chem. Soc.*, 1982, **104**, 2127.
78. (a) H. W. Roesky, H. Djarrah, D. Amirzadeh-Asl and W. S. Sheldrick, *Chem. Ber.*, 1981, **114**, 1554; (b) P. L. Dhingra, P. Lyall and R. D. Verma, *Ind. J. Chem., Sect. A*, 1988, **27**, 323.
79. J. Dahlmann and K. Winsel, *J. Prakt. Chem.*, 1979, **321**, 370.
80. S. K. Shukla, A. Ranjan and A. K. Saxena, *Phosphorus, Sulfur Silicon Relat. Elem.*, 2003, **178**, 785.
81. G. A. Abakumov, A. I. Poddel'sky, E. V. Grunova, V. K. Cherkasov, G. K. Fukin, Y. A. Kurskii and L. G. Abakumova, *Angew. Chem., Int. Ed.*, 2005, **44**, 2767.
82. (a) A. J. Edwards, M. A. Paver, P. R. Raithby, M.-A. Rennie, C. A. Russell and D. S. Wright, *Angew. Chem., Int. Ed.*, 1994, **33**, 1277; (b) M. A. Beswick, N. L. Cromhout, C. N. Harmer, M. A. Paver, P. R. Raithby, M.-A. Rennie, A. Steiner and D. S. Wright, *Inorg. Chem.*, 1997, **36**, 1740; (c) A. Bashall, M. McPartlin, M. A. Beswick, E. A. Harron, A. D. Hopkins, S. J. Kidd, P. R. Raithby, D. S. Wright and A. Steiner, *Chem. Commun.*, 1999, 1145; (d) D. Barr, A. J. Edwards, S. Pullen, M. A. Paver, P. R. Raithby, M.-A. Rennie, C. A. Russell and D. S. Wright, *Angew. Chem., Int. Ed.*, 1994, **33**, 1875; (e) A. J. Edwards, M. A. Beswick, J. R. Galsworthy, M. A. Paver, P. R. Raithby, M.-A. Rennie, C. A. Russell, K. L. Verhorevoort and D. S. Wright, *Inorg. Chim. Acta*, 1996, **248**, 9; (f) A. Bashall, M. A. Beswick, N. Feeder, A. D. Hopkins, S. J. Kidd, M. McPartlin, P. R. Raithby and D. S. Wright, *J. Chem. Soc., Dalton Trans.*, 2000, 1841.
83. (a) M. A. Beswick, E. A. Harron, A. D. Hopkins, P. R. Raithby and D. S. Wright, *J. Chem. Soc., Dalton Trans.*, 1999, 107; (b) M. A. Beswick, M. A. Paver, D. S. Wright, N. Choi, A. D. Hopkins and M. McPartlin, *Chem. Commun.*, 1998, 261.
84. (a) A. Bashall, M. A. Beswick, N. Choi, A. D. Hopkins, S. J. Kidd, Y. G. Lawson, M. E. G. Mosquera, M. McPartlin, P. R. Raithby, A. E. H. Wheatley, J. A. Wood and D. S. Wright, *J. Chem. Soc., Dalton Trans.*, 2000, 479; (b) M. A. Beswick, N. Choi, C. N. Harmer, A. D. Hopkins, M. McPartlin and D. S. Wright, *Science*, 1998, **281**, 1500.
85. M. C. Copsey, J. C. Jeffery, A. P. Leedham, C. A. Russell and J. M. Slattery, *Dalton Trans.*, 2003, 2103.
86. S. C. James, N. C. Norman, A. G. Orpen, M. J. Quayle and U. Weckenmann, *J. Chem. Soc., Dalton Trans.*, 1996, 4159.
87. (a) C. A. Stewart, R. L. Harlow and A. J. Arduengo III, *J. Am. Chem. Soc.*, 1985, **107**, 5543; (b) A. J. Arduengo III, C.A. Stewart, F. Davidson, D. A. Dixon, J. Y. Becker, S. A. Culley and M. B. Mizen, *J. Am. Chem. Soc.*, 1987, **109**, 627.
88. (a) C. A. Stewart and A. J. Arduengo III, *Inorg. Chem.*, 1986, **25**, 3847; (b) A. J. Arduengo III, M. Lattman, H. V. R. Dias, J. C. Calabrese and M. Klein, *J. Am. Chem. Soc.*, 1991, **113**, 1799.
89. (a) M. A. Paver, J. S. Joy and M. B. Hursthouse, *Chem. Commun.*, 2001, 2480; (b) M. A. Paver, J. S. Joy, S. J. Coles, M. B. Hursthouse and J. E. Davies, *Polyhedron*, 2003, **22**, 211; (c) F. García, R. J. Less, V. Naseri, M. McPartlin, J. M. Rawson and D. S. Wright, *Dalton Trans.*, 2008, 997.
90. B. N. Diel, T. L. Hubler and W. G. Ambacher, *Heteroatom Chem.*, 1999, **10**, 423.

Index

Note: Page numbers in *italic* refer to figures or tables.